ORGANIC SYNTHESES

ORGANIC SYNTHESES

Collective Volume 7

A REVISED EDITION OF ANNUAL VOLUMES 60–64

John Wiley & Sons, Inc.
NEW YORK / CHICHESTER / BRISBANE / TORONTO / SINGAPORE

Published by John Wiley & Sons, Inc.

Library of Congress Catalog Card Number: 21-17747
ISBN 0-471-51559-0

Printed in the United States of America

10 9 8 7 6 5 4 3 2 1

PREFACE

Beginning a new tradition for *Organic Syntheses*, collective volumes will be compiled every 5 years instead of the previous 10. Thus Collective Volume VII contains procedures previously published in annual volumes 60–64 (1981–1985) but revised and updated in the light of experience and advances since their first appearance. This new format reflects in part the increased pace of research in organic chemistry and our belief that *Organic Syntheses* should be publishing the most up-to-date and significant procedures for the use of our readership. The Editor is grateful to the submitters for their cooperation in reviewing and updating their procedures. In a few instances the Editor has revised the original title so that each procedure has a title compound.

Through the efforts of Assistant Editor Theodora Greene the nomenclature, presentation of spectroscopic data, and other variable elements of style have been standardized during the preparation of recent annual volumes. She reexamined those volumes published before her association with the enterprise to ensure as much conformity as possible in this compendium.

In accord with past practice, extensive hazard warnings have been included. In addition we are now soliciting from submitters any information needed for special disposal problems. The Board of Editors is discouraging the use of potentially hazardous solvents such as benzene and hexamethylphosphoric triamide (HMPA) by asking submitters to replace them with others. In the particular case of HMPA procedures involving that solvent have been rechecked with a replacement. One of those (p. 326) appears in this volume.

In some instances changes have been made in the experimental procedure reflecting experience since their publication. In the case of the reduction of α-amino acids to amino alcohols, a version employing lithium aluminum hydride has been added to the original version, which used borane–dimethyl sulfide. The conversion of cycloheptanone to azocinone has been improved by extending the reaction time, which removes small amounts of the oxime that had contaminated the desired lactam. In the case of the preparation of the homochiral Wieland–Miescher ketone, some reactions in the procedure described by the original submitters have been included. Because of an unusual combination of circumstances, these submitters had not had the opportunity to review the checker's version of their procedure which appeared in Volume 63.

Following the practice of Collective Volume VI, the table of contents has been arranged alphabetically by title compound (not by the name of the method). Since this listing is probably the least used, this ordering, while often not keeping related procedures adjacent to each other, is likely to have the least effect on users. The concordance listing introduced in Collective Volume VI, which relates the title to the annual volume in which it first appears, is retained in the contents section. If one has a literature citation to an annual volume, the concordance index at the end of the volume allows the reader to find the latest version.

In this volume the Editor has returned to the earlier practice of multiple indices. Where names of title compounds, isolated intermediates, and uncommon reagents appear, they are accompanied by *Chemical Abstracts* registry numbers. In the titles in the text, the *Chemical Abstracts* name, which is usually different, is given below the main title name in brackets. The practice introduced in recent annual volumes, of following each

v

procedure with an appendix of *Chemical Abstracts* registry numbers and names has been dropped in this collective volume to save space, but, as indicated above, this information is retained in the indices.

The editors of *Organic Syntheses* welcome corrections, suggestions, and procedures being submitted for consideration by the Board of Editors. Prospective submitters should consult the section entitled "Submission of Preparations" at the front of one of the latest annual volumes for guidance. Correspondence should be addressed to the current Secretary of the Board of Editors of Organic Syntheses, Dr. Jeremiah P. Freeman, Department of Chemistry, University of Notre Dame, Notre Dame, IN 46556.

The Editor is grateful to the submitters, checkers, and editors of annual volumes 60–64 who made this collective volume possible. He is indebted to previous editors Richard E. Benson, Robert M. Coates, and the late William E. Sheppard for the useful Style Guide for *Organic Syntheses*, and to Theodora Greene, whose careful attention to the detail of this guide and whose skill in the use of *Chemical Abstracts* provides us with accurate nomenclature, registry numbers, and conformity of style. The major burden of this work as well as the preparation of the annual volumes in recent years has fallen to my secretary, Mrs. Myra Martin, whose diligence and passion for thoroughness has left these volumes as error-free as is humanly possible.

JEREMIAH P. FREEMAN

Notre Dame, Indiana
January 1990

CONTENTS

vii

CONTENTS ix

ORGANIC SYNTHESES

REDUCTIVE COUPLING OF CARBONYLS TO ALKENES: ADAMANTYLIDENEADAMANTANE

(Tricyclo[3.3.1.1^{3,7}]decane, tricyclo[3.3.1.1^{3,7}]decylidene-)

Submitted by Michael P. Fleming and John E. McMurry[1]
Checked by Steven R. Villaseñor and Carl R. Johnson

1. Procedure

A 2-L, three-necked flask is thoroughly flamed while being flushed with argon and is then fitted with three rubber septa. Anhydrous titanium trichloride (63.16 g, 0.409 mol) (Notes 1 and 2) is added to the weighed flask in an argon-filled glove bag. The flask is reweighed, and one of the rubber septa is replaced with a dry (12 hr at 120°C) reflux condenser through which a stream of argon is flowing (Note 3). The flask is fitted with a mechanical stirrer equipped with a glass shaft and Teflon paddle (Note 4). Into the flask is syringed 600 mL of 1,2-dimethoxyethane (Note 5), and the remaining rubber septum is exchanged for a glass stopper. Lithium (8.52 g, 1.23 mol) (Note 6) is etched to brilliance in methanol, quickly washed in petroleum ether (Note 7), and cut into small pieces directly into the stirred suspension. The mixture is heated at reflux for 1 hr by an oil bath that is then removed (Note 8). Immediately after the solvent has ceased to reflux, 2-adamantanone (15.36 g, 0.102 mol) (Note 1) is added in one portion, and the mixture is heated at reflux for 18 hr.

Stirring is maintained as the mixture is allowed to cool to room temperature, and 600 mL of petroleum ether (Note 7) is added in 100-mL portions at 5-min intervals (Note 9). The stirrer is disconnected from its motor, and the solution is poured into a sintered-glass funnel containing 50 g of Florisil (approximately 7 cm in depth) (Note 1). The black material remaining in the reaction vessel is washed with eight 50-mL portions of petroleum ether, which are poured into the same pad of Florisil (Notes 10 and 11). The filter pad is then washed with 400 mL of petroleum ether. Removal of the solvent from the combined filtrates by means of a rotary evaporator followed by high vacuum (0.05 mm) gives 12–13 g of a white crystalline solid. This crude product is dissolved in 3.5 L of hot methanol (Note 1), and the volume is reduced to 2 L by boiling. The solution is allowed to slowly cool to room temperature. The colorless needles are vacuum filtered through sintered glass (medium frit) and washed with 50 mL of ice-cold methanol. The crystals are dried under vacuum (0.05 mm Hg) to give 10.3–10.4 g (75–76%) of adamantylideneadamantane, mp 184–186°C; ^{1}H NMR (CDCl$_3$) δ: 1.5–2.1 (br m, 24 H) 2.7–3.1 (br m, 4 H). Concentration of the mother liquor to 350 mL and crystallization as described above yields an additional 1.2–1.5 g (9-11%) of product, mp 182–184°C (Note 12).

2. Notes

1. The following reagents were used as supplied: titanium trichloride from Alfa Products, Morton Thiokol, Inc.; 2-adamantanone from Aldrich Chemical Company, Inc.; methanol from MCB, Inc.; and acetone from Mallinckrodt, Inc.

2. Because of its sensitivity toward oxygen and water, anhydrous titanium trichloride should always be handled under an inert atmosphere. The submitters report that titanium trichloride in bottles that have been opened and resealed undergoes a slow deterioration that causes erratic results in the coupling reaction. This decomposition is frequently detectable by the evolution of white fumes from the titanium trichloride during transfer. If a number of small-scale reactions are to be performed, the use of a Schlenk tube is advisable to extend the useful life of the titanium trichloride.

3. Substitution of nitrogen for argon does not significantly decrease the yield.

4. The coupling reaction may be adversely affected if metallic stirrers are employed. The bore of the stirrer should be water-cooled. Lubricants such as mineral oil are to be avoided since they complicate product isolation.

5. The 1,2-dimethoxyethane was obtained from Aldrich Chemical Company, Inc., and was allowed to stand over molecular sieves (type 4A in 1/16-in. pellet form from Union Carbide Corporation) for several days. Final purification was accomplished by heating at reflux over potassium in a nitrogen atmosphere for at least 10 hr, followed by distillation from potassium. The solvent was used on the same day that it was distilled to minimize the formation of peroxides.

6. Lithium wire (3.2-mm diam, 0.02% sodium) was obtained from Alfa Products, Morton Thiokol, Inc., and was washed in petroleum ether before weighing.

7. Petroleum ether (bp 35–65°C) was obtained from Fisher Scientific Company and was distilled from potassium permanganate.

8. The color of the reaction mixture at this point varies from gray-green to gray-black. The success of the reaction seems to be independent of the color.

9. Addition of petroleum ether causes a viscous black precipitate to cling to the walls of the flask, leaving a milky-white solution that can be conveniently poured into the filter.

10. The black residue, which consists of inorganic salts, titanium, and unreacted lithium, is retained in the reaction vessel, where it is washed with petroleum ether while the mass is manually stirred with the paddle. No problem has been encountered in exposing the black material to the air during the washing procedure.

11. The black residue is destroyed in the following manner. The stirrer motor is reattached, and the flask is flushed with argon. The flask is cooled in an ice-water bath before adding 300 mL of petroleum ether and 300 mL of acetone. As the mixture is vigorously stirred, ca. 10 mL of methanol is added from a dropping funnel. After reaction has begun (an induction period of up to 0.5 hr may occur before gas evolution becomes noticeable), an additional 590 mL of methanol is added dropwise over a 6–10-hr period. Stirring is continued at 0°C until pieces

of lithium can no longer be seen (approximately 1 hr after the addition of the methanol has been completed).

12. The second crop is slightly impure, as shown by its NMR spectrum.

3. Discussion

Adamantylideneadamantane has been prepared by (1) photolysis of 2-adamantylketene dimer,[2] (2) reduction of 4e-chloroadamantylideneadamantane with sodium in liquid ammonia,[3] (3) rearrangement of spiro[adamantane-2,4'-homoadamantan-5'-ol] with Lewis acids,[4,5] (4) reduction of 2,2-dibromoadamantane with magnesium[6] or zinc-copper couple,[7] and (5) treatment of the azine of 2-adamantanone with hydrogen sulfide, followed by oxidation with lead tetraacetate and heating with triphenylphosphine.[8]

The present method is a modification of a previous procedure by the submitters.[9] Handling of lithium in the air is less hazardous and more convenient than that of potassium, which was originally used. Higher yields were obtained when the higher-boiling solvent 1,2-dimethoxyethane was employed rather than tetrahydrofuran. Although the titanium trichloride-lithium system results in slightly lower yields for aliphatic ketones than the corresponding potassium method, the former is considerably more convenient for large-scale reactions. The lithium procedure is applicable to both aromatic and aliphatic aldehydes and ketones (Table I). Reductive coupling of unsymmetrical carbonyl compounds usually results in a mixture of geometric isomers.

Details of the titanium-induced dicarbonyl coupling reaction can be found in a full paper[10] and in a review article.[11]

TABLE I

REACTION OF KETONES AND ALDEHYDES WITH TITANIUM
TRICHLORIDE-LITHIUM IN 1,2-DIMETHOXYETHANE

Carbonyl Compound	Yield of Alkene Product (%)
Acetophenone	94
Benzaldehyde	97
Benzophenone	96
Cholestanone	84
Cyclododecanone	90
Cycloheptanone	85
Cyclohexanone	81
Decanal	59
Hexanal	58 (28 : 72, *cis* : *trans*)

1. Thimann Laboratories, University of California, Santa Cruz, CA 95064. Present address of J. E. M.: Baker Laboratory, Department of Chemistry, Cornell University, Ithaca, NY 14853.
2. Strating, J.; Scharp, J.; Wynberg, H. *Recl. Trav. Chim. Pays-Bas* **1970**, *89*, 23–31.
3. Wieringa, J. H.; Strating, J.; Wynberg, H. *Tetrahedron Lett.* **1970**, 4579–4582.
4. Boelema, E.; Wynberg, H.; Strating, J. *Tetrahedron Lett.* **1971**, 4029–4032.
5. Gill, G. B.; Hands, D. *Tetrahedron Lett.* **1971**, 181–184.

6. Bartlett, P. D.; Ho, M. S. *J. Am. Chem. Soc.* **1974**, *96*, 627–629.
7. Geluk, H. W. *Synthesis*, **1970**, 652–653.
8. Schaap, A. P.; Faler, G. R. *J. Org. Chem.* **1973**, *38*, 3061–3062.
9. McMurry, J. E.; Fleming, M. P. *J. Org. Chem.* **1976**, *41*, 896–897.
10. McMurry, J. E.; Fleming, M. P.; Kees, K. L.; Krepski, L. R. *J. Org. Chem.* **1978**, *43*, 3255.
11. McMurry, J. E. *Acc. Chem. Res.* **1983**, *16*, 405.

SYNTHESIS AND DIELS–ALDER REACTIONS OF 3-ACETYL-2(3*H*)-OXAZOLONE: 6-AMINO-3,4-DIMETHYL-*cis*-3-CYCLOHEXEN-1-OL

[2(3*H*)-Oxazolone, 3-acetyl- and 3-cyclohexen-1-ol, 6-amino-3,4-dimethyl-, *cis*-]

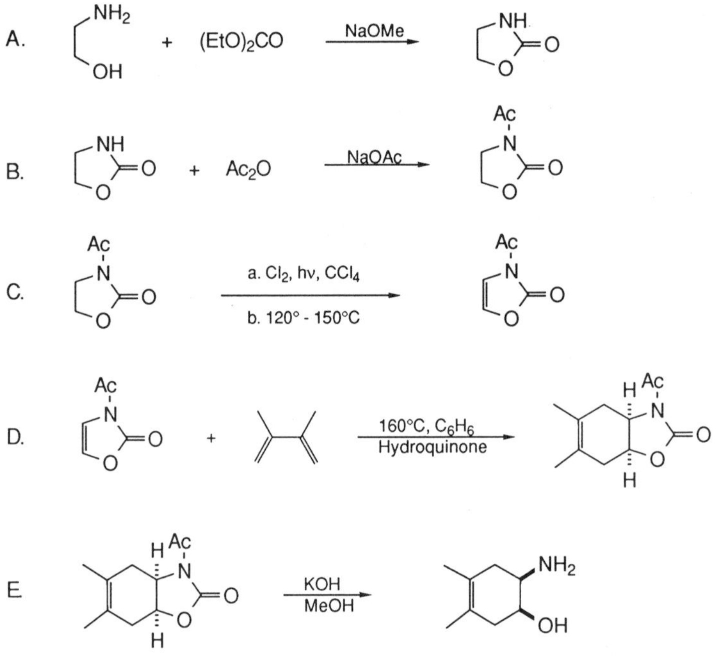

Submitted by KARL-HEINZ SCHOLZ, HANS-GEORG HEINE, and WILLY HARTMANN[1]
Checked by ASHOK B. SHENVI, BRUCE M. MONROE, RICHARD E. BENSON, and BRUCE E. SMART

1. Procedure

A. *2-Oxazolidinone.* A 2-L, three-necked flask equipped with a thermometer, magnetic stirring bar, and a Vigreux column fitted with a distillation head is charged with 305 g (5.0 mol) of freshly distilled 2-aminoethanol, 730 g (6.2 mol) of diethyl carbonate, and 2.5 g (0.05 mol) of sodium methoxide (Note 1). The mixture is stirred and the flask is heated in an oil bath maintained at 125–130°C. Ethanol begins to distill off when the internal temperature reaches 95–100°C. After heating for about 8 hr, the internal temperature reaches 125°C and ethanol ceases to distill (Note 2). The reaction mixture is allowed to cool to 60–70°C and is poured into 1 L of cold chloroform (Note 3). The resulting solution is chilled thoroughly in an ice–water bath and the precipitated product is recovered by filtration. The filtrate is concentrated to 250 mL and chilled to give a second crop. The combined crops are dried in a vacuum oven at 50°C to give 320–339 g (74–78%) of white crystals, mp 86–88°C [lit.[2] mp 87–89°C] (Note 4).

B. *3-Acetyl-2-oxazolidinone.* A 3-L, one-necked flask equipped with a reflux condenser and a magnetic stirring bar is charged with 326 g (3.75 mol) of 2-oxazoli-dinone, 94 g (1.15 mol) of anhydrous sodium acetate, and 1.6 L of acetic anhydride. The stirred solution is refluxed for 3 hr and the acetic anhydride is then removed by distillation at 15–20 mm. The residue is extracted with three 875-mL portions of boiling toluene (Note 5), and the hot toluene extractions are filtered, combined, and concentrated to a total volume of 675 mL. Diethyl ether (675 mL) is added with stirring to the toluene solution and the mixture is chilled in an ice–water bath. The precipitate is removed by filtration and washed with 250 mL of diethyl ether to give 328–403 g (68–83%) of colorless to very light tan crystals, mp 65–67°C [lit.[2] mp 69–70°C] (Note 6). A second crop of 63–24 g (13–5%), mp 65–68°C, is obtained by concentrating the filtrate to 275 mL and chilling it in an ice–water bath.

C. *3-Acetyl 4- and 5-chloro-2-oxazolidinone.* A 3-L, four-necked flask is equipped with a reflux condenser topped with a gas discharge tube, thermometer, fritted-glass inlet tube extending to the bottom of the flask, and a glass sleeve for accepting an ultra-violet (UV) lamp (Note 7). The reaction vessel is charged with 258 g (2.0 mol) of 3-acetyl-2-oxazolidinone, 2 L of carbon tetrachloride, and several boiling chips. The mixture is heated to gentle reflux, the light source is turned on, and 155 g (2.18 mol) of chlorine gas (Note 8) is introduced at such a rate that no chlorine escapes from the condenser (Note 9). After the addition is complete, heating is discontinued and nitrogen is bubbled through the reaction mixture to remove the dissolved hydrogen chloride. The solvent is then removed on a rotary evaporator to give a yellow oil, which consists of a mixture of 3-acetyl 4- and 5-chloro-2-oxazolidinones[3] and is used in Step D without further purification.

D. *3-Acetyl-2(3H)-oxazolone.* The crude mixture of 3-acetyl 4- and 5-chloro-2-oxazolidinone from Step C is placed in a 2-L, three-necked flask equipped with a thermometer, sealed mechanical stirrer, and gas discharge tube. The material is heated to 120°C with stirring, and the temperature is then slowly increased to 150°C and held there until the evolution of gas ceases (Note 10). The cooled, black reaction mixture is distilled at 20 mm. The fractions boiling up to 150°C are collected and redistilled through a 50-cm × 3-cm Vigreux column fitted with a variable take-off head. There is obtained 140–172 g (55–68%) of product, bp 108–112°C (24 mm), which solidifies, mp 35–37°C (Note 11).

E. *4-Acetyl-7,8-dimethyl-2-oxa-4-azabicyclo[4.3.0]non-7-en-3-one.* A solution of 63.5 g (0.5 mol) of 3-acetyl-2(3H)-oxazolone, 27.5 g (0.33 mol) of 2,3-dimethylbutadiene (Note 12), and 2.0 g of hydroquinone in 125 mL of benzene is heated at 160°C under nitrogen in a 360-ml Hastelloy C shaker tube (Note 13). After 12 hr, the pressure vessel is cooled to room temperature, recharged with 27.5 g of 2,3-dimethylbutadiene, and heated another 12 hr at 160°C. The vessel is again cooled, recharged with 27.5 g of 2,3-dimethylbutadiene, and heated at 160°C for a final 12 hr. The resulting mixture is distilled to give 36.4–40.3 g (35–39%) of crude product, bp 115–130°C (0.5 mm), which solidifies (Notes 14 and 15). This material can be recrystallized by adding 36.4 g of melted product to 50 mL of boiling hexane, followed by cooling to give 27.6 g (26%) of crystals, mp 72–78°C (Note 16).

F. *6-Amino-3,4-dimethyl-cis-3-cyclohexen-1-ol.* A solution of 26.1 g (0.125 mol) of 4-acetyl-7,8-dimethyl-2-oxa-4-azabicyclo[4.3.0]non-7-en-3-one and 42.0 g (0.75 mol) of potassium hydroxide in 200 mL of methanol and 100 mL of water is refluxed for 36 hr. The resulting mixture is exhaustively extracted with diethyl ether using a liquid–liquid continuous extraction apparatus. The ethereal extract is concentrated on a rotary evaporator and the residue is taken up in 150 mL of methylene chloride. The resulting solution is dried over anhydrous sodium sulfate, the drying agent is removed by filtration, and the filtrate is concentrated to dryness. The solid residue (16.5 g) is recrystallized from 100 mL of diethyl ether to give 11.2 g (64%) of colorless crystals, mp 63–65.5°C. A second crop of 2.3 g (13%) is obtained by concentrating the mother liquor to 50 mL and chilling in an ice–water bath (Note 17).

2. Notes

1. The checkers obtained 2-aminoethanol, diethyl carbonate, and anhydrous sodium methoxide from the Aldrich Chemical Company.

2. About 625 mL (theoretical: 583 mL) of ethanol is collected. If the reaction is stopped before this volume is collected, the yield is reduced.

3. If the reaction mixture cools below 60°C, the product solidifies in the flask.

4. The product shows the following ^{1}H NMR spectrum (d_6-DMSO) δ: 3.3–3.8 (m, 2 H), 4.2–4.6 (m, 2 H), 6.5–7.5 (br s, 1 H) and is analytically pure. Anal. calcd. for $C_3H_5NO_2$: C, 41.38; H, 5.79; N, 16.09. Found: C, 41.61; H, 5.70; N, 16.06. The submitters report that they obtained pure material, mp 89–90°C, after three recrystallizations from chloroform.

5. This is conveniently done by adding the toluene to the residue in the flask, heating to reflux in an oil bath, and then filtering the hot mixture.

6. This material shows the following ^{1}H NMR spectrum ($CDCl_3$) δ: 2.49 (s, 3 H), 3.8–4.7 (complex m, 4 H) and has acceptable analysis. Anal. calcd. for $C_5H_7NO_3$: C, 46.51; H, 5.46; N, 10.85. Found: C, 46.49; H, 5.40; N, 10.99. The submitters report that they obtained colorless, analytically pure material, mp 69–70°C, after three recrystallizations from benzene.

7. The submitters used a Philips HPK 125-W high-pressure mercury vapor lamp. The sleeve is 2-mm Pyrex glass with an NS45 ground joint. The lamp does not require cooling. The checkers obtained equally good results by shining a Westinghouse 250-W sun lamp on the reaction flask from a distance of 25 cm.

8. The chlorine gas is passed successively through three wash bottles. The center bottle is filled with concentrated sulfuric acid and the other two are left empty to serve as safety traps.

9. The photochlorination takes 4–6 hr. The hydrogen chloride evolved is absorbed in water.

10. Dehydrochlorination begins at about 120°C. The temperature is raised about 10°C/hr to 150°C to avoid vigorous gas evolution. The elimination of hydrogen chloride is complete after 5–6 hr.

11. The submitters report yields of 150–200 g, and that analytically pure material boils at 110°C (24 mm) and melts at 35–37°C after recrystallization from diethyl ether. The material obtained by the checkers showed a satisfactory analysis without further purification. Anal. calcd. for $C_5H_5NO_3$: C, 47.25; H, 3.97; N, 11.02. Found: C, 46.81; H, 4.00; N, 11.21. The material shows the following ^{1}H NMR spectrum (CDCl$_3$) δ: 2.63 (s, 3 H), 6.90 (d, 1 H, $J = 2.5$), 7.30 (d, 1 H, $J = 2.5$); IR (CCl$_4$) cm^{-1}: 1880, 1735 (C=O).

12. The sample of 2,3-dimethylbutadiene was obtained from the Aldrich Chemical Company.

13. The submitters employed a nickel autoclave and noted that product from Step D may contain a small amount of hydrogen chloride or chlorinated material that can adversely affect a stainless-steel pressure vessel. Hastelloy C is a high-nickel alloy.

14. The submitters obtained 48.0 g of product and 33.5 g of recovered starting material, bp 110°C (24 mm). The checkers found that the forerun collected at 108–112°C (24 mm) contained starting material, but it was highly contaminated with unidentified by-products.

15. The checkers obtained erratic results in this step, possibly because of surface effects or trace impurities in the pressure vessel. In two other runs, only 16.8–18.8 g of crude product were obtained. In one case, high-boiling oligomers were formed, but none of the desired product was produced. Impurities in the diene or dienophile did not appear to be the problem since runs that employed recrystallized 3-acetyl-2(3*H*)-oxazolone and redistilled 2,3-dimethylbutadiene also gave variable results.

16. The submitters report pure product with bp 135–137°C (1.2 mm) and mp 78–80°C after recrystallization from chloroform. The checkers found that recrystallization from chloroform gave very poor recovery of product with mp 75–78°C. Material with mp 72–78°C is pure by NMR, mass spectroscopy, and combustion analysis; ^{1}H NMR (CDCl$_3$) δ: 1.70 (s, 6 H), 2.33 (m, 4 H), 2.45 (s, 3 H), 4.40 (d of t, 1 H, $J = 9.0, 4.5$), 4.83 (d of t, 1 H, $J = 9.0, 4.5$); infrared (IR) (KBr) cm^{-1}: 1780, 1690. Mass spectrum m/e calculated: 209.1051. Found: 209.1030. Anal. calcd. for $C_{11}H_{15}NO_3$: C, 63.14; H, 7.23; N, 6.69. Found: C, 63.19; H, 7.10; N, 6.67.

17. The product shows the following ^{1}H NMR spectrum (CDCl$_3$) δ: 1.60 (s, 6 H), 2.13 (br m, 4 H), 2.30 (s, 3 H), 3.00 (m, 1 H), 3.80 (m, 1 H) and is analytically pure. Anal. calcd. for $C_8H_{15}ON$: C, 66.35; H, 10.71; N, 9.92. Found: C, 66.17, H, 10.48; N, 10.26.

3. Discussion

The dienophile, 3-acetyl-2(3H)-oxazolone,[4] is an attractive intermediate for the synthesis of vicinal aminoalcohols with cis configurations. It reacts with 1,3-dienes, even under quite mild conditions, to form (4 + 2) cycloadducts.[5,6] Its high reactivity with deactivated 1,3-dienes is noteworthy. This property is present also in 2(3H)-oxazolone,[4] which can be obtained easily through solvolysis of 3-acetyl-2(3H)-oxazolone in methanol. 3-Acetyl-2(3H)-oxazolone, on UV irradiation in the presence of a sensitizer, combines easily with olefins to form (2 + 2) cycloadducts,[7] the hydrolysis of which leads to the class of cis-2-aminocyclobutanols.

1. Central Division for Research and Development, Main Scientific Laboratory of Bayer AG, D-4150, Krefeld-Uerdingen, West Germany.
2. Homeyer, A. H. U.S. Patent 2399118, 1946, Chem. Abstr. **1946**, 40, 4084.
3. Kunieda, T.; Abe, Y.; Iitaka, Y.; Hirobe, M. J. Org. Chem. **1982**, 47, 4291–4297.
4. Scholz, K.-H.; Heine, H.-G.; Hartmann, W. Justus Liebigs Ann. Chem. **1976**, 1319.
5. Scholz, K.-H; Heine, H.-G.; Hartmann, W. Justus Liebigs Ann. Chem. **1977**, 2027.
6. Deyrup, J. A.; Gingrich, H. L. Tetrahedron Lett. **1977**, 3115.
7. Scholz, K.-H.; Heine, H.-G.; Hartmann, W. Tetrahedron Lett. **1978**, 1467.

ELECTROPHILIC N-AMINATION OF IMIDE SODIUM SALTS WITH O-DIPHENYLPHOSPHINYLHYDROXYLAMINE (DPH): 7-AMINOTHEOPHYLLINE

(1H-Purine-2,6-dione, 7-amino-3,7-dihydro-1,3-dimethyl-)

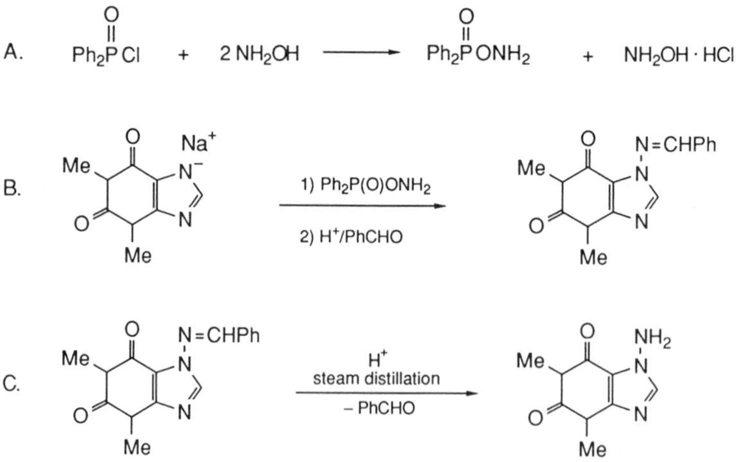

Submitted by W. KLÖTZER, J. STADLWIESER, and J. RANEBURGER[1]
Checked by MICHAEL J. LUZZIO and ANDREW S. KENDE

1. Procedure

A. *O-Diphenylphosphinylhydroxylamine.*[2] A 500-mL, round-bottomed flask, equipped with a reflux condenser, drying tube, an efficient mechanical stirrer, a dropping funnel, and a nitrogen-inlet tube, is charged with 300 mL of anhydrous methylene chloride, 16.5 g (0.5 mol) of hydroxylamine base (Note 1), and 1.0 g of dry sodium bicarbonate. While the suspension is stirred vigorously at −30°C (bath temperature), a solution of 52.06 g (0.22 mol) of diphenylphosphinyl chloride (Note 2) in 70 mL of anhydrous methylene chloride is added under a nitrogen atmosphere at a constant rate within 30 min. The resulting thick suspension is stirred at −30°C for 2 hr and for an additional 2 hr after the cooling bath is removed. The reaction mixture is filtered through a sintered-glass funnel (porosity 3) and the residue is washed with two 80-mL portions of methylene chloride. The methylene chloride is removed from the colorless solid by a stream of air for 2 hr. The dry solid, still on the funnel, is then mixed thoroughly with 200 mL of deionized water. The water is removed by suction. The same operation is performed sequentially with 150 mL of 5% aqueous sodium bicarbonate solution and then with two 150-mL portions of water. This solid, which retains water tenaciously, is dried by suction and by pressing down on the funnel for several hours, followed by drying in a phosphorus pentoxide-charged vacuum desiccator until its weight is constant (24 hr) to give 36 g (70%) of impure *O*-diphenylphosphinylhydroxylamine, mp 120–135°C, with decomposition.

A 500-mL, two-necked flask, equipped with a reflux condenser and a drying tube, is charged with 240 mL of anhydrous ethanol. The solvent is preheated to 70°C and a 12-g portion of this finely powdered dry product is added all at once. The resulting suspension is refluxed for 2–3 min when almost all of the solid has dissolved. The hot solution is filtered as quickly as possible through a sintered-glass funnel (porosity 3) and the filtrate is chilled to 0°C for 30 min. Isolation of the crystalline deposit and washing with 20 mL of ether provides 7.8 g of pure product. Recrystallization of three 12-g portions furnishes 23.4 g (44%) of *O*-diphenylphosphinylhydroxylamine, mp > 140°C, with decomposition (Note 3).

B. *7-Benzylideneaminotheophylline.* A 2000-mL, round-bottomed flask, equipped with an efficient mechanical stirrer, thermometer, and drying tube, is charged with 600 mL of anhydrous *N*-methylpyrrolidone (Note 4) and 20.2 g (0.1 mol) of anhydrous theophylline sodium salt (Note 5). The flask is cooled with an ice–salt bath to 0°C (internal temperature). Then 23.4 g (0.1 mol) of *O*-diphenylphosphinylhydroxylamine is added in three equal portions while the suspension is stirred vigorously. After the ice–salt bath is removed, the resulting viscous suspension is stirred for 6 hr at 20°C.

After the solution is diluted with 1200 mL of water, the pH is adjusted to 1–2 with concentrated hydrochloric acid and the mixture stirred at 5°C for 1 hr. The precipitated diphenylphosphinic acid is isolated by filtration and washed with 50 mL of water (Note 6). The filtrate is placed in a 2000-mL, round-bottomed flask, equipped with a reflux condenser and an efficient mechanical stirrer. A solution of 20 mL of benzaldehyde in 50 mL of ether is added and the mixture is stirred vigorously for 20 min. The precipitate that forms is isolated by filtration and washed sequentially with 50 mL of water and 50 mL of ether to yield 19.6 g (69%) of 7-benzylideneaminotheophylline, mp 207–209°C.[3] An analytical sample may be prepared by recrystallization from ethanol (mp 209°C).

C. *7-Aminotheophylline.* The reaction flask of a steam distillation apparatus is charged with 19.6 g (0.069 mol) of 7-benzylideneaminotheophylline and 100 mL (0.1 mol) of 1 N hydrochloric acid. The suspension is steam-distilled until no more benzaldehyde is detected in the distillate (Note 7). The resulting clear solution in the reaction flask is concentrated by rotary evaporation to a volume of 30 mL, adjusted to pH 10 with concentrated ammonium hydroxide, transferred to a separatory funnel, and extracted with five 60-mL portions of chloroform. The combined chloroform extracts are dried with anhydrous sodium sulfate, filtered, and concentrated to dryness by rotary evaporation. The residue is recrystallized from 75 mL of water to afford 11.3 g (84%) of 7-amino-theophylline, mp 222°C.[3]

2. Notes

1. Hydroxylamine base has been prepared by the method of Lecher and Hofmann.[4] The free base can be stored in a tightly stoppered flask at $-20°C$ for several days. The checkers found it expedient to prepare free hydroxylamine by a modification of the Lecher–Hofmann procedure in which a Schlenk tube under dry N_2 was used to filter the NaCl precipitate and the NH_2OH base was crystallized from the filtrate at $-30°C$, then isolated by inverting the Schlenk apparatus and filtering the product (74% yield from the hydrochloride).

2. Diphenylphosphinyl chloride can be purchased from Aldrich Chemical Company, Inc. or from EGA-Chemie, D-7924 Steinheim, West Germany (an Aldrich Chemical Company). Diphenylphosphinyl chloride can also be prepared by oxygen-mediated oxidation of diphenylchlorophosphine[5] (purchased from Fluka AG, CH-9470 Buchs, Switzerland).

3. The recrystallization should be performed as quickly as possible in portions below 15 g. Prolonged heating in ethanolic solution causes substantial losses. The pure, dry compound can be stored in a tightly stoppered flask at 0°C for at least 6 months without loss of aminating capacity. The submitters report that the pure compound showed no signs of spontaneous decomposition during 4 years of use, except when heated to $> 140°C$, where the compound decomposes with effervescence.

4. *N*-Methylpyrrolidone (purum grade) was purchased from Fluka AG, CH-9470 Buchs, Switzerland, dried over calcium hydride, and vacuum-distilled [bp 78–79°C (12 mm)].

5. The sodium salt of theophylline was obtained as follows: to a solution of 36.34 g (0.2 mol) of theophylline in 120 mL of 50% aqueous ethanol at 80°C was added 50 mL (0.2 mol) of aqueous 4 N sodium hydroxide. Chilling to 0°C, filtration of the precipitate, washing with 50 mL of 96% ethanol, then with 100 mL of ether, and drying in a vacuum desiccator over phosphorus pentoxide provides 28.0 g of the anhydrous salt.

6. The recovered and dried diphenylphosphinic acid, 19.6 g (90%), is ready to be recycled to diphenylphosphinyl chloride.[6,7]

7. Traces of benzaldehyde can be detected with Brady's reagent (2,4-dinitrophenyl-hydrazine sulfate solution) or by its characteristic smell.

3. Discussion

Electrophilic N-aminations of imide salts have been performed with hydroxylamine-O-sulfonic acid (HOSA),[8-10] O-(2,4-dinitrophenyl) hydroxylamine,[11,12] and O-mesitylenesulfonylhydroxylamine (MSH).[11] The use of HOSA is mainly restricted to aqueous reaction media.[8,9] O-(2,4-Dinitrophenyl)hydroxylamine, MSH, and O-diphenylphosphinylhydroxylamine (DPH) can be applied in anhydrous or even nonpolar solvents. O-(2,4-Dinitrophenyl)hydroxylamine and MSH require N-protected hydroxylamine for their preparation.[11,12] MSH has been found to be explosive.[13,14] DPH has the advantage of being prepared directly from unprotected hydroxylamine and seems to have no tendency toward spontaneous decomposition. The possibility of recycling diphenylphosphinic acid may be regarded as a further advantage. The advantage of using unprotected hydroxylamine to prepare DPH is partially negated by the required somewhat delicate preparation and handling of the free hydroxylamine base. The large amount of solvent that is sometimes required because of the low solubility of DPH and the resulting diphenylphosphinic acid salt may be regarded as a disadvantage, too.

O-Diphenylphosphinylhydroxylamine has also been used to aminate carbanions,[15,16] tertiary phosphines, and thio ethers.[2]

TABLE I

N-Amino Compounds from Imide Sodium Salts and DPH[a]

Educt Alkali Salt	Solvent[b]	Product	Yield (%)	Ref.
Imidazole	DMF	1-Aminoimidazole	28	3
2-Nitroimidazole	NMP	1-Amino-2-nitroimidazole	40	3
2-Methyl-4(5)-nitroimidazole	NMP	1-Amino-2-methyl-4-nitroimidazole	30	3
Theobromine	DMF	1-Aminotheobromine	71	3
Theophylline	NMP	7-Aminotheophylline	60	3
Phthalimide	DMF	N-Aminophthalimide	90	3

[a] DPH = O-diphenylphosphinylhydroxylamine.
[b] DMF = anhydrous dimethylformamide; NMP = anhydrous N-methylpyrrolidone.

1. Institut für Organische und Pharmazeutische Chemie, Universität Innsbruck, A-6020 Innsbruck, Innrain 52a, Austria.
2. The method represents a modification of a recently published preparation: Harger, M. J. P. *J. Chem. Soc., Perkin Trans. 1* **1981**, 3284.
3. Klötzer, W.; Baldinger, H.; Karpitschka, E. M.; Knoflach, J. *Synthesis* **1982**, 592.
4. Lecher, H.; Hofmann, J. *Chem. Ber.* **1922**, *55*, 912.
5. Tyssee, D. A.; Bausher, L. P.; Haake, P. *J. Am. Chem. Soc.* **1973**, *95*, 8066.
6. Kreutzkamp, N.; Schindler, H.; *Arch. Pharm. (Weinheim)* **1960**, *293*, 296; *Chem. Abstr.* **1964**, *60*, 4179g.
7. Higgins, W. A.; Vogel, P. W.; Craig, W. G. *J. Am. Chem. Soc.* **1955**, *77*, 1864.
8. Klötzer, W.; Herberz, M. *Monatsh. Chem.* **1965**, *96*, 1731; Klötzer, W. *Monatsh. Chem.* **1966**, *97*, 1117.

9. Broom, A. D.; Robins, R. K. *J. Org. Chem.* **1969**, *34*, 1025.
10. Wallace, R. G. *Aldrichimica Acta* **1980**, *13*, 3.
11. Tamura, Y.; Minamikawa, J; Ikeda, M. *Synthesis* **1977**, 1.
12. Sheradsky, T. *J. Heterocycl. Chem.* **1967**, *4*, 413.
13. Ning, R. Y. *Chem. Eng. News* **1973**, *51*, 36.
14. Taylor, E. C.; Sun, J.-H. *Synthesis* **1980**, 801.
15. Colvin, E. W.; Kirby, G. W.; Wilson, A. C. *Tetrahedron Lett.* **1982**, *23*, 3835.
16. Boche, G.; Bernheim, M.; Schrott, W. *Tetrahedron Lett.* **1982**, *23*, 5399.

ATROPALDEHYDE

(Benzeneacetaldehyde, α-methylene)

A. $C_6H_5CH{=}CH_2$ + $HCCl_3$ + $NaOH$ $\xrightarrow[\text{40-60°C}]{\text{cat.}}$ $C_6H_5CH\!-\!CH_2$ (ring with CCl_2)

B. $C_6H_5CH\!-\!CH_2$ (ring with CCl_2) + $2\,NaOH$ + $2\,C_2H_5OH$ $\xrightarrow[\text{reflux}]{\text{ethanol}}$ H, H $C{=}C$ $CH(OC_2H_5)_2$, C_6H_5

C. H, H $C{=}C$ $CH(OC_2H_5)_2$, C_6H_5 + H_2O $\xrightarrow[\text{4°C to –4°C}]{\text{HCOOH}}$ H, H $C{=}C$ CHO, C_6H_5

Submitted by INGOLF CROSSLAND[1]
Checked by THOMAS J. BLACKLOCK and ANDREW S. KENDE

1. Procedure

A. *1,1-Dichloro-2-phenylcyclopropane.* In a 1-L, three-necked, round-bottomed "Morton" flask (Note 1) equipped with a mechanical stirrer, a thermometer, and a reflux condenser are placed 57 mL (0.50 mol) of styrene, 50 mL of chloroform, 2 g of triethyl-benzylammonium chloride, 25 mL of methylene chloride, and a solution of 77 g of sodium hydroxide in 77 mL of water (Note 2). The mixture is stirred vigorously.

The temperature is allowed to rise to 40°C and then kept between 40 and 45°C by cooling with water (Note 3). After about an hour evolution of heat subsides, and the dark reaction mixture is heated to 55–60°C for an additional hour. The products are transferred to a 1-L separatory funnel with 250 mL of water and shaken. The organic layer is separated and the aqueous phase extracted with 25 mL of petroleum ether (Note 4). The organic fractions are combined, dried over anhydrous magnesium sulfate, filtered, concentrated in vacuo, and distilled through a 20-cm Vigreux column. A forerun at about 50°C (16 mm) consists mainly of styrene. Distillation of the remainder affords 80–82 g (86–88%) of dichlorophenylcyclopropane, bp 118–120°C (16 mm) (Note 5).

B. *Atropaldehyde diethyl acetal.* A mixture of 18.7 g (0.100 mol) of 1,1-dichloro-2-phenylcyclopropane, 16 g (0.40 mol) of sodium hydroxide, and 160 mL of ethanol is placed in a 250-mL flask fitted with a reflux condenser. The mixture is heated under reflux for 24 hr. Some bumping may occur. Water (200 mL) is added, and the mixture is extracted with three 30-mL portions of petroleum ether. The extracts are combined, dried as above with magnesium sulfate, concentrated in vacuo, and distilled through a 20-cm Vigreux column. The acetal begins to distill at about 70°C (0.5 mm), and the product (14–15 g) is collected until the temperature reaches about 100°C. Gas chromatographic analysis of the product shows it to be about 85% pure (yield 58–62%) (Note 6).

C. *Atropaldehyde.* The acetal (15 g), placed in a 100-mL flask fitted with a magnetic stirrer and a thermometer, is cooled to about 4°C in an ice bath. A mixture of 15 mL of formic acid and 4 mL of water is similarly cooled and added in one lot with stirring to the acetal. The temperature drops to about −4°C. The homogenous mixture is stirred for 60 sec and then quenched by adding 15 mL of petroleum ether and 25 mL of water. The mixture is transferred to a separatory funnel and thoroughly shaken. The aqueous phase is extracted with two additional 15-mL portions of petroleum ether, and the combined extracts are dried as above with magnesium sulfate and concentrated in vacuo (Note 7). A mixture of 10 mL each of petroleum ether and diethyl ether is added to the crude aldehyde, and the solution is cooled to about −50°C. After 15 min the colorless crystals are filtered and washed with a few milliliters of the solvent cooled to 0°C. The yield of vacuum-dried product (Note 7) is 5.8–6.8 g (71–83%). Recrystallization from a mixture of 10 mL each of diethyl ether and petroleum ether as above gives 5–6 g, mp 38–40°C (Notes 8 and 9).

2. Notes

1. The checkers used a 1-L, three-necked "Morton" flask[2] containing deep vertical creases for more efficient mixing and temperature control. The sodium hydroxide solution was added to the stirred reaction mixture through the reflux condenser in one portion.

2. The submitters used Merck styrene 99%, stabilized with 4-tert-butylpyrocatechol. Triethylbenzylammonium chloride was prepared by refluxing equimolar amounts of triethylamine and benzyl chloride in ethanol for 2 hr and removing the solvent in vacuo. The salt is commercially available. The other reagents were technical grade.

3. The reaction is exothermic, and it may be a problem to keep the temperature low enough during the first 5 min. A bath with cold water must be kept ready below the flask. The methylene chloride may be added to moderate the reaction.

4. The organic layer is heavier than the aqueous phase. It may be necessary to allow the mixture to stand for an hour before the phases separate.

5. The product shows ^{1}H NMR (60 MHz, CCl$_4$) δ: 1.87 (q, 2), 2.83 (triplet, 1), 7.20 (singlet, 5). The reported bp is 118–119°C (16 mm) or 114°C (13 mm).[3,4] Gas chromatographic analysis indicates the product to be 99% pure, n_D^{23} 1.551.

6. The product shows ^{1}H NMR (90 MHz, CDCl$_3$) δ: 1.18 (triplet J = 7 Hz, 6), 3.58 (multiplet, 4), 5.22 (singlet, 1), 5.54 (singlet, 2), 7.18–7.58 (multiplets, 5). Minor resonance signals near the foot of the triplet revealed the presence of 1-phenyl-2,2-diethoxycyclopropane as the major by-product. Gas chromatographic analyses on a 44-m SCOT column, 150–200°C, indicated that the crude product contained the ketal (10%), the starting material (2%), and an unidentified compound (2%). The submitters have not observed polymerization or other deterioration of the crude acetal when it was stored without special precautions in the laboratory for 2 months at ca. 20°C.

7. The aldehyde must be kept cold; see Note 8. If the solution is cooled much below room temperature, crystallization of the aldehyde may take place and render some of the manipulations difficult. The crystals may be dissolved in diethyl ether.

8. The product shows ^{1}H NMR (90 MHz, CDCl$_3$) δ: 6.11 (singlet, 1), 6.56 (singlet, 1), 7.36 (multiplet, 5), 9.72 (singlet, 1). Mass spectrum (70 eV, m/e, relative intensity): 132 (M$^+$, 51%), 104 (69%), 86 (100%), 78 (13%), 77 (35%). The crystalline aldehyde is unstable at room temperature. When kept for 24 hr in a vacuum-sealed ampoule at 20°C, the crystals slowly deliquesce. The aldehyde may, however, be kept at −6°C for 10 days without any observable deterioration.

9. Procedures B and C work well on a larger scale. Thus atropaldehyde was obtained in 20–26-g quantities from 0.5 mol of styrene (30–39%).

3. Discussion

The method presented here is a simple procedure for the preparation of pure atropaldehyde via its stable acetal, starting from inexpensive chemicals.

The described synthesis of 1,1-dichloro-2-phenylcyclopropane is a slightly modified version of published procedures.[3,4]

Syntheses of acetals of atropaldehyde have been reported previously, but all required either multistep sequences or difficultly accessible starting materials.[5,6] Thus the ethylene glycol acetal has been prepared from 2-phenylpropanal in a three-step procedure.[5] Ring openings of dihalocyclopropanes to give acetals are well known.[7-10] The reaction of 1,1-dichloro-2-phenylcyclopropane with methanolic sodium methoxide has been shown to give 1-phenyl-2,2-dimethoxycyclopropane.[11]

The only described preparatively useful route to atropaldehyde is the hydrolysis of the ethylene glycol acetal mentioned above.[5] The present method is fast and affords labile aldehyde that is pure enough to allow crystallization. A cyclopropene is suggested to be an intermediate in the ring-opening reaction.[12] Nitro-substituted atropaldehydes have been prepared via a Mannich reaction.[13]

Formally the reactions amount to an α-formylation of styrene. The homologous aldehyde may be prepared from propenylbenzene.[10]

1. Institute of Organic Chemistry, The Technical University of Denmark, Building 201, DK 2800 Lyngby, Denmark.
2. Morton, A. A.; Knott, D. M. *Ind. Eng. Chem. Anal. Ed.* **1941**, *13*, 649.
3. Juliá, S.; Ginebreda, A. *Synthesis* **1977**, 682–683.
4. Makosza, M.; Wawrzyniewicz, M. *Tetrahedron Lett.* **1969**, 4659–4662.
5. Elkik, E. *Bull. Soc. Chim. Fr.* **1968**, 283–288.
6. Normant, H. *C. R. Hebd Seances Acad. Sci. Ser. C* **1955**, *240*, 1435; *Chem. Abstr.* **1956**, *50*, 3424g.
7. Skattebøl, S. *J. Org. Chem.* **1966**, *31*, 1554–1559.
8. Nerdel, F.; Buddrus, J.; Brodowski, W.; Hentschel, P.; Klamann, D.; Weyerstahl, P. *Justus Liebigs Ann. Chem.* **1967**, *710*, 36–58.
9. Tobey, S. W.; West, R. *J. Am. Chem. Soc.* **1966**, *88*, 2478–2481.
10. Henseling, K.-O.; Quast, D.; Weyerstahl, P. *Chem. Ber.* **1977**, *110*, 1027–1023.
11. Shields, T. C.; Gardner, P.D. *J. Am. Chem. Soc.* **1967**, *89*, 5425–5428.
12. Crossland, I. *Acta Chem. Scand.* **1987**, *B41*, 310–312.
13. Hengartner, U.; Batcho, A. D.; Blount, J. F.; Leimgruber, W.; Larscheid, M. E.; Scott, J. W. *J. Org. Chem.* **1979**, *44*, 3748–3752.

AZULENE

A.

B.

Submitted by KLAUS HAFNER and KLAUS-PETER MEINHARDT[1]
Checked by STEPHEN G. SENDEROFF and ANDREW S. KENDE

1. Procedure

A 4-L, three-necked, round-bottomed flask equipped with a mechanical stirrer, 500-mL pressure-equalizing dropping funnel, thermometer, and reflux condenser provided with a calcium chloride drying tube is charged with 202.6 g (1.0 mol) of 1-chloro-2,4-dinitrobenzene (Note 1) and 1.2 L of dry pyridine (Note 2). The mixture is heated while it is stirred in a water bath to 80–90°C for 4 hr, during which time a thick yellow precipitate of N-(2,4-dinitrophenyl)pyridinium chloride is formed (Note 3). After cooling to 0°C a solution of 100.0 g (2.22 mol) of dimethylamine in 300 mL of dry pyridine was prechilled to 0°C and added dropwise over a period of about 30 min with stirring. The resulting brownish-red liquid reaction mixture is allowed to warm to room temperature and stirring is continued for 12 hr. The drying tube is replaced by a gas inlet and the system is flushed with dry nitrogen in a hood. Under nitrogen, 70.0 g (1.06 mol) of ice-cold, freshly distilled cyclopentadiene (Note 4) is added, and subsequently 400 mL of 2.5 *M* sodium methoxide solution (Note 5) is slowly added dropwise to the stirred reaction mixture. During addition of the sodium methoxide, the temperature rises to 35–40°C. After the addition is completed, stirring is continued for another 4 hr. The reaction vessel is immersed in an oil bath, the dropping funnel removed, and the flask is fitted with a distillation head. The stirred mixture is cautiously heated under nitrogen (Note 6), and a mixture of pyridine and methanol is distilled off until the temperature of the reaction mixture has increased to 105–110°C (Note 7). After the distillation head is removed and 1 L of dry pyridine added, the black mixture is heated with stirring under a nitrogen atmosphere for 4 days with a bath temperature of 125°C. It is then cooled to 60°C, the reflux condenser is replaced by a distillation head, and pyridine is removed under reduced pressure (Note 8). The gummy black solid residue is removed by a spatula and rinsed with hexanes. It is extracted in a Soxhlet apparatus with 1.5 L of hexanes in several batches. To remove the remaining pyridine, the combined blue hexane rinse and the extraction solutions are carefully washed with two 150-L portions of 10% aqueous hydrochloric acid, then water (Note 9). The organic layer is dried with anhydrous sodium sulfate, the drying agent is removed by filtration, and the solvent is distilled through a 50-cm vacuum-jacketed Vigreux column. The crude azulene is purified by chromatography on activity II alumina (Note 10) with hexane and yields azulene as blue plates, mp 96–97°C, yield 65–75 g (51–59%) (Note 11).

2. Notes

1. Commercial 1-chloro-2,4-dinitrobenzene was obtained from Aldrich Chemical Company, Inc. (Milwaukee) or from Bayer, AG (Leverkusen, FRG) and used directly.

2. Commercial pyridine was dried over potassium hydroxide or calcium hydride and distilled prior to use. The checkers used reagent-grade pyridine (Mallinckrodt AR), which was distilled from KOH and stored over Linde 4A molecular sieves.

3. The reaction mixture should be evenly warmed to 80–90°C within 30 min with efficient mechanical stirring to prevent caking or "hot spots."

4. Dicyclopentadiene, obtained from the Aldrich Chemical Company, Inc. (or E. Merck, Darmstadt, FRG), was cracked just prior to use according to the procedure of Fieser and Williamson,[2] to give the monomer, bp 40–42°C.

5. Sodium methoxide was prepared just prior to use from 23.0 g (1.0 g-atom) of sodium metal and 400 mL of anhydrous methanol (distilled from magnesium turnings), then cooled to room temperature.

6. *Caution!* Dimethylamine is evolved.

7. Approximately 600 mL of distillate will be collected.

8. The blue pyridine distillate is redistilled through a 50-cm vacuum-jacketed Vigreux column (to avoid loss of azulene) until approximately 1.7 L is collected; the residual azulene is combined with the main residues for extraction.

9. A total volume of 2 L of hexane washes results, accompanied by the gradual precipitation of a yellow solid from the hexane washes. The acid-wash procedure frequently leads to emulsions and gummy yellow solid in both phases; back-extraction of the "aqueous" layer with hexane may be necessary.

10. Alumina was purchased from Macherey, Nagel and Co., Düren [Federal Republic of Germany (FRG)]. The checkers employed 650 g of neutral alumina (Fisher, adsorption grade, 80–200 mesh) packed in a 40-cm-high column. Yellow impurities remained on the column, while the blue azulene came off with the hexane solvent front.

11. Further purification of azulene may be achieved by sublimation at reduced pressure, mp 99°C.[3] The checkers found that mechanical losses, particularly as mentioned in Note 9, lead to reduction in yield with reduction in scale (0.1 mol, 39% yield; 0.5 mol, 43% yield; 0.8 mol, 79% yield).

3. Discussion

Azulene has been synthesized by a variety of methods: by dehydrogenation of hydroazulenes,[3] by annelation of a seven-membered ring on a five-membered ring either by ring-closure of vinylogous aminopentafulvenes,[4,5] or by cycloadditions of aminopentafulvenes with activated 1,3-dienes or alkynes,[6] and by annelation of a five-membered ring on a seven-membered ring starting from troponoids or heptafulvenes.[7,3b] Of these, the Ziegler–Hafner synthesis of azulene[4] by thermal cyclization of the 6-(4-methylanilino-1,3-butadienyl) pentafulvene proved to be the most versatile. Azulene is also simply prepared from 6-dimethylaminopentafulvene and thiophene 1,1-dioxide or from 6-acyloxypentafulvenes and 1-diethylaminobutadiene, but with lower yields.[6b,d,e]

The present procedure, based on the Ziegler–Hafner synthesis, is simple and avoids the use of benzidine for the ring closure of the pentafulvene and isolation of the 5-dimethylamino-2,4-pentadienylidenedimethyliminium perchlorate.[8] Other amines were also checked; with N-methylaniline, ring closure of the resulting pentafulvene in pyridine failed; and with diethylamine, a delay in boiling can take place during the reaction.

Substituted azulenes can be prepared in the same manner by the use of substituted cyclopentadienes or substituted pentamethinium salts.

1. Institut für Organische Chemie der Technischen Hochschule, Petersenstr. 22, D-6100 Darmstadt (FRG).
2. Fieser, L. F.; Williamson, K. L. In "Organic Experiments," 3rd ed.; D. C. Heath and Co.: Lexington, MA, 1975; pp 118–120.
3. (a) Treibs, W.; Kirchhof, W.; Ziegenbein, W. *Fortschr. Chem. Forsch.* **1955**, *3*, 334; Keller-Schierlein, W.; Heilbronner, E. In "Non-Benzenoid Aromatic Compounds," Ginsburg, D., Ed.; Interscience: New York, 1959; Chapter 6, p. 277; (b) Nozoe, T.; Ito, S. *Fortschr. Chem. Org. Naturst.* **1961**, *19*, 32.
4. (a) Ziegler, K.; Hafner, K. *Angew. Chem.* **1955**, *67*, 301; (b) Hafner, K. *Liebigs Ann. Chem.* **1957**, *606*, 79.
5. Jutz, J. C. In *Top. Curr. Chem.* **1978**, *73*, 125.
6. (a) Sato, M.; Ebine, S.; Tsunetsugu, J. *Tetrahedron Lett.* **1974**, 2769; (b) Copland, D.; Leaver, D.; Menzies, W. B. *Tetrahedron Lett.* **1977**, 639; (c) Severin, T.; Ipach, I. *Synthesis* **1978**,

592; (d) Mukherjee, D.; Dunn, L. C.; Houk, K. N. *J. Am. Chem. Soc.* **1979**, *101*, 251; Gupta, Y. N.; Mani, S. R.; Houk, K. N. *Tetrahedron Lett.* **1982**, 495.

7. (a) Nozoe, T.; Seto, S.; Matsumura, S.; Murase, Y. *Bull. Chem. Soc. Jpn.* **1962**, *35*, 1179, 1990; (b) Oda, M.; Kitahara, Y.; *J. Chem. Soc., Chem. Commun.* **1969**, 352; (c) Ehntholt, D. J.; Kerber, R. C. *J. Chem. Soc., Chem. Commun.* **1970**, 1451; (d) Yang, P. W.; Yasunami, M.; Takase, K. *Tetrahedron Lett.* **1971**, 4275.

8. (a) König, W.; Regner, W. *Ber. Dtsch. Chem. Ges.* **1930**, *63*, 2823; (b) Malhotra, S. S.; Whiting, M. C. *J. Chem. Soc.* **1960**, 3812.

REDUCTION OF QUINONES WITH HYDRIODIC ACID: BENZ[a]ANTHRACENE

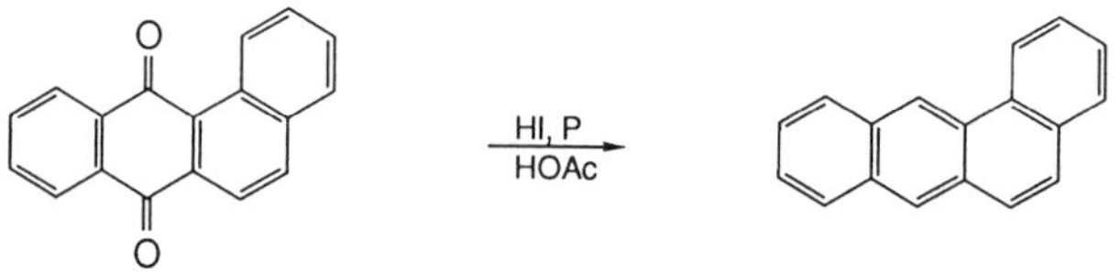

Submitted by MARIA KONIECZNY and RONALD G. HARVEY[1]
Checked by GREGORY A. REED and CARL R. JOHNSON

1. Procedure

Caution! Benzene has been identified as a carcinogen; OSHA has issued emergency standards on its use. All procedures involving benzene should be carried out in a well-ventilated hood, and glove protection is required. Benz[a]anthracene is also a known carcinogen.

A 500-mL, one-necked, round-bottomed flask is equipped with a magnetic stirring bar and an efficient condenser, and charged with 10.3 g (0.04 mol) of benz[a]anthracene-7,12-dione (Note 1), 5 g (0.16 mol) of red phosphorus (Note 2), and 100 mL of glacial acetic acid. The stirred suspension is heated to reflux, and 60 mL of 56% hydriodic acid (ca. 0.44 mol) (Note 3) is introduced through the condenser. The suspension is heated at reflux for 24 hr. The hot reaction mixture is poured into 500 mL of distilled water containing ∼30 g of sodium bisulfite. The suspension is stirred for 16 hr and filtered. The dry filter cake is transferred to a beaker and treated with sufficient hot dichloromethane (∼ 120 mL) to dissolve all of the benz[a]anthracene, and the mixture is filtered once again to remove the residual phosphorus. The volume of the filtrate is reduced to 40 mL. The solution is adsorbed on basic alumina, activity I (Note 4). A 2-cm × 40-cm chromatography column is slurry-packed with ca. 10 g of basic alumina and the benz[a]anthracene adsorbed on alumina is added to the top of the column. Elution with 5% benzene in hexane (occasional rinsing of the column tip with benzene to remove crystallized product may be necessary) and evaporation of the solvent in a rotary evaporator affords 7.7–7.9 g (84–87%) of pure, white benz[a]anthracene, mp 159.5–160°C (Note 5).

2. Notes

1. Benz[a]anthracene-7,12-dione, available from Eastman Organic Chemicals, was used without further purification.

2. Phosphorus, which serves to scavenge the I_2 produced, can be omitted. However, the product tends to retain traces of a yellow impurity that is difficult to remove.

3. The hydriodic acid employed was a 56% aqueous solution preserved with ~0.8% hypophosphorous acid obtained from Fisher Scientific Co. Once a bottle is opened, the contents tend to deteriorate, becoming dark-colored in less than 2 days. However, shelf life can be extended indefinitely if the container is purged with dry nitrogen before resealing.

4. Alumina sufficient to adsorb the complete solution is added, then the solvent is removed under vacuum. While benz[a]anthracene, mp 157–158°C, sufficiently pure for most purposes, can be obtained by crystallization of the crude product from ethanol–water, "filtration" through alumina removes residual, colored impurities, affording a pure, white product.

5. Pure benz[a]anthracene has been reported to melt at 158–159°,[2] 160°,[3] and 167°C.[4] The submitters report a melting point of 164–164.5°C. The submitters conducted this preparation on a scale five times larger and reported yields of up to 95%.

3. Discussion

The synthetic procedure described is based on that reported earlier for the synthesis on a smaller scale of anthracene, benz[a]anthracene, chrysene, dibenz[a, c]anthracene, and phenanthrene[5] in excellent yields from the corresponding quinones.[6] Although reduction of quinones with HI and phosphorus was described in the older literature, relatively drastic conditions were employed and mixtures of polyhydrogenated derivatives were the principal products.[7] The relatively milder experimental procedure employed herein appears generally applicable to the reduction of both *ortho-* and *para-* quinones directly to the fully aromatic polycyclic arenes. The method is apparently inapplicable to quinones having an olefinic bond, such as o-naphthoquinone, since an analogous reaction of the latter provides a product of undetermined structure (unpublished result). As shown previously,[6] phenols and hydroquinones, implicated as intermediates in the reduction of quinones by HI, can also be smoothly deoxygenated to fully aromatic polycyclic arenes under conditions similar to those described herein.

Although previous experience indicates that phosphorus is not essential for these reductions,[6,8] purification of the product is more difficult with its omission. With hydrocarbons sensitive to further reduction, phosphorus can have a deleterious effect through promotion of hydrogenation of the desired product. Whether or not phosphorus should be employed in an individual case will be dictated by experience with the particular compound and by the degree of purity required.

While the reduction of polycyclic quinones to phenols, hydroquinones, dihydrodiols, dihydro arenes, and arenes by a variety of reagents has been described, no entirely satisfactory general method is currently available for reduction directly to the fully aromatic arenes. Reagents previously employed for this purpose include LiAlH$_4$,[9,10] NaBH$_4$,[11] NaBH$_4$-BF$_3$,[11a,12] diborane,[12] aluminum and cyclohexanol,[13] zinc dust distillation,[14] and diphenylsilane.[15] These methods commonly furnish lower yields and are less general than the present procedure.

1. The Ben May Institute, The University of Chicago, 5841 S. Maryland Ave., Hospital Box 24, Chicago, IL 60637.
2. Clar, E. "Polycyclic Hydrocarbons," Vol. 1, Academic Press: New York, 1964, p. 311.
3. Fieser, L. F.; Hershberg, E. B. *J. Am. Chem. Soc.* **1937**, *59*, 2502.
4. I. G. Farbenind, A.-G. Ger. Patent 481819, 1925; 486766, 1925; *Chem. Abstr.* **1930**, *24*, 2139, 1390.
5. Reduction of phenanthrene-9,10-dione with HI in acetic acid afforded 9-hydroxyphenanthrene as sole product, while an analogous reaction without acetic acid furnished phenanthrene essentially quantitatively.
6. Konieczny, M.; Harvey, R. G. *J. Org. Chem.* **1979**, *44*, 4813.
7. According to Clar, "the reduction of quinones with hydriodic acid and red phosphorus at 200°C yields mostly hydrogenated hydrocarbons. These can be dehydrogenated by sublimation over copper at 400°C"; see reference 2, p. 171.
8. Konieczny, M.; Harvey, R. G. *J. Org. Chem.* **1980**, *45*, 1308.
9. Davies, W.; Porter, Q. N. *J. Chem. Soc.* **1957**, 4967.
10. Harvey, R. G.; Goh, S. H.; Cortez, C. *J. Am. Chem. Soc.* **1975**, *97*, 3468; Harvey, R. G.; Fu, P. P. In "Polycyclic Hydrocarbons and Cancer: Environment, Chemistry, and Metabolism," Gelboin, H. V.; Ts'o, P. O. P., Eds.; Academic Press: New York, 1978; Vol. I, Chapter 6, p. 133.
11. (a) Rerick, M. In "Reduction," Augustine, R. L.; Ed.; Marcel Dekker; New York, 1968; p. 39; (b) Criswell, T. R.; Klanderman, B. H. *J. Org. Chem.* **1974**, *39*, 770; (c) Cho, H.; Harvey, R. G. *J. Chem. Soc., Perkin Trans. 1* **1976**, 836.
12. Bapat, D. S.; Subba Rao, B. C.; Unni, M. K.; Venkataraman, K. *Tetrahedron Lett.* **1960**, 15.
13. Coffey, S.; Boyd, V. *J. Chem. Soc.* **1954**, 2468.
14. Reference 2, p. 161.
15. Gilman, H.; Diehl, J. *J. Org. Chem.* **1961**, *26*, 4817.

CONVERSION OF KETONES TO CYANOHYDRINS: BENZOPHENONE CYANOHYDRIN

(Benzeneacetonitrile, α-hydroxy-α-phenyl)

Submitted by PAUL G. GASSMAN and JOHN J. TALLEY[1]
Checked by TOD HOLLER and GEORGE BÜCHI

1. Procedure

A. *O-(Trimethylsilyl)benzophenone cyanohydrin.* A 250-mL, one-necked flask equipped with a reflux condenser, magnetic stirring bar, and drying tube is charged with 22.0 g (0.12 mol) of benzophenone (Note 1), 13.9 g (0.14 mol) of trimethylsilyl cyanide (Note 2), 600 mg (1.9 mmol) of anhydrous zinc iodide (Note 3), and 50 mL of dry methylene chloride (Note 4). The solution is heated at 65°C in an oil bath for 2 hr (Notes 5 and 6). The solvent is removed on a rotary evaporator to yield 36.4–37.9 g of crude product (Note 7), which is used in the next step without purification (Note 8).

B. *Benzophenone cyanohydrin.* To the 250-mL flask, which contains the crude *O*-(trimethylsilyl)benzophenone cyanohydrin, is added 50 mL of tetrahydrofuran (Note 9) and 30 mL of 3 *N* hydrochloric acid. The mixture is heated at 65°C (oil bath temperature) for 1 hr. The solution is poured into a separatory funnel and 30 mL of water is added. The aqueous phase is separated and back-extracted with three 100-mL portions of diethyl ether. The ethereal extracts are combined with the tetrahydrofuran solution and dried over anhydrous magnesium sulfate, filtered, and solvent is removed by evaporation on a rotary evaporator to give a yellow solid. The material is recrystallized from 300 mL of toluene and dried at a pressure of 0.05 mm overnight to give 17.7–18.8 g of white crystals. Concentration of the mother liquors to 100 mL produced a second crop of 1.1–3.8 g for a combined yield of 17.9–21.5 g (79–86%), mp 127–130°C (Note 10).

2. Notes

1. Benzophenone was purchased from Distillation Products (Eastman Organic Chemicals) and used without purification.

2. Trimethylsilyl cyanide was prepared shortly before use according to the procedure of Livinghouse, T. *Org. Synth. Coll. Vol. VII* **1990,** 517.

3. Anhydrous zinc iodide was purchased from Alfa Products, Ventron Corporation, and used without further purification.

4. The use of this solvent is optional. The reaction can be carried out in the absence of solvent without significant change in yield. For certain unhindered ketones the solvent is helpful in dissipating the heat generated in the reaction.

5. For certain unhindered ketones external cooling may be necessary instead of heating due to the exothermicity of the reaction.

6. The checkers followed the reaction by IR spectroscopy and found that the benzo-phenone carbonyl peak (1640 cm^{-1}) disappeared after 2 hr.

7. If purification is desired, it may be achieved by vacuum distillation of this crude product, bp 104°C (0.5 mm).

8. On prolonged standing the crude product appears to undergo some decomposition. Thus it should be used directly in the next step for maximum yield.

9. For many analogs the use of solvent and/or heating is not necessary. However, both solvent and heating are necessary for hindered cyanohydrins, such as the one described in this procedure.

10. The submitters report a melting point of 131–132.5°C. The melting point has been previously reported to be 127–130°C.[2]

3. Discussion

Traditionally cyanohydrins have been prepared by processes that require the establishment of an equilibrium between a ketone and its corresponding cyanohydrin. For many ketones, especially those that are sterically hindered, the position of the equilibrium is unsatisfactory for the effective synthesis of the cyanohydrin. We describe herein a general method for the synthesis of cyanohydrins that does not depend on an equilibrium process. As a result this synthetic procedure can be used to convert a wide variety of dialkyl, diaryl, and arylalkyl ketones into their corresponding cyanohydrins. In addition to the described conversion of benzophenone into its cyanohydrin, acetophenone, fluorenone, *tert*-butyl phenyl ketone, and a wide variety of aliphatic ketones have been converted into cyanohydrins by this general procedure (Table I).[3]

O-(Trimethylsilyl)benzophenone cyanohydrin has been prepared previously by the addition of trimethylsilyl cyanide to benzophenone using an aluminum chloride catalyst.[4] The preparation of cyanohydrin silyl ethers described in the synthesis is based on the general procedure of Evans, Carroll, and Truesdale.[5] Trimethylsilyl cyanide has been prepared also by Zubrick, Dunbar, and Durst.[6] The overall procedure is that of Gassman and Talley.[3]

Benzophenone cyanohydrin has been synthesized previously by the nitrogen dioxide oxidation of 1,2-dicyanotetraphenylethane.[2]

TABLE I

PREPARATION OF CYANOHYDRINS

Product	Yield (%)	mp [bp] (°C)
Cyclohexanone cyanohydrin	90	27–28 [63° (10^{-6} mm)]
Cyclooctanone cyanohydrin	93	115–116
Fluorenone cyanohydrin	98	118.5–120
tert-Butyl phenyl ketone cyanohydrin	99	82–83
p-Chloroacetophenone cyanohydrin	94	91.5–92.5
p-Nitroacetophenone cyanohydrin	89	112–113
p-Acetylbenzonitrile cyanohydrin	95	77.5–78.5
p-Methoxyacetophenone cyanohydrin	96	78–80
p-Methylacetophenone cyanohydrin	97	79.5–80

1. Department of Chemistry, University of Minnesota, Minneapolis, MN 55455.
2. Wittig, G.; Pockels, U. *Chem. Ber.* **1936**, *69*, 790–792.
3. Gassman, P. G.; Talley, J. J. *Tetrahedron Lett.* **1978**, 3773–3776.
4. Liddy, W.; Sundermeyer, W. *Chem. Ber.* **1973**, *106*, 587–593.
5. Evans D. A.; Carroll, G. L.; Truesdale, L. K. *J. Org. Chem.* **1974**, *39*, 914–917.
6. Zubrick, J. W.; Dunbar, B. I.; Durst, H. D. *Tetrahedron Lett.* **1975**, 71–74.

1,3-OXAZEPINES VIA PHOTOISOMERIZATION OF HETEROAROMATIC N-OXIDES: 3,1-BENZOXAZEPINE

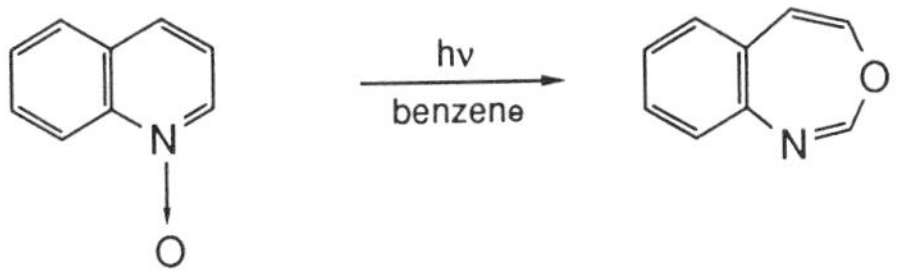

Submitted by ANGELO ALBINI, GIAN FRANCO BETTINETTI, and GIOVANNA MINOLI[1]
Checked by PATRICK MACMANUS and ROBERT M. COATES

1. Procedure

Caution! 3,1-Benzoxazepine is a strong lacrimator and a moderate skin irritant. The preparation should be carried out in a well-ventilated hood. The apparatus should be shielded to avoid exposure to ultraviolet light.

Benzene has been identified as a carcinogen; OSHA has issued emergency standards on its use. All procedures involving benzene should be carried out in a well-ventilated hood, and glove protection is required.

Irradiation is carried out in a round-bottomed, cylindrical, Pyrex vessel (Note 1) equipped with a Pyrex immersion well (Note 2), nitrogen inlet, distillation sidearm, a small sidearm fitted with a rubber septum for removing aliquots, and a magnetic stirring bar. The flask is charged with 12 g (0.066 mol) (Note 3) of quinoline N-oxide dihydrate (Note 4) and 1.3 L of dry benzene (Note 5). The mixture is stirred and heated to boiling with a heating mantle as a slow stream of nitrogen (Note 6) is bubbled into the vessel. Benzene is distilled through the sidearm until the distillate is perfectly clear (Note 7). The lamp (Note 8) is placed in the immersion well, water is circulated through the cooling jacket, and the nitrogen flow is adjusted as necessary to maintain an outward flow of gas while the light yellow solution cools to room temperature (Note 9). The solution is stirred vigorously (Note 10) and irradiated for 2.5–3 hr, at which time the N-oxide is largely consumed (Note 11). The orange solution is transferred to a 1-L, round-bottomed flask, and the benzene is removed by rotary evaporation at room temperature. The red-orange oily residue, which contains some solid, is extracted with three 40-mL portions of dry cyclohexane, the combined cyclohexane extracts are evaporated under reduced pressure at room temperature, and the extraction operation is repeated on the oil thus obtained (Note 12). Evaporation of the combined cyclohexane extracts affords 6.1–6.3 g (63–65%) of crude 3,1-benzoxazepine (Note 13). Bulb-to-bulb distillation in a Kugelrohr apparatus at 0.2 mm with an oven temperature of 80°C affords 4.7–4.8 g (49–50%) of 3,1-benzoxazepine as a pale yellow oil, n_D^{24} 1.6074 (Notes 14 and 15).

2. Notes

1. The apparatus used by the checkers for irradiations at a 9-g scale was 33 cm high and 9 cm in diameter and had a 60/50 ⊤ joint at the top for the immersion well. The joints for the distillation and sampling sidearms were 24/40 and

14/20, respectively. The gas inlet was located about 6 cm from the bottom of the vessel to accommodate the use of a heating mantle. A disk of coarse, sintered glass was sealed into the gas inlet near its point of attachment. The capacity of the vessel with the immersion well in place was ca. 900 mL. The submitters used a similar but flat-bottomed apparatus of 1.2-L capacity. The flat-bottomed vessel facilitates vigorous stirring, but it does not fit as well into the heating mantle and may therefore be somewhat hazardous to use.

2. The checkers used a Vycor immersion well and a Pyrex filter sleeve. The immersion well, 450-W mercury lamp, and the requisite transformer are available from Hanovia Lamp Division, Canrad-Hanovia Inc, 100 Chestnut Street, Newark, NJ 07105.

3. The checkers carried out the irradiation on a 9-g scale in 750 mL of benzene after azeotropic distillation.

4. Quinoline N-oxide dihydrate is supplied by Aldrich Chemical Company, Inc. and EGA Chemie KG, Steinheim/Albuch, Germany. The submitters prepared the compound by the procedure of Hayashi[2] with minor modifications. The water of hydration may be removed under reduced pressure in a drying pistol. However, since the anhydrous N-oxide is very hygroscopic, the submitters have found that it is more expedient to use the dihydrate and remove the water by azeotropic distillation in the irradiation vessel.

5. The checkers dried the benzene by distillation from calcium hydride immediately before use. The submitters report that toluene may be used instead of benzene; however, since the product is not very thermally stable, they advise that the toluene should be evaporated without heating above 35–40°C during the isolation.

6. Other dry gases may be used. The submitters report that the reaction is not quenched by oxygen.

7. A total of ca. 150–300 mL was collected. The distillation time may be reduced by insulating the vessel and sidearm with glass wool.

8. The submitters used a Helios Italquartz 500-W lamp which has emission characteristics similar to those of the Hanovia 450-W medium pressure mercury lamp used by the checkers.

9. The checkers noticed that a thin film of oil that was evidently quinoline N-oxide deposited on the surface of the immersion well and irradiation vessel during cooling.

10. Vigorous stirring is essential for optimum yields. The checkers obtained lower isolated yields (ca. 32–33%) in two runs in which a relatively slow stirring rate was employed. The low yields were probably caused in part by deposition of oil on the surface of the immersion well and the resulting interference with the transmission of UV light into the solution.

11. The submitters emphasize the importance of terminating the irradiation before all of the N-oxide is consumed. Overirradiation gives rise to a more complicated mixture of products from which the product can no longer be isolated by the simple extraction procedure described.

The progress of the irradiation was determined by the checkers by proton NMR analysis. At appropriate intervals 5-mL aliquots were removed, the solvent was evaporated, and hexamethylbenzene was added as an internal standard. The ratio of N-oxide, benzoxazepine, and hexamethylbenzene was determined from integration of the resonances at δ 2.26 (s, 18H), 5.55 (d, 1H, $J = 6$), and 8.46 (d, 1H, $J = 6$), respectively, in chloroform-d. After 2.5–3 hr of irradiation the amount of benzoxazepine present was ca. 60–68% of theoretical and ca. 10% of starting N-oxide remained.

The submitters followed the course of the irradiation by TLC analysis on silica gel with 5% (v/v) methanol in chloroform as developing solvent. Since some by-products have R_1 values coincident with the N-oxide, this spot will not completely disappear and caution must be exercised to avoid overirradiation.

12. This extraction procedure separates most of the carbostyril, that is, 2(1H)-quinolinone, which is formed to the extent of ca. 20% in the irradiation. The submitters have isolated the carbostyril by-product by crystallization of the extraction residue from 95% ethanol in runs carried out to high conversion. Alternatively, the carbostyril may be isolated by chromatography of the crude product on silica gel with 5% methanol-chloroform as eluant. However, the benzoxazepine cannot be obtained by this method since it undergoes hydrolysis during the chromatography.

13. The purity of the crude product is about 90% according to NMR analysis, the remaining material being mostly unchanged N-oxide.

14. The spectral properties of the product are as follows: IR (liquid film) cm^{-1}: 1665 (C=N), 1630 (C=C), 1480 (sharp), 1440 (sharp), 1035 (strong), 765 (strong); ^{1}H NMR (CDCl$_3$) δ: 5.55 (d, 1 H, $J = 6$, CH=CH—O), 5.84 (d, 1 H, $J = 6$, CH=CH—O), 6.44 (s, 1 H, O—CH=N), 6.96 (m, 4 H, aromatic protons).

15. Samples of the product stored in tightly stoppered flasks in a freezer at $-20°$C for several weeks showed no sign of decomposition.

3. Discussion

The preparation of 3,1-benzoxazepines by photochemical isomerization of quinoline N-oxides constitutes a rather general entry into this class of seven-membered heterocycles. Since the structure of the photoisomer of 2-phenylquinoline N-oxide was first recognized as 2-phenyl-3,1-benzoxazepine by Buchardt et al.,[3] the scope of this method for oxidative ring expansion of six-membered heterocyclic N-oxides to 1,3-oxazepines has been extensively explored.[4] For example, irradiation of 2-cyano-, 2-phenyl-, and 2-methoxyquinoline N-oxides affords the corresponding 2-substituted 3,1-benzoxazepines in 70–90% yield.[5] However, isolation of the moisture-sensitive parent compound was only recently accomplished in the submitters' laboratories.[6]

Related 1,3-oxazepines have been obtained from irradiation of many other heterocyclic N-oxides including pyridine N-oxides, isoquinoline N-oxides, quinoxaline N-oxides, quinazoline N-oxides, phenanthridine N-oxides, benzophenazine N-oxides, and acridine N-oxides.[4] However, the reported yields are variable and have generally been higher for phenyl and other aryl-substituted derivatives. This procedure is also satisfactory for 1,3-benzoxazepine (from isoquinoline N-oxide) and, in general, for oxazepines not substituent-stabilized.

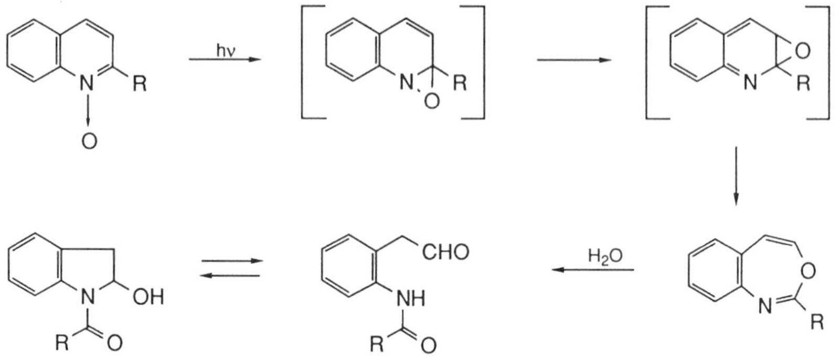

A mechanism involving initial cyclization to an oxaziridine, [1,5] sigmatropic rearrangement to an imino epoxide, and electrocyclic ring opening was originally proposed for the photochemical isomerization.[4] However, since later attempts to detect intermediates by flash photolysis were unsuccessful,[7] ground-state oxaziridines, if formed at all, must have exceedingly short lifetimes. The benzoxazepines undergo facile hydrolysis to o-(N-acylamino)phenylacetaldehydes, which frequently exist as the cyclic carbinol amide tautomers. If water is present during the irradiation from use of the N-oxide hydrate or moist solvent, the hydrolysis products may be isolated instead of the benzoxazepine. Dehydration of the carbinol amide to N-acyl indoles may also occur during irradiation and/or purification of the products. The formation of carbostyrils is sometimes an important competing reaction in the irradiation of quinoline N-oxides, and this by-product is, in fact, formed to the extent of ca. 20% in the present procedure. The use of polar protic solvents such as water or alcohols favors carbostyril formation in contrast to aprotic solvents such as benzene or acetone in which the pathway leading to benzoxazepines usually predominates.

1. Instituto di Chimica Organica dell'Università, v. le Taramelli 10, 27100 Pavia, Italy.
2. Hayashi, E., private communication to Ochiai, E. in "Aromatic Amine N-Oxides," Elsevier Publishing Co.: Amsterdam, 1967, p. 24.
3. Buchardt, O.; Jensen, B.; Larsen, I. K. *Acta Chem. Scand.* **1967**, *21*, 1841–1854.
4. (a) Spence, G. G.; Taylor, E. C.; Buchardt, O. *Chem. Rev.* **1970**, *70*, 231–265; (b) Bellamy, F.; Streith, J. *Heterocycles* **1976**, *4*, 1391–1447.
5. (a) Kaneko, C.; Yamada, S. *Chem. Pharm. Bull.* **1966**, *14*, 555–557; (b) Kaneko, C.; Yamada, S.; Ishikawa, M. *Tetrahedron Lett.* **1966**, 2145–2150; (c) Buchardt, O.; Kumler, P. L.; Lohse, C. *Acta Chem. Scand.* **1969**, *23*, 1155–1167 (d) Albini, A.; Fasani, E.; Dacrema, L. M. *J. Chem, Soc. Perkin Trans. 1* **1980**, 2738–2742.
6. Albini, A.; Bettinetti, G. F.; Minoli, G. *Tetrahedron Lett.* **1979**, 3761–3764.
7. (a) Lohse, C. *J. Chem. Soc., Perkin Trans. II* **1972**, 229–233; (b) Tomer, K. B.; Harrit, N.; Rosenthal, I.; Buchardt, O.; Kumler, P. L.; Creed, D. *J. Am Chem. Soc.* **1973**, *95*, 7402–7406.

OXIDATION OF 5-AMINOTETRAZOLES: BENZYL ISOCYANIDE

(Benzene, isocyanomethyl)

A.

$$C_6H_5CHO + H_2N-\underset{\underset{H}{N-N}}{\overset{N-N}{\|}} \xrightarrow[CH_3OH,\ 50°C]{Et_3N} \left[C_6H_5CH=N-\underset{\underset{H}{N-N}}{\overset{N-N}{\|}} \right] + H_2O$$

$$\downarrow H_2 \mid Pd/C$$

$$C_6H_5CH_2NH-\underset{\underset{H}{N-N}}{\overset{N-N}{\|}}$$

B.

$$C_6H_5CH_2NH-\underset{\underset{H}{N-N}}{\overset{N-N}{\|}} \xrightarrow[H_2O,\ CH_2Cl_2]{NaOBr,\ 0°C} C_6H_5CH_2N{\equiv}C + 2\,N_2 + NaBr + H_2O$$

Submitted by GERHARD HÖFLE and BERND LANGE[1]
Checked by ORVILLE L. CHAPMAN and THOMAS C. HESS

1. Procedure

Caution! This preparation should be conducted in an efficient hood because of the obnoxious odor of the isocyanide.

A. *5-Benzylaminotetrazole.* Freshly distilled benzaldehyde (21.2 g, 0.2 mol) is added in one portion to a warm (50°C) solution of 5-aminotetrazole (17.2 g, 0.2 mol) (Note 1) and triethylamine (20.2 g, 0.2 mol) in 100 mL of absolute methanol. After 15 min the reaction mixture is cooled to room temperature, transferred to an autoclave, and hydrogenated with agitation at room temperature over Pd (10%) on carbon (1 g) for 18 hr at 500 psi (pounds per square inch) of hydrogen. The catalyst is removed by filtration and all volatile material is removed at 60°C under aspirator pressure. The gummy tan solid is triturated with 250 mL of hot water. Aqueous 20% HCl is added until pH 3 is reached. The mixture is cooled to room temperature and the solid collected, washed with water, and dried over-night at room temperature under reduced pressure (100 μ); yield: 27.5 g (80%), mp 183.5–185°C (lit.[2] mp 183°C).

B. *Benzyl isocyanide.* In a 500-mL, round-bottomed flask equipped with a magnetic stirring bar and a pressure-equalizing funnel are placed 5-benzylaminotetrazole (10.5 g, 60 mmol), 100 mL of 10% sodium hydroxide solution, and 70 mL of dichloromethane. The mixture is cooled to 0°C and a solution of NaOBr in water (165 mL, 65 mmol) (Note 2) is added with vigorous stirring over a 15-min period (Note 3). The dichloromethane layer is separated and the aqueous phase extracted with five 50-mL portions of dichloromethane. The combined dichloromethane extracts are dried over anhydrous MgSO₄, the drying agent is removed by filtration, and the dichloromethane is removed by simple distillation. The pressure is then reduced to ~20 mm with an aspirator and benzyl isocyanide is distilled at 98–100°C; yield: 5.91 g (84%) (Notes 4 and 5).

2. Notes

1. 5-Aminotetrazole monohydrate is available from Aldrich Chemical Company, Inc.; it was dehydrated by heating over P_2O_5 at 100°C under reduced pressure (100 μ) for 4 hr.

2. The NaOBr solution was prepared according to a procedure described in *Organic Syntheses*.[3] Bromine [12.6 g (4 mL, 79 mmol)] was added dropwise with vigorous stirring to 150 mL of a 10% NaOH solution at −10°C. Enough 10% NaOH solution was added to the yellow solution to give 200 mL of reagent.

3. During addition of the NaOBr solution the mixture warms to 20°C. The reaction is virtually instantaneous and can be monitored by the liberated nitrogen.

4. The product was pure by IR and NMR spectroscopy. The IR spectrum showed a very strong band at 2150 cm^{-1}, the NMR spectrum a distorted triplet at δ 4.5 (2 H) and a broad singlet at δ 7.3 (5 H).

5. Glassware can be freed from the odor of isocyanide by rinsing with a 1 : 10 mixture of concentrated hydrochloric acid and methanol.

3. Discussion

By this method high yields of isocyanides are obtained by an oxidation process. Since this oxidation can also be performed anodically or with bromine or lead tetraacetate and triethylamine in the absence of water (see Table I),[4] it represents a valuable alternative to other procedures: dehydration reactions,[5-7] the alkylation of silver cyanide[8,9] or the carbylamine (isocyanide) reaction.[10] The starting materials, 5-aminotetrazoles, can be readily obtained by reductive alkylation of 5-aminotetrazole[2] or from monosubstituted thioureas and sodium azide.[11] A limitation of the reaction is that the substituent R must be stable toward oxidation. From a mechanistic point of view the oxidation of 5-aminotetrazoles is a two step process with a pentaazafulvene as an unstable, undetectable intermediate.

TABLE I

PREPARATION OF ISOCYANIDES ($R-N\!=\!C$) BY OXIDATION OF 5-AMINOTETRAZOLES

		Yield (%)		Anodic Oxidation[a]
R	NaOBr[a]	Pb(OAc)$_4$/NEt$_3^b$	Br$_2$/Net$_3^b$	
C_6H_5	92	70	43	39
C_4H_9	75			
$C_6H_5CH_2$	84			48

[a] In 2N sodium hydroxide solution.
[b] In dichloromethane.

Benzyl isocyanide is a useful precursor of compounds containing the α-benzylamino moiety. Substituted styrenes, vinyl isocyanides, 2-oxazolines, 1-pyrrolines, imidazoles, and α-amino acids and ketones can be obtained by metalation of isocyanides with butyllithium[12] or copper salts,[13] and subsequent reaction with various electrophiles.[12]

1. Institut für Organische Chemie, Technische Universität Berlin, D-1000 Berlin 12, Strasse des 17, Juni 135. This work was supported by the Deutsche Forschungsgemeinschaft.

2. Henry, R. H.; Finnegan, W. G. *J. Am. Chem. Soc.* **1954,** *76,* 926–928. The present procedure represents a modification of the procedure described herein.

3. Allen, C. F. H.; Wolf, C. N. *Org. Synth., Coll. Vol. IV* **1963,** 45.

4. Höfle, G.; Lange, B. *Angew. Chem.* **1976,** *88,* 89; *Angew. Chem., Int. Ed. Engl.* **1976,** *15,* 113–114.

5. Ugi, I.; Meyr, R.; Lipinski, M.; Bodesheim, F.; Rosendahl, F. *Org. Synth., Coll. Vol. V* **1973,** 300; Ugi, I.; Meyr, R. *Org. Synth., Coll. Vol. V* **1973,** 1068.

6. Niznik, G. E.; Morrison, W. H., III, Walborsky, H. M. *Org. Synth., Coll. Vol. VI* **1988,** 751.

7. Schuster, R. E.; Scott, J. E.; Casanova, J., Jr. *Org. Synth., Coll. Vol. V* **1973,** 772.

8. Jackson, H. L.; McKusick, B. C. *Org. Synth., Coll. Vol. IV* **1963,** 438.

9. Engemyr, L. B.; Martinsen, A.; Songstad, J. *Acta. Chem. Scand., Ser. A* **1974,** *28,* 255–266.

10. Gokel, G. W.; Widera, R. P.; Weber, W. P. *Org. Synth., Coll. Vol. VI* **1988,** 232.

11. Stolle, R. *J. Prakt. Chem.* **1982,** *134,* 282–309; Garbrecht, W. L.; Herbst, R. M. *J. Org. Chem.* **1953,** *18,* 1269–1282.

12. Hoppe, D. *Angew. Chem., Int. Ed. Engl.* **1974,** *13,* 789–804.

13. Saegusa, T.; Ito, Y.; Kinoshita, H.; Tomita, S. *J. Org. Chem.* **1971,** *36,* 3316–3323.

PEPTIDE SYNTHESIS USING 1-(4-CHLOROPHENYL)-3-(4'-METHYL-1'-PIPERAZINYL)-2-PROPYN-1-ONE AS REAGENT: BENZYLOXYCARBONYL-L-ALANYL-L-CYSTEINE METHYL ESTER AND BENZYLOXYCARBONYL-L-ASPARTYL-(*tert*-BUTYL ESTER)-L-PHENYLALANYL-L-VALINE METHYL ESTER

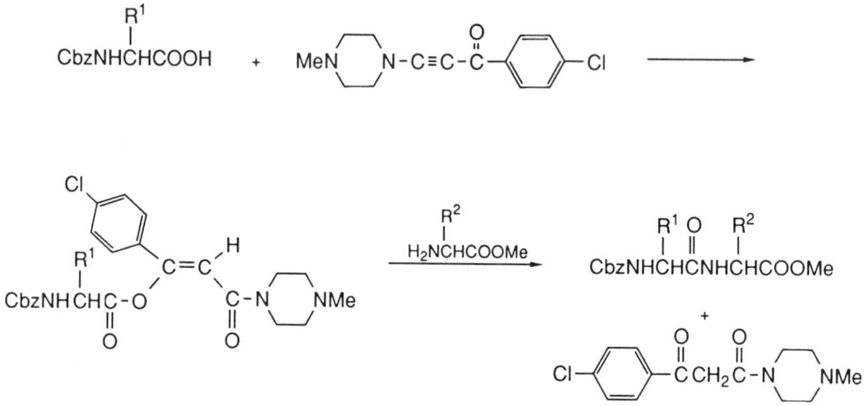

Submitted by H. P. Fahrni, U. Lienhard, and M. Neuenschwander[1]
Checked by David R. Bolin and Gabriel Saucy

1. Procedure

A. *Benzyloxycarbonyl-L-alanyl-L-cysteine methyl ester.* A round-bottomed, three-necked, 100-mL flask is equipped with a magnetic stirring bar, 10-mL dropping funnel, thermometer, and nitrogen bubbler (Note 1). The apparatus is flushed with dry nitrogen and then charged with 446.5 mg (0.002 mol) of benzyloxycarbonyl-L-alanine in 10 mL of dry dichloromethane. The mixture is stirred until solution is complete and then cooled to 0°C. Within 20 min a solution of 525.5 mg (0.002 mol) of 1-(4-chlorophenyl)-3-(4'-methyl-1'-piperazinyl)-2-propyn-1-one (Note 2) in 5 mL of dry dichloromethane is added. Stirring is continued for 1 hr at 0°C and for an additional hour at room temperature (t_1). The mixture is cooled again to 0°C and a suspension of 343.3 mg (0.002 mol) of L-cysteine methyl ester hydrochloride is quickly added, followed by a solution of 202.3 mg (0.002 mol) of N-methylmorpholine in 5 mL of dry dichloromethane. While the nitrogen atmosphere is maintained, the mixture is allowed to warm up and is stirred for 12 hr (t_2) at room temperature. The solvent is removed by rotary evaporation, and the residue is shaken intensively with 30 mL of ethyl acetate and 10 mL of water. The organic layer is extracted two times with 10-mL portions of aqueous 10% citric acid and once with 5 mL of 1 N sodium hydrogen carbonate. The organic phase is dried over sodium sulfate. The solvent is removed by rotary evaporation to leave 647 mg (95%) of the crude pale-yellow dipeptide. Recrystallization from ethyl acetate provides 551 mg (81%) of colorless crystals of benzyloxycarbonyl-L-alanyl-L-cysteine methyl ester, mp 115–117°C; $[\alpha]_D^{20}$ − 26.4° (CH$_3$OH, *c* 1.29) (Note 3).

B. *Benzyloxycarbonyl-L-aspartyl-(tert-butyl ester)-L-phenylalanyl-L-valine methyl ester.* A round-bottomed, three-necked, 100-mL flask is equipped with a 10-mL dropping funnel, thermometer, magnetic stirring bar, and a nitrogen bubbler. The flask is flushed with dry nitrogen and then charged with a solution of 941.1 mg (0.002 mol) of benzyloxycarbonyl-L-aspartyl-(*tert*-butyl ester)-L-phenylalanine (Note 4) in 10 mL of dry dichloromethane. The flask is maintained under a dry nitrogen atmosphere and cooled to 0°C with an ice–salt bath. The mixture is stirred and a solution of 525.5 mg (0.002 mol) of 1-(4-chlorophenyl)-3-(4′-methyl-1′-piperazinyl)-2-propyn-1-one (Note 2) in 5 mL of dry dichloromethane is added during a period of 20 min. Stirring is continued for 1 hr at 0°C and for 5 hr at room temperature (t_1). The mixture is again cooled to 0°C and a suspension of 335.3 mg (0.002 mol) of L-valine methyl ester hydrochloride and 202.3 mg (0.002 mol) of N-methylmorpholine in 5 mL of dichloromethane is added. After 30 min the reaction mixture is allowed to warm up and is stirred overnight (18 hr; t_2) at room temperature. The solvent is removed by rotary evaporation and the residue is shaken intensively with 40 mL of ethyl acetate and 10 mL of water. The organic layer is extracted twice with 10-mL portions of aqueous 10% citric acid and once with 5 mL of 1 N sodium hydrogen carbonate.

The organic phase is dried over sodium sulfate, and solvent is removed by rotary evaporation to leave 1132 mg (97%) of the crude pale-yellow tripeptide. For further purification the crude product is dissolved in ethyl acetate, treated with some activated carbon, and filtered through Celite. Removal of the solvent and crystallization from ethyl acetate/ether/petroleum ether (ca. 2 : 1 : 1) yields 993 mg (85%) of colorless crystals of benzyloxycarbonyl-L-aspartyl-(*tert*-butyl ester)-L-phenylalanyl-L-valine methyl ester; mp 119–120°C (Note 5).

2. Notes

1. Cysteine derivatives are oxidized to cystine by oxygen. The nitrogen atmosphere for preparation of Cbz-alanylcysteine methyl ester is therefore indispensable and is recommended for other cases as well.

2. This reagent is available from Fluka Chemical Corp.

3. The literature[2] value is $[\alpha]_D^{20} - 26.5°$ (CH$_3$OH, c 1.27). The reported[2] mp is 116.5-118°C.

4. Benzyloxycarbonyl-L-aspartyl-(*tert*-butyl ester)-L-phenylalanine dicyclohexyl-amine salt was conveniently prepared by standard procedures.[3] The salt was dissolved in ethyl acetate and extracted three times with aqueous 10% citric acid, and once with water. The organic phase was dried and solvent was removed to leave the dipeptide as an oil.

5. The product has the following physical properties: Specific rotation: $[\alpha]_D^{20} - 36.5°$ (C$_2$H$_5$OH, c 2); IR (KBr), cm^{-1}: 3285, 1732, 1691, 1640, 1531, 1367, 1229, 1158, 1050, 746, 701; ^{1}H NMR (100 MHz, CDCl$_3$) δ: 0.81 (d, 3 H, J = 7) 0.84 (d, 3 H, J = 7), 1.41 (s, 9 H), 1.8–2.3 (m, 2 H), 2.62 (d of d, 1 H, J = 17, J′ = 6), 2.86 (d of d, 1 H, J = 17, J′ = 5), 3.08 (d, 1 H, J = 8), 3.71 (s, 3 H), 4.3–4.8 (m, in total 3 H), 5.11 (s, 2 H), 5.78 (d, 1 H, J = 8), 6.28 (d, 1 H, J = 8), 7.02 (d, 1 H, J = 8), 7.21 (s, 5 H), 7.57 (s, 5 H).

3. Discussion

This procedure illustrates the use of 1-(4-chlorophenyl)-3-(4'-methyl-1'-piperazinyl)-2-propyn-1-one[4] as a reagent for peptide synthesis.[5] The same method also gives amides in excellent yields.[6]

The preparation of Cbz-L-alanylcysteine methyl ester shows the advantage of using, as the amine component, an amino acid with an unprotected sulfhydryl moiety. No problems were encountered with the use of amino acid derivatives with unprotected hydroxyl or sulfhydryl groups as either the amine[5] or carboxyl component.[7] This procedure is based on the pronounced selective reactivity of the enol ester, which is generated by the addition of carboxylic acids to "push–pull acetylenes." Generally, the yields of peptides are good and a broad variety of solvents (e.g., dichloromethane, tetrahydrofuran, acetonitrile, dimethylformamide) may be used, depending on the solubility of the coupling components. It is also possible to change the solvent after the activation step or to isolate the activated components. Normally, however, this is neither necessary nor recommended. Purification of the reaction mixture is simple, since the piperazine by-product is conveniently extracted with an acidic water phase.

The following peptides and further examples have been prepared[5] by the following procedure.

Peptide	t_1^a	t_2^b	Yield (%)^c
Cbz-L-Ala-Gly-OMe	2^d	12	91
Cbz-L-Ala-L-Val-OMe	2^d	24	84
Cbz-L-Ala-L-Phe-OMe	2^d	18	88
Cbz-Gly-L-Phe-Gly-OEt	2	12	90
Cbz-L-Asp(O-t-Bu)-L-Phe-L-Val-OMe	6	18	85
Cbz-L-Ile-L-Ile-OBzl	18	24	75
Cbz-L-Ala-L-Ser-OMe	2^d	72	90
Cbz-L-Ala-L-Tyr-OMe	2^d	72	91
Cbz-L-Ala-L-Cys-OMe	2^d	12	81
Cbz-L-Ala-L-Met-OMe	2^d	15	85
Cbz-L-Ser-Gly-OEt	2	24	81

$^a t_1$: time for activation of the carboxylic component (see procedures).
$^b t_2$: time for coupling (see procedures).
^{c} Yield of pure recrystallized product.
^{d} Stirring 1 hr at 0°C, then 1 hr at 20°C.

During our experiments no side reactions were detected. This is in contrast to peptide synthesis with isoxazolium salts,[8] where some side reactions, one leading to a diacyl amino compound, were observed.[9] In most cases, these side reactions are due to a secondary amino group in the reagent which is impossible in the case of push–pull acetylenes.

Compared with ynamines, which have also been applied to peptide synthesis,[10] push–pull acetylenes are much more selective. They do not show the side reactions observed with ynamines,[11] and the yields are not markedly influenced by the sequence of addition of compounds in the activation step or by excess of acetylene reagent.

A crucial point in peptide synthesis is racemization of the activated amino acid. Three different tests were made to evaluate the degree of racemization. Using the Anderson test peptide[12] Cbz-Gly-Phe-Gly-OEt, no racemization could be detected when the peptide was prepared in dichloromethane, acetonitrile, or tetrahydrofuran. This means racemization is below the detection limit of 1%. Benzylleucylglycine ethyl ester is used in the very sensitive Young test.[13] In this test, designed to exaggerate racemization, we found 5% of racemate, when the solvent was dichloromethane. In the more polar solvent, dimethylformamide, this value rose to 12%. Therefore, racemization is in the same range as that observed for the racemization-resistant azide procedure. The coupling of Cbz-L-aspartyl(O-t-Bu)-L-phenylalanine with valine methyl ester is reported to be very sensitive to racemization.[14] The tripeptide was prepared as described above, and the crude product was hydrolyzed. Gas-layer chromatography (GLC) showed the presence of 2-3% D-phenylalanine. Again, in contrast to the ynamine procedure,[15] racemization seems to be no problem when push–pull acetylenes are used.

So far, the only observable disadvantage of the reagent is the somewhat long reaction time for the coupling of the activated amino acids (or peptides) with the amine component. The increase in reaction time t_2 could be a limiting factor, if longer peptide fragments are to be linked.

1. Institut für Organische Chemie der Universität Bern, CH 3012 Bern, Freiestrasse 3, Switzerland.
2. Inui, T. *Bull Chem. Soc. Jpn.* **1971**, *44*, 2515.
3. Kanaoka, M.; Ishida, T.; Kikuchi, T. *Chem. Pharm. Bull.* **1978**, *26*, 605–610.
4. (a) Lienhard, U.; Fahrni, H.-P.; Neuenschwander, M. *Helv. Chim. Acta* **1978**, *61*, 1609; (b) Gais, H. J.; Hafner, K.; Neuenschwander, M. *Helv. Chim. Acta* **1969**, *52*, 2641.
5. Neuenschwander, M.; Fahrni, H.-P.; Lienhard, U. *Helv. Chim. Acta* **1978**, *61*, 2437.
6. Neuenschwander, M.; Lienhard, U.; Fahrni, H.-P.; Hurni, B. *Helv. Chim. Acta* **1978**, *61*, 2428.
7. Gais, H.-J. *Angew. Chem.* **1978**, *90*, 625; *Angew Chem., Int. Ed. Engl.* **1978**, *17*, 597.
8. (a) Woodward, R. B.; Olofson, R. A. *J. Am. Chem. Soc.* **1961**, *83*, 1007; (b) Woodward, R. B.; Olofson, R. A.; Mayer, H. *Tetrahedron* **1966**, *22*, Suppl. No. 8, 321.
9. Woodman, D. J.; Davidson, A. I. *J. Org. Chem.* **1970**, *35*, 83.
10. Buijle, R.; Viehe, H. G. *Angew. Chem., Intern. Ed. Engl.* **1964**, *3*, 582.
11. Steglich, W.; Höfle, G.; König, W.; Weygand, F. *Chem. Ber.* **1968**, *101*, 308.
12. Anderson, G. W.; Callahan, F. M.; *J. Am. Chem. Soc.* **1958**, *80*, 2902.
13. Williams, M. W.; Young, G. T. *J. Chem. Soc.* **1963**, 881.
14. Sieber, P.; Riniker, B. In "Pept., Proc. Eur. Pept. Symp., 11th," 1971; Nesvadba, H., Ed.; North-Holland: Amsterdam, 1973; pp. 49–53.
15. Weygand, F.; König, W.; Buijle, R.; Viehe, H. G. *Chem. Ber.* **1965**, *98*, 3632.

INDOLES FROM 2-METHYLNITROBENZENES BY CONDENSATION WITH FORMAMIDE ACETALS FOLLOWED BY REDUCTION: 4-BENZYLOXYINDOLE

[1*H*-Indole, 4-(phenylmethoxy)-]

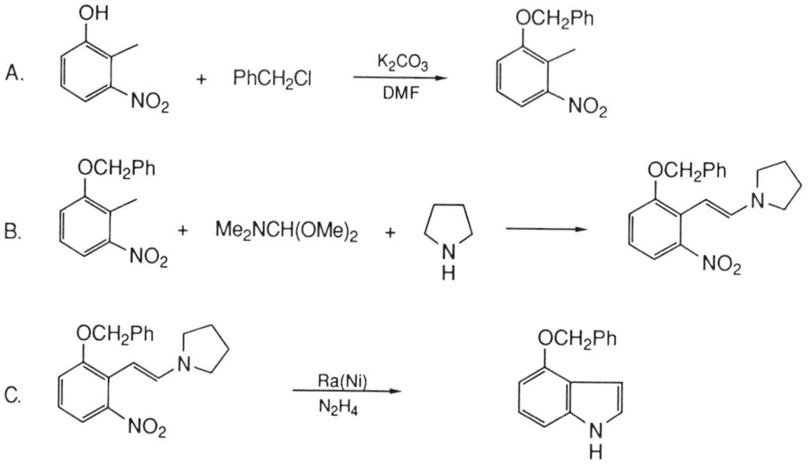

Submitted by ANDREW D. BATCHO[1] and WILLY LEIMGRUBER[2]
Checked by DAVID J. WUSTROW and ANDREW S. KENDE

1. Procedure

A. *6-Benzyloxy-2-nitrotoluene.* A stirred mixture of 124.7 g (0.81 mol) of 2-methyl-3-nitrophenol (Note 1), 113.2 g (0.90 mol) of benzyl chloride, 112.2 g (0.81 mol) of anhydrous potassium carbonate, and 800 mL of dimethylformamide (DMF) is heated at 90°C for 3 hr. Most of the DMF is removed on a rotary evaporator (20 mm) and the oily residue is poured into 400 mL of 1 *N* sodium hydroxide and extracted with ether (3 × 800 mL). The combined extracts are dried (Na$_2$SO$_4$), filtered, and evaporated to give 203.5 g of yellowish solid. Recrystallization from 1 L of methanol cooled to 0°C affords 177.6 (90%) of 6-benzyloxy-2-nitrotoluene as pale-yellow crystals, mp 61–63°C[3] (Note 2).

B. *(E)-6-Benzyloxy-2-nitro-β-pyrrolidinostyrene.* To a solution of 175.4 g (0.72 mol) of 6-benzyloxy-2-nitrotoluene in 400 mL of DMF are added 102.5 g (0.84 mol) of *N,N*-dimethylformamide dimethyl acetal (Note 3) and 59.8 g (0.84 mol) of pyrrolidine. The solution is heated at reflux (110°C) for 3 hr (Note 4) under nitrogen and allowed to cool to room temperature. The volatile components are removed on a rotary evaporator, and the red residue (Note 5) is dissolved in 200 mL of methylene chloride and 1.60 L of methanol. The solution is concentrated to a volume of about 1.40 L on the steam bath and then is cooled to 5°C. Filtration and washing of the filtercake with 200 mL of cold methanol affords 209.8 g of red crystals, mp 87–89°C (Note 6). The mother liquors are evaporated, and the residue is recrystallized from 50 mL of methanol (5°C) to give an additional 12.30 g of red solid, mp 81–83°C (Note 7). Thus the total yield is 222.1 g (95%) of a 15 : 1 mixture of (E)-6-benzyloxy-2-nitro-β-pyrrolidino-styrene (Note 8) and (E)-6-benzyloxy-β-dimethylamino-2-nitrostyrene.

C. *4-Benzyloxyindole.* To a stirred solution of 162.2 g (0.50 mol) of (E)-6-ben-zyloxy-2-nitro-β-pyrrolidinostyrene (Note 9) in 1 L of THF and 1 L of methanol at 30°C under nitrogen is added 10 mL of Raney nickel (Note 10) followed by 44 mL (0.75 mol) of 85% hydrazine hydrate. Vigorous gas evolution is observed. The red color turns to dark brown within 10 min, and the reaction temperature rises to 46°C. An additional 44 mL of 85% hydrazine hydrate is added after 30 min and again 1 hr later. The temper-ature is maintained between 45 and 50°C with a water bath during the reaction and for 2 hr after the last addition. The mixture is cooled to room temperature and the catalyst is removed by filtration through a bed of Celite (Note 11) and is washed several times with methylene chloride. The filtrate is evaporated and the residue dried by evaporating with 500 mL of toluene. The reddish residue (118.5 g), dissolved in ca. 1 L of toluene-cyclohexane (1 : 1), is applied to a column of 500 g of silica gel (70–230-mesh, Merck) prepared in the same solvent. Elution with 6.0 L of toluene–cyclohexane (1 : 1) followed by 3 L of toluene–cyclohexane (1 : 2) affords 108.3 g of white solid, which is crystal-lized from 150 mL of toluene and 480 mL of cyclohexane (Note 12). A total of 107.3 g (96% yield) of 4-benzyloxyindole (Note 13) is obtained in three crops as white prisms, mp 60–62°C (Note 14).

2. Notes

1. 2-Methyl-3-nitrophenol was obtained from Aldrich Chemical Company, Inc.

2. The ^{1}H NMR spectrum is as follows δ: 2.42 (s, 3 H), 5.10 (s, 2 H), 7.13 (m, 3 H), 7.35 (m, 5 H).

3. *N,N*-Dimethylformamide dimethyl acetal was prepared according to a procedure of Bredereck.[4] *N,N*-Dimethylformamide diethyl acetal can also be used. Both the dimethyl and the diethyl acetal are commercially available from Aldrich Chemical Company, Inc.

4. The reaction was followed by TLC on silica gel plates developed with ether–petroleum ether (1 : 1).

5. Since it contained *non*volatile *N*-formylpyrrolidine, direct reduction of the crude material was not attempted.

6. This crop contained 5% 6-benzyloxy-β-dimethylamino-2-nitrostyrene (by NMR). Pure 6-benzyloxy-2-nitro-β-pyrrolidinostyrene melts at 91.5–92.5°C.

7. This crop contained 15% 6-benzyloxy-β-dimethylamino-2-nitrostyrene (by NMR).

8. The ^{1}H NMR spectrum is as follows: δ: 1.8 (m, 4 H), 3.08 (m, 4 H), 5.03 (5, 2 H), 5.20 (d, 1 H, J = 12.2), 6.91 (dd, 2 H, J = 9), 7.25 (m, 6 H), 7.75 (d, 1 H, J = 12.2).

9. This compound may contain varying amounts of 6-benzyloxy-β-dimethylamino-2-nitrostyrene.

10. Raney nickel is commercially available as type #28 from the Davison Chemical Division of W. R. Grace and Co.

11. The catalyst is pyrophoric and should not be sucked dry.

12. The material tenaciously holds hydrocarbons, such as pentane, hexane, and petroleum ether, which cannot be removed even under high vacuum. The solvated crystals show hydrocarbon protons in NMR and exhibit a broad melting point. However, we found that cyclohexane is not retained in the crystals.

13. The ^{1}H NMR spectrum is as follows: δ: 6.65 (m, 2 H), 6.95 (m, 3 H), 7.9 (br s, 1 H), 7.32 (m, 5 H).

14. We could not reproduce the reported[5] melting point of 72–74°C (toluene). The material has the proper microanalysis and is pure by NMR and thin-layer chromatography (TLC).

3. Discussion

Through the years, widespread interest in the synthesis of natural products and their analogs bearing the oxygenated indole nucleus has led to the development of several routes to protected hydroxylated indoles. However, 4-benzyloxyindole was first prepared relatively recently in modest overall yield by the Reissert method, which involves condensation of 6-benzyloxy-2-nitrotoluene with ethyl oxalate, reductive cyclization to the indole-2-carboxylate, hydrolysis to the acid, and decarboxylation.[5]

Although a variety of synthetic methods have been used to prepare indoles,[6,7] many of these lack generality and are somewhat restrictive since they employ conditions, such as acid or strongly basic cyclizations or thermal decarboxylations, which are too harsh for labile substituents. This efficient, two-step procedure[8,9] illustrates a general, simple, and convenient process for preparing a variety of indoles substituted in the carbocyclic ring, as can be seen in Table I. Since many of these examples served to determine the scope of this method, the yields in most cases have not been optimized. In many cases, the starting materials are readily available or can be easily prepared.

As can be seen in Table I, variation of the substituent has a profound effect on the rate of reaction of the o-nitrotoluene derivative with dimethylformamide acetals but has little effect on the yields, which are often almost quantitative. As can be predicted, electron-withdrawing groups accelerate the reaction. To shorten the somewhat lengthy reaction times that are often necessary when electron-donating substituents are present, more reactive aminomethylenating reagents such as pyrrolidine (or piperidine) acetals,[8] aminals,[10] or trisaminomethanes[11] can be employed. Alternatively, as described above, simply adding pyrrolidine to the reaction mixture also generates in situ a very effective aminomethylenating reagent.[12,13] Thus, for example, in the case of 6-benzyloxy-2-nitrotoluene, the reaction with N,N-dimethylformamide dimethyl acetal requires 51 hr versus 3 hr when pyrrolidine is added. Pyrrolidine undergoes exchange reactions with N,N-dimethylformamide acetals to produce an equilibrium mixture of formylpyrrolidine acetal and the mixed pyrrolidine dimethylamine aminal (alkoxydimethylaminopyrrolidinomethane) as well as other trisaminomethane species.[14] (In cases where pyrrolidine reacts with the aromatic substrate, addition of the substrate can be delayed until pyrrolidine exchange is complete.) This mixture of reagents gives rise to condensation

TABLE I

INDOLES FROM 2-METHYLNITROBENZENES BY CONDENSATION WITH N,N-DIMETHYLFORMAMIDE ACETALS AND REDUCTION

Structures: **1** (2-methylnitrobenzene bearing R^1, R^2, R^3, R^4, with CH_3 and NO_2) → **2** (2-(2-pyrrolidinovinyl)nitrobenzene bearing R^1, R^2, R^3, R^4, with NO_2) → **3** (indole bearing R^1, R^2, R^3, R^4, N–H)

	Substituents			Intermediates 2 (mp or bp/mm)	Reaction Time	Purified Yield (%) (Procedure)[a]	Indoles 3 (mp or bp/mm)	Yield (%) (Procedure)[b]	Refs.
R^1	R^2	R^3	R^4						
OCH_2Ph	—	—	—	125°/0.03	22 hr	97(M,E)	52.5–53.5°	80(A)	8, 9
—	OCH_2Ph	—	—	67–68°	51 hr	90(M)	60–62°	70(B)	12
—	—	OCH_2Ph	—	98–99°	29 hr	78(E)	103–105°	45(B)[64][c]	8, 9, 12
—	OCH_2Ph	OCH_2Ph	—	108.5–110°	41 hr	97(M)	118–120°	75(B)[i]	12
—	OCH_2Ph	CH_3	—	99.5–101°	48 hr	86(M)	112–113°	54(B)[j]	8, 9
—	OCH_3	—	—	113–134°	31 hr	87(M)	82–83°	84(B)	12
—	—	OCH_3	—	68.5–70°	16 hr	92(M)	56.5–57.5°	83(A)	8, 9
—	OCH_3	OCH_3	—	152°/0.06	70 hr	64(E)	88–90°	63(A)[62][c]	8, 9, 12
—	OCH_3	OCH_3	—	125–126°	48 hr	68(M)	154–155°	28(A)	8, 9
—	—	—	CH_3	100–101°	8 hr	54(M)	100–110°/0.15	66(A)	22
—	OCH_2O		—	114–116°	18 hr	72(E)	110–111°	50(A)[52][c]	8, 9, 12
Cl	—	—	—	111°/0.03	6 hr	89(E)	90°/0.04	63(B)	8, 9
—	Cl	—	—	81.5–82.5°	7 hr	88(E)	71–72°	78(B)	8, 9
—	—	Cl	—	44–46°	24 hr	57(M)	86.5–88°	52(B)[75][c]	8, 9, 12
—	—	NH_2[d]	—	173–174°	2 hr	82(E)[f]	77.5–78.5°	43(A)	8, 9, 12
CN	—	—	—	66–68°	3 hr	93(M)	116–117°	67(C)	17

TABLE I (Continued)
INDOLES FROM 2-METHYLNITROBENZENES BY CONDENSATION WITH N,N-DIMETHYLFORMAMIDE ACETALS AND REDUCTION

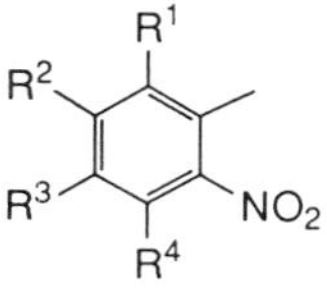

Substituents				Intermediates 2 (mp or bp/mm)	Reaction Time	Purified Yield (%) (Procedure)[a]	Indoles 3 (mp or bp/mm)	Yield (%) (Procedure)[b]	Refs.
R^1	R^2	R^3	R^4						
—	—	CN	—	134–137.5°	2.5 hr	86(E)	128–129°	65(A)	8, 9
—	F	—	—	57.5–59°	3.5 hr	92(E)	46.5–47°	51(B)	8, 9
—	—	F	—	46–47°	22 hr	63(M)	74–75°	80(B)[80][c]	8, 12
CH_3	—	—	—	108°/0.05	24 hr	70(E)	82°/0.4	57(A)	8, 9
—	—	CH_3	—	41.5–43.5°	37 hr	83(M)	29–30.5°	83(A)	23
—	—	—	CH_3	76–76.5	46 hr	40(E)	83–84°	48(A)	8, 9
—	—	$CH(CH_3)_2$	—	138–140°/0.06	42 hr	84(E)	40–41°	51(A)	8, 9
—	—	$CH(OCH_3)_2$	—	67–68°	8 hr	55(E)	62–63.5°	31(A)	8, 9
$COOCH_3$	—	—	—	120–130°/0.2	6 hr	86(M)	63°	82(A)[63][c]	17, 24
$COOC_2H_5$	—	—	—	(Oil)	5 days	93(E)	67–69°	38(D)	17
—	$COOC_2H_5$[e]	—	—	55–56.5°	4.5 hr	70(E)	95–96°	39(A)	8, 9
—	—	—	$COOCH_3$	132–134°	9 hr	88(M)	46–48°	72(A)[g]	12

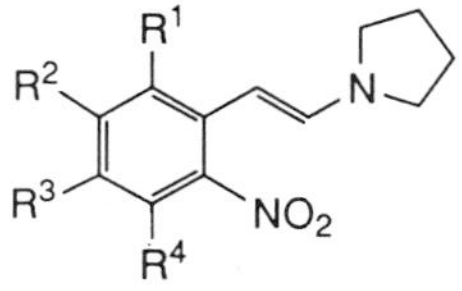

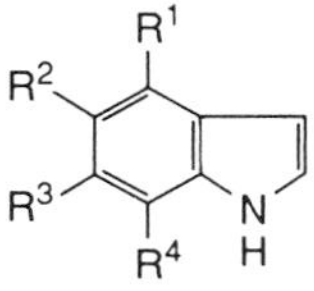

Cl	OCH$_3$	—	—	—	Over-night	—(M)	109–111°	(B)[59]^c	25
—	OCH$_3$	Cl	—	140–141°	Over-night	78(M)	126–128°	46(B)[45]^c	25
—	OCH$_3$	F	—	116–117°	Over-night	64(M)	73–74°	54(B)	25
—	—	Br	—	—	31 hr	—(M)	93°	37(B)^h	26

^{a}(M) = N,N-Dimethylformamide dimethyl acetal; (E) = N,N-dimethylformamide diethyl acetal.

^{b}A = Catalytic hydrogenation in benzene using palladium on charcoal; B = catalytic hydrogenation in benzene using Raney nickel; C = iron in acetic acid; D = stannous chloride.

^{c}Yield in brackets represents overall yield without purification of intermediate 2.

^{d}R^3 = NO$_2$ in compounds 1 and 2.

^{e}R^2 = COOH in compound 1.

^{f}No solvent was used.

^{g}Methanol was the solvent.

^{h}Ethanol was the solvent.

^{i}(M) + pyrrolidine gave a mixture (10 : 1) of pyrrolidine enamine, mp 108–110° (MeOH), and N,N-dimethylenamine (5 hr reflux, 97% yield), which, on reduction (Raney nickel–hydrazine), gave the indole (93% yield).

^{j}(M) + pyrrolidine gave a mixture (9 : 1) of pyrrolidine enamine and N,N-dimethylenamine (5 hr reflux, 95% yield), which, on reduction (Raney nickel–hydrazine), gave the indole (89% yield).

39

products—pyrrolidine enamines—that contain 5–10% of the corresponding *N,N*-dimethylenamines.

The enamine intermediates are usually crystalline, red compounds that can be stored at room temperature for reasonable periods. In cases where the enamines are noncrystalline, it is recommended that the crude product be used directly in the next step, since purification is, in such cases, not practical. Although the more volatile derivatives can be distilled under high vacuum, this entails some risk because of their thermal instability. Moreover, the enamines are not stable to silica gel (TLC or column) chromatography.

Conversion of the intermediate nitroenamine into the indole requires selective reduction of the nitro group. Catalytic hydrogenation results in spontaneous formation of the indole and is generally the mildest and most convenient method of reduction. Although selectivity does vary with the substituent on the aromatic ring, it is generally highly in favor of the nitro group. However, scale-up requires access to large autoclaves or special equipment. To avoid hydrogenolysis of benzyl or chloro functions, Raney nickel is the catalyst of choice. Excellent yields have been obtained using hydrazine and the appropriate catalyst[15] as, in essence, a hydrogenation process which does not require an autoclave or special equipment and can be easily carried out in the laboratory.

Alternative methods of reduction have also been used: sodium dithionite,[16] iron in acetic acid,[17] stannous chloride,[17] and titanium trichloride.[18]

This method has been applied to the preparation of polycyclic indoles[12, 19] and azaindoles[19-21] as well.

1. Research Division, Hoffmann–La Roche, Inc., Nutley, NJ 07110.
2. Deceased.
3. Beer, R. J. S.; Clarke, K.; Khorana, H. G.; Robertson, A. *J. Chem. Soc.* **1948**, 1605.
4. Bredereck, H.; Simchen, G.; Rebsdat, S.; Kantlehner, W.; Horn, P.; Wahl, R.; Grieshaber, P. *Chem. Ber.* **1968**, *101*, 47.
5. Stoll, A.; Troxler, F.; Peyer, J.; Hofmann, A. *Helv. Chim. Acta* **1955**, *38*, 1452.
6. Sundberg, R. J. "The Chemistry of Indoles," Academic Press: New York, 1970.
7. Brown, R. K. In "Indoles," Houlihan, W. J., Ed.; Wiley-Interscience: New York, 1972, Part 1, Chapter 2.
8. Leimgruber, W.; Batcho, A. D. Third International Congress of Heterocyclic Chemistry, Japan, August 23–27, 1971.
9. Batcho, A. D.; Leimgruber, W. U.S. Patent 3732245, 1973; U.S. Patent 3976639, 1976; *Chem Abstr.* **1977**, *86*, 29624.
10. Bredereck, H.; Simchen, G.; Wahl, R. *Chem. Ber.* **1968**, *101*, 4048.
11. Kruse, L. I. *Heterocycles* **1981**, *16*, 1119.
12. Batcho, A. D.; Leimgruber, W., unpublished results.
13. Maehr, H.; Smallheer, J. M. *J. Org. Chem.* **1981**, *46*, 1752.
14. See: Abdulla, R. F.; Brinkmeyer, R. S. *Tetrahedron* **1979**, *35*, 1675.
15. Furst, A.; Berlo, R. C.; Hooton, S. *Chem. Rev.* **1965**, *65*, 51.
16. Garcia, E. E.; Fryer, R. I. *J. Heterocycl. Chem.* **1974**, *11*, 219.
17. Ponticello, G. S.; Baldwin, J. J. *J. Org. Chem.* **1979**, *44*, 4003.
18. Somei, M.; Inoue, S.; Tokutake, S.; Yamada, F.; Kaneko, C. *Chem. Pharm. Bull.* **1981**, *29*, 726.
19. Gassman, P. G.; Schenk, W. N. *J. Org. Chem.* **1977**, *42*, 3240.
20. Prokopov, A. A.; Yakhontov, L. N. *Khim. Geterotsikl. Soedin* **1977**, 1135; *Chem. Abstr.* **1977**, *87*, 201382w.

21. Prokopov, A. A.; Yakhontov, L. N. *Khim. Geterotsikl. Soedin* **1979**, 86; *Chem. Abstr.* **1979**, *90*, 152049u.

22. Benington, F.; Morin, R. D.; Bradley, R. J. *J. Heterocycl. Chem.* **1976**, *13*, 749.

23. Hengartner, U.; Valentine, D., Jr.; Johnson, K. K.; Larscheid, M. E.; Pigott, F.; Scheidl, F.; Scott, J. W.; Sun, R. C.; Townsend, J. M.; Williams, T. H. *J. Org. Chem.* **1979**, *44*, 3741.

24. Kozikowski, A. P.; Ishida, H.; Chen, Y.-Y. *J. Org. Chem.* **1980**, *45*, 3350.

25. Flaugh, M. E.; Crowell, T. A.; Clemens, J. A.; Sawyer, B. D. *J. Med. Chem.* **1979**, *22*, 63.

26. Dellar, G.; Djura, P.; Sargent, M. V. *J. Chem. Soc., Perkin Trans. 1* **1981**, 1679.

CHIRAL MEDIA FOR ASYMMETRIC SOLVENT INDUCTIONS. (S,S)-(+)-1,4-BIS(DIMETHYLAMINO)-2,3-DIMETHOXYBUTANE FROM (R,R)-(+)-DIETHYL TARTRATE

(1,4-Butanediamine, 2,3-dimethoxy-N,N,N',N'-tetramethyl-[S,S]-)

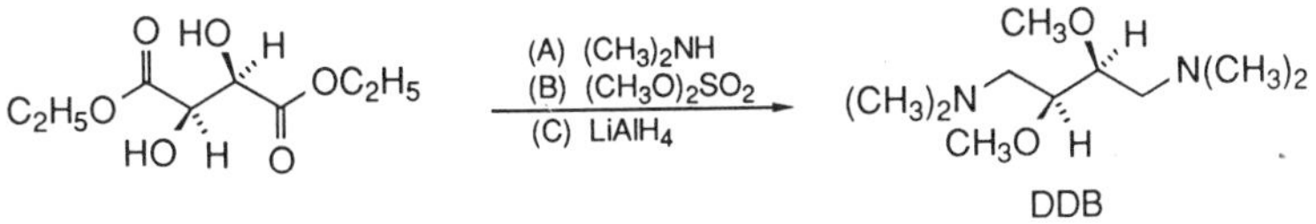

Submitted by Dieter Seebach, Hans-Otto Kalinowski, Werner Langer, Gerhard Crass, and Eva-Maria Wilka[1]

Checked by M. F. Semmelhack and Diane Facciolo

1. Procedure

Caution! Because of the high toxicity and carcinogenicity of dimethyl sulfate (Step B), this step should be carried out in a well-ventilated hood.

A. *(R,R)-(+)-N,N,N',N'-Tetramethyltartaric acid diamide.* Into a mixture of 618 g (3 mol) of diethyl tartrate (Note 1) and 600 mL of freshly distilled methanol in a 2-L Erlenmeyer flask is poured at least 450 mL (7 mol) of liquid, anhydrous, cold (−78°C) dimethylamine (Note 2). The mixture is swirled briefly and then allowed to stand in a hood for 3 days with a drying tube in place. After seeding (Note 3) and cooling in a refrigerator overnight, the massive crystals are collected by suction filtration. The filtrate is concentrated, seeded, and cooled to yield a second crop. The combined crystals are washed with cold methanol (−30°C) and dried under reduced pressure at 70–100°C (oil bath). The diamide thus obtained is sufficiently pure to be used in the following step. The yield is 570–580 g (93–95%). Recrystallization from methanol–ethyl acetate furnishes an analytically pure sample, mp 189–190°C, [α]$_D$ +43° (ethanol, *c* 3.0).

B. *(R,R)-(+)-2,3-Dimethoxy-N,N,N',N'-tetramethylsuccinic acid diamide.* Into a 4-L, three-necked, round-bottomed flask, fitted with a mechanical stirrer, reflux condenser, and stopper, are introduced 240 mL of 50% aqueous sodium hydroxide (3 mol), 1.5 L of methylene chloride, 0.2 g of benzyltriethylammonium chloride (TEBA), and then 260 g (2.06 mol) of dimethyl sulfate (Note 4). The mixture is stirred vigorously (Note 5), and a total of 204 g (1 mol) of the powdered tartaric acid diamide is added in portions at such a rate as to maintain refluxing (Note 6). Stirring is continued for 24 hr without heating, whereupon 1 L of water is added. Separation of the organic phase, extraction of the aqueous layer with three 300-mL portions of methylene chloride, drying of the combined organic solutions over sodium sulfate, and removal of the solvent in a rotary evaporator (bath temperature below 80°C, water aspirator vacuum) furnishes a slightly yellow oil that crystallizes at 25°C and is sufficiently pure for use in the following reduction step. Recrystallization from cyclohexane/benzene yields 220 g (95%, Note 7) of colorless prisms, mp 63.2–63.5°C, $[\alpha]_D$ + 116° (benzene, c 3).

C. *(S,S)-(+)-1,4-Bis(dimethylamino)-2,3-dimethoxybutane (DDB).* A 4-L, three-necked, round-bottomed flask is fitted with a heating jacket, mechanical stirrer, reflux condenser with drying tube, and a stoppered, pressure-equalizing dropping funnel, flushed with nitrogen or argon, and charged with 2.2 L of dry tetrahydrofuran (THF, Note 8) and 60 g (1.6 mol) of lithium aluminum hydride (LiAlH$_4$, Note 9). A mixture of 250 mL of THF and 232 g (1.0 mol) of the diamide is added, with stirring, at a rate sufficient to reach and maintain refluxing. After the addition is completed, the reaction mixture is kept boiling for 2 hr. The flask is immersed in an ice bath, and 60 mL of water, 180 mL of 10% aqueous potassium hydroxide, and again 60 mL of water are added cautiously with very vigorous stirring. The hydrogen gas that is generated is led well above the stirring motor into the hood exhaust. Temporarily, the slurry becomes viscous and difficult to stir; during this period addition has to be made extremely carefully. The pale yellow, completely hydrolyzed slurry is filtered by suction, the filter cake extracted twice by refluxing with THF in a round-bottomed flask, and the combined solutions are concentrated in a rotary evaporator. The residual liquid is distilled through a 20-cm Vigreux column; bp 62–64°C (3 mm), yield 180 g (88%). For use in organometallic reactions, DDB is freshly distilled from LiAlH$_4$, $[\alpha]_D$ + 14.7° (neat), d_4^{20} 0.896 (Note 10).

2. Notes

1. Commercial (R,R)-(+)-diethyl tartrate can be used. The submitters prepared it from (R,R)-(+)-tartaric acid (Firma Benckiser, D-Ludwigshafen, or Firma Boehringer, D-Ingelheim), $[\alpha]_D$ +12.7° (water, c 17): a 4-L, three-necked, round-bottomed flask equipped with a mechanical stirrer, water separator for organic solvents heavier than water, and a stopper is charged with 1.5 kg (10 mol) of tartaric acid, 1.5 L (26 mol) of 96% ethanol, 1 L of chloroform, and 30 g of freshly activated (1 N HCl), highly acidic ion-exchange resin (Lewatit 3333). The stirred mixture is heated at reflux until no more water separates (up to 60 hr). Filtration, evaporation, and vacuum distillation (oil bath temperature must not exceed 145°C, no column, fast distillation) yield 1.85 kg (90%) of the ester, $[\alpha]_D$ +8.16° (neat).

2. Dimethylamine (bp 6°C) is condensed into a 1-L flask cooled to −78°C and fitted with an inlet tube and an opening protected from atmospheric moisture with a silica gel drying tube. It is either taken from a cylinder or freed from 1.5

L of a stirred 40% aqueous solution by heating at 60–80°C with 50 g of potassium hydroxide and leading the amine vapors first through a reflux condenser, then through a 50-cm (2 cm i.d.) drying tube filled with potassium hydroxide pellets, and finally into the receiver flask cooled to −78°C.

3. Sometimes spontaneous crystallization occurs; if it does not, a small amount of the solution is withdrawn and evaporated on a watch glass, and the crystals that are obtained by scratching with a glass rod are used for seeding.

4. Dimethyl sulfate was purchased from Riedel de Haen, D-Seelze-Hannover, and used without purification.

5. Since the reaction mixture becomes very gelatinous on addition of the tartaric acid amide, a powerful motor and a large stirring blade are necessary.

6. Since dimethyl sulfate decomposes rapidly in concentrated alkaline medium, addition of the powdered tartaric acid amide must begin *immediately* after the dimethyl sulfate is introduced. The amide should be added *as fast as possible* (ca. 20–30 min) within the limits of the capacity of the reflux condenser and the mechanical stirrer. The amount of dimethyl sulfate can be increased up to 2.5 equivalents and fresh benzyltriethylammonium chloride can be added toward the end of the addition. With less rapid addition and stirring, the yield drops to 45–55%.

7. The yield obtained by the checkers was 78%.

8. Tetrahydrofuran (THF) was obtained from BASF AG, D-Ludwigshafen, and was distilled twice from potassium hydroxide pellets.

9. Lithium aluminum hydride ($LiAlH_4$) was used as a white powder purchased from the Metallgesellschaft AG, D-Frankfurt.

10. The specific rotation is highly sensitive to the water content of the DDB; only material distilled from $LiAlH_4$ shows this value.

3. Discussion

The three compounds whose syntheses are described in the present procedure have been reported previously by the submitters.[2,3]

The amino ether DDB has been used extensively as a chiral solvent for asymmetric syntheses.[2-8] It is readily available on a large scale in both enantiomeric forms: starting from the unnatural (S,S)-(−)-tartaric acid,[9] (−)-DDB is equally accessible[3] following the procedures described herein.

As demonstrated by the examples listed in Table I, DDB induces chirality in enantioface, enantiotope, and enantiomer differentiating[10] reactions in which it acts as a metal (Li, Mg, Cu, Zn) complexing ligand, as a hydrogen-bond mediating component, and as a base catalyst. It can be used at temperatures as low as −150°C if mixed with appropriate cosolvents.[3] It is readily recovered and separated from products by acid extraction during workup. The enantiomeric excess (e.e.) obtained in this asymmetric induction is generally in the range of 10–20%; in optimized and/or fortuitous cases, optical yields of up to 50% have been obtained. The chemical yields are as high as in conventional achiral solvent systems. An application of DDB is described on p. 000 of this volume.

TABLE I
ASYMMETRIC SYNTHESES WITH (+)-DDB AS A CHIRAL AUXILIARY AGENT[2-8]

Reagents	Conditions (DDB : Reagent, Temp. °C, Solvent)	Product	$[\alpha]_D$ (Solvent, c) (%, e.e)
C_6H_5CHO + Bu_2Mg	2 : 1, −78, ether	$C_6H_5\overset{\text{OH}}{\underset{\|}{C}}HC_4H_9$	−2.5° (C_6H_6, 7.0) (8)
C_6H_5CHO + BuLi	10 : 1, −150, pentane	$C_6H_5\overset{\text{OH}}{\underset{\|}{C}}HC_4H_9$	+7° (neat) (40)
C_6H_5CHO + i-PrLi	4 : 1, −120, pentane	$C_6H_5\overset{\text{OH}}{\underset{\|}{C}}H\text{-}i\text{-Pr}$	+6.1° (ether, 7.5) (14)
C_6H_5CHO + $(C_6H_5S)_3CLi$	10 : 1, −78, pentane	$C_6H_5\overset{\text{OH}}{\underset{\|}{C}}HC(SC_6H_5)_3$	+23° (C_6H_6, 1.03) (12)
C_6H_5CHO + $CH_3\overset{\text{NO}}{\underset{\|}{N}}CH_2Li$	10 : 1, −78, pentane	$C_6H_5\overset{\text{OH}}{\underset{\|}{C}}HCH_2\overset{\text{NO}}{\underset{\|}{N}}CH_3$	+6.5 (CH_2Cl_2, 3.0) (14.8)
C_6H_5CHO + $CH_2{=}\overset{\text{OLi}}{\underset{\|}{C}}{-}O\text{-}t\text{-Bu}$	10 : 1, −78, pentane	$C_6H_5\overset{\text{OH}}{\underset{\|}{C}}HCH_2\overset{O}{\overset{\|}{C}}{-}O\text{-}t\text{-Bu}$	+5.1° (C_6H_6, 11.1)
C_6H_5CHO + $CH_2{=}\overset{\text{OLi}}{\underset{\|}{C}}NMe_2$	10 : 1, −78, pentane	$C_6H_5\overset{\text{OH}}{\underset{\|}{C}}HCH_2\overset{O}{\overset{\|}{C}}NMe_2$	+9.1° (C_6H_6, 12.4) (14)
$(C_6H_5)_2CO$ + $CH_3CH{=}CNMe_2$	10 : 1, −78, pentane	$(C_6H_5)_2\overset{\text{OH}}{\underset{\|}{C}}{-}\overset{\;}{\underset{\underset{CH_3}{\|}}{C}}{-}\overset{O}{\overset{\|}{C}}{-}NMe_2$	+8.5° (C_6H_6, 11.3) (~22)

$C_6H_5COCH_3 + LiAlH_4$ -78, pentane

$$\underset{\underset{C_6H_5CHCH_3}{}}{OH}$$

$+4.7°$ (neat) (11)

$C_6H_5CH{=}NC_6H_5$ + 2-methyl-2-lithio-1,3-dithiane $3:1$, -30, hexane

2-methyl-2-(C_6H_5CH–NH–C_6H_5)-1,3-dithiane

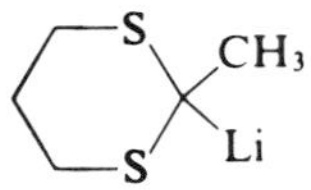

$+15.5°$ (CH_2Cl_2, 18.7)

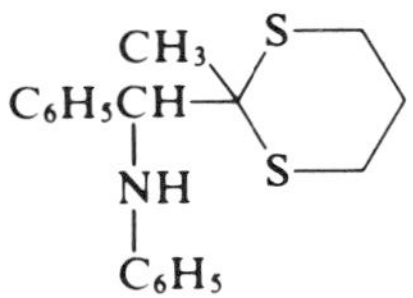

cyclopent-2-enone + Bu_2CuLi $10:1$, -78, ether

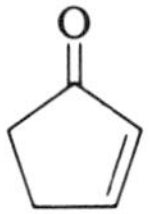

3-Bu-cyclopentanone

$-10°$ (C_6H_6, 5.9)

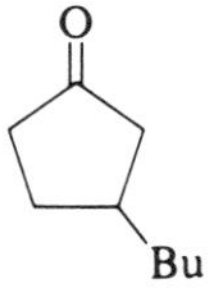

cyclopent-2-enone + Bu_3ZnLi $10:1$, -78, ether

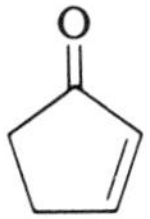

3-Bu-cyclopentanone

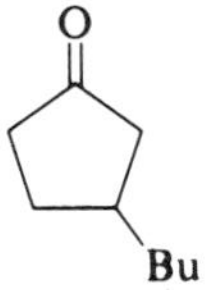

$+6.9°$ (C_6H_6, 6.8)

cyclohex-2-enone + Bu_2CuLi $10:1$, -78, ether

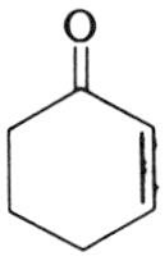

3-Bu-cyclohexanone

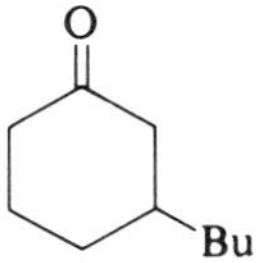

$-0.9°$ (C_6H_6, 11.5)

cyclohex-2-enone + Me_2CuLi $10:1$, -78, ether

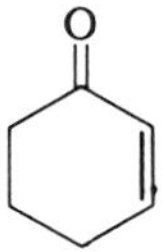

3-Me-cyclohexanone

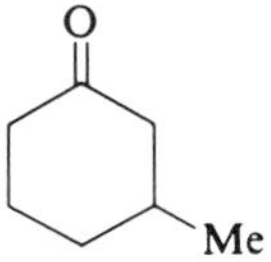

$-1.7°$ (neat) (13.6)

TABLE I (*Continued*)

Reagents	Conditions (DDB : Reagent, Temp. °C, Solvent)	Product	$[\alpha]_D$ (Solvent, c) (%, e.e)
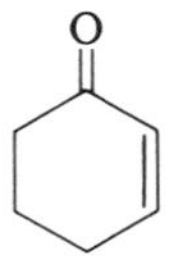 + Me₃ZnLi	10 : 1, −78, ether	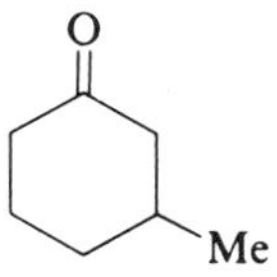	−0.74° (C₆H₆, 6)
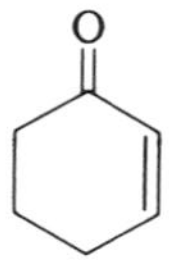 + Bu₃ZnLi	10 : 1, −78, ether	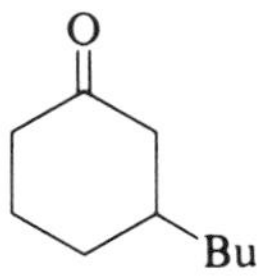	−1.0° (C₆H₆, 4.3)
$CH_3CH{=}CHNO_2$ + BuLi	10 : 1, −78, pentane	$O_2NCH_2CHCHC_4H_9$ $\quad\quad\quad\ CH_3$	+0.9° (C₆H₆, 10.4) (28)
$CH_3CH{=}CHNO_2$ 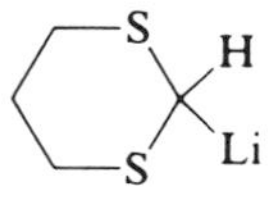 +	10 : 1, −78, pentane	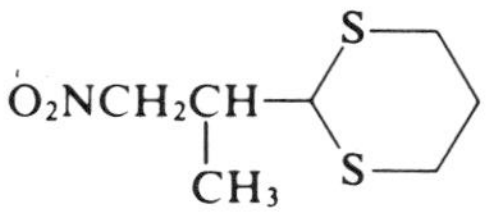	−4.3° (C₆H₆, 7.5) (45)
$CH_3CH{=}CHNO_2$ + $H_2C{=}C(OLi)NMe_2$	10 : 1, −78, pentane		+0.7° (C₆H₆, 5.4) (10)

(CH₃)₃SiCl + [aziridine, Li, N–CH₃] 9 : 1, − 125, pentane → [aziridine, (CH₃)₃Si, N–CH₃] − 27.5° (C₆D₆, 4.9) (12.4)

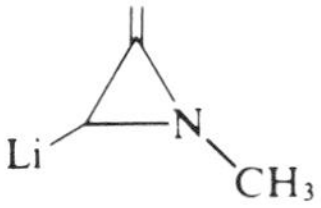

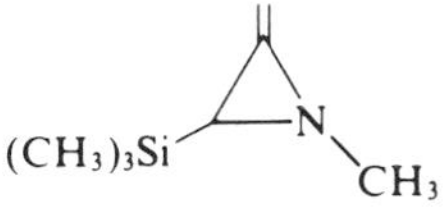

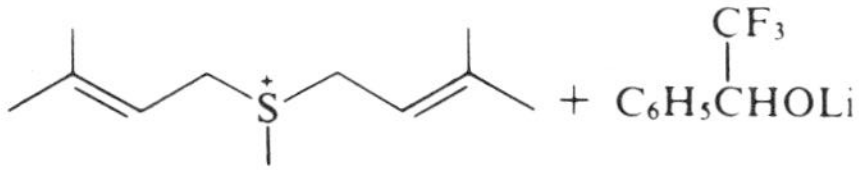

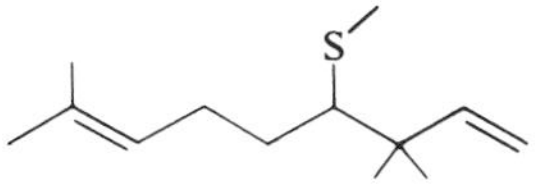

[allyl sulfonium ylide] + C₆H₅CHOLi (CF₃) − 20, THF → [S–CH substituted alkene] + 2.9 ± 0.3° [365 nm] (CHCl₃, 0.62) (12)

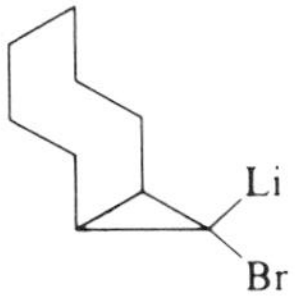

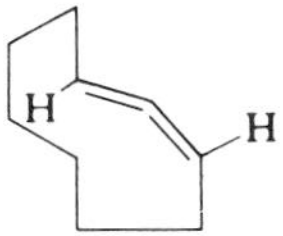

[bicyclic Li, Br] 10 : 1, − 78, pentane → [bicyclic H, H] + 8.2° (neat) (5)

C₆H₅[cyclopropane Br, Br] + ½ BuLi + H₂O 10 : 1, − 100, pentane → [cyclopropane Br, Br, C₆H₅, H] + 1.9° (neat)

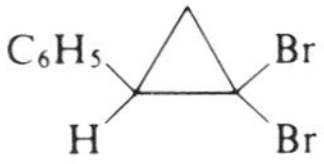

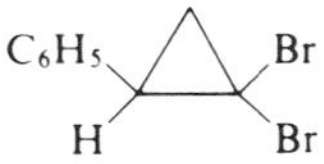

C₆H₅COCH₃ + hν 7.5 : 1, − 72, pentane → [diol: Br OH C₆H₅ / C₆H₅ OH CH₃ ...] + 8.0° (C₂H₅OH, 5.0) (23.5)

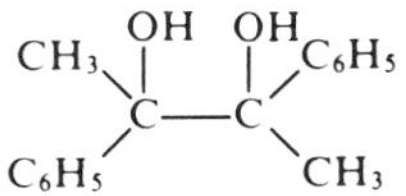

C₆H₅COCH₃ + cathodic. electrochemical reduction 24, methanol → [diol] + 2.2° (C₂H₅OH, 5.0) (6.4)

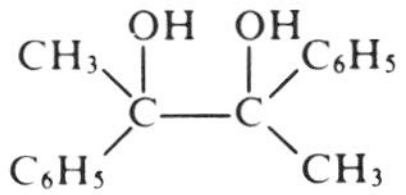

C₆H₅COCH₃ + Li/naphthalene 6 : 1, 25, → [diol] 2.5° (C₂H₅OH, 5.0) (7.3)

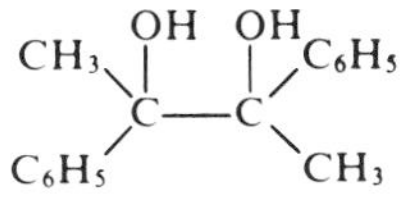

47

TABLE I (*Continued*)

Reagents	Conditions (DDB : Reagent, Temp. °C, Solvent)	Product	$[\alpha]_D$ (Solvent, c) (%, e.e)
C_6H_5CO-t-Bu + $h\nu$	−30, neat DDB		−1.2°[($C_2H_5)_2O$, 5.0]
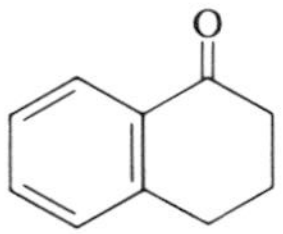 + $h\nu$	−35, pentane	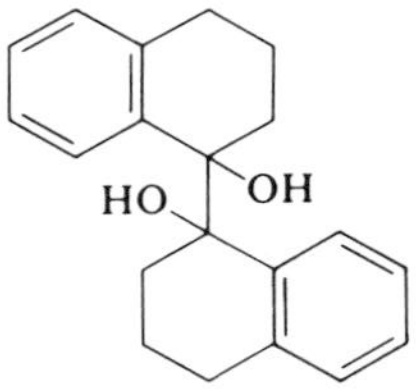	+22.3° (CHCl₃, 5.0)
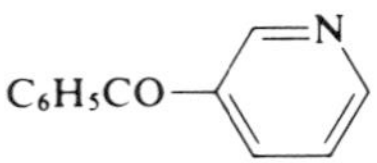 + $h\nu$	−15, neat DDB	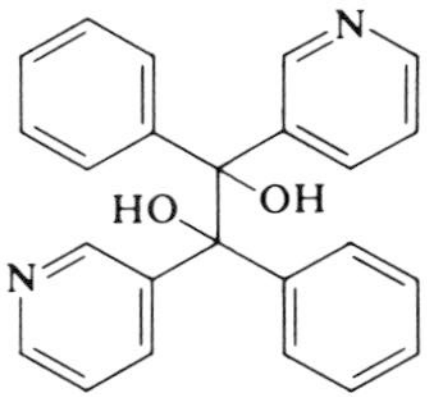	−20.0° (CH₃SOCH₃, 2.0)
C_6H_5CHO + $C_6H_5CH_2COOH$ + AcOAc	6 : 1, −25		diastereomer a +5.0°, diastereomer b +26.0° (C₂H₅OH, 3)

TABLE II

COMPARISON OF TMB[a] WITH DDB[b] USED AS COSOLVENTS IN THE ADDITION OF
BUTYLLITHIUM TO VARIOUS ALDEHYDES AT $-78°C$ IN PENTANE[3]

$$RCHO + C_4H_9Li \rightarrow RCH(OH)C_4H_9$$

R	e.e. (%)		Sense of Rotation, Absolute Configuration
	With TMB[a]	With DDB[b]	
CH_3	1.2	7.5	(+)-S
C_2H_5	8.8	11.5	(+)-S
i-C_3H_7	18.0	19.0	(+)-R
t-C_4H_9	22.8	13.5	(+)-R
$(C_2H_5)_2CH$	20.0	19.0	(+)
c-C_6H_{11}	25.0	22.5	(+)-R
C_6H_5	30.0	19.0	(+)-R
4-CH_3-C_6H_4	32.5	11.5	(+)
2-CH_3-C_6H_4	45.3	10.5	(+)
2,4,6-$(CH_3)_3$-C_6H_2	23.0	2.3	(+)

[a]TMB = (S,S)-(−)-1,2,3,4-tetramethoxybutane.
[b]DDB = (S,S)-(+)-1,4-bis(dimethylamino)-2,3-dimethoxybutane.

Another chiral cosolvent, which is less readily separated from low-boiling and/or water-soluble products and is somewhat less stable toward organolithium reagents, is (S,S)-(−)-1,2,3,4-tetramethoxybutane (TMB).[3] As is shown in Table II, it is a cosolvent that is superior to DDB in differentiating between the enantiotopic faces of aldehydes with organolithium reagents.[3] Finally, the octamethyl-1,4-diamino-2,3-bis(2-aminoethoxy)butane (DEB)[5] can be used in a 2 : 1 ratio with alkyllithium reagents to produce carbinols in even higher enantiomeric yields.

DDB, TMB, and DEB are far superior to other neutral chiral auxiliary agents used in the same reactions.[3, 10-13]

1. Laboratorium für Organische Chemie der Eidgenössischen Technischen Hochschule, ETH-Zentrum, Universitätstrasse 16, CH-8092-Zürich and Institut für Organische Chemie der Justus Liebig-Universität, Giessen, Fachbereich 14, Heinrich-Buff-Ring 58, D-6300-Giessen.

2. Seebach. D.; Dörr, H.; Bastani, B.; Ehrig, V. *Angew. Chem.* **1969,** *81,* 1002; *Angew. Chem., Int. Ed. Engl.* **1969, 8,** 982.

3. Seebach, D.; Kalinowski, H.-O.; Bastani, B.; Crass, G.; Daum, H.; Dörr, H; Du Preez, N. P.; Ehrig, V.; Langer, W.; Nüssler, Ch.; Oei, H.-A.; Schmidt, M. *Helv. Chim. Acta* **1977,** *60,* 301.

4. Seebach, D.; Oei, H.-A. *Angew. Chem.* **1975,** *87,* 629; *Angew. Chem., Int. Ed. Engl.* **1975,** *14,* 634. Seebach. D.; Oei. H.-A.; Daum, H. *Chem. Ber.* **1977,** *110,* 2316.

5. Seebach, D; Langer, W.; Crass, G.; Wilka, E. M.; Hilvert, D.; Brunner, E. *Helv. Chim. Acta* **1979,** *62,* 2695; Seebach, D.; Langer, W. *Helv. Chim. Acta* **1979,** *62,* 1701, 1710.

6. Trost, B. M.; Biddlecom, W. G. *J. Org. Chem.* **1973,** *38,* 3438.

7. Karlsen, S.; Frøyen, P.; Skattebøl, L. *Acta Chem. Scand. Ser. B* **1976,** *30,* 664.

8. Quast, H.; Weise Velez, C. A. *Angew. Chem.* **1978,** *90,* 224; *Angew. Chem., Int. Ed. Engl.* **1978,** *17,* 213.

9. Natural (R,R)-$(+)$-tartaric acid costs DM 142/3 kg (Aldrich Chemical Company, Inc.); the unnatural (S,S)-enantiomer can be purchased from Chemische Fabrik Uetikon, CH-Uetikon, at SFr. 350/kg, 195/kg (as a 100-kg batch), or 70/kg (> 1000-kg batch).
10. Izumi, Y.; Tai, A. "Stereo-Differentiating Reactions," Academic Press: New York, San Francisco, London; Kodansha Ltd. Tokyo, 1977.
11. Morrison, J. D.; Mosher, H. S. "Asymmetric Organic Reactions," Prentice-Hall: Englewood Cliffs, NJ, 1971.
12. Langer, A. W. "Polyamine-Chelated Alkali Metal Compounds," Advances in Chemistry Series, American Chemical Society: Washington, DC, 1974.
13. Nozaki, H.; Aratani, T.; Toraya, T.; Noyori, R. *Tetrahedron*, **1971**, *27*, 905.

CONDENSATION OF DIMETHYL 1,3-ACETONEDICARBOXYLATE WITH 1,2-DICARBONYL COMPOUNDS: *cis*-BICYCLO[3.3.0]OCTANE-3,7-DIONES

[2,5(1*H*,3*H*)-Pentalenedione, tetrahydro-, *cis*-, and 2,5(1*H*,3*H*)-pentalenedione, tetrahydro-, 3a,6a-dimethyl-, *cis*]

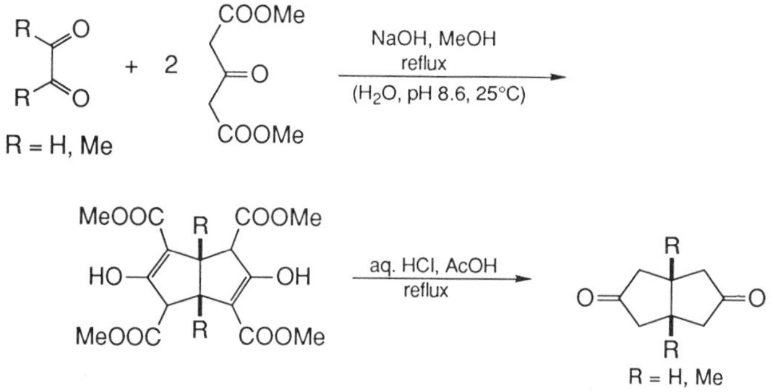

Submitted by STEVEN H. BERTZ,[1] JAMES M. COOK,[2] ALI GAWISH,[2] and ULRICH WEISS[3]
Checked by TODD K. JONES, SCOTT E. DENMARK, S. V. GOVINDAN, and ROBERT M. COATES

1. Procedure

I. Specific Procedure for Glyoxal

A. *Tetramethyl 3,7-dihydroxybicyclo[3.3.0]octa-2,6-diene-2,4,6,8-tetracarboxylate.* A 3-L, three-necked, round-bottomed flask is equipped with a thermometer, mechanical stirrer, pressure-equalizing dropping funnel, reflux condenser (Note 1), and a heating mantle. A solution of 64 g (1.60 mol) of sodium hydroxide (Note 2) in 1.15 L of methanol is prepared in the flask, cooled in an ice bath, and stirred as 273 g (1.57 mol) of dimethyl 1,3-acetonedicarboxylate (Note 3) is added dropwise. The resulting slurry is stirred and heated to reflux, at which point the white salt dissolves. The heating mantle is removed, and the solution is stirred rapidly while 128.5 g of aqueous 40% glyoxal (51.4 g, 0.886 mol) (Notes 3 and 4) is added at a rate sufficient to maintain the internal temperature at 65°C (Note 5). After the addition is completed (40–60 min, Note

6), the mixture is allowed to cool to room temperature and stirred overnight (Note 7). The precipitate is collected by suction filtration, washed with 500 mL of methanol (Note 8), and dried under reduced pressure. The yield of the white to light yellow disodium salt is 197–215 g (58–63%) (Note 9).

A 6-L Erlenmeyer flask equipped with a large magnetic stirring bar (Note 10) is charged with 1 L of chloroform and a solution of the disodium salt (0.46–0.50 mol) in 800 mL of water. The two-phase mixture is stirred rapidly as 2.00 equiv (920–1000 mL, 0.92–1.00 mol) of cold 1 M hydrochloric acid is added. The layers are separated and the aqueous phase is extracted with three 500-mL portions of chloroform. The combined organic layers are washed once with saturated sodium chloride, dried over anhydrous sodium sulfate, and concentrated under reduced pressure by rotary evaporation, keeping the water bath temperature at or below 40°C. Crystallization of the remaining waxy solid from 2 : 1 hexane–ethyl acetate affords 158–171 g (54–59% based on dimethyl 1,3-acetonedicarboxylate) of the tetraester, mp 97–100°C (Note 11).

B. *cis-Bicyclo[3.3.0]octane-3,7-dione.* A 3-L, three-necked, round-bottomed flask equipped with a heating mantle, two reflux condensers, and a magnetic stirrer (Note 12) is charged with 135 g (0.364 mol) of the tetraester, 66 mL of glacial acetic acid, and 600 mL of 1 M hydrochloric acid (Note 13). The mixture is stirred vigorously and heated at reflux for 2.5 hr (Note 14). The solution is cooled in an ice bath and the product is extracted with five 250-mL portions of chloroform. The chloroform extracts are combined, and the solution is concentrated by rotary evaporation (bath temperature at or below 40°C) until most of the acetic acid is removed. The residue is dissolved in 300 mL of fresh chloroform. The solution is washed with 60-mL portions of saturated sodium bicarbonate until the aqueous layer remains basic to litmus paper, dried with anhydrous sodium sulfate, and evaporated cautiously under reduced pressure. The yield is 44–45.5 g (88–90%) of white to light yellow solid, mp 84–85°C (Note 15). The product is sufficiently pure for most purposes; it may be purified by recrystallization from methanol or ethanol and/or by sublimation at 70°C (0.1 mm).

II.   General Procedure Using Aqueous Buffer

A. *Tetramethyl   3,7-dihydroxy-1,5-dimethylbicyclo[3.3.0]octa-2,6-diene-2,4,6,8-tetracarboxylate.* A freshly prepared solution (pH 8.3) of 5.6 g of sodium bicarbonate in 400 mL of water, 70 g (0.40 mol) of dimethyl 1,3-acetonedicarboxylate, and a magnetic stirring bar are placed in a 1-L Erlenmeyer flask. The resulting solution is stirred rapidly as 17.2 g (0.20 mol) of biacetyl (Note 3) is added in one portion. Stirring is continued for 24 hr, during which time white crystals separate. The solid is collected by suction filtration and dried under reduced pressure to afford 60–62 g, mp 155–158°C. The filtrate is cooled in an ice bath, acidified to pH 5 (pHydrion paper) with dilute hydrochloric acid, and extracted with three 100-mL portions of chloroform. The chloroform extracts are combined, washed with saturated sodium chloride, and dried with anhydrous sodium sulfate. Evaporation of the solvent under reduced pressure gives another 2–4 g of crude product (Note 16). Recrystallization from hot methanol gives 58–60 g (73–75%) of the tetraester, mp 155–157°C, in two crops (Note 17).

B. *cis-1,5-Dimethylbicyclo[3.3.0]octane-3,7-dione.* A 1-L, one-necked, round-bottomed flask is equipped with a heating mantle, reflux condenser, and a magnetic stirring bar. The flask is charged with 200 mL of 1 M hydrochloric acid, 40 mL of glacial acetic acid, and 24 g (0.060 mol) of the tetraester from Section IIA. The mixture is stirred vigorously and heated at reflux for 3–6 hr (Note 18). The solution is cooled in an ice bath and the product is isolated as described in Section IB above. The yield of c white to light yellow solid, mp 219–221°C, is 9.5–9.8 g (95–98%). Recrystallization from a minimum amount of hot ethanol affords 7.5–7.7 g (75–77%) of the diketone, mp 222–225°C, in one crop (Note 19).

2. Notes

1. The dropping funnel and the reflux condenser were connected to the same neck using a Claisen adapter.

2. The submitters recommend that a high-purity grade of sodium hydroxide be used. Otherwise insoluble impurities are formed that must be removed by filtration through a sintered-glass Buchner funnel.

3. Dimethyl 1,3-acetonedicarboxylate, aqueous 40% glyoxal, and biacetyl were purchased from Aldrich Chemical Company, Inc. The submitters advise against using glyoxal solution that contains a significant amount of white solid.

4. The submitters report that the yield is decreased by 5% if exactly 0.5 equiv of glyoxal is used. The yield is improved to 75–76% in runs carried out on smaller scale (ca. 0.1 mol of glyoxal).

5. Heating should be resumed if necessary to maintain a temperature of 65°C. The submitters report that lower yields are obtained at lower temperatures (e.g., 37% at 25°C).

6. The submitters caution that the addition time is critical. In one run by the checkers with a 30-min addition time, the yield of the disodium salt was reduced to 49%.

7. Similar yields were obtained by the submitters when the reaction mixture was allowed to cool at room temperature for 2 hr and in an ice bath for another 2 hr.

8. The solid is washed first by allowing methanol to percolate through the filter cake with gentle suction until the brown color is removed. The product is suspended in methanol, filtered, and washed again by the percolation procedure.

9. The submitters obtained 411–430 g (61–64%) from reactions conducted on twice the scale described using a 2-hr addition time. Elemental analyses by the submitters and checkers indicate a variable degree of hydration ($n = 1$–2) for the product. The yield and molar quantities of the disodium salt are calculated assuming a monohydrate, $C_{16}H_{16}O_{10}Na_2 \cdot H_2O$. For further characterization of this salt, see reference 6.

10. The checkers used a mechanical stirrer to achieve more efficient mixing of the layers.

11. The submitters obtained 176–182 g (62–64%) of product, mp 103–105°C, after trituration with a minimum amount of cold methanol. Crystallization was facilitated by scraping the sticky solid with a silver spatula. The reported melting point is 104–107°C.[4] The crude product obtained initially by the checkers was a low-melting solid, mp 70–75°C, that was conveniently transferred and purified by recrystallization from about 1 L of hot 2 : 1 hexane–ethyl acetate. Elemental

analyses of the product by the submitters were within $\pm 0.4\%$ of the theoretical value. The spectral properties of the product are as follows: IR (CHCl$_3$) cm^{-1}: 3025, 2960, 1740, 1673, 1632, 1450, 1438, 1330, 1250, 1200, 1155, 1058; ^{1}H NMR (CDCl$_3$, 200 MHz) δ: 3.64 (apparent t, 2 H, J_{app} = 2.4, two CH), 3.78 (s, 6 H, two OCH$_3$), 3.81 (s, 6 H, two OCH$_3$), 3.87 (apparent t, 2 H, J_{app} = 2.4, two CH), 10.35 (broad s, 2 H, two enolic OH).

12. The volume of the flask should be at least three times larger than the volume of the solution to avoid losses from excessive foaming caused by rapid evolution of carbon dioxide. The checkers used a mechanical stirrer to facilitate stirring of the initially heterogeneous mixture. Two reflux condensers are recommended for efficient venting of the gas evolved.

13. The submitters point out that the disodium salt may be used directly provided that two additional equivalents of 1 *M* hydrochloric acid are employed. Acetic acid may be omitted to simplify the isolation procedure. In this case the reaction mixture remains heterogeneous throughout.

14. Progress of the reaction can be followed by observing the gas evolved through a bubbler connected to the top of the reflux condensers.

15. The product gave satisfactory elemental analyses. Calcd. for C$_8$H$_{10}$O$_2$: C, 69.54; H, 7.30. Found: C, 69.42; H, 7.36. The literature melting point is 84–86°C.[4] The spectral properties of the product are as follows: IR (CHCl$_3$) cm^{-1}: 1738 (C=O), 1405, 1222, 1208, 1175, 792; ^{1}H NMR (CDCl$_3$, 220 MHz) δ: 2.16 (dd, 4 H, J = 4.2, 19.3, H$_A$ of CH$_A$H$_B$), 2.59 (dd, 4 H, J = 8.5, 19.3, H$_B$ of CH$_A$H$_B$), 3.04 (m, 2 H, CH); ^{13}C NMR (CDCl$_3$) δ: 35.5 (d, CH), 42.6 (t, CH$_2$), 217.2 (s, C=O); mass spectrum (70 eV) m/e (relative intensity): 138 (M$^+$, 41), 69 (36), 68 (58), 41 (100), 39 (53).

16. The checkers obtained ca. 19 g of a viscous red oil that on dissolution in ca. 150 mL of methanol deposited 4 g of crude crystalline product.

17. The product gave a satisfactory elemental analysis. Calcd. for C$_{18}$H$_{22}$O$_{10}$: C, 54.28; H, 5.53. Found: C, 54.00; H, 5.57. The melting point reported initially (167–169°C after sublimation)[5] is apparently incorrect. The spectral properties of the product are as follows: IR (KBr) cm^{-1}: 3539, 1742, 1664, 1425, 1340, 1260, 1235, 1070, 1020; ^{1}H NMR (CDCl$_3$) δ: 1.29 (s, 6 H, two CH$_3$), 3.75 (s, 6 H, two CO$_2$CH$_3$), 3.87 (s, 6 H, two CO$_2$CH$_3$), 3.94 (s, 2 H, two CH), 10.62 (br s, 2 H, two OH).

18. The checkers recovered a 1 : 1 mixture of tetraester and diketone from two runs conducted for 2.5 hr. When the reflux time was extended to 6 hr, complete conversion to product was attained. The reaction progress was monitored by gas evolution (Note 14). Some variability of reaction times is probably attributable to differences in stirring efficiency, temperature gradients, and/or particle size of the crystalline starting material.

19. The recrystallized product was analyzed by the checkers. Calcd. for C$_{10}$H$_{14}$O$_2$: C, 72.26; H, 8.49. Found: C, 72.45; H, 8.54. The spectral properties of the product are as follows: IR (KBr) cm^{-1}: 1736, 1390, 1245, 1210, 1180, 1145, 1070; ^{1}H NMR (CDCl$_3$, 360 MHz) δ: 1.22 (s, 6 H, two CH$_3$), 2.36 and 2.39 (AB q, 8 H, J = 18.5, four CH$_2$).

3. Discussion

Bicyclo[3.3.0]octane-3,7-dione has been prepared in five steps from dimethyl malonate and chloral in about 20% overall yield.[4] The direct formation of bicyclo[3.3.0]octane-3,7-diones by the 2 : 1 condensation of acetone-1,3-dicarboxylate and 1,2-dicarbonyl compounds was discovered by Weiss and Edwards[5] and is commonly called the Weiss Reaction.[3] The variation described in Section I has been optimized for large-scale preparation of the parent diketone.[6] It is a good example of what Turner[7] has called a "point reaction," as it is very sensitive to experimental details such as temperature and stirring rate. The aqueous buffer procedure given in Part II is a "plateau reaction"[7] and affords a general method for preparing a variety of angularly-substituted bicyclo[3.3.0]octane-3,7-diones (Table I).[8–15] The parent diketone can also be prepared by the aqueous buffer procedure (Part II), but chromatography is required to purify the product.[13] The mechanism of this novel annulation reaction involves a complex sequence of aldol condensations, dehydrations, and Michael additions,[9,16,17] the order of which

TABLE I

2 : 1 CONDENSATION OF DIMETHYL 1,3-ACETONEDICARBOXYLATE WITH VARIOUS 1,2-DICARBONYL COMPOUNDS

R'—CO—CO—R	MeOOC R' COOMe HO— —OH MeOOC R COOMe	Yield (%)	Ref.
Glyoxal	$R = R' = H$	70	6, 13
Pyruvaldehyde	$R = H, R' = CH_3$	52	5
Phenylglyoxal	$R = H, R' = Ph$	66	9
3-Cyclopentenylglyoxal	$R = H, R' = C_5H_7$	90	15
4-Cycloheptenylglyoxal	$R = H, R' = C_7H_{11}$	70	15
4,5-Dioxopentanoic Acid	$R = H, R' = CH_2CH_2CO_2H$	80	8
Dimethyl 3-(dioxo ethyl)glutarate	$R = H, R' = CH(CH_2CO_2CH_3)_2$	51	16a
Biacetyl	$R = R' = CH_3$	84	5
Bis(cyclopentyl) 1,2-ethane-dione	$R = R' = C_5H_9$	12	15, 16a
2,3-Pentanedione	$R = CH_3, R' = CH_2CH_3$	70	15
2,3-Hexanedione	$R = CH_3, R' = CH_2CH_2CH_3$	64	15
Cyclopentane-1,2-dione	$R = R' = (CH_2)_3$	45	5, 14
Cyclohexane-1,2-dione	$R = R' = (CH_2)_4$	81	10
Cyclooctane-1,2-dione	$R = R' = (CH_2)_6$	80	11
Cyclooct-5-ene-1,2-dione	$R = R' = (CH_2)_2CH=CH(CH_2)_2$	87	12
Cyclododecane-1,2-dione	$R = R' = (CH_2)_{10}$	94	11
1-Phenyl-1,2-propanedione	$R = CH_3, R' = Ph$	68	15
Ninhydrin	$R = R' = C_7H_4O$	60	9
Furil	$R = R' = C_4H_3O$	60	16c
Phenanthrenequinone	$R - R' = 2,2'$-biphenyl	54	15, 16c
Dimethyl 2,3-dioxosuccinate	$R = R' = CO_2CH_3$	14	19g

may be pH-dependent.[17] The isolation of γ-hydroxycyclopentenones in certain cases[9,16] implicates these reactive Michael acceptors as intermediates. A number of other interesting products have been isolated from reaction of glyoxal with acetonedicarboxylate.[17,18] The various bicyclic diketones prepared by this method have served as starting materials for syntheses of polycyclic compounds[19] and natural products.[20] The reaction of dimethyl 1,3-acetonedicarboxylate with cyclic 1,2-diketones is a particularly effective method of propellane synthesis;[5,10,21] for instance, Ginsburg has used it to prepare a propellane with a 40-membered ring.[21c] Routes have been developed to triquinacene[22a] and its derivatives[22b-d] which employ as the key step the chemistry discussed in this procedure. This general synthesis of bicyclo[3.3.0]octane derivatives has also been extended to selected examples of bicyclo[3.3.1]nonane and bicyclo[3.3.2]decane systems by substituting malondialdehyde[23] and *o*-phthalaldehyde,[24] respectively, for glyoxal. Reviews of recent progress by the Cook group are available.[25]

1. AT&T Bell Laboratories, Murray Hill, NJ 07974.

2. Department of Chemistry, University of Wisconsin-Milwaukee, Milwaukee, WI 53201.

3. National Institute of Arthritis, Diabetes, and Digestive and Kidney Diseases, National Institutes of Health, Bethesda, MD 20205. Dr. Weiss died July 11, 1989. His co-authors dedicate this paper in appreciation of his constant generosity.

4. (a) Vossen, G. Dissertation, Bonn, 1910; (b) Yates, P.; Hand, E. S.; French, G. B. *J. Am. Chem. Soc.* **1960**, *82*, 6347-6353.

5. Weiss, U.; Edwards, J. M. *Tetrahedron Lett.* **1968**, 4885-4887.

6. (a) Bertz, S. H. Ph.D. Dissertation, Harvard University, 1978; (b) Bertz, S. H.; Rihs, G.; Woodward, R. B. *Tetrahedron* **1982**, *38*, 63-70.

7. Turner, S. "The Design of Organic Syntheses"; Elsevier: Amsterdam, 1976; p. 21.

8. Oehldrich, J.; Cook, J. M.; Weiss, U. *Tetrahedron Lett.* **1976**, 4549-4552.

9. Yang-Lan, S.; Mueller-Johnson, M.; Oehldrich, J.; Wichman, D.; Cook, J. M.; Weiss, U. *J. Org. Chem.* **1976**, *41*, 4053-4058.

10. Weber, R. W.; Cook, J. M. *Can. J. Chem.* **1978**, *56*, 189-192.

11. Yang, S.; Cook, J. M. *J. Org. Chem.* **1976**, *41*,1903-1907.

12. Gawish, A.; Mitschka, R.; Cook, J. M.; Weiss, U. *Tetrahedron Lett.* **1981**, *22*, 211-214.

13. Oehldrich, J.; M. S. Thesis, University of Wisconsin, Milwaukee, 1976.

14. Deshpande, M. N.; Wehrli, S.; Jawdosiuk, M.; Guy, J. T.; Bennett, D. W.; Cook, J. M.; Depp, M. R.; Weiss, U. *J. Org. Chem.* **1986**, *51*, 2436.

15. Kubiak, G.; Ph.D. Thesis, University of Wisconsin, Milwaukee, 1988.

16. (a) Avasthi, K.; Deshpande, M. N.; Han, W.-C.; Cook, J. M.; Weiss, U. *Tetrahedron Lett.* **1981**, *22*, 3475-3478; (b) Kubiak, G.; Cook, J. M.; Weiss, U. *J. Org. Chem.* **1984**, *49*, 561-564; (c) Kubiak, G.; Cook, J. M.; Weiss, U. *Tetrahedron Lett.* **1985**, *26*, 2163-2166.

17. Bertz, S. H.; Adams, W. O.; Silverton, J. V. *J. Org. Chem.* **1981**, *46*, 2828-2830.

18. (a) Edwards, J. M.; Qureshi, I. H.; Weiss, U.; Akiyama, T.; Silverton, J. V. *J. Org. Chem.* **1973**, *38*, 2919-2920; (b) Rice, K. C.; Weiss, U.; Akiyama, T.; Highet, R. J.; Lee, T.; Silverton, J. V. *Tetrahedron Lett.* **1975**, 3767-3770; (c) Bertz, S. H.; Kouba, J.; Sharpless, N. E. *J. Am. Chem. Soc.* **1983**, *105*, 4116-4117.

19. (a) Mitschka, R.; Cook, J. M.; Weiss, U. *J. Am. Chem. Soc.* **1978**, *100*, 3973-3974; (b) Borden, W. T.; Ravindranathan, T. *J. Org. Chem.* **1971**, *36*, 4125-4127; (c) Askani, R.; Kirsten, R.; Dugall, B. *Tetrahedron* **1981**, *37*, 4437-4444; (d) Paquette, L. A.; Ley, S. V.; Meisinger, R. H.; Russell, R. K.; Oku, M. *J. Am. Chem. Soc.* **1974**, *96*, 5806-5815; (e) Baldwin, J. E.; Kaplan, M. S. *J. Am. Chem. Soc.* **1971**, *93*, 3969-3977; (f) Camps, P.; Iglesias, C.; Lozano, R.; Miranda, M. A.; Rodrígues, M. J. *Tetrahedron Lett.* **1987**, *28*, 1831-1832; (g) Camps, P.; Figueredo, M. *Can. J. Chem.* **1984**, *62*, 1184-1193; (h) Lannoye, G.; Cook, J. M. *Tetrahedron Lett.* **1988**, *29*, 171-174; (i) Venkatachalam, M.; Jawdosiuk,

M.; Deshpande, M.; Cook, J. M . *Tetrahedron Lett.* **1985**, *26*, 2275-2278; (j) Venkatachalam, M.; Wehrli, S.; Kubiak, G.; Cook, J. M.; Weiss, U. *Tetrahedron Lett.* **1986**, *27*, 4111-
4114; (k) Venkatachalam, M.; Wehrli, S.; Kubiak, G.; Cook, J. M.; Weiss, U. *J. Org. Chem.*
1987, *52*, 4110-4115; (l) Lannoye, G.; Sambasivarao, K.; Wehrli, S.; Cook, J. M.; Weiss,
U. *J. Org. Chem.* **1988**, *53*, 2327-2340; (m) Sambasivarao, K.; Kubiak, G.; Lannoye, G.;
Cook, J. M.; *J. Org. Chem.*, **1988**, *53*, 5173-5175.
20. (a) Wrobel, J.; Takahashi, K.; Honkan, V.; Lannoye, G.; Cook, J. M.; Bertz, S. H. *J. Org.
Chem.* **1983**, *48*, 139-141; (b) Coates, R. M.; Shah, S. K.; Mason, R. W. *J. Am. Chem. Soc.*
1982, *104*, 2198-2208; (c) Paquette, L. A.; Han, Y.-K. *J. Am. Chem. Soc.* **1981**, *103*, 1831-
1835; (d) Nicolaou, K. C.; Sipio, W. J.; Magolda, R. L.; Seitz, S.; Barnette, W. E. *J. Chem.
Soc., Chem. Commun.* **1978**, 1067-1068; (e) Shibasaki, M.; Ueda, J.-i., Ikegami, S. *Tetrahedron Lett.* **1979**, 433-436; (f) Sakan, F.; Hashimoto, H.; Ichihara, A.; Shirahama, H.;
Matsumoto, T. *Tetrahedron Lett.* **1971**, 3703-3706; (g) Caille, J. C.; Bellamy, F.; Guiland,
R. *Tetrahedron Lett.* **1984**, *25*, 2345-2346; (h) Piers, E.; Karunaratne. V. *J. Chem. Soc.,
Chem. Commun.* **1984**, 959-960.
21. (a) Ashkenazi, P.; Kettenring, J.; Migdal, S.; Gutman, A. L.; Ginsburg, D. *Helv. Chim. Acta*
1985, *68*, 2033-2036; (b) Natrajan, A.; Ferrara, J. D.; Youngs, W. J.; Sukenik, C. N. *J.
Am. Chem. Soc.* **1987**, *109*, 7477-7483; (c) Ginsburg, D. unpublished results (private communication with U. Weiss, February 8, 1987).
22. (a) Bertz, S. H.; Lannoye, G.; Cook, J. M. *Tetrahedron Lett.* **1985**, *26*, 4695-4698; (b)
Lannoye, G.; Cook, J. M. *Tetrahedron Lett.* **1987**, *28*, 4821-4824; (c) Gupta, A. K.; Cook,
J. M.; Weiss, U. *Tetrahedron Lett.* **1988**, *29*, 2535-2538; (d) Gupta, A. K.; Lannoye, G. S.;
Kubiak, G.; Schkerantz, J.; Wehrli, S.; Cook, J. M. *J. Am. Chem. Soc.* **1989**, *111*, 2169-
2179.
23. Bertz, S. H. *J. Org. Chem.* **1985**, *50*, 3585-3592.
24. Föhlisch, B.; Dukek, U.; Graessle, I.; Novotny, B.; Schupp, E.; Schwaiger, G.; Widmann,
E. *Liebigs Ann. Chem.* **1973**, 1839-1850.
25. (a) Mitschka, R.; Oehldrich, J.; Takahashi, K.; Cook. J. M.; Weiss, U.; Silverton, J. V.
Tetrahedron **1981**, *37*, 4521-4542; (b) Venkatachalam, M.; Deshpande, M. N.; Jawdosiuk,
M.; Kubiak, G.; Wehrli, S.; Cook, J. M.; Weiss, U. *Tetrahedron* **1986**, *42*, 1597-1605.

BIS(2,2,2-TRICHLOROETHYL) AZODICARBOXYLATE

[Diazenedicarboxylic acid, bis(2,2,2-trichloroethyl) ester]

A. $H_2NNH_2 \cdot H_2O$ +   2 $ClCO_2CH_2CCl_3$ $\longrightarrow$ $Cl_3CCH_2O_2CNHNHCO_2CH_2CCl_3$

B. $Cl_3CCH_2O_2CNHNHCO_2CH_2CCl_3$ $\xrightarrow[\substack{HNO_3 \\ (fuming)}]{[O]}$ $Cl_3CCH_2O_2CN{=}NCO_2CH_2CCl_3$

Submitted by R. DANIEL LITTLE and MANUEL G. VENEGAS[1,2]
Checked by SANDY BANKS and ORVILLE L. CHAPMAN

1. Procedure

A. *Bis(2,2,2-trichloroethyl) hydrazodicarboxylate.* In a 500-mL, three-necked flask
equipped with mechanical stirrer, thermometer, and 250-mL and 125-mL dropping

funnels (Note 1) is placed a solution of 13.34 g (0.23 mol) of 64% hydrazine hydrate (Note 2) in 60 mL of 95% ethanol. The reaction flask is cooled in an ice bath and 96 g (0.46 mol) of 2,2,2-trichloroethyl chloroformate (Note 3) is added dropwise so that the temperature is kept below 20°C. During the addition of 1 equiv of the chloroformate, a white precipitate is formed. After exactly one-half of the chloroformate has been added, a solution of 25 g (0.24 mol) of sodium carbonate in 100 mL of water is added dropwise along with the remaining chloroformate. The rate of addition of these two reagents is such that the flow of the chloroformate is faster than that of the sodium carbonate so that there is always an excess of chloroformate present; the temperature is kept below 20°C during the addition. As the second equivalent of chloroformate is added the white precipitate dissolves.

After the addition of the reactants is complete, the reaction is allowed to stir for an additional 30 min while the solution warms to room temperature. The reaction mixture is then transferred to a separatory funnel. The viscous organic bottom layer is separated from the aqueous layer and is dissolved in 200 mL of ether. The reaction vessel is washed with 100 mL of ether, and this ether portion is used to extract further the aqueous layer. The ether layers are combined, dried over magnesium sulfate, and filtered, and the solvent is removed under reduced pressure. The viscous oil is allowed to crystallize in an ice bath (0°C). The crystals are collected on a Büchner funnel, washed with 500 mL of water, and dried in a vacuum desiccator at 0.5 mm for 48 hr. 80.8 g (93%) of white crystalline bis(2,2,2-trichloroethyl) hydrazodicarboxylate (mp 85–89°C) is obtained. This material is sufficiently pure for the next preparation. However, further purification can be achieved using an Abderhalden drying apparatus (refluxing 95% EtOH for 12 hr at 0.05 mm; MgSO₄ desiccant). Material purified in this way melted at 96.5–97.5°C (Notes 4 and 5).

B. *Bis (2,2,2-trichloroethyl) azodicarboxylate. Caution! Large amounts of nitrogen oxides are evolved during the oxidation with fuming nitric acid. Therefore, operations should be conducted in an efficient fume hood.*

In a 500-mL, three-necked flask equipped with mechanical stirrer, thermometer, pressure-equalizing dropping funnel, and gas outlet tube is added 78.55 g (0.21 mol) of bis(2,2,2-trichloroethyl) hydrazodicarboxylate dissolved in 180 mL of chloroform (Note 6). The solution is cooled to 0°C and 53.2 mL (1.26 mol) of fuming nitric acid (Notes 7 and 8) is added so that the temperature of the solution does not rise above 5°C. The reaction mixture is then allowed to warm slowly to room temperature over 4 hr (Note 9). After an additional 2 hr at room temperature, the material is transferred to and shaken in a 1-L separatory funnel half-filled with ice chips. The two layers are allowed to separate and the bottom organic layer is removed. The aqueous layer is extracted with 250 mL of chloroform. The organic layers are combined and washed with 300 mL of water, 300 mL of aqueous 5% sodium bicarbonate, and again with 300 mL of water. The organic layer is dried with magnesium sulfate, filtered, and the solvent is removed under reduced pressure. The yellow crystals that form are collected on a Büchner funnel and washed with pentane. The pentane filtrate is concentrated under reduced pressure to afford more crystalline material which is again collected on a Büchner funnel and washed with more pentane. The cycle is repeated until no more crystals appear after removal of pentane. The yellow crystals so obtained are air dried for 1 hr to afford 59.2 g (75.8%) of bis(2,2,2-trichloroethyl) azodicarboxylate which melts at 108–110°C. Further drying using an Abderhalden drying apparatus (refluxing 95% EtOH for 12 hr at 0.5 mm; MgSO₄ desiccant) affords a compound that melts at 109–110.5°C (Notes 10–12).

2. Notes

1. The thermometer is fitted into one of the necks of the flask so that when it is immersed in the solution, the range between 10 and 20°C is easily visible. A two-necked adapter is used for the dropping funnels.

2. Hydrazine hydrate, 64%, practical grade, was obtained from MCB, Inc.

3. 2,2,2-Trichloroethyl chloroformate (96%) is commercially available from Aldrich Chemical Company, Inc., and is used without further purification.

4. The average yield obtained for five runs performed by three different people was 83%.

5. The spectral properties of bis(2,2,2-trichloroethyl) hydrazodicarboxylate are as follows: ^{1}H NMR (CDCl₃) δ: 4.80 (s, 4 H, CH₂CCl₃), 7.0–7.6 (s, br, 2 H, −NH, the position is concentration-dependent).

6. The solution can be warmed gently without harm to facilitate solution of the hydrazo compound.

7. Mallinckrodt fuming nitric acid (90–95%, d 1.5) was used.

8. The reaction seems to be surprisingly dependent on the amount of nitric acid used. A run with 78.6 g of hydrazo compound and a sixfold excess of nitric acid was quenched after 22 hr and afforded 100% conversion to the desired azo compound (NMR analysis). Another run with 80.0 g of hydrazo compound and a fivefold excess of nitric acid gave only 92% conversion after 25 hr. In another run with 2.0 g of hydrazo compound and a sixfold excess of nitric acid the reaction was complete after 4 hr. In addition, the oxidation was found to be temperature dependent. For example, in a run in which the temperature was maintained between 0 and 5°C for 3 hr and the solution was not allowed to warm to room temperature, only 18% yield was obtained (NMR analysis).

9. The evolution of large amounts of nitrogen oxides was noticed after approximately 1.5 hr (the temperature had reached 13°C).

10. Yields ranged from 76 to 94% (six runs performed by three different people).

11. The NMR spectrum (CDCl₃) for bis(2,2,2-trichloroethyl) azodicarboxylate shows only a singlet at δ 5.05.

12. The title compound is now commercially available from Aldrich Chemical Co., Inc.

3. Discussion

Bis(2,2,2-trichloroethyl) azodicarboxylate has been prepared by oxidation of bis(2,2,2-trichloroethyl) hydrazodicarboxylate with dinitrogen tetroxide.[3]

Bis(2,2,2-trichloroethyl) azodicarboxylate is a yellow crystalline material that is stable indefinitely in a vacuum desiccator stored in the dark. This compound offers a number of important advantages over diethyl and dimethyl azodicarboxylate for the synthesis of azo compounds. Probably the most important advantage is that in contrast to the ethyl and methyl esters, the trichloroethyl ester grouping can be removed under neutral conditions—a requirement when the product of the transformation is acid- or base-labile.[4] Furthermore, in contrast to dimethyl azodicarboxylate and diethyl azodicarboxylate, which have been known to explode when heated and require distillation for purification, bis(2,2,2-trichloroethyl) azodicarboxylate is isolated as a crystalline solid requiring no heating whatsoever. Another advantage is that Diels–Alder cycloadducts with bis(2,2,2-trichloroethyl) azodicarboxylate are often crystalline solids that can be purified by recrystallization. This is in marked contrast to the viscous oils that are often obtained when the commercially available diethyl azodicarboxylate is used. Finally, we have found that Diels–Alder cycloadditions using bis(2,2,2-trichloroethyl) azodicarboxylate often proceed faster and at a temperature lower than that required for the dimethyl and diethyl analogues (e.g., reaction with 6,6-dimethylfulvene and 6-acetoxyfulvene).

1. Department of Chemistry, University of California, Santa Barbara, CA 93106.
2. The authors wish to thank Ahmed Bukhari for the data that he supplied for this publication.
3. Mackay, D; Pilger, C. W.; Wong, L. L. *J. Org. Chem.* **1973,** *38*, 2043.
4. See, for example: (a) Semmelhack, M. F.; Foos, J. S.; Katz, S. *J. Am. Chem. Soc.* **1973,** *95*, 7325; (b) Berson, J. A.; Bushby, R. J.; McBride, J. M.; Tremelling, M. *J. Am. Chem. Soc.* **1971,** *93*, 1554; (c) Toong, Y. C.; Borden, W. T.; Gold, B. *Tetrahedron Lett.* **1975,** 1549; (d) Little, R. D.; Venegas, M. G., *J. Org. Chem.* **1978,** *43*, 2921; (e) Little, R. D.; Carroll, G. L. *Org. Chem.* **1979,** *44*, 4720; (f) Little, R. D.; Muller, G. W. *J. Am. Chem. Soc.,* **1981,** *103*, 2744.

β-HALOACETALS AND KETALS: 2-(2-BROMOETHYL)-1,3-DIOXANE AND 2,5,5-TRIMETHYL-2-(2-BROMOETHYL)-1,3-DIOXANE

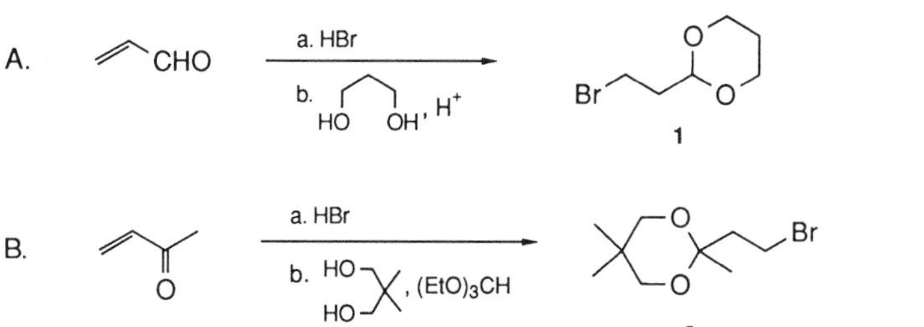

Submitted by J. C. STOWELL, D. R. KEITH, and B. T. KING[1]
Checked by YUMI NAKAGAWA and ROBERT V. STEVENS

1. Procedure

A. *2-(2-Bromoethyl)-1,3-dioxane* (**1**). A 2-L, three-necked flask is equipped with a mechanical stirrer, thermometer, and gas inlet tube. In the flask are placed 750 mL of dichloromethane, 112 g (2.00 mol) of acrolein (Note 1), and 0.10 g of dicinnamalacetone indicator (Note 2) under nitrogen. The yellow solution is cooled to 0–5°C with an ice bath. Gaseous hydrogen bromide (Note 3) is bubbled into the solution with stirring until the indicator becomes deep red (Note 4). The ice bath is removed and 1.0 g of *p*-toluenesulfonic acid monohydrate and 152.2 g (2.00 mol, 144 mL) of 1,3-propanediol (Note 1) are added. The yellow solution is stirred at room temperature for 8 hr and then concentrated with a rotary evaporator. The residual oil is washed with two 250-mL portions of saturated aqueous sodium bicarbonate and dried over anhydrous potassium carbonate. Vacuum distillation through a 30-cm Vigreux column yields 252 g (65%) of **1** as a colorless liquid, bp 72–75°C (2.0 mm), n_D^{20} 1.4809 (Note 5).

B. *2,5,5-Trimethyl-2-(2-bromoethyl)-1,3-dioxane* (**2**). A 1-L, three-necked flask is equipped with a magnetic stirrer and a gas inlet tube. In the flask are placed 700 mL of dichloromethane, 140 g (2.00 mol) of methyl vinyl ketone (Note 6), and 0.010 g of dicinnamalacetone indicator (Note 2). Anhydrous hydrogen bromide (Note 3) is bubbled into the solution with stirring until the indicator changes to deep red (Note 7). The gas inlet tube is removed and 208 g (2.00 mol) of neopentanediol, 296 g (2.00 mol) of triethyl orthoformate, and 0.67 g of *p*-toluenesulfonic acid monohydrate are added to the solution. The flask is stoppered and stirred at room temperature for 1–2 hr and then concentrated by rotary evaporation (Note 8). The concentrated solution is washed twice with saturated sodium bicarbonate solution. *(Caution: There is some foaming.)* The bicarbonate washes are extracted three times with dichloromethane and the combined organic portions dried over anhydrous K_2CO_3. Rotary evaporation followed by vacuum distillation of the residue through a 30-cm Vigreux column yields 256 g (54%) of **2** as a clear, colorless oil, bp 65°C (0.3 mm), n_D^{22} 1.4687 (Note 9).

2. Notes

1. The acrolein, 1,3-propanediol, and cinnamaldehyde were purchased from Aldrich Chemical Company, Inc.

2. The indicator was prepared by the method of Diehl and Einhorn.[2] A solution of 5 g of sodium hydroxide in 50 mL of water and 40 mL of ethanol is prepared in a 250-mL Erlenmeyer flask. To this is added a solution of 1.84 mL (0.025 mol, 1.45 g) of acetone in 6.3 mL (0.050 mol, 6.6 g) of freshly distilled cinnamaldehyde (Note 1). This mixture is stirred thoroughly at room temperature for 30 min. The resulting voluminous yellow precipitate is filtered with suction, washed with 100 mL of water, and dried, affording 6.5 g of 1,9-diphenylnona-1,3,6,8-tetraen-5-one. Recrystallization from 200 mL of hot 95% ethanol gives 3.5 g of yellow crystals, mp 142–143°C (lit.[2] mp 142°C). This indicator is also available from Aldrich Chemical Co.

3. The anhydrous hydrogen bromide was purchased in a lecture bottle from Matheson. A trap is used between the lecture bottle and the gas inlet tube.

4. When the red color persists 5 min after the hydrogen bromide has been turned off, the addition is finished. At this point the proton magnetic resonance spectrum shows only dichloromethane and 3-bromopropanal (60 MHz, CH_2Cl_2) δ: 3.04 (t, 2 H), 3.59 (t, 2 H), 10.67 (s, 1 H).

5. Product **1** has the following spectral characteristics: IR (neat) cm^{-1}: 2980, 2870, 1250, 1150, 1140, 1015; ^1H NMR (90 MHz, CDCl$_3$) δ: 1.38 (d of m, 1 H, one

5-position on dioxane ring), 1.8–2.4 (m, the other 5-position on the dioxane ring), 2.14 (d of t, 2 H, CH_2-C-Br), 3.45 (t, 2 H, CH_2Br), 3.80 (d of t, 2 H, 4, and 6-positions on ring), 4.15 (d of double d, 2 H, 4, and 6-positions on ring), 4.71 (t, 1 H, 2-position on ring; ^{13}C magnetic resonance (22.5 MHz, CDCl$_3$) δ: 100.06, 66.86, 38.08, 27.79, 25.79.

6. The neopentanediol and triethyl orthoformate were purchased from Aldrich Chemical Co., Inc. and used as received. Failure to distill the methyl vinyl ketone, also obtained from Aldrich Chemical Co. Inc., to a clear, colorless liquid before use resulted in difficulty in determining the endpoint of the reaction with HBr. Therefore, the methyl vinyl ketone was distilled prior to use at reduced pressure.

7. When the red color persists 5 min after the hydrogen bromide has been turned off, the addition is finished. At this point the proton magnetic resonance spectrum shows only dichloromethane and 4-bromo-2-butanone (60 MHz, CH_2Cl_2) δ: 2.15 (s, 3 H, CH_3CO), 3.02 (t, 2 H, CH_2CO), 3.52 (t, 2 H, CH_2Br); ^{13}C NMR (22.5 MHz, CDCl$_3$) δ: 25.75, 30.11, 45.91, 205.12. Little or no exotherm is noticed during the hydrogen bromide addition.

8. The reaction can be conveniently monitored by TLC using silica plates and eluting with 1 : 4 ethyl acetate–heptane.

9. Product 2 has the following characteristics: IR (neat liquid) cm^{-1}: 2970, 2880, 1260, 1220, 1125, 1085; ^{1}H NMR (60 MHz, CDCl$_3$) δ: 0.81 (s, 3 H, 5-methyl), 1.01 (s, 3 H, 5-methyl), 1.34 (s, 3 H, 2-methyl), 2.05–2.45 (m, 2 H, CH_2-C-Br), 3.2–3.8 (m, 6 H, CH_2O and CH_2Br); ^{13}C NMR (22.5 MHz, CDCl$_3$) δ: 19.64, 22.24, 22.76, 26.99, 29.72, 43.25, 70.23, 98.26.

3. Discussion

Cyclic β-haloacetals and ketals have been prepared by variations of two basic methods. The most frequently used method involves the combination of an α,β-unsaturated carbonyl compound (acrolein, methyl vinyl ketone, crotonaldehyde, etc.) a diol, and the anhydrous hydrogen halide. All possible sequences of combining these three have been used. In most cases the anhydrous acid was dissolved in the diol and then the carbonyl compound was added slowly.[3,7] Alternatively, the acetals of the α,β-unsaturated carbonyl compounds were prepared and isolated and then the hydrogen halide was added.[8] Finally the hydrogen halide may be added to the α,β-unsaturation followed by acetal formation,[9] and this is the basis of the present procedures.

The second general method is the aluminum halide-catalyzed reaction of acid halides with ethylene to give β-halo ketones, which are subsequently converted to ketals.[10,11]

The preparations are much simplified if a stoichiometric amount of hydrogen halide is added using an indicator to determine the endpoint. We have found that 1,9-diphenylnona-1,3,6,8-tetraen-5-one (dicinnamalacetone)[12] is of appropriate basicity to detect excess anhydrous hydrogen halides in organic solvents including chloroform, dichloromethane, benzene, toluene, acetic acid, and acetone (but not in alcohols). The reaction between the hydrogen halide and the α,β-unsaturated carbonyl compound is fast enough at 0–25°C that the endpoint is readily detected, and the yield-lowering use of excess hydrogen halide or long contact times[13] are avoidable. The intermediate β-halo aldehydes are unstable toward trimerization[14] if they are not diluted by a solvent and therefore should not be isolated but used directly in the next step. β-Bromo ketones darken on isolation and brief storage, so they, too, should be protected directly.

The conversion of the intermediate bromo aldehyde to the dioxane proceeds readily because of a favorable equilibrium position. However, the equilibrium for the reaction of the bromo ketone with the diol is unfavorable and requires removal of the by-product, water. This is done under mild conditions using ethyl orthoformate.[15]

We have chosen to use 1,3-diols because the Grignard reagents derived from the 1,3-dioxanes are thermally stable.[16] This contrasts with the use of ethylene glycol where the resulting β-haloalkyl dioxolanes give Grignard reagents that decompose at 25–35°C.[17-19] Acyclic acetals give insufficient protection to allow preparation of Grignard reagents.[19] The protection of the ketone with 1,3-propanediol is not readily driven to completion, but with neopentanediol the equilibrium lies further toward ketal formation,[20] giving a better yield of more stable ketal.

β-Haloacetals and ketals have recently seen wide use as alkylating agents[10, 21-24] and in the preparation of Grignard reagents. The Grignard reagents have been alkylated,[25] acylated,[16, 26] added to carbonyl groups,[5, 18, 27-32] and used in Michael additions.[33-35] One example also gives a useful Wittig reagent.[9] Subsequent reactions of these products generally require removal of the acetal and ketal groups to regenerate the carbonyl function. This is readily done with aqueous acid in most cases, but not when aldehydes were protected with 1,3-diols because of the high equilibrium stability of the corresponding dioxanes. This problem is readily overcome by first converting to the dimethyl acetal in methanol and then using aqueous acid hydrolysis, or by using other specialized methods.[9, 16]

1. Department of Chemistry, University of New Orleans, New Orleans, LA 70148.
2. Diehl, L.; Einhorn, A. *Chem. Ber.* **1885**, *18*, 2320.
3. Hill, H. S.; Potter, G. J. C. *J. Am. Chem. Soc.* **1929**, *51*, 1509.
4. Faass, U.; Hilgert, H. *Chem. Ber.* **1954**, *87*, 1343.
5. Büchi, G.; Wüest, H. *J. Org. Chem.* **1969**, *34*, 1122.
6. Kriesel, D. C.; Gisvold, O. *J. Pharm. Sci.* **1971**, *60*, 1250.
7. Brown, E.; Dahl, R. *Bull. Soc. Chim. Fr.* **1972**, 4292.
8. Fischer, R. F.; Smith, C. W. *J. Org. Chem.* **1960**, *25*, 319.
9. Stowell, J. C.; Keith, D. R. *Synthesis* **1979**, 132; Dauben, W. G.; Gerdes, J. M.; Bunce, R. A. *J. Org. Chem.* **1984**, *49*, 4293; Baldwin, S. W.; Crimmins, M. T. *J. Am. Chem. Soc.* **1982**, *104*, 1132; Cohen, N.; Banner, B. L.; Lopresti, R. J. *Tetrahedron Lett.* **1980**, *21*, 4163.
10. Hajos, Z. G; Micheli, R. A.; Parrish, D. R.; Oliveto, E. P. *J. Org. Chem.* **1967**, *32*, 3008.
11. Trost, B. M.; Kunz, R. A. *J. Am. Chem. Soc.* **1975**, *97*, 7152.
12. This and other aromatic cross-conjugated ketones give deeply colored crystalline hydrochlorides with the anhydrous acid. See Stobbe, H.; Haertel, R. *Justus Liebigs Ann. Chem.* **1909**, *370*, 99. Furthermore, this indicator has been used to plot acidity in toluene solution: Sanders, W. N.; Berger, J. E. *Anal. Chem.* **1967**, *39*, 1473,
13. Roedig, A. "Methoden der Organischen Chemie" (Houben-Weyl), Vol. V/4, Georg Thieme: Stuttgart, 1960, p. 120.
14. Jacobs, T. L.; Winstein, S.; Linden, G. B.; Seymour, D. *J. Org. Chem.* **1946**, *11*, 223.
15. Marquet, A.; Dvolaitzky, M.; Kagan, H. B.; Mamlok, L.; Ouannes, C.; Jacques, J. *Bull. Soc. Chim. Fr.* **1961**, 1822.
16. Stowell, J. C. *J. Org. Chem.* **1976**, *41*, 560.
17. Eaton, P. E.; Mueller, R. H.; Carlson, G. R.; Cullison, D. A; Cooper, G. F.; Chou, T.-C.; Krebs, E.-P. *J. Am. Chem. Soc.* **1977**, *99*, 2751.

18. Ponaras, A. A. *Tetrahedron Lett.* **1976**, 3105.
19. Feugeas, Cl. *Bull. Soc. Chim. Fr.* **1963**, 2568.
20. Hine, J. "Structural Effects on Equilibria in Organic Chemistry," Wiley-Interscience: New York, 1975, p. 287.
21. Johnson, W. S.; Hughes, L. R.; Kloek, J. A.; Niem, T.; Shenvi, A. *J. Am. Chem. Soc.* **1979**, *101*, 1279.
22. Larchevêque, M.; Cuvigny, T. *Tetrahedron Lett.* **1975**, 3851.
23. Ellison, R. A.; Lukenbach, E. R.; Chiu, C.-W. *Tetrahedron Lett.* **1975**, 499.
24. Lundeen, A. J.; Hoozer, R. V. *J. Am. Chem. Soc.* **1963**, *85*, 2178.
25. Gras, J.-L.; Bertrand, M. *Tetrahedron Lett.* **1979**, 4549.
26. Dodge, J.; Hedges, W.; Timberlake, J. W.; Trefonas, L. M.; Majeste, R. J. *J. Org. Chem.* **1978**, *43*, 3615.
27. Martin, S. F.; Puckette, T. A.; Colapret, J. A. *J. Org. Chem.* **1979**, *44*, 3391.
28. Monti, S. A.; Yang, Y.-L. *J. Org. Chem.* **1979**, *44*, 897.
29. Pattenden, G.; Whybrow, D. *Tetrahedron Lett.* **1979**, 1885.
30. Gotschi, E.; Schneider, F.; Wagner, H.; Bernauer, K.; *Helv. Chim. Acta* **1977**, *60*, 1416.
31. Goldberg, O.; Dreiding, A. S. *Helv. Chim. Acta* **1977**, *60*, 964.
32. Loozen H. J. J. *J. Org. Chem.* **1975**, *40*, 520.
33. Paquette, L. A.; Han, Y. K. *J. Org. Chem.* **1979**, *44*, 4014.
34. Marfat, A.; Helquist, P. *Tetrahedron Lett.* **1978**, 4217.
35. Brattesani, D. N.; Heathcock, C. H. *J. Org. Chem.* **1975**, *40*, 2165.

CONVERSION OF METHYL KETONES INTO TERMINAL ALKYNES: (*E*)-BUTEN-3-YNYL-2,6,6-TRIMETHYL-1-CYCLOHEXENE

[Cyclohexene, 2-(1-buten-3-ynyl)-1,3,3-trimethyl-, (*E*)-]

1) LDA, THF
2) ClPO(OEt)$_2$
3) LDA (2.25 equiv)
4) H$_2$O

Submitted by EI-ICHI NEGISHI, ANTHONY O. KING, and JAMES M. TOUR[1]
Checked by WEYTON W. TAM and ROBERT V. STEVENS

1. Procedure

An oven-dried, 500-mL, two-necked, round-bottomed flask equipped with a magnetic stirring bar, a rubber septum inlet, and an outlet connected to a mercury bubbler is flushed with nitrogen and charged with 100 mL of tetrahydrofuran (THF) (Note 1). To this are added sequentially at 0°C, diisopropylamine (Note 2) (10.6 g, 14,7 mL, 105 mmol) and butyllithium in hexane (Note 3) (2.22 *M*, 47.3 mL, 105 mmol). The reaction mixture is stirred for 30 min and cooled to −78°C. β-Ionone (Note 4) (19.2 g, 20.3 mL, 100 mmol) is slowly added. After stirring the mixture for 1 hr at −78°C, diethyl chlorophosphate (Note 5) (18.1 g, 15.2 mL, 105 mmol) is added, and the reaction mixture is allowed to warm to room temperature over 2–3 hr (reaction mixture A) (Note 6).

Lithium diisopropylamide is prepared in a separate 1-L flask from diisopropylamine (22.8 g, 31.6 mL, 225 mmol), butyllithium in hexane (2.22 M, 101 mL, 225 mmol), and THF (200 mL), as described above. To this is added over approximately 45 min at −78°C reaction mixture A prepared above via a 16-G double-ended needle under a slight pressure of nitrogen. The resulting mixture is allowed to warm to room temperature over 2–3 hr and is quenched with water (200 mL) at 0°C. The organic layer is separated, and the aqueous layer is extracted with pentane (3 × 50 mL). The combined organic layer is treated with ice-cold hydrochloric acid (1 N, 200 mL), water (2 × 100 mL), and saturated aqueous sodium bicarbonate (100 mL) to pH ≥8 (Note 7). After drying over magnesium sulfate, the volatile compounds are evaporated using a rotary evaporator at ca. 20 mm. The residue is distilled at 0.7 mm to provide (E)-buten-3-ynyl-2,6,6-trimethyl-1-cyclohexene (Note 8) in one fraction boiling at 69–73°C (0.7 mm) (Note 9). The yield by isolation has ranged from 12.5 g (72%) to 14.8 g (85%) (Note 10). The purity of the product by GLC is 98%.

2. Notes

1. Tetrahydrofuran available from Aldrich Chemical Company, Inc. was purified by distillation from sodium and benzophenone.

2. The submitters used diisopropylamine (99%) available from Aldrich Chemical Company, Inc. without further purification.

3. The submitters used butyllithium in hexane available from Alfa Products, Morton Thiokol, Inc.

4. The submitters used 98% pure β-ionone available from Aldrich Chemical Company, Inc. without further purification.

5. The submitters used diethyl chlorophosphate available from Aldrich Chemical Company, Inc.

6. Reduced yields of product were obtained by the checkers when reaction time at room temperature was reduced from 2–3 hr to 1½ hr.

7. After extraction with hydrochloric acid, the pentane layer, on addition of 100 mL of water, formed a poorly separating emulsion. Checkers found that, by addition of 100 mL of saturated aqueous sodium bicarbonate to this pentane-water emulsion, two easily separable layers can be formed.

8. The distilled product was found to be slightly yellow, and deepened to orange at room temperature. Storage at −5°C maintained the initial coloration for several weeks.

9. The product displays the following data: n_D^{24} 1.5130; IR (neat) cm^{-1}: 3300 (s), 2920 (s), 2080 (m), 1770 (w), 1630 (w), 1600 (w), 1455 (s), 1380 (m), 1355 (m), 1200 (m), 1030 (m), 960 (s); ^{1}H NMR (CDCl$_3$, TMS) δ: 1.01 (s, 6 H), 1.2–1.8 (m with a singlet at 1.71, 7 H), 1.85–2.15 (m, 2 H), 2.90 (d, 1 H, J = 2), 5.42 (dd, 1 H, J = 17 and 2), 6.67 (d, 1 H, J = 17); ^{13}C NMR (CDCl$_3$, TMS) δ: 19.17, 21.48, 28.75, 33.07, 33.98, 39.59, 77.29, 83.10, 111.36, 131.38, 136.90, 142.33.

10. The GLC trace (SE-30) of the reaction mixture shows essentially one peak (≥98) in the product region. In separate 5–20-mmol scale experiments, the GLC yields observed by using a paraffin internal standard were 90–95%.

TABLE I

CONVERSION OF METHYL KETONES INTO TERMINAL ACETYLENES VIA ENOL
PHOSPHATES

Ketone	Base[a]	Yield of Acetylene (%)	
		GLC	Isolated
β-Ionone	LDA	95	85
Dihydro-β-ionone	LDA	90	85
Acetophenone	LDA	85	80
Pinacolone	LDA	90	78
Cyclohexyl methyl ketone	LDA	85	80
2-Octanone	LDA	23	—
2-Octanone	LTMP	75	—
6-Methyl-5-hepten-2-one	LDA	25	—
6-Methyl-5-hepten-2-one	LTMP	75	61

[a] LDA = lithium diisopropylamide; LTMP = lithium 2,2,6,6-tetramethylpiperidide.

3. Discussion

This procedure is based on a study of conversion of methyl ketones into terminal alkynes.[2] The scope of the procedure may be indicated by the results summarized in Table I.

As can be seen in Table I, lithium diisopropylamide (LDA) is a satisfactory base in cases where the carbon group (R) of a methyl ketone ($RCOCH_3$) either is bulky or does not contain an α-methylene or α-methine group. In the other cases, LDA is relatively ineffective. In such cases, however, the use of lithium 2,2,6,6-tetramethylpiperidide (LTMP) in place of LDA gives satisfactory results. The LTMP procedure appears to be the only documented method that is satisfactory for the conversion of the above-mentioned type.

The submitters have attempted the conversion of β-ionone into the desired dienyne by various known methods. In general, those involving acidic reagents or reaction conditions yielded the desired product in low yields ($< 50\%$) along with by-products, such as isomeric allenes, that appear near the product on GLC traces (SE-30). Such procedures include (a) PCl_5 in benzene, then $NaNH_2$ in NH_3;[3] (b) PCl_5 and 2,6-lutidine, then $NaNH_2$ in NH_3;[4] (c) $POCl_3$ in DMF, then NaOH;[5] and (d) $(CF_3SO_2)_2O$, CCl_4, pyridine, then heat.[6] Also unsatisfactory in the hands of submitters was a method involving the use of hydrazine in triethylamine, then iodine and triethylamine in THF, then methanolic potassium hydroxide.[7] A procedure involving the use of sodium ethoxide, then diethyl chlorophosphate, and finally $NaNH_2$ in NH_3,[8] on the other hand, converted β-ionone into the desired dienyne in $\leq 73\%$ GLC yield. The procedure reported here may be viewed as a modification of the method described above.

1. Department of Chemistry, Purdue University, West Lafayette, IN 47907.
2. Negishi, E.-i.; King, A. O.; Klima, W. L.; Patterson, W.; Silveira, A., Jr. J. Org. Chem. **1980**, *45*, 2526.

3. Sweet, R. S.; Marvel, C. S. *J. Am. Chem. Soc.* **1932**, *54*, 1184.

4. Corey, E. J.; Katzenellenbogen, J. A.; Posner, G. H. *J. Am. Chem. Soc.* **1967**, *89*, 4245.

5. Rosenblum, M.; Brawn, N.; Papenmeier, J.; Applebaum, M. *J. Organomet. Chem.* **1966**, *6*, 173.

6. Hargrove, R. J.; Stang, P. J. *J. Org. Chem.* **1974**, *39*, 581.

7. Krubiner, A. M.; Gottfried, N.; Oliveto, E. P. *J. Org. Chem.* **1969**, *34*, 3502.

8. Craig, J. C.; Moyle, M. *J. Chem. Soc.* **1963**, 3712.

REDUCTIVE CLEAVAGE OF VINYL PHOSPHORODIAMIDATES: 17β-*tert*-BUTOXY-5α-ANDROST-2-ENE

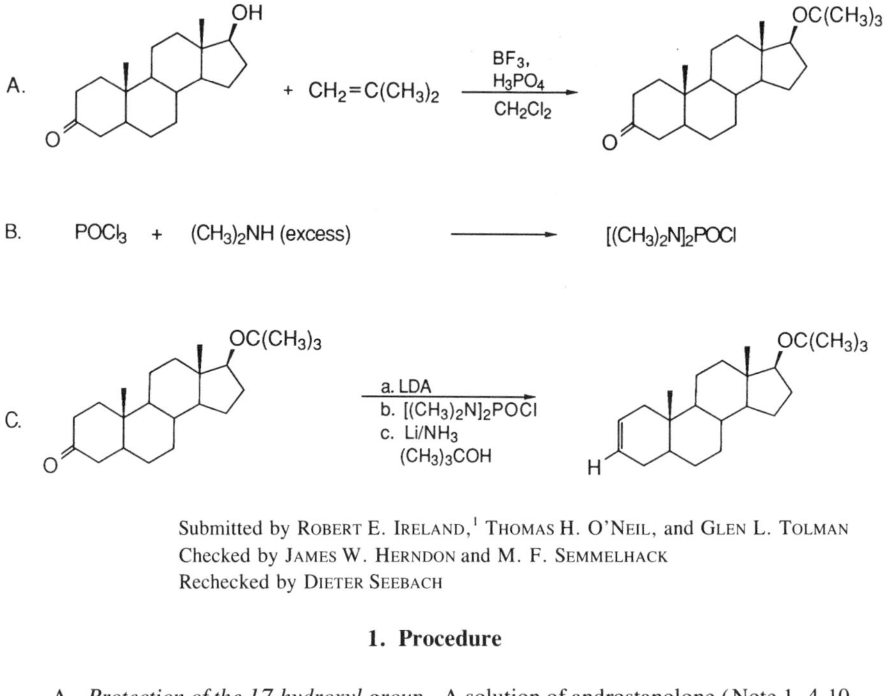

Submitted by ROBERT E. IRELAND,[1] THOMAS H. O'NEIL, and GLEN L. TOLMAN
Checked by JAMES W. HERNDON and M. F. SEMMELHACK
Rechecked by DIETER SEEBACH

1. Procedure

A. *Protection of the 17-hydroxyl group.* A solution of androstanolone (Note 1, 4.10 g, 14 mmol) in 60 mL of dichloromethane in a 250-mL, one-necked, round-bottomed flask equipped with a magnetic stirring bar and a rubber septum bearing two syringe needles (argon inlet and exit) is cooled to −20°C (refrigerated bath). Argon is allowed to pass over the surface of the mixture for 15 min and then boron trifluoride etherate (Note 2, 0.125 mL, 0.90 mmol) is added rapidly, via syringe, followed by anhydrous phosphoric acid (Note 3, 0.053 mL, 1.0 mmol). Isobutene (Note 4) is added as a gas through a large-bore syringe needle until approximately 100 mL has condensed. The steroid precipitates during addition of the isobutene and redissolves as the reaction proceeds. The drying tube is replaced with a stopper, the tightly sealed flask is allowed to warm to 25°C, and the mixture is stirred at this temperature for 4 hr (Note 5). The flask is cooled to 0°C, opened, and warmed to 25°C to allow excess isobutene to evaporate. The residue is poured into 2 N aqueous ammonium hydroxide (100 mL) and ethyl acetate (75 mL) is added. After the layers are vigorously shaken, the aqueous solution is washed with a second portion of ethyl acetate. The combined organic extracts are washed with saturated sodium chloride solution, dried over anhydrous magnesium sulfate, filtered, and concentrated by rotary evaporation. The residue is recrystallized from hexane to give colorless crystals, mp 146–148°C, 4.10 g (86%, Note 6).

B. *Preparation and reductive cleavage of the vinyl phosphorodiamidate.* A dry, 250-mL flask equipped with magnetic stirrer, syringe port (Note 7), and argon outlet is flushed three times with argon. To the flask are added 40 mL of dry tetrahydrofuran (THF) and 1.17 mL (8.4 mmol) of dry diisopropylamine (Note 8). The flask is cooled in an acetone–dry-ice bath while 7.4 mmol of butyllithium in hexane (Note 9) is added dropwise with stirring. After the addition is complete, the solution is allowed to warm for 15 min. The flask is then cooled in an ice–water bath. To this solution is added 1.61 g (4.6 mmol) of 17β-*tert*-butoxy-5α-androstan-3-one in 30 mL of 2 : 1 THF/DMPU (Note 8) solution. The reaction mixture is stirred with ice cooling for 15 min. N,N,N',N'-Tetramethyldiamidophosphorochloridate, 5.83 mL (0.038 mol) (Note 10), is added dropwise with stirring. After 15 min, the bath is removed; the flask is allowed to warm to 25°C and is stirred for an additional 2 hr. The excess reagent is hydrolyzed by slow addition of 30 mL of saturated aqueous sodium bicarbonate solution and stirring for 30 min. After three extractions with 100-mL portions of diethyl ether, the combined organic layers are washed twice with 100 mL of water and 100 mL of saturated sodium chloride solution. The solution is dried over anhydrous magnesium sulfate and the ether is removed under reduced pressure on a rotary evaporator to afford 2.9–3.0 g of a crude yellow solid (Note 11). The crude phosphorodiamidate is dissolved in 40 mL of dry THF and added to a dry, three-necked, 250-mL flask equipped with overhead stirrer, cold finger condenser (acetone–dry ice), argon bubbler, and acetone–dry-ice bath. Dry ammonia is distilled into the flask until the phosphorodiamidate begins to precipitate. The bath is removed and the solution is allowed to warm to reflux. Dry *tert*-butyl alcohol (1.75 mL, Note 12) is added in one portion. To the clear solution is added 1.5 cm of ⅛-in., cleaned lithium wire in 0.3-cm portions. The blue color is maintained (by the addition of lithium wire if necessary) with stirring for 4 hr and then allowing the stirred solution to warm to room temperature overnight. Sodium benzoate is added in 25-mg portions until the blue color is discharged. Ammonium chloride (0.50 g) is added in one portion, the condenser removed, and the ammonia allowed to evaporate. The residue is taken up in 100 mL of diethyl ether and 100 mL of water. The layers are separated and the aqueous phase is extracted with 100 mL of diethyl ether. The combined organic layers are washed with 100 mL of saturated aqueous sodium chloride solution, dried over anhydrous magnesium sulfate, and filtered. The ether is removed under reduced pressure on a rotary evaporator. The crude olefin is filtered through 15 g of silica gel (Note 13) using benzene–ethyl acetate (2 : 1) as eluant, to give an off-white solid that is recrystallized from a minimum amount of absolute ethanol to give, after drying, 1.0 g (67%) (Note 14) of 17β-*tert*-butoxy-5α-androst-2-ene, mp 114–117°C (Note 15).

2. Notes

1. Androstanolone was obtained from Aldrich Chemical Company, Inc., and used without purification. The recheckers used material from Fluka Chemical Corp.

2. Boron trifluoride etherate was distilled before use.

3. Anhydrous phosphoric acid was prepared by slow addition of 5 g of 15% phosphoric acid to 2 g of phosphorus pentoxide.[2]

4. Isobutene, reagent grade, was obtained from Phillips Company.

5. The flask was stoppered with a greased 24/40 ground-glass stopper held in place by rubber bands stretched over appropriately placed wire hooks. The pressure at 25°C was slightly more than 1 atm.

6. The spectral properties are as follows: ^{1}H NMR (CDCl$_3$) δ: 0.74 (s, 3 H), 1.02 (s, 3 H), 1.13 (s, 9 H), 3.36 (m, 1 H); IR (CHCl$_3$) cm^{-1}: 1715 (C=O), 1255, 1205.

7. All solutions were added via glass syringes under rigorously anhydrous conditions.

8. Diisopropylamine and tetrahydro-1,3-dimethyl-2-(1H)-pyrimidinone (dimethyl propylene urea, DMPU) (Fluka Chemical Corp.) were distilled from calcium hydride. The original procedure used a THF–hexamethylphosphoric amide (HMPA) mixture. Because of the suspected carcinogenicity of HMPA, *Organic Syntheses* is recommending its replacement with DMPU whenever possible.

9. Butyllithium in hexane was obtained from Alfa Products, Morton Thiokol, Inc. or Foote Mineral Company. The checkers titrated the solution before[3] use.

10. *N,N,N',N'*-Tetramethyldiamidophosphorochloridate was obtained by the recheckers from Fluka Chemical Corp. If desired, it may be prepared as follows. In a dry, 2-L, three-necked flask equipped with overhead stirrer, thermometer, argon outlet, and pressure-equalizing addition funnel is placed 400 mL of diethyl ether (Note 7). The flask is cooled in an isopropyl alcohol–dry-ice bath while 100 g (2.2 mol) of anhydrous dimethylamine is added in one portion. A solution of 85 g (0.56 mol) of phosphoryl chloride in 200 mL of diethyl ether is added at a rate to maintain the temperature at −35 ± 5°C. The addition time is approximately 1.5 hr. After the addition is complete, the bath is removed, and stirring is continued for 4 hr. The thick white slurry is filtered through a coarse frit and the filter cake is washed with 4000 mL of diethyl ether. The combined filtrates are concentrated under aspirator pressure on a rotary evaporator. Fractional distillation of the concentrate through a 10-cm Vigreux column gives 71–80 g (74–84%) of the *N,N,N',N'*-tetramethyldiamidophosphorochloridate, bp 58.5–59°C (0.6 mm) d 1.126; IR (neat) cm^{-1}: 1470, 1450, 1290, 1230, 980; ^{1}H NMR (neat) δ: 2.69 (d J_{P-H} = 13).

11. Spectral data are as follows: IR (CCl$_4$) cm^{-1}: 1660, 1350, 1215; ^{1}H NMR (CDCl$_3$) δ: 0.69 (s, 3 H, CH_3), 0.78 (s, 3 H, CH_3), 1.11 (s, 9 H, CH_3), 2.60 (d, 3 H, J_{P-H} = 10, N-CH_3), 3.28 (m, 1 H, OCH), 5.12 (m, 1 H, C=CH).

12. *tert*-Butyl alcohol was dried by distillation from calcium hydride.

13. Silica gel 60 (particle size 0.063–0.200 μm) is available from E. Merck, A. G.

14. In the original procedure employing HMPA, the yield was 1.1–1.2 g (71–78%).

15. Spectral data are as follows: ^{1}H NMR (CDCl$_3$) δ: 0.63 (s, 3 H, CH_3), 0.67 (s, 3 H, CH_3), 1.02 (s, 9 H, CH_3), 3.28 (m, 1 H, OCH), 5.43 (m, 2 H, vinyl H); IR (CHCl$_3$): The product was characterized by cleavage of the *tert*-butyl ether (CF$_3$CO$_2$H, CH$_2$Cl$_2$, 0°C) to give 17β-hydroxy-5α-androst-2-ene, mp 161–162°C, lit.[4] mp 163–165°C.

3. Discussion

The reduction of a carbonyl group to an olefin has been accomplished by the Shapiro modification[5] of the Bamford–Stevens reaction and by the hydride reduction of the corresponding enol ether,[6] enol acetate,[7] or enamine.[8] The nickel reduction of the thioketal has also been used successfully.[9]

The lithium/amine reduction of N,N,N',N'-tetramethylphosphorodiamidates is a general method for the cleavage of the C—O bond.[10] In addition to the reductive deoxygenation of carbonyl compounds to generate olefins, the phosphorodiamidates of alcohols are reduced in high yield to give alkanes. Alcohols in which the hydroxyl group is greatly hindered could be unreactive toward N,N,N',N'-tetramethyldiaminophosphorochloridate. In such cases, treatment of the alcohols with butyllithium and N,N-dimethylphosphoramidic dichloride in 1,2-dimethoxyethane and N,N,N',N'-tetramethylethylenediamine followed by addition of dimethylamine gave rise to N,N,N',N'-tetramethylphosphorodiamidates in good yields.[11] Combined in a two-step process (e.g., RCOR' → RCHOHR' → RCH$_2$R'), the method allows the reductive removal of a carbonyl functionality. This two-step process compares favorably with the analogous Wolff–Kishner reduction. Additionally, reduction of the enol phosphorodiamidate by dialkyl cuprate reagents generates a substituted olefin.[12]

The phosphorodiamidate group can also serve as a protecting group for the hydroxyl function, since it is stable to CH$_3$Li, LiAlH$_4$, KOH, and 0.2 N aqueous HCl, but is quantitatively cleaved by butyllithium–TMEDA (tetramethylethylenediamine).[10]

1. The Chemical Laboratories of the California Institute of Technology, Pasadena, CA 91125. Contribution No, 5718. This work was made possible in part by a grant from the National Institutes of Health. Present address: Department of Chemistry, University of Virginia, Charlottesville, VA.
2. Beyeman, H. C.; Heiswolf, G. J. *Recl. Trav. Chim. Pays-Bas* **1965**, *84*, 203.
3. Kofron, W. G.; Baclawski, L. M. *J. Org. Chem.* **1976**, *41*, 1879.
4. Fetizon, M.; Jurion, M.; Nguyen Trong, A. *Org. Prep. Proced, Int.* **1974**, *6*, 31–35.
5. Shapiro, R. H. *Org. React.* **1976**, *23*, 450 and references cited therein.
6. Larson, G. L.; Hernandez, E.; Alonzo, C.; Nieves, I. *Tetrahedron Lett.* **1975**, 4005; Pino, P.; Lorenzi, G. P. *J. Org. Chem.* **1966**, *31*, 329.
7. Cagliot, L.; Caielli, G.; Maina, G. Salva, A. *Gazz. Chim. Ital.* **1962**, *92*, 309.
8. Lewis, J. W.; Lynch, P. P. *Proc. Chem. Soc. (London)* **1963**, 19; Coulter, J. M.; Lewis, J. W.; Lynch, P. P. *Tetrahedron* **1968**, *24*, 4489.
9. Ben-Efraim, D. A.; Sondheimer, F. *Tetrahedron* **1969**, *25*, 2826; Fishman, J.; Torigoe, M.; Guzik, H. *J. Org. Chem.* **1963**, *28*, 1443.
10. Ireland, R. E.; Muchmore, D. C.; Hengarten, U. *J. Am. Chem. Soc.* **1972**, *94*, 5098; Ireland, R. E.; Pfister, G. *Tetrahedron Lett.* **1969**, 2145.
11. Liu, H.-J.; Lee, S. P.; Chan, W. H. *Can. J. Chem.* **1977**, *55*, 3797.
12. Blaszczak, L.; Winkler, J.; O'Kuhn, S. *Tetrahedron Lett.* **1976**, 4405.

tert-BUTOXYCARBONYLATION OF AMINO ACIDS AND THEIR DERIVATIVES: *N-tert*-BUTOXYCARBONYL-L-PHENYLALANINE

(L-Phenylalanine, *N*-[(1,1-dimethylethoxy)carbonyl]-)

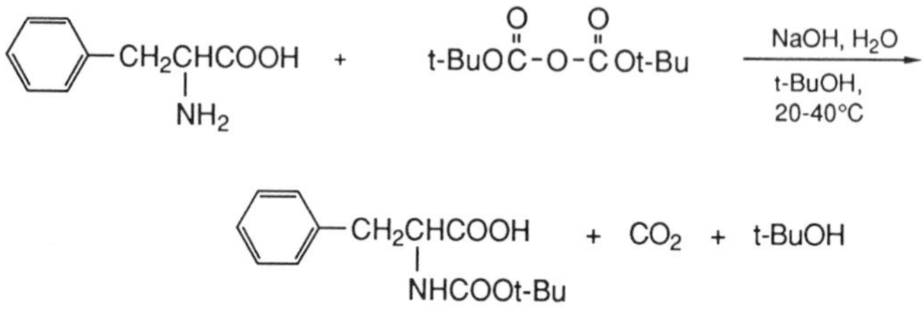

Submitted by OSKAR KELLER, WALTER E. KELLER, GERT VAN LOOK, and GERNOT WERSIN[1]
Checked by THOMAS VON GELDERN, MARK A. SANNER, and CLAYTON H. HEATHCOCK

1. Procedure

A 4-L, four-necked, round-bottomed flask, equipped with an efficient stirrer, a dropping funnel, reflux condenser, and thermometer is charged with a solution of 44 g (1.1 mol) of sodium hydroxide in 1.1 L of water. Stirring is initiated and 165.2 g (1 mol) of L-phenylalanine (Note 1) is added at ambient temperature, and then diluted with 750 mL of *tert*-butyl alcohol (Note 2). To the well-stirred, clear solution (Note 3) is added dropwise within 1 hr, 223 g (1 mol) of di-*tert*-butyl dicarbonate (Note 4). A white precipitate appears during addition of the di-*tert*-butyl dicarbonate. After a short induction period, the temperature rises to about 30–35°C. The reaction is brought to completion by further stirring overnight at room temperature. At this time, the clear solution will have reached a pH of 7.5–8.5. The reaction mixture is extracted two times with 250 mL of pentane, and the organic phase is extracted three times with 100 mL of saturated aqueous sodium bicarbonate solution. The combined aqueous layers are acidified to pH 1–1.5 by careful addition of a solution of 224 g (1.65 mol) of potassium hydrogen sulfate in 1.5 L of water (Note 5). The acidification is accompanied by copious evolution of carbon dioxide. The turbid reaction mixture is then extracted with four 400-mL portions of ethyl ether (Note 6). The combined organic layers are washed two times with 200 mL of water, dried over anhydrous sodium sulfate or magnesium sulfate, and filtered. The solvent is removed under reduced pressure using a rotary evaporator at a bath temperature not exceeding 30°C (Note 7). The yellowish oil that remains is treated with 150 mL of hexane and allowed to stand overnight (Note 8). Within 1 day the following portions of hexane are added with stirring to the partially crystallized product: 2 × 50 mL, 4 × 100 mL, and 1 × 200 mL. The solution is placed in a refrigerator overnight; the white precipitate is collected on a Büchner funnel and washed with cold pentane. The solid is dried under reduced pressure at ambient temperature to constant weight to give a first crop. The mother liquor is evaporated to dryness leaving a yellowish oil, which is treated in the same manner as described above, giving a second crop (Note

9). The total yield of pure white *N-tert*-butoxycarbonyl-L-phenylalanine is 207–230 g (78–87%), mp 86–88°C, $[\alpha]_D^{20}$ + 25.5° (ethanol, *c* 1.0) (Note 10).

2. Notes

1. L-Phenylalanine puriss. from Fluka AG or Tridom Chemical Inc. was used.

2. All the solvents and reagents used were of pure grade and obtained from Fluka AG.

3. At this stage, the reaction mixture has a pH of 12–12.5.

4. Di-*tert*-butyl dicarbonate can be prepared according to Pope, B. M.; Yamamoto, Y.; Tarbell, D. S. *Org. Synth., Coll. Vol. VI* **1988,** 418 or purchased from Fluka AG. Di-*tert*-butyl dicarbonate melts at 22–24°C; this compound can be liquified by immersing the reagent bottle in a water bath with a maximum temperature of 35°C. Commercial material is 97–98% pure; a total of 223 g must be employed.

5. It is recommended that acidification be carried out at a temperature of 0–5°C.

6. Ethyl or isopropyl acetate may also be used as extraction solvents for less lipophilic *N-tert*-butoxycarbonyl amino acids.

7. Evaporation should be performed first at 10–20 mm, then at a pressure less than 1 mm in order to remove the *tert*-butyl alcohol completely. Remaining small quantities of *tert*-butyl alcohol lead to difficulty in crystallization.

8. Seeding or scratching with a glass rod helps to induce crystallization.

9. Normally it is not worthwhile to isolate a third crop, which is of lower purity.

10. *N-tert*-Butoxycarbonyl-L-phenylalanine prepared by this method is obtained in a very pure state. Thin-layer chromatography shows a single spot and a content of less than 0.05% free amino acid. Acylation of lipophilic amino acids with excess di-*tert*-butyl dicarbonate may result to some extent in formation of the corresponding *N-tert*-butoxycarbonyl dipeptide.

3. Discussion

In recent years the *tert*-butoxycarbonyl (Boc) group has achieved a leading role as a protective group for the amino moiety of amino acids in peptide synthesis.[2] At one time the most widely used *tert*-butoxycarbonylating agent was the hazardous[3] and toxic *tert*-butyl azidoformate.[4] Di-*tert*-butyl dicarbonate[5] is a highly reactive and safe reagent of the "ready-to-use" type that reacts under mild conditions with amino acids,[5a,6a–i] peptides,[6j–l] hydrazine and its derivatives,[7] amines,[8a–g] and CH-acidic compounds[8h] in aqueous organic solvent mixtures to form pure derivatives in very good yields. Acylation with di-*tert*-butyl dicarbonate proceeds normally without strict pH control. The procedure given here demonstrates a suitable large-scale and safe preparation of an *N-tert*-butoxycarbonylamino acid with extremely simple experimental operations. Table I shows some other Boc-amino acids and derivatives prepared by this method. *N-tert*-Butoxycarbonyl-L-phenylalanine has also been prepared by acylation of L-phenylalanine with other *tert*-butoxycarbonylating agents: *tert*-butyl 4-nitrophenyl carbonate,[9] *tert*-butyl

TABLE I

Boc-Amino Acids Prepared by Acylation with Di-tert-Butyl Dicarbonate

Boc-Amino Acids[a]	Solvent[b]	Base	Time (hr)[c]	Yield, (%)	mp, (°C)	$[\alpha]_D^{20}$	Remarks
Boc-Ala-OH	A	NaOH	16	92-94	82-83	-25.5 (acetic acid, c 2.0)	pH 8.0[e]
Boc-β-Ala-OH	A	NaOH	16	85-86	76-77		
Boc-Arg-OH	B	—	15	88	159-160 (dec)	-6.8 (acetic acid, c 1.0)	Extraction with n-butyl alcohol
Boc-Arg(NO₂)-OH[d]	B	NaOH	15	82	107	-22.0 (pyridine, c 2.0)	pH 8.5[e]
Boc-Asn-OH	C	NaOH	18	80-81	176 (dec)	-7.2 (dimethylformamide, c 2.0)	5 hr, 45-50°C
Boc-Asp(OBzl)-OH	A	NaOH	16	81-89	101-102	-19.7 (dimethylformamide, c 2.0)	pH 8.0[e]
Boc-Cys(Bzl)-OH	B	NaOH	15	65	86-87	-43.4 (acetic acid, c 1.0)	
(Boc-Cys-OH)₂	D	NaOH	16	85	143-145 (dec)	-115.6 (acetic acid, c 2.0)	
Boc-Gln-OH	E	NaOH	18	76	125 (dec)	-3.4 (ethanol, c 2.0)	pH 8.0[e]
Boc-Glu(OBzl)-OH[f]	B	NaOH	15	86	142-143	+13.2 (methanol, c 1.0)	pH 8.5-9[e]
Boc-Gly-OH	A	NaOH	16	96	87-88		
Boc-His(Boc)-OH	A	KHCO₃	18	75	170 (dec)	+19.5 (chloroform, c 2.0)	
Boc-Ile-OH	A	NaOH	16	78	69-71	+2.8 (acetic acid, c 2.0)	
Boc-Leu-OH[g]	A	NaOH	18	96	85-87	-24.7 (acetic acid, c 2.0)	
Boc-Lys(Boc)-OH[f]	A	NaOH	16	82	138-139	+6.1 (dimethylformamide, c 1.5)	

Compound	Solvent	Base	Time (h)	Yield (%)	mp (°C)	$[\alpha]_D$	Notes
Boc-Lys(CBZ)-OH	A	NaOH	18	96	Oil	−22.8 (methanol, c 1.3)	
Boc-Met-OH	A	NaOH	18	60[h]	50–51	+18.2 (ethanol, c 2.0)	
Boc-Met-OH[f]	A	NaOH	18	85	139–140	−60.6 (acetic acid, c 2.0)	pH 8.5–9[e]
Boc-Pro-OH	A	NaOH	12	95	134–135	−3.6 (acetic acid, c 2.0)	pH 8.5–9[e]
Boc-Ser-OH	A	NaOH	16	66–82	86–88	+19.2 (80% ethanol, c 2.0)	
Boc-Ser(Bzl)-OH	B	NaOH	16	90	62–63	−8.2 (acetic acid, c 1.0)	
Boc-Thr-OH	A	NaOH	16	85	71–73	−18.2 (dimethylformamide, c 1.0)	
Boc-Trp-OH[i]	A	NaOH	16	96	137–138 (dec)		
Boc-Trp-(FOR)-OH[f]	F	Et$_3$N	48	61	158–159[k]	+36.0 (ethanol, c 2.0)	
Boc-Tyr-OH	A	NaOH[l]	24	75	137[m]	+2.6 (acetic acid, c 1.0)	
Boc-Tyr-OH[f]	A	NaOH[l]	24	84	216	+2.6 (acetic acid, c 1.0)	
Boc-Tyr(Bzl)-OH	B	NaOH	18	70	110–111	+27.6 (ethanol, c 1.0)	pH 10.4[e]
Boc-Tyr(2,6-Cl$_2$-Bzl)-OH	A	NaOH	24	48	104 (dec)	+20.6 (ethanol, c 2.0)	
Boc-Val-OH	A	NaOH	16	85	76–78	−7.5 (acetic acid, c 1.0)	

[a] The amino acids used, with the exception of β-alanine and glycine, were of L-configuration. The abbreviations used for amino acids and their protecting substituents concur with E. Wünsch.[2]

[b] Solvent systems: A: tert-butyl alcohol–water; B: dioxane–water; C: dimethylformamide–water; D: methanol–water; E: acetonitrile–water; F: dimethylformamide.

[c] The reaction was generally carried out at room temperature after the exothermic starting period had subsided. Progress of the reaction was monitored by TLC. Reaction times are not optimized.

[d] pH control is necessary.

[e] Crystallizes with 15% solvent (ethyl acetate).

[f] Dicyclohexylamine salt.

73

azidoformate,[10] *tert*-butyl 2,4,5-trichlorophenyl carbonate,[11] *tert*-butyl pentachlorophenyl carbonate,[12] *tert*-butyl 8-quinolyl carbonate,[13] *tert*-butyl chloroformate,[14] *tert*-butyl fluoroformate,[15] *tert*-butyl phenyl carbonate,[16] N-*tert*-butoxycarbonyl-1H-1,2,4-triazole,[17] *tert*-butyl 4,6-dimethylpyrimidyl-2-thiol carbonate,[18] N-*tert*-butoxycarbonyloxyimino-2-phenylacetonitrile,[19] *tert*-butyl α-methoxyvinyl carbonate,[20] *tert*-butyl aminocarbonate (*tert*-butoxycarbonyloxyamine).[21] Recently, a simple method for the 4-dimethylaminopyridine-catalyzed *tert*-butoxycarbonylation of various types of amides has been reported.[22]

1. Fluka AG, Chemische Fabrik, CH-9470 Buchs, Switzerland.
2. Wunsch, E. In "Methoden der Organischen Chemie" (Houben-Weyl), 4th ed.; Thieme: Stuttgart, 1975; Vol. 15.
3. (a) Fenlon, W. J. *Chem. Eng. News* **1976**, *54*, 3; (b) Koppel, H. C. *Chem. Eng. News* **1976**, *54*, 5; (c) Feyen, P. *Angew. Chem.* **1977**, *89*, 119; *Angew. Chem., Int. Ed.* **1977**, *16*, 115; (d) *WARNING: Org. Synth., Coll. Vol. VI* **1988**, 207.
4. (a) Insalaco, M. A.; Tarbell, D. S. *Org. Synth., Coll. Vol. VI* **1988**, 207, 9-12; (b) Carpino, L. A.; Carpino, B. A.; Crowley, P. J.; Giza, C. A.: Terry, P. H. *Org. Synth. Coll. Vol. V* **1973**, 157-159.
5. (a) Pozdnev, V. F. *Khim. Prir. Soedin.* **1974**, 764-767; *Chem. Abstr.* **1975**, *82*, 156690d; (b) Pope, B. M.; Yamamoto, Y.; Tarbell, D. S. *Org. Synth., Coll. Vol. VI* **1988**, 418; (c) Pozdnev, V. F.; Smirnova, E. A.; Podgornova, N. N.: Zentsova, N. K.; Kalei, U. O. *Zhur. Org. Khim.* **1979**, *15*, 106-109; *Chem. Abstr.* **1979**, *90*, 168196a.
6. (a) Fauchère, J. L.; Leukart, O.; Eberle, A.; Schwyzer, R. *Helv. Chim. Acta* **1979**, *62*, 1385-1395; (b) Do, K. W.; Thanei, P.; Caviezel, M.; Schwyzer, R. *Helv. Chim. Acta* **1979**, *62*, 956-964; (c) Hubbuch, A.; Danho, W.; Zahn, H. *Liebigs Ann. Chem.* **1979**, 776-783; (d) Hofmann, K.; Finn, F. M.; Kiso, Y. *J. Am. Chem. Soc.* **1978**, *100*, 3585-3590; (e) Pozdnev, V. F. *Zhur. Obshch. Khim.* **1978**, *48*, 476-477; *Chem. Abstr.* **1978**, *89*, 24739m; (f) Pozdnev, V. F. *Bioorg. Khim.* **1977**, *3*, 1605-1610; *Chem. Abstr.* **1978**, *88*, 62595y; (g) Moroder, L.; Hallett, A.; Wünsch, E.; Keller, O.; Wersin, G. *Hoppe-Seyler's Z. Physiol. Chem.* **1976**, *357*, 1651-1653; *Chem. Abstr.* **1977**, *86*, 107005h; (h) Pozdnev, V. F. *Khim. Prir. Soedin.* **1980**, 379; *Chem. Abstr.* **1980**, *93*, 204996j; (i) Scott, J. W.; Parker, D.; Parrish, D. R. *Synth. Commun.* **1981**, *11*, 303-304; (j) Bullesbach, E. E.; Naithani, V. K. *Hoppe-Seyler's Z. Physiol. Chem.* **1980**, *361*, 723-734; *Chem. Abstr.* **1981**, *94*, 16076f; (k) Gattner, H. G.; Danho, W.; Behn, C.; Zahn, H. *Hoppe-Seyler's Z. Physiol. Chem.* **1980**, *361*, 1135-1138; *Chem. Abstr.* **1980**, *93*, 239908j; (l) Naithani, V. K.; Gattner, H. G. *Hoppe-Seyler's Z. Physiol. Chem.* **1981**, *362*, 685-695; *Chem. Abstr.* **1981**, *95*, 220291x.
7. (a) Pozdnev, V. F. *Zhur. Org. Khim.* **1977**, *13*, 2531-2535; *Chem. Abstr.* **1978**, *88*, 89063h; (b) Dolinskaya, S. I.; Pozdnev, V. F.; Chaman, E. S. *Khim. Prir. Soedin.* **1974**, 266-267; *Chem. Abstr.* **1974**, *81*, 49994q.
8. Pirkle, W. H.; Simmons, K. A.; Boeder, C. W. *J. Org. Chem.* **1979**, *44*, 4891-4896; (b) Pozdnev, V. F. *Khim. Prir. Soedin.* **1980**, 408-409; *Chem. Abstr.* **1980**, *93*, 186700b; (c) Smith, M. A.; Weinstein, B.; Greene, F. D. *J. Org. Chem.* **1980**, *45*, 4597-4602; (d) Muchowski, J. M.; Venuti, M. C. *J. Org. Chem.* **1980**, *45*, 4798-4801; (e) Hofmann, K.; Finn, F. M.; Kiso, Y. *J. Am. Chem. Soc.* **1978**, *100*, 3585-3590; (f) Jones, N. F.; Kumar, A.; Sutherland, I. O. *Chem. Commun.* **1981**, 990-992; (g) Ernest, I. *Helv. Chim. Acta* **1980**, *63*, 201-213; (h) Rachon, J.; Schollkopf, U. *Liebigs Ann. Chem.* **1981**, 99-102.
9. Anderson, G. W.; McGregor, A. C. *J. Am. Chem. Soc.* **1957**, *79*, 6180-6183.
10. (a) Schwyzer, R.; Sieber, P.; Kappeler, H. *Helv. Chim. Acta* **1959**, *42*, 2622-2624; (b) Schnabel, E. *Liebigs Ann. Chem.* **1967**, *702*, 188-196; (c) Ali, A.; Fahrenholz, F.; Weinstein, B. *Angew. Chem.* **1972**, *84*, 259; *Angew. Chem., Int. Ed.* **1972**, *11*, 289.

11. Broadbent, W.; Morley, J. S.; Stone, B. E. *J. Chem. Soc. (C)* **1967**, 2632–2636.
12. Fujino, M.; Hatanaka, C. *Chem. Pharm. Bull.* **1967**, *15*, 2015–2016.
13. Rzeszotarska, B.; Wiejak, S. *Liebigs Ann. Chem.* **1968**, *716*, 216–218.
14. (a) Sakakibara, S.; Honda, I.; Takada, K.; Miyoshi, M.; Ohnishi, T.; Okumura, K. *Bull. Chem. Soc. Jpn.* **1969**, *42*, 809–811; (b) Klengel, H.; Schumacher, K. J.; Losse, G. *Z. Chem.* **1973**, *13*, 221–222; *Chem. Abstr.* **1973**, *79*, 137466g.
15. Schnabel, E.; Herzog, H.; Hoffmann, P.; Klauke, E.; Ugi, I. *Liebigs Ann. Chem.* **1969**, *716*, 175–185.
16. Ragnarsson, U.; Karlsson, S. M.; Sandberg, B. E.; Larsson, L. E. *Org. Synth., Coll. Vol. VI* **1988**, 203.
17. Bram, G. *Tetrahedron Lett.* **1973**, 469–472.
18. Nagasawa, T.; Kuriowa, K.; Narita, K.; Isowa, Y. *Bull. Chem. Soc. Jpn.* **1973**, *46*, 1269–1277.
19. Itoh, M.; Hagiwara, D.; Kamiya, T. *Bull. Chem. Soc. Jpn.* **1977**, *50*, 718–721.
20. Kita, Y.; Haruta, J.; Yasuda, H.; Fukunaga, K.; Shirouchi, Y.; Tamura, Y. *J. Org. Chem.* **1982**, *47*, 2697–2700.
21. Harris, R. B.; Wilson, I. B. *Int. J. Pept. Prot. Res.* **1984**, *23*, 55–60.
22. Grehn, L.; Gunnarsson, K.; Ragnarsson, U. *Acta Chem. Scand.* **1986**, *B40*, 745–750.

N-tert-BUTOXYCARBONYL-L-PHENYLALANINE

(L-Phenylalanine, *N*[(1,1-dimethylethoxy)carbonyl)]-)

Submitted by WILLIAM J. PALEVEDA, FREDERICK W. HOLLY, and DANIEL F. VEBER[1]
Checked by MARK A. SANNER, THOMAS VON GELDERN, and CLAYTON H. HEATHCOCK

1. Procedure

To a stirred mixture of 16.51 g (0.1 mol) of L-phenylalanine in 60 mL of water and 60 mL of peroxide-free dioxane (Note 1) is added 21 mL of triethylamine. To the resulting solution is added 27.1 g (0.11 mol) of 2-(*tert*-butoxycarbonyloxyimino)-2-phenylacetonitrile (Note 2). Solution is obtained during the first hour of stirring. After 3 hr (Note 3) the solution is diluted with 150 mL of water. The resulting turbid solution is extracted with at least four 200-mL portions of ethyl ether (Note 4). The aqueous layer is then acidified to pH 2.5 with cold 2.5 *N* hydrochloric acid to yield an oily layer. The mixture is extracted with three 100-mL portions of methylene chloride. The combined organic extracts are dried with anhydrous sodium sulfate. After filtration of the sodium sulfate, the filtrate is evaporated under reduced pressure at a bath temperature of 30°C. Hexane is added to the thick oil to turbidity. Crystallization occurs after cooling and stirring the mixture for a short time. More hexane is added in portions until no further crystallization occurs. A total of 200 mL of hexane is required. The mixture is allowed to stand for 1 hr. The white crystalline solid is collected by filtration, washed with three 100-mL portions of hexane, and dried under reduced pressure to yield 21.4–22.0 g (80–83%) of *tert*-butoxycarbonyl-L-phenylalanine, mp 86–88°C, $[\alpha]_D^{20}$ −3.6° (HOAc, c 1), $[\alpha]_{546}^{20}$ 29.9° [EtOH, c 1) (Note 5).

2. Notes

1. Peroxides are removed from dioxane by its passage through a column of neutral alumina.[2]

2. 2-(*tert*-Butoxycarbonyloxyimino)-2-phenylacetonitrile is obtained from Aldrich Chemical Company, Inc., under the trademark "BOC-ON."

3. The reaction is allowed to continue until TLC (Whatman K1F, ethyl acetate–pyridine–acetic acid–water, 10:5:1:3) shows that the unprotected amino acid (R_f 0.4) is no longer present, as evidenced by negative ninhydrin spray.

4. It is imperative that all the by-product is removed at this point; otherwise it will contaminate the product, making crystallization difficult. Each ether extract is spotted on a Whatman K1F plate and the plate viewed under UV light to ascertain that all of the by-product has been extracted. The checkers found that six or seven ether extractions were required to remove the by-product completely.

5. The literature gives melting points ranging from 79–80°C to 84–86°C; the optical rotation is reported as $[\alpha]_D^{25}$ −0.8° (HOAc, c 4.957), $[\alpha]_D^{20}$ −4.8° (HOAc, c 1), $[\alpha]_{546}^{20}$ 30° (EtOH, c 1).

The spectral properties of *tert*-butoxycarbonyl-L-phenylalanine are as follows: ^{1}H NMR (CD$_3$OD) δ: 1.36 (s, 9 H, *t*-butyl), 2.87 (dd, 1 H, J = 14.9, H$_\beta$), 3.16 (dd, 1 H, J = 14.6, H$_\beta$), 4.36 (dd, 1 H, J = 9.6, H$_\alpha$), 7.26 (s, 5 H, phenyl). In CDCl$_3$ solution, both carbamate rotamers may be seen in the ^{1}H NMR spectrum.

3. Discussion

Various reagents have been used for the introduction of the *tert*-butoxycarbonyl group, including *tert*-butyl *p*-nitrophenyl carbonate,[3] *tert*-butyl azidoformate[4] (no longer commercially available because of its toxic and potentially explosive nature), *tert*-butyl 2,4,5-trichlorophenyl carbonate,[5] di-*tert*-butyl dicarbonate,[6] and the reagent described herein, 2-(*tert*-butoxycarbonyloxyimino)-2-phenylacetonitrile.[7] Using the same reagent, the crystalline BOC derivatives of the following amino acids have been prepared in these laboratories in the indicated yields: 7-aminoheptanoic acid (88%), DL-tyrosine (96%), 6-fluoro-DL-tryptophan (87%), 5-methyl-DL-tryptophan (95%), 5-bromo-DL-tryptophan (94%), 5-methoxy-DL-tryptophan (67%), 1-methyl-DL-tryptophan (82%), and 5-fluoro-DL-tryptophan (62%).

1. Merck Sharp & Dohme Research Laboratories, West Point, PA 19486.
2. Stewart, J. M.; Young, J. D. "Solid Phase Peptide Synthesis," W. H. Freeman & Co.: San Francisco, 1969, p 31.
3. Anderson, G. W.; McGregor, A. C. *J. Am. Chem. Soc.* **1957,** *79,* 6180.
4. Schwyzer, R.; Sieber, P.; Kappeler, H. *Helv. Chim. Acta* **1959,** *42,* 2622.
5. Broadbent, W.; Morley, J. S.; Stone, B. E. *J. Chem. Soc. C* **1967,** 2632.
6. Moroder, L.; Hallett, A.; Wünsch, E.; Keller, O.; Wersin, G. *Hoppe-Seyler's Z. Physiol. Chem.* **1976,** *357,* 1651.
7. Itoh, M.; Hagiwara, D.; Kamiya, T. *Tetrahedron Lett.* **1975,** 4393; *Bull. Chem. Soc. Jpn.* **1977,** *50,* 718.

GENERATION AND REACTIONS OF ALKENYLLITHIUM REAGENTS: 2-BUTYLBORNENE

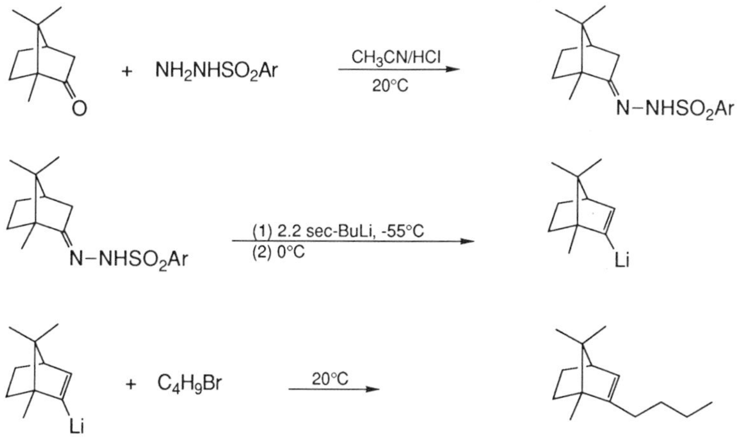

Submitted by A. Richard Chamberlin, Ellen L. Liotta, and F. Thomas Bond[1]
Checked by Hiroko Masamune and Robert V. Stevens

1. Procedure

A. *d-Camphor 2,4,6-triisopropylbenzenesulfonylhydrazone.* In a 500-mL Erlenmeyer flask equipped with a magnetic stirring bar is placed 66.0 g (0.22 mol) of 2,4,6-triisopropylbenzenesulfonylhydrazide (Note 1), 30.4 g (0.20 mol) of *d*-camphor (Note 2), 100 mL of freshly distilled acetonitrile, and 20.0 mL (0.24 mol) of concentrated hydrochloric acid. The resulting solution is stirred overnight while a granular solid precipitates. The white crystals are cooled at −10°C for 4 hr and collected by suction filtration, dissolved in 175 mL of dichloromethane, filtered to remove a small amount of insoluble material, and concentrated under reduced pressure on a rotary evaporator to give 60.8–63.4 g (70–73%) of a white solid, mp 196–199°C (dec) (Note 3).

B. *2-Butylbornene.* A 1-L, three-necked flask is equipped with a 250-mL addition funnel (sealed with a rubber septum), a mechanical stirrer, and a rubber septum. The system is vented (via a hypodermic needle inserted through the addition funnel septum) through a mineral oil bubbler, and the apparatus is flame-dried while it is flushed with prepurified nitrogen introduced through the septum of the flask. The flask is charged with 40.0 g (0.092 mol) of *d*-camphor 2,4,6-triisopropylbenzenesulfonylhydrazone, resealed, and again flushed with nitrogen. Hexane, 200 mL, (Note 4), and 200 mL of tetramethylethylenediamine (Note 5) are added, and the stirred solution, under an atmosphere of nitrogen, is cooled to approximately −55°C with an ethanol–water(2:1)/dry ice bath. Using a Luer-Lok syringe, 158 mL (0.20 mol) of 1.29 *M* *sec*-butyllithium (Note 6) is transferred to the addition funnel. The solution is stirred rapidly and the *sec*-butyllithium added over a period of 15–20 min. The resulting orange solution is stirred for 2 hr, and the cold bath removed. After 20 min the flask is immersed in an ice bath until nitrogen evolution ceases (approximately 10 min).

To this stirred solution of 2-lithiobornene is added, via syringe, 15.2 g (0.11 mol) of butyl bromide (Note 7) over a 1-min period. The ice bath is then removed, and the reaction mixture is stirred at room temperature overnight. The mixture is poured into 500 mL of water. The layers are separated and the aqueous layer extracted with two 100-mL portions of ether. The combined organic extracts are washed with five 200-mL portions of water, one 50-mL portion of 1 *N* hydrochloric acid, and two 200-mL portions of water. The solution is dried over anhydrous magnesium sulfate, filtered, and concentrated on a rotary evaporator at aspirator pressure and room temperature. Distillation of the residual yellow liquid through a 20-cm Vigreux column affords 8.9–9.4 g (50–53%) of product as a colorless liquid, bp 57–59°C (0.5 mm), n_D^{25} 1.4664, $[\alpha]_D^{25}$ − 10.7° (*c* MeOH, 0.0747) (Note 8).

2. Notes

1. The submitters used material prepared following a literature procedure.[2]
2. *d*-Camphor was purchased from Eastman Kodak Co., $[\alpha]_D^{25}$ + 39.5°.
3. The ¹H NMR spectrum is as follows: δ: 0.61 (s, 3 H), 0.86 (s, 6 H), 1.26 (overlapping doublets, *J* = 6.7, 18 H), 1.4–2.2 (m, 7 H), 2.90 (septuplet, *J* = 7, 1 H), 4.20 (septuplet, *J* = 7, 2 H), 7.15 (s, 2 H).
4. MCB, Inc. reagent-grade hexane was distilled from lithium aluminum hydride.
5. This compound was purchased from Aldrich Chemical Company, Inc., and distilled from lithium aluminum hydride.
6. The *sec*-butyllithium was purchased from Alfa Products, Morton Thiokol, Inc. and standardized by double titration or diphenylacetic acid titration. Other alkyllithium bases such as butyllithium and methyllithium cannot be substituted for the stronger *sec*-butyllithium since larger amounts of bornylene are formed because of incomplete dianion formation. Careful attention must be paid to stoichiometry in this reaction; failure to do so also results in increasing the amount of bornylene formed.

Even under ideal conditions the NMR of crude product shows 20–30% borny-lene, which, however, is easily separated from the desired product during distil-lation as a "forerun" which sublimes into the vacuum pump trap.

7. Analytical reagent material was purchased from Mallinckrodt, Inc., and distilled from calcium hydride.

8. The ^{1}H NMR spectrum ($CDCl_3$) is as follows: δ 0.74 (s, 3 H), 0.76 (s, 3 H), 0.94 (s, 3 H), 0.7–1.0 (broad m, 7 H), 1.4 (m, 4 H), 1.9 (m, 2 H), 2.19 ("t", $J = 4$, 1 H), 5.51 (m, 1 H).

3. Discussion

The sequence described here illustrates a general procedure for converting ketones into alkylated olefins:

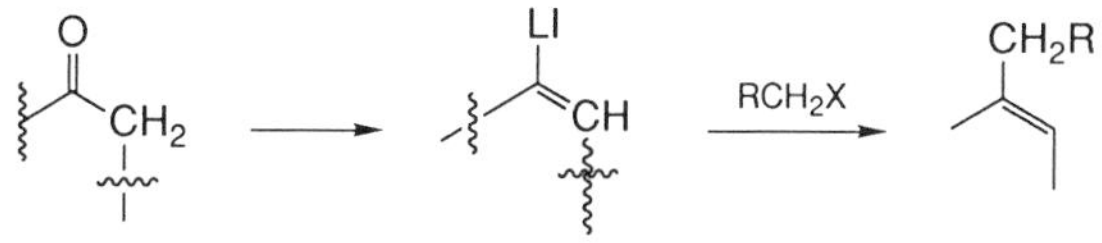

It is a modification of the Shapiro olefin synthesis[3] that allows the alkenyl anion inter-mediate to be trapped with primary halides and other electrophiles. Use of triisopropyl-benzenesulfonylhydrazones as the vinyllithium precursor[4] is an improvement over previously[4] used toluenesulfonylhydrazones,[5,6] which can be employed in the sequence provided excess *sec*-butyllithium (typically 4.5 equiv) and alkyl halide (3.0 equiv) are used. Methyl ketones (e.g., acetone, acetophenone, 2-octanone) can also be used and can be converted into their dianions using 2.2 equiv of the weaker base, butyllithium. The conditions described above, with the slight modifications noted, have been used for a variety of ketones as shown in Table I.

The submitters have found that the hexane–tetramethylethylenediamine solvent system described above, which is required for toluenesulfonylhydrazones, may be replaced with tetrahydrofuran when triisopropylbenzenesulfonylhydrazones are used, provided that the electrophilic reagent is added to the alkenyllithium species as soon as it is formed (as indicated by cessation of nitrogen evolution).

Primary alkyl bromides react well in this sequence except for particularly reactive compounds (e.g., methyl bromide, allyl bromide) that give the vinyl halide by metal-halogen exchange. Secondary halides, as expected, suffer from elimination as a side reaction. Other electrophiles have been used successfully including D_2O, aldehydes and ketones, dimethylformamide,[4,7] chlorotrimethylsilane,[4,8] 1,2-dibromoethane,[4] and carbon dioxide. Such sequences allow for relatively straightforward preparation of deuterated olefins, allylic alcohols, α,β-unsaturated aldehydes, alkenylsilanes, alkenyl bromides, and α,β-unsaturated acids. The major advantages of this route to alkenylli-thium reagents[9] lie in the availability of the ketone precursors and the regiospecificity

TABLE I
Ketone to Butylalkene Conversions

Ketone	Product
	a
	a,b
	a
	a,c
	d

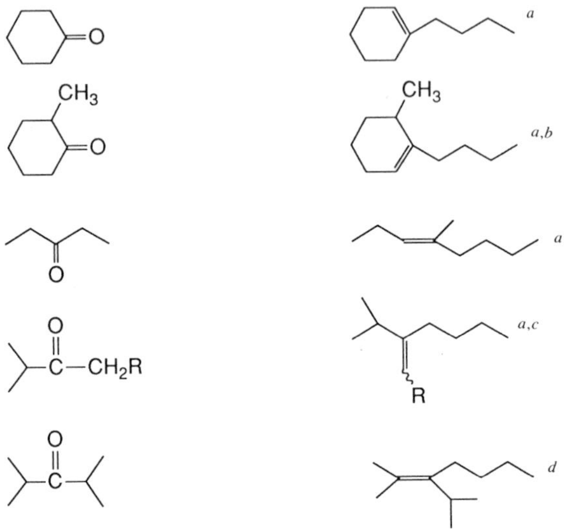

[a] *sec*-Butyllithium is added at $-8°C$.
[b] Approximately 2% of the isomeric 1-butyl-2-methylcyclohexene is formed.
[c] A mixture of (Z) and (E) isomers is formed.
[d] Tertiary hydrogen removal is slower. *sec*-Butyllithium (3.0 equiv) is added at $-78°C$; the solution is immediately warmed to room temperature and stirred for 1–2 hr before butyl bromide (2.0 equiv) is added.

of the Shapiro reaction.[3,10] There are numerous alternative routes to trisubstituted olefins.[11]

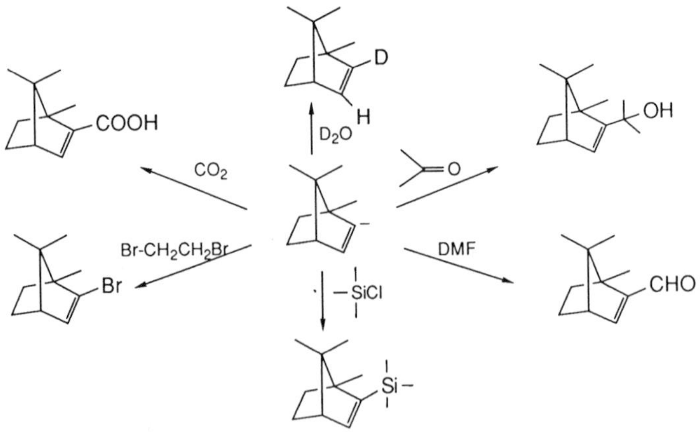

1. Department of Chemistry, University of California, San Diego, La Jolla, CA 92093.
2. Cusack, N. J.; Reese, C. B.; Risins, A. C.; Roozpeikar, B. *Tetrahedron*, **1976**, *32,* 2157.
3. Shapiro, R. H. *Org. Reactions*, **1975**, *23,* 405; Shapiro, R. H.; Duncan, J. H. *Org. Synth. Coll. Vol. VI* **1988**, 172.
4. Chamberlin, A. R.; Stemke, J. E.; Bond, F. T. *J. Org. Chem.*, **1978**, *43,* 147.
5. Shapiro, R. H.; Lipton, M. F.; Kolonko, K. J.; Buswell, R. L.; Capuan, L. A. *Tetrahedron Lett.*, **1976**, 1811, Burgstahler, A. W.; Boger, D. L.; Naik, N. C. *Tetrahedron*, **1976**, *32,* 309.
6. Stemke, J. E.; Chamberlin, A. R.; Bond, F. T. *Tetrahedron Lett.*, **1976**, 2947.
7. Traas, P. C.; Boelens, H.; Takken, H. J. *Tetrahedron Lett.*, **1976**, 2287.
8. Chan, T. H.; Baldassare, A.; Massuda, D. *Synthesis*, **1976**, 801; Taylor, R. C.; Degenhardt, C. R.; Melega, W. P.; Paquette, L. A. *Tetrahedron Lett.*, **1977**, 159.
9. Neuman, H.; Seebach, D. *Chem. Ber.*, **1978**, *111,* 2785, report a detailed study of generation of vinyllithium reagents from vinyl bromides, and reactions thereof.
10. Kolonko, K. J.; Shapiro, R. H. *J. Org. Chem.*, **1978**, *43,* 1404; Lipton, M.; Shapiro, R. H. *J. Org. Chem.*, **1978**, *43,* 1409.
11. Reucroft, J.; Sammes, P. G. *Quart. Rev.*, **1971**, *25,* 135; Faulkner, D. J. *Synthesis*, **1971**, 175.

PREPARATION OF THIOL ESTERS: S-*tert*-BUTYL CYCLOHEXANECARBOTHIOATE AND S-*tert*-BUTYL 3α,7α,12α-TRIHYDROXY-5β-CHOLANE-24-THIOATE

[Cyclohexanecarbothioic acid, S-(1,1-dimethylethyl)ester and cholane-24-thioic acid, 3,7,12-trihydroxy-S-(1,1-dimethylethyl)ester, (3α,5β,7α,12α)]

A. $(CH_3)_3CSH$ + $TlOC_2H_5$ $\xrightarrow{\text{benzene}}$ $TlSC(CH_3)_3$ + C_2H_5OH

B.

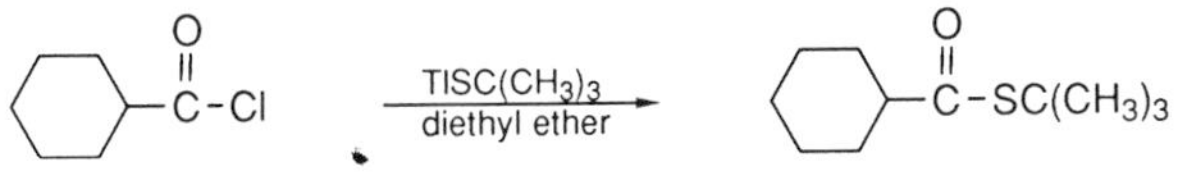

C.

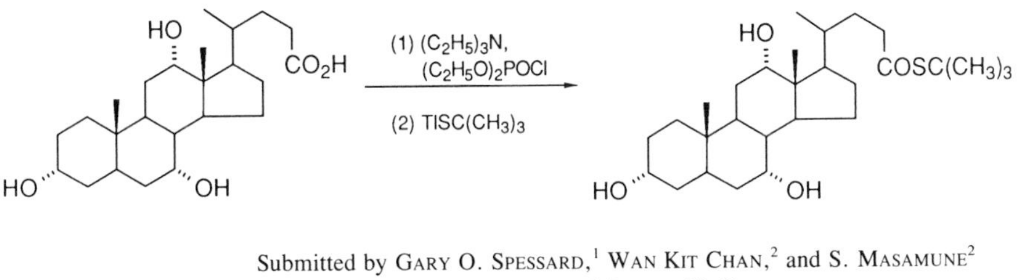

Submitted by GARY O. SPESSARD,[1] WAN KIT CHAN,[2] and S. MASAMUNE[2]
Checked by TRINA KITTREDGE and ROBERT V. STEVENS

1. Procedure

Caution! Thallium compounds are very toxic. However, they may be safely handled if prudent laboratory practices are followed. Rubber gloves and laboratory coats should be worn, and reactions should be carried out in an efficient hood. Thallium wastes should be collected and disposed of separately.[3]

A. *Thallium(I) 2-methylpropane-2-thiolate.* A 500-mL, round-bottomed flask equipped with a magnetic stirring bar and a pressure-equalizing dropping funnel to which a nitrogen inlet adapter is attached is charged with 47.2 g (0.189 mol) of thallium(I) ethoxide (Note 1) and 200 mL of anhydrous benzene (Note 2). Over a period of 15 min 19.2 g (24 mL, 0.213 mol) of 2-methylpropane-2-thiol (Note 1) is added. The reaction mixture is stirred under a nitrogen atmosphere for 1 hr and the resulting precipitate is collected by filtration. After washing with three 100-mL portions of anhydrous pentane (Note 3), 48.5–51.2 g (90–95%) of the product is obtained as bright yellow crystals, mp 165–170°C dec (Note 4). This material is sufficiently pure for use in the following steps.

B. *S-tert-Butyl cyclohexanecarbothioate.* A solution of 4.38 g (0.030 mol) of cyclo-hexanecarboxylic acid chloride (Note 5) in 150 mL of ether (Note 6) is placed in a dry, 500-mL, round-bottomed flask equipped with a magnetic stirring bar and a gas inlet. The system is flushed with nitrogen and the solution is cooled in an ice bath. Stirring is initiated and 8.82 g (0.031 mol) of the thallium(I) 2-methylpropane-2-thiolate prepared in Step A is added. After the resulting milky suspension is stirred for 2 hr at room temperature, the fine precipitate is removed by filtration through Celite (Note 7) and washed thoroughly with four 50-mL portions of ether. The combined filtrate and washings are concentrated on a rotary evaporator to give a pale-yellow oil, which is distilled under reduced pressure through a 5-cm Vigreux column. After separation of a forerun, 5.36–5.44 g (90–91%) of the colorless thiol ester is collected, bp 100°C (7 mm) (Note 8).

C. *S-tert-Butyl ester from cholic acid.* A dry, 250-mL, one-necked, round-bottomed flask is equipped with a magnetic stirring bar and a nitrogen inlet adapter; the system is purged with, and maintained under, dry nitrogen. After 4.90 g (0.0120 mol) of cholic acid (Note 9), 1.33 g (0.0131 mol) of triethylamine (Note 10), and 60 mL of dry tetra-hydrofuran (THF, Note 11) are placed in the flask, a stoppered, pressure-equalizing dropping funnel charged with a solution of 2.18 g (0.0127 mol) of diethyl phosphor-ochloridate (Note 9) in 30 mL of dry THF is attached to the top of the nitrogen inlet adapter (see Figure 1). The solution is added to the stirred reaction mixture over a period of 5 min and stirring is continued for 3.5 hr at room temperature. The dropping funnel is removed, and the reaction mixture is taken up into a dry, 100-mL syringe and trans-ferred to a dry filtering apparatus. This apparatus is shown in Figure 2. The glass-fritted filter funnel of medium porosity with a built-in vacuum adapter is connected to the middle neck of a 500-mL, three-necked, round-bottomed flask. A calcium chloride drying tube is connected to the vacuum adapter and a nitrogen inlet adapter is attached to the top of the filter funnel. The precipitated triethylamine hydrochloride is now removed from the reaction mixture by stoppering the nitrogen inlet adapter and using the positive

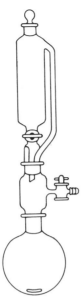

Figure 1

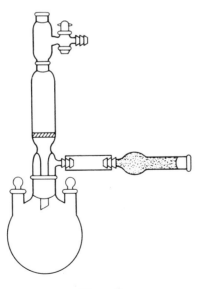

Figure 2

nitrogen pressure to force the solution through the glass frit. Dry tetrahydrofuran, 40 mL, is used to rinse the original reaction flask. The stopper of the nitrogen inlet adapter (Figure 2) is removed, and this washing is transferred via the same syringe to the filtering apparatus and forced through the filter in the same manner described above. One of the stoppers of the three-necked flask is replaced by a nitrogen inlet adapter and the filter funnel is replaced by a mechanical stirrer. As the filtrate is stirred at room temperature, the remaining stopper is removed and 3.90 g (0.0133 mol) of thallium(I) 2-methylpropane-2-thiolate is added. After the addition is complete, the neck is restoppered, and the resulting mixture is vigorously stirred under nitrogen at room temperature overnight. The precipitate is removed by suction filtration through Celite filter aid (Note 7) and washed with three 30-mL portions of THF. The filtrate and washings are combined and concentrated under reduced pressure, and the resulting residue is dissolved in 160 mL of ethyl acetate. This solution is washed with two 100-mL portions of aqueous 5% $NaHCO_3$, then with 50 mL of aqueous saturated NaCl, and finally is dried over anhydrous Na_2SO_4. The solvent is removed by rotary evaporator to afford a white, gummy paste which crystallizes upon trituration with 20 mL of acetonitrile. The crystals are collected by suction filtration to afford 4.2 g of crude product. Recrystallization from 90 mL of hot acetonitrile provides 3.5 g of the thiol ester as small white needles, mp 166–167°C (Note 12). A second crop of 0.5 g, mp 165–166°C, can be obtained upon concentration of the mother liquor to approximately 30 mL, for a combined yield of 70%.

2. Notes

1. Thallium(I) ethoxide and 2-methylpropane-2-thiol were purchased from Aldrich Chemical Company, Inc.

2. Benzene, reagent grade, was purified and dried by first removing the benzene–water azeotrope by simple distillation and then collecting the remaining liquid under an atmosphere of nitrogen.

3. Dry pentane was obtained by allowing practical grade pentane to be shaken with and then distilled from concentrated sulfuric acid.

4. The product should be stored in a dark bottle under an atmosphere of argon to prevent discoloration and possible decomposition.

5. Cyclohexanecarboxylic acid chloride may be prepared in the following way: a pressure-equalizing addition funnel fitted with a nitrogen inlet tube is attached to a 500-mL, round-bottomed flask equipped with a magnetic stirring bar and also charged with 12.8 g (0.100 mol) of cyclohexanecarboxylic acid (purchased from Aldrich Chemical Company, Inc.) and 250 mL of anhydrous ether. (Anhydrous benzene may also be used.) The ethereal solution is cooled to ice-bath temperature and 25.4 g (0.200 mol) of oxalyl chloride (purchased from Aldrich Chemical Company, Inc.) is added over a period of 20 min. Under nitrogen, the resulting solution is stirred for 26 hr before it is concentrated on a rotary evaporator to afford a pale-yellow oil. Distillation of the oil yields 13.5 g (92%) of cyclohexanecarboxylic acid chloride as a clear, colorless liquid, bp 75°C (30 mm); IR (liquid film) cm^{-1}: 1800 (strong).

6. Anhydrous ether was obtained from Mallinckrodt Inc. and used without further purification.

7. Celite (C-211), purchased from Fisher Scientific Company, was washed thoroughly with ether.

8. The spectral characteristics of the product are as follows: IR (liquid film) cm^{-1}: 1675 (strong); ^{1}H NMR (neat) δ: 1.42 [s, 9 H, C$(CH_3)_3$], 1.0–2.0 (m, 10 H, all CH_2 in cyclohexane portion), 2.3 (m, 1 H, CH).

9. Cholic acid and diethyl phosphorochloridate were obtained from Aldrich Chemical Company, Inc.

10. Triethylamine was purchased from Eastman Organic Chemicals.

11. Tetrahydrofuran, reagent grade, was refluxed over and distilled from lithium aluminum hydride immediately prior to use (see Org. Synth., Coll. Vol. V **1973**, 976 for warning).

12. The spectral properties of the product are as follows: IR (CHCl$_3$) cm^{-1}: 3600 (sharp, weak), 3430 (broad, medium), 1675 (strong), no absorption at 1700.

3. Discussion

Methods available before 1971 for the preparation of thiol esters are briefly summarized in a review article.[4] Since then, several newer techniques have been developed to meet a certain set of criteria required for recent synthetic operations. This development may be summarized as follows. Whenever an acid chloride is available, the reaction of the Tl(I) salt of a thiolate of virtually any kind, including alkane-, benzene-, 2-benzothiazoline-, and 2-pyridinethiol, proceeds efficiently and near-quantitatively. However, if selective thiol ester formation in the presence of hydroxy or other functional groups in the same molecule is required, three main procedures are available. First, reaction of an acid (**1**), with a dialkyl or diphenyl phosphorochloridate affords the anhydride (**2**)

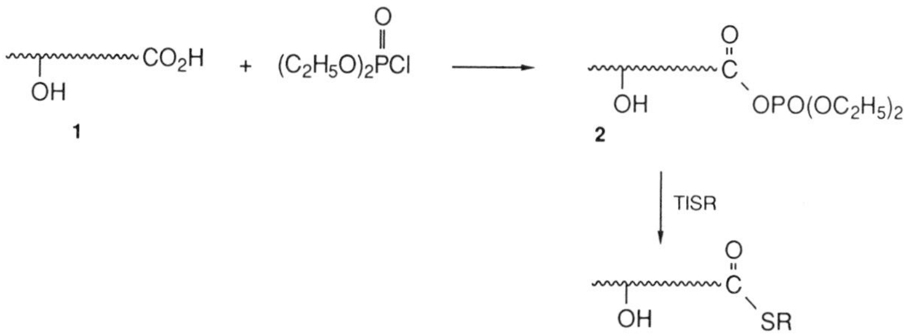

(with the hydroxy groups intact) which is subsequently converted to the thiol ester.[5] This method can be applied to any type of thiol and a variety of hydroxy acids (except for β-hydroxy acids[6]). A mixed anhydride method using ethyl chloroformate and pyridine also effects selective thiol ester formation in many cases.[7] Second, the imidazolide of an acid that is prepared from **1** and N,N-carbonyldiimidazole reacts efficiently with relatively acidic thiols such as benzenethiol to yield the thiol ester.[6,8] Third, use of a disulfide and triphenylphosphine effects the selective formation of thiol esters, but this technique is applicable only to relatively reactive disulfides such as those derived from 2-benzothiazole-, 2-pyridinethiol,[9, 10] and 4-tert-butyl-N-isopropylimidazole-2-thiol.[11]

Other methods that can be used to prepare thiol esters from carboxylic acids include the use of aryl thiocyanates,[12] thiopyridyl chloroformate,[13] 2-fluoro-N-methylpyridinium tosylate,[14] 1-hydroxybenzotriazole,[15] and boron thiolate.[16] Direct conversion of O-esters to S-esters can also be effected via aluminum and boron reagents.[17] However, the applicability of these[12-17] and other methods,[18] including the carboxyl group activation by means of 4-dimethylaminopyridine (DMAP) and dicyclohexylcarbodiimide (DCC),[18d] to the selective thiol ester formation discussed above has not been clearly defined.

Thiol esters have recently been utilized, with and without activation, for the preparation of O-esters for lactones, in particular, in macrolide syntheses. The accompanying procedure illustrates this conversion.[19]

1. Department of Chemistry, Saint Olaf College, Northfield, MI 55057.
2. Department of Chemistry, University of Alberta, Edmonton, Alberta, Canada, T6G 2G2. The present address of S. Masamune is the Department of Chemistry, Massachusetts Institute of Technology, Cambridge, MA 02139.
3. Taylor, E. C.; Robey, R. L.; Johnson, D. K.; McKillip, A. *Org. Synth., Coll. Vol. VI* **1988,** 791.
4. Field, L. *Synthesis* **1972,** 106.
5. Masamune, S.; Kamata, S.; Diakur, J.; Sugihara, Y.; Bates, G. S. *Can J. Chem.* **1975,** *53,* 3693. See also Yamada, S.; Yokoyama, Y.; Shioiri, T. *Chem. Pharm. Bull.* **1977,** *25,* 2423.
6. (a) Masamune, S.; Hayase, Y.; Chan, W. K.; Sobczak, R. L. *J. Am. Chem. Soc.* **1976,** *98,* 7874; (b) Gais, H.-J. *Angew. Chem., Int. Ed. Engl.* **1977,** *16,* 244. The preparation of an alkanethiol ester requires a trace amount of the corresponding sodium thiolate.
7. Masamune, S.; Hirama, M.; Mori, S.; Ali, Sk. A.; Garvey, D. S. *J. Am. Chem. Soc.* **1981,** *103,* 1568; Ref. 9.
8. Bates, G. S.; Diakur, J.; Masamune, S. *Tetrahedron Lett.* **1976,** 4423.
9. (a) Mukaiyama, T.; Matsueda, R.; Maruyama, H. *Bull. Chem. Soc. Jpn.* **1970,** *43,* 1271; (b) Mukaiyama, T. *Synth. Commun.* **1972,** *2,* 243.
10. Masamune, S.; Souto-Bachiller, F.; Hayase, Y. unpublished results.
11. Corey, E. J.; Brunelle, D. J. *Tetrahedron Lett.* **1976,** 3409.
12. Grieco, P. A.; Yokoyama, Y.; Williams, E. *J. Org. Chem.* **1978,** *43,* 1283.
13. Corey, E. J.; Clark, D. A. *Tetrahedron Lett.* **1979,** 2875.
14. Watanabe, Y.; Shoda, S.; Mukaiyama, T. *Chem. Lett.* **1976,** 741.
15. Horiki, K. *Synth. Commun.* **1977,** *7,* 251.
16. Pelter, A.; Levitt, T. E.; Smith, K. *J. Chem. Soc., Perkin Trans. I,* **1977,** 1672.
17. (a) Warwel, S.; Ahlfaenger, B. *Chem.-Ztg.* **1977,** *101,* 103; (b) Cohen, T.; Gapinski, R. E. *Tetrahedron Lett.* **1978,** 4319; (c) Hatch, R. P.; Weinreb, S. M. *J. Org. Chem.* **1977,** *42,* 3960.
18. (a) Kunieda, T.; Abe, Y.; Hirobe, M. *Chem. Lett.* **1981,** 1427; (b) Khim, S.; Yang, S. *Chem. Lett.* **1981,** 133; (c) Kim, H. J.; Lee, S. P.; Chan, W. H. *Synth. Commun.* **1979,** *9,* 91; (d) Neiss, B.; Steglich, W. *Angew. Chem. Int. Ed. Engl.* **1978,** *17,* 522.
19. Chan, W. K.; Spessard, G. O.; Masamune, S. *Org. Synth., Coll. Vol. VII* **1990,** 87.

PREPARATION OF *O*-ESTERS FROM THE CORRESPONDING THIOL ESTERS: *tert*-BUTYL CYCLOHEXANECARBOXYLATE

(Cyclohexanecarboxylic acid, 1,1-dimethylethyl ester)

Submitted by WAN KIT CHAN,[1] S. MASAMUNE,[1] and GARY O. SPESSARD[2]
Checked by TRINA KITTREDGE and ROBERT V. STEVENS

1. Procedure

A 500-mL, round-bottomed flask equipped with a magnetic stirring bar is flushed with nitrogen. The flask is then charged with 5.56 g (0.028 mol) of *S-tert*-butyl cyclohexanecarbothioate (Note 1), 5.55 g (0.075 mol) of *tert*-butyl alcohol, and 250 mL of anhydrous acetonitrile (Note 2). The mixture is stirred vigorously and 23.7 g (0.056 mol) of mercury(II) trifluoroacetate (Note 3) is added in one portion. The resulting mixture is stirred vigorously for 45 min and then concentrated to approximately 50–75 mL on a rotary evaporator (Note 4). To this concentrated mixture is added 250 mL of hexane and the orange solid that forms is removed by filtration. The filter pad is then washed with 50 mL of hexane. The filtrate and washings are combined and washed with a 50 mL portion of aqueous saturated sodium chloride, dried over anhydrous sodium sulfate, and concentrated on a rotary evaporator to give a pale-yellow liquid (Note 4).

The crude product is purified by passing it through a column (4.5 cm × 30 cm) of neutral alumina (Note 5) using chloroform as eluant. The desired product moves with the solvent front, and the first 300–350 mL of eluant contains all of the product. Removal of the solvent gives 4.96 g. The product, which contains a small amount of *tert*-butyl alcohol, can be further purified by distillation through a short-path apparatus to give 4.6 g (90%) of pure *O*-ester, bp 91°C (25 mm) (Notes 4 and 6).

2. Notes

1. This compound is prepared according to *Org. Synth., Coll. Vol. VII* **1990**, 81.

2. Acetonitrile, obtained from J. T. Baker Chemical Co., was refluxed overnight with phosphorus pentoxide and then distilled under nitrogen onto freshly activated Linde 4A molecular sieves. The acetonitrile was stored over the molecular sieves for 24 hr before use.

3. Although mercury(II) trifluoroacetate may be obtained commercially, the submitters recommend that it be freshly prepared. A mixture of red mercury(II) oxide (108.3 g, 0.5 mol) (obtained from BDH Chemicals Ltd.) and freshly distilled trifluoroacetic acid (137.0 g, 1.2 mol) (purchased from J. T. Baker Chemical Co.) was heated at 80°C for 30 min. The excess trifluoroacetic acid and the water formed in the reaction were removed under reduced pressure. The white crystal-

line residue was then dried (50°C, 0.01 mm) for 48 hr to give a quantitative yield of product.

4. The temperature of the water bath was kept below 28°C during evaporation of the acetonitrile.

5. Woelm neutral alumina, activity grade 1, (300 g) was used. The column was packed using hexane.

6. The spectral properties of the product are as follows: IR (neat) cm^{-1}: 1735 (strong); ^{1}H NMR (CDCl$_3$) δ: 1.38 [singlet, 9 H, C(CH$_3$)$_3$] 1.0–2.4 (multiplet, 11 H, cyclohexane protons).

3. Discussion

In recent years much attention has been directed toward efficient ester (and lactone) formation in connection with the synthesis of naturally occurring macrolides.[3,4] Four principal methods for such a reaction have emerged from these studies:

Method 1. Use of a thiophilic metal ion to activate an alkane- or arenethiol ester for nucleophilic displacement by an alcohol is applicable to both ester and lactone formation.[5]

Method 2. Corey's "double activation" method for lactone formation is patterned after Mukaiyama's procedure for peptide formation and involves refluxing a solution of the 2-pyridinethiol ester of a hydroxy acid in a high-boiling solvent for a prolonged period of time.[6]

Method 3. Gerlach's modification of Method 2 uses AgClO$_4$ or AgBF$_4$ to catalyze the cyclization.[7]

Method 4. Mitsunobu's method uses a combination of diethyl azodicarboxylate and triphenylphosphine as a condensing agent.[8]

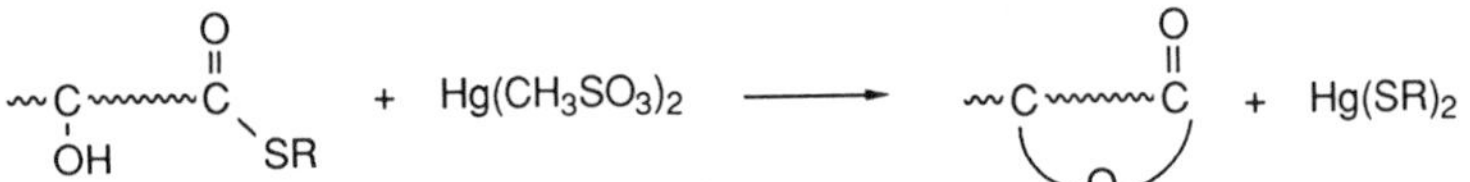

Method 1

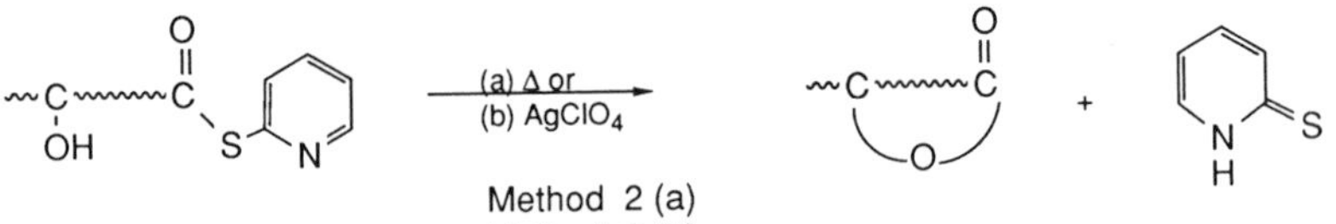

Method 2 (a)
Method 3 (b)

Method 1 offers some distinct advantages. First, an ester such as the 1,1-dimethylethylthiol ester serves as an excellent protective group, surviving both relatively mild alkaline and acid conditions, and has been used successfully in the synthesis of many macrolide natural products.[4b,g,j,n,o,q] Second, reaction of a metal ion such as Hg(II) with the thiol ester formally creates a highly reactive trivalent sulfur species, and thus ester (and lactone) formation proceeds very rapidly at room temperature or below. More importantly, bulky substituents or double bonds located near the reaction centers (i.e., near the hydroxy and acyl groups) do not impede the reaction (see Table I).[4j] Thus *tert*-

TABLE I

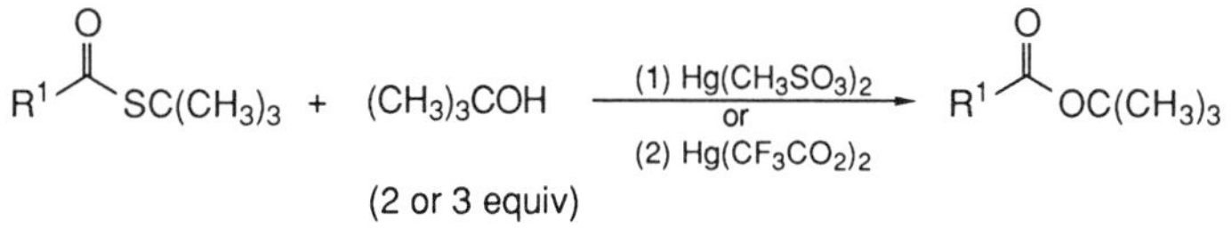

$$R^1-\overset{\displaystyle O}{\overset{\|}{C}}-SC(CH_3)_3 \; + \; (CH_3)_3COH \; \xrightarrow[\substack{or \\ (2)\ Hg(CF_3CO_2)_2}]{(1)\ Hg(CH_3SO_3)_2} \; R^1-\overset{\displaystyle O}{\overset{\|}{C}}-OC(CH_3)_3$$

(2 or 3 equiv)

R^1	Reagent	Buffer	Yield (%)[a]
c-C_6H_{11}—	1 or 2	Na_2HPO_4 (or none)	100
$(CH_3)_3C$—	1	 (epoxide)—CH_2Cl	90
(E)—$CH_3CH{=}CH$—	1 or 2	 (epoxide)—CH_2Cl	85
(Z)—$CH_3CO_2C(CH_3){=}CH$—	1	Na_2HPO_4	90

[a]The yields are estimated by GC analysis.

butyl pivalate and *tert*-butyl crotonate are prepared in excellent yields. In the absence of alcohols, *tert*-butyl cyclohexanecarbothioate reacts with $Hg(CF_3CO_2)_2$ to form cyclohexanecarboxylic trifluoroacetic anhydride. Reaction of this anhydride with *tert*-butyl alcohol to give the ester, however, proceeds ca. 10 times more slowly than the Hg(II)-catalyzed ester formation described above.[9] The intermediacy of the corresponding ketene has been eliminated by use of an appropriately deuterated compound[9] and pivalic acid. Thus the metal-catalyzed ester formation appears to proceed for the most part through coordination of the alcohol to the metal, as shown in a possible intermediate (**1**), followed by collapse into the ester and mercuric salts with retention of stereochemistry at the carbon atom alpha to the carboxy group.

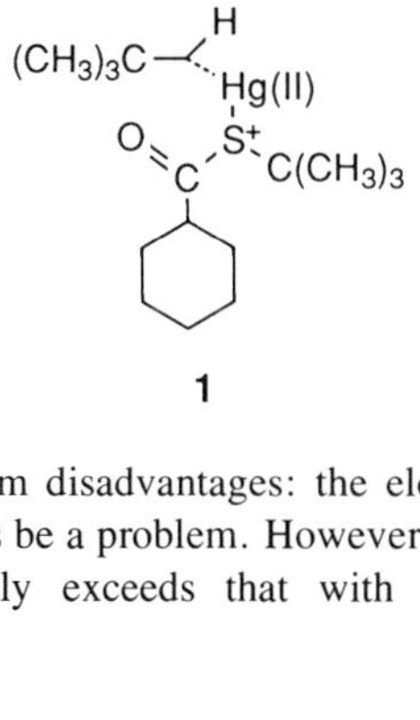

1

This method is not free from disadvantages: the electrophilicity of Hg(II) toward reactive alkenes may sometimes be a problem. However, in most cases the reactivity of Hg(II) with sulfur significantly exceeds that with ordinary or electron-deficient

(C=C—C=O) double bonds, and other combinations of thiol esters and thiophilic metals may be used to overcome this problem. The more acidic the reacting thiol, the less thiophilic is the metal needed to effect the reaction, and in some cases Cu(I), Cu(II), and Ag(I) are superior to Hg(II). For example, the combination of $Ag(I)CF_3CO_2$ [but not $Ag(I)ClO_4$ or $Ag(I)BF_4$] and a benzenethiol ester is very efficient for ester formation. The presence of electron-withdrawing groups such as the C=C bond and protected hydroxy groups somewhat retards the ester formation. A few examples are shown in Table II.[4j] All these observations appear to conform with the hard and soft acid and base principle of Pearson. Further, it is clear that Gerlach's report of the use of Ag(I) to activate 2-pyridinethiol esters (Method 3) is fully in accord with this trend.

Corey's "double activation" procedure (Method 2) does not use an external reagent to activate the functional group, but effects cyclization by heating a solution of the 2-pyridinethiol ester of a hydroxy acid for a prolonged period. Several pieces of evidence point to the intermediacy of 2 in this lactonization.[10] If one accepts this intermediate, it follows that a hydroxy(2-pyridinethiol)ester, heavily substituted near the reaction centers (i.e., near the hydroxyl and acyl groups), would encounter a high energy barrier in the process leading to 2. This inference has been confirmed by measuring the approximate rates of reaction of 2-pyridine- and 2-benzothiazolethiol esters of cyclohexanecarboxylic acid with primary, secondary, and tertiary alcohols.[3a,9] This steric retardation of the reaction may constitute a major drawback to Method 2. A marked improvement has been made, however, and the latest version of this method[11] involves use of the 1-methyl- or 1-isopropyl-4-*tert*-butylimidazole-2-thiol ester of the hydroxy acid which undergoes cyclization ~ 100 times faster than the corresponding 2-pyridinethiol ester. This improved method has been used in syntheses of erythronolide A[4m] and B.[4k]

2

Lactonization of aliphatic hydroxy acids proceeds with the aid of two reagents, diethyl azodicarboxylate and triphenylphosphine. This fourth procedure has been selected as a method of choice for the final cyclization to yield vermiculine[4i] and pyrenophorin.[4h]

Several other methods to effect ester and lactone formation are now available. Mukaiyama uses 2-chloro-*N*-methylpyridinium iodide and its derivatives as a condensing agent.[12] Staab's imidazole method,[13] successfully utilized in a synthesis of pyrenophorin[4c] and a model study for erythronolide B,[14] requires a catalytic amount of strong base, and thus is applicable only to compounds stable under such conditions. The mixed anhydride of a hydroxycarboxylic acid and 2,4,6-trichlorobenzoic acid is efficiently cyclized to provide the corresponding lactone.[14,15] Similarly, the use of a reactive phosphoric acid anhydride intermediate is equally effective.[16] Some other methods for carboxyl activation, using dibutyltin oxide,[17] distannoxane,[18] *N*,*N*,*N'*,*N'*-tetramethylchloroformamidinium chloride,[14] and the combination of dicyclohexylcarbodiimide (DCC), 4-dimethylaminopyridine (DMAP), and DMAP hydrochloride,[20] have also appeared.

TABLE II

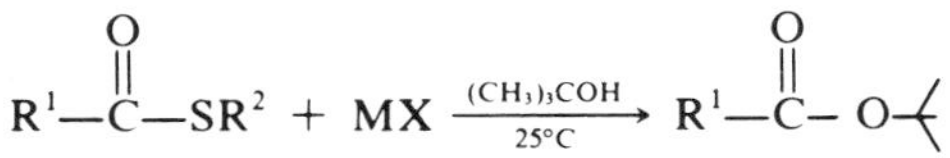

$$R^1-\overset{\overset{\displaystyle O}{\|}}{C}-SR^2 + MX \xrightarrow[25°C]{(CH_3)_3COH} R^1-\overset{\overset{\displaystyle O}{\|}}{C}-O-\!\!\!<$$

R¹	R²	MX	Solvent	Time	Yield (%)
c-C₆H₁₁—	—C₆H₅	Cu(CF₃SO₃)	C₆H₆/THF	10 min	95
c-C₆H₁₁—	![benzothiazole ring]	Ag(CF₃CO₂)	C₆H₆	10 min	100
(E)-CH₃CH=CH—	—C₆H₅	Cu(CF₃SO₃)	C₆H₆/THF	5 hr	80
(E)-C₆H₅CH=CH—	—C₆H₅	Cu(CF₃SO₃)₂	CH₃CN	1.5 hr	24
		Ag(CF₃CO₂)	C₆H₆(Δ)	1.5 hr	100
		AgBF₄	C₆H₆(Δ)	1 hr	<5
C₆H₅—	![dihydrobenzothiazole ring]	Ag(CF₃CO₂)	C₆H₆(Δ)	1.5 hr	100
		Cu(CF₃SO₃)	C₆H₆/THF	5 hr	90
		Cu(CF₃SO₃)₂	CH₃CN	30 min	100

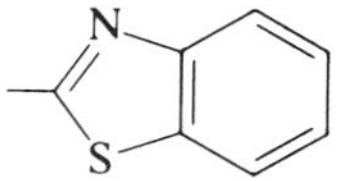

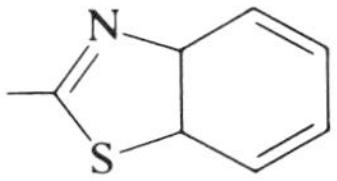

The preceding discussion is a summary of the lactonization methods known at present; newer methods continue to be explored. The selection of a method for an individual case depends to a large extent on the structure and functionalities of the substrate.

1. Department of Chemistry, University of Alberta, Edmonton, Alberta, Canada, T6G 2G2. The present address of S. Masamune is the Department of Chemistry, Massachusetts Institute of Technology, Cambridge, MA 02139.
2. Department of Chemistry, Saint Olaf College, Northfield, MI 55057.
3. For recent reviews, see (a) Masamune, S.; Bates, G. S.; Corcoran, J. W. *Angew. Chem., Int. Ed. Engl.* **1977**, *16*, 585; (b) Nicolaou, K.C. *Tetrahedron* **1977**, *33*, 683; (c) Back, T. G. *Tetrahedron* **1977**, *33*, 3041.
4. (a) Masamune, S.; Kim, C. U.; Wilson, K. E.; Spessard, G. O.; Georghiou, P. E.; Bates, G. S. *J. Am. Chem. Soc.* **1975**, *97*, 3512; (b) Masamune, S.; Yamamoto, H; Kamata, S.; Fukuzawa, A. *J. Am. Chem. Soc.* **1975**, *97*, 3513; (c) Colvin, E. W.; Purcell, T. A.; Raphael, R. A. *J. Chem. Soc., Perkin Trans. 1* **1976**, 1718; (d) Corey, E. J.; Nicolaou, K. C.; Toru, T. *J. Am. Chem. Soc.* **1975**, *97*, 2287; (e) Corey, E. J.; Nicolaou, K. C.; Melvin, Jr., L. S. *J. Am. Chem. Soc.* **1975**, *97*, 653 and 654; (f) Corey, E. J.; Ulrich, P.; Fitzpatrick, J. M. *J. Am. Chem. Soc.* **1976**, *98*, 222; (g) Masamune, S.; Hayase, Y.; Chan, W. K.; Sobczak, R. L. *J. Am. Chem. Soc.* **1976**, *98*, 7874; (h) Gerlach, H.; Oertle, K.; Thalmann, A. *Helv. Chim. Acta* **1977**, *60*, 2860; (i) Fukuyama, Y.; Kirkemo, C. L.; White, J. D. *J. Am. Chem. Soc.* **1977**, *99*, 646; (j) Masamune, S.; Hayase, Y.; Schilling, W.; Cahn, W. K.; Bates, G. S. *J. Am. Chem. Soc.* **1977**, *99*, 6756; (k) Corey, E. J.; Kim, S.; Yoo, S.; Nicolaou, K. C.; Melvin, Jr., L. S.; Brunelle, D. J.; Falck, J. R.; Trybulski, E. J.; Lett, R.; Sheldrake, P. W. *J. Am. Chem. Soc.* **1978**, *100*, 4620; (l) Inanaga, J.; Katsuki, T.; Takimoto, S.; Ouchida, S.; Inoue, K.; Nakano, A.; Okukado, N.; Yamaguchi, M. *Chem. Lett.* **1979**, 1021; (m) Corey, E. J.; Hopkins, P. B.; Kim, S.; Yoo, S.; Nambiar, K. P.; Falck, J. R. *J. Am. Chem. Soc.* **1979**, *101*, 7134; (n) Tatsuta, K.; Amemiya, Y.; Maniwa, S.; Kinoshita, M. *Tetrahedron Lett.* **1980**, *20*, 2837; (o) Huang, J.; Meinwald, J. *J. Am. Chem. Soc.* **1981**, *103*, 861; (p) Tatsuta, K.; Amemiya, Y.; Kanemura, Y.; Kinoshita, M. *Tetrahedron Lett.* **1981**, *21*, 3997; (q) Masamune, S.; Hirama, M.; Mori, S.; Ali, Sk. A.; Garvey, D. S. *J. Am. Chem. Soc.* **1981**, *103*, 1568; (r) Woodward, R. B. et al. *J. Am. Chem. Soc.* **1981**, *103*, 3213.
5. Masamune, S.; Kamata, S.; Schilling, W. *J. Am. Chem. Soc.* **1975**, *97*, 3515.
6. Corey, E. J.; Nicolaou, K. C. *J. Am. Chem. Soc.* **1974**, *96*, 5614.
7. Gerlach, H.; Thalmann, A. *Helv. Chim. Acta* **1974**, *57*, 2661; Thalmann, A.; Oertle, K.; Gerlach, H. *Org. Synth., Coll. Vol. VII* **1990**, 470.
8. Kurihara, T.; Nakajima, Y.; Mitsunobu, O. *Tetrahedron Lett.* **1976**, 2455.
9. Masamune, S.; Chan, W. K.; Sugihara, Y. unpublished results.
10. Corey, E. J.; Brunelle, D. J.; Stork, P. J. *Tetrahedron Lett.* **1976**, 3405.
11. Corey, E. J.; Brunelle, D. J. *Tetrahedron Lett.* **1976**, 3409.
12. (a) Mukaiyama, T.; Usui, M.; Saigo, K. *Chem. Lett.* **1976**, 49; (b) Mukaiyama, T.; Narasaka, K.; Kikuchi, K. *Chem. Lett.* **1977**, 441; (c) Narasaka, K.; Masui, T.; Mukaiyama, T. *Chem. Lett.* **1977**, 763; (d) Narasaka, K.; Yamaguchi, M.; Mukaiyama, T. *Chem. Lett.* **1977**, 959; (e) Narasaka, K.; Maruyama, K.; Mukaiyama, T. *Chem. Lett.* **1978**, 885.
13. Staab, H. A.; Mannerschreck, A. *Chem. Ber.* **1962**, *95*, 1284.
14. White, J. D.; Lodwig, S. N.; Trammell, G. G.; Fleming M. P. *Tetrahedron Lett.* **1974**, 3263.
15. (a) Inanaga, J.; Hirata, K.; Saeki, H.; Katsuki, T.; Yamaguchi, M. *Bull. Chem. Soc. Jpn.* **1979**, *52*, 1989; (b) Yamaguchi, M. *Yuki Gosei Kagaku Kyokeishi* **1980**, *38*, 22; *Chem. Abstr.* **1980**, *93*, 25345z.
16. Kaiho, T.; Masamune, S.; Toyoda, T. *J. Org. Chem.* **1982**, *47*, 1612.
17. (a) Steliou, K.; Szczygielska-Nowosielska, A.; Favre, A.; Poupart, M.-A.; Hanessian, S. *J.*

Am. Chem. Soc. **1980,** *102,* 7578; (b) Steliou, K.; Poupart, M.-A. *J. Am. Chem.* **1983,** *105,* 7130.

18. Otera, J.; Yano, T.; Himeo, Y.; Nogaki, H. *Tetrahedron Lett.* **1986,** *27,* 4501.
19. Fujisawa, T.; Mori, T.; Fukumoto, K.; Sato, T. *Chem. Lett.* **1982,** 1891.
20. Boden, E. P.; Keck, G. E. *J. Org. Chem.* **1985,** *50,* 2394.

ESTERIFICATION OF CARBOXYLIC ACIDS WITH DICYCLOHEXYLCARBODIIMIDE/ 4-DIMETHYLAMINOPYRIDINE: *tert*-BUTYL ETHYL FUMARATE

[(*E*)-2-Butenedioic acid, ethyl 1,1-dimethylethyl ester]

Submitted by B. Neises and Wolfgang Steglich[1]
Checked by Cheryl Stubbs and Robert V. Stevens

1. Procedure

Caution! Dicyclohexylcarbodiimide is a potent allergen and should be handled with gloves.

A 500-mL, one-necked flask equipped with a calcium chloride drying tube is charged with 28.83 g (0.20 mol) of monoethyl fumarate (Note 1), 200 mL of dry dichloromethane (Note 2), 44.47 g (0.60 mol) of *tert*-butyl alcohol (Note 3), and 2.00 g (0.16 mol) of 4-dimethylaminopyridine (Note 4). The solution is stirred and cooled in an ice bath to 0°C while 45.59 g (0.22 mol) of dicyclohexylcarbodiimide (Note 5) is added over a 5-min period. After a further 5 min at 0°C the ice bath is removed and the dark-brown reaction mixture is stirred for 3 hr at room temperature. The dicyclohexylurea that has precipitated is removed by filtration through a fritted Büchner funnel (G3), and the filtrate is washed with two 50-mL portions of 0.5 *N* hydrochloric acid (Note 6) and two 50-mL portions of saturated sodium bicarbonate solution. During this procedure some additional dicyclohexylurea is precipitated, which is removed by filtration of both layers to facilitate their separation. The organic solution is dried over anhydrous sodium sulfate and concentrated with a rotary evaporator. The concentrate is distilled under reduced pressure, affording, after a small forerun, 30.5–32.5 (76–81%) of *tert*-butyl ethyl fumarate, bp 105–107°C (12 mm) (Note 7).

2. Notes

1. Monoethyl fumarate was purchased from Ega-Chemie, D-7924 Steinheim, Germany.

2. Dichloromethane was freshly distilled over P_4O_{10}.

3. *tert*-Butyl alcohol was purchased from E. Merck, D-6100 Darmstadt, Germany, and used without further purification.

4. 4-Dimethylaminopyridine was obtained from Schering AG, D-1000 Berlin, Germany. 4-Pyrrolidinopyridine, which is equally well suited as a catalyst in this reaction may be purchased from Ega-Chemie, D-7924 Steinheim, Germany.

5. Dicyclohexylcarbodiimide was freshly distilled with a Kugelrohr apparatus (Büchi GKR-50), bp 135–140°C (0.5 mm). It may be either added in crystalline form or dissolved in 50 mL of dry dichloromethane.

6. For esters more sensitive to acids, the use of concentrated aqueous citric acid solution is advisable.

7. The proton magnetic resonance spectrum of the product in chloroform-d shows the following absorptions: δ 1.30 [t, 3 H, J = 7.5, CH_3CH_2), 1.50 (s, 9 H, $C(CH_3)_3$], 4.23 (q, 2 H, J = 7.5, CH_3CH_2), 6.77 (s, 2 H, CH=CH). *tert*-Butyl ethyl fumarate may be easily converted into ethyl fumarate by alkaline hydrolysis.

3. Discussion

This procedure offers a convenient method for the esterification of carboxylic acids with alcohols[2-4] and thiols[2] under mild conditions. Its success depends on the high efficiency of 4-dialkylaminopyridines as nucleophilic catalysts in group transfer reactions.[5] The esterification proceeds without the need of a preformed, activated carboxylic acid derivative, at room temperature, under nonacidic, mildly basic conditions. In addition to dichloromethane other aprotic solvents of comparable polarity such as diethyl ether, tetrahydrofuran, and acetonitrile can be used. The reaction can be applied to a wide variety of acids and alcohols, including polyols,[2,4,6] α-hydroxycarboxylic acid esters,[7] and even very acid labile alcohols such as vitamin A.[8] It has also been used for the esterification of urethane-protected α-amino acids with polymeric supports carrying hydroxy groups.[9] In this case, however, some racemization of the amino acid is observed because of 2-alkoxyoxazolin-5-one formation.[10] Racemization can be decreased by shortening the coupling time[10] or completely avoided by working with N-(p-nitrophenylsulfenyl)amino acids.[11]

With increasing steric hindrance, the rate of esterification is decreased and the formation of N-acylureas may become a serious side reaction. This is indicated by the decrease in yield in the esterification of 2,5-cyclohexadiene-1-carboxylic acid with different alcohols: MeOH (95%), EtOH (84%), i-PrOH (75%), c-$C_6H_{11}OH$ (65%), t-BuOH (65%).[12] Diminished acidity because of the influence of electron-donating substituents in aromatic carboxylic acids can also lead to low yields.

The dicyclohexylcarbodiimide/4-dialkylaminopyridine method is also well suited to the synthesis of a wide variety of thiol esters.[2,13]

4-Dimethylaminopyridine also catalyzes the formation of esters and thiol esters in the reaction of mixed carboxylic anhydrides[14] or 2,4,6-trinitrophenyl esters[15] with alcohols and thiols. 1-Acyl-4-benzylidene-1,4-dihydropyridines have been introduced recently as promising reagents for the synthesis of sterically hindered esters.[16] The current methods available for ester and thiol ester formation have been reviewed recently by Haslam.[17]

1. Institut für Organische Chemie und Biochemie der Universität Bonn, Gerhard-Domagk-Strasse 1, D-5300 Bonn, Germany.
2. Neises, B.; Steglich, W. *Angew. Chem., Int. Ed. Engl.* **1978,** *17,* 522; *Angew. Chem.* **1978,** *90,* 556.
3. Hassner, A.; Alexanian, V. *Tetrahedron Lett.* **1978,** 4475.
4. Ziegler, F. E.; Berger, G. D. *Synth. Commun.* **1979,** *9,* 539.
5. Steglich, W.; Höfle, G. *Angew. Chem., Inter. Ed. Engl.* **1969,** *8,* 981; *Angew. Chem.* **1969,** *81,* 1001; Höfle, G.; Steglich, W. *Synthesis* **1972,** 619; Höfle, G.; Steglich, W.; Vorbrüggen, H. *Angew. Chem., Inter. Ed. Engl.* **1978,** *17,* 569; *Angew. Chem.* **1978,** *90,* 602; Hassner, A.; Krepski, L. R.; Alexanian, V. *Tetrahedron* **1978,** *34,* 2069.
6. Žinić, M.; Bosnić-Kašnar, B.; Kolbah, D. *Tetrahedron Lett.* **1980,** 1365.
7. Gilon, Ch.; Klausner, Y.; Hassner, A. *Tetrahedron Lett.* **1979,** 3811.
8. Niewöhner, U.; Steglich, W., unpublished results.
9. Wang, S.-S.; Yang, C. C.; Kulesha, I. D.; Sonenberg, M.; Merrifield, R. B. *Int. J. Pept. Protein Res.* **1974,** *6,* 103; Wang, S.-S.; Kulesha, I. D. *J. Org. Chem.* **1975,** *40,* 1227; Chang, C.-D.; Meienhofer, J. *Int. J. Pept. Protein Res.* **1978,** *11,* 246.
10. Atherton, E.; Benoiton, N. L.; Brown, E.; Sheppard, R. C.; Williams, B. J. *J. Chem. Soc., Chem. Commun.* **1981,** 336.
11. Neises, B.; Andries, Th.; Steglich, W., publication in preparation.
12. Neises, B.; Steglich, W., unpublished results.
13. Neises, B.; Steglich, W.; van Ree, T. *S. Afr. J. Chem.* **1981,** *34,* 58.
14. Inanaga, J.; Hirata, K.; Saeki, H.; Katsuki, T.; Yamaguchi, M.; *Bull. Chem. Soc. Jpn.* **1979,** *52,* 1989; Kawanami, Y.; Dainobu, Y.; Inanaga, J.; Katsuki, T.; Yamaguchi, M. *Bull. Chem. Soc. Jpn.* **1981,** *54,* 943.
15. Kim, S.; Yang, S. *Synth. Commun.* **1981,** *11,* 121; Kim, S.; Yang, S. *Chem. Lett.* **1981,** 133.
16. Anders, E.; Will, W. *Synthesis* **1980,** 485.
17. Haslam, E. *Tetrahedron* **1980,** *36,* 2409.

ACYLOIN CONDENSATION BY THIAZOLIUM ION CATALYSIS: BUTYROIN

(4-Octanone, 5-hydroxy-)

Submitted by H. STETTER and H. KUHLMANN[1]
Checked by SHARBIL J. FIRSAN and ROBERT M. COATES

1. Procedure

A 500-mL, three-necked, round-bottomed flask is equipped with a mechanical stirrer, a short gas inlet tube, and an efficient reflux condenser fitted with a potassium hydroxide drying tube. The flask is charged with 13.4 g (0.05 mol) of 3-benzyl-5-(2-hydroxyethyl)-4-methyl-1,3-thiazolium chloride (Note 1), 72.1 g (1.0 mol) of butyraldehyde (Note 2), 30.3 g (0.3 mol) of triethylamine (Note 2), and 300 mL of absolute ethanol. A slow stream of nitrogen (Note 3) is begun, and the mixture is stirred and heated in an oil bath at 80°C. After 1.5 hr the reaction mixture is cooled to room temperature and concentrated by rotary evaporation. The residual yellow liquid is poured into 500 mL of water contained in a separatory funnel, and the flask is rinsed with 150 mL of dichloromethane which is then used to extract the aqueous mixture. The aqueous layer is extracted with a second 150-mL portion of dichloromethane. The combined organic phases are washed with 300 mL of saturated sodium bicarbonate and with 300 mL of water. The dichloromethane is removed by rotary evaporation under slightly diminished pressure. Distillation through a 20-cm Vigreux column gives 51–54 g (71–74%) of product as a colorless to light-yellow liquid, n_D^{20} 1.4309, bp 90–92°C (13–14 mm) (Notes 4 and 5).

2. Notes

1. The catalyst, 3-benzyl-5-(2-hydroxyethyl)-4-methyl-1,3-thiazolium chloride, is supplied by Fluka AG, Buchs, Switzerland, and by Tridom Chemical, Inc., Hauppauge, New York. The thiazolium salt may also be prepared as described below[2] by benzylation of 5-(2-hydroxyethyl)-4-methyl-1,3-thiazole, which is commercially available from E. Merck, Darmstadt, West Germany, and Columbia Organic Chemicals Co., Inc., Columbia, SC. The acetonitrile used by the checkers was dried over Linde 3A molecular sieves[3] and distilled under nitrogen, bp 77–78°C. The same yield of thiazolium salt was obtained by the checkers when benzyl chloride and acetonitrile from commercial sources were used without purification.

    A 250-mL, three-necked, round-bottomed flask is equipped with a mechanical stirrer, a reflux condenser fitted with a drying tube, and a stopper. The flask is charged with 14.3 g (0.1 mol) of 5-(2-hydroxyethyl)-4-methyl-1,3-thiazole, 12.7 g (0.1 mol) of freshly distilled benzyl chloride, and 50 mL of dry acetonitrile. The mixture is heated at reflux for 24 hr and cooled to room temperature. Crystallization is induced by scratching or seeding. The solid is collected by suction filtration, washed colorless with two 50-mL portions of acetonitrile, and dried partially in the air. Drying is completed under reduced pressure by gentle rotation on a rotary evaporator heated with a water bath at about 90°C. The yield of thiazolium salt, mp 141–143°C, is 18.2–19.6 g (67–73%).

2. Butyraldehyde is supplied by Aldrich Chemical Co., Inc. and Eastman Organic Chemicals. The aldehyde was freshly distilled before use. Triethylamine was dried over potassium hydroxide pellets and distilled.

3. The submitters recommend that the nitrogen stream be passed through a bubbler and that the flow rate be adjusted to ca. one bubble per second. If the nitrogen flow is too fast, some of the butyraldehyde will be swept out of the flask.

4. The procedure may be conducted on a larger scale, in which case the proportion of catalyst and base are reduced. The submitters report that they obtained 169 g (78%) of butyroin from 216.3 g (3.0 mol) of butyraldehyde, 26.8 (0.1 mol) of thiazolium catalyst, 60.6 g (0.6 mol) of triethylamine, and 600 mL of absolute ethanol. Although the scale may be increased further, appropriate precautions should be taken to control the reaction. For example, the aldehyde may be added in portions or the flask may be cooled initially.

5. The product obtained by the checkers boiled at 86–87.5°C (15–16 mm). A boiling point of 85–87°C (12–13 mm) and an index of refraction n_D^{20} 1.4325 have been recorded for butyroin.[4] The product exhibits the following spectral characteristics: IR (neat) cm^{-1}: 3505 and 1704; ^{1}H NMR (CCl$_4$) δ: 0.94 (unsymmetrical t, 6 H, 2 CH$_3$), 1.18–1.56 (m, 4 H, 2 CH$_2$), 1.64 (sextet, 2 H, J = 7, CH_2CH$_2$C=O), 2.41 (t, 2 H, J = 7, CH$_2$C=O), 3.31 (s, 1 H, OH), 3.98 (m, 1 H, CHOH).

3. Discussion

This procedure is representative of a new general method for the preparation of noncyclic acyloins by thiazolium-catalyzed dimerization of aldehydes in the presence of weak bases (Table I).[5] The advantages of this method over the classical reductive coupling of esters[6] or the modern variation, in which the intermediate enediolate is

TABLE I

ACYLOINS PREPARED BY THIAZOLIUM ION-CATALYZED
CONDENSATION OF ALDEHYDES[5]

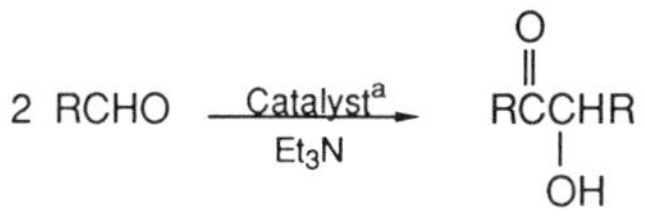

$$2\ RCHO \xrightarrow[\text{Et}_3\text{N}]{\text{Catalyst}^a} \underset{\overset{|}{\text{OH}}}{R\overset{\overset{\text{O}}{\|}}{C}CHR}$$

R	Yield (%)	Bp or mp (°C)
C$_4$H$_9$	79	83 (2.2 mm)
C$_5$H$_{11}$	81	90 (1.5 mm)
C$_7$H$_{15}$	83[b]	39
C$_9$H$_{19}$	85[b]	53
C$_{11}$H$_{23}$	83[b]	62
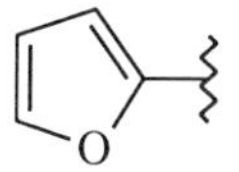	80[c,d]	136

[a]3-Benzyl-5-(2-hydroxyethyl)-4-methyl-1,3-thiazolium chloride.
[b]The product was isolated by pouring the ethanolic solution into well-stirred, ice-cold water, filtering, and recrystallizing from aqueous ethanol. The solutions should be ice-cold for the isolation of the low-melting acyloins. The products may also be isolated by extraction as described for butyroin.
[c]In this case furoin crystallized from the ethanolic solution on cooling.
[d]The following somewhat simpler procedure may also be used. A solution of 13.4 g (0.05 mol) of catalyst, 96.1 g (1.0 mol) of 2-furaldehyde, 300 mL of absolute ethanol, and 30.3 g (0.3 mol) of triethylamine is stirred at room temperature for 12 hr. The product (84.5 g, 88%) crystallizes directly from solution and is isolated by filtration.

trapped by silylation,[4,7] are the simplicity of the procedure, the inexpensive materials used, and the purity of the products obtained. For volatile aldehydes such as acetaldehyde and propionaldehyde the reaction is conducted without solvent in a small, heated autoclave. With the exception of furoin the preparation of benzoins from aromatic aldehydes is best carried out with a different thiazolium catalyst bearing an N-methyl or N-ethyl substituent, instead of the N-benzyl group.[5] Benzoins have usually been prepared by cyanide-catalyzed condensation of aromatic and heterocyclic aldehydes.[8,9,10] Unsymmetrical acyloins may be obtained by thiazolium-catalyzed cross-condensation of two different aldehydes.[11] The thiazolium ion-catalyzed cyclization of 1,5-dialdehydes to cyclic acyloins has been reported.[12]

Although the catalysts of the dimerization of aldehydes to acyloins by thiazolium ion has been known for some time,[13] the development of procedures using anhydrous solvents which give satisfactory yields of acyloins on a preparative scale was first realized in the submitters' laboratories.[5] The mechanism proposed by Breslow[13a] for the thiazolium ion-catalyzed reactions is similar to the Lapworth mechanism[14] for the benzoin condensation with a thiazolium ylide replacing the cyanide ion. Similar mechanisms are involved in many important enzyme-catalyzed transformations that require thiamine as a cofactor. The combination of thiazolium salts and weak bases has also been utilized to catalyze the conjugate addition of aldehydes to electron-deficient double bonds.[2]

Butyroin has been prepared by reductive condensation of ethyl butyrate with sodium in xylene,[6b] or with sodium in the presence of chlorotrimethylsilane,[7] and by reduction of 4,5-octanedione with sodium 1-benzyl-3-carbamoyl-1,4-dihydropyridine-4-sulfinate in the presence of magnesium chloride,[15] or with thiophenol in the presence of iron polyphthalocyanine as electron transfer agent.[16] This acyloin has also been obtained by oxidation of (E)-4-octene with potassium permanganate[17] and by reaction of propylmagnesium bromide with nickel tetracarbonyl.[18]

Acyloins are useful starting materials for the preparation of a wide variety of heterocycles (e.g., oxazoles[19] and imidazoles[20]) and carbocyclic compounds (e.g., phenols[21]). Acyloins lead to 1,2-diols by reduction, and to 1,2-diketones by mild oxidation.

1. Institut für Organische Chemie der Rheinisch-Westfälischen Technischen Hochschule Aachen, West Germany.
2. (a) Stetter, H. Angew. Chem., Int. Ed. Engl. 1976, 15, 639–647; (b) Stetter, H. Angew. Chem. 1976, 88, 695–704.
3. Burfield, D. R.; Lee, K.-H.; Smithers, R. H. J. Org. Chem. 1977, 42, 3060–3065.
4. Schräpler, U.; Rühlmann, K. Chem. Ber. 1963, 96, 2780–2785.
5. Stetter, H.; Rämsch, R. Y.; Kuhlmann, H. Synthesis 1976, 733–735.
6. (a) McElvain, S. M. Org. React. 1949, 4, 256–268; (b) Snell, J. M.; McElvain, S. M. Org. Synth., Coll. Vol. II 1943, 114–116; and references cited therein.
7. Rühlmann, K. Synthesis 1971, 236–253; Bloomfield, J. J.; Owsley, D. C.; Nelke, J. M. Org. React. 1976, 23, 259–403; Bloomfield, J. J.; Nelke, J. M. Org. Synth., Coll. Vol. VI 1988, 167.
8. Ide, W. S.; Buck, J. S. Org. React. 1949, 4, 269–304.
9. Adams, R.; Marvel, C. S. Org. Synth., Coll. Vol. I 1941, 94–95.
10. Hartman, W. W.; Dickey, J. B. J. Am. Chem. Soc. 1933, 55, 1228–1229.
11. Stetter, H.; Dämbkes, G. Synthesis 1977, 403–404.
12. Cookson, R. C.; Lane, R. M. J. Chem. Soc., Chem. Commun. 1976, 804–805.

13. (a) Breslow, R. *J. Am. Chem. Soc.* **1958**, *80*, 3719–3726; (b) Tagaki, W.; Hara, H. *J. Chem. Soc., Chem. Commun.* **1973**, 891; (c) Schilling, C. L., Jr.; Mulvaney, J. E. *Macromolecules* **1968**, *1*, 452–455.
14. Kuebrich, J. P.; Schowen, R. L.; Wang, M.-S.; Lupes, M. E. *J. Am. Chem. Soc.* **1971**, *93*, 1214–1220.
15. Inoue, H.; Sonoda, I.; Imoto, E. *Bull. Chem. Soc. Jpn.* **1979**, *52*, 1237–1238.
16. Inoue, H.; Nagata, T.; Hata, H.; Imoto, E. *Bull. Chem. Soc. Jpn.* **1979**, *52*, 469–473.
17. Srinivasan, N. S.; Lee, D. G. *Synthesis* **1979**, 520–521.
18. Benton, F. L.; Voss, Sr. M. C.; McCusker, P. A. *J. Am. Chem. Soc.* **1945**, *67*, 82–83.
19. Lakhan, R.; Ternai, B. In "Advances in Heterocyclic Chemistry," Katritzky, A. R.; Boulton, A. J., Eds.; Academic Press: New York, 1974; Vol. 17, pp. 99–211.
20. Grimmett, M. R. In "Advances in Heterocyclic Chemistry," Katritzky, A. R.; Boulton, A. J., Eds.; Academic Press: New York, 1970; Vol. 12, pp. 103–183.
21. Egli, C.; Helali, S. E.; Hardegger, E. *Helv. Chim. Acta* **1975**, *58*, 104–110.

(S)-(+)-γ-BUTYROLACTONE-γ-CARBOXYLIC ACID

[2-Furancarboxylic acid, tetrahydro-5-oxo-, (S)-]

Submitted by OLIVIER H. GRINGORE and FRANCIS P. ROUESSAC[1]
Checked by MATTHEW F. SCHLECHT, HOWARD DROSSMAN, and CLAYTON H. HEATHCOCK

1. Procedure

Caution! This procedure should be conducted in a well-ventilated hood to avoid inhalation of poisonous NO₂ vapors. To protect the operator the distillation must be carried out with the usual precautions associated with vacuum distillation.

A 6-L Erlenmeyer flask which contains a large magnetic stirring bar is charged with 294 g (2 mol) of L-glutamic acid (Note 1) and 2 L of distilled water. The suspension is stirred vigorously while solutions of 168 g (2.4 mol) of sodium nitrite in 1.2 L of water and 1.2 L of aqueous 2 N sulfuric acid are added simultaneously from separatory funnels (Note 2). After the addition is complete (Note 3), the solution is stirred at room temperature for an additional 15 hr. The water is then removed by heating below 50°C under reduced pressure with a rotary evaporator (Note 4). The resulting pasty solid is triturated with 500 mL of boiling acetone and the hot solution is filtered and set aside to cool. This operation is repeated four times (Notes 5 and 6). Removal of solvent with a rotary evaporator affords 312 g of crude (+)-γ-butyrolactone-γ-carboxylic acid as a slightly yellow oil (Notes 7 and 8).

A 250-mL, round-bottomed flask is equipped with a magnetic stirring bar and charged with 100 g of the foregoing crude lactone acid (Note 9). The flask is fitted with a Claisen distillation apparatus and connected to a vacuum pump (Notes 10 and 11). The flask is gradually heated with an oil bath (160°C) until gas evolution ceases (Note 12). At this point the oil bath is removed and the black, viscous oil is distilled with the use of a flame (Note 13). The product, 58 g (70%), is collected as a colorless oil at 146–154°C (0.03 mm). The distillate crystallizes in the receiver, mp 66–68°C (Notes 14 and 15).

2. Notes

1. This material was purchased from the Aldrich Chemical Company Inc., $[\alpha]_D^{23}$ +29° (6 N HCl, c 1).

2. The addition requires about 30 min. During addition the reaction mixture should warm to 30–35°C and smooth evolution of NO_2 and N_2 should occur. If the solutions of $NaNO_2$ and H_2SO_4 are added too rapidly, more gas appears to be generated and a reduction in yield occurs.

3. At this point the reaction mixture is clear and colorless. Residual brown gas usually remains in the flask.

4. If a conventional aspirator pump is employed, concentration can require several days. The checkers employed a rotary evaporator that was evacuated to approximately 3 mm by a vacuum pump. Two traps, one cooled in an ice–salt bath and the other in an acetone–dry-ice bath, were inserted between the rotary evaporator and the vacuum pump. In this way, the reaction mixture can be concentrated to a paste in about 16–20 hr.

5. Repetitive extraction may also be performed in a flask heated with a water bath to 65°C; acetone is removed by decantation. Ethyl acetate has also been used for the extraction.[2]

6. The checkers found that a higher recovery is obtained if the pasty solid is vigorously agitated during trituration with five 750-mL portions of boiling acetone.

7. The crude yield reported is in excess of the theoretical yield (260 g). The checkers obtained crude yields of 243–259 g, probably because water was more efficiently removed in the concentration step.

8. Although this material is sufficiently pure for some applications, it is advisable to purify it further before use. Distillation[2] and crystallization[3] have been described. The submitters recommend purification by the distillation procedure given. By direct crystallization of 101 g of crude lactone acid from ether–petroleum ether, the checkers obtained 36.5 g (35%) of material, mp 72–74°C.

9. If the distillation is carried out on a larger scale, the yield is lower.

10. The submitters recommend a short-path distillation apparatus with large sections (i.e., wide bore) since the distillate partially crystallizes in the condenser during the distillation. It is important that the distillation apparatus have a Claisen head because the viscous material tends to bump.

11. The vacuum pump should be protected by a soda–lime trap.

12. During this heating period the system pressure should rise from 0.03 to 0.5 mm and the crude lactone acid should become black. When gas evolution ceases, the pressure decreases to its initial value.

13. Distillation should be carried out briskly. If a simple bunsen burner with a low flame is used, distillation requires several hours. The checkers used a hot flame, about 13 cm in length, from a gas–air torch. In this way, the distillation requires only about 15 min. Distillation is discontinued when colored vapors appear.

14. The checkers distilled crude lactone acid obtained in approximately quantitative yield (259 g). When this material was used, distillation of 100-g portions gave 64.3–66.4 g (65–66% yield).

15. The submitters report that recrystallization from ethyl acetate–petroleum ether raises the melting point to 73°C. The product obtained is analytically pure, $[\alpha]_D^{21}$ +16° (EtOH, c 2). When the checkers used ethyl acetate–petroleum ether they often obtained an oily product.

The spectrum of the lactone acid is as follows: ^{1}H NMR (CD$_3$COCD$_3$) δ: 2.1–2.9 (m, 4 H), 4.85–5.15 (m, 1 H), 5.1 (s, 1 H, COOH).

3. Discussion

The (S)-$(+)$-γ-butyrolactone-γ-carboxylic acid is a useful intermediate for the synthesis of pheromones,[4] natural lignans,[5] and other derivatives.[6] In the same manner, but starting with D-glutamic acid, the (R)-$(-)$-lactone acid may be prepared.[7] Lactonization occurs with full retention of configuration at the chiral center.[8,9] Recently, authors have described an efficient method that allows the formation of derivatives of the (R)-$(-)$-lactone from the more available (S)-$(+)$ counterpart.[10]

The procedure is a detailed description of the Austin–Howard preparation.[2] The mechanism presumably involves anchimeric assistance of the carboxy group in decomposition of an intermediate diazonium ion, leading to a labile α-lactone:[4]

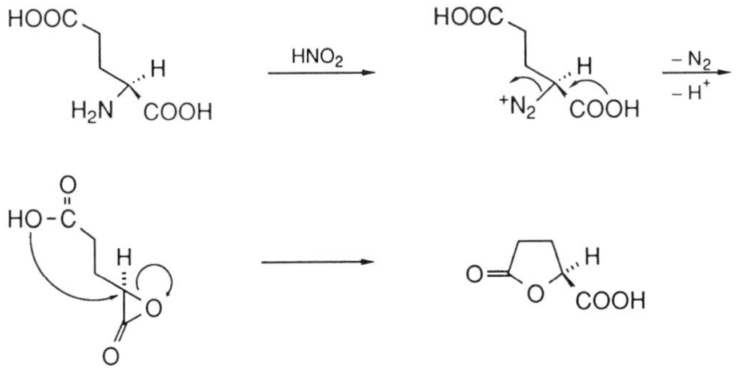

The title compound has also been prepared[11] using hydrochloric acid instead of sulfuric acid, and ethyl acetate instead of acetone. In the hands of the submitters, this procedure gave a lower yield.

1. Laboratoire de Synthèse Organique, Faculté des Sciences, Université du Maine, F-72017 Le Mans.
2. Austin, A. T.; Howard, J. *J. Chem. Soc.* **1961**, 3593.

3. Ravid, U.; Silverstein, R. M.; Smith, L. R. *Tetrahedron* **1978**, *34*, 1449.
4. For a review see Smith, L. R.; Williams, H. J. *J. Chem. Ed.* **1979**, *56*, 696.
5. Tomioka, K.; Koga, K. *Tetrahedron Lett.* **1979**, 3315; Tomioka, K.; Ishiguro, T.; Koga, K. *J. Chem. Soc., Chem. Commun.* **1979**, 652; Tomioka, K.; Ishiguro, T.; Koga, K. *Tetrahedron Lett.* **1980**, 2973; Robin, J. P.; Gringore, O.; Brown, E. *Tetrahedron Lett.* **1980**, 2709.
6. Koga, K.; Taniguchi, M.; Yamada, S.-i. *Tetrahedron Lett.* **1971**, 263.
7. Eguchi, C.; Kakuta, A. *Bull. Chem. Soc. Jpn.* **1974**, *47*, 1704.
8. Cervinka, O.; Hub, L. *Collect. Czech. Chem. Commun.* **1968**, *33*, 2927.
9. Brewster, P.; Hiron, F.; Hughes, E. D.; Ingold, C. K.; Rao, P. A. D. S. *Nature (London)* **1950**, *166*, 178.
10. Takano, S.; Yonaga, M.; Ogasawara, K. *Synthesis* **1981**, 265.
11. Taniguchi, M.; Koga, K.; Yamada, S. *Tetrahedron* **1974**, *30*, 3547; Iwaki, S.; Marumo, S.; Saito, T.; Yamada, M.; Katagiri, K. *J. Am. Chem. Soc.* **1974**, *96*, 7842.

3-(1-HYDROXYBUTYL)-1-METHYLPYRROLE AND 3-BUTYROYL-1-METHYLPYRROLE

[1*H*-Pyrrole-3-methanol, 1-methyl-α-propyl- and 1-Butanone, 1-(1-methyl-1*H*-pyrrol-3-yl)-]

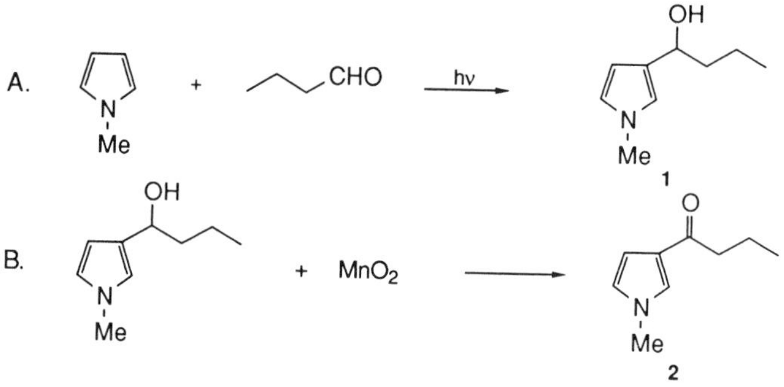

Submitted by H. M. GILOW and G. JONES, II[1]
Checked by STEVEN M. PITZENBERGER, RICHARD A. HAYES, and ORVILLE L. CHAPMAN

1. Procedure

A. *3-(1-Hydroxybutyl)-1-methylpyrrole* (**1**). A photochemical quartz immersion well (220 mm length) (Note 1) equipped with 450-W Hanovia medium-pressure mercury lamp and a Vycor filter, cooled with water, is used. To a 125-mL Pyrex reaction vessel (230 mm long, 64 mm i.d.) equipped with a gas inlet and outlet is added 60 mL (55 g, 0.676 mol) of 1-methylpyrrole (Note 2) and 65 mL (54 g, 0.936 mol) of butyraldehyde (Note 3). Dry nitrogen is slowly bubbled through the solution during 48 hr of photolysis (Note 4).

The solution is concentrated under reduced pressure. The remaining oil is distilled under reduced pressure using a simple distillation apparatus. After a small forerun, 27 g (0.179 mol, 26% yield) (Note 5) of **1** is collected, as a light yellow oil, bp 90–94°C/0.05 mm (Note 6). Further purification is accomplished by a second distillation under reduced pressure, bp 90.2°C/0.05 mm (Notes 7 and 8).

B. *3-Butyroyl-1-methylpyrrole* (**2**). A 100-mL, one-necked, round-bottomed flask is fitted with an efficient reflux condenser and arranged for magnetic stirring and heating. The flask is charged with 50 mL of pentane (Note 9) and 2.0 g (13 mmol) of **1** (Note 10). To the rapidly stirred solution is added 16 g (180 mmol) of activated manganese-(IV) oxide (Note 11) in small portions over 5 min. The solution is heated at reflux for 18 hr and then an additional 8 g (90 mmol) of activated manganese(IV) oxide is added in portions (Note 12). After being heated at reflux for 24 hr, the reaction mixture is filtered through a 2-cm Celite filter pad. The filtered manganese oxides are thoroughly washed with about 200–300 mL of dichloromethane. Evaporation of solvent from the combined filtrates leaves 1.4–1.6 g of a light yellow oil. Bulb-to-bulb distillation at 100°C/0.1 mm (Note 13) gives 1.27–1.40 g (8.4–9.3 mmol, 64–71% yield) of an oil (**2**) (Note 14).

2. Notes

1. The photochemical quartz immersion well was obtained from Ace Glass Inc.
2. 1-Methylpyrrole was obtained from Aldrich Chemical Company, Inc. and distilled before use, bp 112–112.5°C.
3. Butyraldehyde was obtained from Aldrich Chemical Company, Inc. and distilled before use, bp 74.5–75.5°C. It is important that a freshly distilled sample, free of trimer, be used, or the final product will be contaminated with trimer.
4. When the reaction mixture was monitored by GLC (500-mm × 3.2-mm column, packed with 5% OV 101 on chromosorb G, HP, 100/120 mesh) most of the product was formed in the first 24 hr of photolysis, as shown by the following profile:

Time of Photolysis (hr)	% of Alcohol (Based on Starting Pyrrole)
2	4
19	18
24	23
48	25

5. The checkers found that the distillate contained 15–30% butyraldehyde (as monitored by NMR), which depended on the efficiency of the distillation. A 10-cm column packed with glass helices was the most efficient, but the yield of distilled product dropped drastically.
6. The susceptibility of 3-(1-hydroxybutyl)-1-methylpyrrole to air oxidation and decomposition with acid requires that prolonged storage be done in tightly capped containers in a refrigerator.

7. The spectral properties of 3-(1-hydroxybutyl)-1-methylpyrrole are as follows: ^{1}H NMR (CDCl$_3$) δ: 0.90 (t, 3 H, CH$_3$—C), 1.10–1.80 (m, 4 H, —CH$_2$CH$_2$—), 2.88 (s, 1 H, H—O—), 3.51 (s, 3 H, CH$_3$N—), 4.50 (t, 1 H, HC—), 5.9 (t, 1 H, 4-pyrrole) and 6.41 (d, 2 H, 2,5-pyrrole). IR (neat)cm^{-1}: 3400 (H—O stretch) and 1175 (C=O stretch).

8. The reaction can also be carried out using smaller amounts of 1-methylpyrrole (0.113 mol), butyraldehyde (0.113 mol), and a solvent (245 mL acetonitrile, ACS grade) in a somewhat larger reaction vessel. After 17 hr of photolysis, and after removal of the volatile material and distillation of the remaining oil under reduced pressure, 4–5 g of the alcohol is isolated.

9. The submitters used dichloromethane. The checkers found that use of pentane[2] resulted in increased yields for the oxidation.

10. When the alcohol (1) is contaminated with small amounts of butyraldehyde, oxidation proceeds with a much lower yield of product.

11. Activated manganese(IV) oxide was purchased from Alfa Products, Morton Thiokol, Inc.

12. Progress of the reaction can be monitored by taking an aliquot of the reaction and filtering it, removing the solvent in a vacuum, dissolving the residual oil in carbon tetrachloride, and observing the ^{1}H NMR spectrum. Relative integration of the proton resonances of the pyrrole 2-position (6.1 ppm for the alcohol and 7.2 ppm for the ketone) gives an indication of the percent conversion. The checkers found only 77% conversion after the first reflux period. A higher conversion, 90–97%, was achieved after a second addition of activated manganese(IV) oxide and subsequent heating at reflux.

13. The submitters used a short-path simple distillation apparatus, bp 85–87°C (0.2 mm).

14. The following spectral properties were recorded for 3-butyroyl-1-methylpyrrole, 2: ^{1}H NMR (CDCl$_3$, 200 MHz) δ: 0.97 (t, 3 H), 1.72 (sextet, 2 H), 2.68 (t, 2 H), 3.68 (s, 3 H), 6.6 (m, 2 H, 4,5-pyrrole), 7.23 (t, 1 H, 2-pyrrole); IR (neat) cm^{-1}: 1660 (C=O stretch); MS (70 eV) m/e (rel. int.): 151 (8.6, M$^+$), 123 (1.6), 108 (34), 28 (100). The submitters reported the following spectral data: ^{1}H NMR (CDCl$_3$) δ: 0.95 (t, 3 H), 1.65 (sextet, 2 H), 2.65 (t, 2 H), 3.63 (s, 3 H), 6.47 (m, 2 H), 7.15 (m, 1 H); IR (neat)cm^{-1}: 1700.

3. Discussion

This procedure provides a method for functionalizing the pyrrole ring in the 3-position, normally a difficult synthetic step when conventional electrophilic substitution is used.[3] The technique has been extended to addition of several aldehydes and acetone and to a number of pyrroles.[4] The generality includes photoaddition to imidazoles that are substituted in the 4-position. Pyrrole photoadduct alcohols are readily dehydrated to 3-alkenylpyrroles or oxidized to 3-acyl derivatives.

The precedent is strong for the involvement of oxetanes as intermediates in carbonyl additions to pyrroles.[5-7] NMR evidence has been obtained for an oxetane adduct of acetone and N-methylpyrrole.[4] The initial photoadduct was shown to rearrange readily on workup to the 3-(hydroxyalkyl)pyrrole derivative.

Oxidation of the 3-(hydroxyalkyl)pyrrole derivative gives a pure 3-acylpyrrole derivative that is difficult to obtain by direct substitution in the pyrrole ring. Acylation of pyrrole yields 1- and/or 2-acetylpyrrole, whereas acylation of 1-methylpyrrole forms both 2- and 3-acetyl-1-methylpyrrole, the latter in smaller amount.[3] When a similar procedure was used, 3-(1-hydroxyethyl)-1-methylpyrrole was converted to 3-acetyl-1-methylpyrrole in 76% yield.[4] Recently the decarbonylation of 1-methyl-4-acetyl-2-pyrrolaldehyde was used as a method to prepare 3-acetyl-1-methylpyrrole.[8]

1. Department of Chemistry, Boston University, Boston, MA 02215 (H. M. G. on leave from Southwestern at Memphis, Memphis, TN 38112). This work was supported by the donors of the Petroleum Research Fund, administered by the American Chemical Society.
2. Corey, E. J.; Gilman, N. W.; Ganem, B. E. *J. Am. Chem. Soc.* **1968**, *90*, 5616–5617.
3. Jones, R. A.; Bean, G. P. "The Chemistry of Pyrroles"; Academic Press: New York, 1977.
4. Jones, II, G.; Gilow, H. M. *J. Org. Chem.* **1979**, *44*, 2949.
5. Arnold, D. R. *Adv. Photochem.* **1968**, *6*, 301.
6. Rivas, C.; Bolivar, R. A. *J. Heterocycl. Chem.* **1976**, *13*, 1037.
7. Nakano, T.; Rivas, C.; Perez, C.: Larrauri, J. M. *J. Heterocycl. Chem.* **1976**, *13*, 173.
8. Private communication with Professor H. J. Anderson.

REDUCTIVE ARYLATION OF ELECTRON-DEFICIENT OLEFINS: 4-(4-CHLOROPHENYL)BUTAN-2-ONE

[2-Butanone, 4-(4-chlorophenyl)-]

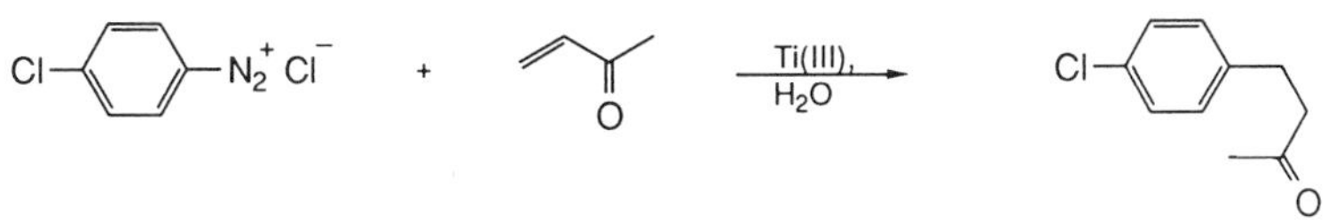

Submitted by ATTILIO CITTERIO
Checked by ROBERT HAESSIG, LEO WIDLER, and DIETER SEEBACH

1. Procedure

Caution! Like all vinyl monomers, 3-buten-2-one is toxic and the preparation should be carried out in a well-ventilated hood.

A 500-mL, four-necked, round-bottomed flask equipped with a magnetic stirring bar, a thermometer, a gas inlet, an externally cooled, pressure-equalizing dropping funnel (Note 1), and a gas bubbler is charged with 15% aqueous titanium trichloride (92 mL, 0.109 mol) (Note 2). N,N-Dimethylformamide (Note 3) (70 mL) is added during 45 min with stirring and cooling (ice-bath; 0–5°C) while nitrogen is bubbled through the solution. Freshly distilled 3-buten-2-one (5.7 mL, 0.066 mol) is added at 0–5°C by syringe. The nitrogen flow is stopped, and 4-chlorobenzenediazonium chloride solution (0.044 mol) (Note 4) is added dropwise at 0–5°C from the dropping funnel. After 2–3 min, nitrogen evolution commences, and the rate of addition is adjusted so that 1–2 bubbles per second are vented through the bubbler. Nitrogen evolution continues for 20 min after the addition is complete (1.5 hr). The ice bath is removed and the solution stirred for 1 hr at room temperature. Ether, 50 mL, is added with stirring, and the organic phase is separated. The aqueous phase is extracted with ether (3 × 50 mL) and the combined organic extracts are washed with 3% aqueous Na_2CO_3 (2 × 30 mL) and water, dried over magnesium sulfate, and concentrated under reduced pressure. The residue is distilled to give 5.2–6.0 g (65–75% yield) of 4-(4-chlorophenyl)butan-2-one as a pale-yellow liquid, bp 90–91°C (0.5 mm) (Note 5).

2. Notes

1. The checkers used a dropping funnel with temperature-control jacket [Normag N 8055, Otto Fritz GmbH, Normschliff-Aufbaugeräte (Normag), D-6238 Hofheim am Taunus].

2. The 15% titanium trichloride solution was purchased from Carlo Erba Chemicals or from Merck & Company, Inc., but can also be prepared by dissolving metallic titanium in 20% aqueous hydrochloric acid[2] or by dissolving solid titanium trichloride in 1 M aqueous hydrochloric acid. Titanium(III) sulfate (from BDH Chemicals Ltd.) can also be used. All titanium(III) solutions were titrated with aqueous cerium(IV) sulfate prior to use.

3. N,N-Dimethylformamide from Carlo Erba Chemicals, from Fluka AG or from Merck & Company, Inc. was used as received. Other solvents (e.g., acetone, acetic acid, acetonitrile) can also be used.

4. The 4-chlorobenzenediazonium chloride solution is prepared as follows: finely powdered 4-chloroaniline (5.65 g, 0.044 mol) is suspended in 18 mL of 24% aqueous hydrochloric acid and cooled to 0°C. Sodium nitrite (3.2 g, 0.046 mol) in water (7 mL) is added dropwise during 45 min at 0–5°C to give a pale-yellow solution of the diazonium salt.

5. The physical properties of the product are as follows: n_D^{25} 1.5251; IR (liquid film) cm^{-1}: 1715; ^1H NMR ($CDCl_3$) δ: 2.0 (s, 3 H), 2.6–2.8 (m, 4 H), 6.8–7.3 (m, 4 H); mass spectrum m/e: 182 (M); semicarbazone, mp 165°C (164–165.5°C[4]). GLC analysis (glass capillary column, 20 m, pluronic L-64, program: 120–200°C at 5°C/min): >99% pure.

3. Discussion

This synthesis is only one example of a wide range of reactions that involve aryl (or alkyl) radical addition to electron-deficient double bonds resulting in reduction.[3,5,6] The corresponding oxidative reaction using aryl radicals is the well-known Meerwein reaction,[7] which uses copper(II) salts.

General arylation reactions are summarized by the following equations and some specific examples are presented in Table I.

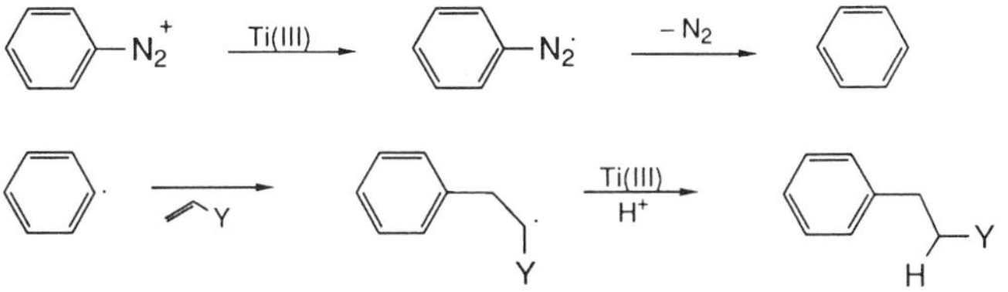

Homolytic cleavage of diazonium salts to produce aryl radicals is induced by titanium(III) salt, which is also effective in reducing the α-carbonylalkyl radical adduct to olefins, telomerization of methyl vinyl ketone, and dimerization of the adduct radicals. The reaction can be used with other electron-deficient olefins, but telomerization or dimerization are important side reactions.

Other limitations of the reaction are related to the regioselectivity of the aryl radical addition to double bond, which is determined mainly by steric and radical delocalization effects.[8] Thus, methyl vinyl ketone gives the best results, and lower yields are observed when bulky substituents are present in the β-position of the alkene. However, the method represents complete positional selectivity because only the β-adduct radicals give reductive arylation products whereas the α-adduct radicals add to diazonium salts, because of the different nucleophilic character of the alkyl radical adduct.[8,9]

TABLE I

REDUCTIVE ARYLATION OF ELECTRON-DEFICIENT OLEFINS BY ARENEDIAZONIUM SALTS INDUCED BY TITANIUM(III) SALTS

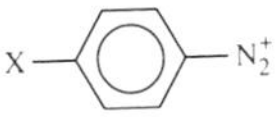 X	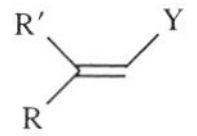 R	R'	Y	Yield (%)[a]
4-OCH$_3$	H	H	COCH$_3$	65
H	H	H	COCH$_3$	75
4-Br	H	H	COCH$_3$	68
4-COCH$_3$	H	H	COCH$_3$	72
4-Cl	H	H	CHO	63
4-Cl	CH$_3$	H	COCH$_3$	44
4-Cl	CH(CH$_3$)$_2$	H	COCH$_3$	28
4-Cl	C(CH$_3$)$_3$	H	COCH$_3$	14
4-Cl	CH$_3$	CH$_3$	COCH$_3$	12
4-Cl	Ph	H	COCH$_3$	18
4-Cl	H	H	CN	25[b]
4-Cl	H	H	COOH	33[b]
4-Cl	H	H	COOEt	32[b]

[a]From the diazonium salt.
[b]Telomers are formed; the reactions are carried out with twice the amount of titanium(III) salt.

The product described here, 4-(4-chlorophenyl)butan-2-one, was previously prepared in the following ways: (a) by reduction of the corresponding benzalacetone,[10] (b) by catalyzed decarbonylation of 4-chlorophenylacetaldehyde by $HFe(CO)_4$ in the presence of 2,4-pentanedione,[11] (c) by reaction of 4-chlorobenzyl chloride with 2,4-pentanedione under basic catalysis (K_2CO_3 in EtOH),[4] (d) by reaction of 4-chlorobenzyl chloride with ethyl 3-oxobutanoate under basic catalysis (LiOH),[12] and (e) by reaction of 3-(4-chloro-phenyl)-propanoic acid with methyllithium.[13]

1. Istituto di Chimica del Politecnico, Piazza L. da Vinci 32, I-20133 Milano, Italy.
2. Moszner, N.; Hartmann, M.; *Macromol. Chem.* **1979**, *180*, 97.
3. Citterio, A.; Vismara, E. *Synthesis* **1980**, 291; Citterio, A., Cominelli, A.; Bonavoglia, F. *Synthesis* **1986**, 308.
4. Boatman, S.; Harris, T. M.; Hauser, C. R. *J. Org. Chem.* **1965**, *30*, 3321.
5. Citterio, A.; Minisci, F.; Serravalle, M. *J. Chem. Res.* **1981**. *Synop., 198, Miniprint*, 2174.
6. Citterio, A.; Vismara, E. *Synthesis* **1980**, 751.
7. Rondestvedt, C. S., Jr. *Org. React.* **1976**, 24, 225; Rondestvedt, C. S., Jr. *Org. React.* **1960**, *11*, 189; Doyle, M. P.; Siegfried, B.; Elliott, R. C.; Dellaria, J. F. *J. Org. Chem.* **1977**, *42*, 2431.
8. Citterio, A.; Minisci, F.; Vismara, E. *J. Org. Chem.* **1982**, *47*, 81.
9. Citterio, A.; Ramperti, M.; Vismara, E.; *J. Heterocycl. Chem.* **1980**, *18*, 763.
10. Malinowski, S. *Rocz. Chem.* **1955**, *29*, 37; *Chem. Abstr.* **1956**, *50*, 3292R.
11. Yamashita, M.; Watanabe, Y.; Mitsud, T.; Takegami, Y. *Bull. Chem. Soc. Jpn.* **1978**, *51*, 835.
12. Nakai, S. *Jpn. Kokai* **1977**, *77*, 25, 709; *Chem. Abstr.* **1977**, *87*, 52781w.
13. Krubsack, A. J.; Sehgal, R.; Loong, W. A.; Slack, W. E. *J. Org. Chem.* **1975**, *40*, 3179.

SYNTHESIS OF α,β-UNSATURATED NITRILES FROM ACETONITRILE: CYCLOHEXYLIDENEACETONITRILE AND CINNAMONITRILE

[Acetonitrile, cyclohexylidene- and 2-propenenitrile, 3-phenyl, (*E*)- and (*Z*)-]

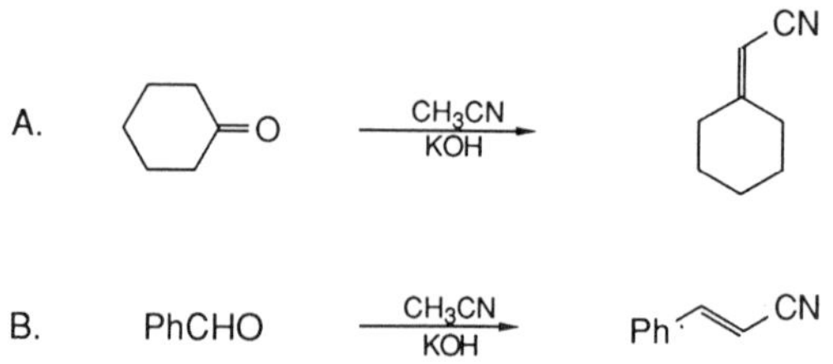

Submitted by STEPHEN A. DiBIASE, JAMES R. BEADLE, and GEORGE W. GOKEL[1]
Checked by YUMI NAKAGAWA and ROBERT V. STEVENS

1. Procedure

A. *Cyclohexylideneacetonitrile.* A 1-L three-necked, round-bottomed flask equipped with a reflux condenser, mechanical stirrer, and addition funnel is charged with potassium hydroxide (85% pellets, 33.0 g, 0.5 mol, Note 1) and acetonitrile (250 mL, Notes 2 and 3). The mixture is brought to reflux and a solution of cyclohexanone (49 g, 0.5 mol, Note 4) in acetonitrile (100 mL) is added over a period of 0.5–1.0 hr. Heating at reflux is continued for 2 hr (Note 5) after the addition is complete and the hot solution is then poured onto cracked ice (600 g). The resulting binary mixture is separated and the aqueous phase is extracted with ether (3 × 200 mL). The combined organic extracts are evaporated under reduced pressure, or may be placed in a 2-L Erlenmeyer flask containing several boiling chips and the volume reduced on a steam bath (internal temperature ca. 50°C). The resulting sweet-smelling, yellow to yellow–orange oil is transferred to a 1- or 2-L, three-necked, round-bottomed flask (depending on whether internal or external steam generation is used) and steam distilled (bp 81–99°C, Note 6). The distillate is extracted with three to five 200-mL portions of ether until the aqueous phase is clear (Note 7). The ether phase is washed with brine (2 × 100 mL), dried over sodium sulfate, and evaporated under reduced pressure to give a pale yellow oil (29–36 g, 48–60%) that consists of a mixture of isomers (α,β 80–83%, β,γ 17–20%, Note 8).

Isolation of the pure α,β-isomer. A 250-mL Erlenmeyer flask equipped with a magnetic stirring bar is charged with the isomeric nitriles (20 g, 0.165 mol), prepared in Section A above, and carbon tetrachloride (20 mL). A solution of bromine in carbon tetrachloride (1/9, v/v, ca. 25–30 mL) is added dropwise until the color of excess bromine persists. The reaction vessel is cooled in an ice bath for 30 min and filtered by gravity and the solvent evaporated under reduced pressure. The crude oil is distilled at reduced pressure (bp 40–42°C/0.15 mm) to give a colorless liquid (11–15 g, 55–75%) that is the pure α,β-isomer (Notes 9 and 10).

B. *Preparation of E- and Z-Cinnamonitrile.* A 1-L, three-necked, round-bottomed flask equipped with a mechanical stirrer, reflux condenser, and addition funnel is charged with potassium hydroxide pellets (33 g, 0.5 mol, Note 1) and acetonitrile (400 mL, Note 2). The mixture is brought to reflux under nitrogen and a solution of benzaldehyde (53 g, 0.5 mol, Note 4) in acetonitrile (100 mL) is added in a stream (1–2 min). After addition, stirring is continued for 10 min (Note 5) and the hot solution is then poured onto 500 g of cracked ice in a 1-L beaker. After being cooled for a few minutes, the two-phase mixture is transferred to a 2-L, three-necked flask and steam-distilled (Note 11). The distillate is transferred to a separatory funnel, and the upper aqueous phase is separated and then extracted with two 500-mL portions of ether (Note 7). The combined organic material is dried briefly over Na_2SO_4 and the ether evaporated to yield pure cinnamonitrile (20–29 g, 31–45%) as a pale-yellow oil (E/Z ratio ca. 5.5; Note 12).

2. Notes

1. Potassium hydroxide (85% pellets, AR grade) should be as fresh as possible (see Note 5).

2. Acetonitrile (99%) was obtained from Aldrich Chemical Company, Inc. and may be used without purification.

3. The yield of product is dependent on concentration. An increase in the amount of acetonitrile in Step A to ca. 1000 mL increases the yield of the isomer mixture to 65–75% without affecting isomer distribution. Further dilution to ca. 5000 mL increases the yield to 80–85%.

4. Cyclohexanone and benzaldehyde were purchased from either Aldrich Chemical Company, Inc. or Eastman Organic Chemicals and used without additional purification.

5. The reaction time depends on the quality of the potassium hydroxide employed. An induction period is often observed when older potassium hydroxide samples are used, possibly because surface formation of carbonates reduces the solubility of the salt in acetonitrile. An attempt was made to monitor the cinnamonitrile reaction by GLC, following loss of starting material. Although formation of the product was observed and reached a maximum, the starting material peak never completely disappeared. Prolonged reaction times (>2 hr) resulted in failure to isolate any of the desired product. Reaction times of less than 30 min gave the expected yields. Undissolved potassium hydroxide was observed in the reaction vessel when these reactions were terminated. At a column temperature of 150°C and a gas flow rate of ca. 60 mL/min (5-ft × 0.25-in. column, 10% SE-30 on firebrick), the retention times are as follows: cyclohexylideneacetonitrile and isomer, 2.8 min; Z-cinnamonitrile, 3.0 min; E-cinnamonitrile, 3.7 min). The reaction may also be monitored by a 2,4-dinitrophenylhydrazone spot test.

6. Distillation may be conducted using an apparatus designed for either internal or external steam generation. The first 1000-mL portion of distillate contains ca. 35 g of product. An additional 500 mL of distillate yields less than 1 g. Vacuum distillation gave product in 22% yield.

7. To facilitate phase separation, solid sodium chloride was added to the aqueous layer.

8. The product thus obtained is of high purity. The trace of color may be removed by distillation at reduced pressure (bp 50°C/0.5 mm).

9. Bromination can be monitored by ^{1}H NMR in CCl_4. The vinyl protons are observed at 5.08 (α,β-isomer) and 5.65 ppm.

10. The ^{1}H NMR spectra (in CCl_4) for the two isomers are as follows. Cyclohexylideneacetonitrile: δ 1.25-2.0 (m, 6 H), 2.0-2.8 (m, 4 H, methylene protons), 5.08 (m, 1 H, olefin); 2-(1-cyclohexenyl)acetonitrile: 1.25-2.0 (m, 4 H), 2.0-2.8 [m, 4 H, $-(CH_2)_4-$], 3.05 (pseudo-s, 2 H, $-CH_2CN$), 5.65 (m, 1 H, olefin).

11. Steam distillation may be conducted using apparatus designed either for internal or external steam generation. Using internally generated steam, 2.5 L of distillate was collected. The last 500 mL contained less than 1 g of product.

12. Isomer distribution and purity were assessed by GLC (see Note 5). The ^{1}H NMR spectra (in CCl_4) for the pure isomers are as follows: E-isomer: δ 5.71 [d, 1 H, $J = 17$, ArCH=CH—CN; 7.44 (d, 1 H, $J = 17$, ArCH=CHCN), 7.3 (pseudo-s, 5 H, aromatic protons). Z-Isomer: δ 5.31 (d, 1 H, $J = 12$, ArCH=CHCN), 6.98 (d, 1 H, $J = 12$, ArCH=CHCN), 7.3 (pseudo-s, 5 H, aromatic protons).

3. Discussion

Introduction of the two-carbon fragment is a cornerstone of synthetic methodology and many of the condensation reactions frequently used have been known for decades, if not for a century. Examples include the malonic ester[2] and acetoacetic ester[3] reactions, the Perkin[4] condensation, and the Doebner–Knoevenagel[5] reaction. Addition of the cyanomethyl group has been accomplished by a variety of methods,[6] mostly circuitous, and is certainly not in the group of classical reactions named above. The direct approach is found in a recent application of lithio trimethylsilylacetonitrile,[7] but the difference in expense and convenience between using this method and a mixture of potassium hydroxide and acetonitrile is manifest.

The direct synthesis of α,β-unsaturated nitriles can be accomplished by treating the appropriate carbonyl compound with potassium hydroxide in acetonitrile.[8] In order for direct condensation to succeed, acetonitrile must be deprotonated by the relatively weak base potassium hydroxide and the carbanion thus formed must add to the carbonyl. The cyanohydrin is presumably dehydrated to leave the α,β-unsaturated compound, which may or may not isomerize in the medium. We have run this reaction with a large number of carbonyl compounds[8] and have found that it is most successful for aromatic aldehydes (36–86%) and other nonenolizable carbonyl compounds such as benzophenone (84%). Yields are also acceptable for most cyclic ketones with six or more carbons in the ring (e.g., 2-methylcyclohexanone, 78%; cis-octalone, 80%; cycloheptanone, 78%; cyclooctanone, 66%, cyclododecanone, 45%), and for aliphatic ketones having three or more carbons bonded on each side (e.g., diethyl ketone, 35%; di-n-propyl ketone, 65%, di-n-butyl ketone 65%). Ketones that are sterically hindered (camphor) or highly enolized (cyclopentanone) are not useful substrates in this reaction.

We present here examples of this condensation with an aromatic aldehyde and a cyclic ketone. Both of these examples are useful because, although other methods are available for their preparation, problems often attend these syntheses. In the synthesis of cyclohexylideneacetonitrile, for example, the standard method[9] results exclusively in the β,γ-isomer and none of the α,β-isomer. In Part A of this procedure, cyclohexanone is condensed with acetonitrile to give predominantly the conjugated isomer (80–83%), which is then separated from the nonconjugated isomer by selective bromination.

The procedures presented here are simple, inexpensive, and may be used on a large scale. The use of potassium hydroxide in this reaction may, however, prove incompatible with certain base-sensitive functional groups.

1. Department of Chemistry, University of Maryland, College Park, MD 20742. Present address: Department of Chemistry, University of Miami, Coral Gables, FL 33124.
2. House, H. O. In ''Modern Synthetic Reactions,'' 2nd ed.; W. A. Benjamin: Menlo Park, CA, 1972; pp. 510–518.
3. Claisen, L.; Lowman, O. *Ber.* **1887**, *20*, 651.
4. Perkin, W. H. *J. Chem. Soc.* **1868**, *21*, 181.
5. Knoevenagel, E. *Ber.* **1898**, *31*, 2596; Doebner, O. *Ber.* **1900**, *33*, 2140.
6. For preparations of cinnamonitrile, for example, see: Ghosez, J. *Bull. Soc. Chem. Belg.* **1932**, *41*, 477 and references cited therein; Schiemenz, G. P.; Engelhard, H. *Chem. Ber.* **1962**, 95, 195; Krüger, C. *J. Organometal. Chem.* **1967**, *9*, 125; Crouse, D. N.; Seebach, D. *Chem. Ber.* **1968**, *101*, 3113; Pattieson, I.; Wade, K.; Wyatt, B. K. *J. Chem. Soc. (A)* **1968**, 837; Kametani, T.; Yamaki, K.; Ogasawara, K. *Yakugaku Zasshi* **1969**, 89, 154; *Chem. Abstr.* **1969**, *70*, 106342y; Uchida, A.; Saito, S.; Matsuda, S. *Bull. Chem. Soc. Jpn.* **1969**, *42*, 2989; Vishnyakova, T. P.; Koridze, A. A. *Zh. Obshch. Khim.* **1969**, *39*, 210; *Chem. Abstr.* **1969**, *70*, 96901p; Uchida, A.; Doyama, A.; Matsuda, S. *Bull. Chem. Soc. Jpn.* **1970**, *43*, 963; Louvar, J. J.; Sparks, A. K. Ger. Off., 2 041 563; *Chem. Abstr.* **1971**, *75*, P19989z; Das, R.;

Wilkie, C. A. *J. Am. Chem. Soc.* **1972**, *94*, 4555; Takahashi, K.; Sasaki, K.; Tanabe, H.; Yamada, Y.; Iida, H. *Nippon Kagaku Kaishi* **1973**, 2347; *Chem. Abstr.* **1974**, *80*, 70499v; Krause, J. G.; Shaikh, S. *Synthesis* **1975**, 502.

7. Matsuda, I.; Murata, S.; Ishii, Y. *J. Chem. Soc., Perkin Trans. 1* **1979**, 26. See also: Tanaka, K.; Ono, N.; Kubo, Y.; Kaji, A. *Synthesis* **1979**, 890.

8. DiBiase, S. A.; Lipisko, B. A.; Haag, A.; Wolak, R. A.; Gokel, G. W. *J. Org. Chem.* **1979**, *44*, 4640.

9. Cope, A. C.; D'Addieco, A. A.; Whayte, D. E.; Glickman, S. A. *Org. Synth., Coll. Vol. IV* **1963**, 234.

1,2-CYCLOBUTANEDIONE

Submitted by J. M. Denis, J. Champion, and J. M. Conia[1]
Checked by Robert V. Stevens and Steven R. Angle

1. Procedure

Caution! This reaction should be carried out in a dark hood to prevent the photoinduced polymerization of the dione.

A 1-L, three-necked, round-bottomed flask equipped with a 500-mL dropping funnel, a nitrogen-inlet tube, a mechanical stirrer, a low-temperature thermometer, and a calcium chloride drying tube is charged with 172 g (0.75 mol) of 1,2-bis(trimethylsiloxy) cyclobutene (Note 1) and 375 mL of anhydrous pentane (Note 2). A dry nitrogen atmosphere is maintained in the system (Note 3) and the solution is cooled to $-60°C$ by means of a dry ice-methanol bath. Then a solution of 120 g (0.75 mol) of bromine and 375 mL of anhydrous pentane is added dropwise with stirring over a period of 2 hr. When the addition is complete, the mixture must be heated to 40°C for 2 hr (Note 4) and concentrated by removing ca. 550 mL of solvent under reduced pressure (15 mm) at room temperature. To isolate the dione the residue is cooled to $-60°C$ by immersion of the flask in a dry ice–methanol bath. The crystallized product is *quickly* filtered through a hermetically sealed (Note 3), 250-mL sintered-glass funnel (porosity 3), a dry nitrogen pressure being used to push down solvent. The yellow solid is washed with stirring with eight 25-mL portions of anhydrous pentane cooled to $-60°C$ and sucked as dry as possible on the filter. The pentane used in the washing is concentrated under reduced pressure (15 mm) at room temperature to ca. 15 mL, then the flask is again cooled to $-60°C$. The crystallized product is washed with four 15-mL portions of anhydrous pentane cooled to $-60°C$ as before. The two batches of crystals are of approximately equal purity. The yield of 1,2-cyclobutanedione is 42–46 g (70–73%), mp 65°C (Note 5).

2. Notes

1. The 1,2-bis(trimethylsiloxy)cyclobutene is prepared according to the procedure of Bloomfield.[2]

2. Pentane is distilled prior to use (bp 36°C/760 mm) and stored over sodium wire.

3. *Caution! Moisture must be avoided to prevent the ring contraction of the dione into 1-hydroxycyclopropanecarboxylic acid.*[2]

4. This heating is necessary to complete the reaction.[2]

5. The yellow product shows in the ^{1}H NMR (CCl$_4$) a single signal at 2.98 ppm indicating its high degree of purity. It can be sublimed under vacuum (15 mm) at room temperature, mp 65°C; IR (CCl$_4$) cm^{-1}: 1778 and 1810; UV (hexane) nm max (ϵ): 407 (4), 423 (8), 436 (10.5), 453 (19), 463 (17), 489 (42), and 500.5 (28). The dione can be stored at 0°C in a hermetically sealed flask in the dark for months.

3. Discussion

This method of preparation of the 1,2-cyclobutanedione is an adaptation of that independently described by Denis and Conia[3] and by Heine.[4] Acyloins, 1,2-cyclobutanediols, imidazole, thioimidazole, and amino- and cyanofuran derivatives are readily available[5,6] from bis(trimethylsiloxy)alkenes.

The bis(trimethylsiloxy)alkene bromination procedure is a large-scale preparation that gives excellent yields of cyclic and acyclic 1,2-diones; however, when enolizable 1,2-diketones are produced, some complications can be encountered.[7,8]

1. Laboratoire des Carbocycles, Universite de Paris-Sud, 91405 ORSAY, France.
2. Bloomfield, J. J.; Nelke, J. M. *Org. Synth. Coll. Vol. VI*, **1988**, 167.
3. Denis, J. M.; Conia, J. M. *Tetrahedron Lett.* **1971**, 2845–2846.
4. Heine, H. G. *Chem. Ber.* **1971**, *104*, 2869–2872.
5. Ruhlmann, K. *Synthesis* **1971**, 236–253.
6. Conia, J. M.; Barnier, J. P. *Tetrahedron Lett.* **1971**, 4981–4984.
7. Strating, J.; Reiffers, S.; Wynberg, H. *Synthesis* **1971**, 209–211.
8. Strating, J.; Reiffers, S.; Wynberg, H. *Synthesis* **1971**, 211–212.

CYCLOBUTANONE

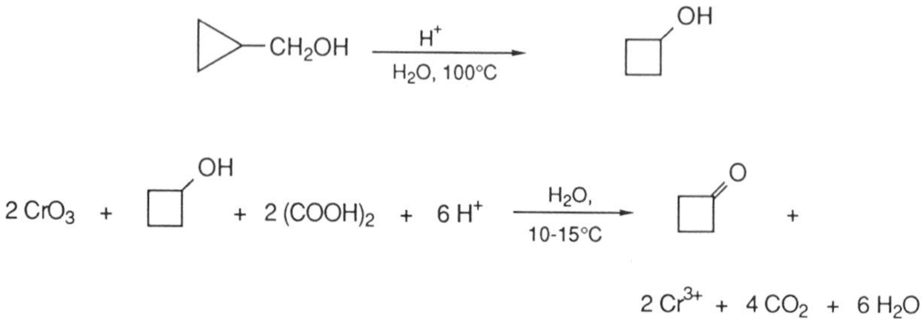

Submitted by MIROSLAV KRUMPOLC and JAN ROCEK[1]
Checked by D. SEEBACH, R. DAMMANN, F. LEHR, and M. POHMAKOTR

1. Procedure

In a 2-L, three-necked, round-bottomed flask equipped with a reflux condenser are placed 250 mL of water, 48 mL (ca. 0.55 mol) of concentrated hydrochloric acid, and 49.5 g (0.65 mol) of cyclopropylcarbinol (Note 1); the reaction mixture is refluxed for ca. 100 min. The formation of cyclobutanol can be observed nearly instantaneously, as this alcohol is only partially soluble in water and soon separates (Note 2). The flask is then immersed in an ice bath equipped with a mechanical stirrer, a thermometer, and a dropping funnel (using a three-way adapter, parallel sidearm), and the reflux condenser is replaced by an ethanol–dry ice trap connected to a U-tube immersed in an ethanol–dry ice bath to ensure condensation of the very volatile cyclobutanone. The flask is charged with an additional 48 mL (ca. 0.55 mol) of concentrated hydrochloric acid in 200 mL of water and 440 g (3.5 mol) of oxalic acid dihydrate (Note 1). The heterogeneous mixture is stirred for ca. 15 min to saturate the solution with oxalic acid. A solution of 162 g (1.62 mol) of chromium trioxide in 250 mL of water is added dropwise with stirring at such a rate that the temperature of the reaction mixture is kept between $10°C$ and $15°C$ (NaCl–ice bath, $-5°C$ to $-10°C$) and the generation of carbon dioxide remains gentle. The reduction of each drop of chromic acid is practically instantaneous. As the addition of the reagent proceeds (1.5–2 hr), oxalic acid gradually dissolves and a dark-blue solution containing chromium(III) salts results (Note 3). Just before the end of the oxidation (ca. 10 mL of the chromic acid solution left), the cyclobutanone (with traces of cyclobutanol) trapped in the U-tube (a few milliliters) is added to the reaction mixture. After the oxidation is completed, the ice bath is removed and stirring is continued for ca. 1 hr to bring the reaction mixture to room temperature and to reduce the amount of carbon dioxide in the solution.

The reaction mixture is poured into a 2-L separatory funnel and extracted with four 200-mL portions of methylene chloride (Note 4). The extracts (the lower phase) are combined, dried over anhydrous magnesium sulfate containing a small amount of anhydrous potassium carbonate (to remove traces of hydrochloric acid), and filtered, and the filtrate is concentrated by distillation through a vacuum-insulated silvered column (20-cm length, 1-cm i.d.) packed with glass helices (size 2.3 mm, Lab Glass, Inc.) and equipped with an adjustable stillhead, until the pot temperature rises to 80°C (Note 5). The crude product is then transferred to a 100-mL flask and distilled through the same column (reflux ratio 10 : 1) to give 14–16 g (0.20–0.23 mol), 31–35% overall yield (based on pure cyclopropyl carbinol) of cyclobutanone, bp 98–99°C, d_4^{25} 0.926, n_D^{25} 1.4190 (Note 6). The product is sufficiently pure (98–99%) for most purposes (Notes 5, 7, 8, and 9).

2. Notes

1. The following compounds were used as supplied: cyclopropylcarbinol (Aldrich Chemical Company, Inc. or Fluka AG, 95% pure), hydrochloric acid (Fisher Reagent, 36.5–38%), chromium trioxide (Fisher Certified), oxalic acid dihydrate (Fisher Certified), methylene chloride (Fisher Certified).

2. At this point cyclopropylcarbinol has been completely converted into a mixture of products containing ca. 80% cyclobutanol, 8% 3-butene-1-ol, and several additional products observable by GLC analysis in varying amounts. About 95–97% pure cyclobutanol (60–65% yield) can be obtained if the reaction mixture is neutralized with sodium hydroxide and sodium bicarbonate, saturated with magnesium sulfate, extracted with ether, and fractionally distilled on an efficient distillation column. The remaining impurities are extremely difficult to remove.

3. Oxalic acid is used in excess to ensure a rapid oxidation of the alcohol and to destroy the excess chromic acid when the cooxidation process is over. Part of the oxalic acid is consumed by chromium(III) to form oxalatochromium(III) complexes.

4. As cyclobutanone is considerably soluble in water, a thorough and vigorous agitation is recommended to ensure good extraction of the aqueous layer by methylene chloride. Oxalic acid is insoluble in this solvent.

5. The checkers used a silvered, vacuum-insulated column 30 cm in length with 1.5-cm i.d., filled with 4-mm × 4-mm helices; distillation of CH_2Cl_2 was first done from a 250-mL, two-necked flask with dropping funnel from which the dried extraction solution was continuously added. When ca. 50-mL total volume of solution remained (bath temperature ca. 90°C), it was transferred into a 100-mL, one-necked flask. Eight fractions of the cyclobutanone were collected at a 15–20 : 1 reflux ratio: bp/g/% purity of ketone (by VPC): 80–90/1.17/37, 90–95/4.3/53, 95–96/1.71/99.5, 96–97/1.41/—, 96–97.5/1.2/99.9, 97.5–98/3.95/99.9, 98/3.76/100, 98/1.78/99.8. The $n_D^{20.5}$ of fraction 7 was 1.4210.

6. The reported physical constants of cyclobutanone[2] are bp 99–100°C, d_4^{24} 0.924, n_D^{25} 1.4188.

7. Gas-liquid chromatography [⅛-in. × 6-ft, 10% diethylene glycol succinate (LAC-728) column, 70°C] of cyclobutanone (99.2% pure) revealed the presence of small amounts of methylene chloride (0.6%) and cyclobutanol (0.2%). No cleavage product, 4-hydroxybutyraldehyde, was found. The traces of water, detected by NMR spectroscopy using CD_3COCD_3 as a solvent, can be removed by drying over molecular sieves.

8. ^{1}H NMR (CCl$_4$) δ: 1.98, degenerate quintet (2 H, J = 8 Hz); 3.03, t (4 H, J = 8 Hz). IR (liquid film on KBr plates) cm^{-1}: 1783 (strong, C=O).

9. If the preparation of cyclobutanone from cyclopropylcarbinol is carried out in two steps, with cyclobutanol isolated first, somewhat higher yields can be achieved (70–80% based on cyclobutanol, 45–50% overall yield, purity 98–99%).

3. Discussion

Cyclobutanone has been prepared (1) by pyrolysis of 1-hydroxycyclobutane-1-carboxylic acid[3] (15% yield), (2) by reaction of diazomethane with ketene[4-6] (36% overall yield based on precursors used for the generation of both components[6]), (3) from pentaerythritol, the final step being the oxidative degradation of methylenecyclobutane[7,8] (30–45% overall yield), (4) by oxidation of cyclobutanol with chromic acid–pyridine complex in pyridine[9] (no yield is given), (5) by oxidative cleavage of 5,9-dithiaspiro[3.5]nonane, prepared via 2-(ω-chloropropyl)-1,3-dithiane[10,11] from 1,3-propanedithiol[12] (40% overall yield), (6) via solvolytic cyclization of 3-butyn-1-ol[13,14] (30% yield), (7) by epoxidation of methylenecyclopropane followed by ring expansion of resulting oxaspiropentane[15-17] (28% overall yield), (8) from 1,3-dibromopropane and methyl methylthiomethyl sulfoxide via cyclobutanone dimethyl dithioacetal S-oxide[18] (75% overall yield), and (9) from 4-chlorobutyraldehyde cyanohydrin, the final step being hydrolysis of cyclobutanone cyanohydrin[19] (45% overall yield).

The present procedure offers a simple and fast (2–3 days are required) preparation of pure cyclobutanone from cyclopropylcarbinol. The synthesis is carried out in one operation, without isolating the intermediate cyclobutanol. The first reaction, acid-catalyzed rearrangement of cyclopropylcarbinol, has been described by Caserio, Graham, and Roberts.[9] The novel feature is the preparation of cyclobutanone from cyclobutanol in the presence of oxalic acid. It is based on rapid cooxidation of two substrates proceeding via a three-electron oxidation–reduction mechanism[20,21] in which chromium(VI) is reduced directly to chromium(III). In the absence of oxalic acid the chromic acid oxidation of cyclobutanol gives along with cyclobutanone ca. 30–40% of 4-hydroxybutyraldehyde,[2] as the alcohol undergoes extensive carbon–carbon cleavage by chromium(IV).[2,21,22] The participation of oxalic acid in the reaction process serves to suppress the formation of a chromium(IV) intermediate; the only by-product formed is carbon dioxide.

Cyclobutanone is a versatile starting material used for numerous synthetic and theoretical studies in the chemistry of small rings. The preparation of this compound by the cooxidation process illustrates the synthetic utilization of three-electron oxidation–reduction reactions.

1. Department of Chemistry, University of Illinois at Chicago, Chicago, IL 60680.
2. Roček, J.; Radkowsky, A. E. *J. Am. Chem. Soc.* **1973**, *95*, 7123–7132.
3. Demjanow, N. J.; Dojarenko, M. *Chem. Ber.* **1922**, *55*, 2737–2742.
4. Lipp, P.; Köster, R. *Chem. Ber.* **1931**, *64*, 2823–2825.
5. Kaarsemaker, S.; Coops, J. *Recl. Trav. Chim. Pays-Bas* **1951**, *70*, 1033–1041.
6. Machinskaya, I. V.; Smirnova, G. P.; Barkhash, V. A. *J. Gen. Chem. USSR* **1961**, *31*, 2390–2393; *Chem. Abstr.* **1962**, *56*, 12751g.
7. Roberts, J. D.; Sauer, C. W. *J. Am. Chem. Soc.* **1949**, *71*, 3925–3929.

8. Conia, J. M.; Leriverend, P.; Ripoll, J. L. *Bull. Soc. Chim. Fr.* **1961,** 1803–1804.
9. Caserio, M. C.; Graham, W. H.; Roberts, J. D. *Tetrahedron* **1960,** *11,* 171–182.
10. Seebach, D.; Jones, N. R.; Corey, E. J. *J. Org. Chem.* **1968,** *33,* 300–305.
11. Seebach, D.; Beck, A. K. *Org. Synth., Coll. Vol. VI* **1988,** 316.
12. Corey, E. J.; Seebach, D. *Org. Synth., Coll. Vol. VI* **1988,** 556.
13. Hummel, K.; Hanack, M. *Liebigs Ann. Chem.* **1971,** *746,* 211–213.
14. Hanack, M.; Dehesch, T.; Hummel, K.; Nierth, A. *Org. Synth., Coll. Vol. VI* **1988,** 324.
15. Salaün, J. R.; Conia, J. M. *J. Chem. Soc., Chem. Commun.* **1971,** 1579–1580.
16. Salaün, J. R.; Champion, J.; Conia, J. M. *Org. Synth., Coll. Vol. VI* **1988,** 320.
17. Aue, D. H.; Meshishnek, M. J.; Shellhamer, D. F. *Tetrahedron Lett.* **1973,** 4799–4802.
18. Ogura, K.; Yamashita, M.; Suzuki, M.; Tsuchihashi, G. *Tetrahedron Lett.* **1974,** 3653–3656.
19. Stork, G.; Depezay, J. C.; d'Angelo, J. *Tetrahedron Lett.* **1975,** 389–392.
20. Hasan, F.; Roček, J. *J. Am. Chem. Soc.* **1972,** *94,* 3181–3187.
21. Hasan, F.; Roček, J. *J. Am. Chem. Soc.* **1974,** *96,* 534–539.
22. Wiberg, K. B.; Mukherjee, S. K. *J. Am. Chem. Soc.* **1974,** *96,* 6647–6651.

CYCLOBUTENE

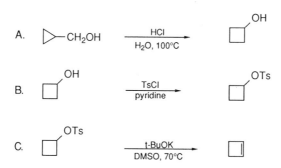

Submitted by J. SALAÜN and A. FADEL[1]
Checked by LAWRENCE R. MCGEE and BRUCE E. SMART

1. Procedure

A. *Cyclobutanol.* A 1-L, three-necked, round-bottomed flask equipped with a reflux condenser and a magnetic stirring bar is charged with 600 mL of water, 57.5 mL (ca. 0.68 mol) of concentrated hydrochloric acid, and 57.7 g (0.80 mol) of cyclopropylcarbinol (Note 1). The reaction mixture is stirred and refluxed for 3 hr. Cyclobutanol is only partially soluble in water and soon separates. The reaction mixture is allowed to cool to room temperature, and the flask is then immersed in an ice bath. To the cold, stirred mixture is added 24 g (0.6 mol) of sodium hydroxide pellets, followed by 6.7 g (0.08 mol) of sodium bicarbonate to complete the neutralization. The mixture is saturated with sodium chloride and extracted for 30 hr with diethyl ether using a liquid–liquid continuous extraction apparatus. The ethereal extract is dried over anhydrous sodium sulfate and the drying agent is removed by filtration. The bulk of the solvent is distilled from the filtrate to give 55.0 g of residual liquid containing 88% cyclobutanol and 12% 3-buten-1-ol by gas chromatography (Note 2). The crude product is carefully distilled through spinning band columns to give 32.8 g (57%) of cyclobutanol, bp 122–124°C. Gas chromatographic analysis of the product shows it to be 95% pure (Notes 2–4).

B. *Cyclobutyl tosylate.* A 500-mL, three-necked, round-bottomed flask fitted with a stirrer and a theromometer is charged with 200 mL of pyridine (Note 5) and 32.3 g (0.448 mol) of cyclobutanol. The solution is stirred and chilled to 0°C, and then 89.8 g (0.471 mol) of *p*-toluenesulfonyl chloride (Note 6) is added in portions over a 20-min period. The reaction mixture is allowed to warm to room temperature and is stirred for 16 hr. The mixture is recooled to 0°C, and poured into 260 mL of concentrated hydrochloric acid in 800 mL of ice water. The mixture is extracted with three 300-mL portions of ether and the combined ethereal extracts are dried over anhydrous magnesium sulfate. The drying agent is removed by filtration and the filtrate is concentrated on a rotary evaporator. The residue is held under high vacuum (0.03 mm) at room temperature for 3 hr to give 93.3 g (92%) of cyclobutyl tosylate as a pale-yellow oil (Note 7).

C. *Cyclobutene.* A 500-mL, two-necked, round-bottomed flask is fitted with a 100-mL dropping funnel equipped with an argon-inlet tube, a magnetic stirring bar, and a water-cooled condenser. The outlet of the condenser is attached to an all-glass transfer manifold. Two weighed traps fitted with gastight stopcocks and immersed in dry ice–acetone baths are attached to the manifold. A calcium chloride drying tube is attached to the exit of the second trap. While the system is continuously purged with a slow stream of argon, the flask is charged with 33.6 g (0.30 mol) of potassium *tert*-butoxide and 120 mL of anhydrous dimethyl sulfoxide (Note 8), and a solution of 25.6 g (0.113 mol) of cyclobutyl tosylate in 30 mL of anhydrous dimethyl sulfoxide is placed in the dropping funnel. The potassium *tert*-butoxide suspension is stirred vigorously and heated to 70°C. The cyclobutyl tosylate solution is then added dropwise over a 10-min period (Note 9). After the addition is completed, the reaction mixture is stirred at 70°C for an additional 2 hr. The manifold system is closed off from the reaction vessel and the material collected in the first trap is slowly warmed. The product distills at ca. 2°C into the second dry-ice-cooled trap to give 4.3–5.1 g (70–84%) of cyclobutene [lit.[2] bp 2°C] (Notes 10–12).

2. Notes

1. The checkers obtained cyclopropylcarbinol from the Aldrich Chemical Company, Inc. It can be readily prepared by the reduction of cyclopropanecarboxylic acid with lithium aluminum hydride.[3]

2. A 25-m × 0.3-mm HP Ultra Silicone capillary column at 70°C with 30–psi helium head pressure was used for the chromatographic analysis: retention times of 3-buten-1-ol and cyclobutanol are 1.19 and 1.35 min, respectively. The submitters used a 3-m × 0.3-cm 20 *M* carbowax column at 90°C/8 psi hydrogen and reported retention times of 13 and 20 min for 3-buten-1-ol and cyclobutanol, respectively.

3. The crude product was first distilled on a 50-cm × 0.8-cm spinning band column (reflux ratio 10 : 1) to give 19.6 g of cyclobutanol, bp 124°C. The forerun

fractions, bp 66–123°C (23.0 g), were combined and redistilled on a 30-cm × 0.8-cm spinning band column (reflux ratio 25 : 1) to give an additional 13.2 g of cyclobutanol, bp 122–123°C. The major by-product, 3-buten-1-ol, boils at 112–114°C. Gas chromatographic analysis of the combined product fractions indicates a mixture of 95% cyclobutanol/3-buten-1-ol (99.7%/0.3%) and 5% unidentified compounds.

4. Cyclobutanol shows the following ^{1}H NMR spectrum (CDCl$_3$): δ: 1.1–2.4 (m, 6 H), 4.16 (quintet, 1 H, J = 7.5), 4.54 (s, 1 H, OH).

5. The pyridine was distilled from calcium hydride and stored over potassium hydroxide.

6. The p-toluenesulfonyl chloride was obtained from the Aldrich Chemical Company, Inc., and was recrystallized from hexane prior to use.

7. The product is >99% pure by NMR and shows the following spectrum: ^{1}H NMR (CDCl$_3$) δ: 1.1–2.3 (m, 6 H), 2.47 (s, 3 H), 4.77 (quintet, 1 H, J = 7.5), 7.32 (d, 2 H, J = 9.0), 7.79 (d, 2 H, J = 9.0).

8. Potassium tert-butoxide was obtained from the Aldrich Chemical Company, Inc. The dimethyl sulfoxide was distilled from calcium hydride and stored under argon.

9. The reaction mixture turns green, then blue indigo, and finally dark pink during the addition.

10. The submitters report obtaining 5.2 g of 99.2% pure cyclobutene. The product obtained by the checkers was pure by NMR spectroscopy and shows the following ^{1}H NMR spectrum (CDCl$_3$) δ: 2.55 (s, 4 H), 6.00 (s, 2 H).

11. The cyclobutene can be converted to 1,2-dibromocyclobutane by distilling 4.3–7.3 g (0.079–0.135 mol) of cyclobutene into 100 mL of pentane chilled to −40°C, followed by adding a solution of 15.5–32.0 g (0.097–0.200 mol) of bromine in 30 mL of pentane. After the usual workup with aqueous sodium thiosulfate and distillation, 14.3–25.3 g (84–87.5%) of pure 1,2-dibromocyclobutane, bp 60°C (6 mm), is obtained. It shows the following ^{1}H NMR spectrum (CDCl$_3$) δ: 1.90–3.03 (m, 4 H), 4.27–4.70 (m, 2 H). The 1,2-dibromocyclobutane can be conveniently converted back to cyclobutene by debromination with zinc in ethanol.[4]

12. Cyclobutene can be prepared on a larger scale (25–37 g) simply by scaling up the reactants.

3. Discussion

Cyclobutene has been prepared (1) by pyrolysis of cyclobutyldimethylamine oxide[4-6] and cyclobutyltrimethylammonium hydroxide[4,6,7] (50–73% yield), which were prepared in eight steps from malonate esters (2.0–2.1% overall yield of cyclobutene contaminated with 1,3-butadiene), (2) by pyrolysis of the products of cycloaddition of dimethyl acetylenedicarboxylate with cyclooctatriene[8-10] (30–32% overall yield) or with cyclooctatetraene[11-13] (34–39% overall yield), (3) by photolysis of butadiene leading to cyclobutene (30% yield) and bicyclo[1.1.0]butane (5% yield),[14,15] (4) by oxidation of cyclobutylcarboxylic acid[16] with lead tetraacetate[17] (67% yield) (11.8% overall yield),

(5) by fragmentation of 1,2-cyclobutyl thiocarbonate with trialkyl phosphite[18] (68% yield based on cis-1,2-dihydroxycyclobutane), (6) by ring expansion of cyclopropylcarbene[19] (7) from cyclobutylidene[20] and (8) by base-induced ring expansion of cyclopropylmethyl tosylate with potassium tert-butoxide in dimethyl sulfoxide leading to a 1 : 1 mixture of cyclobutene and methylenecyclopropane.[21] None of these methods appears to be practical.

The present procedure offers in good yields a simple and ready preparation of pure cyclobutene from the easily available cyclopropylcarbinol. The product is free of the impurities (e.g., 1,3-butadiene, bicyclobutane, methylenecyclopropane) usually obtained with the various methods so far reported. The procedure described for the synthesis of cyclobutanol is patterned after the acid-catalyzed rearrangement of cyclopropylcarbinol reported by Roberts[22] and Roček.[23]

1. Laboratoire des Carbocycles, U.A. 478, Bât. 420, Université de Paris-Sud, 91405 Orsay, France.
2. *Beilsteins Handbuch der Organischen Chemie*, E III, 5 (1), 170.
3. McCloskey, C. M.; Coleman, G. H. *Org. Synth., Coll. Vol. III* **1955**, 221.
4. Heisig, G. B. *J. Am. Chem. Soc.* **1941**, 63, 1698.
5. Willstätter, R.; von Schmaedel, W. *Chem. Ber.* **1905**, 38, 1992.
6. Roberts, J. D.; Sauer, C. W. *J. Am. Chem. Soc.* **1949**, 71, 3925.
7. Ripoll, J.-L.; Conia, J.-M. *Bull. Soc. Chim. Fr.* **1965**, 2755.
8. Cope, A. C.; Hochstein, F. A. *J. Am. Chem. Soc.* **1950**, 72, 2515.
9. Cope, A. C.; Haven, A. C., Jr.; Ramp, F. L.; Trumbull, E. R. *J. Am. Chem. Soc.* **1952**, 74, 4867.
10. Normant, H.; Maitte, P. *Bull. Soc. Chim. Fr.* **1960**, 1424.
11. Anderson, A. G., Jr.; Fagerburg, D. R. *Tetrahedron* **1973**, 29, 2973.
12. Avram, M.; Nenitzescu, C. D.; Marica, E. *Chem. Ber.* **1957**, 90, 1857.
13. Reppe, W.; Schlichting, O.; Klager, K.; Toepel, T. *Liebigs Ann., Chem.* **1948**, 560, 1.
14. Srinivasan, R. *J. Am. Chem. Soc.* **1963**, 85, 4045.
15. Srinivasan, R. U.S. Patent 3427241, 1969; *Chem. Abstr.* **1969**, 70, 67763q.
16. Heisig, G. B.; Stodola, F. H. *Org. Synth., Coll. Vol. III* **1955**, 213.
17. Bacha, J. D.; Kochi, J. K. *Tetrahedron* **1968**, 24, 2215.
18. Hartmann, W.; Fischler, H.-M.; Heine, H.-G. *Tetrahedron Lett.* **1972**, 853.
19. Friedman, L.; Shechter, H. *J. Am. Chem. Soc.* **1960**, 82, 1002.
20. (a) Schoeller, W. W. *J. Org. Chem.* **1980**, 45, 2161; (b) Schoeller, W. W. *J. Am. Chem. Soc.* **1979**, 101, 4811.
21. Dolbier, W. R., Jr.; Alonso, J. H. *J. Chem. Soc., Chem. Commun.* **1973**, 394.
22. Caserio, M. C.; Graham, W. H.; Roberts, J. D. *Tetrahedron* **1960**, 11, 171.
23. Krumpolc, M.; Roček, J. *Org. Synth., Coll. Vol. VII* **1990**, 114.

DEOXYGENATION OF EPOXIDES WITH LOWER VALENT TUNGSTEN HALIDES: *trans*-CYCLODODECENE

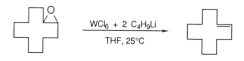

Submitted by MARTHA A. UMBREIT and K. BARRY SHARPLESS[1]
Checked by RONALD F. SIELOFF, MICHAEL F. REINHARD, and CARL R. JOHNSON

1. Procedure

Caution! Concentrated butyllithium may ignite spontaneously on exposure to air or moisture. Manipulations with this reagent should be performed with great care.

A dry, 1-L, three-necked flask equipped with a thermometer, a mechanical stirrer, and a rubber septum is flushed with nitrogen (admitted through a hypodermic needle in the septum). A nitrogen atmosphere is maintained throughout the subsequent reaction. The flask is charged with 420 mL of tetrahydrofuran (Note 1), and the solvent is cooled to −62°C in an acetone–dry ice bath. Tungsten hexachloride (60g, 0.15 mol) (Note 2) is then introduced. While the cold suspension is stirred, 31 mL (0.30 mol) of 90% butyllithium in hexane (Note 3) is added slowly from a hypodermic syringe. The rate of addition (complete in ca. 5 min) is such that the temperature remains below −15°C. The resulting mixture is allowed to warm slowly to room temperature with stirring. The green-brown viscous suspension becomes a dark-brown homogeneous solution that eventually turns green at room temperature. Because the reaction with the epoxide is exothermic, the flask is momentarily returned to the acetone–dry ice bath while 14.8 g (0.081 mol) of *trans*-cyclododecene oxide (Notes 4 and 5) is introduced with a hypodermic syringe. The cooling bath is again removed and the reaction mixture is allowed to stir at room temperature. After 30 min (Note 6) the mixture is poured into 600 mL of an aqueous solution that is 1.5 M in sodium tartrate and 2 M in sodium hydroxide (Note 7), and extracted with 240 mL of hexane. The organic phase is washed with a mixture of 160 mL of water and 80 mL of aqueous saturated sodium chloride, dried over anhydrous magnesium sulfate, and concentrated under reduced pressure with a rotary evaporator. The residual liquid is distilled under reduced pressure affording 10.5–10.8 g (78–82%) of cyclododecene as a colorless liquid, bp 92–98°C (4.1 mm) (Note 8), 92% *trans* and 8% *cis* by analysis on a ¼-in. × 6-ft GLC column packed with 10% $AgNO_3$ and 5% Carbowax 20 M on 80–100 mesh Chromosorb W at 110°C.

2. Notes

1. Reagent-grade tetrahydrofuran was freshly distilled from sodium and benzo-phenone and maintained under nitrogen. In small-scale experiments (1 mmol of tungsten hexachloride in 10 mL of solvent) anhydrous ether was equally effective, but did not give a homogeneous reaction solution.
2. Tungsten hexachloride, purchased from Pressure Chemical Company, was used without further purification. The dark blue-black crystals were pulverized in a dry

box or glove bag under a nitrogen atmosphere prior to use. Upon exposure to air or moisture yellow or orange oxides form. Slight contamination from these products does not interfere with the deoxygenation.

3. A suspension of 90% butyllithium in hexane was purchased from Alfa Products, Morton Thiokol, Inc. and was not standardized. The suspension was shaken to obtain uniform density before it was taken up into the syringe. For smaller-scale reactions the submitters report that 15% (1.6 M) butyllithium in hexane or methyllithium in ether is convenient and effective.

4. *trans*-Cyclododecene oxide, purchased from Aldrich Chemical Company, Inc., was used without further purification. The purity of the cyclododecene oxide sold by Aldrich varies, but it is usually > 95% trans; in this case it was 98% trans and 2% cis by analysis on a ⅛-in. × 6-ft. GLC column packed with 3% OV-17 on 80–100-mesh Gas-Chrom Q at 140°C. It is not necessary to wait until the brown solution becomes green before adding the epoxide. After epoxide addition the solution is dark green and appears homogeneous.

5. Molar ratios of tungsten reagent to epoxide of less than 1.5 : 1 resulted in incomplete reaction, while ratios greater than 3 : 1 did not improve, and in some cases actually diminished, the yield of alkene. Ratios of ca. 2 : 1 proved generally effective for a variety of epoxides. Molar ratios of alkyllithium to tungsten hexachloride of less than 2 : 1 also gave incomplete reaction; ratios of 3 : 1 or 4 : 1 are believed to give rise to different reduced tungsten species, which may be used in other reductions.

6. The reaction may be monitored by quenching small aliquots in aqueous 20% sodium hydroxide, extracting into hexane, and analyzing by gas chromatography.

7. Aqueous alkali alone, unless in huge excess, produces an emulsion. The addition of a chelating agent such as tartrate permits a clean separation of phases in a workup of reasonable dimension. A minimum of 6 mol of tartrate and 6 mol of hydroxide per mole of tungsten hexachloride used is adequate to suppress emulsions.

8. An IR spectrum of the product was identical to that of an authentic sample of *trans*-cyclododecene. The ^{1}H NMR spectrum of the product was as follows: δ 1.4 (m, 16 H); 2.2 (m, 4 H, $-CH_2-CH=CH-CH_2$); 5.5 (m, $-CH=CH-$).

3. Discussion

This procedure illustrates a general, one-step method to deoxygenate di- or trisubstituted epoxides to olefins in high yield and with high retention of stereochemistry.[2] Reductions are usually complete in less than 1 hr at room temperature or below. In certain cases yields and stereochemical retention have been maximized by using 3 equiv of alkyllithium for each equivalent of WCl$_6$, or by adding LiI.[2] The reagent is compatible with ethers and esters. It has been used to reductively couple aldehydes and ketones, but considerably longer reaction times and excess reagent are required for appreciable coupling.

Chlorohydrin salts are reduced by the reagent at elevated temperatures and extended reaction times, with complete loss of stereochemistry. Unlike the more highly substituted epoxides, terminal and unsubstituted cyclohexene epoxides appear to proceed, at least in part, via such intermediates, and must be refluxed for several hours to obtain olefins.

Epoxides have been converted to olefins stereoselectively and in good yield by preparation of the iodohydrins, which are then reduced with stannous chloride in the presence of phosphoryl chloride and pyridine.[3] A mild, stereoselective epoxide reduction can be achieved with sodium (cyclopentadienyl) dicarbonylferrate, after several in situ steps; however, the large steric demands of this reagent limit its use to terminal or very accessible epoxides.[4] Olefins of inverted stereochemistry have been obtained by the reaction of epoxides with lithium diphenylphosphide and methyl iodide, followed by cis elimination of the resulting betaine.[5] The reduced tungsten reagent complements these methods by reducing the more highly substituted epoxides with retention of stereochemistry. It should be especially useful when iodohydrin formation is sterically impeded or when conditions for the stereospecific iodohydrin reduction are objectionable.

Deslongchamps[6] and Masamune[7] have both encountered molecules in which the epoxide moiety was so severely shielded on the backside that any trans addition (e.g., iodohydrin formation) was inconceivable. Reduction with the tungsten reagent gave excellent yields of olefin in both cases. Parker employed the tungsten reagent to selectively reduce the trisubstituted epoxides of the trisepoxide of humulene, in effect functionalizing the least reactive double bond of the parent triene.[8] Masamune and Parker found that other standard reagents for epoxide reduction failed in these cases; Deslongchamps did not try other methods.

A variety of reducing metals,[3,9] chromous salts,[10] and lower valent iron[11] and titanium[12] salts convert epoxides to olefins in one step, but yields are usually low or moderate and stereochemistry is largely or completely lost. Routes involving thionocarbonates,[13] episulfides,[14] and episelenides[15] have also been used to convert epoxides to olefins. Epoxides activated by adjacent carbonyl, ester, or hydroxy groups have been reduced by special methods.[16]

1. Department of Chemistry, Massachusetts Institute of Technology, Cambridge, MA 02139.
2. Sharpless, K. B.; Umbreit, M. A.; Nieh, M. T.; Flood, T. C. *J. Am. Chem. Soc.* **1972**, *94*, 6538-6540.
3. Cornforth, J. W.; Cornforth, R. H.; Mathew, K. K. *J. Chem. Soc.* **1959**, 112-127.
4. Giering, W. P.; Rosenblum, M.; Tancrede, J. *J. Am. Chem. Soc.* **1972**, *94*, 7170-7172.
5. Vedejs, E.; Fuchs, P. L. *J. Am. Chem. Soc.* **1971**, *93*, 4070-4072.
6. Bélanger, A.; Berney, D. J. F.; Borschberg, H.-J.; Brousseau, R.; Doutheau, A.; Durand, R.; Katayama, H.; LaPalme, R.; Leturc, D. M.; Liao, C.-C.; MacLachlan, F. N.; Maffrand, J.-P.; Marazza, F.; Martino, R.; Moreau, C.; Saint-Laurent, L.; Saintonge, R.; Soucy, P. Ruest, L.; Deslongchamps, P. *Can. J. Chem.* **1979**, *57*, 3348-3354.
7. Masamune, S.; Massachusetts Institute of Technology, private communication; see also Suda, M. Ph.D. Dissertation, University of Alberta, 1975, p. 54; Souto-Bachillar, F. A. Ph.D. Dissertation, University of Alberta, 1978, p. 63.
8. Sattar, A.; Forrester, J.; Moir, M.; Roberts, J. S.; Parker, W. *Tetrahedron Lett.* **1976**, 1403-1406.
9. Kupchan, S. M.; Maruyama, M. *J. Org. Chem.* **1971**, *36*, 1187-1191; Bertini, F.; Grasselli, P.; Zubiani, G. *J. Chem. Soc. Chem. Commun.* **1970**, 144; Sharpless, K. B. *J. Chem. Soc. Chem. Commun.* **1970**, 1450-1451.
10. Kochi, J. K.; Singleton, D. M.; Andrews, L. J. *Tetrahedron* **1968**, *24*, 3503-3515.
11. Fugisawa, T.; Sugimoto, K.; Ohta, H. *Chem. Lett.* **1974**, 883-886.
12. McMurray, J. E.; Fleming, M. P. *J. Org. Chem.* **1975**, *40*, 2555-2556.

13. Corey, E. J.; Carey, F. A.; Winter, R. *J. Am. Chem. Soc.* **1965**, *87*, 934–935.
14. Schuetz, R. D.; Jacobs, R. L. *J. Org. Chem.* **1961**, *26*, 3467–3471.
15. Clive, D. L. J.; Denyer, C. *J. Chem. Soc. Chem. Commun.* **1973**, 253; Chan, T. H.; Finken-bine, J. R. *Tetrahedron Lett.* **1974**, 2091–2094; Behan, J. M.; Johnstone, R. A. W.; Wright, M. J. *J. Chem. Soc., Perkin Trans. 1* **1975**, 1216–1217; Krief, A.; VanEnde, D. *Tetrahedron Lett.* **1975**, 2709–2712.
16. Cole, W.; Julian, P. L. *J. Org. Chem.* **1954**, *19*, 131–138; Dowd, P.; Kang, K. *J. Chem. Soc. Chem. Commun.* **1974**, 384–385; Halsworth, A. S.; Henbest, H. B. *J. Chem. Soc.* **1960**, 3571–3575.

MERCAPTANS FROM THIOKETALS: CYCLODODECYL MERCAPTAN

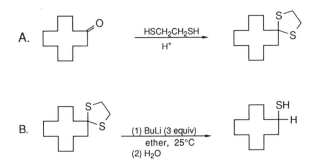

Submitted by S. R. WILSON[1] and G. M. GEORGIADIS
Checked by E. VEDEJS, P. C. CONRAD, and M. W. BECK

1. Procedure

Caution! This procedure should be carried out in an efficient hood to prevent exposure to alkane thiols.

A. *1,4-Dithiaspiro[4.11]hexadecane.* A mixture of 46.5 g (0.26 mol) of cyclodo-decanone (Note 1), 24.1 g (21.5 mL, 0.26 mol) of 1,2-ethanedithiol (Note 1), and 0.75 g (0.004 mol) of *p*-toluenesulfonic acid monohydrate (Note 2), in 200 mL of toluene (Note 3) is placed in a 500-mL, three-necked reaction flask equipped for reflux under a water separator.[2] The mixture is heated at reflux for several hours until the theoretical amount of water (0.26 mol = 4.6 mL) has collected in the Dean-Stark trap. The reaction mixture is cooled and transferred to a separatory funnel. The mixture is washed with water, the toluene is removed on a rotary evaporator, and the residue is placed under reduced pressure (<0.1 mm) for several hours to remove traces of solvent. Approximately 66 g (99%) of a white solid is recovered (0.26 mol, mp 84–86°C). The crude material is pure by GLC and TLC, and is used in the next step with no further purification.

B. *Cyclododecyl mercaptan.* In a 1-L, three-necked, round-bottomed flask equipped with a mechanical stirrer and nitrogen inlet and outlet stopcocks are placed 25.8 g (0.10 mol) of 1,4-dithiaspiro[4.11]hexadecane and 300 mL of ether, freshly distilled from sodium. The mixture is purged with nitrogen, cooled to 0°C with an ice bath, and 125 mL (0.30 mol, 2.4 M in hexane) of butyllithium is added by syringe (Notes 4, 5) under a slow flow of nitrogen. The light-yellow mixture is then allowed to warm to room temperature and stirred overnight with nitrogen stopcocks closed (Note 6). The reaction mixture is cooled to 0°C and 50 mL of water is added slowly and very carefully (Note 7). The resulting light brown solution is poured into 200 mL of water in a separatory funnel and, after shaking, the organic layer is separated. The solution is dried over $MgSO_4$, concentrated (aspirator), and distilled through a 10-cm Vigreux column at 103–108°C (1 mm) to give 17.2–17.9 g (86–90%) of pure cyclododecyl mercaptan (Notes 8, 9). A small forerun, bp < 95°C, (ca. 2 mL) is discarded.

2. Notes

1. The submitters used cyclododecanone and 1,2-ethanedithiol obtained from Aldrich Chemical Company, Inc.
2. The submitters used *p*-toluenesulfonic acid monohydrate from MCB, Inc.
3. The submitters used benzene in place of toluene.
4. The submitters used butyllithium from Alfa Products, Ventron Corporation.
5. The reaction also occurs well with only 2 mol of butyllithium, but traces of starting material remain.
6. The reaction is complete in about 6 hr.
7. *Caution! Quenching of excess butyllithium is exothermic.*
8. By GLC analysis, the distilled cyclododecyl mercaptan is >95% pure. Sometimes the product is pale pink.
9. The distilled cyclododecyl mercaptan has the following spectral data: ^{1}H NMR (CCl_4) δ: 1.1 (d, 1 H, J = 6, S-*H*), 1.32 (broad s, 20 H), 1.64–1.82 (m, 2 H), 2.81 (m, 1 H, C*H*SH); IR (neat, μ) 3.4, 6.82, 6.94. Anal. calcd. for $C_{12}H_{24}S$: C, 71.93; H, 12.07; S, 16.00. Found: C, 71.83; H, 12.19; S, 16.03.

3. Discussion

Mercaptans are generally prepared by displacement reactions.[3] However, secondary or hindered mercaptans are more difficult to obtain. The dithiolane cleavage reaction[4] is a convenient "in situ" generation of thioketones which are known to be reduced[5] with butyllithium to secondary mercaptans by β-hydrogen transfer. Table I shows a number of mercaptans prepared from *saturated* thioketals in 78–90% yields. The aryl example gives lower yields partly because of ring metalation.

TABLE I

MERCAPTANS FROM ETHYLENE THIOKETAL CLEAVAGE / REDUCTION

Ketone Thioketal	Bp/mp (°C)	Yield (%)
Cyclododecanone	103–108 (1 mm)	90
4-*tert*-Butylcyclohexanone	~ 100 (0.5 mm)	90[a]
2-Adamantanone	mp 139–142	79
4-Heptanone	127–135 (760)	81
Acetophenone	70–75 (0.5 mm)	36[b]
Cyclohexanone	130–140 (760)	78[c]
Estrone	mp 170–175	90[d]
Pregnenolone	mp 108–113	65[d]
Undecan-5-one	110–120 (0.3 mm)	93

[a]Axial: equatorial ratio, 2 : 1.
[b]By extraction into KOH (purity = 85–93%).
[c]Distillation could not cleanly separate thiol from octane (formed from the butyllithium).
[d]Mixture of isomers.

1. Department of Chemistry, Indiana University, Bloomington, IN 47405. Present address: Department of Chemistry, New York University, New York, NY 10003.
2. Jones, R. H.; Lukes, G. E.; Basher, J. T. U.S. Patent No. 2,690,988 (1954); *Chem. Abstr.* **1955,** *49,* 9868d.
3. Klayman, D. L.; Shine, R. J.; Bower, J. D. *J. Org. Chem.* **1972,** *37,* 1532.
4. Wilson, S. R.; Georgiadis, G. M.; Khatri, H. N.; Bartmess, J. E. *J. Am. Chem. Soc.* **1980,** *102,* 3577.
5. Rautenstrauch, V. *Helv. Chim. Acta* **1974,** *57,* 496; Ohno, A.; Yamabe, T.; Nagata, S. *Bull. Chem. Soc. Jpn.* **1975,** *48,* 3718.

EPOXIDATION OF OLEFINS BY HYDROGEN PEROXIDE–ACETONITRILE: *cis*-CYCLOOCTENE OXIDE

(*cis*-9-Oxabicyclo[6.1.0]nonane)

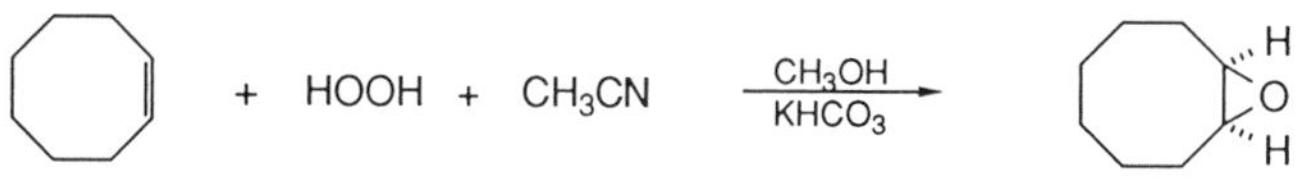

Submitted by R. D. BACH and J. W. KNIGHT[1]
Checked by K. W. FOWLER and G. BÜCHI

1. Procedure

Caution! Organic-soluble peroxides may be explosive (Note 4)!

In a 5-L, three-necked, round-bottomed flask fitted with a mechanical overhead stirrer, addition funnel, and thermometer are placed 484 g (4.4 mol) of *cis*-cyclooctene, 3 L of reagent methanol (Note 1), 330 g (8.04 mol) of acetonitrile, and 77 g (0.77 mol) of potassium bicarbonate (Note 2). To the resulting heterogeneous mixture is added dropwise 522 g (4.6 mol) of 30% hydrogen peroxide with cooling at a rate that maintains the temperature of the reaction at 25–35°C (Note 3). Following the addition of hydrogen peroxide, the ice bath is removed and the reaction mixture is allowed to stir at room temperature overnight. The reaction mixture is divided in half, and each portion is diluted with 500 mL of a saturated sodium chloride solution. Each portion is then extracted with four 500-mL portions of methylene chloride (Note 4). The organic phases are combined, dried over magnesium sulfate, and concentrated at reduced pressure by rotary evaporation. Short-path distillation of the crude product (Note 5) under reduced pressure gives 333–337 g (60–61%) of *cis*-cyclooctene oxide, bp 85–87°C (20 mm), as a white solid, mp 53–56°C (Note 6).

2. Notes

1. Omission of the methanol resulted in substantially reduced yields.
2. The reaction does not proceed well when sodium bicarbonate is used as the base.
3. The reaction is exothermic and caution should be exercised to keep the reaction temperature from rising. The time required for complete addition of the hydrogen peroxide is ca. 2–3 hr. The temperature is maintained at 25–35°C by employing an ice-water bath. When the hydrogen peroxide was added too rapidly, the reaction temperature rose until the solvents refluxed.
4. To check for organic-soluble peroxides, add several milliliters of the methylene chloride solution to a solution containing ca. 1 mg of sodium dichromate, 1 mL of water, and 1 drop of dilute sulfuric acid. A blue color in the organic layer is a positive test for perchromate ion. The checkers found that the combined organic phases exhibited a positive test and therefore stirred them overnight with a solution of 100 g of sodium metabisulfite in 500 mL of water prior to drying.
5. Heat from an IR lamp or heat gun must be applied to the condenser to keep the product from solidifying. The distillation pot should not be taken to dryness because of the possibility of the presence of organic peroxides.
6. The crude product may be used in many cases without further purification. Sublimation of the distilled oxirane affords the product as white needles, mp 56–57°C. The checkers obtained a broader melting point of the distillate, but the product was pure by analytical VPC.

3. Discussion

cis-Cyclooctene oxide has been prepared from *cis*-cyclooctene by the action of perbenzoic acid,[2] hydrogen peroxide,[3] molybdenum hexacarbonyl and *tert*-butyl hydroperoxide,[4] peracetic acid,[5] chromic acid,[6] and polymer-supported peracids.[7]

Oxiranes are typically formed by the action of a peracid such as *m*-chloroperbenzoic acid[8] on an alkene.[9] The present method has the advantage of being useful for both large- and small-scale reactions. The actual epoxidizing agent is generated in situ from the addition of hydrogen peroxide to a nitrile, forming a peroxyimidic acid.[10] This procedure is an adaptation of the method of Payne that utilized an intermediate peroxyimidic acid derived from the reaction of hydrogen peroxide with acetonitrile[11a] and benzonitrile.[11b] The alkaline hydrogen peroxide-benzonitrile system has more recently been used with steroids,[12] and in the total synthesis of prostaglandin $F_{2\alpha}$.[13] The present method does not require the separation of benzamide from the product. In addition, the reagents are inexpensive and the method is convenient and safe since it does not require large-scale preparation and handling of an organic peracid. Recently, it has been shown that substitution of trichloroacetonitrile for acetonitrile produces an even more reactive reagent.[14]

This epoxide has been found to be particularly useful in the laboratory in the large-scale preparation of *trans*-cyclooctene using the procedure of Whitham.[15] *trans*-Cyclooctane-1,2-diol is obtained from *cis*-cyclooctene oxide on treatment with sodium acetate in acetic acid and alkaline hydrolysis of the intermediate *trans*-2-acetoxycyclooctanol. The *trans* diol is converted to its benzaldehyde acetal, which on treatment with butyllithium affords *trans*-cyclooctene in a stereospecific manner.

1. Department of Chemistry, Wayne State University, Detroit, MI 48202.
2. Godchot, M.; Cauquil, G. *C. R.* **1931,** *192*, 962.
3. Treibs, W. *Chem. Ber.* **1939,** *72*, 7–10.
4. Sheng, M. N. *Synthesis*, **1972,** 194–195.
5. Cope, A. C.; Fenton, S. W.; Spencer, C. F. *J. Am. Chem. Soc.* **1952,** *74*, 5884–5888.
6. Cope, A. C.; Kinter, M. R.; Keller, R. T. *J. Am. Chem. Soc.* **1952,** *76*, 2757–2760.
7. Harrison, C. R.; Hodge, P. *J. Chem. Soc. Chem. Commun.* **1974,** 1009–1010; Harrison, C. R.; Hodge, P. *J. Chem. Soc., Perkin Trans. 1* **1976,** 605–609.
8. Fieser, M.; Fieser, L. F. "Reagents for Organic Synthesis"; Wiley-Interscience: New York, 1967; Vol. 1, p. 135.
9. Swern, D. *Org. React.* **1953,** *7*, 378–433; Harrison, I. T.; Harrison, S. "Compendium of Organic Synthetic Methods"; Wiley-Interscience: New York, 1971; Vol. 1, pp 325–326; Harrison, I. T.; Harrison, S. "Compendium of Organic Synthetic Methods"; Wiley-Interscience: New York, 1974; Vol. 2, pp 134–135.
10. Wiberg, K. B. *J. Am. Chem. Soc.* **1953,** *75*, 3961–3964.
11. Payne, G. B.; Deming, P. H.; Williams, P. H. *J. Org. Chem.* **1961,** *26*, 659–663; Payne, G. B. *Tetrahedron* **1962,** *18*, 763–765.
12. Ballantine, J. D.; Sykes, P. J. *J. Chem. Soc. C* **1970,** 731–735.
13. Woodward, R. B.; Gosteli, J.; Ernest, I.; Friary, R. J.; Nestler, G.; Raman, H.; Sitrin, R.; Suter, Ch.; Whitesell, J. K. *J. Am. Chem. Soc.* **1973,** *95*, 6853–6858.
14. Arias, L. A.; Adkins, S.; Nacel, C. S.; Bach, R. D. *J. Org. Chem.* **1983,** *48*, 888.
15. Hines, J. N.; Peagram, M. J.; Thomas, E. J.; Whitham, G. H. *J. Chem. Soc., Perkin Trans. 1* **1973,** 2332–2337.

CYCLOPROPANECARBOXALDEHYDE

A.

B.

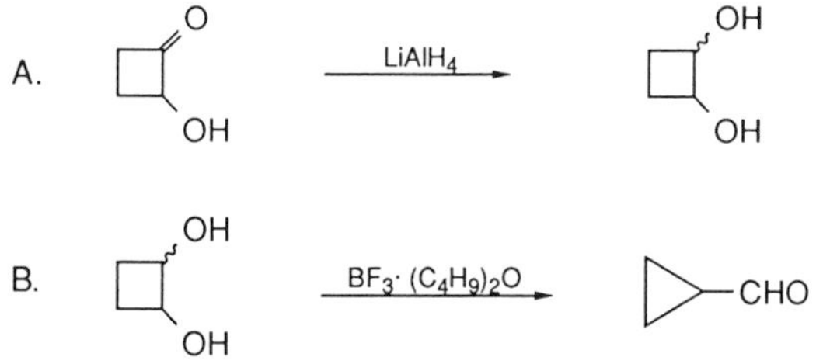

Submitted by J. P. BARNIER, J. CHAMPION, and J. M. CONIA[1]
Checked by ROBERT V. STEVENS and STEVEN R. ANGLE

1. Procedure

A. *1,2-Cyclobutanediol.* A 2-L, three-necked, round-bottomed flask fitted with a 200-mL dropping funnel, a mechanical stirrer, a nitrogen-inlet tube, and a reflux condenser equipped with a calcium chloride drying tube is charged with 6.2 g (0.16 mol) of lithium aluminum hydride (Notes 1 and 2) and 200 mL of anhydrous diethyl ether (Note 3). The dropping funnel is charged with 42 g (0.48 mol) of 2-hydroxycyclobutanone (Note 4) and 150 mL of anhydrous diethyl ether. While the suspension of lithium aluminum hydride is gently stirred under a nitrogen atmosphere, the solution of 2-hydroxycyclobutanone is added dropwise at a rate maintaining a gentle reflux. When the addition is complete, the mixture is heated at reflux for 1 hr. After the mixture has returned to room temperature, 200 mL of anhydrous diethyl ether is added. The gray reaction mixture is hydrolyzed by addition, in small parts, of a sufficient amount of wet sodium sulfate (Note 5). The reaction mixture is filtered through a sintered-glass funnel (porosity 3). The organic layer is decanted and dried over sodium sulfate. The solid is extracted with anhydrous tetrahydrofuran (Note 6) by means of a Soxhlet apparatus over a period of 24 hr. The combined organic layers are concentrated by distillation of the solvent with a rotary evaporator. The yield of crude *cis-* and *trans-*1,2-cyclobutanediol (ca. 50:50) is 34–40 g (80–95%) (Note 7).

B. *Cyclopropanecarboxaldehyde.* A 50-mL distilling flask equipped with a receiver cooled to −20°C with a dry ice–methanol bath is charged with 34 g (0.39 mol) of a crude mixture of both *cis-* and *trans-*1,2-cyclobutanediol and 10 μL of boron trifluoride butyl etherate (Note 8). The mixture is heated to 230°C with a metal bath. Drops of liquid appear on the condenser, and the aldehyde and water distill into the receiver. The temperature of the distillate oscillates between 50 and 100°C. Each time the distillation stops, 5–10 μL of boron trifluoride butyl etherate is added to the distilling flask (Note 9). The distillate is transferred into a separatory funnel and sodium chloride is added. The organic layer is decanted and the aqueous layer is extracted three times with 25-mL portions of methylene chloride. The combined organic solution is dried over sodium sulfate, and the solvent is removed by distillation through a 15-cm, helix-packed, vacuum-insulated column. The residue consists of practically pure cyclopropanecarbox-aldehyde, 17.5–21.6 g (65–80%) (Note 10).

2. Notes

1. Lithium aluminum hydride is available from Prolabo—France or Alfa Products, Morton Thiokol, Inc.

2. On one occasion the checkers found it necessary to add more lithium aluminum hydride (0.3 g) for complete reaction.

3. The checkers used diethyl ether distilled from sodium–benzophenone.

4. The checkers prepared 2-hydroxycyclobutanone by the Bloomfield procedure.[2] The submitters prepared it by their aqueous hydrolysis procedure.[3] This procedure was checked also and proceeds in quantitative yield as described.[3]

5. Sodium sulfate is mixed with water to form a thick slurry. It is added to the reaction mixture with vigorous stirring to obtain a good dispersion of the slurry in the medium. The added amount of wet sodium sulfate is sufficient when the reaction effervescence ceases and the gray color of the mixture turns to yellow–white.

6. Tetrahydrofuran is purified by distillation from lithium aluminum hydride after 48-hr refluxing over potassium hydroxide (see *Org. Synth. Coll. Vol. V* **1973,** 976 for hazard warning).

7. The crude 1,2-cyclobutanediol is dried by azeotropic distillation with anhydrous benzene. The product is a mixture of *cis* and *trans* isomers (ca. 50:50) readily separable by gas chromatography on a 12-ft column containing 20% silicone SE 30 on Chromosorb W at 140°C. *cis*-1,2-Cyclobutanediol (mp 12–13°C): IR (CCl$_4$) cm^{-1}: hydroxyl absorption at 3625 and 3580; ^{1}H NMR (CCl$_4$) δ: multiplet at 1.98, multiplet at 4.20, and a singlet at 4.51 in a ratio 4:2:2, respectively. *trans*-1,2-Cyclobutanediol (mp 72°C): IR (CCl$_4$) cm^{-1}: hydroxyl absorption at 3620: ^{1}H NMR (CD$_3$COCD$_3$) δ: multiplet between 0.9 and 2.2, multiplet between 3.6 and 4.0, and singlet at 3.6 in a ratio of 4:2:2, respectively.

8. Boron trifluoride butyl etherate, purchased from Fluka AG, is chosen for its convenient boiling point.

9. In a typical run 10 μL of boron trifluoride butyl etherate is added every 10–15 min over a period of 3–4 hr.

10. The crude product is more than 99% pure as shown by gas chromatography; IR (CCl$_4$) cm^{-1}: carbonyl absorption at 1730; ^{1}H NMR (CCl$_4$) δ: doublet at 8.93, multiplet between 1.5 and 2.2, and a multiplet between 1.02 and 1.75 in the ratio 1:1:4, respectively. The product has bp 95–98°C (760 mm).

3. Discussion

This method of preparation of cyclopropanecarboxaldehyde is an adaptation of that given by J. M. Conia and J. P. Barnier.[3] The various methods so far reported, which involve in the last step oxidation,[4] reduction,[5] or hydrolysis[6] of a suitable cyclopropane derivative, are tedious or require expensive starting materials. The other routes involve the direct ring contraction of cyclobutane derivatives into cyclopropanecarboxaldehyde starting from cyclobutene oxide[7] or from 2-bromo- or 2-tosyloxycyclobutanol.[8] The present procedure uses a particularly easy ring contraction, that of 1,2-cyclobutanediol, and it involves cheap, easily available starting materials. This method can be applied to symmetrical dialkylcyclobutane-1,2-diols, but it gives a mixture of two cyclopropyl carbonyl compounds from unsymmetrical diols.

1. Laboratoire des Carbocycles, Universite de Paris-Sud, 91405 ORSAY, France.
2. Bloomfield, J. J.; Nelke, J. M. *Org. Synth., Coll. Vol. VI* **1988**, 167.
3. Conia, J. M.; Barnier, J. P. *Tetrahedron Lett.* **1971**, 4981–4984.
4. Venus-Danilova, E. D.; Kazimirova, V. F. *Zh. Obshch. Khim.* **1938**, *8*, 1438; *Chem. Abstr.* **1939**, *33*, 4204[1]; Lee, C. C.; Bhardjaw, I. S. *Can. J. Chem.* **1963**, *41*, 1031–1033; Shuikina, Z. I. *Zh. Obshch, Khim.* **1937**, *7*, 933; *Chem. Abstr.* **1937**, *31*, 5332[3]; Brewster Young, L.; Trahanowsky, W. S. *J. Org. Chem.* **1967**, *32*, 2349–2350.
5. Smith, L. I.; Rogier, E. R. *J. Am. Chem. Soc.* **1951**, *73*, 4047–4049; Brown, H. C.; Garg, C. P. *J. Am. Chem. Soc.* **1964**, *86*, 1085–1089; Brown, H. C.; Subba Rao, B. C. *J. Am. Chem. Soc.* **1958**, *80*, 5377–5380; Brown H. C.; Tsukamoto, A. *J. Am. Chem. Soc.* **1961**, *83*, 4549–4552; Brown, H. C.; Tsukamoto, A. *J. Am. Chem. Soc.* **1964**, *86*, 1089–1095.
6. Meyers, A. I.; Adickes, H. W.; Politzer, I. R.; Beverung, W. N. *J. Am. Chem. Soc.* **1969**, *91*, 765–767.
7. Ripoll, J. L.; Conia, J. M. *Bull. Soc. Chim. Fr.* **1965**, 2755–2762.
8. Salaun, J.; Conia, J. M. *J. Chem. Soc. Chem. Commun.* **1970**, 1358–1359.

CYCLOPROPANONE ETHYL HEMIACETAL FROM ETHYL 3-CHLOROPROPANOATE

(Cyclopropanol, 1-ethoxy-)

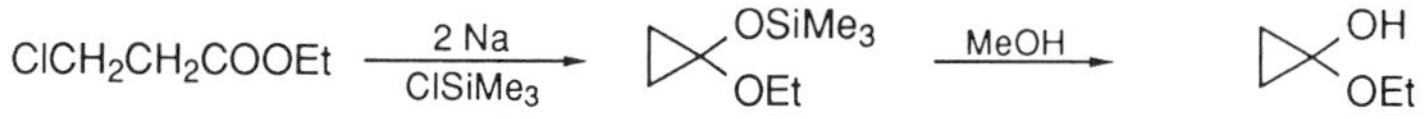

Submitted by J. SALAÜN and J. MARGUERITE[1]
Checked by STEVEN D. YOUNG, SYUN-ICHI KIYOOKA, and CLAYTON H. HEATHCOCK

1. Procedure

A. *1-Ethoxy-1-(trimethylsilyloxy)cyclopropane.* A 1-L, three-necked, round-bottomed flask is fitted with an efficient mechanical stirrer (Note 1), a reflux condenser provided with a calcium chloride tube, and a 500-mL pressure-equalizing dropping funnel equipped with a nitrogen inlet at the top. The flask is flushed with dry nitrogen, and 500 mL of anhydrous toluene (Note 2) and 52.9 g (2.3 g-atom) of sodium cut in small pieces (Note 3) are introduced. The mixture is brought to reflux by means of a heating mantle and the sodium is finely pulverized by vigorous stirring. Heating and stirring are stopped (Note 4), and the mixture is allowed to cool to room temperature. Toluene is removed under nitrogen pressure by means of a double-ended needle and replaced by 500 mL of anhydrous diethyl ether (Notes 5 and 6). At this point, 108.5 g (1 mol) of chlorotrimethylsilane (Note 7) is added to the flask. To the mixture, 136.58 g (1 mol) of ethyl 3-chloropropanoate is added dropwise with stirring at a rate sufficient to maintain a gentle reflux over a 3-hr period (Note 8). When about 0.3 mol of chloro ester has been added, a deep-blue precipitate appears (Note 9). When the addition is over, the reaction mixture is heated at reflux for 30 min. The contents of the flask are cooled and filtered through a sintered-glass funnel under a stream of dry nitrogen (Note 10). The precipitate is washed twice with 100 mL of anhydrous diethyl ether.

The colorless filtrate is transferred to a distilling flask and the solvent is distilled through a 25-cm vacuum-jacketed Vigreux column, and the residue is distilled under reduced pressure. After a small forerun (1–2 g), 1-ethoxy-1-(trimethylsilyloxy)cyclopropane is obtained at 43–45°C (12 mm) as a colorless liquid, 106 g (61%) (Note 11).

B. *Cyclopropanone ethyl hemiacetal.* Into a 500-mL Erlenmeyer flask fitted with a magnetic stirring bar is placed 250 mL of reagent-grade methanol. Freshly distilled 1-ethoxy-1-(trimethylsilyloxy)cyclopropane (100 g, 0.56 mol) is added all at once to the methanol and the solution is stirred overnight (12 hr) at room temperature (Note 12). An aliquot (50 mL) of the solution is concentrated by slow evaporation of methanol with a rotary evaporator at room temperature (Note 13) and formation of the methanolysis product is checked by NMR examination of the residue (Note 14). When the reaction is complete (Note 15), the solution is concentrated by removal of the methanol (Note 16). Distillation of the residue through a 20-cm helix-packed, vacuum-insulated column under reduced pressure gives 52 g (89%) of 1-ethoxycyclopropanol, bp 60°C (20 mm) (Notes 14 and 17), which contains trace amounts of 1-methoxycyclopropanol (Notes 18 and 19).

2. Notes

1. An efficient stirrer is used at a spinning rate sufficient to disperse the molten sodium into small beads of a diameter of approximately 0.1 mm. The checkers found it necessary to use a mechanical stirrer equipped with a nichrome wire "beater" rather than a Teflon paddle. If the sodium sand particles are too large, the final product will be contaminated with starting chloro ester, from which it is very difficult to separate.

2. Toluene is freshly distilled from phosphorus pentoxide into the reaction flask.

3. Sodium pieces are washed in dry pentane or toluene to remove oil.

4. It is essential that stirring be discontinued before cooling is begun to prevent the molten sodium from coalescing into one gigantic lump.

5. Diethyl ether is dried by molecular sieves and distilled from lithium aluminum hydride.

6. To remove the toluene completely, the finely divided sodium is washed under nitrogen with anhydrous diethyl ether (3 × 50 mL).

7. Chlorotrimethylsilane, obtained from Aldrich Chemical Co. or Prolabo (France), is distilled from quinoline or calcium hydride.

8. For the acyloin condensation of diesters it has been recommended that the diester and chlorotrimethylsilane be added together to the sodium dispersion;[2] no difference has been noted with our procedure.

9. The deep-blue color seems to be indicative of a satisfactory reduction. When the color is yellow–green, the yield is usually poor.

10. *Caution! Because of the pyrophoric nature of finely divided alkali metal residues or production of free acid (HCl) from the chlorosilane, the products are sensitive to moisture. Unreacted sodium is destroyed by careful addition of ethanol to the residual solid.*

11. The yield varies from 60 to 85%, bp 50–52°C (18 mm); 60–62°C (35 mm); 66–68°C (40 mm); the proton magnetic resonance (PMR) spectrum (CCl_4 solution, $HCCl_3$ external reference) shows absorption at δ: 0.08 (s, 9 H), 0.70 (m, 4 H), 1.05 (t, 3 H, $J = 7.11$) and 3.55 (q, 2 H, $J = 7.11$); the IR spectrum (CCl_4) exhibits absorption at 3090 and 3010 (cyclopropane), 1250, 845, and 758 cm^{-1} ($-Si[CH_3]_3$).

12. After the solution is stirred for 5–10 min, the clear solution becomes slightly turbid for a few minutes and then turns clear again. When these changes are not observed, methanolysis has not occurred.

13. If some 1-ethoxy-1-(trimethylsilyloxy)cyclopropane is still present, it will be lost by too rapid evaporation of methanol.

14. The product has the following spectral properties: IR (CCl_4): 3600 and 3400 (hydroxyl), 3010 and 3090 cm^{-1} (cyclopropyl); ^{1}H NMR (CCl_4) δ: 0.84 (s, 4 H), 1.18 (t, 3 H, $J = 7.11$), 3.73 (q, 2 H, $J = 7.11$) and 4.75 (m, 1 H).

15. Lack of NMR absorption around δ 0.08 shows that the trimethylsilyloxy group has been completely removed.

16. If the reaction is not complete, as shown by the presence of a singlet around δ 0.08, a spatula tip full of pyridinium p-toluenesulfonate[3] is added and the mixture is stirred for 4 hr. Methanol is then removed, and the residue is dissolved in 200 mL of diethyl ether. The solution is washed with saturated sodium chloride until neutral, dried over anhydrous sodium sulfate, and concentrated. Addition of a drop of HCl or of chlorotrimethylsilane is also effective to complete the reaction. Then, the hydrochloric acid is removed with methanol. (Thus, it is not necessary to wash with saturated sodium chloride until neutral.)

17. The yield varies from 78 to 95%, bp 51°C (12 mm), 64°C (25 mm), 75°C (46 mm).

18. On standing with methanol at 25°C for 1 week, 65% of 1-ethoxycyclopropanol is converted into 1-methoxycyclopropanol; conversion appears to be complete after 15 days.[4] The spectral properties of the 1-methoxycyclopropanol are: IR (CCl_4): 3600 and 3400 (hydroxyl), 3010 and 3090 cm^{-1} (cyclopropyl); ^{1}H NMR (CCl_4) δ: 0.85 (s, 4 H) and 3.40 (s, 3 H).

19. Cyclopropanone hemiacetal can be kept unaltered for several months at 0°C in the refrigerator. On heating above 100°C or on standing in acidic solvents, it undergoes ring opening to give ethyl propionate.

3. Discussion

Cyclopropanone ethyl hemiacetal was first synthesized by the reaction of ketene and diazomethane in ether at $-78\,^{\circ}C$ in the presence of ethanol.[4] The yield is low (43%) and the reaction is hazardous, especially when a large-scale reaction is required. The method described in this procedure for the preparation of cyclopropanone ethyl hemiacetal from ethyl 3-chloropropanoate is an adaptation of that described previously;[5] the procedure described for the synthesis of 1-ethoxy-1-(trimethylsilyloxy)cyclopropane is patterned after the method reported by Rühlmann.[6]

Cyclopropanone ethyl hemiacetal is a molecule of considerable interest since its reactions appear to involve the formation of the labile cyclopropanone.[7] It readily undergoes nucleophilic addition of Grignard reagents,[4,5] azides,[4] and amines[8] to provide 1-substituted cyclopropanols in high yields. It has been reported that upon treatment with an equimolar amount of methylmagnesium iodide, the cyclopropanone ethyl hemiacetal is converted into iodomagnesium 1-ethoxycyclopropylate,[9] which can react with hydrides, organometallic reagents, cyanide carbanion, and phosphorus ylides[10] to provide useful synthons. The preparation of some challenging 2,3-disubstituted cyclopentanones including a total synthesis of the 11-deoxyprostaglandin has been reported from the cyclopropanone hemiacetal.[11] The ready availability of this compound should lead to other synthetic applications. For a recent review dealing with the chemistry of the cyclopropanone hemiacetals, see reference 12.

On the other hand, silylated cyclopropanols, such as 1-ethoxy 1-(trimethylsilyloxy)cyclopropane, work well as homoenolate anion precursors. They undergo ring opening reactions with a variety of metal halides ($TiCl_4$, $GaCl_3$, $SbCl_5$, $ZnCl_2$, $HgCl_2$ . . .). Thus, in the presence of suitable catalysts, the zinc homoenolates of alkyl propionates undergo a variety of carbon-carbon bond forming reactions with a very high degree of chemoselectivity.[13]

1. Laboratoire des Carbocycles, ERA n° 316, Bat. 420, Université de Paris-Sud, 91405 Orsay, France.
2. Bloomfield, J. J.; Owsley, D. C.; Nelke, J. M. *Org. React.* **1976**, *23*, 306; Bloomfield, J. J.; Nelke, J. M. *Org. Synth., Coll. Vol. VI* **1988**, 167.
3. Miyashita, M.; Yoshikoshi, A.; Grieco, P. A. *J. Org. Chem.* **1977**, *42*, 3772.
4. Wasserman, H. H.; Clagett, D. C. *J. Am. Chem. Soc.* **1966**, *88*, 5368; Wasserman, H. H.; Cochoy, R. E.; Baird, M. S. *J. Am. Chem. Soc.* **1969**, *91*, 2375.
5. Salaün, J. *J. Org. Chem.* **1976**, *41*, 1237; **1977**, *42*, 28.
6. Rühlmann, K. *Synthesis* **1971**, 236.
7. Turro, N. J. *Acc. Chem. Res.* **1969**, *2*, 25; Shaafsma, S. E.; Steinberg, H.; de Boer, T. J. *Recl. Trav. Chim. Pays-Bas* **1966**, *85*, 1170.
8. Wasserman, H. H.; Adickes, H. W.; de Ochoa, D. Espejo *J. Am. Chem. Soc.* **1971**, *93*, 5586.
9. Brown, H. C.; Rao, C. G. *J. Org. Chem.* **1978**, *43*, 3602.
10. Salaün, J.; Fadel, A. *Tetrahedron Lett.* **1979**, 4375; Salaün, J.; Bennani, F.; Compain, J. C.; Fadel, A.; Ollivier, J. *J. Org. Chem.* **1980**, *45*, 4129.
11. Salaün, J.; Ollivier, J. *Nouv. J. Chim.* **1981**, *5*, 587.
12. Salaün, J. *Chem. Rev.* **1983**, *83*, 619; Salaün, J. *Top. Curr. Chem.* **1988**, *144*, 1.
13. Nakamura, E.; Aoki, S.; Sekiya, K.; Oshino, H.; Kuwajima, I. *J. Am. Chem. Soc.* **1987**, *109*, 8056 and references cited therein.

CYCLOUNDECANECARBOXYLIC ACID

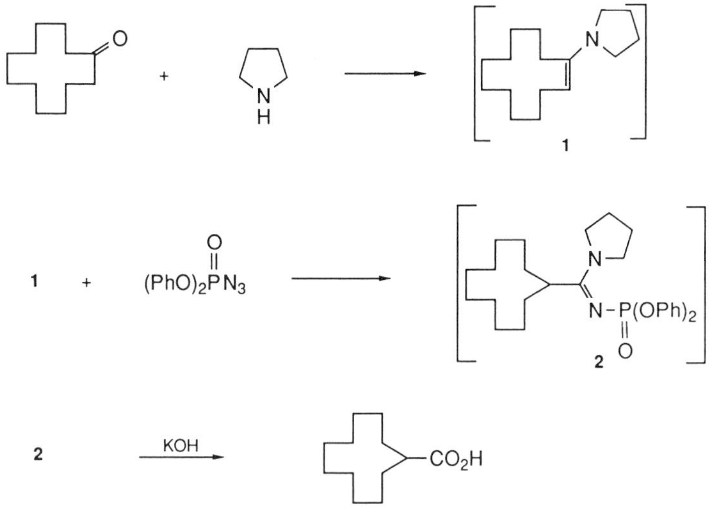

Submitted by Yasumasa Hamada and Takayuki Shioiri[1]
Checked by M. F. Semmelhack and E. Spiess

1. Procedure

To a 300-mL, round-bottomed flask fitted with a water separator (Note 1) that contains 15 g of Linde 4A molecular sieve $1/16$-in. pellets and is filled with toluene, are added 7.3 g (0.04 mol) of cyclododecanone, 11.4 g (0.16 mol) of pyrrolidine, 100 mL of toluene, and 0.57 g (0.004 mol) of boron trifluoride etherate. The solution is heated under reflux for 20 hr. The water separator is replaced by a distillation head, and about 90 mL of the toluene is removed by distillation at atmospheric pressure. The residue containing 1-(N-pyrrolidino)-1-cyclododecene (**1**) is used in the next step without further purification (Note 2).

The crude enamine (**1**) is dissolved in 20 mL of toluene, and the solution is transferred (Note 3) to a 100-mL, three-necked flask equipped with a magnetic stirring bar, a 50-mL dropping funnel, reflux condenser protected with a calcium chloride tube, and a thermometer immersed in the solution. A solution of 13.2 g (0.048 mol) of diphenyl phosphorazidate (Note 4; *Warning*) in 20 mL of toluene is added with stirring during 30 min while the reaction temperature is maintained at about 25°C. The mixture is stirred for 4 hr at 25°C and heated at reflux for 1 hr. The mixture is transferred to a 300-mL, round-bottomed flask and most of the toluene is removed under reduced pressure to yield 23.7 g of a reddish-brown oil, **2** (Note 5).

Ethylene glycol (200 mL) and 40 g (0.71 mol) of potassium hydroxide are added to the residual oil. The mixture is heated at reflux for 24 hr and then concentrated at 80–115°C (25 mm) (bath temperature ca. 190°C) until 100 mL of distillate is collected.

The residue is dissolved in 300 mL of water and cooled to room temperature. Carbon dioxide is introduced as a gas until the pH of the solution reaches 9. The mixture is washed with three 80-mL portions of diethyl ether (Note 6). The aqueous layer is acidified with about 53 mL of concentrated hydrochloric acid and extracted with four 80-mL portions of benzene. The combined benzene extracts are washed with 50 mL of water and dried over anhydrous sodium sulfate. The solvent is removed under reduced pressure to give 4.5–5.5 g of a black–brown oil. Distillation of the oil at 110–115°C (0.1 mm) yields 3.5–3.8 g (40–48%) of cycloundecanecarboxylic acid as a colorless oil.

2. Notes

1. The apparatus described in *Organic Syntheses*[2] is satisfactory.
2. Pure 1-(*N*-pyrrolidino)-1-cyclododecene, bp 144°C (1.5 mm), may be isolated by distillation through a Vigreux column.
3. The original flask used for the enamine formation can be used after the attachment of a Y-shaped tube fitted with a dropping funnel and a reflux condenser protected with a tube packed with a drying agent such as anhydrous calcium chloride.
4. Diphenyl phosphorazidate is prepared by the action of sodium azide with diphenyl phosphorochloridate[3]. It is also available from Aldrich Chemical Co. and was used after purification by distillation at 134–136°C (0.2 mm). *Warning: Diphenyl phosphorazidate may produce explosive hydrogen azide when it is in contact with moisture for a long time. When diphenyl phosphorazidate, which has been stored for a long time, is used, it should be washed with saturated aqueous sodium bicarbonate and dried over sodium sulfate before distillation.*
5. Purification of 1 g of the crude oil was made by column chromatography using 50 g of Merck silica gel with 0.063–0.200-mm particles (catalog No. 7734) in a column 2.2-cm × 40-cm and 1 : 1 (v/v) ethyl acetate-hexane as eluant to give pure diphenyl (cycloundecyl-1-pyrrolidinylmethylene)phosphoramidate (2) as a colorless oil, 632 mg (78%). When a Merck precoated silica gel F254 thin layer plate, layer thickness 0.25 mm, is developed with 1 : 1 (v/v) ethyl acetate-hexane and visualized with ultraviolet light, the phosphoramidate appears at R_f 0.3. Thus the crude oil contained about 15 g of the phosphoramidate.
6. This procedure is designed primarily to remove phenol.

3. Discussion

Cycloundecanecarboxylic acid has been prepared by the bromination of cyclododecanone followed by the Favorskii rearrangement of 2-bromocyclododecanone.[4]

The present preparation illustrates a general and convenient method for ring contraction of cyclic ketones.[5] The first step is the usual procedure for the preparation of enamines. The second step involves 1,3-dipolar cycloaddition of diphenyl phosphorazidate to an enamine followed by ring contraction with evolution of nitrogen. Ethyl acetate and tetrahydrofuran can be used as a solvent in place of toluene. Pyrrolidine enamines from various cyclic ketones smoothly undergo the reaction under similar reaction conditions. Diphenyl (cycloalkyl-1-pyrrolidinylmethylene)phosphoramidates with 5, 6, 7, and 15 members in the ring have been prepared in yields of 68–76%.

The third step is hydrolysis of the *N*-phosphorylated amidines, which is carried out by either acid or alkali depending on the substrate.

Similar reaction sequences can be used successfully to convert alkyl aryl ketones to α-arylalkanoic acids.[6]

1. Faculty of Pharmaceutical Sciences, Nagoya City University, Nagoya 467, Japan.
2. Natelson, S.; Gottfried, S. *Org. Synth., Coll. Vol. III* **1955**, 381.
3. Shioiri, T.; Yamada, S. *Org. Synth., Coll. Vol VII* **1990**, 206.
4. Ziegenbein, W. *Chem. Ber.* **1961**, *94*, 2989.
5. Yamada, S.; Hamada, Y.; Ninomiya, K.; Shioiri, T. *Tetrahedron Lett.* **1976**, 4749.
6. Shioiri, T.; Kawai, N. *J. Org. Chem.* **1978**, *43*, 2936.

A GENERAL SYNTHETIC METHOD FOR THE PREPARATION OF METHYL KETONES FROM TERMINAL OLEFINS: 2-DECANONE

$$C_8H_{17}CH=CH_2 \xrightarrow[CuCl,\,O_2]{PdCl_2} C_8H_{17}\overset{\overset{\displaystyle O}{\|}}{C}CH_3$$

Submitted by JIRO TSUJI, HIDEO NAGASHIMA, and HISAO NEMOTO[1]
Checked by EDWIN VEDEJS, J. GEGNER, and T. K. MALLMAN

1. Procedure

A 100-mL, three-necked, round-bottomed flask is fitted with a magnetic stirrer and a pressure-equalizing dropping funnel containing 1-decene (4.2 g, 30 mmol). The flask is charged with a mixture of palladium chloride (0.53 g, 3 mmol), cuprous chloride (2.97 g, 30 mmol) (Note 1), and aqueous dimethylformamide (DMF/H_2O = 7 : 1, 24 mL). With the other outlets securely stoppered and wired down, an oxygen-filled balloon (Note 2) is placed over one neck, and the flask is stirred at room temperature to allow oxygen uptake (Note 3). After 1 hr, 1-decene (4.2 g, 30 mmol) (Note 4) is added over 10 min (Note 5) using the dropping funnel, and the solution is stirred vigorously at room temperature under an oxygen balloon (Note 6). The color of the solution turns from green to black within 15 min and returns gradually to green. After 24 hr, the mixture is poured into cold 3 *N* hydrochloric acid (100 mL) and extracted with five 50-mL portions of ether. The extracts are combined and washed successively with 50 mL of saturated sodium bicarbonate solution, 50 mL of brine, and then dried over anhydrous magnesium sulfate. After filtration, the solvent is removed by evaporation and the residue is distilled using a 15-cm Vigreux column to give 2-decanone as a colorless oil (3.0–3.4 g, 65–73%, bp 43–50°C/1 mm (Notes 7 and 8).

2. Notes

1. Cupric chloride can be used, but it tends to chlorinate the products and cuprous chloride is preferable; reagent-grade dimethylformamide (DMF) was distilled before use.

2. The balloon was purchased at a toy shop; its inflated volume was approximately 500 mL.

3. The initial black solution gradually turns green by oxygen absorption.

4. The sample of 1-decene was obtained from the Aldrich Chemical Company, Inc. and distilled before use.

5. In cases where the alkene is soluble, up to 30% of the aqueous DMF can be mixed with the alkene to facilitate controlled addition. With 1-decene, DMF forms a two-phase mixture.

6. The reaction is slightly exothermic.

7. The first fraction (bp 30-40°C) contains decenes that are formed by palladium-catalyzed isomerization of 1-decene (indicated by a broad signal at δ 5.2-5.5 in the ^{1}H NMR spectrum).

8. The spectral properties of 2-decanone are as follows: ^{1}H NMR (CCl$_4$) δ: 0.7-1.8 (15 H, complex), 2.02 (3 H, s), 2.37 (2 H, t, J = 7); IR (neat) 1722 cm^{-1}.

3. Discussion

Methyl ketones are important intermediates for the synthesis of methyl alkyl carbinols, annulation reagents, and cyclic compounds. A common synthetic method for the preparation of methyl ketones is the alkylation of acetone derivatives, but the method suffers limitations such as low yields and lack of regioselectivity. Preparation of methyl ketones from olefins and acetylenes using mercury compounds is a better method. For example, hydration of terminal acetylenes using HgSO$_4$ gives methyl ketones cleanly.[2] Oxymercuration of 1-olefins and subsequent oxidation with chromic oxide is another method.[3] Preparation of an epoxide from a 1-olefin and its rearrangement catalyzed by a cobalt catalyst to give methyl ketones has been reported briefly.[4]

Compared with these methods, the palladium-catalyzed oxidation of 1-olefins described here is more convenient and practical. The industrial method of ethylene oxidation to acetaldehyde using PdCl$_2$–CuCl$_2$–O$_2$ is the original reaction of this type.[5] The oxidation of various olefins has been carried out.[6-9]

Use of DMF as a solvent for the oxidation of 1-olefins has been reported by Clement and Selwitz.[6] The method requires only a catalytic amount of PdCl$_2$ and gives satisfactory yields under mild conditions. A small amount of olefin migration product is the only noticeable contaminant in the cases reported. The procedure can be applied satisfactorily to various 1-olefins with other functional groups. This useful synthetic method for the preparation of methyl ketones has been applied extensively in the syntheses of natural products such as steroids,[10] macrolides,[11,12] dihydrojasmone,[13] and muscone.[14] A comprehensive review article on the palladium-catalyzed oxidation of olefins has been published.[15]

1. Department of Chemical Engineering, Tokyo Institute of Technology, Meguro, Tokyo 152, Japan.
2. Thomas, R. J.; Campbell, K. N.; Hennion, G. F. *J. Am. Chem. Soc.* **1938**, *60*, 718.
3. Rogers, H. R.; McDermott, J. X.; Whitesides, G. M. *J. Org. Chem.* **1975**, *40*, 3577.
4. Eisenmann, J. L. *J. Org. Chem.* **1962**, *27*, 2706.
5. Smidt, J.; Hafner, W.; Jira, R.; Sedlmeier, J.; Sieber, R.; Ruttinger, R.; Kojer, H. *Angew. Chem.* **1959**, *71*, 176.

6. Clement, W. H.; Selwitz, C. M. *J. Org. Chem.* **1964,** *29,* 241.
7. Fahey, D. R.; Zuech, E. A. *J. Org. Chem.* **1974,** *39,* 3276.
8. Lloyd, W. G.; Luberoff, B. J. *J. Org. Chem.* **1969,** *34,* 3949.
9. Tsuji, J.; Shimizu, I.; Yamamoto, K. *Tetrahedron Lett.* **1976,** 2975.
10. Tsuji, J.; Shimizu, I.; Suzuki, H.; Naito, Y. *J. Am. Chem. Soc.,* **1979,** *101,* 5070.
11. Takahashi, T.; Kasuga, K.; Takahashi, M.; Tsuji, J. *J. Am. Chem. Soc.* **1979,** *101,* 5072.
12. Takahashi, T.; Hashiguchi, S.; Kasuga, K.; Tsuji, J. *J. Am. Chem. Soc.* **1978,** *100,* 7424.
13. Tsuji, J.; Kobayashi, Y.; Shimizu, I. *Tetrahedron Lett.* **1979,** 39.
14. Tsuji, J.; Yamakawa, T.; Mandai, T. *Tetrahedron Lett.* **1979,** 3741.
15. Tsuji, J. *Synthesis* **1984,** 369.

DEOXYGENATION OF SECONDARY ALCOHOLS: 3-DEOXY-1,2 : 5,6-DI-*O*-ISOPROPYLIDENE-α-D-*ribo*-HEXOFURANOSE

[α-D-*ribo*-Hexofuranose, 3-deoxy-1,2 : 5,6-bis-*O*-(1-methylethylidene)-]

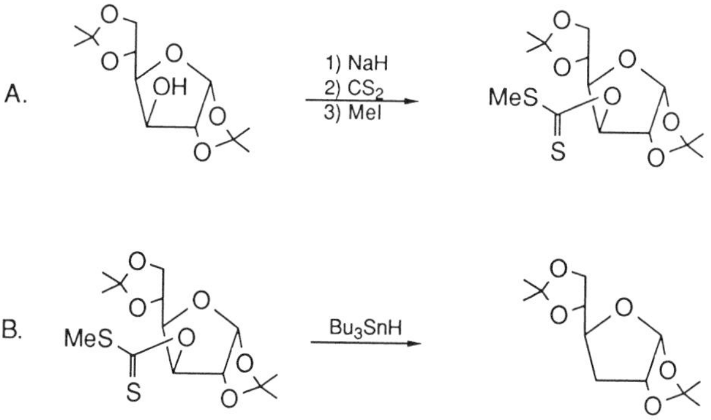

Submitted by S. IACONO and JAMES R. RASMUSSEN[1]
Checked by PETER J. CARD and BRUCE E. SMART

1. Procedure

Caution! Carbon disulfide, iodomethane, and tributyltin hydride are poisonous and should be handled in a well-ventilated hood.

A. *1,2 : 5,6-Di-O-isopropylidene-3-O-(S-methyldithiocarbonate)-α-D-glucofuranose.* A 1-L, three-necked, round-bottomed flask equipped with a magnetic stirring bar, nitrogen-inlet adapter, pressure-equalizing addition funnel, and stopper is charged with 26.0 g (0.10 mol) of 1,2 : 5,6-di-*O*-isopropylidene-α-D-glucofuranose, 25 mg of imidazole (Note 1), and 400 mL of anhydrous tetrahydrofuran (Note 2). The reaction vessel is flushed with nitrogen and a nitrogen atmosphere is maintained during the ensuing steps. Over a 5-min period, 7.2 g (0.150 mol) of a 50% sodium hydride dispersion (Note 3) is added. Vigorous gas evolution is observed. After the reaction mixture is stirred for 20 min, 22.8 g (0.30 mol) of carbon disulfide is added all at once. Stirring is continued for 30 min, after which time 25.3 g (0.177 mol) of iodomethane is added in a single portion. The reaction mixture is stirred for another 15 min, and 5.0 mL of glacial acetic acid is added dropwise to destroy excess sodium hydride. The solution is filtered (Note 4) and the filtrate is concentrated on a rotary evaporator. The semisolid residue is extracted with three 100-mL portions of ether, and the combined ether extracts are washed with two 100-mL portions of saturated sodium bicarbonate solution and two 100-mL portions of water. The ethereal solution is dried over anhydrous magnesium sulfate, the drying agent is removed by filtration, and the solvent is removed by rotary evaporation. The product is dried further at 0.05 mm overnight. The resulting orange syrup is distilled (Kugelrohr) to give 32.2–33.0 g (92–94%) of product, bp 153–160°C (0.5–1.0 mm) (Note 5).

B. *3-Deoxy-1,2 : 5,6-di-O-isopropylidene-α-D-ribo-hexofuranose.* A   dry,   1-L, round-bottomed flask is equipped with a magnetic stirring bar and a reflux condenser to which a nitrogen inlet is attached. The apparatus is charged with 500 mL of anhydrous toluene (Note 6), 24.7 g (0.085 mol) of tributyltin hydride (Note 7) and 19.25 g (0.055 mol) of 1,2 : 5,6-di-O-isopropylidene-3-O-(S-methyldithiocarbonate)-α-D-glucofura-nose. The reaction mixture is heated at reflux under a nitrogen atmosphere until TLC analysis indicates the disappearance of starting materials (4–7 hr) (Note 8). During this time the reaction solution changes from deep yellow to nearly colorless. The toluene is removed on a rotary evaporator to yield a thick, oily residue that is partitioned between 250-mL portions of petroleum ether and acetonitrile. The acetonitrile layer is separated and washed with three 100-mL portions of petroleum ether and is then concentrated on a rotary evaporator. The residual yellow oil is taken up in hexane–ethyl acetate (10 : 1) and filtered through a pad of silica gel (Note 9). The filtrate is concentrated and the residual oil is distilled to give 10.0 g (75%) of product as a colorless syrup, bp 72–73°C (0.2 mm); n_D^{25} 1.4474 (Note 10).

2. Notes

1. 1,2 : 5,6-Di-O-isopropylidene-α-D-glucofuranose and imidazole were purchased from Aldrich Chemical Company, Inc. and used without further purification. Alternatively, the glucofuranose starting material can be prepared by standard methods from D-glucose.[2]

2. Reagent-grade tetrahydrofuran was freshly distilled from a purple solution of sodium and benzophenone.

3. Sodium hydride, a 50% dispersion in mineral oil, was purchased from Alfa Products, Morton Thiokol, Inc. It is not necessary to remove the mineral oil before conducting the reaction.

4. The collected salts should be disposed of carefully by first rinsing with isopropyl alcohol to ensure that no sodium hydride remains.

5. The submitters report pure product with bp 135–136°C (0.07 mm). The material obtained by the checkers is pure by NMR analysis. It shows ^{1}H NMR (CDCl$_3$) δ: 1.35 (s, 6 H), 1.42 (s, 3 H), 1.55 (s, 3 H), 2.60 (s, 3 H), 3.90–4.40 (m, 4 H), 4.68 (d, 1 H), 5.85–6.0 (m, 2 H).

6. Reagent-grade toluene was dried by distilling the toluene–water azeotrope and then cooling the remaining liquid under an atmosphere of nitrogen.

7. Tributyltin hydride was purchased from Aldrich Chemical Company, Inc. and stored under nitrogen at 4°C.

8. An E. Merck Silica Gel 60 F-254 0.25-mm plate was used for the TLC analysis.

9. Silica Woelm TSC, obtained from Woelm Pharma, was used.

10. The product is pure by NMR and TLC analyses and shows ^{1}H NMR (CDCl$_3$) δ: 1.27 (s, 3 H), 1.31 (s, 3 H), 1.38 (s, 3 H), 1.46 (s, 3 H), 1.60–1.90 (m, 1 H), 2.05–2.30 (dd, 1 H), 3.65–4.25 (m, 4 H), 4.71 (t, 1 H), 5.77 (d, 1 H).

3. Discussion

This procedure illustrates a simple, general method for the deoxygenation of secondary hydroxyl groups. It is particularly useful for reducing hindered alcohols. The method was first described by Barton and McCombie,[3] who have reviewed a number of other examples.[4]

A variety of thiocarbonyl derivatives, in addition to xanthate esters, undergo reductive homolytic cleavage when treated with tributyltin hydride. These include thiobenzoates,[3] thiocarbonylimidazolides,[3,5] and phenyl thionocarbonate esters.[6] The S-methyl xanthate ester is a particularly convenient intermediate to prepare because of its ease of formation and the low cost of the reagents. Its use is precluded, however, by the presence of base-labile protecting groups and, in such cases, the thiocarbonylimidazolide or phenyl thionocarbonate ester will generally prove satisfactory. Additional methods for the radical deoxygenation of alcohols are described in a review by Hartwig.[7]

The tributyltin hydride reduction usually proceeds without complications. The most common byproduct is starting alcohol, which is postulated to be derived from a mixed thioacetal.[3] Use of the phenyl thionocarbonate ester has been reported to minimize this side reaction in cases where it is a problem.[6]

3-Deoxy-1,2 : 5,6-di-O-isopropylidene-α-D-*ribo*-hexofuranose has been prepared by a variety of other methods, the most widely used of which is the Raney nickel reduction of the 3-S-[(methylthio)carbonyl]-3-thioglucofuranose derivative.[8]

1. Department of Chemistry, Cornell University, Ithaca, NY 14853. Present address: Genzyme Corporation, 75 Kneeland Street, Boston, MA 02111.
2. Schmidt, O. T. "Methods in Carbohydrate Chemistry," Whistler, R. L.; Wolfrom, M. L., Eds.; Academic Press: New York, 1963; Vol. 2, pp. 318–325.
3. Barton, D. H. R.; McCombie, S. W. *J. Chem. Soc., Perkin Trans. 1* **1975**, 1574–1585.
4. Barton, D. H. R.; Motherwell, W. B. *Pure Appl. Chem.* **1981**, *53*, 15–31.
5. Rasmussen, J. R.; Slinger, C. J.; Kordish, R. J.; Newman-Evans, D. D. *J. Org. Chem.* **1981**, *46*, 4843–4846.
6. Robins, M. J.; Wilson, J. S.; Hansske, F. *J. Am. Chem. Soc.* **1983**, *105*, 4059–4065.
7. Hartwig, W. *Tetrahedron* **1983**, *39*, 2609.
8. Hedgley, E. J.; Overend, W. G.; Rennie, R. A. C. *J. Chem. Soc.* **1963**, 4701–4711.

DI-*tert*-BUTYL METHYLENEMALONATE

[Propanedioic acid, methylene-, bis(1,1-dimethylethyl)ester]

$$CH_2(COO\text{-}t\text{-}Bu)_2 \quad + \quad (CH_2O)_x \quad \xrightarrow[\text{AcOH}]{\substack{\text{AcOK} \\ \text{Cu(OAc)}_2 \cdot H_2O}} \quad CH_2\text{=}C(COO\text{-}t\text{-}Bu)_2$$

Submitted by PALOMA BALLESTEROS and BRYAN W. ROBERTS[1]
Checked by DOREEN L. WELLER and JAMES D. WHITE

1. Procedure

Caution! This reaction should be carried out in an efficient hood to prevent exposure to formaldehyde and acetic acid.

A 250-mL, one-necked, round-bottomed flask is equipped with a magnetic stirrer and a reflux condenser protected by a calcium chloride drying tube. Into the flask are placed 30.0 g (0.14 mol) of di-*tert*-butyl malonate (Note 1), 8.4 g (0.28 mol) paraformaldehyde (Note 2), 1.4 g (0.014 mol) of potassium acetate, 1.4 g (0.007 mol) of cupric acetate monohydrate, and 70 mL of glacial acetic acid. The resulting green–white suspension is placed in an oil bath preheated to 90–100°C and stirred for 2 hr (Note 3). The reaction mixture is allowed to cool to room temperature, and the reflux condenser is replaced with a short-path distillation apparatus, the vacuum outlet of which is connected in sequence to a trap cooled in acetone–dry ice, a potassium hydroxide trap, another trap cooled in acetone–dry ice, and a vacuum pump. The receiving flask is cooled in acetone–dry ice, and the system is evacuated over approximately 1 hr to remove acetic acid and other volatile material (Note 4). The bath temperature is increased to 40–50°C for 15 min and then is rapidly raised to 140–150°C to drive over crude product, which is collected over a boiling-point range of 60–82°C (Note 5). When distillate ceases to come over, the bath temperature is increased to 170°C and distillate is collected over the same boiling-point range until the reaction mixture turns from blue to green–brown. The total amount of crude product collected is 24.3 g. This material is dissolved in 50 mL of ether and washed with saturated aqueous sodium bicarbonate solution (4 × 20 mL) and water (25 mL). The combined aqueous fractions are extracted with 50 mL of ether, and the combined ether extracts are dried over magnesium sulfate for 10 min (Note 6). Filtration and evaporation on a rotary evaporator give 20.0 g of crude product, which is distilled through an 8-cm Vigreux column. The di-*tert*-butyl methylenemalonate is collected at 60–67°C/0.1 mm and weighs 15.3 g (48%) (Note 7). The product is somewhat unstable and should be stored in Pyrex in the refrigerator.

2. Notes

1. Di-*tert*-butyl malonate was prepared according to the procedure of Johnson; see *Org. Synth., Coll. Vol. IV* **1963,** 261.

2. Paraformaldehyde was obtained from Aldrich Chemical Company, Inc., and stored in a desiccator over phosphorus pentoxide.

3. After approximately 25 min, the suspension dissolves and the reaction mixture becomes blue–green.

4. At the beginning of the evaporation, the pressure is controlled to minimize bumping of the vigorously boiling mixture.

5. During this operation the pressure varies between 0.3 and 1.5 mm. As the temperature is raised, the reaction mixture turns blue and gas evolution is observed.

6. The procedure can be interrupted at this point and the ether extracts dried over magnesium sulfate overnight in the refrigerator.

7. The bath temperature should not exceed 100°C in order to prevent contamination of the product with the bis(hydroxymethyl) derivative of di-*tert*-butyl malonate. The product exhibits single peaks in the ^{1}H NMR spectrum ($CDCl_3$, 250 MHz) at 1.51 and 6.25 ppm and contains approximately 6% of di-*tert*-butylmalonate as indicated by a peak at 1.47 ppm. Contamination by the bis(hydroxymethyl) derivative is indicated by a peak at 1.48 ppm.

3. Discussion

Methylenemalonate esters are potentially useful activated alkenes that can serve as electrophilic partners in the Michael and cycloaddition reactions and, in the process, introduce a *gem*-diester functionality for further synthetic transformation. The simple esters, however, have a marked propensity toward spontaneous polymerization and, as a consequence, have been used only sparingly in the Michael reaction,[2] the Diels–Alder reaction,[3] [2 + 2] cycloaddition,[4] and [3 + 2] cycloaddition.[5] The recently prepared di-*tert*-butyl analog[6] is advantageous in being longer lived and suitable for conventional synthetic operations, and in introducing a readily cleaved diester moiety. In its most useful application thus far, the compound has been found to react under mild conditions with enamines with no added catalyst or with enol ethers and acetates under Lewis acid catalysis to give either cyclobutanes or Michael adducts, depending on alkene structure.[7]

Di-*tert*-butyl methylenemalonate was originally prepared by phenylsulfenylation of di-*tert*-butyl methylmalonate and thermal elimination of the related sulfoxide.[6] Because methylenemalonate esters are customarily prepared by Knoevenagel-type condensation of malonic esters with formaldehyde equivalents, the considerably more convenient procedure described herein was subsequently adapted from Bachman and Tanner's study using paraformaldehyde under metal-ion catalysis.[3a] The approximately 6% di-*tert*-butyl malonate accompanying the product has presented no interference in the aforementioned reactions with nucleophilic alkenes under neutral or acidic conditions, but its presence should be taken into consideration in other applications.

1. Department of Chemistry, University of Pennsylvania, Philadelphia, PA 19104.

2. (a) Meerwein, H.; Schürmann, W. *Ann. Chem.* **1913,** *398,* 196; (b) Welch, K. N. *J. Chem. Soc.* **1930,** 257; (c) Baum, K.; Guest, A. M. *Synthesis* **1979,** 311; (d) Pelter, A.; Rao, J. M. *Tetrahedron Lett.* **1981,** *22,* 797.

3. (a) Bachman, G. B.; Tanner, H. A. *J. Org. Chem.* **1939,** *4,* 493; (b) Levina, R. Ya.; Godovikov, N. N. *Zhur. Obshchei Khim.* **1955,** *25,* 986; *Chem. Abstr.* **1956,** *50,* 3458c; (c) Marx, J. N.; Bombach, E. J. *Tetrahedron Lett.* **1977,** 2391.

4. Komiya, Z.; Nishida, S. *J. Org. Chem.* **1983,** *48,* 1500.

5. (a) Christl, M.; Huisgen, R.; Sustmann, R. *Chem. Ber.* **1973**, *106*, 3275; (b) Cutler, A.; Ehntholt, D.; Giering, W. P.; Lennon, P.; Raghu, S.; Rosan, A.; Rosenblum, M.; Tancrede, J.; Wells, D. *J. Am. Chem. Soc.* **1976**, *98*, 3495; (c) Abram, T. S.; Baker, R.; Exon, C. M. *Tetrahedron Lett.* **1979**, 4103; (d) Ali, S. A.; Senaratne, P. A.; Illig, C. R.; Meckler, H.; Tufariello, J. J. *Tetrahedron Lett.* **1979**, 4167; (e) Mishchenko, A. I.; Prosyanik, A. V.; Pleshkova, A. P.; Isobaev, M. D.; Markov, V. I.; Kostyanovskii, R. G. *Izv. Akad. Nauk SSSR, Ser. Khim.* **1979**, 131; *Chem. Abstr.* **1979**, *90*, 186844q; (f) Prosyanik, A. V.; Mishchenko, A. I.; Zaichenko, N. L.; Zorin, Ya. Z.; Kostyanovskii, R. G.; Markov, V. I. *Khim. Geterotsikl. Soedin.* **1979**, 599; *Chem. Abstr.* **1979**, *91*, 175249u.
6. Ballesteros, P.; Roberts, B. W.; Wong, J. *J. Org. Chem.* **1983**, *48*, 3603.
7. Ballesteros, P.; Baar, M. R.; Roberts, B. W., manuscript in preparation.

2,6-DI-*tert*-BUTYL-4-METHYLPYRIDINE

[Pyridine, 2,6-bis(1,1-dimethylethyl)-3-methyl-]

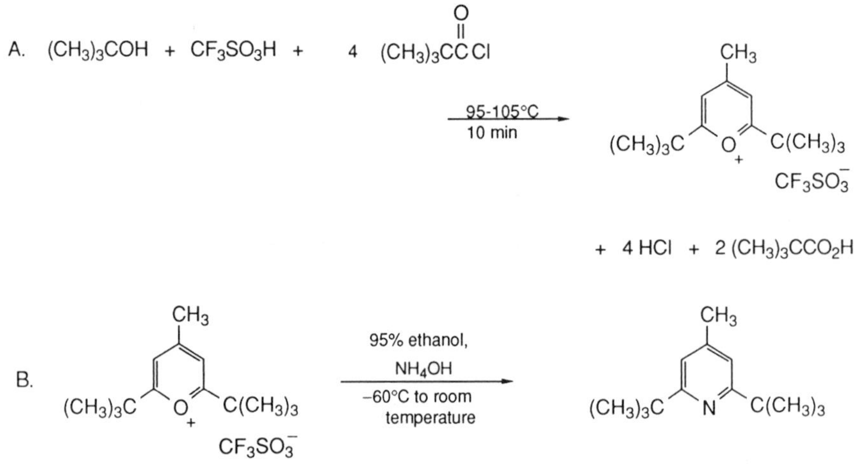

A. $(CH_3)_3COH$ + CF_3SO_3H + 4 $(CH_3)_3\overset{\overset{\displaystyle O}{\|}}{C}CCl$

$\xrightarrow[\text{10 min}]{\text{95-105°C}}$

+ 4 HCl + 2 $(CH_3)_3CCO_2H$

B.

$\xrightarrow[\substack{\text{−60°C to room} \\ \text{temperature}}]{\substack{\text{95% ethanol,} \\ \text{NH}_4\text{OH}}}$

Submitted by ALBERT G. ANDERSON[1] and PETER J. STANG[2]
Checked by MARK T. DuPRIEST and GEORGE BÜCHI

1. Procedure

Caution! The reaction described in Step A should be conducted in a hood, since some carbon monoxide is generated by partial decarbonylation of pivaloyl chloride.

A. *2,6-Di-*tert-*butyl-4-methylpyrylium trifluoromethanesulfonate* (**1**). The center neck of a 5-L, three-necked, round-bottomed flask equipped with a thermometer port, magnetic stirrer bar coated with Teflon, and heating mantle is fitted with a 125-mL pressure-equalizing dropping funnel. The two side necks are fitted with 7 cm (diam) × 27 cm dry ice condensers vented through oil-filled bubblers into traps containing aqueous 1 *N* sodium hydroxide (Note 1). A thermometer is placed in the thermometer port to extend nearly to the bottom of the flask without contacting the stirrer bar. The apparatus is purged with dry nitrogen, the nitrogen flow is stopped, and to the flask are added 300 g (2.5 mol) of pivaloyl chloride and 46 g (0.62 mol) of anhydrous *tert*-butyl alcohol (Note 2). The condensers are charged with isopropyl alcohol-dry ice and, with stirring, the reaction mixture is warmed to 85°C. Heating is discontinued, and the mantle is allowed to remain in place. Then 187.5 g (109 mL, 1.25 mol) of trifluoromethanesulfonic acid is added with stirring during a period of 2–3 min (Note 3). After the addition is complete, the temperature is maintained at 95–105°C for 10 min with the heating mantle. The mantle is removed, and the brown reaction mixture is first allowed to spontaneously cool to 50°C and finally cooled to -10°C with an isopropyl alcohol-dry-ice bath. On addition of 1 L of cold diethyl ether (Note 4), a precipitate forms immediately and is collected, washed with three 300-mL portions of diethyl ether, and air-dried on a medium-porosity fritted-glass filter to give 118–137 g (53–62%) of light tan 2,6-di-*tert*-butyl-4-methylpyrylium trifluoromethanesulfonate, mp 153–164°C (Note 5).

B. *2,6-Di-tert-butyl-4-methylpyridine* (2). To 119–128 g (0.33–0.36 mol) of crude pyrylium salt in a 5-L, three-necked, round-bottomed flask equipped with a mechanical stirrer is added 2 L of 95% ethanol. The mixture is cooled to -60°C with an isopropyl alcohol-dry ice bath and to the fine slurry is added, with stirring, in one portion 1 L of concentrated (d 0.90) ammonium hydroxide also cooled to -60°C. The yellow reaction mixture is held at -60°C for 30 min, then allowed to warm to -40°C and maintained at that temperature for 2 hr, during which time the slurry dissolves. The mixture is then allowed to spontaneously warm to room temperature (Note 6). The reaction mixture is divided into two portions. Each portion is poured into a 4-L separatory funnel, 1 L of aqueous 10% sodium hydroxide is added, and the mixture is extracted with four 250-mL portions of pentane (Note 7). The extracts from both portions are combined and washed with 100 mL of saturated aqueous sodium chloride solution. The pentane is removed on the rotary evaporator (Note 8), leaving a residual light-yellow oil that is dissolved in 150 mL of pentane and introduced slowly during 20–30 min onto the top of a 40 × 4.5 cm water-jacketed chromatography column (Note 9) containing 300 g of activated alumina (Note 10). After the solution has been added, the column is filled with pentane and a 1-L constant-pressure addition funnel, also filled with pentane, is fitted to the top of the column to provide a slight head pressure. The elution is completed in ca. 90 min. All the pyridine is obtained in the first 2 L of eluant (Note 11). The pentane is removed on the rotary evaporator to yield 62.7–66.3 g (90–93%) of a colorless, odorless oil that solidifies on cooling or standing, mp 31–32°C (Note 12).

2. Notes

1. Since gas evolution at the onset of the reaction is quite vigorous, the operator should check to see that the passage of gas is unobstructed. Gas entering the sodium hydroxide trap should be passed over the solution, not bubbled through it, to guard against the possibility of sodium hydroxide solution being drawn back into the reaction flask.

2. *tert*-Butyl alcohol was obtained from Eastman Organic Chemicals and was used as received. The checkers also used *tert*-butyl alcohol freshly distilled from

potassium with equal results. Pivaloyl chloride was obtained from Aldrich Chemical Company, Inc.

3. Trifluoromethanesulfonic acid FC-24 was obtained directly from the manufacturer, Minnesota Mining and Manufacturing Co. (3M), 15 Henderson Dr., West Caldwell, NJ 07006. Adherence to both the time and temperature during the addition is critical for best results.

4. The diethyl ether is conveniently cooled to $-10\,°C$ by addition of some dry ice.

5. The crude salt darkens somewhat on standing because of further polymerization of impurities, but this does not affect the preparation of the base. It may be further purified by two recrystallizations at $-30\,°C$ from isopropyl alcohol (8.7 mL/g) to give colorless needles (94% recovery), mp 168–169°C. The salt is not hygroscopic and may be stored indefinitely at room temperature. It is characterized by NMR [$(CD_3)_2SO$] δ: 1.45 (s, 18 H), 2.72 (s, 3 H), 8.10 (s, 2 H).

6. The submitters state that the reaction may be monitored by the formation of a brilliant-yellow intermediate that fades on completion of the reaction.[3] The checkers found it most convenient to remove the cold bath and allow the reaction to stir overnight at room temperature. If the reaction is worked up before completion, a yellow impurity is formed which cannot be removed by subsequent chromatography.

7. Phillips Petroleum Company pentane was used as received. Other brands required distillation to remove small amounts of higher-boiling compounds.

8. Ethanol should not be removed by distillation or use of a rotary evaporator since considerable amounts of product codistil with the ethanol.

9. The water-jacketed chromatography column shown in Figure 1 is useful when low-boiling solvents or heat-sensitive compounds are chromatographed. Considerable heat is generated when the pentane solution containing the pyridine is introduced onto the column. This may cause boiling of the pentane and separation of the alumina. A flow of cold water through the jacket avoids separation of the alumina. The column was packed by slowly adding the alumina to the column half filled with pentane.

10. Aluminum oxide, activated, acidic, was obtained from Aldrich Chemical Company, Inc., and used as received.

11. The progress of the elution may be monitored by occasionally spotting a fluorescent TLC plate and examining the plate under short-wave UV light; the pyridine appears as a dark-blue spot. Attempts to completely remove colored impurities by distillation, acid-base extraction, or activated charcoal were unsuccessful.

12. Additional physical constants are bp 148–153°C (95 mm), 223°C (760 mm); $HPtCl_6$ salt mp 213–214°C (decomp), CF_3SO_3H salt mp 202.5–203.5°C (from CH_2Cl_2); NMR (CCl_4) δ: 1.29 (s, 18 H), 2.25 (s, 3 H), 6.73 (s, 2 H); pK_a in 50% ethanol: 4.41[4] vs. 4.38 for pyridine in the same solvent.[5] Approximately 0.1% of an impurity, identified by gas chromatography–mass spectrum as 2,6-di-*tert*-butyl-4-neopentylpyridine, is also present; this arises by acid-catalyzed dimerization of isobutylene generated in situ during formation of the pyrylium salt.

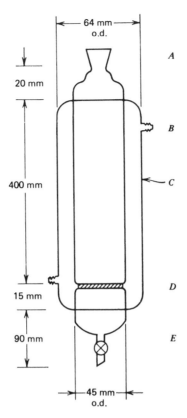

Figure 1. Not drawn to scale. (*A*) 29/26 joint; (*B*) hose connection; (*C*) water jacket; (*D*) 40 mm Kimflow fritted disk, size 40-C (coarse), Lab Glass LG28280; (*E*) Teflon Stopcock 2-mm plug bore, Lab Glass LG9605T.

3. Discussion

2,6-Di-*tert*-butyl-4-methylpyrylium salts previously have been prepared in yields of 4–40% starting with the chloride or anhydride of pivalic acid and employing various counterions, such as ClO_4^-, $FeCl_4^-$, or $AlCl_4^-$.[6] A more recent multistep preparation yields 33% of the perchlorate.[7] The pyrylium salt has been used to prepare pyrylotrimethinecyanine compounds.[8] 2,6-Di-*tert*-butyl-4-methylpyridine has been prepared in 44% yield by treating 4-picoline with a 10 molar excess of *tert*-butyl lithium[4] or by an anionic condensation reaction.[9] The present procedure is essentially that of Anderson and Stang.[3]

The pyrylium trifluoromethanesulfonate salt is nonexplosive. The resulting pyridine possesses the ability to distinguish between Lewis and Brønsted acids.[3,5] It will not react with metal cations[10] or BF_3. The combination of 2,6-di-*tert*-butyl-4-methylpyridine and methyl trifluoromethanesulfonate results in improved yields under very mild conditions of methylated carbohydrates containing acid- or base-labile groups.[11] This pyridine was used as a hindered base in the synthesis of an antigenic bacterial hexasaccharide from *Salmonella newington*.[12] The base has also found use in silylation studies.[13,14] The hindered pyridine makes possible Friedel-Crafts alkylation of aromatic rings under basic conditions.[15,16] Substitution of pyridine by the hindered base results in substantially improved yields of a variety of vinyl esters.[16] The use of the sterically hindered base was essential for the preparation of 1-(ethynyl)vinyl trifluoromethanesulfonates.[17] After use the protonated base can be economically recovered in greater than 95% yield by addition of the pyridinium salt to a two-phase mixture of aqueous 50% sodium hydroxide and pentane followed by elution of the pentane solution through an unactivated silica gel column.[17]

Recently, 2,6,-di-*tert*-butyl-4-methylpyridine was incorporated into a polymer[18] via the 4-methyl group and the resulting polymer bound 2,6-di-*tert*-butylpyridine used in vinyl triflate synthesis. It has also been shown[20] that unlike 2,6-dimethyl and 2,4,6-trimethylpyridine, the 2,6-di-*tert*-butyl-4-methylpyridine does not react with triflic anhydride, ($CF_3SO_2)_2O$, and this might account for the advantageous use of this nonnucleophilic base when preparing reactive triflates.[21]

Because of its ease of preparation and ready availability in quantity 2,6-di-*tert*-butyl-4-methylpyridine continues to be the base of choice in a variety of applications when nonnucleophilic basic reaction conditions are required.[22]

1. Central Research & Development Department, Experimental Station, E. I. du Pont de Nemours & Company, Wilmington, DE 19898.
2. Department of Chemistry, University of Utah, Salt Lake City, UT 84112.
3. Anderson, A. G.; Stang, P. J. *J. Org. Chem.* **1976**, *41*, 3034–3036.
4. Deutsch, E.; Cheung, N. K. V. *J. Org. Chem.* **1973**, *38*, 1123–1126.
5. Brown, H. C.; Kanner, B. *J. Am. Chem. Soc.* **1966**, *88*, 986–992.
6. Balaban, A. T.; Nenitzescu, C. D. *Justus Liebigs Ann. Chem.* **1959**, *625*, 66–73, 74–88; Dimroth, K.; Mach, W. *Angew. Chem. Int. Ed. Engl.* **1968**, *7*, 460–461.
7. Balaban, A. T. *Org. Prep. Proced. Int.* **1977**, *9*, 125–130.
8. Balaban, A. T. *Tetrahedron Lett.* **1978**, 599–600.
9. Bates, R. B.; Gordon, B., III; Keller, P. C.; Rund, J. V.; Mills, N. S. *J. Org. Chem.* **1980**, *45*, 168–169.
10. Balaban, A. T. *Tetrahedron Lett.* **1978**, 5055–5056.
11. Arnarp, J.; Kenne, L.; Lindberg, B.; Lönngren, J. *Carbohydr. Res.* **1975**, *44*, C5–C7.
12. Kochetkov, N. K.; Dmitriev, B. A.; Nikolaev, A. V.; Bairomova, N. E.; Shashkov, A. S. *Bio. Org. Khim.* **1979**, *5*, 64; *Chem. Abstr.* **1979**, *90*, 168858.
13. Barton. T. J.; Tully, C. R. *J. Org. Chem.* **1978**, *43*, 3649–3653.
14. Miller, R. D.; McKean, D. R. *Tetrahedron Lett.* **1979**, 2305–2308.
15. Stang, P. J.; Anderson, A. G. *J. Am. Chem. Soc.* **1978**, *100*, 1520–1525.
16. Forbus, T. R., Jr.; Martin, J. C. *J. Org. Chem.* **1979**, *44*, 313–314.
17. Stang, P. J.; Fisk, T. E. *Synthesis* **1979**, 438–440.
18. Wright, M. E.; Pulley, S. R. *J. Org. Chem.*, **1987**, *52*, 1623.
19. Wright, M. E.; Pulley, S. R. *J. Org. Chem.*, **1987**, *52*, 5036.
20. Binkley, R. W.; Ambrose, M. G. *J. Org. Chem.*, **1983**, *48*, 1776.
21. Stang, P. J.; Hanack, M.; Subramanian, L. R. *Synthesis*, **1982**, 85.
22. Auchter, G.; Hanack, M. *Chem. Ber.* **1982**, *115*, 3402; Ambrose, M. G.; Binkley, R. W. *J. Org. Chem.*, **1983**, *48*, 674; Hoffmann, R. V.; Kumar, A. *J. Org. Chem.*, **1984**, *49*, 4011;

Martinez, A. G.; Rios, I. E.; Barcina, J. O.; Hernando, M. M. *Chem. Ber.*, **1984**, *117*, 982; Schoenleber, R. W.; Kim, Y.; Rapoport, H. *J. Am. Chem. Soc.*, **1984**, *106*, 2645; Petersen, J. S.; Fels, G.; Rapoport, H. *J. Am. Chem. Soc.*, **1984**, *106*, 4539; Harder, I. O.; Hanack, M.; *Chem. Ber.*, **1985**, *118*, 2974; Taylor, S. L.; Forbus, T. R., Jr.; Martin, J. C. *Org. Synth. Coll. Vol. VII* **1990**, 506; Forbus, T. R., Jr.; Taylor, S. L.; Martin, J. C. *J. Org. Chem.*, **1987**, *52*, 4156.

α-AMINO ACETALS: 2,2-DIETHOXY-2-(4-PYRIDYL)ETHYLAMINE

(4-Pyridineethanamine, β,β-diethoxy)

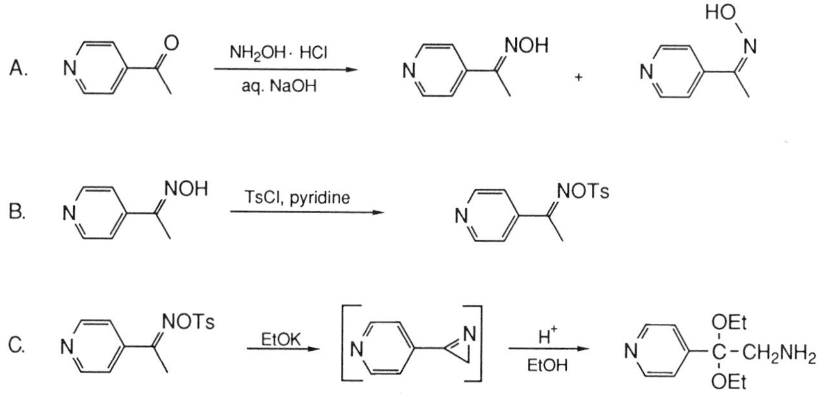

Submitted by John L. LaMattina and R. T. Suleske[1]
Checked by Paul Hebeisen and Andrew S. Kende

1. Procedure

A. *4-Acetylpyridine oxime.* Hydroxylamine hydrochloride (25.0 g, 0.36 mol) (Note 1) is dissolved in 50 mL of water, and the solution is added to 70 mL of 20% aqueous sodium hydroxide in a 500-mL Erlenmeyer flask. To this magnetically stirred solution is added at one time 4-acetylpyridine (36.3 g, 0.30 mol) (Note 2); a precipitate forms rapidly. The reaction mixture is stirred at 0–5°C for 2 hr; then the precipitate is collected by suction filtration and washed with 500 mL of cold water.

The product (mp 122–146°C, 33–36 g, 81–88%) can be shown from its ^{1}H NMR spectrum (Note 3) to be a 5:1 mixture of the *E*- and *Z*-isomers of 4-acetylpyridine oxime. To obtain pure *E*-isomer (Note 4), the product is recrystallized twice as follows. The crude product is dissolved in 600 mL of hot water in a 2-L Erlenmeyer flask, the hot solution is decanted from any undissolved residue, and the supernatant liquid is allowed to cool slowly to 30°C during 2–3 hr by placing the flask on a cork ring. The precipitate is collected at this temperature by suction filtration. A second crystallization by the same procedure yields pure *E*-oxime, which is dried under reduced pressure over

Drierite to constant weight. The yield of *E*-4-acetylpyridine oxime, mp 154–157°C (Note 5), is 27.1–28.3 g (66–69%).

B. *4-Acetylpyridine oxime tosylate.* Pure *E*-oxime (27.1 g, 0.20 mol) and *p*-toluenesulfonyl chloride (47.9 g, 0.22 mol) (Note 6) are added to 100 mL of anhydrous pyridine (Note 7) in a 1-L, round-bottomed flask fitted with a drying tube and a large magnetic stirring bar. The reaction mixture is stirred at 25°C for 24 hr; a precipitate of pyridine hydrochloride forms. A 500-mL portion of ice water is added with continued stirring. the initial precipitate dissolves, and a voluminous white precipitate soon forms. This is collected by suction filtration, washed with three 150-mL portions of cold water and dried under reduced pressure and over Drierite to constant weight. The yield of pure tosylate, mp 79–81°C (Note 8), is 55.1 g (95%).

C. *2,2-Diethoxy-2-(4-pyridyl)ethylamine.* To a 2-L, round-bottomed flask containing 80 mL of absolute ethanol (Note 9) and fitted with a magnetic stirrer and a reflux condenser with a drying tube, potassium metal (7.60 g, 0.19 mol) is slowly added (Note 10). When the metal has dissolved, the solution is cooled to 0–5°C and (*E*)-4-acetyl-pyridine tosylate (55.1 g, 0.19 mol) dissolved (with gentle warming) in 320 mL of absolute ethanol is added over 15 min through a dropping funnel to the stirred solution at 0–5°C. During this period a precipitate of potassium *p*-toluenesulfonate forms. The temperature of the stirred mixture is allowed to rise to room temperature for 1 hr. The mixture is diluted with 1 L of anhydrous ether and filtered by suction. The precipitate is quickly washed with 150 mL of anhydrous ether. The ether filtrates are combined, and hydrogen chloride gas is bubbled through the ether solution for 15 min. A precipitate forms immediately. The precipitate is collected by suction filtration, washed with three 170-mL portions of anhydrous ether, and dried briefly under reduced pressure. The dihydrochloride thus obtained is dissolved in 200 mL of water, and powdered sodium carbonate is added until the mixture reaches a pH of >10. The mixture is extracted four times with 125-mL portions of chloroform. The combined chloroform extracts are dried over anhydrous magnesium sulfate and concentrated at reduced pressure to an oil. This orange–red oil is distilled at 0.2 mm to yield 29.7 g (74.5%) of the amine as a colorless oil, bp 93–95°C (Note 11).

2. Notes

1. Hydroxylamine hydrochloride 97% (mp 155–157°C), available from Aldrich Chemical Company, Inc. or Fisher Scientific Company, is suitable for use without further purification.

2. 4-Acetylpyridine (98%) from Aldrich Chemical Company, Inc. was distilled under reduced pressure (bp 103–104°C/14–16 mm) prior to use.

3. In dimethyl sulfoxide-d_6, the *E* and *Z* isomers show OH proton resonances at δ 11.65 and 10.97, respectively.

4. Use of the isomer mixture prevents isolation of oxime tosylate in crystalline form at the next step and leads to reduced overall yield of pure amine.

5. The lit.[2] melting point for the oxime is 158°C.

6. *p*-Toluenesulfonyl chloride was purified prior to use by the procedure of L. Fieser and M. Fieser in "Reagents for Organic Synthesis."[3]

7. Pyridine AR (Mallinckrodt, Inc.) was used directly.

8. The lit.[2] mp for this compound is 80°C.

9. Ethanol was dried by reflux over magnesium ribbon.

10. For the safe handling and disposal of potassium metal, see *Org. Synth., Coll. Vol. IV* **1963**, 134.

11. This compound has the following 90-MHz ^{1}H NMR spectrum (CDCl$_3$) δ: 0.75 (br s, 2 H, NH$_2$), 1.20 [t, 6 H, $J = 7$, (CH$_3$CH$_2$O)$_2$], 2.97 (s, 2 H, CH$_2$NH$_2$), 3.41 [m, 4 H, (OCH$_2$CH$_3$)$_2$], 7.37 (d of d, 2 H, $J = 2$, 4.5, pyridine H$_3$ and H$_5$), 8.58 (d of d, 2 H, $J = 2$, 4.5, pyridine H$_2$ and H$_6$). Anal. calcd. for C$_{11}$H$_{18}$N$_2$O$_2$: C, 62.83; H, 8.63; N, 13.32. Found: C, 62.63; H, 8.52; N, 13.20.

3. Discussion

α-Amino ketones are useful intermediates for the preparation of a variety of heterocycles, including imidazoles,[4] oxazoles,[5] and pyrazines.[6] Unfortunately, pyrazine formation can be a complicating side reaction because of the tendency of α-amino ketones to dimerize. One way to avoid this problem is to generate these intermediates in a protected form, specifically, as α-amino acetals.[7] Such derivatives allow one to manipulate the amino moiety as desired. The acetal can then be hydrolyzed at the appropriate interval to complete the synthesis.

α-Amino acetals can be prepared from α,α-dialkoxynitriles either via catalytic hydrogenation[8] or by reaction with organometallic reagents.[9] However, these methods are limited by the availability of the appropriate starting material. The procedure here offers a more simple approach that involves the Neber rearrangement. Although this reaction is generally used to prepare α-amino ketones, use of an anhydrous ethanol medium readily results in acetal formation. A summary of other α-amino acetals prepared using this procedure appears in Table I.

This reaction, like all Neber rearrangements, is limited by the availability of the appropriate oxime tosylate.[10] Substrates in which the aryl group contains an electron-donating function are unstable, since they have a propensity to undergo Beckmann rearrangement. However, this difficulty can be resolved by subsequent conversion of the α-amino acetals. For example, catalytic hydrogenation of 2,2-diethoxy-2-(p-bromo-phenyl)ethylamine yields the known parent compound, 2,2-diethoxy-2-phenylethyl-amine; (these two α-amino acetals readily undergo hydrolysis and should be protected from moisture). Other approaches to α-aminoacetals include reduction of α-azidoace-tals,[11] reaction of α-haloketimines with alcohol,[12] and reaction of *N,N*-dichloroamines with sodium ethoxide.[13]

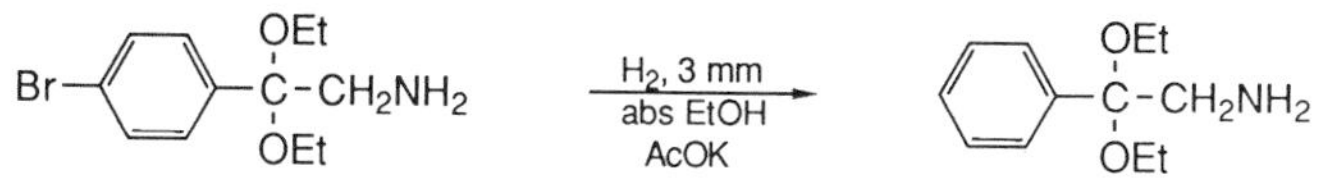

TABLE I
Preparation of α-Amino Acetals

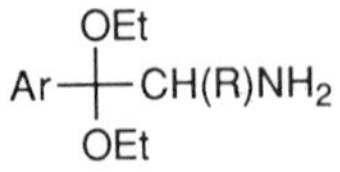

$$Ar\!-\!\!\begin{array}{c}OEt\\|\\|\\OEt\end{array}\!\!CH(R)NH_2$$

Ar	R	Yield (%)	bp (°C/mm)	mp (°C), HCl salt
2-Pyridyl	H	58	82/0.2	150 (dec)[a]
3-Pyridyl	H	53	84/0.2	187–188[a]
4-Pyridyl	CH_3^b	40	98/0.2	129–130[a]
$4\text{-}O_2N\text{-}C_6H_4$	H	78	—[c]	116 (dec)
$4\text{-}Br\text{-}C_6H_4$	H	92	—[d]	

[a] Dihydrochloride salt.
[b] For the preparation of this material in Step C, gaseous HCl is bubbled into the ethereal filtrate for 3 hr. Presumably the longer reaction time is necessary for steric reasons.
[c] This material decomposes on distillation and is purified by column chromatography (silica gel–chloroform).
[d] This material decomposes on distillation and hydrolyzes when chromatographed on silica gel. However, ^{1}H-NMR analysis indicates that it is >95% pure.

1. Central Research, Pfizer Inc., Groton, CT 06340.
2. Cymerman-Craig, J.; Willis, D. *J. Chem. Soc.* **1955**, 4315.
3. Fieser, L. F.; Fieser, M. "Reagents for Organic Synthesis," Vol. 1; Wiley: New York; 1967, pp 1179–1180.
4. Bredereck, H.; Theilig, G. *Chem. Ber.* **1953**, *86*, 88.
5. Kondrat'eva G. Y.; Huang, C.-H. *Zh. Obshch. Khim.* **1962**, *32*, 2348; *J. Gen. Chem. USSR* **1962**, *32*, 2315.
6. Cheeseman, G. W. H.; Werstuik, E. S. G. *Adv. Heterocycl. Chem.* **1972**, *14*, 122.
7. According to IUPAC Nomenclature of Organic Chemistry, Rule C-331.1: "Compounds containing the group —C— are termed acetals (the name ketal is abandoned)."
8. Erickson, J.; Montgomery, W.; Rorso, O. *J. Am. Chem. Soc.* **1955**, *77*, 6640.
9. Chastrette, M.; Axiotis, G. P. *Synthesis* **1980**, 889.
10. Oxley, P.; Short, N. F. *J. Chem. Soc.* **1948**, 1514.
11. Higgins, S. D.; Thomas, C. B. *J. Chem. Soc., Perkin Trans. I* **1983**, 1483.
12. deKimpe, N.; Verhe, R.; deBuyck, L.; Moens, L.; Sulmon, P.; Schamp, N. *Synthesis* **1982**, 765.
13. Coffen, D. L.: Hengartner, U.; Katonak, D. A.; Mulligan, M. E.; Burdick, D. C.; Olson, G. L; Todaro, L. J. *J. Org. Chem.* **1984**, *49*, 5109.

DIASTEREOSELECTIVE α-ALKYLATION OF β-HYDROXYCARBOXYLIC ESTERS THROUGH ALKOXIDE ENOLATES: DIETHYL (2*S*, 3*R*)-(+)-3-ALLYL-2-HYDROXYSUCCINATE FROM DIETHYL (*S*)-(−)-MALATE

(Butanedioic acid, 2-hydroxy-3-(2-propenyl)-, diethyl ester, [*S*-(*R*,*S*)])

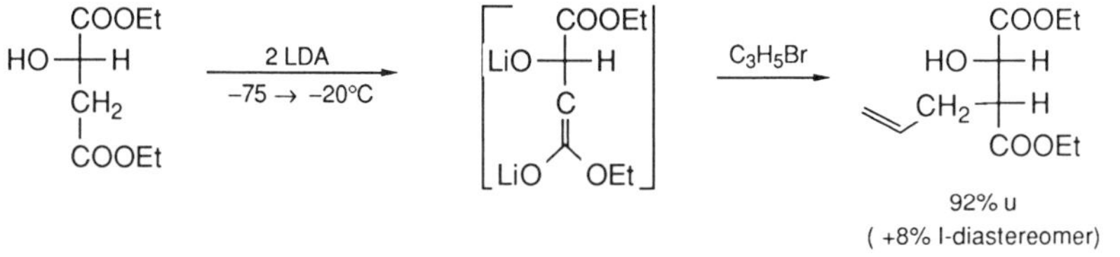

92% u
(+8% l-diastereomer)

Submitted by DIETER SEEBACH, JOHANNES AEBI, and DANIEL WASMUTH[1]
Checked by BRIAN MAXWELL and CLAYTON H. HEATHCOCK

1. Procedure

A 500-mL, three-necked flask containing a magnetic stirring bar is equipped with a 100-mL pressure-equalizing and serum-capped dropping funnel, a three-way stopcock, and a low-temperature thermometer (Note 1). The dry apparatus is filled with argon and kept under an inert gas pressure of ca. 100 mm against the atmosphere until the aqueous workup (Note 2); see Figure 1.

The flask is charged through serum cap B with 17 mL (120 mmol) of diisopropyl-amine (Note 3) and 200 mL of tetrahydrofuran (THF) (Note 4), using syringe techniques. It is cooled to −75°C in a dry ice bath. With stirring, exactly 100 mmol of butyllithium (hexane solution) (Note 5) is introduced from the dropping funnel (Note 6) within 10 min, followed after 0.5 hr, by a mixture of 9.51 g (50 mmol) of (−)-diethyl (*S*)-malate (Note 7) and 5 mL of THF, which is added dropwise through cap B at such a rate that the temperature does not rise above −60°C. The addition takes approximately 10 min (Note 8). The dry ice cooling bath is replaced by an ice–salt bath (ca. −15°C) in which the contents of the flask warm to −20°C within 0.5 hr. The solution is stirred at −20° ± 2°C for 0.5 hr and then is cooled to −75°C.

To the solution of the alkoxide enolate thus prepared is added by syringe within 5 min 10.7 mL (124 mmol) of neat 3-bromo-1-propene (Note 9) at such a rate that the temperature of the reaction mixture does not rise above −70°C. Stirring is continued, first for 2 hr at −75°C, and then overnight while the temperature rises to −5°C (Note 10).

The reaction mixture is quenched by adding a solution of 12 g (200 mmol) of glacial acetic acid in 20 mL of diethyl ether at −50°C and is then poured into a 1-L separatory funnel containing 500 mL of ether and 70 mL of water. The organic layer is washed successively with 40 mL each of saturated sodium bicarbonate and sodium chloride solution, and the aqueous phases are extracted with two 200-mL portions of ether. The combined ethereal solutions are dried by vigorous stirring with dry MgSO$_4$ for 15 min. Removal of the solvent first with a rotary evaporator at a bath temperature no higher than 35°C and then at room temperature under oil pump vacuum (0.1 mm) furnishes

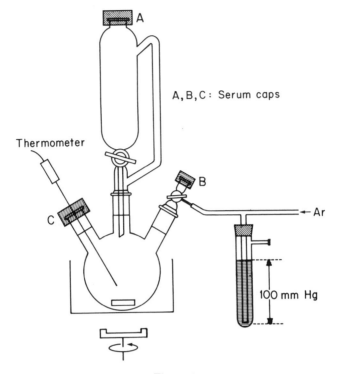

Figure 1

10.4 g of a yellow oil consisting, according to capillary gas chromatography (GC) (Note 11), of 81.3% of the desired allylated (2S,3R) product (73.5% yield), 8.5% of the (2S,3S) diastereoisomer (90.5% ds^2), and 6.3% of the starting diethyl malate (Note 12).

The product is purified by flash chromatography (Notes 13–15). A flash column of 7-cm diameter is charged with 450 g of silica gel (Kieselgel 60, Merck, Korngrösse 0.040–0.063 mm, 230–400 mesh ASTM) and 10.4 g of the crude product. A 1 : 1 mixture of ether and pentane is used for elution, with a running rate of 5-cm column length per minute (pressure 1.25 atm). After a 200-mL forerun, 33-mL fractions are collected. No attempt is made to separate the two diastereoisomers; fractions 22–40 are combined to give 8.0 g (70%) of pure allylated product [ratio of diastereoisomers 92 : 8 (Note 11)], after removal of the solvent; $[\alpha]_D^{20}$ +11.2° (chloroform, c 2.23) (Note 16).

2. Notes

1. A Pt-100 thermometer (Testoterm KG, Lenzkirch, Germany) was used by the submitters. This is preferred to a conventional thermometer because it is more accurate and more convenient to read. Careful temperature control is essential for the present procedure. Unless stated otherwise, all temperatures given are

those of the reaction mixture. The checkers found that a $+30$ to $-100°C$ alcohol thermometer is satisfactory.

2. The glass components of the apparatus are dried overnight in a $170°C$ oven and allowed to cool in a desiccator over a drying agent before assembly. The apparatus is filled with argon by evacuating and pressurizing several times through the three-way stopcock, as described previously.[3]

3. Diisopropylamine was freshly distilled from calcium hydride.

4. Tetrahydrofuran (THF) was first distilled under an inert atmosphere from KOH and then from the blue solution obtained with potassium and benzophenone, as described previously.[3] [However, see warning notice, *Org. Synth., Coll. Vol. V* **1973,** 976–977.]

5. Before use, the commercial 1.6 M solution of butyllithium in hexane was titrated acidimetrically using diphenylacetic acid as an indicator.[4]

6. The dropping funnel was calibrated before use in this procedure. With standard graduated dropping funnels and syringes, the submitters noticed up to 10% deviation from true volumes! Syringe techniques were applied; the dropping funnel was rinsed with ca. 5 mL of dry THF.

7. Commercial (S)-$(-)$-malic acid was esterified under standard conditions, following a procedure by Fischer and Speier.[5] The freshly distilled ester employed by the submitters had an $[\alpha]_D^{20}$ $-10.5°$ (neat) ($d_{20}^4 = 1.128$ g/cm^3), which corresponds to an optical purity of 100%.[6]

8. The flask, in which the ester/THF mixture was prepared, and the syringe are rinsed with a total of ca. 5 mL of dry THF.

9. Commercial allyl bromide was distilled before use.

10. The submitters used a 2-L Dewar cylinder holding, besides the flask, ca. 1 L of ethanol as a cooling liquid. If no excess dry ice was present at the beginning of the warm-up period, it took ca. 12 hr to reach $-5°C$.

11. GLC-analysis were performed using the following column and conditions: 0.3-mm $\times$ 20-m glass capillary column Pluronic L 64, program $120°C$, (3 min), $10°C$/min up to $200°C$, temperature of injector and detector $200°C$, carrier gas: hydrogen (1.3 atm).

12. A total of ca. 4% of four minor side products with longer retention times is also present.

13. This is the fastest method, although it consumes large amounts of solvent and of silica gel. The procedure is that of Still et al.[7] Conventional chromatography is also possible but is more time-consuming.

14. Kugelrohr distillation does not separate the starting material, diethyl malate. Distillation through a 30-cm Vigreux column (silvered vacuum jacket) leads to loss of material (only 40% yield, diastereoisomer ratio 90 : 10, free of starting material).

15. Hydrolysis of the crude product yields, after recrystallization, pure (2S,3R)-3-allyl-2-hydroxysuccinic acid, mp 96.0–$97.5°C$, $[\alpha]_D^{20}$ $+14.7°$ (acetone, c 1.69).

16. The boiling point is 77–78°C (0.07 mm). Previously, a specific rotation of $[\alpha]_D^{25}$ +11.9° (chloroform, c 1.77) was reported.[8a] The ^{13}C NMR spectrum ($CDCl_3$) of the (2S,3R) isomer shows the following signals δ (off-resonance multiplicity, assignment): 14.12 (q, $CO_2CH_2CH_3$), 32.21 (t, C(3)CH_2), 48.25 [d, C(3)], 60.86 and 61.81 (2 t, $CO_2CH_2CH_3$), 70.36 [d, C(2)], 117.78 [t, C(3)CH_2CH=CH_2], 134.94 [d, C(3)CH_2CH=CH_2], 171.92 and 173.48 (2 s, $CO_2CH_2CH_3$).

3. Discussion

The compound described here had not been known prior to our first synthesis of it.[8] Generally, aldol derivatives of this configuration are prepared by the addition of E enolates of esters to aldehydes,[9,10] **1 → 2** in Scheme 1.

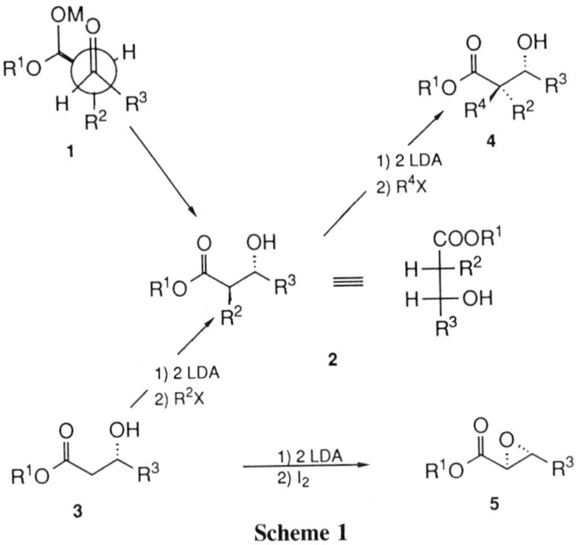

Scheme 1

The method of preparing α-branched β-hydroxy esters by alkylation of dianion derivatives of the parent compounds was first discovered by Herrmann and Schlessinger.[11] It is highly diastereoselective[12] and applicable without racemization to optically active derivatives, as first demonstrated independently by Fráter with β-hydroxybutanoate[13] and by us with malate[8,14] (see **3 → 2** and **3 → 5** in Scheme 1). In the meantime, many applications have been published.[15,16] A related method of preparing derivatives belonging to the same diastereoisomeric series is the alkylation of β-lactone enolates.[17]

Examples of alkylation of malic esters are listed in Table I, together with those of double alkylation, which can also be achieved, see **2 → 4** in Scheme 1. Since the (S) and the (R) forms of malic acid are both readily available,[18] the enantiomers of all structures shown in Table I can be prepared as well. The method is also applicable to β-hydroxy γ-lactones of type **6,** the alkylations of which lead[19] to derivatives of opposite configuration **8,** see **6 → 7** in Scheme 2. Finally, the dilithio derivative **9** of di-*tert*-butyl N-formylaspartate is alkylated (→ **10**; see Scheme 2)[20] with the same relative topicity,[21] **ul,** as the malate dianion derivative (Table I).

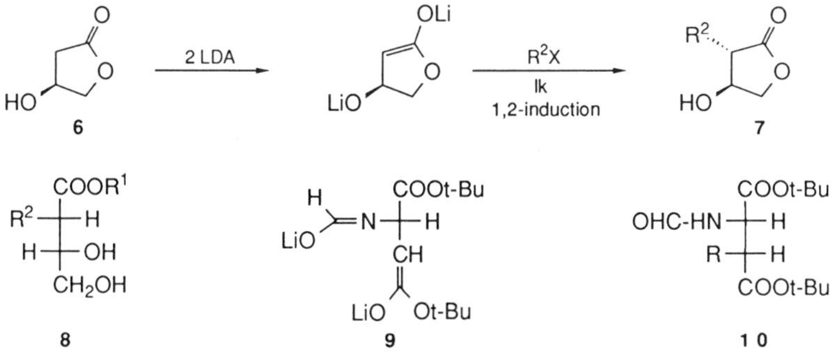

Scheme 2

Structures: **6**, **7**, **8**, **9**, **10**

TABLE I

PRODUCTS OF MONO- AND DIALKYLATION WITH RELATIVE TOPICITY ul^a OF (S)-MALIC ESTERS THROUGH ALKOXIDE ENOLATES. THE RATIOS OF DIASTEREOISOMERS (SEE % ds) WAS DETERMINED BY ^{1}H OR ^{13}C NMR SPECTROSCOPY OR BY GC ANALYSIS.

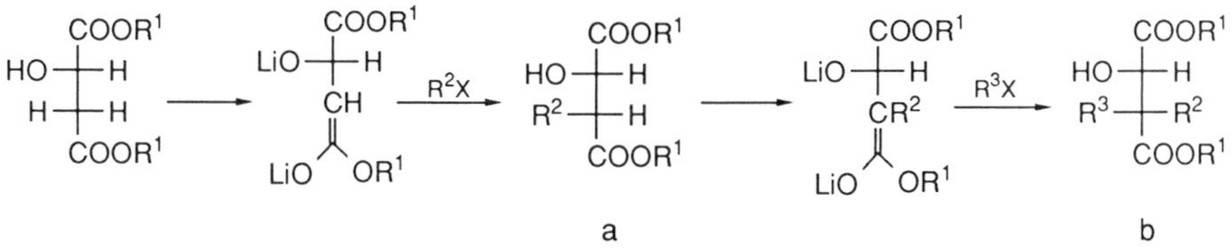

Product	R^1	R^2	R^3	Yield (%)	% ds[b]	Ref.
				(Malate → a)		
a	CH_3	CH_3	—	65	91	8a,14b
	CH_3	$C(OH)(CH_3)_2$	—	55	75	8a
	C_2H_5	CH_3	—	88	91	8a
	C_2H_5	$CH_2C_6H_5$	—	48	91	8a
	C_2H_5	I	—	80	67	8a
	CH_3	$CH_2CH_2NO_2$	—	31	85	14a
	CH_3	C_2H_5	—	64	90	14b
	CH_3	$CH_2CH=CH_2$	—	63	93	8b,14c
				(a → b)		
b	CH_3	CH_3	CH_3	94	—	14b
	CH_3	CH_3	C_2H_5	36	95	14b
	CH_3	C_2H_5	CH_3		72	14b
	CH_3	CH_3	CD_3	92	89	14b
	CH_3	CH_3	$^{13}CH_3$	81	88	14b
	CH_3	CH_3	$CH_2CH=CH_2$	74	95	14c
	CH_3	CH_3	H	>98	67	14c

[a] See reference 21.
[b] See reference 2.

In Table II, a series of useful chiral building blocks is shown, which are accessible through alkylations of malic acid derivatives; the table also contains some natural products that were synthesized from such building blocks.

The alkylation of doubly deprotonated β-hydroxy esters, an example of which is described in the procedure above, is not just a useful alternative to the diastereoselective aldol-type addition, but can supply enantiomerically pure products from appropriate precursors, and it can be used for the preparation of α,α-disubstituted derivatives (see **4** in Scheme 1). These were hitherto not available stereoselectively from enolates of α-branched esters and aldehydes.

TABLE II
CHIRAL, NONRACEMIC BUILDING BLOCKS AND NATURAL PRODUCTS SYNTHESIZED
THROUGH ALKYLATION OF MALIC ACID DERIVATIVES[a]

Products and Intermediates from (S)-Malic Acid

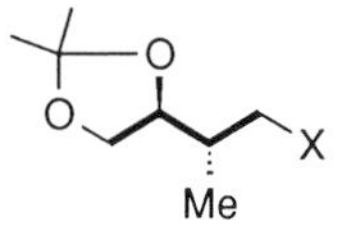

Me
ref. 15b, 22

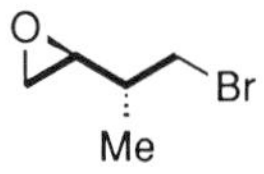

Me
ref. 8b

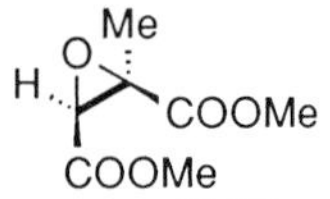

Me
COOMe
COOMe
ref. 14c

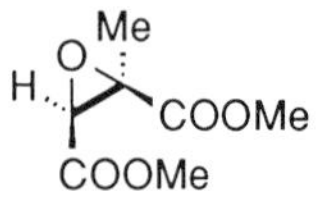

Me
COOMe
COOMe
ref. 14c

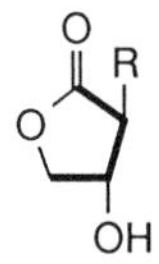

R
OH
R = Me, Et, higher alkyl
allyl, benzyl; ref. 19

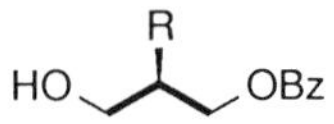

R
HO   OBz
R = Me, allyl, benzyl
(by oxidative degradation
after alkylation); ref. 8c

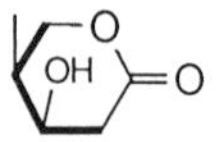

OH
ref. 23

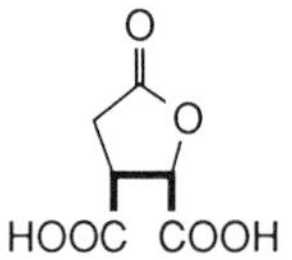

HOOC COOH
(+)-isocitric acid
ref. 8b

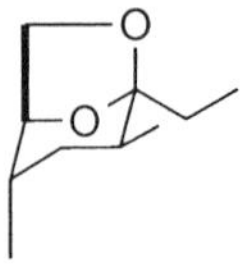

(-)-δ-multistratin
ref. 15b, 22

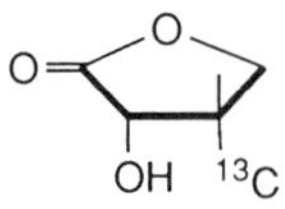

OH   ¹³C
(+)-pantolactone
ref. 14b

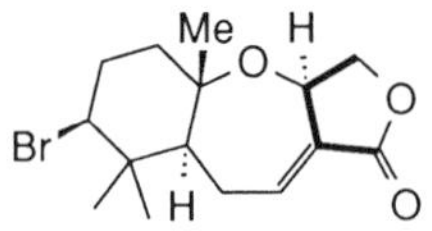

Me   H
Br
H
(-)-aplysistatin
ref. 19c

[a]The four-carbon unit of the structure derived from malic acid is indicated by heavy lines.

1. Laboratorium für Organische Chemie, ETH-Zentrum, Universitätstrasse 16, CH-8092 Zürich, Switzerland.
2. Seebach, D.; Naef, R. *Helv. Chim. Acta* **1981**, *64*, 2704.
3. Seebach, D.; Beck, A. K. *Org. Synth., Coll. Vol. VI* **1988**, 316, 869; Seebach, D.; Hidber, A. *Org. Synth., Coll. Vol. VII* **1990**, 447.
4. Kofron, W. G.; Baclawski, L. M. *J. Org. Chem.* **1976**, *41*, 1879.
5. Fischer, E.; Speier, A. *Chem. Ber.* **1895**, *28*, 3252.
6. Kuhn, R.; Wagner-Jauregg, T. *Chem. Ber.* **1928**, *61*, 504.
7. Still, W. C.; Kahn, M.; Mitra, A. *J. Org. Chem.* **1978**, *43*, 2923.
8. (a) Seebach, D.; Wasmuth, D. *Helv. Chim. Acta* **1980**, *63*, 197; (b) Aebi, J. D. Master's Thesis, ETH Zürich, 1981; (c) Aebi, J. D.; Sutter, M. A.; Wasmuth, D.; Seebach, D. *Liebigs Ann. Chem.* **1983**, 2114; Aebi, J. D.; Sutter, M. A.; Wasmuth, D.; Seebach, D. *Liebigs Ann. Chem.* **1984**, 407.
9. Heathcock, C. H. *Science* **1981**, *214*, 395; Heathcock, C. H. "Asymmetric Synthesis," Morrison, J. D., Ed.; Academic Press: New York, 1983, Vol. 3.
10. Evans, D. A.; Nelson, J. V.; Taber, T. R. *Top. Stereochem.* **1982**, *13*, 1.
11. Herrmann, J. L.; Schlessinger, R. H. *Tetrahedron Lett.* **1973**, 2429.
12. Kraus, G. A.; Taschner, M. J. *Tetrahedron Lett.* **1977**, 4575.
13. Fráter, G. *Helv. Chim. Acta* **1979**, *62*, 2825, 2829; Fráter, G. *Helv. Chim. Acta* **1980**, *63*, 1383; Fráter, G. *Tetrahedron Lett.* **1981**, *22*, 425.
14. (a) Züger, M.; Weller, Th.; Seebach, D. *Helv. Chim. Acta* **1980**, *63*, 2005; (b) Wasmuth, D.; Arigoni, D.; Seebach, D. *Helv. Chim. Acta* **1982**, *65*, 344, 620; (c) Wasmuth, D. Ph.D. Thesis No. 7033, ETH Zürich, 1982.
15. (a) Hoffmann, R. W.; Ladner, W.; Steinbach, K.; Massa, W.; Schmidt, R.; Snatzke, G. *Chem. Ber.* **1981**, *114*, 2786; (b) Hoffmann, R. W.; Helbig, W. *Chem. Ber.* **1981**, *114*, 2802.
16. Kramer, A.; Pfander, H. *Helv. Chim. Acta* **1982**, *65*, 293.
17. Mulzer, J.; Kerkmann, T. *Angew. Chem.* **1980**, *92*, 470; Mulzer, J.; Kerkmann, T. *Angew. Chem., Int. Ed. Engl.* **1980**, *19*, 466.
18. Hungerbühler, E.; Seebach, D.; Wasmuth, D. *Helv. Chim. Acta* **1981**, *64*, 1467.
19. (a) Shieh, H.-M.; Prestwich, G. D. *J. Org. Chem.* **1981**, *46*, 4319; (b) Chamberlin, A. R.; Dezube, M. *Tetrahedron Lett.* **1982**, *23*, 3055; (c) Shieh, H.-M.; Prestwich, G. D. *Tetrahedron Lett.* **1982**, *23*, 4643.
20. Seebach, D.; Wasmuth, D. *Angew. Chem.* **1981**, *93*, 1007; Seebach, D.; Wasmuth, D. *Angew. Chem., Int. Ed. Engl.* **1981**, *20*, 971.
21. Seebach, D.; Prelog, V. *Angew. Chem.* **1982**, *94*, 696; Seebach, D.; Prelog, V. *Angew. Chem. Int. Ed. Engl.* **1982**, *21*, 654.
22. Mori, K.; Iwasawa, H. *Tetrahedron*, **1980**, *36*, 87.
23. Lawson, K., unpublished results, ETH Zürich, 1982.

DIETHYL [(2-TETRAHYDROPYRANYLOXY)METHYL]PHOSPHONATE

(Phosphonic acid, [(tetrahydro-2*H*-pyran-2-yl)oxy]methyl-, diethyl ester)

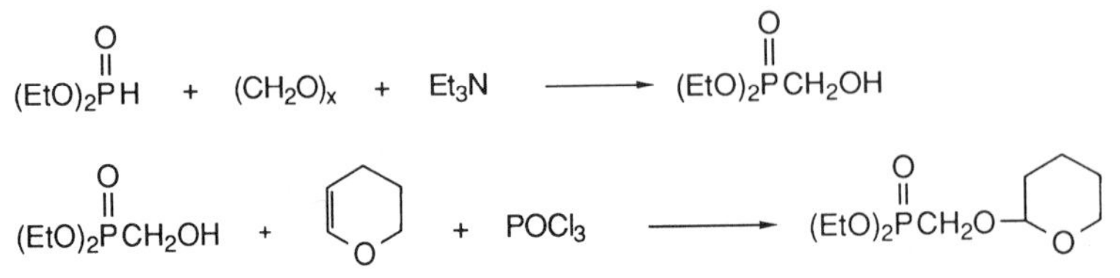

$$(EtO)_2 \overset{O}{\underset{\|}{P}} H \ + \ (CH_2O)_x \ + \ Et_3N \ \longrightarrow \ (EtO)_2 \overset{O}{\underset{\|}{P}} CH_2OH$$

$$(EtO)_2 \overset{O}{\underset{\|}{P}} CH_2OH \ + \ \text{[dihydropyran]} \ + \ POCl_3 \ \longrightarrow \ (EtO)_2 \overset{O}{\underset{\|}{P}} CH_2O\text{[tetrahydropyranyl]}$$

Submitted by ARTHUR F. KLUGE[1]
Checked by RONALDO A. PILLI, KENNETH S. KIRSHENBAUM, CLAYTON H. HEATHCOCK, and K. BARRY SHARPLESS

1. Procedure

A. *Diethyl hydroxymethylphosphonate.* To a 250-mL, round-bottomed flask equipped with a magnetic stirring bar and an efficient reflux condenser are added 69 g (64.4 mL, 0.5 mol) of diethyl phosphite (Note 1), 15 g (0.5 mol) of paraformaldehyde, and 5.1 g (0.05 mol) of triethylamine. The mixture is placed in an oil bath preheated to 100–120°C. The temperature is increased to 120–130°C, and the mixture is stirred at this temperature for 4 hr. The stirring bar is removed, the flask is transferred to a rotary evaporator, and most of the triethylamine is removed by heating under reduced pressure of ca. 15 mm and with a bath temperature of ca. 80°C. Kugelrohr distillation at 125°C (0.05 mm) (Note 2) gives 41.4–54.9 g (49–65%) of material of sufficient purity for the next step (Notes 3 and 4).

B. *Diethyl [(2-tetrahydropyranyloxy)methyl]phosphonate.* A mixture of 33.63 g (0.2 mol) of diethyl hydroxymethylphosphonate, 21 g (0.25 mol) of dihydropyran, and 150 mL of diethyl ether is placed in a stoppered flask, and 20 drops of phosphorus oxychloride is added while the contents are swirled manually. After 3 hr at room temperature the reaction is monitored by TLC (Note 5). The mixture is diluted with diethyl ether, transferred into a separatory funnel, and shaken successively with 100 mL of saturated sodium bicarbonate solution, 100 mL of water, and 100 mL of saturated sodium chloride solution. The ether solution is dried over $MgSO_4$, filtered, and the ether is removed with a rotary evaporator. Kugelrohr distillation of the residue (110°C, 0.05 mm) gives 42.4–46.9 g (84–93%) of material of sufficient purity for use in homologation reactions (Notes 6 and 7).

2. Notes

1. Diethyl phosphite, paraformaldehyde, and triethylamine were obtained from Aldrich Chemical Company, Inc. Dihydropyran was obtained from MCB, Inc.

2. Attempted isolation of diethyl hydroxymethylphosphonate by standard vacuum-distillation technique is accompanied by extensive decomposition. The use of

Kugelrohr apparatus allows the isolation to be accomplished at a lower temperature, and therefore the product is obtained in higher yield. Alternatively, the checkers found that distillation using a 2-in. wiped-film molecular still (Pope Scientific, Inc.) significantly raised product yields, especially when the reaction was performed on a larger scale (Notes 3 and 6).

3. The checkers found that reactions run on up to four times the present scale and rectified using a molecular still (wall temperature 110–120°C, 0.10 mm) gave yields of 89–94%. *Warning: On this larger scale (i.e., four times the present scale) a brief runaway was experienced and some material that escaped from the condenser was caught in a trap; however, the yield was still excellent (94%).*

4. On TLC [silica, visualization with 1.5% phosphomolybdic acid spray and heating] the product has a retardation factor of ca. 0.1 with ethyl acetate development and ca. 0.3 with methanol–dichloromethane [5 : 95] development. The ^{1}H NMR spectrum (CDCl$_3$) is as follows δ: 1.31 (t, 6 H, $J = 6.8$), 3.87 (d, 2 H, $J = 7$), 4.13 (m, 4 H), 5.34 (br s, 1 H, OH).

5. Five drops of reaction mixture is added to a mixture of 20 drops of diethyl ether and 1 drop of triethylamine. On TLC (Note 4) the product has $R_f \sim 0.4$ with ethyl acetate development. If TLC indicates the presence of diethyl hydroxymethylphosphonate an additional 5 g of dihydropyran and 10 drops of phosphorus oxychloride are added. The reaction is checked by TLC for completeness after 1 hr and is worked up at that time.

6. The checkers found that reactions run on up to nine times the present scale could be effected with only a small reduction in yield. Molecular still distillation (wall temperature 105–115°C, 0.10 mm) gave yields of 81–83%.

7. GLC analysis (0.5 × 200 cm 3% OV-17, 170°C, helium flow = 30 mL/min) shows the product with a retention time of 5 min and a purity greater than 97%. The ^{1}H NMR spectrum (CDCl$_3$) is as follows δ: 1.35 (t, 6 H, $J = 7$), 1.4–1.9 (m, 6 H), 3.4–4.45 (m, 8 H), 4.7 (m, 1 H).

3. Discussion

Diethyl [(2-tetrahydropyranyloxy)methyl]phosphonate is useful in the Wittig–Horner synthesis of enol ethers, which are intermediates in one-carbon homologations of carbonyl compounds.[2] This procedure is an adaptation of a general method for preparing dialkyl hydroxymethylphosphonates.[3] An *O*-tetrahydropyranyl derivative also has been prepared from dibutyl hydroxymethylphosphonate, and diethyl hydroxymethylphosphonate has been *O*-silylated with *tert*-butylchlorodimethylsilane and imidazole.[2] Another useful congener in this series has been prepared by an Arbuzov reaction of methoxyethoxymethyl (MEM) chloride and triethyl phosphite.[2]

1. Institute of Organic Chemistry, Syntex Research, Palo Alto, CA 94304.
2. Kluge, A. F.; Clousdale, I. S. *J. Org. Chem.* **1979**, *44*, 4847.
3. Zaripov, R. K.; Abramov, V. S. *Tr. Khim.-Met. Inst., Akad. Nauk Kaz. S.S.R.* **1969**, *5*, 50; *Chem. Abstr.* **1970**, *72*, 21745y.

N-(2,4-DIFORMYL-5-HYDROXYPHENYL)ACETAMIDE

[Acetamide, N-(2,4-diformyl-5-hydroxyphenyl)-]

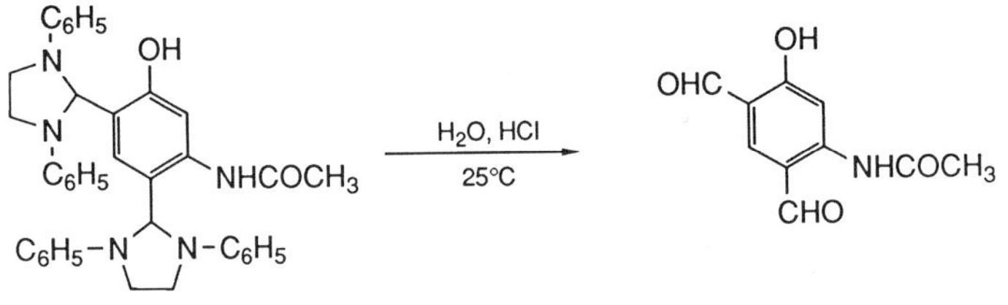

Submitted by Jürgen Hocker and Henning Giesecke[1]
Checked by G. Saucy, Harvey Gurien, and Gerald Kaplan

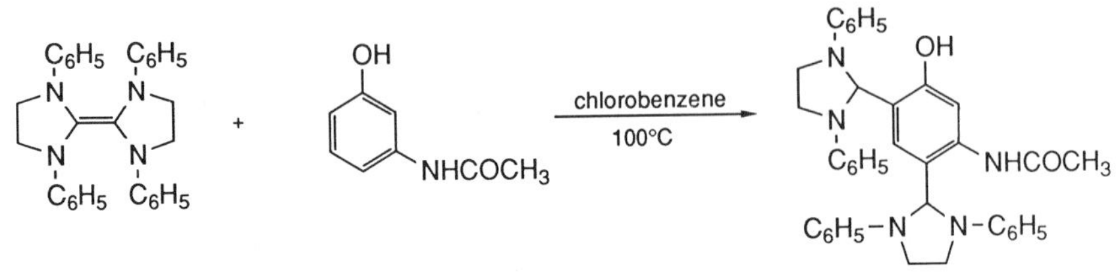

1. Procedure

A. *N-[2,4-Bis(1,3-diphenylimidazolidin-2-yl)-5-hydroxyphenyl]acetamide.* A 500-mL, three-necked flask equipped with nitrogen inlet, mechanical stirrer, reflux condenser, and thermometer is charged with 88.8 g (0.2 mol) of 1,1′,3,3′-tetraphenyl-2,2′-biimidazolidinylidene (Note 1) and 30.2 g (0.2 mol) of 3-acetamidophenol (Note 2). The system is flushed with and maintained under nitrogen, and 100 mL of chlorobenzene (Note 3) is added. The suspension is stirred at 100°C for 6 hr (Note 4). 2-Propanol (250 mL) is added to the hot mixture, which is then allowed to cool to room temperature. A pale-yellow solid precipitates which is filtered and washed with 200 mL of 2-propanol. Drying in vacuum affords 84.6–93.2 g (71–78%) of *N*-[2,4-bis(1,3-diphenylimidazolidin-2-yl)-5-hydroxyphenyl]acetamide, mp 254–256°C (Note 5).

B. *N-(2,4-Diformyl-5-hydroxyphenyl)acetamide.* A suspension of 95 g of the phenol (prepared as described under Section A) (Note 6) in 1 L of aqueous 10% hydrochloric acid is stirred for 2 hr at room temperature. The colorless solid (Note 7) is filtered by suction and twice suspended in water at 40°C. Final filtration is followed with a water wash (1 L, 40°C), and the solid is sucked down on the filter. Crystallization from 800 mL of acetonitrile affords 21.8–22.3 g (66–67.5%) of *N*-(2,4-diformyl-5-hydroxyphenyl)acetamide, mp 215–217°C (Note 8).

2. Notes

1. This material was prepared by the procedure of H. W. Wanzlick, *Org. Synth., Coll. Vol. V* **1973**, 115.

2. The checkers purchased 3-acetamidophenol from Aldrich Chemical Company, Inc.

3. The chlorobenzene was dried by azeotropic distillation.

4. The submitters checked the reaction progress by adding chloroform to an aliquot of the reaction mixture and observing it under long-wavelength UV light (350 mm). A bright fluorescence indicated incomplete reaction. The checkers found that further heating did not eliminate the fluorescence or improve the yield. Adding 1-g quantities of the phenol resulted in quenching of the fluorescence but a lower yield.

5. The product showed the following spectroscopic properties: ^{1}H NMR (d_7-DMF) δ: 1.92 (s, CH_3), 3.5–4.1 (m, 8 H, NCH_2), 6.05 (s, 1 H), 6.36 (s, 1 H), 6.5–7.5 (m, 23 H), 9.13 (s, 1 H).

6. The phenol should be ground in a mortar to eliminate lumps.

7. The checkers always obtained pale-yellow material.

8. The product had the following spectroscopic properties: IR (KBr) cm^{-1}: 1659, 1644, 1620; ^{1}H NMR (d_7-DMF) δ: 2.24 (s, CH_3), 8.22 (s, 1 H), 8.24 (d, $J <$ 1 Hz, 1 H), 9.87 (d, $J <$ 1 Hz, 1 H), 10.16 (s, 1 H), 11.35, 12.05 (2s, OH, NH).

3. Discussion

The reaction of 1,1′,3,3′-tetraphenyl-2,2′-biimidazolidinylidene with phenols illustrated in this procedure is a general method for the preparation of phenolaldehydes.[2] Table I gives the aldehydes that have been prepared by conditions similar to those described here.

Nitrogen-containing heterocyclic compounds such as indoles and imidazoles are also formylated by the electron-rich olefin. 3-Methylimidazole-5-carboxaldehyde can be

TABLE I

Aldehydes from Phenols and 1,1′,3,3′-Tetraphenyl-2,2′-Biimidazolidinylidene

Aldehyde	Yield (%)	
	Imidazolidin-2-yl-phenol	Hydrolysis
4-Hydroxybenzaldehyde	55	88
4-Hydroxy-3,5-dimethylbenzaldehyde	58	80
3-Cyclohexyl-5-methylsalicylaldehyde	43	80
4-Dimethylamino-5-methylsalicylaldehyde	65	54
4-Dimethylaminosalicylaldehyde	52	90
Resorcinol-2,4,6-tricarboxaldehyde	43	96
4-Hydroxy-5-methoxyisophthalaldehyde	55	81
4-Hydroxy-6-methoxyisophthalaldehyde	57	87
Pyrogallol-4,6-dicarboxaldehyde	82	50
8-Hydroxyquinoline-7-carboxaldehyde	76	69

prepared from 2-methylimidazole (yield 83%) and 2-phenylindole-3-carboxaldehyde from 2-phenylindole (yield 64%).

The formylation of phenols with the electron-rich olefin to give imidazolidin-2-yl-phenols is very selective and avoids mixtures of *o*- and *p*-isomers, which are frequently obtained by methods commonly employed for the synthesis of phenolaldehydes. Para substitution of the cyclic aminal group in the phenol is preferred. If the *p*-position is blocked or sterically hindered, the reaction proceeds via the *ortho*-aminals to salicylal-dehydes. Incorporation of more than one aldehyde group in the benzene nucleus is often achieved with hydroxy- and aminophenols.

Reaction of phenols bearing strong electron-withdrawing substituents such as dichlo-rophenols and nitrophenols with 1,1',3,3'-tetraphenyl-2,2'-biimidazolidinylidene results in poor yields. Oxidation of the electron-rich olefin by nitro compounds is also possible.

1. Wissenschaftliches Hauptlaboratorium der BAYER AG, D-5090 Leverkusen-Bayerwerk, Germany.
2. Hocker, J.; Merten, R. *Angew. Chem.* **1972**, *84*, 1022–1031; *Angew. Chem. Int. Ed. Engl.* **1972**, *11*, 964–973; Hocker, J.; Giesecke, H.; Merten, R. *Angew. Chem.* **1976**, *88*, 151; *Angew. Chem. Int. Ed. Engl.* **1976**, *15*, 169–170; Giesecke, G.; Hocker, J. *Justus Liebigs Ann. Chem.* **1978**, 345–361,

STEREOCONTROLLED IODOLACTONIZATION OF ACYCLIC OLEFINIC ACIDS: THE *trans* AND *cis* ISOMERS OF 4,5-DIHYDRO-5-IODOMETHYL-4-PHENYL-2(3*H*)-FURANONE

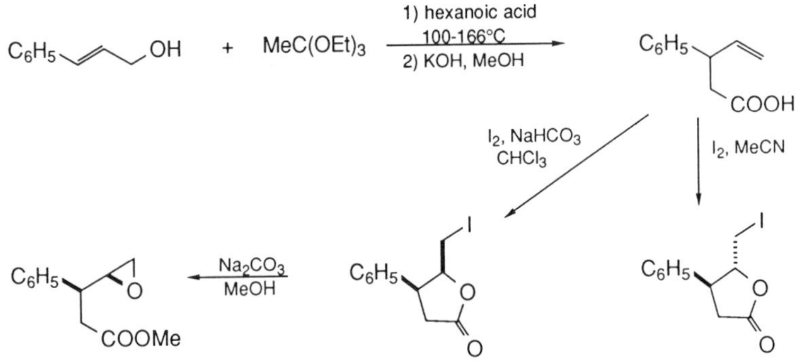

Submitted by F. Bermejo Gonzalez and Paul A. Bartlett[1]
Checked by Pauline J. Sanfilippo and Andrew S. Kende

1. Procedure

A. *3-Phenyl-4-pentenoic acid.* A mixture of 33.7 g (0.25 mol) of cinnamyl alcohol (Note 1), 46.1 mL (0.25 mol) of triethyl orthoacetate (Note 1), and 0.19 mL

(1.5 mmol) of hexanoic acid (Note 2) is placed in a 250-mL, round-bottomed flask equipped with a thermometer, Claisen head, and condenser. The solution is heated in an oil bath with distillation of ethanol. After 3 hr, distillation of ethanol slows and another 0.1-mL portion of hexanoic acid is added. Additional portions (0.1 mL) of the catalyst are added again at 3.5 and 4.5 hr. After 6 hr, a total of 27 mL of ethanol, out of a theoretical 29.2 mL, has been collected, and GC analysis (Note 3) indicates that no cinnamyl alcohol remains. Over this 6-hr period the internal temperature rises from 100 to 166°C.

The solution is allowed to cool, and 19.7 g (0.35 mol) of potassium hydroxide in 25 mL of water and 75 mL of methanol is added. The mixture is heated under reflux for 1 hr under nitrogen. After the alkaline solution is allowed to cool to room temperature, it is washed with ether and acidified with concentrated HCl. The acidic solution is extracted with three 50-mL portions of ether, and the organic layer is dried over $MgSO_4$, filtered, and concentrated under reduced pressure. The yield of crude 3-phenyl-4-pentenoic acid is 38–39 g (86–88%). This material is essentially pure by NMR analysis and can be used directly as starting material for the following iodolactonization reactions. The acid can be further purified by crystallization from hexane (86% recovery in two crops) to give product melting at 44–46°C.

B. *Thermodynamically controlled formation of the trans (4RS,5SR) isomer of 4,5-dihydro-5-iodomethyl-4-phenyl-2(3H)-furanone.* In a 500-mL, round-bottomed flask equipped with a mechanical stirrer (Note 4) and immersed in an ice bath is placed a solution of 10 g (0.057 mol) of 3-phenyl-4-pentenoic acid in 200 mL of acetonitrile (Note 5). Solid iodine (44.5 g, 0.18 mol) (Note 6) is added, and the mixture is protected from light and stirred at 0°C under nitrogen for 24 hr. The mixture is partitioned between 100 mL of ether and 100 mL of saturated aq $NaHCO_3$. The organic layer is washed with 10% aqueous $Na_2S_2O_3$ until colorless, and with water and brine. It is then dried over $MgSO_4$, the solvent is removed at reduced pressure, and the crude trans iodolactone is obtained as a thick oil; weight 14.5–15.6 g (85–91%). NMR analysis indicates that the trans : cis ratio is at least 95 : 5 (Note 7).

C. *Kinetically-controlled formation of the cis (4RS,5RS) isomer of 4,5-dihydro-5-iodomethyl-4-phenyl-2(3H)-furanone.* A mixture of 10 g (0.057 mol) of 3-phenyl-4-pentenoic acid, 9.1 g (0.11 mol) of $NaHCO_3$ and 200 mL of water is placed in a 1000-mL round-bottomed flask equipped with a mechanical stirrer (Note 4) and stirred until a homogeneous solution is obtained. Chloroform (200 mL) is added, the mixture is cooled in an ice bath, and 28.4 g (0.112 mol) (Note 8) of iodine is added. The mixture is stirred at 0°C for 6 hr, the layers are separated, and the organic phase is washed with 10% aqueous $Na_2S_2O_3$ until colorless, and with water and brine. The organic layer is dried over $MgSO_4$, the solvent is removed under reduced pressure, and the crude *cis*-iodolactone is obtained as a semisolid: weight 15.5–16.3 g (91–95%), mp 75–90°C (Note 9). Direct recrystallization of this material from diisopropyl ether (Note 10) affords 9.0–9.5 g (52–55%) of material, mp 103–104°C, with a cis : trans ratio of 15–16 : 1. Further recrystallization from diisopropyl ether gives (in two crops) 8.3–8.9 g (48–52%) of product, mp 104–105°C, with a purity of ≥98%. Additional product can be obtained from the mother liquors.

2. Notes

1. The reagents employed were obtained from Aldrich Chemical Company, Inc. and used as received.

2. Propionic acid may also be used as catalyst; however, its boiling point (141°C) is below that of the reaction temperature at the end of the reaction. The use of *o*-nitrophenol as catalyst resulted in a lower yield in this case.

3. Gas chromatographic analysis was performed on a Varian model 940 gas chromatograph equipped with a 6-in. × ⅛-in. column of 5% OV-101 on Gas-Chrom Q at a column temperature of 155°C.

4. A magnetic stirrer is not recommended because the mixture becomes very thick.

5. Mallinckrodt Inc. analytical reagent-grade acetonitrile was used.

6. Two equivalents of iodine are required by the stoichiometry of the reaction, because of formation of HI_3. In our experience, however, the reaction does not proceed to completion without additional reagent.

7. The crude trans isomer obtained in this way is of suitable purity for conversion to the epoxy ester, as described below (see Discussion). It may be further purified by column chromatography; however, attempted vacuum distillation leads to considerable decomposition.

8. Two equivalents of iodine are required because of formation of NaI_3.

9. NMR analysis of the crude iodolactone indicates that the cis : trans ratio is about 3.4 : 1.

10. The recrystallization is performed using ca. 15 mL of diisopropyl ether per gram of crude product. Dichloromethane, 1–2%, is also added to the hot mixture to effect complete solution before cooling.

3. Discussion

Iodolactonization has become a useful reaction for the stereocontrolled introduction of chiral centers in both cyclic[2] and acyclic[3-9] systems. Depending on the reaction conditions, the cyclization can be carried out under either kinetic or thermodynamic control.[3, 10] The contrast between the stereochemical course of the two procedures is not always as dramatic as with 3-phenyl-4-pentenoic acid, as illustrated by the examples in Table I.[11] A procedure for obtaining high 1,3-asymmetric induction in formation of valerolactones by cyclization of *N*,*N*-dimethylamides has been reported,[12] and the tendency for an oxygen substituent in the allylic position to control the stereochemistry of electrophilic cyclization has been addressed.[13]

Conversion of iodolactones into the corresponding epoxy esters is often one of the major steps in their utilization for the purposes of stereocontrol.[3-7, 14] Methanolysis of the cis isomer of 4,5-dihydro-5-iodomethyl-4-phenyl-2(3*H*)-furanone to methyl (3*RS*,4*RS*)-4,5-epoxy-3-phenylpentanoate is a representative procedure for this transformation.

A mixture of 5.0 g (16.6 mmol) of (4*RS*,5*RS*)-4,5-dihydro-5-iodomethyl-4-phenyl-2(3*H*)-furanone (cis isomer), 75 mL of methanol, and 1.8 g (17.0 mmol) of finely powdered, anhydrous Na_2CO_3 is placed in a 250-mL, round-bottomed flask equipped

TABLE I
Iodolactonization of Olefinic Acids

	Trans : Cis Ratio [Yield (%)]	
Substrate	"Thermodynamic Control"	"Kinetic Control"
3-Methyl-4-pentenoic acid	10 : 1 (84)	1 : 3 (82)
4-Methyl-5-hexenoic acid	10 : 1 (77)	1 : 2.3 (83)
3-Methyl-5-hexenoic acid	1 : 6 (81)	1 : 3 (97)
2-Methyl-5-hexenoic acid	1.1 : 1 (68)	1.8 : 1 (78)
(2 RS,4 SR)-2,4-Dimethyl-5-hexenoic acid	20 : 1 (89)	3.5 : 1

with a mechanical stirrer and heated under nitrogen at reflux for 8 hr. The resulting solution is concentrated under reduced pressure to a volume of 50 mL and partitioned between 100 mL of water and 100 mL of ether. The organic layer is washed with water and brine, dried over $MgSO_4$, and evaporated to give 3.1 g (91%) of crude product. This material, which shows only minor impurities by NMR spectroscopy, can be further purified by chromatography (silica gel/1 : 1 ether : hexane) (98% recovery) and bulb-to-bulb distillation (78°C/0.045 mm) (82% recovery).

1. Department of Chemistry, University of California, Berkeley, CA 94720.
2. Dowle, M. D.; Davies, D. I. *Chem. Soc. Rev.* **1979**, *171*, and references cited therein.
3. Bartlett, P. A.; Myerson, J. *J. Am. Chem. Soc.* **1978**, *100*, 3950.
4. Chamberlin, A. R.; Dezube, M.; Dussault, P. *Tetrahedron Lett.* **1981**, *22*, 4611.
5. Bartlett, P. A.; Myerson, J. *J. Org. Chem.* **1979**, *44*, 1625.
6. Collum, D. B.; McDonald, J. H., III; Still, W. C. *J. Am. Chem. Soc.* **1980**, *102*, 2118.
7. Mori, K.; Umemura, T. *Tetrahedron Lett.* **1982**, *23*, 3391.
8. Williams, D. R.; Barner, B. A.; Nishitani, K.; Phillips, J. G. *J. Am. Chem. Soc.* **1982**, *104*, 4708.
9. Bartlett, P. A. *Tetrahedron* **1980**, *36*, 2.
10. Barnett, W. E.; Sohn, W. H. *J. Chem. Soc., Chem. Commun.* **1972**, 472; *Tetrahedron Lett.* **1972**, 1777.
11. Myerson, J., Ph.D. Dissertation, University of California, Berkeley, 1980.
12. Tamaru, Y.; Mizutani, M.; Furukawa, Y.; Kawamura, S.-I.; Yoshida, Z.-i.; Yanagi, K.; Kazunori, Y.; Minobe, M. *J. Am. Chem. Soc.* **1984**, *106*, 1079–1085.
13. Chamberlin, A. R.; Mulholland, R. L.; Kahn, S. D.; Hehre, W. J. *J. Am. Chem. Soc.* **1987**, *109*, 672–679.
14. House, H. O.; Carlson, R. G.; Muller, H.; Noltes, A. W.; Slater, C. D. *J. Am. Chem. Soc.* **1962**, *84*, 2614; Takano, S.; Hirama, M.; Ogasawara, K. *J. Org. Chem.* **1980**, *45*, 3729; Still, W. C.; Galynker, I. *J. Am. Chem. Soc.* **1982**, *104*, 1774.

OZONOLYTIC CLEAVAGE OF CYCLOHEXENE TO TERMINALLY DIFFERENTIATED PRODUCTS: METHYL 6-OXOHEXANOATE, 6,6-DIMETHOXYHEXANAL, METHYL 6,6-DIMETHOXYHEXANOATE

(Hexanoic acid, 6-oxo-, methyl ester; hexanal, 6,6-dimethoxy-; hexanoic acid, 6,6-dimethoxy-, methyl ester)

A. ⬡ 1) O₃, MeOH, CH₂Cl₂
       2) TsOH
       3) NaHCO₃ ────────▶ (MeO)₂CH⌇⌇⌇CHO
       4) Me₂S

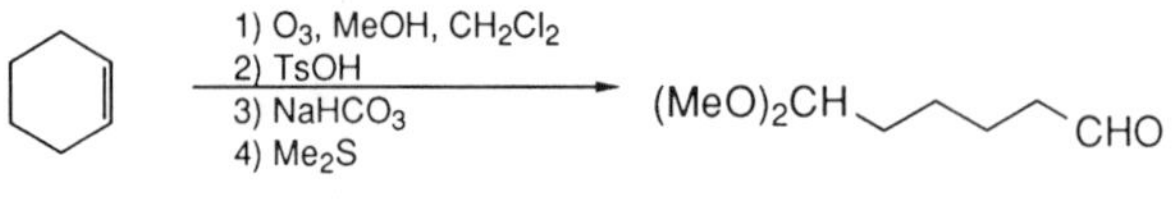

B. ⬡ 1) O₃, MeOH, CH₂Cl₂
       2) Ac₂O, Et₃N ────────▶ MeOOC⌇⌇⌇CHO

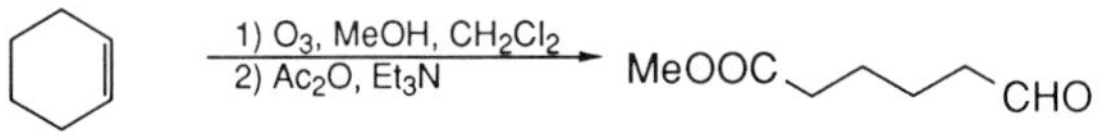

C. ⬡ 1) O₃, MeOH, CH₂Cl₂
       2) TsOH
       3) Ac₂O, Et₃N ────────▶ MeOOC⌇⌇⌇CH(OMe)₂

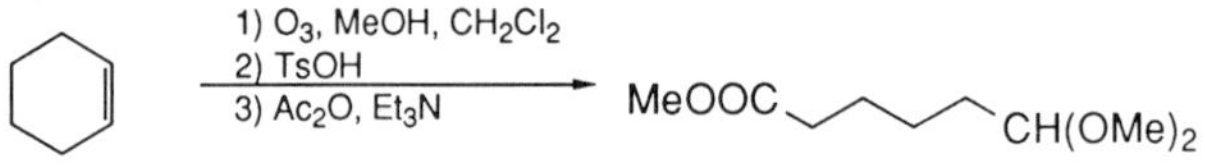

Submitted by RONALD E. CLAUS and STUART L. SCHREIBER[1]
Checked by NAKCHEOL JEONG and MARTIN F. SEMMELHACK

1. Procedure

Caution! The ozonolysis reaction produces peroxidic intermediates that can present a potential explosion hazard. Accordingly, it is recommended that the following experiments be carried out in a hood and behind a safety shield.

A. A 500 mL, three-necked, round-bottomed flask is fitted with a glass tube to admit ozone, a calcium chloride drying tube, a glass stopper, and a magnetic stirring bar and is charged with 6.161 g of cyclohexene (0.075 mol), 250 mL of dichloromethane, and 50 mL of methanol (Note 1). The flask is cooled to ca. −78°C (2-propanol–dry ice), and ozone (Note 2) is bubbled through the solution with stirring. When the solution turns blue, ozone addition is stopped. Nitrogen is passed through the solution until the blue color is discharged (Note 3) and then the cold bath is removed. The drying tube and ozone inlet are replaced with a stopper and rubber septum, and 1.215 g of *p*-toluenesulfonic acid (TsOH) (10% w/w) (Note 4) is added. The solution is allowed to warm to room temperature as it stirs under an atmosphere of nitrogen for 90 min. Anhydrous sodium bicarbonate (2.147 g, 4 mol-equiv) is added to the flask and the mixture is stirred for 15 min, and then 12 mL of dimethyl sulfide (0.150 mol) (Note 5) is added. After being stirred for 12 hr, the heterogeneous mixture is concentrated to approximately 50 mL by rotary evaporation. Dichloromethane (100 mL) is added and the mixture is washed with 75 mL of water (Note 6). The aqueous layer is extracted with two more 100-mL portions of dichloromethane, and the combined organic layers are washed with 100 mL of water. After extracting the aqueous layer with 100 mL of dichloromethane, the organic layers are dried over anhydrous magnesium sulfate, filtered, and concentrated by rotary evaporation. Short-path distillation of the crude product (Note 7) gives 8.2–8.4 g of 6,6-dimethoxyhexanal, 68–70%, bp 80–82°C/1.75 mm (Notes 8 and 9).

B. A round-bottomed flask equipped as in Step A is charged with 6.161 g of cyclohexene (0.075 mol), 250 mL of dichloromethane, 50 mL of methanol, and 2.0 g of anhydrous sodium bicarbonate (Notes 1 and 10). After the apparatus is cooled to ca. −78°C, ozone (Note 2) is bubbled through the solution as it is stirred. Ozone addition is stopped when the solution turns blue. Nitrogen is passed through the solution until the blue color is discharged (Note 3) and then the cold bath is removed. The solution is filtered into a 1-L, round-bottomed flask and 80 mL of benzene is added. The volume is reduced to approximately 50 mL by rotary evaporation (Note 11). After dilution with 225 mL of dichloromethane the flask is cooled to 0°C and 16 mL of triethylamine (0.113 mol) and 21.24 mL of acetic anhydride (0.225 mol) are added via syringe (Note 12), and the solution is stirred under a nitrogen atmosphere for 15 min. The ice bath is removed and stirring is continued for 4 hr. The solution is washed with 150-mL portions of aqueous 0.1 N hydrochloric acid, aqueous 10% sodium hydroxide, and water. The organic layer is dried over anhydrous magnesium sulfate and filtered, and the solvent is removed by rotary evaporation. Short-path distillation of the crude product yields methyl 6-oxohexanoate, (7.0–7.8 g, 65–72%), bp 83–86°C/1.5 mm (Note 13).

C. Cyclohexene, 6.161 g (0.075 mol), is stirred with ozone in dichloromethane and methanol, as above. The resulting solution is treated with p-toluenesulfonic acid and subsequently neutralized with sodium bicarbonate, as described in Procedure A. The solution is filtered into a 1-L, round-bottomed flask, 80 mL of benzene is added, and the volume is reduced to approximately 50 mL by rotary evaporation (Note 11). Dilution with dichloromethane, treatment with triethylamine and acetic anhydride, and workup as described in Procedure B followed by short-path distillation provides methyl (6,6-dimethoxy)hexanoate, (11.2–11.8 g, 78–83%), bp 87–91°C/1.5 mm (Note 14).

2. Notes

1. Cyclohexene was purchased from Aldrich Chemical Company, Inc. and used without purification. Dichloromethane was distilled from calcium hydride. Methanol was distilled from magnesium. The methanol–dichloromethane solvent combination may be replaced with 100% methanol (250 mL) with comparable results.

2. Ozone was produced by a Welsbach Corporation Ozonator, style T-709, with the voltage set at 100 V and oxygen pressure at 7 psi to give approximately 2% ozone concentration. The input oxygen was passed through a column of Hammond Drierite to ensure dryness.

3. The blue color indicates that cleavage of the olefin is complete. Excess ozone is removed to prevent overoxidation

4. Although the ozonolysis product exists in oligomeric form, the amount of acid used was calculated by assuming a theoretical yield of the corresponding monomeric aldehyde–methoxy hydroperoxide. p-Toluenesulfonic acid monohydrate, purchased from Aldrich Chemical Company, Inc., was not purified further.

5. The solution is neutralized to prevent bisacetal formation on subsequent reduction. Dimethyl sulfide was purchased from Aldrich Chemical Company, Inc. and used without purification.

6. An aqueous workup facilitates the removal of dimethyl sulfoxide produced by the reduction of the peroxide.

7. Typically 12.4–13.0 g of crude product is obtained after solvent removal. Material of this quality is satisfactory for most subsequent reactions.

8. The distilled product is similar in purity to the crude material. A small amount of dimethyl sulfoxide and minor impurities remain. Purification of the crude product by flash chromatography (1 : 1 ether : hexanes) affords 6,6-dimethoxyhexanal that is pure by ^{1}H and ^{13}C NMR in 90–95% yield.

9. The following spectral properties of the product were observed: ^{1}H NMR (CDCl$_3$), δ: 1.4–1.7 (m, 6 H), 2.4 (t, 2 H, $J = 7$), 3.3 (s, 6 H), 4.3 (t, 1 H, $J = 5.3$), 9.7 (t, 1 H, $J = 2.5$). ^{13}C NMR (CDCl$_3$), δ: 21.4, 23.7, 31.8, 43.2, 52.1, 103.9, 201.6. IR (film), cm^{-1}: 2700, 1720, 1100. MS, m/e (rel. %): 113(95), 57(100).

10. Sodium bicarbonate serves to buffer the solution and prevent acetal formation.

11. Benzene is added to facilitate the removal of methanol. Although an aqueous wash will remove the methanol, azeotropic removal with benzene is simpler and provides a slightly higher yield.

12. Triethylamine purchased from Aldrich Chemical Company, Inc., was distilled from calcium hydride. Acetic anhydride as supplied by Mallinckrodt, Inc. was distilled from phosphorus pentoxide.

13. The following spectral properties were observed: ^{1}H NMR (CDCl$_3$), δ: 1.5–1.7 (m, 4 H), 2.2–2.4 (m, 4 H), 3.6 (s, 3 H), 9.7 (t, 1 H, $J = 2.5$). ^{13}C NMR (CDCl$_3$), δ: 21.1, 24.0, 33.2, 42.9, 51.0, 173.1, 201.4. IR (film), cm^{-1}: 2700, 1720, 1150. MS: m/e (rel. %): 159(1), 29(3), 75(100).

14. The following spectral properties were observed: ^{1}H NMR (CDCl$_3$), δ: 1.0–1.6 (m, 6 H), 2.15 (t, 2 H, $J = 8$), 3.2 (s, 6 H), 3.6 (s, 3 H), 4.25 (t, 1 H, $J = 5.5$). ^{13}C NMR (CDCl$_3$), δ: 23.7, 24.3, 31.8, 33.4, 50.7, 52.0, 103.9, 173.1. IR (film), cm^{-1}: 1735, 1050, MS: m/e (rel. %): 159(10), 127(30), 75(100).

3. Discussion

This procedure illustrates a recently published method for the ozonolytic cleavage of cycloalkenes to terminally differentiated products.[2] Other examples of the unsymmetrical cleavage of olefins have been reported.[3] In addition, the title compounds have been prepared by other routes. Methyl 6-oxohexanoate has been synthesized from the acid chloride of the half-ester of adipic acid.[4] It has also been prepared from ε-caprolactone by methanolysis followed by oxidation.[4] Lead tetraacetate treatment of 2-hydroxycyclohexanone in methanol and subsequent acidification produces methyl 6,6-dimethoxyhexanoate.[5] A three-step route from cyclohexanone enol acetate (ozonolysis in methanol and reaction with dimethyl sulfide, then with trimethyl orthoformate) has been reported.[6] 6,6-Dimethoxyhexanal has been made by a multistep route.[7]

The present method utilizes commercially available cycloalkenes and proceeds under mild conditions to provide synthetically useful products. The method was shown to be general in the series of cycloalkenes investigated. Yields range from moderate (cyclopentene) to excellent (higher homologues).

The ozonolytic cleavage of cycloalkenes in the presence of methanol produces a chain with an aldehyde and a methoxy hydroperoxide group at the termini.[8] The unsymmetrical ozonolysis product is manipulated in several ways. Dehydration of the methoxy hydroperoxide group affords an ester (Step B). Alternatively, the aldehyde moiety is protected as an acetal. Under these conditions, the methoxy hydroperoxide is reduced[9] (Procedure A) or dehydrated (Step C).

1. Department of Chemistry, Yale University, New Haven, CT 06520. Present address for SLS: Department of Chemistry, Harvard University, Cambridge, MA 02138.
2. Schreiber, S. L.; Claus, R. E.; Reagan, J. *Tetrahedron Lett.* **1982**, *23*, 3867.
3. Sato, T.; Maemoto, K.; Kohda, A. *J. Chem. Soc., Chem. Commun.* **1981**, 1116; Odinokov, V. N.; Tolstikov, G. A.; Galeyeva, R. I.; Karagapoltseva, T. A. *Tetrahedron Lett.* **1982**, *23*, 1371 [See reference 6: French Patent 2309506 (1976); *RZh Khim* **1977**, *24H*, 68]; Bailey, P. S.; Erickson, R. E. *Org. Synth., Coll. Vol. V* **1973**, 489, 493; Bailey, P. S.; Bath, S. S.; Dobinson, F.; Garcia-Sharp, F. J.; Johnson C. D. *J. Org. Chem.* **1964**, *29*, 697; Besten, I. E. D.; Kinstle, T. H. *J. Am. Chem. Soc.* **1980**, *102*, 5968; Trost, B. M.; Ochiai, M.; McDougal, P. G. *J. Am. Chem. Soc.* **1978**, *100*, 7103.
4. Vasilevskis, J.; Gualtieri, J. A.; Hutchings, S. D.; West, R. C.; Scott, J. W.; Parrish, D. R.; Bizzarro, F. T.; Field, G. F. *J. Am. Chem. Soc.* **1978**, *100*, 7423.
5. Hurd, C. D.; Saunders, W. H., Jr. *J. Am. Chem. Soc.* **1952**, *74*, 5324.
6. Büchi, G.; Wuest, H. *Helv. Chim. Acta* **1979**, *62*, 2661.
7. Boeckman, R. K., Jr.; Blum, D. M.; Arthur, S. D. *J. Am. Chem. Soc.* **1979**, *101*, 5060.
8. Bailey, P. S. "Ozonation in Organic Chemistry," Academic Press: New York, 1978; Vol. 1.
9. Pappas, J. J.; Keaveney, W. P.; Gancher, E.; Berger, M. *Tetrahedron Lett.* **1966**, 4273.

PALLADIUM–PHOSPHINE-COMPLEX-CATALYZED REACTION OF ORGANOLITHIUM COMPOUNDS AND ALKENYL HALIDES: (Z)-β-[2-(N,N-DIMETHYLAMINO)PHENYL]STYRENE

[Benzenamine, N,N-dimethyl-2-(2-phenylethenyl)-, (Z)-]

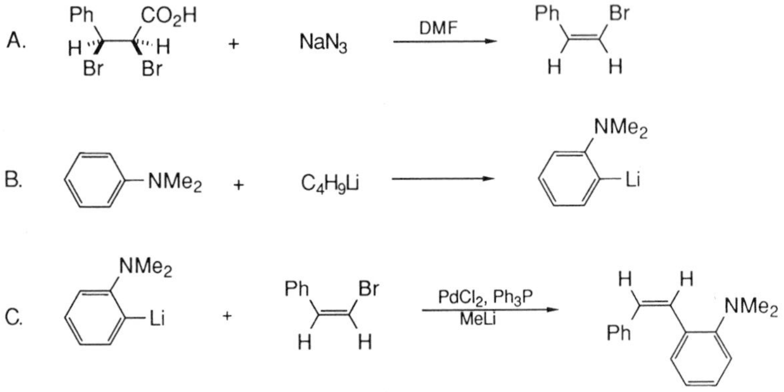

A.

B.

C.

Submitted by SHUN-ICHI MURAHASHI, TAKESHI NAOTA, and YOSHIO TANIGAWA[1]
Checked by JOSEPH FORTUNAK and IAN FLEMING

1. Procedure

Caution! The reaction described in Section (Step) A (Note 1) should be carried out in a well-ventilated hood because bromine is toxic.

A. *(Z)-β-Bromostyrene.* In a 1-L, round-bottomed flask equipped with a magnetic stirring bar are placed 30.8 g (0.100 mol) of *erythro-α,β-dibromo-β-phenylpropionic acid* (Note 1), 13.0 g (0.200 mol) of sodium azide (Note 2), and 500 mL of dry *N,N-dimethylformamide* (Note 3). The reaction mixture is stirred at room temperature for 8 hr and poured into a mixture of 300 mL of ether and 300 mL of water. The organic layer is separated, washed with three 100-mL portions of water, dried over magnesium sulfate, and filtered. After evaporation of the filtrate with a rotary evaporator, the residual liquid is distilled under reduced pressure, giving 13.5–13.9 g (74–76%) of (Z)-β-bromostyrene, bp 54–56°C (1.5 mm) (Note 4).

B. *2-(N,N-Dimethylamino)phenyllithium.* A 100-mL, three-necked, round-bottomed flask equipped with a magnetic stirring bar and a reflux condenser connected to a nitrogen-inlet tube is capped with serum stoppers and flushed with nitrogen. The flask is charged with 18.2 g (0.150 mol) of *N,N-dimethylaniline* (Note 5) and 33.4 mL (0.050 mol) of a 1.50 *M* solution of butyllithium in hexane (Note 6). While a continuous positive nitrogen pressure is maintained, the solution is heated at reflux (in a 90–95°C bath) with stirring for 20 hr and then cooled to room temperature (Note 7).

C. *(Z)-β-[2-(N,N-Dimethylamino)phenyl]styrene.* A 1-L, three-necked, round-bottomed flask equipped with a magnetic stirring bar, a reflux condenser connected to a nitrogen-inlet tube, and a 300-mL, pressure-equalizing dropping funnel is capped with serum stoppers. The flask is flushed with nitrogen and charged with 0.433 g (0.0025 mol) of palladium chloride (Note 8), 2.62 g (0.010 mol) of triphenylphosphine (Note 9), and 300 mL of benzene (Note 10). While a continuous positive nitrogen pressure is maintained, the mixture is stirred at gentle reflux for 30 min, and then 4.25 mL (0.0060 mol) of a 1.41 M solution of methyllithium in ether (Note 11) is added with a syringe. After an additional 10 min at reflux, 9.15 g (0.050 mol) of (Z)-β-bromostyrene as prepared in Section A is added in one portion with a syringe, and the mixture is heated at reflux for 10 min. The solution of 2-(N,N-dimethylamino)phenyllithium prepared as described in Section B is transferred to the dropping funnel with a syringe and diluted by adding 150 mL of benzene (Notes 10 and 12). The resulting solution is then added dropwise to the mixture with stirring at reflux over a period of 30 min (Note 13). After additional stirring for 10 min, the resulting red solution is cooled to room temperature with the help of an ice bath and quenched by adding 100 mL of saturated aqueous ammonium chloride. The organic layer is separated, washed successively with 100 mL of water and 100 mL of saturated aqueous sodium chloride, and then dried over magnesium sulfate and filtered. The solvent is evaporated with a rotary evaporator, and the residue is distilled under reduced pressure to give a forerun (ca. 11 g) of excess N,N-dimethylaniline, bp 31–51°C (1 mm), followed by 7.4–7.5 g (66–67%) of (Z)-β-[2-(N,N-dimethylamino)phenyl]styrene, bp 90.0–92.0°C (0.035 mm), 82–84°C (0.01 mm), as a pale-yellow liquid (Note 14).

2. Notes

1. *erythro-α,β*-Dibromo-β-phenylpropionic acid is prepared from *trans*-cinnamic acid (mp 133–134°C) (Nakarai Chemicals, Japan) by the method used for ethyl α,β-dibromo-β-phenylpropionate (Abbott T. W.; Althousen, D. *Org. Synth., Coll. Vol. II* **1943,** 270) in 83% yield, mp 199–200°C. The checkers used benzene (400 mL per mol) in place of the carbon tetrachloride, because the mixture was then easier to stir and the reaction was more reproducible. The yield before purification was 89% (mp 174–191°C); the yield after recrystallization was 81% (mp 198–199°C). Crude material could be used without appreciable loss of yield.

2. Sodium azide from Wako Pure Chemical Ind., Japan, was used without purification.

3. N,N-Dimethylformamide is distilled over calcium hydride.

4. Gas chromatographic analysis of the distillate (10% PEG-20M on 60–80-mesh, Celite 545 AW, 1-m × 4-mm column, column temperature 100–220°C, injection temperature 200°C) shows that the product is 100% isomerically pure. The spectral properties of the (Z)-β-bromostyrene are as follows: IR (neat) cm^{-1}: strong absorptions at 3095, 3040, 1620, 1500, 1450, 1333, 1032, 930, 920, 830, 770, and 700; ^{1}H NMR (CHCl$_3$) δ: 6.43 (doublet, 1 H, $J = 8$, PhCH=C), 7.08 (doublet, 1 H, $J = 8$, PhC=CHBr), 7.22–7.85 (multiplet, 5 H, aromatic). The checkers also purified the residual oil before distillation by filtration in 250 mL of pentane through three times its weight of silica gel (70–230-mesh) followed by evaporation. The yield before distillation was then reproducibly

84%, distillation was avoided, and the next step proceeded with undiminished yield.

5. N,N-Dimethylaniline from Nakarai Chemicals was dried over calcium hydride and freshly distilled. Three molar equivalents of N,N-dimethylaniline are used to achieve complete conversion of the butyllithium, because in the present particular case free butyllithium, if present, causes the isomerization of the (Z)-alkene to the (E)-isomer.

6. A solution of butyllithium in hexane was obtained from Aldrich Chemical Company, Inc. Before use the solution is titrated with a 1 M solution of 2-butanol in xylene according to the procedure of Watson and Eastham[2] (see Gall, M.; House, H. O. Org. Synth. Coll. Vol. VI 1988, 121) with 2,2'-biquinoline as indicator.

7. The resulting cloudy, yellowish orange solution should be used within 3–4 hr.

8. Palladium chloride from Inuisho Precious Metal Company, Japan, was used without purification.

9. Triphenylphosphine from Nakarai Chemicals, Japan, was used without purification.

10. Benzene is distilled over benzophenone ketyl and stored under a nitrogen atmosphere.

11. A solution of methyllithium in ether is prepared from lithium wire and methyl bromide according to the literature procedure[3] and titrated by the same method as Note 6. The checkers used 1.1 M methyllithium from Aldrich Chemical Co., Inc.

12. Without the dilution, (Z)-1,4-diphenyl-1-buten-3-yne is detected, apparently formed from the cross-coupling with phenylacetylide, derived from lithiation of β-bromostyrene, followed by E2cB elimination or Fritsch–Butlenberg–Wiechell-type rearrangement.

13. Prolonged reaction time causes the isomerization of (Z)-β-[2-(N,N-dimethylamino)phenyl]styrene to the (E)-isomer.

14. Gas chromatographic analysis of the product (5% silicone SE 30 on 80–100-mesh Chromosorb W AB, 0.5-m × 4-mm column, column temperature 100–250°C, injection temperature 180°C) shows that the product is at least 98% (Z)-isomer. The spectral properties of the (Z)-alkene are as follows; IR (neat) cm^{-1}: strong absorptions at 3070, 3025, 2950, 2870, 2835, 2780, 1600, 1490, 1450, 1320, 1190, 1160, 1140, 1100, 1050, 950, 780, 750, and 690; ^{1}H NMR (CCl$_4$) δ: 2.76 (singlet, 6 H, CH$_3$—N), 6.38 (doublet, 1 H, J = 12.3, PhC≡CH), 6.63 (doublet, 1 H, J = 12.3, PhCH=C), 6.50–7.30 (multiplet, 9 H, aromatic).

3. Discussion

The starting materials, (Z)-β-bromostyrene[4] and 2-(N,N-dimethylamino)phenyllithium[5] have been prepared in satisfactory yields by known procedures after slight modifications. The azide procedure[4] gives higher stereospecificity than the earlier procedure using sodium bicarbonate.[6]

This procedure illustrates a general method for the preparation of alkenes from the palladium(0)-catalyzed reaction of vinyl halides with organolithium compounds,[7] which can be prepared by various methods, including direct regioselective lithiation of hydrocarbons.[8] The method is simple and has been used to prepare a variety of alkenes stereoselectively. Similar stoichiometric organocopper reactions sometimes proceed in a nonstereoselective manner[9] and in low yields.[10] Nickel catalysts can be used efficiently for the reaction of alkenyl halides with Grignard reagents but not with organolithium

TABLE I

PALLADIUM-CATALYZED REACTION OF ORGANOLITHIUM COMPOUNDS AND ALKENYL HALIDES[a]

Halides	RLi	Products	Yield (%)[b]
(Z)-C_6H_5CH=CHBr	CH_3Li	(Z)C_6H_5CH=CHCH_3	90
	C_4H_9Li	(Z)C_6H_5CH=CHC_4H_9	62
	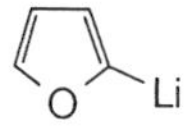	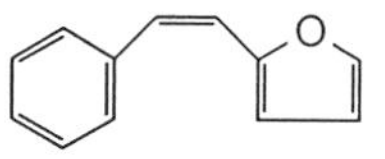	85
	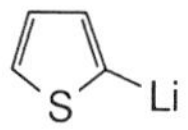	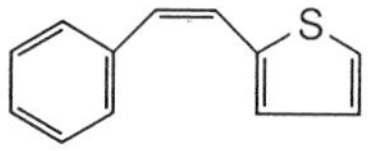	94[c]
	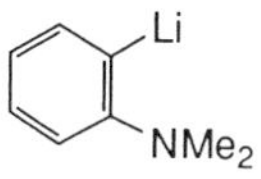	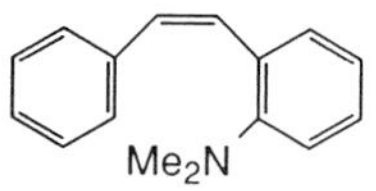	87 (66-67)[d]
	C_6H_5SLi	(Z)-C_6H_5CH=CHSC_6H_5	95[c]
	C_2H_5SLi	(Z)-C_6H_5CH=CHSC_2H_5	93[c]
(E)-C_6H_5CH=CHBr	CH_3Li	(E)-C_6H_5CH=CHCH_3	88
	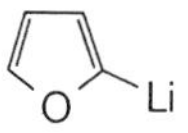	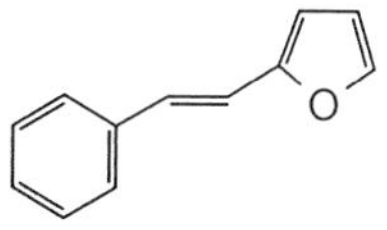	85
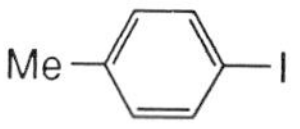 Me—〈 〉—I	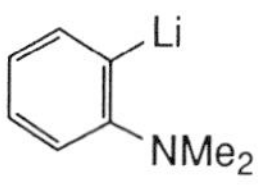	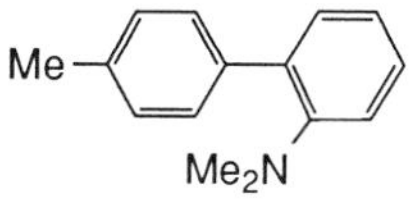 Me—〈 〉—〈 〉Me₂N	(89)[d]

[a] The reaction was carried out on a 1.0–1.5-mmol scale.
[b] Determined by gas chromatography.
[c] Tetrakis-(triphenylphosphine)palladium was used.
[d] Isolated yield.

compounds.[11] Highly reactive zerovalent palladium catalyst can be directly generated in situ from $PdCl_2$–PPh_3–CH_3Li. Tetrakis(triphenylphosphine)palladium can be used alternatively. Grignard reagents undergo the reaction as well with vinyl halides. Organolithium compounds require the limited reaction conditions under which the elimination of alkenyl halides producing lithium acetylides is slower than the cross-coupling reaction.[7] The choice of benzene as a solvent and the dilution of the solution satisfy the above conditions. The palladium-catalyzed alkylation of aryl halides with organolithium compounds proceeds efficiently without such difficulty.[7] Similar reactions with lithium thiolates give the corresponding alkenyl sulfides.[7] Representative reactions of organolithium compounds are shown in Table I.

1. Department of Chemistry, Faculty of Engineering Science, Osaka University, Machikaneyama, Toyonaka, Osaka, Japan.
2. Watson, S. C.; Eastham, J. F. *J. Organometal. Chem.* **1967**, *9*, 165.
3. Wittig, G.; Hesse, A. *Org. Synth., Coll. Vol. VI* **1988**, 901.
4. L'abbé, G.; Miller, M. J.; Hassner, A. *Chem. Ind. (London)* **1970**, 1321.
5. Lepley, A. R.; Khan, W. A.; Giumanini, A. B.; Giumanini, A. G. *J. Org. Chem.* **1966**, *31*, 2047.
6. Cristol, S. J.; Norris, W. P. *J. Am. Chem. Soc.* **1953**, *75*, 2645.
7. Murahashi, S.-I.; Yamamura, M.; Yanagisawa, K.; Mita, N.; Kondo, K. *J. Org. Chem.* **1979**, *44*, 2408; Yamamura, M.; Moritani, I.; Murahashi, S.-I. *J. Organometal. Chem.* **1975**, *91*, C39.
8. Wakefield, B. J. "The Chemistry of Organolithium Compounds," Pergamon: Oxford, 1974; Gschwend, H. W.; Rodriguez, H. R. *Org. React.* **1979**, *26*, 1.
9. Worm, A. T.; Brewster, J. H. *J. Org. Chem.* **1970**, *35*, 1715.
10. Posner, G. H. *Org. React.* **1975**, *22*, 253.
11. Tamao, K.; Sumitani, K.; Kiso, Y.; Zembayashi, M.; Fujioka, A.; Kodama, S.; Nakajima, I.; Minato, A.; Kumada, M. *Bull. Chem. Soc. Jpn.* **1976**, *49*, 1958 and references therein; Kumada, M.; Tamao, K.; Sumitani, K. *Org. Synth., Coll. Vol. VI* **1988**, 407.

COPPER(I)-CATALYZED PHOTOCYCLOADDITION: 3,3-DIMETHYL-cis-BICYCLO[3.2.0]HEPTAN-2-ONE

[Bicyclo[3.2.0]heptan-2-one, 3,3-dimethyl-]

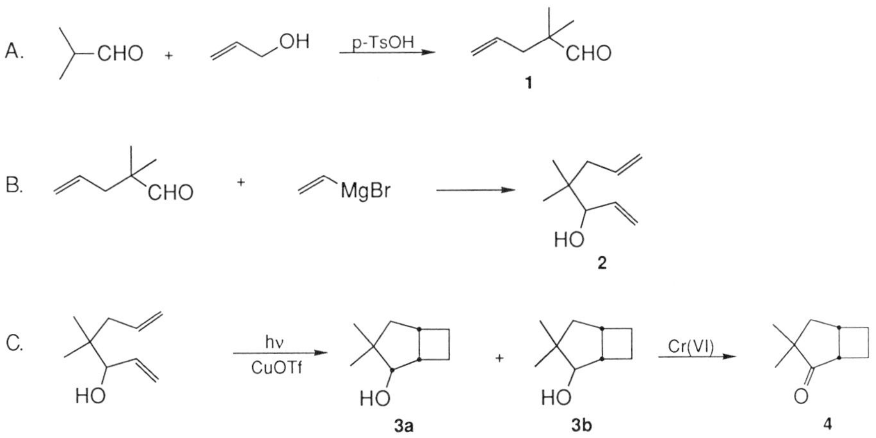

Submitted by ROBERT G. SALOMON and SUBRATA GHOSH[1]
Checked by DANIEL K. JACKSON and RICHARD E. BENSON

1. Procedure

A. *2,2-Dimethyl-4-pentenal.* In a 500-mL, one-necked, round-bottomed flask that contains a magnetic stirring bar are placed 108 g (1.5 mol) of isobutyraldehyde (Note 1), 58 g (1.0 mol) of allyl alcohol (Note 1), 230 mL of *p*-cymene (Note 1), and 0.4 g (2 mmol) of *p*-toluenesulfonic acid monohydrate (Note 1). The mixture is heated with a mantle with stirring for 32 hr under a 50-cm fractionating column packed with 6-mm glass beads and topped by a Dean–Stark trap. The reaction mixture is then distilled through the packed column. The fraction that boils at 120–126°C is collected. The yield is 86.0–87.3 g (77–78%) of 2,2-dimethyl-4-pentenal (**1**) as a clear, colorless oil, n_{D}^{25} 1.4216 (Note 2).

B. *4,4-Dimethyl-1,6-heptadien-3-ol.* In a 1-L, three-necked, round-bottomed flask, fitted with mechanical stirrer, 500-mL pressure-equalizing addition funnel, and condenser topped with a gas inlet for maintaining an atmosphere of dry nitrogen, is placed 17 g (0.7 g-atom) of magnesium turnings (Note 3). The system is flushed with nitrogen, and methanol maintained at −20°C is circulated through the condenser (Note 4). From a solution of 70 g (0.65 mol) of vinyl bromide (Note 5) in 400 mL of tetrahydrofuran (Note 6), a 50-mL quantity is added by means of the addition funnel, and the resulting mixture is stirred mechanically. After a few minutes an exothermic reaction ensues, which subsides after several minutes of vigorous boiling (Note 7). The remainder of the vinyl bromide solution is added at such a rate as to maintain a gentle reflux. After stirring at room temperature for 12 hr, the resulting mixture is cooled with an ice–water bath, and 62 g (0.55 mol) of 2,2-dimethyl-4-pentenal (**1**) is added dropwise over 25–30 min through the addition funnel, which is then rinsed with 10 mL of dry tetrahydrofuran. The resulting mixture is stirred for 1 hr at 23°C and then poured into a mixture of 1 kg of ice, 200 mL of concentrated hydrochloric acid, and 400 mL of water. The resulting mixture is extracted with three 500-mL portions of ether. The combined extracts are washed successively with 400 mL of water, 400 mL of saturated aqueous sodium bicarbonate, and 400 mL of saturated aqueous sodium chloride and then dried over anhydrous sodium sulfate. The drying agent is removed by filtration and the ether is removed with a rotary evaporator. Distillation of the product through a 15-cm vacuum-jacketed Vigreux column gives 63.4–63.8 g (82–83% yield) of 4,4-dimethyl-1,6-heptadien-3-ol (**2**), bp 76–79°C (20 mm), n_D^{24} 1.4562 (Note 8).

C. *3,3-Dimethyl-cis-bicyclo[3.2.0]heptan-2-ol.* A 25-mL test tube is charged with 0.3–0.4 g (0.6–0.8 mmol) of bis(copper trifluoromethanesulfonate)benzene complex (Note 9) and sealed with a rubber septum under an atmosphere of dry nitrogen. A solution of 5 mL (4.3 g, 0.031 mol) of 4,4-dimethyl-1,6-heptadien-3-ol in 10 mL of ether (Note 10) is added by means of a syringe. The resulting solution is poured (Note 11) into a nitrogen-flushed Pyrex 250-mL annular reactor fitted with a magnetic stirrer, an internal concentric water-jacketed quartz immersion well, and a water-cooled reflux condenser topped with a gas inlet for maintaining an atmosphere of dry nitrogen. An additional 20 mL (17.4 g, 0.124 mol) of the hydroxydiene **2** in 200 mL of dry ether is added. The resulting solution is stirred and irradiated for 15 hr with a 450-W medium-pressure Hanovia mercury arc (Note 12) that is suspended in the immersion well. At the end of that time an opaque film of copper is wiped from the immersion well, and irradiation is then continued for an additional 7 hr. The resulting solution is shaken in a separatory funnel with a mixture of 100 g of ice and 100 mL of concentrated aqueous ammonium hydroxide. The organic phase is separated and the aqueous phase is extracted with 100 mL of ether. The organic phases are combined and washed with 100 mL of saturated aqueous sodium chloride and dried over anhydrous sodium sulfate. The solvent is removed by distillation using a rotary evaporator and the product is distilled through a 15-cm vacuum-jacketed Vigreux column to give 19.0–19.9 g (88–92% yield) of 3,3-dimethyl-*cis*-bicyclo[3.2.0]heptan-2-ol (**3**), bp 80–84°C (12 mm), n_D^{25} 1.4761–1.4783 (Note 13).

D. *3,3-Dimethyl-cis-bicyclo[3.2.0]heptan-2-one.* In a 500-mL Erlenmeyer flask containing a magnetic stirring bar is placed 35.1 g (0.25 mol) of 3,3-dimethyl-*cis*-2-bicyclo[3.2.0]heptanol and 200 mL of acetone (Note 14). The solution is cooled with an ice–water bath while 100 mL of 2.7 M Jones reagent (Note 15) is added in small portions over 15 min with vigorous stirring (Note 16). The ice–water bath is removed and the reaction mixture is stirred at 5–20°C for 2 hr. Then 400 mL of saturated aqueous sodium chloride is added, and the resulting mixture is extracted with three 500-mL portions of ether. The extractions are combined and washed successively with 400 mL of saturated aqueous sodium chloride and 200 mL of saturated aqueous sodium bicarbonate. The solvent is removed by means of a rotary evaporator, and the resulting product is transferred to a separatory funnel and separated from the water. The aqueous layer is extracted with 50 mL of ether. The product and the ether layer are combined and dried over anhydrous sodium sulfate. The ether is removed by distillation using a rotary evaporator and the product is distilled through a 15-cm vacuum-jacketed Vigreux column to give 28.7–32.2 g (83–93% yield) of 3,3-dimethyl-*cis*-bicyclo[3.2.0]heptan-2-one (**4**), bp 72–75°C (12 mm), n_D^{20} 1.4622 (Note 17).

2. Notes

1. Isobutyraldehyde, allyl alcohol, *p*-cymene, and *p*-toluenesulfonic acid monohydrate were purchased from Aldrich Chemical Company, Inc., and used as received.

2. The submitters state that the distilled product was about 97% pure as shown by GLC analysis on a 6.4-mm × 1.4-m column packed with 15% FFAP on Chromosorb W, 60–80 mesh, and operated at 140°C. The retention time is about 1.40 min. Two minor impurities with retention times of about 0.95 and 1.15 min were detected in roughly equal amounts. The product has the following spectral properties: IR (neat) cm^{-1}: 2965 (m), 2925 (m), 1725 (vs), 1465 (m), and 915 (m), together with numerous weaker absorption bands; ^{1}H NMR (CDCl$_3$) δ: 1.04 (s, 6 H), 2.22 (d, 2 H, J = 7.0), 4.9–5.3 (m, 2 H), 5.4–6.2 (m, 1 H), 9.40 (s, 1 H).

3. Reagent available from Fisher Scientific Company was used.

4. A Neslab ULT-80 refrigerated circulating bath was used. Alternatively, a Dewar condenser cooled with acetone–dry ice can be used.

5. Vinyl bromide, available from Aldrich Chemical Company, Inc., was used as received.

6. Tetrahydrofuran, anhydrous, 99.9% (water content <0.006%), was purchased from Aldrich Chemical Company, Inc., and used as received. The vinyl bromide solution was prepared in a 500-mL, round-bottomed flask fitted with a glass stopper. The stoppered flask containing the tetrahydrofuran was chilled to about 5°C and weighed. The vinyl bromide, also chilled to about 5°C, was rapidly poured into the tetrahydrofuran until the desired amount had been added. The flask was stoppered, the contents mixed by shaking, allowed to warm to about 16°C, and then added to the pressure-equalizing addition funnel.

7. The checkers found it necessary to initiate the reaction with a crystal of iodine.

8. The submitters state that the purity of the product is greater than 98% by gas chromatographic analysis on a 6.4-mm × 1.4-m column packed with 15% FFAP on Chromosorb W, 60–80 mesh, and operated at 140°C. The retention time is about 4.7 min. An impurity with a retention time of about 2.9 min was detected. The product has the following spectral properties: ^{1}H NMR (CDCl$_3$) δ: 0.84 (s, 3 H), 0.88, (s, 3 H), 1.80–2.30 (m, 2 H), 2.69 (s, 1 H), 3.78 (d, 1 H, J = 6), 4.87–5.33 (m, 4 H), 5.57–6.13 (m, 2 H).

9. The copper complex is available from Strem Chemicals, Inc., under the name "cuprous triflate" (benzene complex). The checkers recommend handling the material in a dry box because of its high sensitivity to moisture and air.

10. Anhydrous ether was distilled from lithium aluminum hydride under dry nitrogen immediately before use.

11. The submitters state that the copper(I) triflate is quite air-stable in solution in the presence of the allylic alcohol.

12. The checkers recommend the use of a relatively new arc lamp. Substantially higher conversions were obtained with a new lamp because of an apparent bathochromic shift in the frequency of the light emitted as the lamp ages, thus lessening the intensity of light in the important absorption region for the reaction.

13. The submitters state that the purity of the product is greater than 97% by GLC analysis on a 6.4-mm × 1.4-m column packed with 15% FFAP on Chromosorb W, 60–80 mesh, and operated at 140°C. The retention time is about 8.0 min. The only impurity is unreacted diene with a retention time of about 4.7 min. The product is an epimeric mixture. TLC analysis by the submitters on 0.25-mm silica gel with 20% ethyl acetate in hexane shows major (>90%) and minor (<10%) epimers with R_f 0.32 and 0.23, respectively. The epimers are separable by column chromatography on silica gel with ethyl acetate–hexane mixtures as eluting solvents. A 3.1-g portion of the distilled isomer mixture was chromatographed by the checkers on 475 g of silica gel (Silica Woelm TSC—activity III/30 mm) using 5% ethyl acetate–hexane as eluent. The elution proceeded as follows: 1520 mL, nil; 1440 mL, 2.7 g of endo isomer; 1400 mL, nil; 2010 mL, 0.20 g of exo isomer. Analysis of the ^{1}H NMR spectrum of the distilled product confirms that the reaction is greater than 90% stereoselective in favor of the endo epimer. The major epimer, 3,3-dimethyl-*endo-cis*-bicyclo[3.2.0]heptan-2-ol, has the following spectral properties: ^{1}H NMR (CDCl$_3$) δ: 0.81 (s, 3 H), 1.13 (s, 3 H), 1.4–3.2 (m, 9 H), 3.66 (d, 1 H), J = 6.7); ^{13}CMR (CDCl$_3$) δ: 16.3 (t), 22.5 (q, CH$_3$), 26.0 (t), 27.8 (q, CH$_3$), 36.0 (d), 42.8 (d), 45.5 (t), 45.8 (s, C-3), 80.9 (d, C-2). The minor epimer, 3,3-dimethyl-*exo-cis*-bicyclo[3.2.0]heptan-2-ol, has the following spectral properties: ^{1}H NMR (CDCl$_3$) δ: 0.77 (s, 3 H), 1.08 (s, 3 H), 1.2–2.9 (m, 9 H), 3.80 (d, 1 H, J = 4.6); ^{13}CMR (CDCl$_3$) δ: 20.7 (q, CH$_3$), 24.0 (t), 26.5 (q, CH$_3$), 27.1 (t), 34.6 (d), 45.6 (s, C-3), 45.8 (d), 46.7 (t), 88.7 (d, C-2).

14. Certified ACS-grade acetone purchased from Fisher Scientific Company was used as received.

15. Eisenbraun, E. J. *Org. Synth., Coll. Vol. V* **1973**, 310–314.

16. Initially a gummy green precipitate is formed that is difficult to stir magnetically. Eventually, however, the inorganic by-products become more fluid. The use of a mechanical stirrer may be desirable.

17. The submitters state that the distilled product is <98% pure by GLC on a 6.4-mm × 1.4-m column packed with 15% FFAP on Chromosorb W, 60–80 mesh, operated at 140°C. The relative retention time is 2.3 versus an unidentified impurity at 1.0. The distilled product has the following spectral properties: IR (neat) cm^{-1}: 2960 (vs), 2940 (vs) and 1735 (vs), and other weaker bands. ^{1}H NMR (CCl$_4$) δ: 0.92 (s, 3 H), 1.12 (s, 3 H), 1.4–3.0 (m, 8 H); ^{13}CMR (CDCl$_3$), δ: 22.7, 24.1, 25.6, 26.4, 31.0, 43.9, 44.2, 48.4, 224.7.

3. Discussion

This procedure illustrates a general method for the preparation of 2-hydroxybicyclo[3.2.0]heptanes by copper(I)-catalyzed photobicyclization of 3-hydroxy-1,6-

heptadienes[2] and a general route to the requisite dienes from allyl alcohols by conversion to 4-pentenals and treatment of the latter with vinyl Grignard reagents.

Compound 1, 2,2-dimethyl-4-pentenal, has been prepared by the Claisen rearrangement route[3] described above and by reaction of isobutyraldehyde with allyl chloride in the presence of aqueous sodium hydroxide and a phase-transfer catalyst.[4] Both routes are applicable to the synthesis of a variety of substituted 4-pentenals.

cis-Bicyclo[3.2.0]heptan-2-ols have been prepared by reduction[5] of the corresponding cis-bicyclo[3.2.0]-heptan-2-ones, which have been prepared by photocycloaddition of alkenes with 2-cyclopentenones.[6] The synthetic strategy of the present procedure is complementary.

1. Department of Chemistry, Case Western Reserve University, Cleveland, OH 44106.
2. Salomon, R. G.; Coughlin, D. J.; Easler, E. M. *J. Am. Chem. Soc.* **1979**, *101*, 3961–3963.
3. Brannock, K. C. *J. Am. Chem. Soc.* **1959**, *81*, 3379–3383.
4. Dietl, H. K.; Brannock, K. C. *Tetrahedron Lett.* **1973**, 1273–1275.
5. Svensson, T. *Chem. Scr.* **1973**, *3*, 171–175.
6. For reviews, see the following: Eaton, P. E. *Acc. Chem. Res.* **1968**, *1*, 50–57; Bauslaugh, P. G. *Synthesis* **1970**, 287–300; de Mayo, P. *Acc. Chem. Res.* **1971**, *4*, 41–47.

ANODIC OXIDATION OF ACIDS: DIMETHYL DECANEDIOATE

(Decanedioic acid, dimethyl ester)

$$2 \ CH_3OOC(CH_2)_4COOH \ \xrightarrow[\text{CH}_3\text{OH, CH}_3\text{ONa}]{\text{electricity}} \ 2 \ CH_3OOC(CH_2)_8COOCH_3$$

$$+ \ 2\,CO_2$$

Submitted by D. A. WHITE[1]
Checked by RONALD F. SIELOFF and CARL R. JOHNSON

1. Procedure

The electrode assembly (Note 1) is constructed; the platinum electrodes are positioned vertically, parallel, and about 5 mm apart by careful bending of the lower platinum wire connections.

To a 500-mL, round-bottomed flask having a central 34/45 standard taper neck and two 24/40 standard taper necks are added 120 g (0.75 mol) of methyl hydrogen hexanedioate (Note 2), 250 mL of methanol, 4.1 g (0.075 mol) of sodium methoxide, 10 mL of pyridine, and a magnetic stirring bar. The electrode assembly is inserted so that the platinum electrodes are immersed in the solution. A thermometer and reflux water condenser are attached. The mixture is heated to 60°C and stirred until the sodium methoxide dissolves.

The mixture is electrolyzed with a constant current of 1.1 A (Note 3) until GLC analysis (Note 4) of the solution shows the absence of the peak due to methyl hydrogen hexanedioate. This requires about 23 hr; electrolysis is continued for an additional 2.5 hr at the same current (Note 5). Throughout the electrolysis the reaction mixture is maintained at 62–65°C by passage of the current. After about 1 hr of electrolysis, when conditions are stabilized, the reaction may be left unattended.

The yellow reaction mixture is allowed to cool and then is acidified with 20 mL of glacial acetic acid. The acidified solution is transferred with methanol washing to a 1-L round-bottomed flask and evaporated (70–80°C, 12 mm) to dryness. The solid residue is stirred with 500 mL of diethyl ether for 1 hr. Undissolved solids are removed by filtration and the residue is washed twice with 100-mL portions of diethyl ether. The combined filtrate and washings are washed with aqueous sodium carbonate until neutral and then three times with 200-mL portions of water. The ether solution is dried over anhydrous calcium sulfate. Filtration and evaporation of the ether afford a yellow oil that is distilled under reduced pressure through a Vigreux column (30 × 2.5 cm). This gives 6–7 g of dimethyl hexanedioate (bp 69–71°C, 0.02 mm), 3–4 g of a mixed fraction (bp 72–105°C, 0.02 mm) (Note 6), and 60–61 g (70–71%) of dimethyl decanedioate (bp 105–107°C, 0.02 mm). The dimethyl decanedioate crystallizes on standing at room temperature, mp 26.5–27.2°C (Notes 7 and 8).

2. Notes

1. The electrode assembly shown (Figure 1) is fairly versatile and has been used by the submitter in flasks with electrolyte volumes of ca. 40 mL to 4 L. Additionally, the platinum electrodes may be replaced by other electrodes that fit directly into the thermometer adaptor, e.g., commercially available ¼ in. graphite or stainless-steel rods. In the present example the electrodes are positioned vertically and are of opposite polarity. In other cases they may be positioned horizontally (parallel to a mercury cathode), and both are anodic.

   The thermometer adaptors are available commercially (Ace Glass Company, Vineland, NJ). The platinum wire to platinum electrode connection was made by laying the wire on the foil, heating both parts to red heat with an oxygen–natural-gas flame, and forcing the two together with a sharp hammer blow.

2. Methyl hydrogen hexanedioate (adipic acid, monomethyl ester) was obtained from Aldrich Chemical Company, Inc.

3. The cell voltage, initially 25 V, increased slowly to 27 V. A power supply operating in a constant current mode was used to supply the current. [Since there is little change in cell voltage, a constant voltage power supply capable of delivering 24 V (e.g., two automobile batteries) could also be used.] The cell voltage is important only in connection with the amount of heat generated in the solution by the passage of current. This depends on the product of current and voltage. The voltage should not be so high that the reflux cannot cope with the heat generated or so low that the reflux temperature is not attained. It should be in the range 20–30 V. If the voltage should not fall within this range, the electrode separation should be adjusted. Decreasing the separation decreases the cell voltage. However, contact of the electrodes should not, of course, occur.

4. The column used was a 6 ft × ⅛ in. stainless-steel column packed with 5% OV17 on 100–120-mesh Chromosorb W. The column temperature was increased at 10°C per minute from 60 to 260°C.

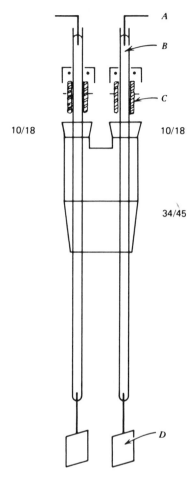

Figure 1. Electrode assembly: (*A*) platinum wire; (*B*) mercury-filled 6-mm-o.d. glass tubing; (*C*) "thermometer" adapter, 10/18 standard taper joint, Teflon; (*D*) platinum electrodes 25 mm × 30 mm.

5. The additional current passed is to allow for the conversion of that portion of the starting material that is converted to the sodium salt by the added sodium methoxide and that is not detected by gas chromatography.

6. The quantity of ester obtained (0.26 mol) theoretically requires 0.52*F*. Actual current passed was 28 A-hr [1.05*F* (Faradays)], corresponding to a current efficiency of 50%.

7. The melting point was determined by remelting the product and allowing it to cool with a thermometer inserted into it. Occasional stirring with the thermometer was necessary to prevent supercooling.

8. The product shows ^{1}H NMR (CDCl$_3$) δ: 1.2–1.8 (complex, 12, internal methylenes), 2.28 (4, α-CH$_2$), 3.65 (6, OCH$_3$).

3. Discussion

The present preparation is based on that of dimethyl tetradecanedioate,[2] with the inclusion of some pyridine into the electrolyte, which has been shown[3] to be effective in preventing anode coating. The reaction used is an example of the Crum Brown-Walker reaction[4] (anodic oxidation of half-esters of α,ω-dicarboxylic acids), which is itself an example of the Kolbe reaction (anodic oxidation of carboxylic acids). The latter is a very general reaction, and its scope and mechanism have been reviewed recently.[5,6] The particular example detailed here has some commercial interest and has been extensively examined. For example, the effects of current density (optimum 0.1–0.4 A/cm^2), degree of neutralization (optimum < 10%), nature of the base used for neutralization, and the nature of the solvent used have been examined.[7,8] These optimized results were obtained[7,8] in a specially constructed cell with an electrode gap of ca. 0.15 mm. To obtain reasonable results with the larger electrode gap used in this example (ca. 5 mm; if the gap is made any smaller, inadvertent contact of the electrodes and short circuiting become a strong possibility) and maintain the cell voltage fairly low (within range of inexpensive power supplies), the current density used (0.14 A/cm^2) was kept toward the low end of the desirable range and the degree of neutralization (10%) at the highest value consistent with a good yield. In a small-gap cell a current efficiency of 66% together with a chemical yield of 85% has been obtained[7] in the same solvent and electrolyte system.

In addition to the route described here, dimethyl decanedioate has been prepared by esterification of decanedioic acid with methanol[9–13] or diazomethane,[14] hydrogenation of dimethyl 2,5,8-decatrienedioate,[15] and by thermal decomposition of bis(1-methoxy-1-cyclopentyl)peroxide.[16]

The present procedure offers an alternative electrochemical setup to accomplish the Kolbe electrolysis of half esters to that reported earlier for the preparation of dimethyl octadecanedioate.[17] In the present case the apparatus offers general versatility and electrode coating is prevented by an additive (pyridine). In the earlier case periodic current reversal was necessary.

1. Corporate Research Department, Monsanto Company, St. Louis, MO 63166.
2. Anderson, J. D.; Baizer, M. M.; Petrovich, J. P. *J. Org. Chem.* **1966**, *31*, 3890–3897.
3. Ross, S. D.; Finkelstein, M. *J. Org. Chem.* **1969**, *34*, 2923–2927.
4. Crum Brown, A.; Walker, J. *Justus Liebigs Ann. Chem.* **1891**, *261*, 107.
5. Eberson, L. In "Organic Electrochemistry," Baizer, M. M., Ed.; Dekker: New York, 1973; p. 469.
6. Utley, J. H. P. In "Techniques of Chemistry," Weinberg, N. L., Ed.; Wiley: New York, 1974; Vol. 5, Part 1, p. 793.
7. Beck, F.; Himmele, W.; Haufe, J.; Brunold, A. U.S. Patent 3625430, 1972; see Ger. Offen. 1802865; *Chem. Abstr.* **1971**, *74*, 37822.
8. Haufe, J.; Beck, F. *Chem. Ing. Tech.* **1970**, *42*, 170–175.
9. Vowinkel, E. *Chem. Ber.* **1967**, *100*, 16–22.
10. Desseigne, G. *Mem. Poudres* **1960**, *42*, 223–227; *Chem. Abstr.* **1961**, *55*, 15344.

11. Droeger, J. W.; Sowa, F. J.; Nieuwland, J. A. *Proc. Indiana Acad. Sci.* **1937**, *46*, 115–117; *Chem. Abstr.* **1938**, *32*, 500.
12. Verkade, P. E.; Coops, J.; Hartmen, H. *Recl. Trav. Chim. Pays-Bas* **1926**, *45*, 585–606.
13. Neison, E. *J. Chem. Soc.* **1876**, *29*, 314.
14. Hückel, W., *Justus Liebigs Ann. Chem.* **1925**, *441*, 1–48.
15. Mettalia, J.; Specht, E. H. French Patent 1507657, 1967; *Chem. Abstr.* **1969**, *70*, 19597.
16. Roedel, M. J. U.S. Patent 2601224, 1952; *Chem. Abstr.* **1953**, *47*, 4363.
17. Swann, S., Jr.; Garrison, S., Jr. *Org. Synth., Coll. Vol. V* **1973**, 463–467.

(2*SR*,3*RS*)-2,4-DIMETHYL-3-HYDROXYPENTANOIC ACID

[Pentanoic acid, 3-hydroxy-2,4-dimethyl-, (*R**,*S**)-(±)-]

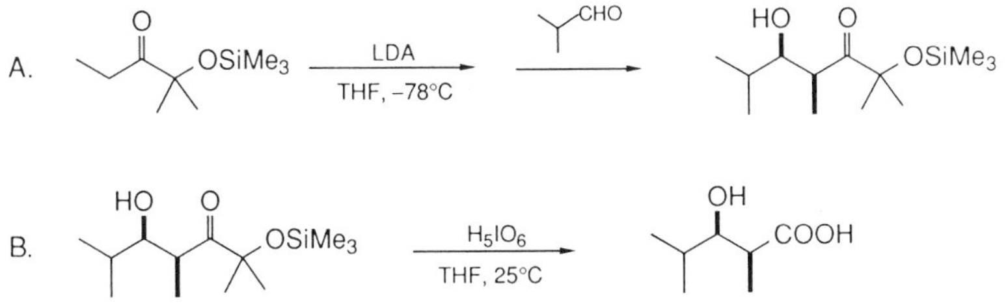

Submitted by B. Bal, C. T. Buse, K. Smith, and Clayton H. Heathcock[1]
Checked by Joseph R. Flisak, Stan S. Hall, Hugh W. Thompson, and Gabriel Saucy

1. Procedure

A. *5-Hydroxy-2,4,6-trimethyl-2-(trimethylsiloxy)heptan-3-one.* A dry, 1-L, four-necked (including a thermometer well), round-bottomed flask equipped with an efficient mechanical stirrer, thermometer, graduated 250-mL pressure-equalizing addition funnel sealed with a rubber septum, and a nitrogen inlet is charged with 125 mL of dry tetrahydrofuran (Note 1) and 31 mL (0.22 mol) of diisopropylamine (Note 2). The stirrer is started and 137 mL (0.20 mol) of 1.5 *M* butyllithium in hexane is transferred to the addition funnel by means of a 16-gauge cannula and argon pressure (Note 3). The reaction flask and its contents are cooled to below −5°C by immersion in a dry ice–acetone bath that is maintained at −10 to −15°C by the occasional addition of dry ice. The butyllithium is added dropwise over a period of 20 min. After the addition is complete 10 mL of dry tetrahydrofuran is added to the addition funnel with a syringe to rinse the walls of the funnel, and the rinse is then added to the pale-yellow solution. After the addition is complete, the solution is stirred for an additional 15 min and is then cooled to below −70°C (dry ice–acetone bath). While the reaction solution is cooling, a solution of 37.7 g (0.20 mol) of 2-methyl-2-(trimethylsiloxy)pentan-3-one (Note 4) in 10 mL of dry tetrahydrofuran is introduced through the septum into the addition funnel. When the lithium diisopropylamide (LDA) solution has cooled to below −70°C, the ketone is slowly added to the solution over a period of 20–25 min to ensure that the reaction temperature is maintained below −70°C. After the addition is complete 10 mL of dry tetrahydrofuran is added to rinse the walls of the addition funnel, the rinse is added, and the stirred reaction solution is maintained below −70°C for an additional 30–40 min. During this time the addition funnel is charged through the septum with a solution of 14.4 g (0.20 mol) of 2-methylpropanal (Note 5) in 10 mL of dry tetrahydrofuran. The aldehyde solution is added dropwise to the vigorously stirring yellow enolate solution at −70°C over a 15-min period, and then the addition funnel is again rinsed with 10 mL of dry tetrahydrofuran and the rinse added to the reaction mixture. After 10–15 min 200 mL of a saturated aqueous ammonium chloride solution is added to the vigorously stirring, −70°C reaction mixture. At this point stirring is discontinued, the cooling bath is removed, and the partially frozen mixture is allowed to warm to room temperature. The contents of the reaction flask are introduced into a 2-L separatory funnel, 200 mL of ether is added to the flask, and the ether rinse is then transferred to the separatory funnel. The layers are shaken and then separated, and the aqueous phase is extracted again with 200 mL of ether. The combined organic phase is washed with 200 mL of water and 200 mL of saturated brine and then dried over magnesium sulfate. After removal of the drying agent by filtration the solvents are removed with a rotary evaporator at aspirator pressure to give 52.1–52.4 g of a pale-yellow oil that is a 63 : 37 mixture of the expected product and the starting material. Most of the starting material is then selectively removed by stirring (magnetic stirring bar) at 25°C at reduced pressure (vacuum pump, 0.1–0.08 mm) for 19 hr to yield 35.2 g of a 90 : 10 mixture (31.7 g, 61%), which is used without further purification for Step B (Notes 6 and 7).

B. *(2SR,3RS)-2,4-Dimethyl-3-hydroxypentanoic acid.* A dry, 500-mL, three-necked, round-bottomed flask equipped with a mechanical stirrer, thermometer, and a nitrogen inlet is flushed with nitrogen, charged with 12.5 g (55 mmol) of periodic acid (Note 8) and 150 mL of dry tetrahydrofuran (Note 1), and then sealed with a stoppered, 25-mL, pressure-equalizing addition funnel. The solution is stirred vigorously and cooled to 0–5°C with an ice–salt bath. During this time the addition funnel is charged with a solution of 12.0 g of 5-hydroxy-2,4,6-trimethyl-2-(trimethylsiloxy)heptan-3-one (10.8 g, 41 mmol of ketone from a 90 : 10 mixture from Part A) in 10 mL of dry tetrahydrofuran, which is then rapidly introduced (1 min) to the cold, stirring solution. After the addition is complete 5 mL of dry tetrahydrofuran is added to the addition funnel to rinse the walls of the funnel, and this rinse is then added to the reaction solution. The cooling bath is removed after 15 min and stirring is continued for 1.5 hr, during which time a white precipitate forms. In the meantime, 52 g (0.5 mol) of sodium bisulfite is mixed with 100 mL of distilled water in a 500-mL filtering flask with a side hose connection and cooled to 0–5°C with an ice–salt bath. The reaction mixture is filtered directly through filter paper with suction into the cold slurry of sodium bisulfite. The residue is rinsed with 50 mL of dry ether (Note 9), which is added to the filter funnel and drawn by suction into the yellow solution. After magnetic stirring of the cold mixture for 20 min the contents of the flask are introduced into a 500-mL separatory funnel and the layers are separated. The aqueous layer (pH 4.3) is extracted twice with 100 mL of ether and the combined yellow organic layer is washed with 125 mL of distilled water and separated (the pH of the wash is 2.6–3.5). The organic layer is dried over magnesium sulfate for 1 hr and filtered to remove the drying agent, and the solvents are removed with a rotary evaporator at aspirator pressure. Distillation of the dark-yellow oil affords

4.9–5.4 g (82–89%) of (2*SR*,3*RS*)-2,4-dimethyl-3-hydroxypentanoic acid, bp 85–89°C (0.01 mm), as a viscous, yellow–green liquid (Note 10). Crystallization from hexane using decolorizing carbon provides 4.6–5.0 g (77–83%) of pure hydroxy acid, mp 75–76°C, as white crystals (Note 11).

2. Notes

1. Tetrahydrofuran was distilled under a nitrogen atmosphere from sodium–benzophenone immediately prior to use.

2. Diisopropylamine was distilled, bp 85°C, under a nitrogen atmosphere from calcium hydride immediately prior to use.

3. *Caution! Concentrated butyllithium may ignite spontaneously on exposure to air or moisture. Manipulations with this reagent should be performed with care.* The submitters used fresh butyllithium from Foote Mineral Company, Johnsonville, Tennessee. The checkers used fresh butyllithium, 1.6 *M* in hexane under argon, from Aldrich Chemical Company, Inc. The butyllithium solutions may be standardized;[2] however, both the submitters and the checkers chose to use fresh reagents and forego the titration. Stainless-steel cannulas with deflected points (double-tip syringe needles) are available from Ace Glass, Inc. and Aldrich Chemical Company, Inc.

4. 2-Methyl-2-(trimethylsiloxy)pentan-3-one was prepared by the method of Young, Buse, and Heathcock, *Org. Synth., Coll. Vol. VII* **1990**, 381.

5. 2-Methylpropanal was freshly distilled; bp 64–65°C.

6. The submitters report that the starting material can be removed within 4 hr to give 38–42 g of the 90 : 10 mixture if the concentration is continued with the rotary evaporator rather than a stationary flask at 0.5–0.1 mm.

7. The ^{1}H NMR (200 MHz, CDCl$_3$) spectrum of the product (taken from a spectrum of a 90 : 10 mixture) is as follows δ: 0.18 (s, 9 H), 0.89 (d, 3 H, J = 6.7), 1.02 (d, 3 H, J = 6.5), 1.08 (d, 3 H, J = 7.1), 1.36 (s, 3 H), 1.37 (s, 3 H), 1.68 (d of septets, 1 H, J = 8.4 and 6.6), 2.95 (d, 1 H, J = 2.6, OH), 3.42 (dt, 1 H, J = 8.5, 2.6, and 2.6), 3.59 (dq, 1 H, J = 7.0 and 2.6). The IR spectrum (film) of a 93 : 7 mixture shows absorptions at 1700 and 3600–3300 cm^{-1}.

8. Fresh periodic acid was obtained from Aldrich Chemical Company, Inc. and stored in a desiccator.

9. Reagent-grade diethyl ether from a freshly opened container was used without further drying.

10. The checkers discovered that the desired hydroxy acid is sensitive to strong acid and heat. Early runs of Step B by the checkers using the original conditions recommended by the submitters involved stirring the bisulfite slurry at room temperature for 3–4 hr, simple partitioning without an aqueous backwash, and drying of the bisulfite oxidation mixture, and distillation of the product at 0.8 mm reduced pressure. These runs consistently resulted in acid-catalyzed transformation in either the workup or in distillation and led to mixtures contaminated with 2-methylpropanal produced by a retroaldol reaction, as well as with other

unsaturated materials. Distillation of one of these runs, which had used a crude 58 : 42 mixture from Step A as starting material, afforded 13.4 g (53%) of α, γ, γ-trimethylbutyrolactone, bp 85–105°C (0.8 mm), as a yellow–green liquid by dehydration–lactonization. Crystallization from hexane provided 10.1 g (40%) of pure lactone, mp 50–51°C, as white crystals. The lactone had the following spectral properties: ^{1}H NMR (200 MHz, CDCl$_3$) δ: 1.28 (d, 3 H, J = 7.1), 1.38 (s, 3 H), 1.46 (s, 3 H), 1.71 (superficial t, 1 H, J = ca. 12), 2.30 (dd, 1 H, J = 12.6 and 8.9), 2.83 (16-line m; 1 H, J = 11.2, 8.9, and 7.1); ^{13}C NMR (50 MHz, CDCl$_3$) δ: 15.6, 27.0, 29.0, 35.6, 43.5, 81.8, 179.1; IR (CCl$_4$) cm^{-1}: 1780; mass spectrum, m/z (relative intensity): 129 (M$^+$ + 1, 1), 113 (33), 84 (16), 69 (30), 59 (34), 43 (100).

11. The hydroxy acid showed the following spectral properties: ^{1}H NMR (200 MHz, CDCl$_3$) δ: 0.89 (d, 3 H, J = 6.6), 1.02 (d, 3 H, J = 6.6), 1.21 (d, 3 H, J = 7.1), 1.71 (octet, 1 H, J = 6.7), 2.71 (dq, 1 H, J = 7.3 and 3.4), 3.64 (dd, 1 H, J = 8.1 and 3.4), 6.7 (br s, 2 H, OH and CO$_2$H); IR (CCl$_4$) cm^{-1}: 1700, 3600–2500.

3. Discussion

The stereochemistry of the aldol addition reaction has been actively investigated in recent years, and several methods for achieving high stereoselectivity have been developed.[3] One of these utilizes the preformed lithium enolates of compounds such 1.[4] Compound **1** gives a single enolate, which has the Z-configuration. This enolate reacts with aldehydes to give β-hydroxy ketones **(2)** with high stereoselectivity:

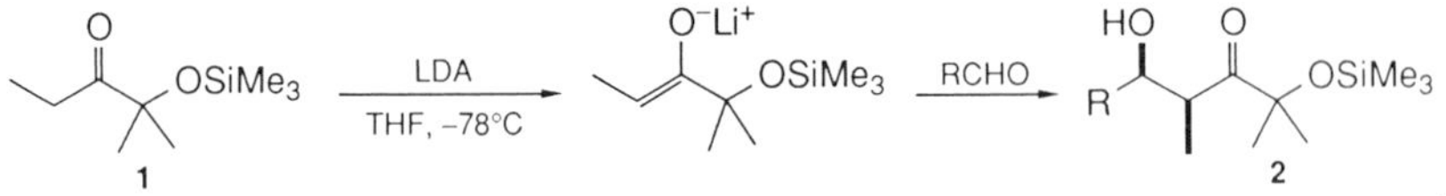

$$(1)$$

Compounds **2** may be directly cleaved with periodic acid to obtain β-hydroxy acids; for example:[4,5]

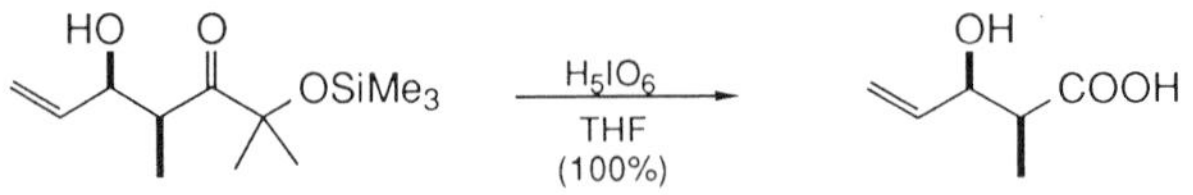

$$(2)$$

Alternatively, the carbonyl group may be reduced, the silyl group hydrolyzed, and the resulting vicinal diol cleaved with buffered sodium periodate to obtain the β-hydroxy aldehyde; for instance:[6,7]

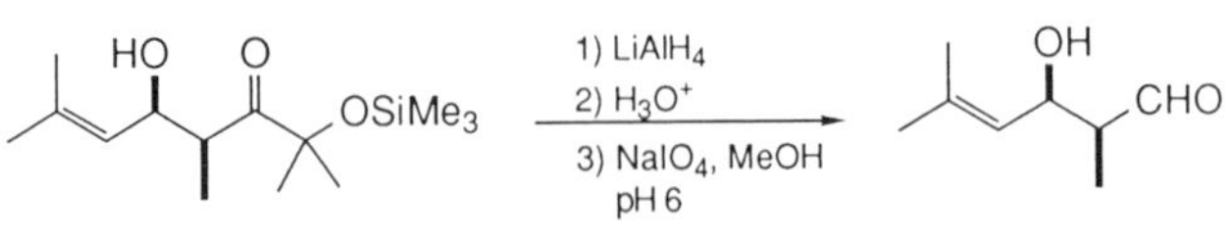

$$(3)$$

Finally, the hydroxy group may be protected as the tetrahydropyranyl ether, an aryl or alkyllithium reagent added to the carbonyl, and the resulting vicinal diol cleaved to obtain the corresponding β-hydroxy ketone; for example:[8]

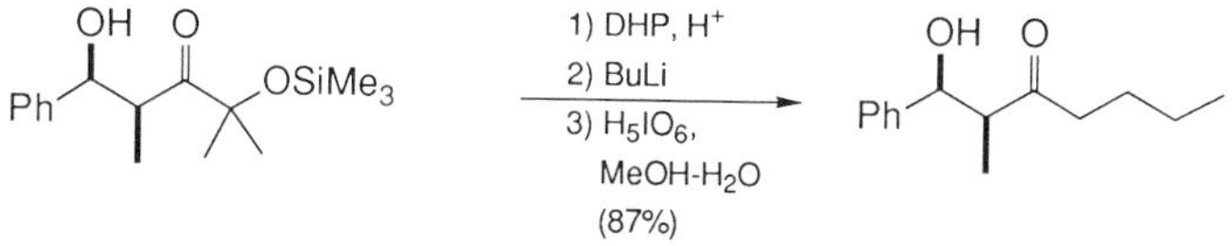

$$
\begin{array}{ccc}
\text{[structure: Ph–CH(OH)–CH(CH}_3\text{)–C(O)–C(CH}_3)_2\text{–OSiMe}_3] &
\begin{array}{l}
\text{1) DHP, H}^+ \\
\text{2) BuLi} \\
\text{3) H}_5\text{IO}_6, \\
\quad \text{MeOH-H}_2\text{O} \\
\text{(87\%)}
\end{array} &
\text{[structure: Ph–CH(OH)–CH(CH}_3\text{)–C(O)–CH}_2\text{CH}_2\text{CH}_2\text{CH}_3]
\end{array} \qquad (4)
$$

Selected examples of the addition of ketone **1** to a variety of aldehydes are collected in Table I.

TABLE I
CONDENSATION OF KETONE **1** WITH ALDEHYDES (Equation 1)

Aldehyde	β-Hydroxy Ketone		β-Hydroxy Acid		Ref.
	Yield (%)	mp	Yield (%)	mp	
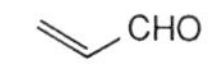 [CHO structure]	80	oil	100	oil	5, 9
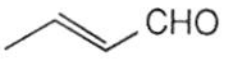 [CHO structure]	97	oil	62	oil	9
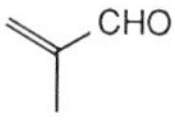 [CHO structure]	99	oil	97	oil	9
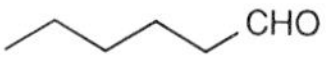 [CHO structure]	43	oil	–	–	7
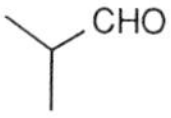 [CHO structure]	93	oil	61	73-75°C	4
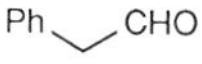 Ph⌣CHO	51	oil	76	119-120°C	4
PhCHO	78	oil	87	oil	4
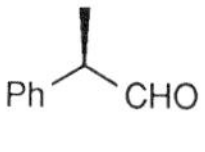 Ph⌣CHO	100[a]	oil	65	134-135°C[d]	4
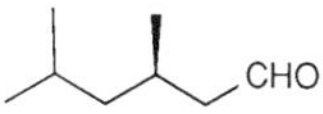 [CHO structure]	61[b]	oil	–	–	7
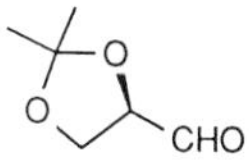 [dioxolane CHO structure]	75[c]	oil	–	–	6

[a] This is a 4 : 1 mixture of Cram : anti-Cram isomers.   [b] Major isomer.
[c] This is a 15 : 1 mixture of Cram : anti-Cram isomers.
[d] This is a 1.3 : 1 mixture of Cram : anti-Cram isomers.

1. Department of Chemistry, University of California, Berkeley, CA 94720.
2. (a) Jones, R. G.; Gilman, H. In "Organic Reactions"; Adams, R., Ed.; Wiley: New York, 1951; Vol. 6, p. 353; (b) Kofron, W. G.; Baclawski, L. M. *J. Org. Chem.* **1976,** *41*, 1879.
3. (a) Heathcock, C. H. In "Comprehensive Carbanion Chemistry," Durst, T.; Buncel, E., Eds.; Elsevier: New York, 1984; Vol. II; (b) Evans, D. A.; Nelson, J. V.; Taber, T. R. In "Topics in Stereochemistry," Eliel, E. L.; Allinger, N. L.; Wilen, S. H., Eds.; Wiley: New York, 1982; Vol. 13; (c) Heathcock, C. H. In "Asymmetric Synthesis," Morrison, J. D., Ed.; Academic Press, Inc.: New York, 1984; Vol. 3.
4. Heathcock, C. H.; Buse, C. T.; Kleschick, W. A.; Pirrung, M. C.; Sohn, J. E.; Lampe, J. *J. Org. Chem.* **1980,** *45*, 1066.
5. Heathcock, C. H.; Jarvi, E. T. *Tetrahedron Lett.* **1982,** *23*, 2825.
6. Heathcock, C. H.; Young, S. D.; Hagen, J. P.; Pirrung, M. C.; White, C. T.; VanDerveer, D. *J. Org. Chem.* **1980,** *45*, 3846.
7. Heathcock, C. H.; Young, S. D., unpublished results.
8. White, C. T.; Heathcock, C. H. *J. Org. Chem.* **1981,** *46*, 191.
9. Heathcock, C. H.; Finkelstein, B. L.; Jarvi, E. T.; Radel, P. A.; Hadley, C. R. *J. Org. Chem.* **1988,** *53*, 1922.

(2SR,3SR)-2,4-DIMETHYL-3-HYDROXYPENTANOIC ACID

(Pentanoic acid, 3-hydroxy-2,4-dimethyl-, (R*,R*)-)

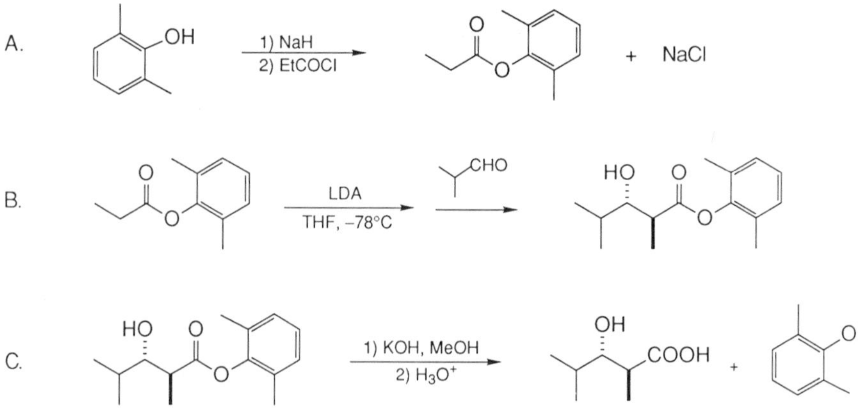

Submitted by STEPHEN H. MONTGOMERY, MICHAEL C. PIRRUNG, and CLAYTON H. HEATHCOCK[1]
Checked by PAULINE J. SANFILIPPO and ANDREW S. KENDE

1. Procedure

A. *2,6-Dimethylphenyl propanoate.* To a 2-L, three-necked, round-bottomed flask is added 26.4 g (0.55 mol) of a 50% dispersion of sodium hydride in mineral oil (Note

1). The sodium hydride is washed several times by decantation with dry hexane and is then covered with 1 L of dry ether (Note 2). The flask is immersed in an ice bath and equipped with a dropping funnel, a mechanical stirrer, and a reflux condenser. A solution of 61.1 g (0.50 mol) of 2,6-dimethylphenol (Note 3) in 150 mL of dry ether is added dropwise over a 10-min period and the mixture is stirred for 5 min, during which time hydrogen evolution ceases. The cold solution is stirred continuously while a solution of 48 mL (50.9 g, 0.55 mol) of propanoyl chloride (Note 1) in 100 mL of dry ether is added dropwise over a 30-min period. After stirring for an additional hour the reaction mixture is poured into a 2-L separatory funnel containing 200 mL of water. The mixture is shaken vigorously and the ether layer is separated and washed successively with 200 mL of aqueous 10% sodium hydroxide, 200 mL of water, and 200 mL of 4% hydrochloric acid, then dried over magnesium sulfate. The ether is removed with a rotary evaporator and the residue distilled through a short, indented Claisen apparatus to obtain 85–86 g (96–97%) of 2,6-dimethylphenyl propanoate, bp 60–65°C (0.05 mm) (Note 4).

B. *2',6'-Dimethylphenyl* *(2SR,3SR)-2,4-dimethyl-3-hydroxypentanoate.* The reaction is carried out in a 2-L, three-necked, round-bottomed flask equipped with an efficient mechanical stirrer, a thermometer, and a 500-mL, pressure-equalizing dropping funnel. The dropping funnel is marked to hold 325 mL and is topped with a rubber septum pierced with a syringe needle attached to a source of dry nitrogen. The flask is charged with 300 mL of dry tetrahydrofuran (Note 2) and 69 mL (0.49 mol) of diisopropylamine (Note 1). Butyllithium (325 mL, 0.49 mol, 1.5 M in hexane) (Note 5) is transferred into the addition funnel with a cannula. The reaction flask and its contents are cooled to below $-5°C$ by immersion in a bath of dry ice and isopropyl alcohol, which is maintained at -10 to $-15°C$ by periodic additions of dry ice. The butyllithium is added dropwise at such a rate as to maintain the temperature of the reaction mixture in the range 0 to $-5°C$. After the addition is complete the mixture is stirred for an additional 15 min and is then cooled to $-70°C$. While the reaction mixture is cooling, the septum is briefly removed and a solution of 85 g (0.48 mol) of 2,6-dimethylphenyl propanoate in 100 mL of dry tetrahydrofuran is added to the addition funnel, the septum is replaced, and nitrogen is passed through the apparatus in a slow stream for 5 min. The ester is then added to the lithium diisopropylamide solution at such a rate that the temperature of the reaction mixture does not exceed $-65°C$. The total addition time is 30–40 min. After the addition is complete the reaction mixture is kept at $-70°C$ for an additional hour, during which time the dropping funnel is charged with a solution of 35.3 g (0.49 mol) of 2-methylpropanal (Note 1) in 100 mL of dry tetrahydrofuran. The aldehyde solution is added dropwise to the vigorously stirring enolate solution at such a rate as to maintain a reaction temperature of less than $-65°C$. After the addition is complete the reaction mixture is kept at $-70°C$ for an additional 30 min. To the vigorously stirring solution is added 500 mL of saturated aqueous ammonium chloride. At this point stirring is discontinued, the cooling bath is removed, and the partially frozen mixture is allowed to warm to room temperature. The contents of the reaction flask are introduced into a large separatory funnel and diluted with 500 mL of ether. The layers are separated, and the organic phase is washed with 300 mL of water and 300 mL of saturated brine and then dried over magnesium sulfate. After removal of the drying agent the solvents are removed with a rotary evaporator to give 112–120 g of an oily semisolid, which is a 7 : 2 mixture of the β-hydroxy ester and 2,6-dimethylphenyl propanoate. This material may be crystallized from ether–hexane to provide 70 g (60%) of pure β-hydroxy ester, mp 75.5–76°C (Note 6). However, it is not necessary to purify the crude product before hydrolysis to the β-hydroxy acid (Note 7).

C. *(2SR,3SR)-2,4-Dimethyl-3-hydroxypentanoic acid.* The crude product from the foregoing preparation (112–120 g) is dissolved in 500 mL of methanol and placed in a 2-L Erlenmeyer flask. A solution of 112 g (2 mol) of potassium hydroxide in a mixture of 500 mL of water and 500 mL of methanol is added with stirring, whereupon the reaction mixture warms to about 40°C. After stirring for 15 min crushed dry ice is added in portions to the vigorously stirring mixture until the pH is 7–8. The resulting solution is concentrated to a volume of about 500 mL with a rotary evaporator and extracted with two 300-mL portions of methylene chloride, which are discarded. The aqueous phase is then acidified to pH 1–2 by addition of 75 mL of concentrated hydrochloric acid (vigorous evolution of CO_2) and extracted with two 500-mL portions of methylene chloride. The combined organic extracts are washed with 200 mL of saturated brine and dried over magnesium sulfate. After removal of the drying agent the solvent is removed with a rotary evaporator to obtain 36–53 g of (2SR,3SR)-2,4-dimethyl-3-hydroxypentanoic acid as a semisolid. Crystallization from hexane provides 30–43 g (41–60% overall yield) of pure hydroxy acid, mp 76–79°C (Note 8).

2. Notes

1. Sodium hydride was obtained from Ventron Corporation, Morton Thiokol Inc. 2,6-Dimethylphenol and propanoyl chloride were obtained from Aldrich Chemical Company, Inc. and used without further purification. Diisopropylamine was distilled from calcium hydride prior to use. 2-Methylpropanal was distilled prior to use.

2. Reagent-grade diethyl ether from a freshly opened container was used without further purification. Reagent-grade tetrahydrofuran was dried over sodium before use.

3. 2,6-Dimethylphenol is a corrosive, poisonous substance that is readily absorbed through the skin. All reactions should be carried out in an efficient hood, and appropriate protective apparel should be used.

4. The IR spectrum (neat) shows an absorption at 1755 cm^{-1}. The ^{1}H NMR spectrum (CDCl$_3$) is as follows δ: 1.27 (t, 3 H, $J = 7$), 2.13 (s, 6 H), 2.55 (q, 2 H, $J = 7$), 6.90 (s, 3 H).

5. Butyllithium was obtained from Foote Mineral Company, Johnsonville, Tennessee. It may be standardized by a double-titration procedure.[2]

6. The IR spectrum (neat) has absorptions at 3500 and 1750 cm^{-1}. The ^{1}H NMR spectrum is as follows δ: 1.00 (d, 3 H, $J = 7$), 1.07 (d, 3 H, $J = 7$), 1.40 (d, 3 H, $J = 7$), 2.20 (s, 6 H), 2.93 (quintet 1 H, $J = 7$), 3.50 (m, 2 H), 7.03 (s, 3 H).

7. The checkers found that hydrolysis of once-crystallized aldol (mp 74–75°C) gives a hydroxy acid that crystallizes readily from hexane, for an overall two-step yield

of 32%. Hydrolysis of the crude aldol product gives the hydroxy acid as an oil that crystallizes with difficulty, for an overall two-step yield of 45%.

8. The IR spectrum (neat) has absorptions at 3500, 3300–2500, and 1695 cm^{-1}. The ^{1}H NMR spectrum is as follows δ: 0.93 (d, 3 H J = 7) 0.99 (d, 3 H, J = 7), 1.24 (d, 3 H, J = 7), 1.81 (octet, 1 H, J = 6), 2.69 (quintet, 1 H, J = 7), 3.44 (t, 1 H, J = 5.6), 7.4 (br s, 2 H, OH).

3. Discussion

A number of methods have been developed for accomplishing aldol addition reactions in a stereoselective manner.[3] The preformed lithium enolates of alkyl esters normally react with aldehydes to give mixtures of the two diastereomeric β-hydroxy esters:[4]

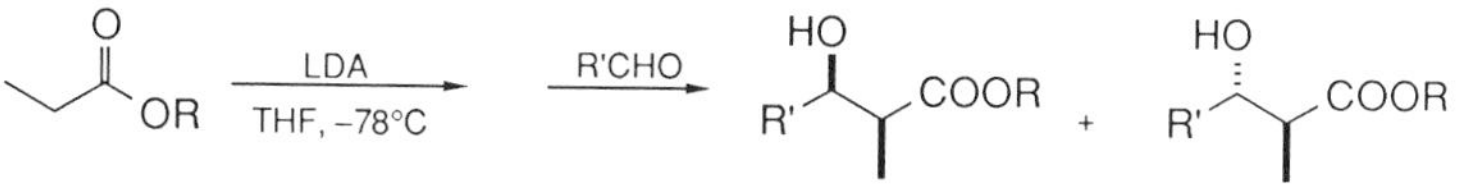

(1)

However, the enolates derived from certain aryl esters add to aldehydes to give largely one stereoisomeric product:[5]

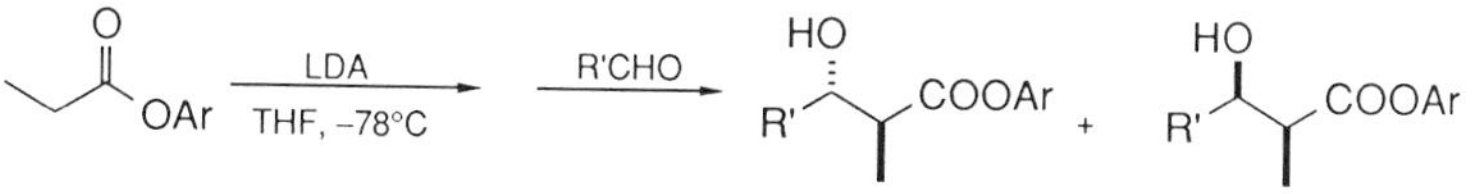

(2)

The aryl groups that have been investigated are 2,6-dimethylphenyl (DMP), 2,6-di-*tert*-butyl-4-methylphenyl (BHT), and 2,6-di-*tert*-butyl-4-methoxyphenyl (DBHA). Selected examples are shown in Table I. The most convenient reagents, because of the ease of their further manipulations, are the DMP esters. With aliphatic aldehydes branched at the α-carbon, the DMP esters give essentially one diastereomeric product, β-hydroxy ester **1**. With aromatic and α-unbranched aliphatic aldehydes, the DMP esters give predominantly, but not entirely, one isomer. In these cases the BHT or DBHA esters may be used. Acrolein gives a mixture of **1** and **2** even with the BHT and DBHA esters.

Aryl esters of other acids show similar stereoselectivity; examples are as follows:

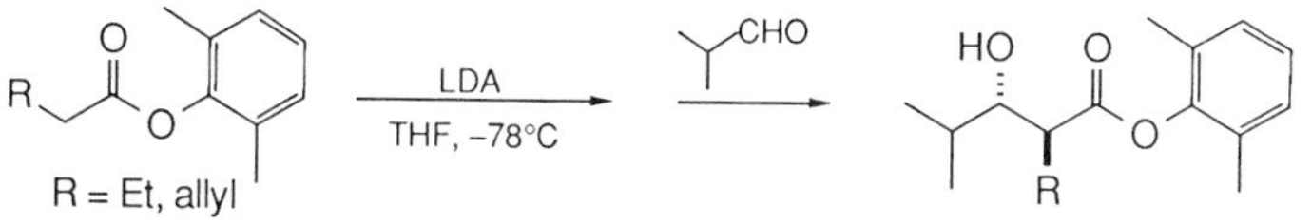

(3)

In addition, the BHT esters of *O*-benzyllactic acid condense with aldehydes to give diastereomerically homogeneous adducts:[6]

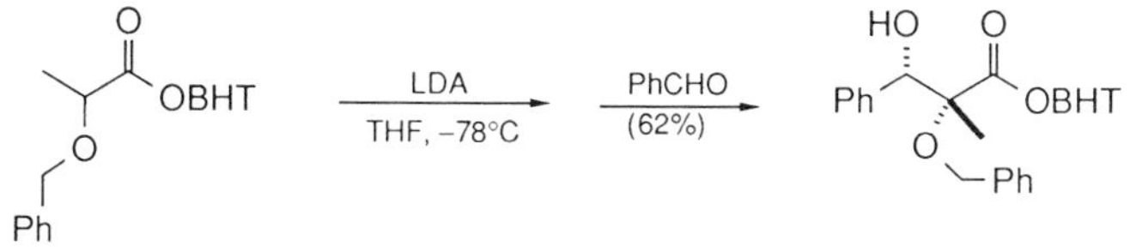

(4)

TABLE I

CONDENSATION OF ARYL ESTERS WITH ALDEHYDES

Ar Ester	R	1 : 2	mp (°C)	Yield (%)[a]
DMP	C_6H_5-	72	88/12	62–63[c]
DMP	$n\text{-}C_5H_{11}$	70	86/14	Oil
DMP	$i\text{-}C_3H_7-$	78	>98/2	76
DMP	$t\text{-}C_4H_9-$	82	>98/2	70–71
DMP	$C_6H_5(CH_3)CH-$	81	>98/2	Oil[d]
BHT	$CH_2=CH-$	88	85/15	64–67[e]
BHT	$CH_2=C(CH_3)-$	88	>98/2	70–71
BHT	C_6H_5-	96	>98/2	Oil
BHT	$i\text{-}C_3H_7-$	100[b]	>98/2	105–106
BHT	$C_6H_5(CH_3)CH-$	100[b]	>98/2	Oil[d]
DBHA	$CH_2=CH-$	90	87/13	65–72[f]
DBHA	C_6H_5-	75	>98/2	59–61
DBHA	$n\text{-}C_5H_{11}-$	70	>98/2	Oil
DBHA	$i\text{-}C_3H_7-$	79	>98/2	91–93
DBHA	$t\text{-}C_4H_9-$	77	>98/2	88–89

[a] All reactions were carried out on a 1-mmol scale. Unless otherwise noted, yields are for high-performance liquid chromatography (HPLC)-purified product. On a larger scale, such as is given in this procedure, yields are somewhat lower.
[b] This is the yield of crude product; these products were not purified by chromatography.
[c] Melting point given is that of the major diastereomer (1).
[d] Mixture of Cram's rule and anti-Cram's rule diastereomers: ratio = 4 : 1.
[e] Melting point given is for a 95 : 5 mixture of 1 : 2.
[f] Melting point given is for a 90 : 10 mixture of 1 : 2.

1. Department of Chemistry, University of California, Berkeley, CA 94720.
2. Jones, R. G.; Gilman, H. In "Organic Reactions," Adams, R., Ed.; Wiley: New York, 1951; Vol. 6, p 353.
3. (a) Heathcock, C. H. In "Comprehensive Carbanion Chemistry," Durst, T.; Buncel, E., Eds.; Elsevier: New York, 1984; Vol. II; (b) Evans, D. A.; Nelson, J. V.; Taber, T. R. In "Topics in Stereochemistry," Eliel, E. L.; Allinger, N. L.; Wilen, S. H., Eds.; Wiley: 1982; Vol. 13; (c) Heathcock, C. H. In "Asymmetric Synthesis," Morrison, J. D., Ed.; Academic Press, Inc.: New York, 1984; Vol. 3.
4. Heathcock, C. H.; Buse, C. T.; Kleschick, W. A.; Pirrung, M. C.; Sohn, J. E.; Lampe, J. J. Org. Chem. 1980, 45, 1066.
5. (a) Heathcock, C. H.; Pirrung, M. C. J. Org. Chem. 1980, 45, 1727; (b) Heathcock, C. H.; Pirrung, M. C.; Montgomery, S. H.; Lampe, J. Tetrahedron 1981, 37, 4087.
6. Heathcock, C. H.; Hagen, J. P.; Jarvi, E. T.; Pirrung, M. C.; Young, S. D. J. Am. Chem. Soc. 1981, 103, 4972.

1,3-DIMETHYLIMIDAZOLE-2-THIONE

(2*H*-Imidazole-2-thione, 1,3-dihydro-1,3-dimethyl-)

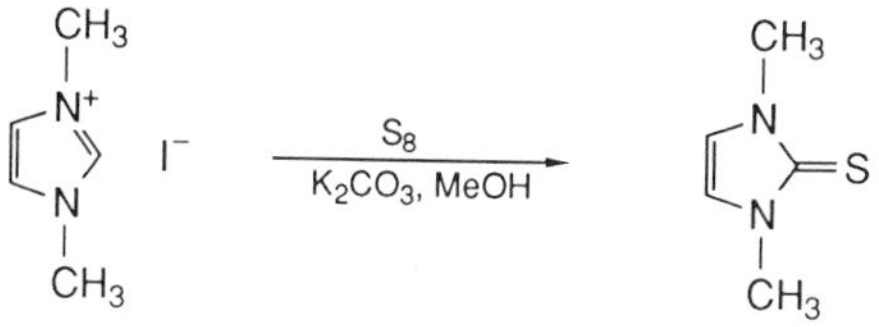

Submitted by BRIAN L. BENAC,[1a] EDWARD M. BURGESS,[2] and ANTHONY J. ARDUENGO, III[1b]
Checked by DAVID R. BRITTELLI, JOSEPH BURIAK, JR., and BRUCE E. SMART

1. Procedure

In a dry, 500-mL, round-bottomed flask, equipped with a magnetic stirrer and a drying tube are placed 44.8 g (0.20 mol) of 1,3-dimethylimidazolium iodide (Note 1), 35.0 g (0.25 mol) of anhydrous potassium carbonate, 6.5 g (0.20 mol) of sulfur (Note 2), and 300 mL of methanol (Note 3). The mixture is stirred for 40 hr at room temperature. The cloudy yellow mixture is filtered through a pad of Celite (Note 4) and the filter cake is washed with 80 mL of dichloromethane. The combined mother liquor and wash is evaporated to dryness on a rotary evaporator. The orange residue is dissolved in 500 mL of hot water and the hot solution is filtered to remove insoluble impurities. The aqueous filtrate is reheated and the product crystallizes on cooling. The white needles are collected by filtration, washed with 50 mL of cold water and air dried for 1 hr. The mother liquor is concentrated to yield a second crop of crystals to give a total of 15–16 g (58–62%) of pure 1,3-dimethylimidazole-2-thione, mp 182–183.5°C (Note 5).

2. Notes

1. The imidazolium iodide salt is conveniently prepared by the following procedure. A 500-mL, three-necked, round-bottomed flask equipped with a dropping funnel, thermometer, water-cooled condenser, and magnetic stirrer is charged with 200 mL of anhydrous methylene chloride and 82.1 g (1.0 mol) of 1-methylimidazole (from the Aldrich Chemical Company, Inc.). The solution is cooled and maintained at 5°C while 143.0 g (1.01 mol) of iodomethane in 75 mL of anhydrous methylene chloride is added dropwise over a 30-min period. When the addition is completed, the cooling bath is removed and the reaction mixture is stirred for 30 min at room temperature. Methylene chloride is removed on a rotary evaporator to yield 213.6–216.7 g (95–97%) of 1,3-dimethylimidazolium iodide, mp 81–83°C, ^{1}H NMR (d_6-DMSO) δ: 3.89 (s, 6 H), 7.73 (s, 2 H), 9.16 (s, 1 H). The submitters report the following spectral data: ^{1}H NMR (d_6-DMSO) δ: 4.08 (s, 6 H), 7.75 (s, 2 H), 9.86 (s, 1 H); ^{13}C NMR (d_6-DMSO) δ: 36.10 (s), 123.04 (s), 136.69 (s). The submitters report that the bromide and methyl sulfate salts of the 1,3-dimethylimidazolium cation gave similar yields in the thione synthesis.

2. Lac (precipitated) sulfur gives the best results. The checkers found that with sublimed sulfur (Fisher Scientific Company, laboratory grade) the yield of thione product is 12.5–12.8 g (49–50%). Lac sulfur is prepared by boiling a suspension of 33 g of calcium oxide and 50 g of sublimed sulfur (Fisher Scientific Company) in 200 mL of water for 30 min, then filtering the hot solution and acidifying the clear filtrate to pH 5 with hydrochloric acid. The precipitated sulfur is collected, washed with water, and dried in a vacuum desiccator.

3. ACS-grade methanol from the Fisher Scientific Company was used without further purification. The submitters report that attempts to use ethanol or water as solvents were unsuccessful.

4. The reaction mixture has a distinct odor of sulfur and should be handled in a hood. The product is odorless.

5. The submitters report a melting point of 182–184°C for material that was recrystallized from water or sublimed under reduced pressure. The product shows the following ^{1}H NMR spectrum (CDCl$_3$) δ: 3.58 (s, 6 H), 6.71 (s, 2 H). The submitters report the following spectral data: ^{1}H NMR (CDCl$_3$) δ: 3.6 (s, 6 H), 6.68 (s, 2 H); ^{13}C NMR (d_6-DMSO) δ: 34.34 (s), 117.82 (s), 161.87 (s); IR (CHCl$_3$) cm^{-1}: 2940 (C—H), 1450, and 1380.

3. Discussion

1,3-Dimethylimidazole-2-thione was first reported by Ansell, Forkey, and Moore,[3] who studied the X-ray crystal structure of this thione. No detailed synthesis of the thione has appeared in the chemical literature. This unusual thione has been used as a precursor to unusual thione ylides,[4,5] tricoordinate sulfuranes[6] and as a desulfurizing agent for a thiirane.[5] The thione also has remarkable antioxidant properties.[7] Compared to tetramethylthiourea, 1,3-dimethylimidazole-2-thione is remarkably resistant to desulfurization.

This procedure has been used to synthesize a variety of 1,3-dialkylimidazole 2-thiones. Other imidazole-2-chalcogenones (Se, Te) can be synthesized by similar procedures.

1. (a) Department of Chemistry, The Roger Adams Laboratory, University of Illinois-Urbana, IL 61801. (b) Present address: E. I. du Pont de Nemours & Co., Central Research and Development Department, E328/304, Wilmington, DE 19898.
2. Department of Chemistry, Georgia Institute of Technology, Atlanta, GA 30332.
3. Ansell, G. B.; Forkey, D. M.; Moore, D. W. *J. Chem. Soc., Chem. Commun.* **1970,** 56.
4. Arduengo, A. J.; Burgess, E. M. *J. Am. Chem. Soc.* **1976,** *98,* 5020.
5. Janulis, E. P., Jr.; Arduengo, A. J., III *J. Am. Chem. Soc.* **1983,** *105,* 3563.
6. Arduengo, A. J.; Burgess, E. M. *J. Am. Chem. Soc.* **1977,** *99,* 2376.
7. Arduengo, A. J.; Burgess, E. M., unpublished results.

β-DIMETHYLAMINOMETHYLENATION: *N,N*-DIMETHYL-*N'*-*p*-TOLYLFORMAMIDINE

[Methanimidamide, *N,N*-dimethyl-*N'*-(4-methylphenyl)-]

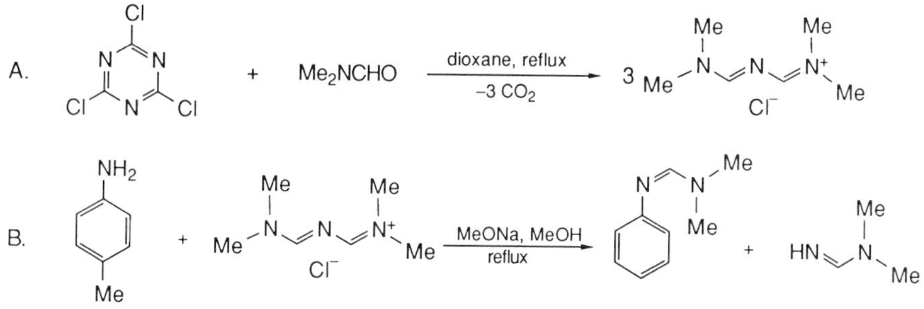

Submitted by JOHN T. GUPTON and STEVEN A. ANDREWS[1]
Checked by T. V. RAJANBABU and BRUCE E. SMART

1. Procedure

Caution! Cyanuric chloride is a lachrymator and causes burns on contact with the skin. All operations with this reagent should be carried out in a well-ventilated hood.

A. *[3-(Dimethylamino)-2-azaprop-2-en-1-ylidene]dimethylammonium chloride.* A 1-L, one-necked, round-bottomed flask is equipped with a Claisen adapter, mechanical stirrer, reflux condenser, and mineral oil bubbler (Note 1). The flask is charged with cyanuric chloride (73.8 g, 0.4 mol) (Note 2), *N,N*-dimethylformamide (175.4 g, 2.4 mol) (Note 3), and 1,4-dioxane (100 mL) (Note 4). The resulting solution is stirred and heated (at approximately 85°C) for 2–3 hr while a considerable amount of carbon dioxide is evolved (Note 5). When gas evolution is minimal, the reaction mixture is allowed to cool to room temperature; the product rapidly solidifies. The flask that contains the solid product is connected to an isopropyl alcohol–dry ice trap and the solvent is removed by evacuating the system to approximately 0.05 mm pressure. The crude product weighs 186–187 g (95%) and melts at 95–103°C (Notes 6, 7, 8).

B. *N,N-Dimethyl-N'-p-tolylformamidine.* A 250-mL, three-necked, round-bottomed flask equipped with a reflux condenser and a magnetic stirring bar coated with Teflon is placed under a positive nitrogen pressure and charged with 100 mL of methanol (Note 9). Sodium metal (1.4 g, 0.06 mol) (Note 10) is then added in small portions. After all of the sodium has reacted, *p*-toluidine (6.4 g, 0.06 mol) (Note 11) is added and the resulting solution is stirred for 5 min. The iminium salt (10.6 g, 0.065 mol) produced in Step A is added in one portion and the resulting mixture is refluxed with stirring overnight. The reaction mixture is cooled to room temperature and the solvent is removed on a rotary evaporator. The residue is taken up in chloroform (100 mL) and extracted twice with a saturated, aqueous solution of sodium bicarbonate (2 × 30 mL). The chloroform phase is dried over anhydrous magnesium sulfate and filtered, and the solvent is removed on a rotary evaporator. The residual dark-brown liquid is distilled using a Kugelrohr apparatus (Note 12); the major fraction boils at 85–100°C (oven temperature), 0.4 mm, and yields 9.1–9.2 g (94–95%) of a pale-yellow liquid (Notes 13 and 14).

2. Notes

1. The bubbler is connected to the condenser to monitor carbon dioxide evolution.

2. Cyanuric chloride was purchased from Aldrich Chemical Co., Inc. and was used without additional purification.

3. N,N-Dimethylformamide was purchased from Aldrich Chemical Co., Inc. and was dried over Linde 3A molecular sieves prior to use.

4. The 1,4-dioxane was reagent-grade and obtained from Fisher Scientific Corp. It was dried over Linde 3A molecular sieves prior to use.

5. The reaction becomes very exothermic with substantial evolution of carbon dioxide within 30–45 min after heating is initiated. It may be necessary to cool the mixture with ice water if the evolution of gas becomes too vigorous.

6. The checkers obtained material free of N,N-dimethylformamide after drying for at least 18 hr. The checkers found variable melting points that depended on the rate of heating. The submitters obtained 195 g (99%) of product which melted at 81–83°C after drying overnight at 1–6 mm of pressure and indicated that the product may contain a small amount of N,N-dimethylformamide but is suitable for use without additional purification. [3-(Dimethylamino)-2-azaprop-2-en-1-ylidene]dimethylammonium chloride is reported to melt at 101–103°C.[2]

7. The product is very hygroscopic and should be handled under a moisture-free environment. If the iminium salt is kept dry, it will have a substantial shelf life. The submitters recommend storing the product in a desiccator over anhydrous calcium sulfate.

8. The product has the following spectral characteristics: IR (CHCl$_3$) cm^{-1}: 1610 (C=N); ^{1}H NMR (CDCl$_3$) δ: 3.27 (s, 6 H, 2 CH$_3$), 3.43 (s, 6 H, 2 CH$_3$), 9.57 (s, 2 H, −CH=N).

9. The methanol that was used was reagent-grade and was dried over Linde 3A molecular sieves.

10. Sodium metal was obtained from Fisher Scientific Corp.

11. p-Toluidine was reagent-grade and was obtained from the Eastman Chemical Co.

12. The Kugelrohr apparatus was obtained from the Aldrich Chemical Co.

13. The submitters obtained 8.3–9.1 g (86–94%) boiling at 85–107°C, 0.4 mm. The reported boiling point of N,N-dimethyl-N'-p-tolylformamidine is 163°C (30 mm).[3] A gas chromatographic analysis of the product using a ¼-in. × 10-ft column packed with 5% Carbowax 20 M supported on 80–100-mesh chromosorb N exhibited a single peak with a retention time of 4.8 min at an oven temperature of 220°C with a flow rate of 60 cm^3/min. The checkers redistilled the product to obtain colorless material, bp 69.5°C (0.2 mm), which was analyzed. Anal.

calcd. for $C_{10}H_{14}N_2$: C, 74.03; H, 8.70; N, 17.27. Found C, 73.57; H, 8.51; N, 17.50

14. The product has the following spectral characteristics: IR (neat) cm^{-1}: 3030 (aromatic CH), 1635 (C=N), 1600 (C=C), ^{1}H NMR (CDCl$_3$) δ: 2.23 (s, 3 H, aromatic CH$_3$), 2.87 [s, 6 H, −N(CH$_3$)$_2$], 6.83 (d, 2 H, J = 8, aromatic CH), 7.06 (d, 2 H, J = 8, aromatic CH), 7.43 (s, 1 H, −CH=N−).

3. Discussion

[3-(Dimethylamino)-2-azaprop-2-en-1-ylidene]dimethylammonium chloride ("Gold's reagent"),[4] the preparation of which is described in Step A of the procedure, is a general β-dimethylaminomethylenating agent that reacts successfully with amines (Equation 1) to produce amidines,[5] with ketones (Equation 2) to produce enaminones,[6] and with amides (Equation 3) to produce acylamidines.[7]

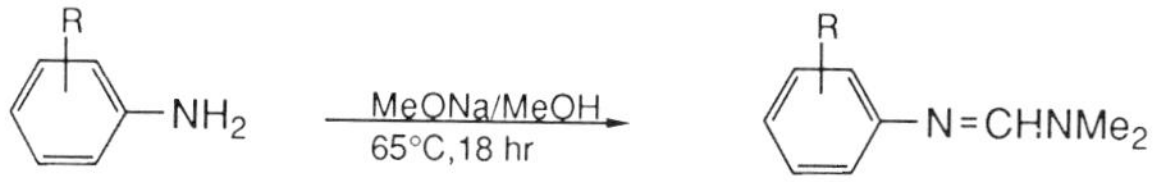

97% (R = 2-Me)
84% (R = 4-NO$_2$)
86% (R = 4-Br)

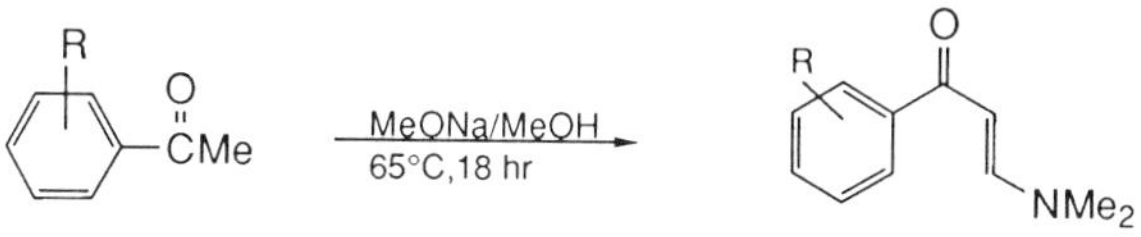

(1)

83% (R = H)
56% (R = 4-NO$_2$)
74% (R = 4-Br)

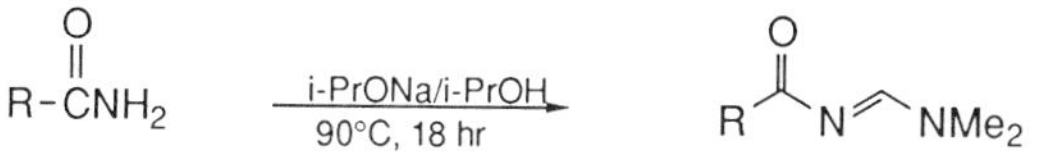

(2)

91% (R = Ph)
81% (R = 4-NO$_2$-C$_6$H$_4$)
89% (R = 3-pyridyl)

(3)

All reactions proceed in high yield and under mild conditions to produce relatively pure products. The most effective β-dimethylamino methylenating agents currently available are the formamide acetals,[8] some of which are available commercially.[9] They are, however, expensive, moisture- and heat-sensitive and require potent, mutagenic alkylating agents for their preparation. Under some circumstances they also necessitate high reaction temperatures and long reaction times. Alternatively, Gold's reagent is prepared in a single step, and in nearly quantitative yield, without purification, from inexpensive raw materials. The reaction of Gold's reagent with an amine or other substrate can be carried out at relatively low temperatures (65–90°C) and moderate reaction times (12-24 hr).

The significance of the aminomethylenated amines, ketones, and amides as important compounds and reaction intermediates is well-documented[5-7] and the use of Gold's reagent, therefore, provides an efficient, economical, and clean method[10] for obtaining such substances.

1. Department of Chemistry, University of Central Florida, P.O. Box 25000, Orlando, FL 32816.
2. Gold, H. *Angew. Chem.* **1960**, *72*, 956; *Chem. Abstr.* **1962**, *57*, 4542d.
3. Meerwein, H.; Florian, W.; Schon, N.; Stopp, F. *Justus Liebigs Ann. Chem.* **1961**, *641*, 1.
4. We have named this compound "Gold's Reagent" to simplify its common usage nomenclature.
5. Patai, S. "The Chemistry of Amidines and Imidates"; Wiley: New York, 1975, Chapter 7.
6. For a recent review of the synthetic importance of enaminones, see Greenhill, J. V. *Chem. Soc. Rev.* **1977**, *6*, 277.
7. (a) Lin, Y.; Lang, S. A.; Lovell, M. F.; Perkinson, N. A. *J. Org. Chem.* **1979**, *44*, 4160; (b) Lin, Y.; Lang, S. A.; Petty, S. R. *J. Org. Chem.* **1980**, *45*, 3750.
8. Abdulla, R.; Brinkmeyer, R. *Tetrahedron Lett.* **1979**, 1675; Simchen, G. In "Advances in Organic Chemistry: Methods and Results," Bohme, H.; Viehe, H., Eds.; Wiley: New York, 1979; Vol. 9, Pt. 2, pp. 393–526.
9. Aldrich Chemical Co., Inc. Milwaukee, Wisconsin.
10. Gupton, J. *Aldrichimica Acta* **1986**, *19*, 43.

1,6-DIMETHYLTRICYCLO[4.1.0.0^{2,7}]HEPT-3-ENE

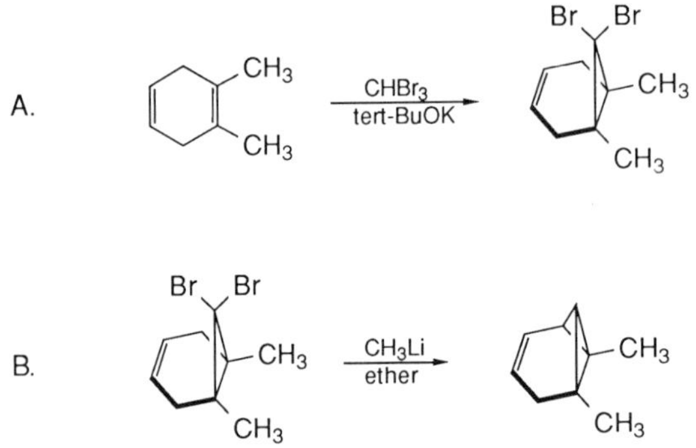

A.

B.

Submitted by R. T. Taylor and L. A. Paquette[1]
Checked by David A. Cortes and M. F. Semmelhack

1. Procedure

A. *7,7-Dibromo-1,6-dimethylbicyclo[4.1.0]hept-3-ene.* Into a 3-L, three-necked flask equipped with an overhead stirrer, 1-L addition funnel, and reflux condenser capped with a nitrogen-inlet tube are introduced 44.8 g (0.4 mol) of powdered potassium *tert*-butoxide (Note 1) and 1 L of olefin-free petroleum ether (bp 35–55°C; Note 2). To this stirred mixture is added a solution containing 38.0 g (0.35 mol) of 1,2-dimethyl-1,4-

cyclohexadiene (Note 3) in 200 mL of the same solvent. With external cooling from an ice bath and under nitrogen, 102.4 g (0.4 mol) of bromoform in 400 mL of petroleum ether is added dropwise during 1 hr. The ice bath is removed and the resultant slurry is stirred at room temperature under nitrogen for 6 hr. Water (500 mL) is added and the mixture is poured into a 3-L separatory funnel containing 300 mL of benzene. The organic layer is washed with four 500-mL portions of water, dried over anhydrous magnesium sulfate, and concentrated on a rotary evaporator (Note 4). Further evacuation at 0.5 mm produces a solid that is recrystallized from ether–petroleum ether (1 : 3) to afford 55–62 g (56.5–63.5%) of colorless solid, mp 95–98 °C (Note 5).

B. *1,6-Dimethyltricyclo[4,1.0.0^{2,7}]hept-3-ene*. A solution of 20.95 g (0.075 mol) of 7,7-dibromo-1,6-dimethylbicyclo[4.1.0]hept-3-ene in 500 mL of anhydrous ether is placed in a 1-L, three-necked flask equipped with a magnetic stirring bar, reflux condenser, addition funnel, and nitrogen-inlet tube. With stirring under nitrogen and external cooling in an ice bath, 50 mL of 1.6 *M* ethereal methyllithium (Note 6) in 70 mL of ether (0.08 mol) is introduced by dropwise addition during 30 min. The ice bath is removed and the mixture is stirred at room temperature for 1 hr. After 100 mL of water has been cautiously introduced, the mixture is transferred to a separatory funnel and the organic layer is separated. This solution is washed with water (3 × 100 mL), dried over anhydrous sodium sulfate (Note 7), and carefully concentrated by slow distillation through a 40-cm Vigreux column at atmospheric pressure, heating at <60 °C (Note 8). The residual liquid is distilled through a short, unpacked column to give 4.2–4.4 g (46–49%) of colorless oil, bp 48–49 °C (23 mm) (Note 9). Under the proper conditions, this hydrocarbon can be stored for 2 weeks at −5 °C without deterioration.

2. Notes

1. Potassium *tert*-butoxide can be obtained commercially from MSA Research Corporation, Callery, Pennsylvania. The checkers used a sample from Aldrich Chemical Company, Inc.

2. A liter of technical grade petroleum ether was treated in a separatory funnel with 200 mL of concentrated sulfuric acid, washed with water, and dried over anhydrous magnesium sulfate.

3. This diene was prepared by the procedure of Paquette and Barrett;[2] satisfactory results can be realized with material of 70–85% purity (15–30% contamination by *o*-xylene) since the aromatic impurity does not react subsequently and is easily removed.

4. Any residual *o*-xylene should be removed prior to crystallization because the dibromide is exceedingly soluble in aromatic solvents.

5. Further recrystallization is not necessary, but pure crystals, mp 107–108 °C, can be obtained in the manner described by Vogel and co-workers.[3]

6. The ethereal methyllithium solutions were purchased from Alfa Inorganics. The concentration of methyllithium in such solutions may be conveniently determined by a procedure described elsewhere[4,5] in which the lithium reagent is titrated with *sec*-butyl alcohol, utilizing the charge transfer complex formed from bipyridyl or *o*-phenanthroline and the lithium reagent as indicator.

7. Anhydrous magnesium sulfate is too acidic for this purpose and promotes rearrangement of the hydrocarbon.

8. All glassware that is to contain the cyclized product should be washed in base and dried (where necessary) prior to use.

9. The checkers found bp 55–56°C/30 mm. Attempted distillation at ca. 50 mm (bp 75°C) led to significant rearrangement to a dimethylcycloheptatriene. The product exhibits the following ^{1}H NMR spectrum (CDCl$_3$) δ:1.08 (s, 3 H, CH$_3$), 1.33 (d, 1 H, $J = 2$, methine C—H), 1.52 (s, 3 H, CH$_3$), 2.15–1.80 (m, 3 H, allylic methylene and methine), 5.50–5.15 (m, 1 H, olefinic C—H), 6.10–5.70 (m, 1 H, olefinic C—H).

3. Discussion

The tricyclo[4.1.0.0$^{2.7}$]hept-3-ene ring system, with its conjugated bicyclobutane ring and double bond and its isomeric relationship to cycloheptatriene, has recently commanded attention as a precursor of yet more highly strained molecules. However, the preparation of the parent hydrocarbon by reaction of 7,7-dibromo-3-norcarene with methyllithium at 0°C, first reported by Klummp and Vrielink,[6] does not proceed in yields above 1–5%.[6,7] Placement of a single methyl group at a ring juncture position of the transient norcarenylidene intermediate is, however, adequate to promote efficient ring closure through C—H alpha insertion.[7,8] The procedure described above is exemplary. Although two alternative routes to tricyclo[4.1.0.0$^{2.7}$]hept-3-enes are currently available,[6,9] alkyllithium-promoted cyclization of readily available 7,7-dibromobicyclo[4.1.0]hept-3-enes constitutes the most direct and efficient approach. In addition, this procedure illustrates an entirely general method for converting norcarane derivatives to endo,endo-1,3-bridged bicyclobutanes.[10–12]

Exposure of tricyclo[4.1.0.0$^{2.7}$]hept-3-enes to catalytic amounts of Ag$^+$ leads instantaneously and quantitatively to cycloheptatriene derivatives.[7] Promise of their usefulness as synthetic intermediates is growing rapidly.[13,14]

1. Department of Chemistry, The Ohio State University, Columbus, OH 43210.
2. Paquette, L. A.; Barrett, J. H. *Org. Synth., Coll. Vol. V* **1973**, 467.
3. Vogel, E.; Wiedemann, W.; Roth, H. D.; Eimer, J.; Gunther, H. *Liebigs Ann. Chem.* **1972**, *759*, 1.
4. Voskiul, W.; Arens, J. F. *Org. Synth., Coll. Vol. V* **1973**, 211.
5. Watson, S. C.; Eastham, J. F. *J. Organomet. Chem.* **1967**, *9*, 165.
6. Klummp, G. W.; Vrielink, J. J. *Tetrahedron Lett.* **1972**, 539.
7. Taylor, R. T.; Paquette, L. A. *Tetrahedron Lett.* **1976**, 2741.
8. Paquette, L. A.; Taylor, R. T. *J. Am. Chem. Soc.* **1977**, *99*, 5708.
9. Christl, M.; Bruntrup, G. *Angew. Chem., Int. Ed. Engl.* **1974**, *13*, 208.
10. Moore, W. R.; Ward, H. R.; Merritt, R. F. *J. Am. Chem. Soc.* **1961**, *83*, 2019; Moore, W. R.; King, B. J. *J. Org. Chem.* **1971**, *36*, 1877.
11. Reinarz, R. B.; Fonken, G. J. *Tetrahedron Lett.* **1973**, 4013.
12. Paquette, L. A.; Wilson, S. E.; Henzel, R. P.; Allen, Jr., G. R. *J. Am. Chem. Soc.* **1972**, *94*, 7761; Paquette, L. A.; Zon, G.; *J. Am. Chem. Soc.* **1974**, *96*, 203; Paquette, L. A.; Zon, G.; Taylor, R. T. *J. Org. Chem.* **1974**, *39*, 2677.
13. Christl, M.; Lechner, M. *Angew. Chem., Int. Ed. Engl.* **1975**, *14*, 765.
14. Paquette, L. A.; Taylor, R. T. *Tetrahedron Lett.* **1976**, 2745; Taylor, R. T.; Paquette, L. A. *J. Org. Chem.* **1978**, *43*, 242.

PREPARATION OF CHLOROPHENYLDIAZIRINE AND THERMAL GENERATION OF CHLOROPHENYL CARBENE: 1,2-DIPHENYL-3-METHYLCYCLOPROPENE

[Benzene, 1,1′-(3-methyl-1-cyclopropene-1,2-diyl)-bis-]

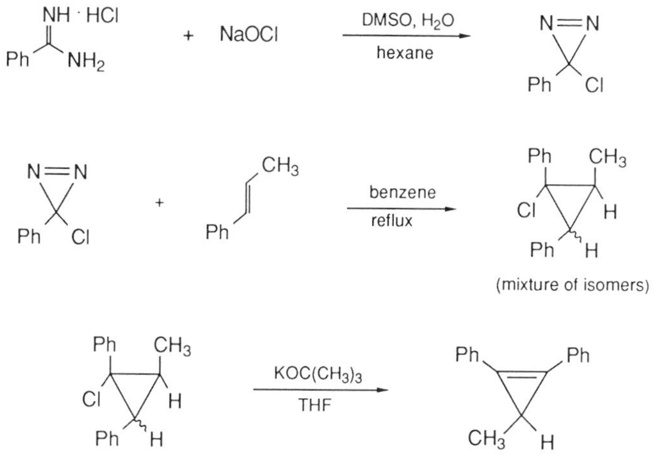

(mixture of isomers)

Submitted by ALBERT PADWA, MITCHELL J. PULWER, and THOMAS J. BLACKLOCK[1]
Checked by M. F. SEMMELHACK and A. ZASK

1. Procedure

Caution! Phenylchlorodiazirine is highly explosive (Note 6). It should always be handled with adequate shielding and normal protective equipment such as face shield and leather gloves.

Benzene has been identified as a carcinogen; OSHA has issued emergency standards on its use. All procedures involving benzene should be carried out in a well-ventilated hood, and glove protection is required.

A 3-L, three-necked, round-bottomed flask equipped with a high-speed mechanical stirrer and a 250-mL pressure-equalized dropping funnel is charged with 22.5 g (0.143 mol) of benzamidine hydrochloride (Note 1), 37.5 g (0.62 mol) of sodium chloride, 300 mL of hexane, and 400 mL of dimethyl sulfoxide. The flask is cooled at 0°C in an ice–salt bath (Note 2), and a mixture containing 15.5 g (2.65 mol) of sodium chloride in 1.2 L of aqueous 5.25% sodium hypochlorite solution (Note 3) is added with vigorous stirring over a 15-min period using the dropping funnel. After the addition is complete, stirring is continued for 15 min. At this time the organic phase is separated and the aqueous phase is extracted three times with 75-mL portions of ether. The ethereal extracts are combined with the organic phase and the mixture is washed successively four times with 125-mL portions of water and once with a 125-mL portion of saturated aqueous sodium chloride. The mixture is dried over anhydrous magnesium sulfate and concentrated to a volume of approximately 75 mL at 25°C under aspirator vacuum. The mixture is filtered through a 3-cm × 14-cm column of silica gel (Note 4) and eluted with 200 mL of anhydrous benzene (Note 5) into a 1-L, single-necked, round-bottomed flask. Concentration under reduced pressure at room temperature to a volume of approximately 50 mL yields a yellow solution (Note 6). The flask is equipped with a magnetic stirrer, heating mantle, and reflux condenser protected from the atmosphere by a calcium chloride drying tube and then charged with 600 mL of anhydrous benzene (Note 5) and 7.49 g (0.0634 mol) of *trans*-β-methylstyrene.

The reaction mixture is then heated at reflux for 3.5 hr and allowed to cool to 25°C. The benzene is removed at 25°C on a rotary evaporator to afford a dark-brown oil. This was diluted with 50 mL of ether, filtered through a 3-cm × 14-cm column of silica gel (Note 4), and eluted with an additional 175 mL of ether. Removal of the ether under reduced pressure yields ca. 21 g of an oily orange solid that consists of a mixture of diastereomeric 1-chloro-1,2-diphenyl-3-methylcyclopropanes (Note 7).

The crude mixture is transferred to a 1-L, one-necked flask equipped with a magnetic stirrer and a calcium chloride drying tube. To the flask is added 450 mL of anhydrous tetrahydrofuran (Note 8), and the mixture is cooled to −78°C in a dry ice–acetone bath. The drying tube is removed for a brief period, and 28.5 g (0.25 mol) of potassium *tert*-butoxide (Note 9) is quickly added in one portion using a powder funnel. The reaction is stirred for 1 hr at −78°C, warmed to 0°C, stirred for 3 hr, and then allowed to warm to room temperature and stirred for 12 hr. At the end of this time 60 mL of water is added slowly to the reaction mixture. The reaction mixture is concentrated on a rotary evaporator at 25°C to a volume of ca. 150 mL. The mixture is taken up in 160 mL of ether and washed successively with six 60-mL portions of water and one 60-mL portion of saturated aqueous sodium chloride. The ether layer is dried over anhydrous magnesium sulfate, and the solvent is removed under reduced pressure at room temperature. The resulting dark-brown oil is chromatographed through a 4-cm × 41-cm column of silica gel (Note 10) with eluting with hexane. Removal of the solvent under reduced pressure at 25°C affords 10.5–11.5 g (80–88%) of 1,2-diphenyl-3-methylcyclopropene as a pale yellow oil (Note 11).

2. Notes

1. Commercial benzamidine hydrochloride may be used without further purification.
2. Dimethylsulfoxide solidifies on the walls of the container but quickly dissolves on addition of the sodium hypochlorite solution.
3. Any commercial laundry bleach containing 5.25% by weight sodium hypochlorite is suitable.
4. Silica gel of 60–200-mesh was used (35 g).
5. Reagent-grade benzene was distilled from calcium hydride. The first 10% of the distillate was discarded.
6. Alternatively, to isolate the pure phenylchlorodiazirine, distillation through a 3-cm Vigreux column at 25°C (0.1 mm) affords 21.0–23.2 g (48–53%) of a pale-yellow oil; IR (cm^{-1}): 3067, 2967, 1706, 1567, 1490, 1437, 1332, 1258, 1200, 1081, 1013, 1001, 905, 758, 692. Foaming may occur during distillation as the residual solvent is removed under high vacuum. A water bath is employed to heat the distillation pot, and at no time should the pot temperature be allowed to

rise above 35°C. The distillation receiving flask should be immersed in a cold bath at $-60°C$ to avoid loss of phenylchlorodiazirine. Phenylchlorodiazirine is reputedly highly explosive and can be detonated by shock and/or elevated temperature.[2] The authors have encountered one such explosion due to a malfunctioning water bath thermostat that allowed the pot temperature to rise to 80°C. At this point the distillation mixture detonated. *In the pure form, phenylchlorodiazirine is considerately more shock-sensitive than nitroglycerine.*[3] Diluted with cyclohexane or benzene, it is not shock-sensitive.

7. The ^{1}H NMR spectrum shows the intermediate chlorocyclopropane to consist of a mixture of two stereoisomers in a 2 : 3 ratio resulting from the nonregio-specific addition of phenylchlorocarbene to *trans*-β-methylstyrene. Crystallization from hexane produced the major isomer as long white needles, mp 98–99°C, which was identified as (S)-1-chloro-(S,S)-1,2-diphenyl-(R)-3-methyl-cyclopropane; ^{1}H NMR (CDCl$_3$, 100 MHz) δ: 1.6 (d, 3 H, J = 6.0 Hz), 2.02 (dq, 1 H, J = 7.5 Hz and J = 6.0 Hz), 2.44 (d, 1 H, J = 7.5 Hz), and 6.6–7.2 (m, 10 H); IR (KBr) cm^{-1}: 3012, 2941, 2899, 2857, 1603, 1580, 1493, 1445, 1383, 1081, 1042, 987, 909, 853, 758, 751, 694. Successive crystallizations to remove the major isomer produced the minor component as a colorless oil identified as (S)-1-chloro-(S,R)-1,2-diphenyl-(S)-3-methylcyclopropane; ^{1}H NMR (CDCl$_3$, 100 MHz) δ: 0.96 (d, 3 H, J = 6.0 Hz), 2.02 (dq, 1 H, J = 7.5 Hz and J = 6.0 Hz), 2.44 (d, 1 H, J = 7.5 Hz), and 7.2–7.6 (m, 10 H); IR (neat, cm^{-1}): 3021, 2941, 2924, 1603, 1493, 1449, 1170, 1079, 1044, 1033, 913, 855, 758, 692.

8. Reagent-grade tetrahydrofuran was distilled from lithium aluminum hydride prior to use.

9. Commercial-grade potassium *tert*-butoxide (available from MSA Research Corporation, Evans City, PA 16033) was used without further purification.

10. Silica gel (200 g, 60–200 mesh) was used as the adsorbent. The eluant was monitored by thin-layer chromatography with collection of only the first eluted component.

11. Distillation of 1,2-diphenyl-3-methylcyclopropene should not be attempted, as much decomposition occurs. The product is characterized by ^{1}H NMR (CDCl$_3$, 100 MHz) δ: 1.36 (d, 3 H, J = 5.0 Hz), 2.18 (q, 1 H, J = 5.0 Hz), and 7.1–7.8 (m, 10 H); IR (neat) cm^{-1}: 3049, 3012, 2907, 1815, 1603, 1490, 1370, 1348, 1087, 1073, 1030, 990, 755, 738, 685. 1,2-Diphenyl-3-methylcyclopropene decomposes slowly at 25°C.

3. Discussion

The formation of aryl-substituted cyclopropenes by the addition of phenylchlorocarbene to olefins followed by dehydrohalogenation is a general reaction. The reagent phenylchlorodiazirine decomposes readily to produce phenylchlorocarbene in high yield.[4-7] Phenylchlorocarbene adds to many olefins to give halocyclopropanes, which can easily eliminate hydrogen chloride on treatment with base. The reaction of phenyl-chlorodiazirine with acetylenes produces cyclopropenyl chlorides, which can readily be converted to the corresponding biscyclopropenyl ethers on treatment with aqueous alcohol.[4,8,9] 1,2-Diphenyl-3-methylcyclopropene has been used to prepare a wide assortment of 1,2-diphenyl-3,3-disubstituted cyclopropenes.[10,11]

1. Chemistry Department, Emory University, Atlanta, GA 30322.
2. Graham, W. H. *J. Am. Chem. Soc.* **1965**, *87*, 4396–4397.
3. DeBoer, C. D. *J. Chem. Soc., Chem. Commun.* **1972**, 377–378.
4. Padwa, A.; Eastman, D. *J. Org. Chem.* **1969**, *34*, 2728–2732.
5. Frey, H. M.; Stevens, I. D. R. *J. Chem. Soc.* **1965**, 1700–1706.
6. Mitch, R. A. *J. Am. Chem. Soc.* **1965**, *87*, 758–761.
7. Moss, R. A. *Tetrahedron Lett.* **1967**, 4905–4909.
8. Breslow, R.; Chang, H. W. *J. Am. Chem. Soc.* **1961**, *83*, 2367–2375.
9. Farnum, D. G.; Burr, M. *J. Am. Chem. Soc.* **1960**, *82*, 2651.
10. Johnson, R. W.; Widlanski, T.; Breslow, R. *Tetrahedron Lett.* **1976**, 4685–4686.
11. Padwa, A.; Blacklock, T. J.; Getman, D.; Hatanaka, N.; Loza, R. *J. Org. Chem.* **1978**, *43*, 1481–1492.

DIPHENYL PHOSPHORAZIDATE

(Phosphorazidic acid, diphenyl ester)

$$(PhO)_2 \overset{\overset{O}{\|}}{P} Cl \quad + \quad NaN_3 \quad \longrightarrow \quad (PhO)_2 \overset{\overset{O}{\|}}{P} N_3$$

Submitted by Takayuki Shioiri[1] and Shun-ichi Yamada[2]
Checked by Christina Bodurow and M. F. Semmelhack

1. Procedure

A mixture of 56.8 g (0.21 mol) of diphenyl phosphorochloridate (Note 1), 16.3 g (0.25 mol) of sodium azide, and 300 mL of anhydrous acetone (Note 2) in a 500-mL round-bottomed flask fitted with a calcium chloride tube is stirred at 20–25 °C for 21 hr. The lachrymatory mixture is filtered in a hood, and the filtrate is concentrated under reduced pressure. The residue is distilled through a short Vigreux column (Note 3). The yield of diphenyl phosphorazidate, bp 134–136 °C (0.2 mm), is 49–52 g (84–89%) (Note 4).

2. Notes

1. Diphenyl phosphorochloridate (diphenyl chlorophosphate), from Aldrich Chemical Company, Inc., was used after purification by distillation at 165–168 °C (5 mm).
2. Commercial acetone was dried over anhydrous potassium carbonate and distilled.
3. The bath temperature should be kept below 200 °C to minimize decomposition of diphenyl phosphorazidate.[3]
4. Diphenyl phosphorazidate is a colorless nonexplosive oil that can be kept for a long time without decomposition if it is protected against light[3] and moisture.

3. Discussion

The procedure described is essentially that of Shioiri and Yamada.[4] Diphenyl phosphorazidate is a useful and versatile reagent in organic synthesis.[5] It has been used for racemization-free peptide syntheses,[4,6,7] thiol ester synthesis,[8] a modified Curtius reaction,[6,9,10] C-acylation of active methylene compounds,[11] esterification of an α-substituted carboxylic acid,[12] formation of diketopiperazines,[13] an alkyl azide synthesis,[14] phosphorylation of alcohols and amines,[15] and polymerization of amino acids and peptides.[16] Furthermore, diphenyl phosphorazidate acts as a nitrene source[3] and as a 1,3-dipole.[17,18] An example of the ring contraction of cyclic ketones to form cycloalkanecarboxylic acids is presented on page 135 in this volume.

1. Faculty of Pharmaceutical Sciences, Nagoya City University, Nagoya 467, Japan.
2. Faculty of Pharmaceutical Sciences, Josai University, Saitama 350-02, Japan.
3. Breslow, R.; Feiring, A.; Herman, F. J. Am. Chem. Soc. **1974,** *96,* 5937.
4. Shioiri, T.; Yamada, S. Chem. Pharm. Bull. **1974,** *22,* 849. No spectral and/or analytical data appear to have been reported for diphenyl phosphorazidate. The checkers recorded the following spectral data: ^{1}H NMR (CDCl$_3$), δ: 7.0–7.3 (br, s, C$_6$H$_5$—); IR (neat) cm^{-1}: 3060 (w, C—H), 2170 (s, —N$_3$), 1590 (m), 1490 (s, arene C=C), 1270 (m, P=O), 960 (s, P—O—aryl).
5. For a review, see (a) Shioiri, T.; Yamada, S. *Yuki Gosei Kagaku Kyokai Shi (J. Synth. Org. Chem. Japan)* **1973,** *31,* 666; Chem. Abstr. **1974,** *81,* 13781a; (b) Shioiri, T. *Nagoya Shiritsu Daigaku Yakugakubu Kenkyu Nempo (Ann. Rept. Pharm. Nagoya City University)* **1977,** *25,* 1.
6. Shioiri, T.; Ninomiya, K.; Yamada, S. J. Am. Chem. Soc. **1972,** *94,* 6203.
7. Yamada, S.; Ikota, N.; Shioiri, T.; Tachibana, S. J. Am. Chem. Soc. **1975,** *97,* 7174.
8. Yokoyama, Y.; Shioiri, T.; Yamada, S. Chem. Pharm. Bull. **1977,** *25,* 2423.
9. Ninomiya, K.; Shioiri, T.; Yamada, S. Tetrahedron **1974,** *30,* 2151.
10. Ninomiya, K.; Shioiri, T.; Yamada, S. Chem. Pharm. Bull. **1974,** *22,* 1398.
11. Hamada, Y.; Shioiri, T. Tetrahedron Lett. **1982,** *23,* 235. For a review, see Shioiri, T.; Hamada, Y. Heterocycles **1988,** *28,* 1035.
12. Ninomiya, K.; Shioiri, T.; Yamada, S. Chem. Pharm. Bull. **1974,** *22,* 1795.
13. Yamada, S.; Yokoyama, Y.; Shioiri, T. Experientia **1976,** *32,* 398.
14. Lal, B.; Pramanik, B. N.; Manhas, M. S.; Bose, A. K. Tetrahedron Lett. **1977,** 1977.
15. Cremlyn, R. J. W. Aust. J. Chem. **1973,** *26,* 1591.
16. Nishi, N.; Nakajima, B.; Hasebe, N.; Noguchi, J. Int. J. Biol. Macromol. **1980,** *2,* 53.
17. Yamada, S.; Hamada, Y.; Ninomiya, K.; Shioiri, T. Tetrahedron Lett. **1976,** 4749; see p. 135 in this volume.
18. Shioiri, T.; Kawai, N. J. Org. Chem. **1978,** *43,* 2936.

THE STORK–DANHEISER KINETIC ALKYLATION PROCEDURE: 3-ETHOXY-6-METHYL-2-CYCLOHEXEN-1-ONE

(2-Cyclohexen-1-one, 3-ethoxy-6-methyl-)

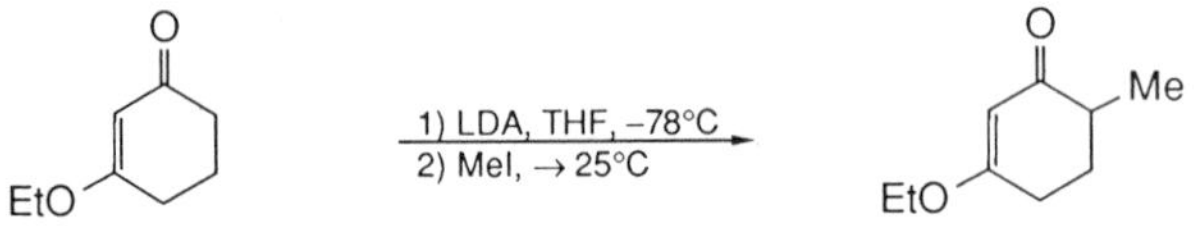

$$\text{EtO} \quad \xrightarrow[\text{2) MeI, } \rightarrow 25°C]{\text{1) LDA, THF, } -78°C} \quad \text{EtO} \quad \text{Me}$$

Submitted by Andrew S. Kende and Pawel Fludzinski[1]
Checked by P. Wovkulich, F. Barcelos, and Gabriel Saucy

1. Procedure

A dry, 2-L, three-necked, round-bottomed flask is equipped with a magnetic stirrer and two 500-mL pressure-equalizing dropping funnels. One of the dropping funnels is fitted with a rubber septum and the air in the system is replaced with dry nitrogen (Note 1). The flask is charged with 400 mL of anhydrous tetrahydrofuran (Note 2) and 51.6 g (71.5 mL, 0.51 mol) of anhydrous diisopropylamine (Note 3). The flask is cooled to 0°C with an ice bath. A 1.7 M hexane solution of butyllithium (288 mL, 0.49 mol) is added dropwise with stirring over a 30-min period. The resulting lithium diisopropyl-amide is cooled to −78°C with a dry ice–acetone bath (Note 4). A solution of 53.9 g (0.385 mol) of 3-ethoxy-2-cyclohexen-1-one (Note 5) in 250 mL of anhydrous tetra-hydrofuran is added dropwise with stirring at −78°C over a 1-hr period. The solution is stirred at −78°C for 30 min followed by the rapid addition of 114 g (50 mL, 0.80 mol) of methyl iodide (Note 6). After 5 min, the cooling bath is removed and the mixture is allowed to warm to room temperature and is stirred overnight. The reaction is quenched with 300 mL of water and the organic phase is separated. The aqueous phase is extracted four times with 75 mL of diethyl ether. The organic phases are combined and washed twice with 150 mL of water, once with 150 mL of brine, and dried over magnesium sulfate. Solvent removal on a rotary evaporator followed by distillation at reduced pressure affords 54–55 g (91–93%) of 3-ethoxy-6-methyl-2-cyclohexen-1-one as a colorless oil, bp 131–133°C (15 mm) (Notes 7 and 8).

2. Notes

1. The procedure described above is accomplished by alternately evacuating and filling the funnel with dry nitrogen two times; an oil bubbler is used to maintain a slight positive pressure throughout the reaction.

2. Tetrahydrofuran is freshly distilled from sodium and benzophenone.

3. Diisopropylamine is distilled from calcium hydride.

4. The flask is cooled with the dry ice–acetone bath for 1 hr before the next addition to ensure complete cooling of the solution.

5. *Org. Synth., Coll. Vol. V* **1973**, 539.

6. Methyl iodide was obtained from Eastman Organic Chemicals and used directly from a fresh bottle.

7. Spectroscopic data for 3-ethoxy-6-methyl-2-cyclohexen-1-one are as follows: ^{1}H NMR (CDCl$_3$) δ: 1.16 (d, 3 H, J = 7), 1.36 (t, 3 H, J = 6), 1.6–2.6 (m, 5 H), 3.92 (q, 2 H, J = 6), 5.32 (s, 1 H); IR (neat, cm^{-1}): 1670, 1600.

8. In the procedure as originally submitted, the authors used 1 equiv of base and distilled the product through a short-path distillation apparatus with 75–80% yields. The checkers used excess lithium diisopropylamide (suggested by Professor Clayton Heathcock) as specified in this procedure, and distilled the product through a 15-cm Vigreux column to afford 1.7–1.9 g of forerun (97–98.5% pure by GC) and 54.1–55.3 g (91.4–93.4% yield) of main fraction. The short-path distillation is probably quite adequate.

3. Discussion

The Stork-Danheiser[2] alkylation of 3-alkoxy-2-cyclohexenones under conditions of kinetic enolate formation at the 6-position has enjoyed extensive application in alicyclic synthesis. Such kinetic enolates have served as nucleophiles for a number of alkylations,[3-24] aldol condensations,[25-27] and Michael additions.[28,29] Reductive transposition of the resulting products to 4-substituted cyclohexenones has likewise found synthetic application.[30-33]

1. Department of Chemistry, University of Rochester, Rochester, NY 14627.
2. Stork, G.; Danheiser, R. *J. Org. Chem.* **1973,** *38,* 1775.
3. Pirrung, M. C. *J. Am. Chem. Soc.* **1979,** *101,* 7130.
4. Johnson, W. S.; McCarry, B. E.; Markezich, R. L.; Boots, S. G. *J. Am. Chem. Soc.* **1980,** *102,* 352.
5. Fried, J.; Mitra, D. K.; Nagarajan, M.; Mehrotra, M. M. *J. Med. Chem.* **1980,** *23,* 235.
6. Macdonald, T. L.; Mahalingam, S. *J. Am. Chem. Soc.* **1980,** *102,* 2113.
7. Stork, G.; Boeckmann, R. K., Jr.; Taber, D. F.; Still, W. C.; Singh, J. *J. Am. Chem. Soc.* **1979,** *101,* 7107.
8. Piers, E.; Zbozny, M.; Wigfield, D. C. *Can. J. Chem.* **1979,** *57,* 1064.
9. Rosenberger, M.; McDougal, P.; Saucy, G.; Bahr, J. *Pure Appl. Chem.* **1979,** *51,* 871.
10. Kende, A. S.; Benechie, M.; Curran, D. P.; Fludzinski, P.; Swenson, W.; Clardy, J. *Tetrahedron Lett.* **1979,** 4513.
11. Cory, R. M.; Chan, D. M. T.; McLaren, F. R.; Rasmussen, M. H.; Renneboog, R. M. *Tetrahedron Lett.* **1979,** 4133.
12. Mellor, M.; Otieno, D. A.; Pattenden, G. *J. Chem. Soc., Chem. Commun.* **1978,** 138.
13. Mellor, M.; Pattenden, G. *Synth. Comm.* **1979,** *9,* 1.
14. Smith, A. B., III; Scarborough, R. M., Jr. *Tetrahedron Lett.* **1978,** 4193.
15. Stork, G.; Taber, D. F.; Marx, M. *Tetrahedron Lett.* **1978,** 2445.
16. Debal, A.; Cuvigny, T.; Larchevêque, M. *Tetrahedron Lett.* **1977,** 3187.
17. Saxton, J. E.; Smith, A. J.; Lawton, G. *Tetrahedron Lett.* **1975,** 4161.
18. Lawton, G.; Saxton, J. E.; Smith, A. J. *Tetrahedron* **1977,** *33,* 1641.
19. Laguerre, M.; Dunogues, J.; Duffaut, N.; Calas, R. *J. Organomet. Chem.* **1976,** *120,* 319.
20. deGroot, A.; Jansen, B. J. M. *Recl. Trav. Chim. Pays-Bas* **1976,** *95,* 81.
21. Gawley, R. E. *Synthesis* **1976,** 777.
22. Dalgaard, L.; Lawesson, S. O. *Acta Chem. Scand., Ser. B* **1974,** *28,* 1077.

23. Stork, G.; Danheiser, R.; Ganem, B. *J. Am. Chem. Soc.* **1973,** *95,* 3414.
24. Thompson, H. W.; Huegi, B. S. *J. Chem. Soc., Chem. Commun.* **1973,** 636.
25. Stork, G.; Kraus, G. A.; Garcia, G. A. *J. Org. Chem.* **1974,** *39,* 3459.
26. Stork, G.; Kraus, G. A. *J. Am. Chem. Soc.* **1976,** *98,* 2351.
27. Torii, S.; Okamoto, T.; Kadono, S. *Chem. Lett.* **1977,** 495.
28. Quesada, M. L.; Schlessinger, R. H.; Parsons, W. H. *J. Org. Chem.* **1978,** *43,* 3968.
29. Posner, G. H.; Brunelle, D. J. *J. Chem. Soc., Chem. Commun.* **1973,** 907.
30. Kieczykowski, G. R.; Quesada, M. L.; Schlessinger, R. H. *J. Am. Chem. Soc.* **1980,** *102,* 782.
31. Kieczykowski, G. R.; Schlessinger, R. H. *J. Am. Chem. Soc.* **1978,** *100,* 1938.
32. Wenkert, E.; Goodwin, T. E. *Synth. Comm.* **1977,** *7,* 409.
33. Quesada, M. L.; Schlessinger, R. H. *Synth. Comm.* **1976,** *6,* 555.

ETHYL α-(BROMOMETHYL)ACRYLATE

[2-Propenoic acid, 2-(bromomethyl)-,ethyl ester)]

A. $(HOCH_2)_2C(CO_2C_2H_5)$ $\xrightarrow[85\text{-}90°C]{48\% \text{ HBr}}$ $CH_2 = C - CO_2H$
                                                                      CH_2Br

B. $CH_2 = C - CO_2H$ $\xrightarrow[reflux]{C_2H_5OH, H^+}$ $CH_2 = C - CO_2C_2H_5$
          CH_2Br CH_2Br

Submitted by K. Ramarajan, K. Ramalingam, D. J. O'Donnell, and K. D. Berlin[1]
Checked by H. S. Shou, E. Tsou, R. A. Hayes, and Orville L. Chapman

1. Procedure

Caution: Ethyl α-(bromomethyl) acrylate is a potent vesicant and lachrymator and should be handled with care. All operations should be carried out in an efficiently ventilated hood in order to avoid contact.

A. *α-(Bromomethyl)acrylic acid.* A 500-mL, three-necked, round-bottomed flask is equipped with a magnetic stirrer, fraction collector, cold-finger condenser, and two thermometers. Into the flask are placed 55.0 g (0.25 mol) of diethyl bis(hydroxymethyl)malonate (Note 1) and 142 mL (1.25 mol) of 47–49% hydrobromic acid (Note 2). The mixture is then heated and the temperature of the liquid maintained between 85 and 90°C. A mixture of ethyl bromide and water distills during the course of 1.5–2 hr. The residue is then boiled for 10 hr, maintaining the temperature between 85–90°C (Note 3). At the end of this period, the mixture is concentrated on a rotary evaporator at 65–70°C (10–15 mm). About 100 mL of water is removed. The residue is cooled in the refrigerator overnight. Crystals of α-(bromomethyl)acrylic acid are filtered in the cold (Note 4) to give, after drying (Note 5), 17.9 g (43%) of acid, mp 71–73°C (Note 6).

B. *Ethyl α-(bromomethyl)acrylate.* In a nitrogen-flushed, 1-L, round-bottomed flask equipped with a magnetic stirrer, Dean-Stark trap, and condenser are placed 42.0 g

(0.25 mol) of α-(bromomethyl)acrylic acid and 300 mL of benzene. Approximately 50 mL of a binary azeotrope of benzene and water is distilled (Note 7). The Dean-Stark trap is removed and 100 mL of absolute ethanol (Note 8) and 1 mL of concentrated sulfuric acid are added slowly. The contents of the flask are boiled in a nitrogen atmosphere for 36 hr, the condensate being passed through 100 g of molecular sieves (Linde 3A) before being returned to the flask. About 125 mL of a mixture of benzene and ethanol is removed from the reaction mixture by distillation (at 67°C). Then 100 mL of benzene is added and another 125 mL of benzene-ethanol mixture distilled (67–75°C). The residue is poured into 200 mL of water and neutralized with solid sodium bicarbonate (ca. 10–15 g) until CO_2 evolution ceases. The resulting solution is extracted with three 75-mL portions of ether, and the combined extracts are dried over anhydrous sodium sulfate for 3 hr. The ether is removed under reduced pressure in a rotary evaporator, and crude ester distilled to give a fraction at 39–40°C (0.9 mm) that weighs 33–34 g (71%). The ester is of high purity, as evidenced by spectral analysis (Note 9).

2. Notes

1. The checkers prepared this ester on a 0.7-mol scale by a modification of the previously published method.[2] The modification was effected as follows. The ethereal extract from the formaldehyde-diethyl malonate reaction, after drying over sodium sulfate for 3 hr, was concentrated in a rotary evaporator and the residue was stored in a refrigerator overnight. The crude ester was obtained as white crystals, mp 47–50°C; yield 85.6%. The checkers found that the ester prepared in this manner gave superior yields of the acrylic acid.

2. The submitters reported that the use of excess hydrobromic acid resulted in the formation of a mixture of dibromoisobutyric acid and α-(bromomethyl)acrylic acid as evidenced by NMR analysis.

3. Temperatures higher that 85–90°C gave a mixture of dibromoisobutyric acid and α-(bromomethyl)acrylic acid.

4. This was done at 4°C to improve the yield; otherwise considerable amounts of α-(bromomethyl)acrylic acid remain in solution.

5. The compound was air-dried for 3 days at room temperature.

6. The product was almost pure. It could be recrystallized from Skelly-solve-B (bp 60–80°C) and further purified by sublimation, mp 73–75°C (Anal. calcd. for $C_4H_5BrO_2$: C, 29.12; H, 3.05. Found: C, 29.07; H, 3.10). IR (KBr) cm^{-1}: 1689 (C=O), 1626 (C=CH_2), ^{1}H NMR (CDCl_3) δ: 4.18 (s, 2 H, H_a), 6.09 (s, 1 H, H_b), 6.49 (s, 1 H, H_c).

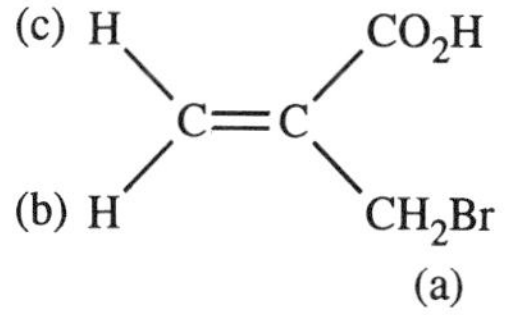

7. There was only about 1 mL of water in the distillate.

8. Absolute alcohol was prepared by boiling commercial absolute alcohol over magnesium turnings for 4 hr in a nitrogen atmosphere.

9. The spectral properties of ethyl α-(bromomethyl)acrylate are as follows: ^{1}H NMR (CDCl$_3$) δ: 1.26–1.40 (t, 3 H, H$_a$), 4.16–4.38 (quintet, 2 H, H$_b$), 4.19 (s, 2 H, H$_c$), 5.96 (s, 1 H, H$_d$), 6.32 (s, 1 H, H$_e$).

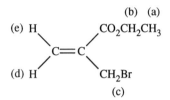

3. Discussion

The procedure described here is a modification of that of Ferris.^{3} The overall yield has been increased from 17 to 30% by making changes as indicated in Notes 2 and 3. In addition, the number of stages in the preparation of ethyl α-(bromomethyl)acrylate from diethyl malonate has been reduced from four to three.

Ethyl α-(bromomethyl)acrylate has proved to be an excellent reagent for conversion of aldehydes and ketones, both acyclic and cyclic, into the corresponding α-methylene-γ-butyrolactone derivatives$^{4-9}$ in a Reformatsky type reaction. The yield was excellent in the case of several spiro α-methylene-γ-butyrolactones.^{10} Synthetic α-methylene-γ-butyrolactone derivatives have been shown to possess antitumor activity.^{5,6,7,11,12,13} Ethyl α-(bromomethyl)acrylate has also proven of value in the synthesis of alkylated products of enol ethers of cyclohexane-1,3-dione.^{14}

1. Department of Chemistry, Oklahoma State University, Stillwater, OK 74078. We gratefully acknowledge partial support of this work by a grant from the National Cancer Institute, CA 14343.

2. Block, Jr., P. *Org. Synth.*, *Coll. Vol. V* **1973**, 381–383.

3. Ferris, A. F. *J. Org. Chem.* **1955**, *20*, 780.

4. Ohler, E.; Reininger, K; Schmidt, V. *Angew. Chem.*, *Int. Ed. Engl.* **1970**, *9*, 456.

5. Howie, G. A.; Stamos, I. K.; Cassady, J. M. *J. Med. Chem.* **1976**, *19*, 309.

6. Stamos, I. K.; Howie, G. A.; Manni, P. E.; Haws, W. J.; Bryn, S. R.; Cassady, J. M. *J. Org. Chem.* **1977**, *42*, 1703.

7. Rosowsky, A.; Papathanasopoulos, N.; Lazarus, H.; Foley, G. E.; Modest, E. J. *J. Org. Chem.* **1974**, *17*, 672.

8. Grieco, P. A. *Synthesis*, **1975**, 67.

9. Gammill, R. B.; Wilson, C. A.; Bryson, T. A. *Synth. Commun.* **1975**, *5*, 245.

10. Ramalingam, K.; Berlin, K. D. *Org. Prep. Proc. Int.* **1977**, *9*, 16.

11. Grieco, P. A.; Noguez, J. A.; Masaki, Yukio; Hiroi, K.; Nishizawa, M. *J. Med. Chem.* **1977**, *20*, 71.

12. Wege, P. M.; Clark, R. D.; Heathcock, C. H. *J. Org. Chem.* **1976**, *41*, 3144.

13. Hoffmann, H. M. R.; Rabe J. *Angew. Chem. Int. Ed. Engl.* **1985**, *24*, 94.

14. deGroot, A.; Jansen, B. J. M. *Recl. Trav. Chim. Pays-Bas* **1976**, *95*, 81.

ALIPHATIC AND AROMATIC β-KETO ESTERS FROM MONOETHYL MALONATE: ETHYL 2-BUTYRYLACETATE

(Pentanoic acid, 4-methyl-3-oxo-, ethyl ester)

$$HO_2CCH_2CO_2C_2H_5 \xrightarrow[\text{(2) }(CH_3)_2CHCOCl]{\text{(1) BuLi, THF}} (CH_3)_2CHCOCH_2CO_2C_2H_5$$

Submitted by W. Wierenga and H. I. Skulnick[1]
Checked by Stefan Blarer, Daniel Wasmuth, and Dieter Seebach

1. Procedure

Ethyl 2-butyrylacetate. In a 1-L, three-necked, round-bottomed flask fitted with a mechanical stirrer, dry nitrogen inlet, and thermometer is placed 19.8 g (0.150 mol) of monoethyl malonate (Note 1), 350 mL of dry tetrahydrofuran (THF, Note 2), and 5 mg of 2,2'-bipyridyl. The solution is cooled to approximately $-70°C$ (in an isopropyl alcohol-dry ice bath) and a 1.6 M solution of butyllithium in hexane is added from a dropping funnel while the temperature is allowed to rise to approximately $-10°C$. Sufficient butyllithium is added (approx. 190 mL) until a pink color persists for several minutes (Note 3). The heterogeneous solution is recooled to $-65°C$ and 7.90 mL (7.98 g, 75 mmol) of isobutyryl chloride (Note 4) is added dropwise over 5 min. The reaction solution is stirred for another 5 min (Note 5) and then poured into a separatory funnel containing 500 mL of ether and 300 mL of cold, 1 N hydrochloric acid (Note 6). The funnel is shaken, the layers are separated, and the organic phase is washed with two 150-mL portions of saturated aqueous sodium bicarbonate, followed by 150 mL of water, and dried over anhydrous sodium sulfate. Removal of the solvents under reduced pressure leaves 11.70 g (98%) of ethyl 2-butyrylacetate (Note 7). The crude product can be distilled at 70–74°C (7 mm) (80% yield, 96% purity by GLC).

2. Notes

1. The potassium salt of monoethyl malonate, available from the Aldrich Chemical Company, Inc., can be used after neutralization. Direct use of the potassium salt with only 1 equiv. of butyllithium gave substantially lower yields. Alternatively, monoethyl malonate can be conveniently prepared in high yield from diethyl malonate.[2]

2. For smaller-scale reactions, THF was dried and used directly by distillation from sodium–benzophenone, or first from KOH and then from $LiAlH_4$. The checkers used only dry THF for the present, large-scale procedure as well.

3. Initially, butyllithium can be added rapidly (20 mL/min) while the cooling bath is removed. A slightly exothermic reaction is noted. Toward the end of the reaction, dropwise addition should be used; the pink color will form and then dissipate. The checkers found it more convenient to use the calculated amount of a freshly titrated[3] solution of butyllithium.

4. Isobutyryl chloride was used as purchased from Aldrich Chemical Company, Inc. or Fluka AG.

TABLE I

REACTION OF ACID CHLORIDES WITH DILITHIO MONOETHYL MALONATE

$$RCOCl \rightarrow RCOCH_2CO_2C_2H_5$$

R	Reaction Time (min)/ Temperature (°C)	Yield (%)[a]
$CH_3CH_2CH_2$	5/−65	95
$PhCH_2$	5/−65	99
Ph	30/−65	97
$4\text{-}CH_3OC_6H_4$	60/−65	90
$4\text{-}ClC_6H_4$	30/−65	96
$2\text{-}ClC_6H_4$	30/−65	95
$2\text{-}C_{10}H_7$	30/−65	95
3-Furyl	15/−65, 60 to 0	97
2-Pyrazinyl	15/−65, 60 to 0	91

[a]The purity of all products isolated is higher than 90% as determined by GLC or ^{1}H NMR. The only contaminants appear to be hydrocarbons including n-octane.

5. Reaction times and temperatures vary, depending on the substrate acid chloride (see Table I).

6. For acid chlorides that contain a basic nitrogen, the aqueous phase is adjusted to approximately pH 7 by limiting the concentration of the hydrochloric acid.

7. Gas chromatographic analysis using a 3-ft, 3% OV-17 column at 90°C indicated a purity of 92% (retention time was 3.2 min) with GC-mass spectrometric identification showing M^+ m/e 158 (27%) and the base peak (100%) at m/e 113 ($C_6H_9O_2$). The ^{1}H NMR spectrum of undistilled material indicates impurities with resonances in the aliphatic region (δ : 1.5–1.0). The checkers recommend distillation of the crude product.

3. Discussion

Since the β-keto ester group is often a key moiety in organic syntheses, a general and efficient route to these 1,3-dicarbonyl compounds is highly desirable. We feel that the one-pot preparation from monoethyl malonate described here[4] represents an attractive alternative to previous methods[5] because of the following characteristics: (1) the reaction is general, as demonstrated by the diversity of examples in Table I; (2) the starting materials, (monoethyl malonate and the acid chlorides) are readily available and inexpensive; (3) the yields are high and therefore omission of purification is possible in many instances; and finally (4) the reaction is simple and easy to scale up.

The optimum ratio for high yields of β-ketoester is 1.7 (monoethyl malonate : acid chloride). A nonstoichiometric reaction for optimum yield is not a serious drawback in this case since the reagent in excess is the inexpensive dilithio monoethyl malonate. Our results show that lowering the ratio also lowers the yield, whereas an increase in the ratio beyond 1.7 has little effect.

1. Department of Experimental Chemistry, The Upjohn Company, Kalamazoo, MI 49001.

2. Strube, R. E. *Org. Synth. Coll. Vol. IV* **1963**, 417.

3. Kofrom, W. G.; Baclawski, L. M. *J. Org. Chem.* **1976**, *41*, 1879.

4. Wierenga, W.; Skulnick, H. I.; *J. Org. Chem.* **1979**, *44*, 310.

5. Early work dating back to the acetoacetic ester condensation is reviewed by (a) Hauser, C. R.; Hudson, B. E. *Org. React.* **1942**, *1*, 266; (b) Hauser, C. R.; Swamer, F. W.; Adams, J. T. *Org. React.* **1954**, *8*, 59; and (c) House, H. O. "Modern Synthetic Reactions," Benjamin: Menlo Park, CA, 1972; Chapter 11. Examples of more recent approaches that do not employ malonates are (d) Krapcho, A. P.; Diamanti, J.; Cayen, C.; Bingham, R. *Org. Synth., Coll. Vol. V* **1973**, 198; (e) Huckin, S. N.; Weiler, L. *J. Am. Chem. Soc.* **1974**, *96*, 1082; (f) Fréon, P.; Tatibouët, F. *C. R. Hebd. Seances Acad. Sci., Ser. C* **1957**, *244*, 2399; (g) Wasserman, H. H.; Wentland, S. H. *J. Chem. Soc., Chem. Commun.* **1970**, 1; (h) Rathke, M. W.; Sullivan, D. F. *Tetrahedron Lett.* **1973**, 1297; and (i) Rathke, M. W.; Deitch, J. *Tetrahedron Lett.* **1971**, 2953. Examples which exploit malonate as a starting material include (j) Breslow, D. S.; Baumgarten, E.; Hauser, C. R. *J. Am. Chem. Soc.* **1944**, *66*, 1286; (k) Ireland, R. E.; Marshall, J. A. *J. Am. Chem. Soc.* **1959**, *81*, 2907; (l) Pichat, L.; Beaucourt, J.-P. *Synthesis* **1973**, 537; (m) Bram, G.; Vilkas, M., *Bull. Soc. Chim. Fr.* **1964**, 9451 (n) Schmidt, U.; Schwochau, M. *Monatsh. Chem.* **1967**, *98*, 1492; (o) Pollet, P.; Gelin, S. *Synthesis* **1978**, 142; (p) Clezy, P. S.; Fookes, C. J. R. *Aust. J. Chem.* **1977**, *30*, 1799; (q) Van der Baan, J. L.; Barnick, J. W. F. K.; Bickelhaupt, F. *Tetrahedron* **1978**, *34*, 223; and (r) Oikawa, Y.; Sugano, K.; Yonemitsu, O. *J. Org. Chem.* **1978**, *43*, 2087.

YEAST REDUCTION OF ETHYL ACETOACETATE: (S)-(+)-ETHYL 3-HYDROXYBUTANOATE

[Butanoic acid, 3-hydroxy-, ethyl ester, (S)]

$$\text{ca. 93% S(+)} \qquad + \qquad \text{ca. 7% R(-)}$$

Submitted by DIETER SEEBACH, MARIUS A. SUTTER, ROLAND H. WEBER, and MAX F. ZÜGER[1]
Checked by TERRY ROSEN and CLAYTON H. HEATHCOCK

1. Procedure

A 4-L, three-necked, round-bottomed flask equipped with mechanical stirrer, bubble counter, and a stopper is charged with 1.6 L of tap water, 300 g of sucrose (Note 1), and 200 g of baker's yeast (Note 2), which are added with stirring in this order. The mixture is stirred for 1 hr at about 30°C, 20.0 g (0.154 mol) of ethyl acetoacetate (Note 3) is added, and the fermenting suspension (Note 4) is stirred for another 24 hr at room temperature. A warm (ca. 40°C) solution of 200 g of sucrose (Note 1) in 1 L of tap water is then added, followed 1 hr later by an additional 20.0 g (0.154 mol) of ethyl acetoacetate (Note 3). Stirring is continued for 50–60 hr at room temperature. When the reaction is complete by gas chromatographic analysis (Note 5), the mixture is worked up by first adding 80 g of Celite and filtering through a sintered-glass funnel (porosity

4, 17-cm diam). After the filtrate is washed with 200 mL of water, it is saturated with sodium chloride and extracted with five 500-mL portions of ethyl ether (Note 6). The combined ether extracts are dried over magnesium sulfate, filtered, and concentrated with a rotary evaporator at 35°C bath temperature to a volume of 50–80 mL. This residue is fractionally distilled at a pressure of 12 mm through a 10-cm Vigreux column, and the fraction boiling at 71–73°C (12 mm) is collected to give 24–31 g (59–76%) of (S)-$(+)$-ethyl 3-hydroxybutanoate (Notes 7 and 8); the specific rotation $[\alpha]_D^{25} + 37.2°$ (chloroform, c 1.3) corresponds to an enantiomeric excess of 85% (Note 9).

The enantiomeric excess may be enhanced by several crystallizations of the 3,5-dinitrobenzoate derivative (Note 10) or else by using "starved" yeast (Note 11).

2. Notes

1. Commercially available sugar (sucrose) from a grocery store is used.

2. Commercially available baker's yeast can be used. The submitters used baker's yeast from E. Klipfel & Co. AG, CH-4310 Rheinfelden (Switzerland). The checkers used Fleischmann's yeast (cubes), obtained from a supermarket, or Red Star Baker's yeast (Universal Food Corporation), obtained from a bakery. The optical rotation of the final product was essentially the same for runs in which the two brands were employed.

3. Ethyl acetoacetate is freshly distilled before use (bp 65°C/12 mm).

4. One to two bubbles per second of CO_2 are developed.

5. A small sample (ca. 1 mL) is removed from the mixture and extracted with ethyl ether. The ether solution is analyzed for remaining ethyl acetoacetate by capillary gas chromatography: 0.3-mm × 20-m glass capillary column Carbowax 20 M, oven temperature 100°C, carrier gas: hydrogen (0.4 atm); retention time of ethyl acetoacetate: 450 sec, of (S)-$(+)$-ethyl 3-hydroxybutanoate: 610 sec. It is important that all the starting material be consumed. If small mounts of ethyl acetoacetate are detected, 100 g of sucrose is added and the mixture is stirred for a further period of 2 days. The checkers detected the presence of residual ethyl acetoacetate by TLC on 250-μm silica gel plates with 1 : 1 ether : hexane as eluant. Plates are developed by dipping the dried plate into a solution of 10% vanillin and 5% sulfuric acid in 95% ethanol and then gently warming over a hot plate; ethyl acetoacetate appears as an intense blue spot with R_f 0.45.

6. In the case of emulsions, addition of methanol may be helpful. The very fine and stable emulsion that still remains is included with the aqueous phase.

7. The spectral properties of (S)-$(+)$-ethyl 3-hydroxybutanoate are as follows: IR[2a] (film) cm^{-1}: 3440, 2980, 1730, 1375, 1300, 1180, 1030; ^{1}H NMR[2b] (CCl_4) δ: 1.15 (d, 3 H, J = 6.5, CH_3), 1.28 (t, 3 H, J = 7 Hz, CH_3), 2.35 (d, 2 H, J = 6.5, CH_2CO), 3.15 (s, 1 H, OH), 4.05 (q, 2 H, J = 7, CH_2O), 4.15 (m, 1 H, CHOH).

8. This ester should be stored in a refrigerator as there has been some indication that it may undergo a transesterification–oligomerization upon standing at room temperature.

9. The specific rotation $[\alpha]_D^{25}$ varies from $+35.5°$ to $+38°$ (82–87% enantiomeric excess). The enantiomeric purity can also be checked by formation of the ester with (R)-$(+)$-1-methoxy-1-trifluoromethylphenylacetyl (MTPA) chloride.[3] The ^{19}F NMR chemical shifts of the diastereomeric esters are 6.13 (R,R) and 6.01 (R,S) ppm downfield of external trifluoroacetic acid.

10. The procedure of enriching the (S)-$(+)$-enantiomer to 100% enantiomeric excess by the previously described crystallization method is tedious.[4] It provides optically pure ethyl (S)-$(+)$-3-$(3',5'$-dinitrobenzoyloxy)butanoate of $[\alpha]_D^{25}$ $+26.3°$ (chloroform, c 2), which after cleavage gives enantiomerically pure (S)-$(+)$-ethyl 3-hydroxybutanoate of $[\alpha]_D^{25}$ $+43.5°$ (chloroform, c 1.0). This optically pure compound has recently become commercially available from Fluka AG, CH-9470 Buchs (Switzerland), but it is very expensive. After submission and checking of this procedure, it was shown[5] that the ee of the product can be increased to $>95\%$ by working under aerobic conditions and by adding the keto ester more slowly; see also Note 11.

11. The analysis of the published procedures for reductions of β-keto esters by baker's yeast indicated[6] that aerobic conditions,[5] the presence of 5–15% ethanol in the medium,[5,7] and "aging" of the yeast[5] might be important for high selectivity. The optimum conditions—"starving" the yeast for at least 4 days in 5% aqueous ethanol aerobically—lead to an activation of the enzyme(s)[8] producing the S-enantiomer of ethyl-3-hydroxybutanoate.

   The procedure[6] was as follows. A suspension of 125 g of baker's yeast in 1000 mL of H_2O/EtOH (95 : 5) was shaken (120 rpm) at 30°C in a 2-L Erlenmeyer flask with indentations for 4 days. After the addition of 5 g (38 mmol) of ethyl acetoacetate the reaction was followed by GLC. When the reaction had reached completion (2–3 days), the mixture was centrifuged and the supernatant was extracted continuously with ether (4 days). The organic layer was dried over magnesium sulfate, filtered, and concentrated with a rotary evaporator at 35°C bath temperature. The crude product was purified by bulb to bulb distillation to give ethyl 3-hydroxybutanoate, 3.54 g (70%), as a colorless liquid with an optical purity of 94% e.e. (enantiomeric excess).

3. Discussion

3-Hydroxybutanoic acid in both enantiomeric forms has been obtained by resolution of the racemic mixture.[9] Hydrogenation of methyl acetoacetate using a Raney nickel catalyst that had been treated with tartaric acid resulted in methyl 3-hydroxybutanoate with an enantiomeric excess of 83–88%.[10] Most recently it was found that enantiomerically pure $(R$-$)$ or $(S$-$)$-ethyl 3-hydroxybutanoate is available by enantioselective hydrogenation with a chiral homogeneous ruthenium catalyst.[11] Furthermore, optically active 3-hydroxybutanoic acid has been obtained in good chemical and optical yield by condensation of chiral α-sulfinyl ester enolates with aldehydes followed by desulfurization.[12] $(R$-$)$-$(-)$-Ethyl 3-hydroxybutanoate in 100% enantiomeric excess resulted from depolymerization of poly-(R)-3-hydroxybutanoate, an intracellular storage product of *Alcaligenes eutrophus H 16*.[13] The method presented in the Seebach–Züger paper[13] is easy to perform. The (S)-$(+)$-ethyl 3-hydroxybutanoate obtained may be enriched to 100%

enantiomeric excess by crystallization of its 3,5-dinitrobenzoate derivative, followed by alcoholysis.[4]

Optically active ethyl 3-hydroxybutanoate is a very useful chiral building block for natural product synthesis. Some applications are shown in Table I. Alkylation of doubly deprotonated ethyl 3-hydroxybutanoate gives branched structures of the following type:[14,15]

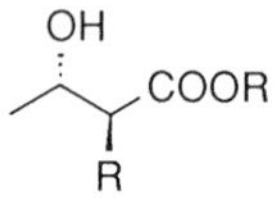

The yeast reduction is not limited to ethyl acetoacetate. It has been applied to other β-keto esters, α-keto esters, α-keto alcohols, α-keto phosphates, and some ketones (Table II). The reductions show a high degree of stereoselectivity. The absolute configuration

TABLE I
NATURAL PRODUCTS FROM (S)- OR (R)-ETHYL 3-HYDROXYBUTANOATE[a]

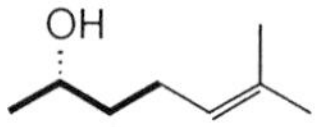

(S)-(+)-Sulcatol[13]

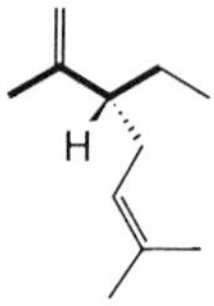

(R)-(–)-Lavandulol[14]

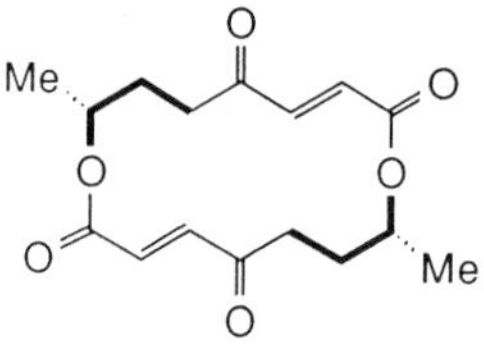

(R, R)-Pyrenophorin[15]

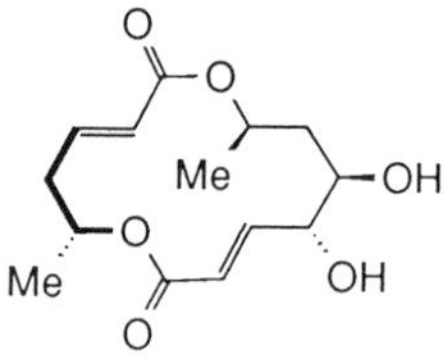

Colletodiol[16]

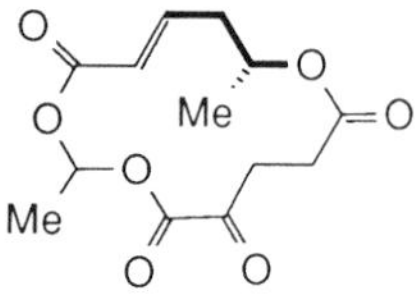

(R,R)-(–)-Grahamimycin A$_1$[16]

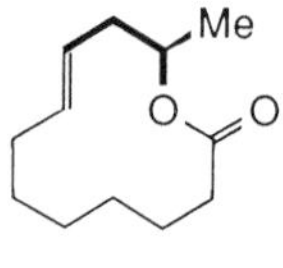

(R)-(+)-Recifeiolide[17]

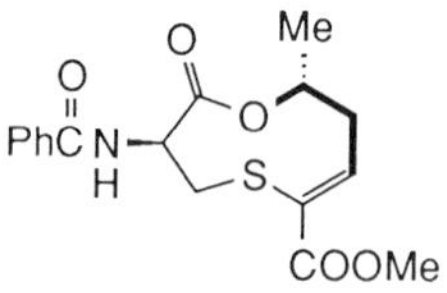

Griseoviridin precursor[19]

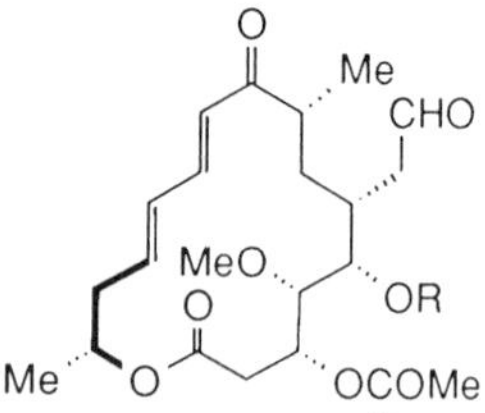

Carbomycin B[19]

[a]The skeleton of ethyl 3-hydroxybutanoate is indicated by heavy lines.

TABLE II

ENANTIOSELECTIVE PREPARATION OF ALCOHOLS FROM THE CORRESPONDING KETONE
BY YEAST REDUCTION

Substrate	Product	Yield (%)	Enantiomeric Excess (%)	Ref.
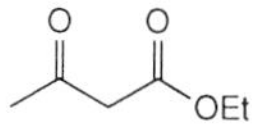	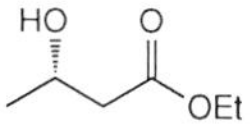	57-67	84-87	25, 26, 27
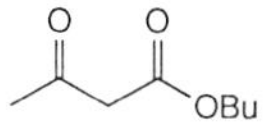	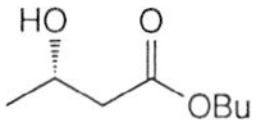	58	90	13
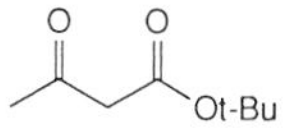	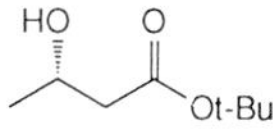	61	85	13
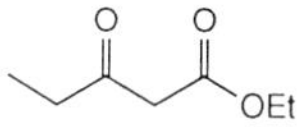	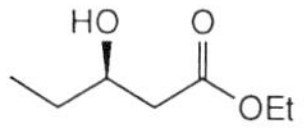	67	40	14
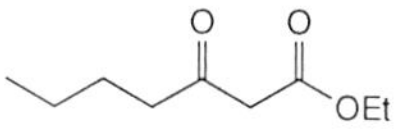	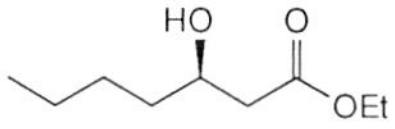		>90	14
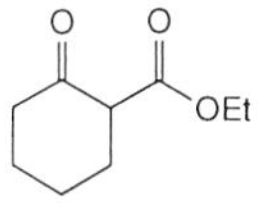	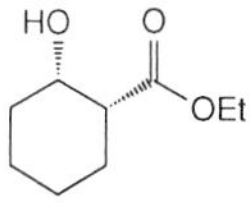	65	86	26, 28
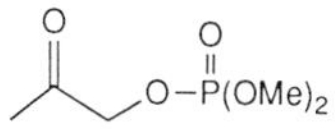	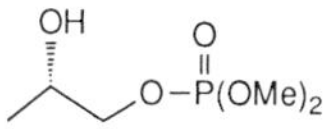	57	74	29
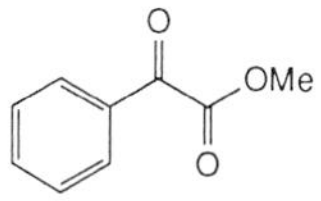	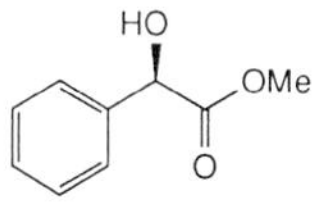	59	>97	26
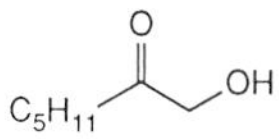	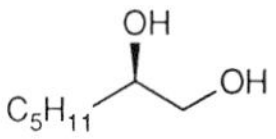	56	100	30
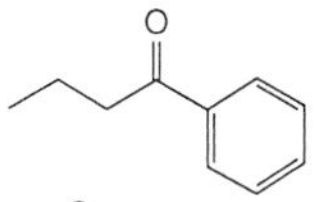	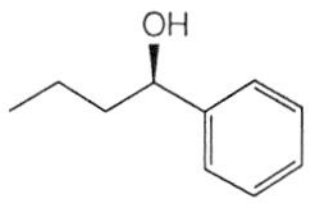	45	85-87	17
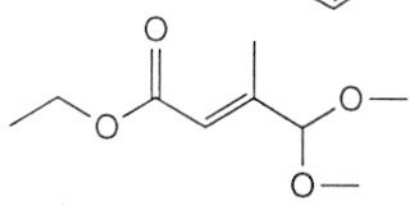	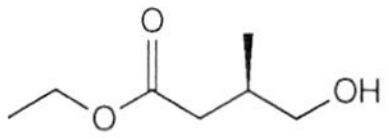	34	>97	31

of the product obtained by reduction of a carbonyl group containing a large group L and a small group S to the alcohol may be determined by application of Prelog's rule.[16,17]

1. Laboratorium für Organische Chemie, ETH-Zentrum, Universitätsstrasse 16, CH-8092 Zürich, Switzerland.
2. See the Sadtler Standard Spectra; (a) no. 17507; (b) no. 4253 M.
3. Dale, J. A; Dull, D. L.; Mosher, H. S. *J. Org. Chem.* **1969**, *34*, 2543-2549.
4. Hungerbühler, E.; Seebach, D.; Wasmuth, D. *Helv. Chim. Acta* **1981**, *64*, 1467-1487.
5. Wipf, B.; Kupfer, E.; Bertazzi, R; Leuenberger, H. G. W. *Helv. Chim. Acta* **1983**, *66*, 485-488.
6. Ehrler, J.; Giovannini, F.; Lamatsch, B.; Seebach, D. *Chimia* **1986**, *40*, 172-173.
7. Zhou, B.; Gopalan, A. S.; van Middlesworth, F.; Sieh, W.-R.; Sih, C. J. *J. Am. Chem. Soc.* **1983**, *105*, 5925-5926; see also Sih, C. J.; Chen, C.-S. *Angew. Chem.* **1984**, *96*, 556-565; *Angew. Chem. Int. Ed. Engl.* **1984**, *23*, 570-578.
8. Gopalan, A. S.; Sieh, W.-R.; Sih, C. J. *J. Am. Chem. Soc.* **1985**, *107*, 2993-2994.
9. Clarke, H. T. *J. Org. Chem.* **1959**, *24*, 1610-1611.
10. Harada, T.; Izumi, Y. *Chem. Lett.* **1978**, 1195-1196.
11. Akutagawa, S.; Kitamura, M.; Kumobayashi, H.; Noyori, R.; Ohkuma, T.; Sayo, N.; Takaya, M. *J. Am. Chem. Soc.* **1987**, *109*, 5856-5858.
12. Mioskowski, C.; Solladié, G. *J. Chem. Soc. Chem. Commun.* **1977**, 162-163.
13. Seebach, D.; Züger, M. *Helv. Chim. Acta* **1982**, *65*, 495-503.
14. Fráter, G. *Helv. Chim. Acta* **1979**, *62*, 2825-2828, 2829-2832.
15. Züger, M. F.; Weller, T.; Seebach, D. *Helv. Chim. Acta* **1980**, *63*, 2005-2009.
16. Prelog, V. *Pure Appl. Chem.* **1964**, *9*, 119-130.
17. MacLeod, R.; Prosser, H.; Fikentscher, L.; Lanyi, J.; Mosher, H. S. *Biochemistry* **1964**, *3*, 838-846.
18. Mori, K. *Tetrahedron* **1981**, *37*, 1341-1342.
19. Kramer, A.; Pfander, H. *Helv. Chim. Acta* **1982**, *65*, 293-301.
20. Seuring, B.; Seebach, D. *Liebigs Ann. Chem.* **1978**, 2044-2073; Seebach, D.; Seuring, B.; Kalinowski, H.-O.; Lubosch, W.; Renger, B. *Angew. Chem.* **1977**, *89*, 270-271; *Angew. Chem. Int. Ed. Engl.* **1977**, *16*, 264-265.
21. Amstutz, R.; Hungerbühler, E.; Seebach, D. *Helv. Chim. Acta* **1981**, *64*, 1796-1799; Hungerbühler, E. Dissertation No. 6862, ETH Zurich, **1981**; Seidel, W.; Seebach, D. *Tetrahedron Lett.* **1982**, *23*, 159-162.
22. Gerlach, H.; Oertle, K.; Thalmann, A. *Helv. Chim. Acta* **1976**, 755-760.
23. Meyers, A. I.; Amos, R. A. *J. Am. Chem. Soc.* **1980**, *102*, 870-872.
24. Nicolaou, K. C.; Pavia, M. R.; Seitz, S. P. *J. Am. Chem. Soc.* **1981**, *103*, 1224-1226.
25. Lemieux, R. U.; Giguere, J. *Can. J. Chem.* **1951**, *29*, 678-690.
26. Deol, B. S.; Ridley, D. D.; Simpson, G. W. *Aust. J. Chem.* **1976**, *29*, 2459-2467.
27. Seuring, B.; Seebach, D. *Helv. Chim. Acta* **1977**, *60*, 1175-1181.
28. Fráter, G. *Helv. Chim. Acta* **1980**, *63*, 1383-1390.
29. Pondaven-Raphalen, A.; Sturtz, G. *Bull. Soc. Chim. Fr.* **1978**, Pt. II, 215-229.
30. Barry, J.; Kagan, H. B. *Synthesis* **1981**, 453-455.
31. Leuenberger, H. G. W.; Boguth, W.; Barner, R.; Schmid, M.; Zell, R. *Helv. Chim. Acta* **1979**, *62*, 455-463.

ETHYL 4-HYDROXYCROTONATE

[2-Butenoic acid, 4-hydroxy-, ethyl ester, (*E*)-]

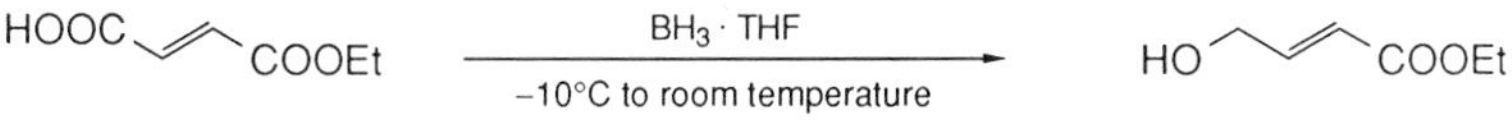

$$\text{HOOC} \diagdown \diagup \text{COOEt} \quad \xrightarrow[\text{−10°C to room temperature}]{\text{BH}_3 \cdot \text{THF}} \quad \text{HO} \diagdown \diagup \text{COOEt}$$

Submitted by ANDREW S. KENDE and PAWEL FLUDZINSKI[1]
Checked by CYNTHIA MCCLURE and EDWIN VEDEJS

1. Procedure

A dry, 2-L, one-necked, round-bottomed flask is equipped with a 1-L pressure-equalizing funnel and a large magnetic stirring bar. The system is flame-dried under an internal atmosphere of dry nitrogen (Note 1). The flask is charged with 300 mL of anhydrous tetrahydrofuran (Note 2) and 100 g of monoethyl fumarate. The solution is then stirred under nitrogen and brought to about $-5°C$ using an ice–salt/methanol bath ($-10°C$) (Note 3). A 1 *M* solution of 700 mL (0.70 mol) of borane–tetrahydrofuran complex (Note 4) is *cautiously* added dropwise (rapid H_2 evolution occurs) with rigorous temperature control to avoid an exothermic reaction. The ice–salt bath is maintained in position throughout the 90 min of addition. The stirred reaction mixture is then gradually allowed to warm to room temperature over the next 8–10 hr. The reaction is carefully quenched at room temperature by dropwise addition of 1 : 1 water : acetic acid (ca. 20 mL) with stirring until no more gas evolution occurs. The reaction is concentrated at room temperature and water pump pressure to a slurry by removal of most of the tetrahydrofuran. The slurry is carefully poured over a 20-min period into 300 mL of ice-cold, saturated sodium bicarbonate solution with mechanical stirring to avoid precipitation of solids, and the product is extracted with 300 mL of ethyl acetate. The aqueous layer is again extracted with 100 mL of ethyl acetate. The organic layers are combined, washed once with 200 mL of saturated sodium bicarbonate, then dried well with anhydrous magnesium sulfate.

Solvent removal at reduced pressure gives 61 g (67% yield) of essentially pure ethyl hydroxycrotonate (Note 5).

An analytical sample may be prepared by quick distillation (or Kugelrohr distillation) at 117–120°C (15 mm), but there is significant loss of material because of decomposition in the distillation pot. From 1 g of product, 0.72 g of pure material is obtained in this way, and recovery decreases as scale of distillation increases.

2. Notes

1. This is accomplished by passing a stream of dry nitrogen through the reaction vessel. During the reaction, a slight positive pressure of nitrogen is maintained throughout the apparatus.
2. The tetrahydrofuran is freshly distilled from sodium and benzophenone.[2]
3. The flask is cooled with the ice–salt/methanol bath for 30 min before the next addition to insure complete cooling of the solution.

4. Borane-tetrahydrofuran is commercially available from Aldrich Chemical Company, Inc. When a fresh bottle is used, titration is not necessary.

5. ^{1}H NMR data for ethyl 4-hydroxycrotonate are as follows (100 MHz, CDCl$_3$): δ 1.30 (t, 3 H, J = 7), 3.58 (br s, 1 H), 4.17 (q, 2 H, J = 7), 4.30 (m, 2 H), 6.03 (dt, 1 H, J = 16), 6.98 (dt, 1 H, J = 16).

3. Discussion

Ethyl (or methyl) 4-hydroxycrotonate has previously been prepared in 51% yield by silver oxide-assisted solvolysis of methyl 4-bromocrotonate,[3] or in 94% yield by reaction of glycolaldehyde with (carbomethoxymethylene) triphenylphosphorane.[4] Both procedures require very expensive starting materials or reagents. Several multistep procedures for preparing the title compound have also been reported.[5] The procedure described above represents a convenient one-step alternative for preparing ethyl 4-hydroxycrotonate, requiring inexpensive starting materials and reagents. This procedure relies on the selective reduction of a carboxylic acid in the presence of a carboxylic ester with borane, which is well documented.[6]

Ethyl 4-hydroxycrotonate has proved to be a valuable intermediate in synthetic chemistry. It has been used in alkaloid synthesis[3] or as a dipolarophile in dipolar cycloadditions.[7] Furthermore, ethyl 4-hydroxycrotonate can be readily oxidized to ethyl 4-oxocrotonate,[4] which has also served as a valuable precursor in synthesis.[8]

1. Department of Chemistry, University of Rochester, Rochester, NY 14627.
2. Perrin, D. D.; Armarego, W. L. F.; Perrin, D. R. "Purification of Laboratory Chemicals," 2nd ed.; Pergamon Press: New York, 1980.
3. Tufariello, J. J.; Tette, J. P. *J. Org. Chem.* **1975,**40, 3866. See also Rambaud, R. *Bull. Soc. Chim. Fr.* **1934,** 1317.
4. Witiak, D. T.; Tomita, K.; Patch, R. J. *J. Med. Chem.* **1981,** 24, 788.
5. (a) Ducher, S.; Journou, M. N. *Ann. Chim.* **1973,** 8, 359; (b) Laporte, J. F.; Rambaud, R. *Bull. Soc. Chim. Fr.* **1969,** 1340; (c) McClure, J. D. *J. Org. Chem.* **1967,** 32, 3888; (d) Kato, T.; Kimura, H. *Chem. Pharm. Bull.* **1977,** 25, 2692.
6. Walker, E. R. H. *Chem. Soc. Rev.* **1976,** 5, 23.
7. Padwa, A.; Ku, H. *J. Org. Chem.* **1979,** 44, 255.
8. (a) Naf, F.; Decorzant, R.; Thommen, W. *Helv. Chim. Acta* **1979,** 62, 114; (b) Devos, M. J.; Hevesi, L.; Bayet, P.; Krief, A. *Tetrahedron Lett.* **1976,** 3911.

OSMIUM-CATALYZED VICINAL OXYAMINATION OF OLEFINS BY N-CHLORO-N-ARGENTOCARBAMATES: ETHYL *threo*-[1-(2-HYDROXY-1,2-DIPHENYLETHYL)]CARBAMATE

[Carbamic acid (2-hydroxy-1,2-diphenylethyl),-ethyl ester, ($R*R*$)-]

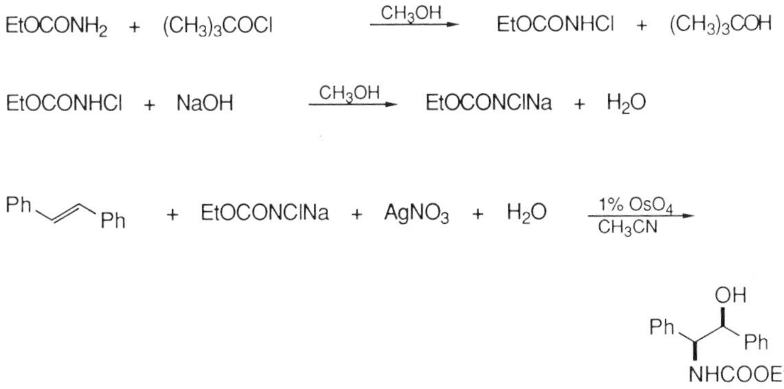

$$EtOCONH_2 + (CH_3)_3COCl \xrightarrow{CH_3OH} EtOCONHCl + (CH_3)_3COH$$

$$EtOCONHCl + NaOH \xrightarrow{CH_3OH} EtOCONClNa + H_2O$$

$$Ph\diagup\!\!\!\diagdown Ph + EtOCONClNa + AgNO_3 + H_2O \xrightarrow[CH_3CN]{1\% OsO_4}$$

Ph–CH(NHCOOEt)–CH(OH)–Ph

Submitted by Eugenio Herranz and K. Barry Sharpless[1]
Checked by Steven D. Young and Clayton H. Heathcock

1. Procedure

A 1-L, one-necked, round-bottomed flask is equipped with a magnetic stirring bar and a 100-mL addition funnel. The flask is placed in an ice bath and charged with 13.36 g (0.15 mol) of ethyl carbamate (Note 1) and 100 mL of reagent-grade methanol. Vigorous stirring is begun and to the ice-cold solution is carefully added 16.9 mL (16.2 g, 0.15 mol) of *tert*-butyl hypochlorite (Note 2). Fifteen minutes after the addition of the *tert*-butyl hypochlorite is complete, a methanolic solution (75 mL) of sodium hydroxide (6.43 g, 0.158 mol) is added dropwise over a period of several minutes (Note 3). After addition of the sodium hydroxide is complete, the ice bath is removed and stirring is continued for a further 10 min. The solvent is removed using a rotary evaporator (bath $< 60°C$) to give the crude ethyl N-chloro-N-sodiocarbamate as a white solid (Note 4). Addition of 400 mL of reagent-grade acetonitrile and 26.33 g (0.1 mol) of silver nitrate (Note 5) results in the gradual appearance of a brown suspension. The solution is stirred for 5 min at room temperature; 18.23 g (0.1 mol) of (*E*)-stilbene (Note 6), 10 mL (~ 1.0 mmol) of a solution of OsO₄ in *tert*-butyl alcohol (Note 7), and 8.1 mL (0.45 mol) of water are then added. The milky brown suspension that results is stirred for 18 hr at room temperature. Filtration of the reaction mixture through a Celite mat on a sintered-glass funnel gives a yellow–brown solution (Note 8). The filtrate is refluxed for 3 hr with 200 mL of 5% aqueous sodium sulfite (Note 9). The resulting mixture is concentrated at aspirator pressure using a rotary evaporator until acetonitrile no longer distills. The residue, which is primarily aqueous, is extracted with two 60-mL portions of methylene chloride (Note 10). The organic phase is dried (MgSO₄) and concentrated to give 24.6 g of crude product as a pale-yellow solid. Crystallization from

50 mL of hot toluene affords 18.6–19.8 g (66–69%) of almost pure ethyl *threo*-1-(2-hydroxy-1,2-diphenylethyl)carbamate, mp 120–122°C (Note 11). Concentration of the mother liquors yields an additional 0.6–0.8 g of the hydroxy carbamate (Note 12).

2. Notes

1. Ethyl carbamate was obtained from the Aldrich Chemical Company, Inc.

2. *tert*-Butyl hypochlorite was obtained from Frinton Laboratories.

3. A 5% excess of sodium hydroxide was used to make sure that the *N*-chloro-*N*-sodiocarbamate was in a basic environment. The sodium hydroxide was obtained from J. T. Baker Chemical Company; it was 97.9% pure.

4. To remove the last traces of methanol the crude *N*-chloro-*N*-sodiocarbamate is placed under high vacuum (0.1 mm) for 15 min. Slightly higher yields of final product are obtained if the crude *N*-chloro-*N*-sodiocarbamate is purified by trituration with ether.

5. Silver nitrate was obtained from Apache Chemicals Inc.

6. (*E*)-Stilbene was used as obtained from Aldrich Chemical Company. The olefin should be added in small portions to avoid overheating of the reaction mixture.

7. Osmium tetroxide was supplied by Matthey-Bishop, Inc. in 1-g amounts in sealed glass ampuls. The procedure that we describe below should be followed to prepare the osmium tetroxide catalyst solution. Work in a well-ventilated hood. One ampul is scored in the middle, broken open, and the two halves are dropped into a clean, brown bottle containing 39.8 mL of reagent grade *tert*-butyl alcohol and 0.20 mL of 70 or 90% *tert*-butyl hydroperoxide (Aldrich). The bottle is capped (use caps with Teflon liners) and then swirled to ensure dissolution of the OsO$_4$. These solutions are stored in the hood at room temperature and seem to be very stable.

8. In this way the silver salts (AgCl) are removed from the reaction mixture. The precipitate is washed twice with 20-mL portions of acetonitrile.

9. The purpose of this sulfite treatment is to reduce and thereby remove the small amount of osmium that is bound to the organic products.

10. If an emulsion forms, addition of Celite and subsequent filtration through a sintered-glass funnel gives a clear separation of the two phases. The checkers found that extraction with three 100-mL portions of methylene chloride avoids emulsion formation.

11. Crystallization occurs at room temperature over a period of ca. 12 hr. The crystals are washed once with 15 mL of toluene or 50 mL of petroleum ether (bp 40–60°C). The product is quite pure. A product of higher purity, however, mp 122–123.5°C, can be obtained by a second crystallization from toluene.

12. After 24 hr at high vacuum (0.1 mm), some crystals appear. Addition of 15 mL of ether, filtration, and washing with 10 mL of ether gives more product, mp 110–121°C. The checkers found that a higher overall yield was obtained if the mother liquors from the first recrystallization were dissolved in 50 mL of boiling diethyl ether. The solution is then brought to cloudiness by addition of petroleum

ether (bp 40–60°C). When this mixture is stored at 0°C overnight, brown crystals are deposited. Recrystallization of this material from 10 mL of hot toluene provides an additional 2.25–3.51 g of hydroxy carbamate, mp 114–117°C.

3. Discussion

This new procedure[2] for vicinal, cis addition of an oxygen and a nitrogen to an olefinic bond constitutes a major improvement over earlier methods,[3,4] since the nitrogen is introduced bearing an easily removed protecting group. Although the procedure described here employs ethyl carbamate, both *tert*-butyl carbamate and benzyl carbamate can also be used. In fact, in most cases, higher yields are realized in oxyaminations using the latter carbamates.

N-Chloro-*N*-argentocarbamates are generated *in situ* by reaction of the corresponding *N*-chlorosodiocarbamates with silver nitrate in acetonitrile. The *N*-chlorosodiocarbamates are prepared from the carbamates according to the method of Campbell and Johnson.[5] There are conflicting statements in the literature about the stability of these *N*-chlorosodiocarbamates.[6] On one occasion, when EtOCONNaCl was prepared by the submitters on a 250-mmol scale, it decomposed rapidly (but not explosively), turning dark and releasing heat and gases. However, this same chloramine salt has been prepared on a 100-mmol scale without incident. The submitters have found that acidic conditions (which lead to contamination by the *N*-chlorocarbamate) are responsible for the spontaneous decomposition of these salts at room temperature. A simple modification of Campbell's procedure for preparing *N*-chloro-*N*-sodiocarbamates avoids this problem. By adding 5% more sodium hydroxide than the calculated amount, it is assured that all the *N*-chlorocarbamate in the reaction mixture is neutralized. No spontaneous decomposition has occurred in the batches of *N*-chloro-*N*-sodiocarbamates prepared in this way.

The regioselectivity of this new procedure toward terminal olefins is considerably better than that realized with the earlier catalytic oxyamination procedures based on chloramine-T[3]. However, the catalytic procedure cannot compete with the regiospecificity exhibited by the stoichiometric *tert*-alkyl imido osmium reagents.[3]

This new catalytic procedure shows a different range of reactivity when compared with the chloramine-T based procedures, being very effective for mono- and 1,2-disubstituted olefins, especially electron-deficient olefins such as dimethyl fumarate and (*E*)-stilbene. However, when the steric hindrance of the olefin increases (trisubstituted olefins), the oxyamination reaction proceeds slowly and affords mixtures of products. Very recently we have been able to oxyaminate trisubstituted olefins (2-methyl-2-heptene, 1-methylcyclohexene, 1-phenylcyclohexene, 3-methyl-2-cyclohexenone) using other *N*-chloro-*N*-metallocarbamates in conjunction with the addition of tetraethylammonium acetate (Et$_4$NOAc).[7]

1. Department of Chemistry, Stanford University, Stanford, CA 94305. Present address: Department of Chemistry, Massachusetts Institute of Technology, Cambridge Mass., 02139.
2. Herranz, E.; Biller, S. A.; Sharpless, K. B. *J. Am. Chem. Soc.* **1978,** *100*, 3596.
3. Sharpless, K. B.; Patrick, D. W.; Truesdale, L. K.; Biller, S. A. *J. Am. Chem. Soc.* **1975,** *97*, 2305; Chong, A. O.: Oshima, K.; Sharpless, K. B. *J. Am. Chem. Soc.* **1977,** *99*, 3420; Patrick, W.; Truesdale, L. K.; Biller, S. A.; Sharpless, K. B. *J. Org. Chem.* **1978,** *43*, 2628.

4. Chong, A. O.; Oshima, K.; Sharpless, K. B. *J. Org. Chem.* **1976**, *41*, 177; Herranz, E.; Sharpless, K. B. *J. Org. Chem.* **1978**, *43*, 2544; Herranz, E.; Sharpless, K. B.; *Org. Synth.*, *Coll. Vol. VII* **1990**, 375.

5. Campbell, M. M.; Johnson, G. *Chem. Rev.* **1978**, *78*, 65.

6. Saika, D.; Swern, D. *J. Org. Chem.* **1968**, *33*, 4548; Chabrier, P. *C. R. Hebd. Seances Acad. Sci. Ser. C* **1942**, *214*, 362.

7. Sharpless, K. B.; Herranz, E. *J. Org. Chem.* **1980**, *45*, 2710-2712.

ETHYL ISOCROTONATE

[Ethyl (Z)-crotonate]

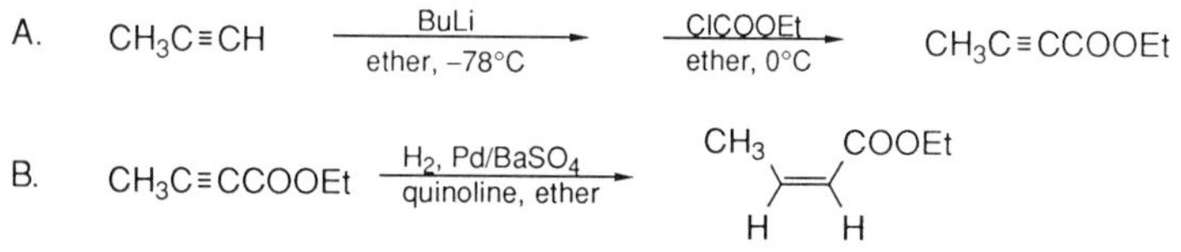

A. $\quad CH_3C \equiv CH \quad \xrightarrow[\text{ether, }-78°C]{\text{BuLi}} \quad \xrightarrow[\text{ether, }0°C]{\text{ClCOOEt}} \quad CH_3C \equiv CCOOEt$

B. $\quad CH_3C \equiv CCOOEt \quad \xrightarrow[\text{quinoline, ether}]{H_2, Pd/BaSO_4} \quad$

$$\begin{array}{cc} CH_3 & COOEt \\ \diagdown & \diagup \\ / & \diagdown \\ H & H \end{array}$$

Submitted by MICHAEL J. TASCHNER, TERRY ROSEN, and CLAYTON H. HEATHCOCK[1]
Checked by JUDY BOLTON and IAN FLEMING

1. Procedure

A. *Ethyl tetrolate.* A 3-L, three-necked, round-bottomed flask is equipped with an overhead mechanical stirrer and charged with 1000 mL of anhydrous ether (Note 1). One neck is fitted with a gas inlet joint connected to a nitrogen line equipped with a mineral oil bubbler. The second neck is fitted with a low-temperature thermometer, and the third is closed with a rubber serum cap after nitrogen has been passed through the flask for a few minutes. The flask is then immersed in a dry ice–acetone bath. While the ether is cooling, 85 mL (60.0 g, 1.50 mol) of propyne (Note 2) is condensed into a flask (Note 3). The stopper is removed briefly, the cold (−78°C) propyne is poured into the flask through a powder funnel inserted into the neck, and stirring is commenced. The stopper is replaced, and 667 mL of a 1.5 M solution of butyllithium in hexane is introduced by syringe at such a rate that the internal temperature does not exceed −65°C (Notes 4–6). During the butyllithium addition a copious white precipitate appears. The slurry is stirred at −78°C for 30 min, and 134 mL (152 g, 1.4 mol) of ethyl chloro-formate (Note 7) is added. At this point, the acetone–dry ice bath is replaced by an ice bath and the reaction mixture is stirred overnight. During this time the ice bath will melt and the reaction mixture should eventually reach room temperature. The mixture is poured onto 400 g of crushed ice, the layers are separated, and the aqueous phase is extracted with two 200-mL portions of ether. The ether solutions are combined, washed with brine and dried over anhydrous MgSO_4. After filtration, the ether is removed with a rotary evaporator (Note 8). The residue is distilled at aspirator pressure to obtain 107–108 g (95–97%) of ethyl tetrolate, bp 60–64°C (20 mm) [lit.[2] 105°C (90 mm)] (Note 9).

B. *Ethyl isocrotonate.* An oven-dried, 500-mL hydrogenation flask equipped with a sidearm fitted with a rubber serum cap and a magnetic stirring bar is charged with 0.4 g of 5% palladium on barium sulfate (Note 10), 0.4 g of quinoline, and 200 mL of anhydrous ether. The flask is attached to an atmospheric-pressure hydrogenation apparatus (Note 11) and flushed with hydrogen. Ethyl tetrolate (23.2 mL, 22.4 g, 0.2 mol) is introduced into the hydrogenation flask with a syringe, and stirring is commenced. The progress of the reaction may be monitored by the uptake of hydrogen (theoretical $\cong$ 4500 mL), by gas chromatography, or by removing aliquots that are concentrated and analyzed by ^{1}H NMR, monitoring the disappearance of the methyl singlet at δ 1.95; a total hydrogenation time of 10–15 hr is required (Note 12). After hydrogenation is complete, the catalyst is removed by filtration of the reaction mixture through a Celite pad. The ether is removed with a rotary evaporator (Note 8) to obtain 21.1–22.4 g (93–98%) of ethyl isocrotonate as a light-yellow liquid. This material contains traces of quinoline but is of a purity suitable for many uses (Notes 6 and 13). The quinoline may be removed, if desired, by washing the ether solution with 1 *M* aqueous acetic acid, followed by aqueous sodium carbonate, or by distillation at atmospheric pressure, bp 128–132°C [lit.[3] bp 129–130.5°C] (Note 14).

2. Notes

1. Although stirring can be done with a large magnetic stirring bar, the reaction mixture becomes rather thick as the 1-lithiopropyne is formed, and effective stirring is difficult. The checkers found that the yield in this step is only 77% when a magnetic stirrer is used.

2. Methylacetylene (technical grade) from Linde Division of the Union Carbide Corporation was employed. The checkers used Matheson Lecture bottles.

3. The propyne is passed directly from the tank or lecture bottle to a cold-finger condenser filled with a slush of isopropyl alcohol and dry ice. The condenser is attached to a 200-mL, three-necked flask equipped with a gas inlet adapter and a glass stopper. The flask has been previously calibrated to hold 85 mL of liquid.

4. Alternatively, the butyllithium solution may be forced into the reaction flask by means of an 18-gauge cannula inserted through the serum cap.

5. Butyllithium was obtained from Foote Mineral Co., Johnsonville, Tennessee. It may be standardized by a double titration procedure.[4]

6. If care is not taken in the formation of 1-lithiopropyne, the final product can be contaminated with as much as 10% of an impurity, which is presumed to be ethyl pentanoate. This impurity has a GLC retention time on conventional packed columns that is quite similar to that of ethyl (*E*)-crotonate. The by-product presumably results from the presence of butyllithium when the ethyl chloroformate is added. The submitters have not observed the formation of this product when care was taken to maintain the reaction temperature below −65°C during addition of the butyllithium to the propyne.

7. Ethyl chloroformate (practical grade) was obtained from MCB, Inc., Cincinnati, Ohio 45212, and used without purification.

8. It is important that the rotary evaporator bath be kept at 5–10°C, or some of the product will be lost by evaporation.

9. The IR spectrum (neat) has absorptions at 2250, 1700, and 1260 cm^{-1}. The ^{1}H NMR spectrum (CDCl$_3$) is as follows δ: 1.23 (t, 3 H, J = 7), 1.95 (s, 3 H), 4.07 (q, 2 H, J = 7).

10. The catalyst was obtained from The American Platinum Works, Newark, NJ.

11. The submitters employed an apparatus similar to that described by Wiberg.[5]

12. The hydrogenation can also be carried out without special apparatus by the following method. The ether solution is placed in a 500-mL, three-necked flask fitted with a fritted-gas inlet tube, a rubber serum cap, an oil bubbler, and a magnetic stirring bar. The catalyst, quinoline, and ethyl tetrolate are introduced, and the reaction flask is cooled in an ice bath. Hydrogen is bubbled through the cold solution at such a rate as to maintain atmospheric pressure in the flask as evidenced by the oil bubbler. When using this technique, it is necessary to monitor the course of hydrogenation by GLC or ^{1}H NMR. However, the rate of hydrogenation decreases rather abruptly after one molar equivalent has been absorbed, and there is little danger of overhydrogenation.

13. Capillary GLC analysis (12 m, cross-linked methyl silicone, programmed, 45°C, 3°C/min, retention time of ethyl (Z)-crotonate, 2.5 min). Ethyl (E)-crotonate has a retention time of 2.95 min under the same conditions. Careful quantitative analysis reveals that the ratio of Z and E isomers is reproducibly in the range 58 : 1 to 59 : 1.

14. The IR spectrum (neat) has absorptions at 3040, 1710, 1640, 1175, 1025, and 810 cm^{-1}. The ^{1}H NMR spectrum is as follows (CDCl$_3$) δ: 1.23 (t, 3 H, J = 7), 2.05 (dd, 3 H, J = 2, 7), 4.03 (q, 2 H, J = 7), 5.62 (dq, 1 H, J = 12, 2), 6.19 (dq, 1 H, J = 12, 7).

3. Discussion

A previous *Organic Syntheses* procedure for the preparation of isocrotonic acid involves the stereospecific Favorskii rearrangement of 1,3-dibromo-2-butanone.[6] However, the procedure is rather laborious and, in our hands, gives only a modest overall yield of acid. Isocrotonic acid has also been prepared by carbonation of *cis*-propenyllithium[7] and by sodium amalgam reduction of β-chloroisocrotonic acid.[8] The present procedure for semihydrogenation of ethyl tetrolate is based on early work of Bourguel[9] and of Allan, Jones, and Whiting.[10] The procedure for acylation of propyne is general and may be employed for the preparation of other α, β-acetylenic esters.[11]

1. Department of Chemistry, University of California, Berkeley, CA 94720.
2. In "Handbook of Chemistry and Physics," 52nd ed.; Weast, R. C., Ed.; Chemical Rubber Company: Cleveland, 1971; p. C-227.
3. von Auwers, K. *Liebigs Ann. Chem.* **1923**, *432*, 46.
4. Jones, R. G.; Gilman, H. *Org. React.* **1951**, *6*, 353.
5. Wiberg, K. B. "Laboratory Technique in Organic Chemistry"; McGraw-Hill: New York, 1960; p. 228.
6. Rappe, C. *Org. Synth., Coll. Vol. VI* **1988**, 711.

7. Seyferth, D.; Vaughan, L. G. *J. Am. Chem. Soc.* **1964**, *86*, 883.

8. Plisov, A. K.; Bogatsky, A. V. *Zhur. Obshchei Khim.* **1957**, *27*, 360; *Chem. Abstr.* **1957**, *51*, 15401c.

9. Bourguel, M. *Bull. Soc. Chim. Fr. (4)* **1929**, *45*, 1067.

10. Allan, J. L. H.; Jones, E. R. H.; Whiting, M. C. *J. Chem. Soc.* **1955**, 1862.

11. Brandsma, L. "Preparative Acetylene Chemistry"; Elsevier: Amsterdam, 1971; p. 80.

THE *C*-ARYLATION OF β-DICARBONYL COMPOUNDS: ETHYL 1-(*p*-METHOXYPHENYL)-2-OXOCYCLOHEXANECARBOXYLATE

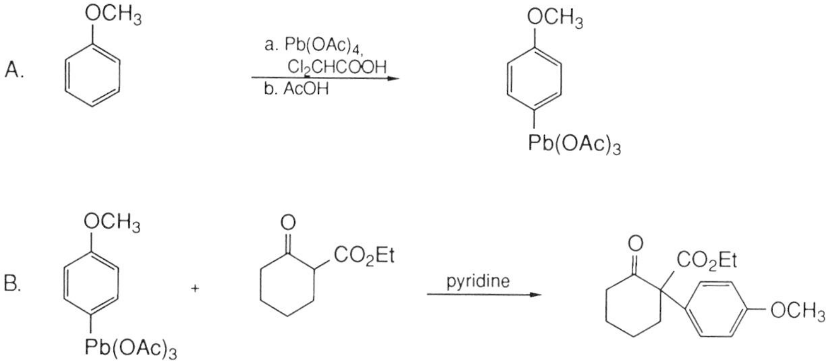

Submitted by ROBERT P. KOZYROD and JOHN T. PINHEY[1]
Checked by M. F. SEMMELHACK and DAVID ZIERING

1. Procedure

A. *p-Methoxyphenyllead triacetate.* A 1-L Erlenmeyer flask, equipped with a magnetic stirring bar, is charged with 50 g (0.11 mol) of lead tetraacetate (Note 1), chloroform (200 mL), and 140 g (1.09 mol) of dichloroacetic acid (Note 2). To this solution is added 16 g (0.15 mol) of anisole (Note 3), and the mixture is stirred at 25°C until lead tetraacetate can no longer be detected (Note 4). The reaction mixture is washed with water (2 × 250 mL) and the chloroform solution is treated with 1.5 L of hexane (Note 5). The yellow precipitate (44 g) is collected by suction filtration and stirred with a mixture of glacial acetic acid (250 mL) and chloroform (200 mL) for 1 hr.

The chloroform solution is washed with water (2 × 250 mL) and stirred with glacial acetic acid (250 mL) for 1 hr (Note 6). The solution that results is washed with water (2 × 250 mL), and the chloroform phase is treated with 1.5 L of hexane and kept at 2°C for 48 hr. The material that precipitates is collected and dried at 0.1 mm in a desiccator (calcium chloride) for 5 hr to give *p*-methoxyphenyllead triacetate (20–22 g, 35–40%) as pale-yellow crystals, mp 138–139°C (Note 7). The product may be kept for at least 3 weeks if stored at 2°C in a sealed container.

B. *Ethyl 1-(p-methoxyphenyl)-2-oxocyclohexanecarboxylate.* A 250-mL, one-necked, round-bottomed flask, equipped with a magnetic stirring bar, is charged with 22.2 g (45 mmol) of p-methoxyphenyllead triacetate, 10.8 g (135 mmol) of pyridine (Note 8), and 70 mL of chloroform (Note 9). To this solution is added 7.0 g (41 mmol) of ethyl 2-oxocyclohexanecarboxylate (Note 10), a calcium chloride drying tube is put in place, and the mixture is stirred at 40°C (Note 11).

After 24 hr, the reaction mixture is diluted with chloroform (80 mL), and washed with water (150 mL) and 3 M sulfuric acid (2 × 150 mL). The water and sulfuric acid washings are each washed (Note 12) with 100 mL of chloroform. The combined chloroform extracts are washed with water (2 × 250 mL), dried with magnesium sulfate, and the solvent removed to give an orange–colored oil (10.3 g) that slowly crystallizes on standing. Crystallization from hexane (Note 5) gives 9.4 g (82%) of ethyl 1-(p-methoxyphenyl)-2-oxocyclohexanecarboxylate, mp 49–50°C.

2. Notes

1. Lead tetraacetate from Merck & Company, Inc. was used. Acetic acid was removed from the reagent at 0.1 mm for 24 hr, in the dark, in a desiccator containing potassium hydroxide pellets.

2. Dichloroacetic acid from Merck & Company, Inc. was used without further purification.

3. Anisole from Fluka AG was distilled before use.

4. A few drops of reaction mixture were shaken with water. A brown precipitate of PbO_2 indicates the presence of unreacted lead tetraacetate. For the quantities given, a reaction time of 1 hr at 15–20°C is adequate.

5. Hexanes, bp 60–69°C, certified by Fisher Scientific Company were used.

6. A second metathesis with glacial acetic acid is carried out to ensure complete conversion of the oligomer into the product.

7. It has been found that the yield of product is generally higher when the reaction is performed on a smaller scale. Reactions carried out on approximately one-third of the above scale have given yields of approximately 60%.

8. Pyridine from Merck & Company, Inc. was distilled and stored over potassium hydroxide pellets.

9. Chloroform was dried over calcium chloride and distilled prior to use.

10. The ethyl 2-oxocyclohexanecarboxylate used was Fluka AG practical grade and was distilled (bp 106–108°C/12 mm) before use.

11. The submitters report that after approximately 1 hr some lead(II) acetate is deposited as an orange–red gum that may temporarily restrict the motion of the stirring bar; this was not observed by the checkers. The material generally crystallizes after a short period as a white solid.

12. These washings are extracted separately in order to minimize formation of solid lead(II) sulfate.

3. Discussion

The procedure described here serves to illustrate a new, general method for effecting the α-arylation of β-dicarbonyl compounds by means of an aryllead triacetate under very mild conditions. Although the first synthesis of an aryllead triacetate was reported relatively recently, a wide range of these compounds can now be readily prepared.[2] The most direct route to these compounds is plumbation of an aromatic compound with lead tetraacetate, and in the procedure reported here p-methoxyphenyllead triacetate has been prepared in this way. It may also be obtained by reaction of the diarylmercury[3] or the corresponding aryltrialkylstannane[4] with lead tetraacetate; the latter provides a convenient and very general route to aryllead triacetates.

The first synthesis of p-methoxyphenyllead triacetate by direct plumbation was reported by Harvey and Norman,[5] who obtained the compound in 24% yield by heating anisole and lead tetraacetate in acetic acid at 80°C for 4 days. Recently it has been found[2] that a much faster reaction and higher yield of aryllead compounds can be achieved by use of a haloacetic acid in place of acetic acid, and this has allowed the synthesis of a greater range of aryllead triacetates by direct plumbation. The improved reaction rate is presumably due to an increase in electrophilicity of lead when acetate is exchanged for a more electron-withdrawing ligand. The choice of the haloacetic acid depends on the reactivity of the aromatic substrate; thus, while dichloroacetic acid has been found best for the plumbation of anisole, trichloroacetic acid is preferred in the case of toluene and biphenyl.[2]

Aryllead tricarboxylates have been shown to be intermediates in two new routes to phenols,[6,7] and to have considerable potential as reagents for the C-arylation of carbon acids that are more acidic than diethyl malonate. A study of their reactions with β-diketones,[8] β-keto esters,[9] α-hydroxymethylene ketones,[10] derivatives of Meldrum's acid and barbituric acid,[11] and nitroalkanes[12] has established that such compounds, which contain only one replaceable hydrogen, undergo smooth arylation in high yield under the conditions outlined in this procedure. Compounds that contain two replaceable hydrogens are less predictable in their behavior. When a 1 : 1 ratio of substrate to aryllead compound is used, dimedone gave only diarylated product in high yield, while ethyl acetoacetate gave both mono- and diarylated products in only moderate yield.

Recently it has been shown that triphenylbismuth carbonate[13] and pentaphenylbismuth[14] can be used to achieve a similar arylation of β-dicarbonyl compounds. These reagents also react under very mild conditions and yields are generally high. Prior to the introduction of the organolead and organobismuth reagents, the most promising procedure for arylation of β-dicarbonyl compounds involved reaction of the enolate anion with a diaryliodonium salt, usually at 80–100°C.[15] Although only a limited range of substrates has been examined, it would appear that yields are only moderate, and in the case of dimedone a mixture of mono-, di-, and O-arylated products is produced. A further method, which has obvious limitations, involves the copper-catalyzed substitution of bromine in 2-bromobenzoic acids by the enolate anion of a β-dicarbonyl compound.[16]

1. Department of Organic Chemistry, University of Sydney, Sydney, N.S.W. 2006, Australia.
2. Bell, H. C.; Kalman, J. R.; Pinhey J. T.; Sternhell, S. *Aust. J. Chem.* **1979**, *32*, 1521 and references cited therein.

3. Criegee, R.; Dimroth, P.; Schempf, R. *Chem. Ber.* **1957**, *90*, 1337.
4. Kozyrod, R. P.; Morgan, J.; Pinhey, J. T. *Aust. J. Chem.* **1980**, *38*, 1147; Ackland, D. J.; Pinhey, J. T. *J. Chem. Soc., Perkin 1* **1987**, 2689; Ackland, D. J.; Pinhey, J. T. *J. Chem. Soc., Perkin 1* **1987**, 2695.
5. Harvey, D. R.; Norman, R. O. C. *J. Chem. Soc.* **1964**, 4860.
6. Campbell, J. R.; Kalman, J. R.; Pinhey, J. T.; Sternhell, S. *Tetrahedron Lett.* **1972**, 1763.
7. Kalman, J. R.; Pinhey, J. T.; Sternhell, S. *Tetrahedron Lett.* **1972**, 5369.
8. Pinhey, J. T.; Rowe, B. A. *Aust. J. Chem.* **1979**, *32*, 1561.
9. Pinhey, J. T.; Rowe, B. A. *Aust. J. Chem.* **1980**, *33*, 113.
10. Pinhey, J. T.; Rowe, B. A. *Aust. J. Chem.* **1983**, *36*, 789.
11. Pinhey, J. T.; Rowe, B. A. *Tetrahedron Lett.* **1980**, 965; Kopinski, R. P.; Pinhey, J. T. *Aust. J. Chem.* **1984**, *37*, 1245.
12. Kozyrod, R. P.; Pinhey, J. T. *Aust. J. Chem.* **1985**, *38*, 713.
13. Barton, D. H. R.; Lester, D. J.; Motherwell, W. B.; Barros Papoula, M. T. *J. Chem. Soc., Chem. Commun.* **1980**, 246.
14. Barton, D. H. R.; Blazejewski, J.-C.; Charpiot, B.; Lester, D. J.; Motherwell, W. B.; Barros Papoula, M. T. *J. Chem. Soc., Chem. Commun.* **1980**, 827.
15. Beringer, F. M.; Forgione, P. S. *J. Org. Chem.* **1963**, *28*, 714 and references cited therein.
16. McKillop, A.; Rao, D. P. *Synthesis* **1977**, 759 and references cited therein.

α-ALLENIC ESTERS FROM α-PHOSPHORANYLIDENE ESTERS AND ACID CHLORIDES: ETHYL 2,3-PENTADIENOATE

(2,3-Pentadienoic acid, ethyl ester)

Submitted by ROBERT W. LANG[1a] and HANS-JÜRGEN HANSEN[1b]
Checked by WILLIAM F. BURGOYNE and ROBERT M. COATES

1. Procedure

A 1-L, three-necked, round-bottomed flask is equipped with a nitrogen inlet, a 250-mL, pressure-equalizing dropping funnel fitted with a gas outlet, and a Teflon-coated magnetic stirring bar. The flask is charged with 300 mL of dichloromethane (Note 1) and 34.8 g (0.10 mol) of ethyl (triphenylphosphoranylidene)acetate (Note 2) and flushed with nitrogen. The yellow solution is stirred at 25°C as a solution of 10.1 g (0.10 mol) of triethylamine (Note 3) in 100 mL of dichloromethane is added dropwise over 5 min. After 10 min, 9.25 g (0.10 mol) of propionyl chloride (Note 4) in 100 mL of dichloromethane is added dropwise to the vigorously stirred solution over 15 min (Note 5). Stirring is continued for an additional 0.5 hr (Note 6), after which the clear, yellow-tinted mixture is evaporated on a rotary evaporator at reduced pressure using a water bath maintained at 25°C (Note 7). A 500-mL portion of pentane (Note 8) is added to the semisolid residue, and the slurry is allowed to stand for 2 hr while it is shaken periodically to facilitate solidification and to complete the extraction of the product. The precipitate is removed by filtration through a coarse, sintered-glass Büchner funnel, and the filter cake is washed with a 50-mL portion of pentane. The filtrates are combined and concentrated at reduced pressure to approximately one-fourth of the original volume using a water bath maintained at 25°C. The mixture is filtered again to remove triphenylphosphine oxide, and the remaining solvent is then evaporated. Rapid distillation of the residual liquid in a short-path distillation apparatus under reduced pressure (Note 9) affords a small forerun amounting to 0.5 mL or less and 7.8–8.1 g (62–64%) of ethyl 2,3-pentadienoate, bp 57–59°C (12–14 mm) (Notes 10 and 11).

2. Notes

1. Dichloromethane was purified by percolation through Woelm activity grade 1 basic alumina and stored under nitrogen.

2. Ethyl (triphenylphosphoranylidene)acetate is available from Fluka AG and Tridom Chemical Inc. under the name (ethoxycarbonylmethylene)triphenyl-phosphorane and from Aldrich Chemical Company, Inc. under the name (carbethoxymethylene)triphenylphosphorane. The reagent may be prepared from triphenylphosphine and ethyl bromoacetate by the following procedure.[2]

   A 1-L, two-necked, round-bottomed flask fitted with a dropping funnel and a mechanical stirrer is charged with 131.0 g (0.5 mol) of triphenylphosphine (Fluka AG, purum) and 250 mL of benzene (Merck, pro analysi). The solution is stirred vigorously while 83.5 g (0.5 mol) of ethyl bromoacetate (Fluka AG, practical-grade) is added dropwise at a rate that maintains the reaction mixture at, or slightly above, room temperature. After a total of 2 hr the reaction is complete and the colorless phosphonium salt is filtered. The salt is washed with 300 mL of cold benzene and 200 mL of pentane and then dissolved in 3 L of water at room temperature. Some further organic impurities are removed by extraction with ether, after which 2 drops of 2% alcoholic phenolphthalein are added. The aqueous solution is stirred vigorously and cooled in an ice bath as 2 M aqueous sodium hydroxide is added slowly until the pink endpoint is reached (pH 8–10). The crystalline phosphorane is collected by filtration, washed thoroughly with cold water, and dried, first with a rotary evaporator under reduced pressure at 60°C and then overnight in a drying oven at 180 mm and 70°C. The white to cream-colored crop of ethyl (triphenylphosphoranyli-dene)acetate, mp 124–126°C, weighs 150–156 g (86–90%) and may be used for the preparation of α-allenic esters without further purification.

3. Triethylamine was supplied by Fluka AG and Aldrich Chemical Company, Inc.

4. Propionyl chloride was purchased from Fluka AG and Aldrich Chemical Company, Inc. and was freshly distilled at 78–80°C (760 mm) prior to use.

5. The checkers maintained the temperature of the reaction mixture at ca. 25°C by cooling with a water bath during the addition of propionyl chloride.

6. The progress of the reaction may be followed by analytical thin-layer chroma-tography on alumina. The submitters used polygram precoated plastic sheets (Alox N/UV$_{254}$) purchased from Macherey-Nagel, Inc. The plates were devel-oped with 1 : 1 hexane : ether and stained with basic permanganate. The retar-dation factor of the product is 0.56.

7. For the isolation of relatively volatile α-allenic esters such as ethyl 2,3-penta-dienoate, the submitters recommend that the rotary evaporation be carried out with cooling in an ice bath. When this precaution was taken, the submitters obtained 8.5–9.5 g (67–75%) of product after distillation.

8. The checkers dried the pentane over sodium wire prior to use.

9. The checkers stirred the distilling liquid rapidly with a magnetic stirrer and maintained a bath temperature of 75–85°C throughout the distillation.

10. Ethyl 2,3-pentadienoate has the following spectral properties: IR (thin film) cm^{-1}: 1965, 1720, 1410, 1250, 1025, 865, 790; ^{1}H NMR (CCl$_4$) δ: 1.26 (t, 3 H, J = 7, OCH$_2$CH$_3$), 1.78 (m, 3 H, CH$_3$), 4.11 (q, 2 H, J = 7, OCH$_2$CH$_3$), 5.28–5.68 (m, 2 H, at C-2 and C-4).

11. On 0.01-mol scale the yield of ethyl 2,3-pentadienoate is 0.79–0.93 g (64–74%). The product was purified by bulb-to-bulb distillation with a Kugelrohr apparatus at 12–14 mm with an oven temperature at 75–85°C.

3. Discussion

The acylation of Wittig reagents provides the most convenient means for the preparation of allenes substituted with various electron-withdrawing substituents.[3] The preparation of α-allenic esters has been accomplished by the reaction of resonance-stabilized phosphoranes with isolable ketenes[4-9] and ketene itself[10] and with acid chlorides in the presence of a second equivalent of the phosphorane.[5] The disadvantages of the first method are the necessity of preparing the ketene and the fact that the highly reactive monosubstituted ketenes evidently cannot be used. The second method fails when the α-carbon of the phosphorane is unsubstituted.[11]

TABLE I

PREPARATION OF α-ALLENIC ESTERS BY THE WITTIG REACTION[12]

R^1	R^2	R^3	R^4	Solvent	Procedure^a	Yield (%)
CH$_3$	H	H	H	CH$_2$Cl$_2$	A	40
C$_2$H$_5$	H	(CH$_3$)$_3$C	H	CH$_3$CN	B	55
CH$_3$	H	C$_6$H$_5$	H	CH$_3$CN	B	23
C$_2$H$_5$	CH$_3$	H	H	CH$_2$Cl$_2$	A	59
C$_2$H$_5$	CH$_3$	CH$_3$	H	CH$_2$Cl$_2$	A	74
C$_2$H$_5$	CH$_3$	CH$_3$	CH$_3$	CH$_2$Cl$_2$	B	39
CH$_3$	CH$_3$	(CH$_3$)$_3$C	H	CH$_3$CN	B	66
C$_2$H$_5$	CH$_3$	C$_6$H$_5$	H	CH$_2$Cl$_2$	A	70

^{a}The reaction times varied from 10 min to 18 hr. Procedure: A—the corresponding phosphonium salt was used with the addition of 2 mol of triethylamine; B—the corresponding phosphorane was used with the addition of 1 mol of triethylamine.

The present procedure affords a general method for preparing α-allenic esters (Table I) that avoids the limitations of the previous methods.[12] Thus, α-allenic esters unsubstituted at C-2 are now available in generally satisfactory yields. Ethyl 2,3-pentadienoate, the title compound, had not been prepared prior to the development of this procedure by the submitters. The mild conditions (i.e., room temperature for relatively short times), avoid the base-catalyzed isomerization of the conjugated allenes to acetylenes.[13] The corresponding phosphonium salts may also be used directly in the reaction provided two equivalents of triethylamine are employed, obviating the lengthy process for drying the phosphorane.[14] Dichloromethane and acetonitrile have been used as solvents for the reaction.[12] The α-allenic esters are usually obtained in analytically pure form after bulb-to-bulb distillation. They may also be purified by column chromatography on alumina with 9 : 1 hexane : ether as eluant.[14]

The submitters have shown that these reactions proceed by dehydrochlorination of the acid chloride to the ketene, which is then trapped by reaction with the phosphorane. The resulting betaine decomposes to the allenic ester via an oxaphosphetane. In contrast, the reaction of acid chlorides with 2 equiv of phosphoranes involves initial acylation of the phosphorane followed by proton elimination from the phosphonium salt.[5]

1. (a) Zentrale Forschungslaboratorien, CIBA-Geigy AG, Postfach CH-4002, Basel, Switzerland; (b) Zentrale Forschungseinheiten, F. Hoffmann-La Roche & Co. AG, Postfach, CH-4002 Basel, Switzerland. Work done at the Institute of Organic Chemistry, University of Fribourg, CH-1700 Fribourg, Pérolles, and supported by the Swiss National Science Foundation.

2. Isler, O.; Gutmann, H.; Montavon, M.; Rüegg, R.; Ryser, G.; Zeller, P. *Helv. Chim. Acta* **1957**, *40*, 1242–1249.

3. Traylor, D. R. "The Chemistry of Allenes," *Chem. Rev.* **1967**, *67*, 317–359; Sandler, S. R.; Karo, W. "Organic Functional Group Preparations"; Academic Press: New York, 1971; Vol. 2, pp. 1–40; Murray, M. "Allene bzw. Kumulene" in "Methoden der Organischen Chemie" (Houben-Weyl), 4th ed., Müller, E., Ed.; Thieme: Stuttgart, 1977; Vol. 5/2a, pp. 963–1076; "The Chemistry of Allenes," Landor, S. R., Ed.; Academic Press: London, 1982; Vols. 1–3; Schuster, H. F.; Coppola, G. M.; "Allenes in Organic Synthesis," J. Wiley & Sons, New York, 1984.

4. Wittig, G.; Haag, A. *Chem. Ber.* **1963**, *96*, 1535–1543.

5. Bestmann, H.-J.; Hartung, H. *Chem. Ber.* **1966**, *99*, 1198–1207.

6. Aksnes, G.; Frøyen, P. *Acta Chem. Scand.* **1968**, *22*, 2347–2352.

7. Andrews, S. D.; Day, A. C.; Inwood, R. N. *J. Chem. Soc. (C)* **1969**, 2443–2449.

8. Orlov, V. Y.; Lebedev, S. A.; Ponomarev, S. V.; Lutsenko, I. F. *J. Gen. Chem. U.S.S.R.* (Engl. transl.) **1975**, *45*, 696.

9. Gompper, R.; Wolf, U. *Liebigs Ann. Chem.* **1979**, 1388–1405, 1406–1425.

10. Hamlet, Z.; Barker, W. D. *Synthesis* **1970**, 543–544.

11. Bestmann, H.-J.; Arnason, B. *Chem. Ber.* **1962**, *95*, 1513–1527; Bestmann, H.-J. *Angew. Chem.* **1965**, *77*, 651–666.

12. Lang, R. W.; Hansen, H.-J. *Helv. Chim. Acta* **1980**, *63*, 438–455.

13. Cymerman Craig, J.; Moyle, M. *J. Chem. Soc.* **1963**, 5356–5360; Busby, R. J. *Quart. Rev. Chem. Soc.* **1970**, *24*, 585–600.

14. Lang, R. W.; Hansen, H.-J. *Helv. Chim. Acta* **1979**, *62*, 1025–1039.

ADDITION OF AN ETHYLCOPPER COMPLEX TO 1-OCTYNE: (E)-5-ETHYL-1,4-UNDECADIENE

[1,4-Undecadiene, 5-ethyl-, (E)-]

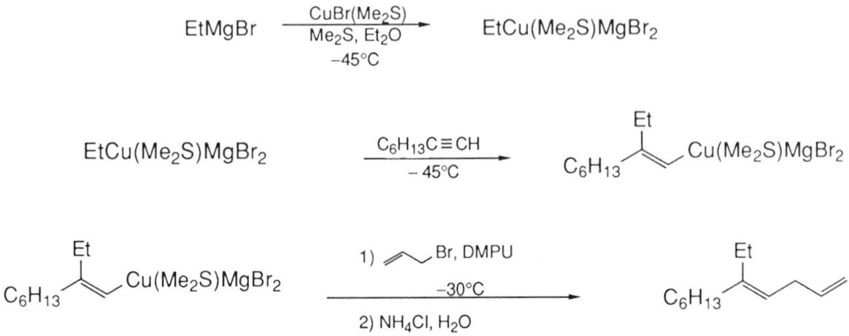

$$EtMgBr \xrightarrow[\substack{Me_2S,\ Et_2O \\ -45°C}]{CuBr(Me_2S)} EtCu(Me_2S)MgBr_2$$

$$EtCu(Me_2S)MgBr_2 \xrightarrow[-45°C]{C_6H_{13}C\equiv CH} \underset{C_6H_{13}}{\overset{Et}{\diagup}}Cu(Me_2S)MgBr_2$$

$$\underset{C_6H_{13}}{\overset{Et}{\diagup}}Cu(Me_2S)MgBr_2 \xrightarrow[\substack{-30°C \\ 2)\ NH_4Cl,\ H_2O}]{1)\ \diagdown Br,\ DMPU} \underset{C_6H_{13}}{\overset{Et}{\diagup}}\diagdown\diagup\diagdown$$

Submitted by Ramnath S. Iyer and Paul Helquist[1]
Checked by Brian H. Johnston and Andrew S. Kende
Rechecked by Ronald C. Newbold and Andrew S. Kende

1. Procedure

Caution! This experiment should be performed in an efficient fume hood because of the unpleasant odor of dimethyl sulfide.

A dry, 1-L, one-necked, round-bottomed flask is equipped with a Teflon-coated magnetic stirring bar and a three-way stopcock bearing a rubber septum (Note 1), and the flask is charged with 25.2 g (0.123 mol) of the dimethyl sulfide complex of cuprous bromide (Note 2). An argon or nitrogen (Note 3) atmosphere is established in the flask by repeated cycles of evacuation with an oil pump and refilling with the inert gas. Through use of a syringe or cannula, 150 mL of diethyl ether (Note 4) and 120 mL of dimethyl sulfide (Note 4) are added. After the mixture is stirred for a few minutes at 25°C, the resulting clear and colorless solution is cooled to −45°C (Notes 5 and 6). A 2.73 M solution (45.0 mL, 0.123 mol) of ethylmagnesium bromide in ether (Note 7) is added dropwise with a syringe or cannula over a period of 10 min. The suspension of yellow–orange solid is stirred at −45°C for 2 hr, and 1-octyne (16.0 mL, 0.109 mol; Note 4) is added with a syringe or cannula over a period of 2 min. After the solution is stirred at −45°C for 2 hr, it is cooled to −78°C (Note 5) and maintained at this temperature during the successive additions of N,N'-dimethylpropyleneurea, DMPU (27.7 mL, 0.223 mol; Note 8) and allyl bromide (11.4 mL, 0.131 mol; Note 4). The mixture is immediately warmed to −30°C and stirred at −30°C for 12 hr; it is warmed to 0°C and quenched by the addition of 30 mL of a saturated, aqueous solution of ammonium chloride adjusted to pH 8 with ammonium hydroxide. The mixture is stirred at 25°C in the air for 1.5 hr (Note 9) and is then shaken in a separatory funnel with a mixture of additional diethyl ether (50 mL) and water (50 mL). The dark-blue aqueous layer is drawn off and the organic layer is washed with additional 50-mL portions of the ammonium chloride solution (pH 8) until the washings are colorless. The organic layer is washed separately with water (50 mL) and saturated aqueous sodium chloride solution (50 mL), dried over anhydrous magnesium sulfate, filtered, and concentrated by rotary evaporation at 25°C (25 min). The residue consists of 15.9 g of yellow oil that is purified by distillation under reduced pressure through a 15-cm Vigreux column to give 13.9 g (71%) (Note 10) of (E)-5-ethyl-1,4-undecadiene as a clear, colorless liquid, bp 56°C (0.70 mm; Note 11).

2. Notes

1. The stopcock is constructed as shown below so that a source of inert gas and vacuum may be attached to the horizontal tubulation and liquid reagents and solutions may be transferred into the reaction flask with a syringe needle or cannula inserted through a rubber septum placed over the end of the vertical tubulation. In order to avoid air leaks through the septum into the reaction flask when reagents are not being added, the stopcock is normally turned to close off the vertical tubulation, but to leave the flask open to the argon source.

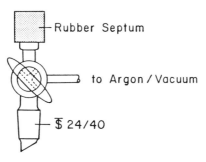

2. This complex is prepared from cuprous bromide and dimethyl sulfide according to the procedure of House.[2] The complex must be pure white. Slightly impure samples will produce pinkish solutions that are unsatisfactory for this procedure. Normally, the complex is dark red when it is first prepared, but the required state of purity can be achieved by two or three recrystallizations under a nitrogen atmosphere as described by House.[2] We have found, however, that if the initially formed complex has a distinctly green appearance, it cannot be purified satisfactorily. Others have also been concerned about this matter of purification.[3]

3. The checkers found that the yields were diminished when the reaction was run under prepurified nitrogen, but they did obtain the reported yields using argon. The submitters, however, have never experienced difficulty using the prepurified nitrogen available in their laboratory.

4. Commercially obtained materials were purified before use as described below. Diethyl ether was distilled from a dark-blue or dark-purple solution of sodium benzophenone radical anion or dianion under nitrogen. This solution was obtained by dissolving 10 g of benzophenone in 1 L of commercial anhydrous ether, adding 10 g of freshly pressed sodium wire, and heating the mixture at reflux under nitrogen until the characteristic blue or purple color developed. Dimethyl sulfide (Aldrich Chemical Company, Inc.), 1-octyne (Chemical Samples

Company or Albany International Chemicals), and allyl bromide (Columbia Organics) were distilled under nitrogen at atmospheric pressure. Dimethylpropyleneurea (DMPU) (Aldrich Chemical Company, Inc.) was distilled at aspirator pressure from calcium hydride.

5. Constant temperatures were maintained by using dry ice–acetone ($-78°C$) or dry ice–acetone/carbon tetrachloride baths (-25 to $-45°C$; the temperature tends toward the upper part of this range as the amount of acetone used is decreased) or more conveniently through the use of an acetone bath equipped with a Neslab CryoCool Model CC-100F low-temperature unit, a Cole-Parmer Versa-Therm Model 2158 temperature controller, and a 500-W immersible heating coil. The temperature of the alkenylcopper solution must not be allowed to exceed $-15°C$; above this temperature rapid coupling to give a diene occurs.

6. When the solution of the cuprous bromide complex is cooled, a portion of the reagent may precipitate, but this behavior does not affect the overall results of the experiment.

7. The ethylmagnesium bromide solution was obtained from Alfa Products, Morton Thiokol, Inc. and was titrated before use by the method of Watson and Eastham.[4]

8. (a) Checkers observed the freezing of some material on addition of DMPU and allyl bromide, making it difficult to maintain stirring. Vigorous, frequent shaking was performed to ensure good mixing. (b) The submitters used hexamethylphosphoric triamide, HMPA, in the original submission, but the Board of Editors of *Organic Syntheses* decided that, in view of the suspected carcinogenicity of this solvent, all procedures using it would be rechecked with DMPU. See Note 10 for the effect of this substitution on the yield.

9. Stirring the mixture in the air simplifies the workup procedure because cuprous complexes are oxidized to cupric compounds that are highly soluble in water or the aqueous ammonia workup medium of this experiment.

10. This yield is approximately 14% lower than that reported using HMPA; however, this result is consistent with those reported by Seebach[5] for the substitution of DMPU.

11. The spectral characteristics of the final product are as follows: IR (neat) cm^{-1}: 3080, 2960, 2915, 2800, 1660, 1640, 1465, 1175, 940, and 910; H^1 NMR (80 MHz, CDCl$_3$) δ: 0.78–2.13 (several overlapping m, 18 H, other saturated C$-$H's), 2.63 (t, 2 H, $J = 6.5$, C$=$CH$-$CH$_2$$-CH=CH_2$), 4.70–5.20 (m, 3 H, other alkenyl C$-$H's), 5.50–6.05 (m, 1 H, alkenyl C$-$H); mass spectrum (70 eV) m/e (relative intensity) 181.1 (M + 1, 1.3), 180.1 (M, 9.2).

3. Discussion

The procedure described above provides an approach to trisubstituted alkenes, compounds that are very common among natural products and that serve as key intermediates in the synthesis of other types of compounds.[6] The methods that have been developed for the preparation of trisubstituted alkenes are far too numerous to discuss to any significant extent here, but they have been the subject of previous review articles.[7] Very briefly, however, a large portion of the available methods may be divided among the following categories:[8] (1) elimination or cleavage reactions of organic halides and other compounds bearing leaving groups; (2) carbonyl condensation reactions of phosphonium ylides and other carbanionic or at least nucleophilic organic intermediates; (3) cleavage or rearrangements of other systems; (4) substitution reactions of alkenyl halides and related compounds; (5) reactions of various allylmetal and other allylic systems; (6) reactions of various alkenylmetal species; and (7) addition reactions to acetylenes, 1,3-dienes, and allenes.

The reaction of organocopper reagents with simple, unactivated acetylenes, an example of the last class of methods in the preceding list, serves as the basis of the procedure described here. This addition reaction was first reported by Normant in 1971[9] and has been investigated extensively since that time by not only Normant,[10] but also by Vermeer,[11] Helquist,[8] Levy,[12] and others.[3b, 13] An important modification[8] of the reaction that has been incorporated in the present preparation is the use of dimethyl sulfide as a ligand and cosolvent, which permits much higher yields and a broader range of applicability than the originally reported procedure.[9]

The specific example reported here is one of several preparations that have been developed using the same general reaction sequence. A brief summary of other representative cases is given in Table I.[8, 14] Notice that the alkenylcopper intermediates react with a variety of electrophilic reagents in addition to alkyl halides. It is especially

TABLE I
TRISUBSTITUTED ALKENES FROM ADDITION OF GRIGNARD-DERIVED ALKYLCOPPER COMPLEXES TO ACETYLENES, FOLLOWED BY REACTION WITH ELECTROPHILIC REAGENTS

Grignard Reagent	Acetylene	Electrophile	Product	Overall Yield (%)[a,b]
MeMgBr	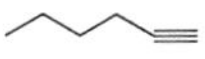	 Br	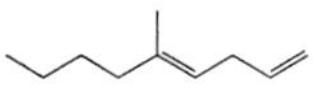	84
MeMgBr	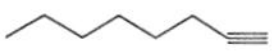	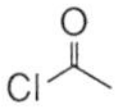 Cl (acetyl chloride)	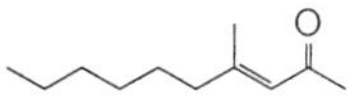	65
EtMgBr	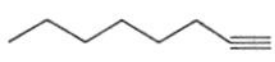	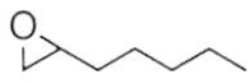 (epoxide)	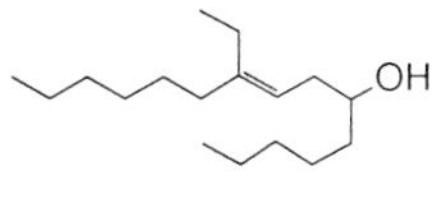 OH	94
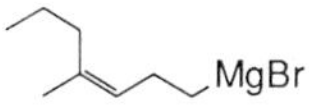 (allyl) MgBr		CO$_2$	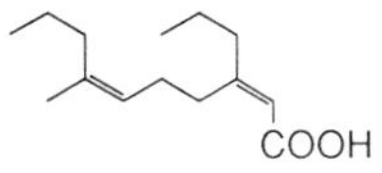 COOH	50
 MgBr		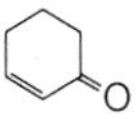 (cyclohexenone) O	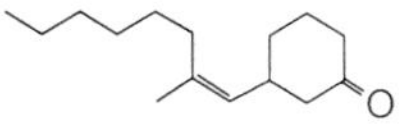 O	73

[a] Since yields are sensitive to traces of oxygen during the reactions, the use of an argon atmosphere is strongly recommended.
[b] See refs. 7, 13.

noteworthy that the overall sequences leading to trisubstituted alkenes have been shown to proceed with greater than 99.9% stereoselectivity.[8] This unusually high degree of control of alkene configuration is of great value in natural products synthesis. The overall stereochemistry is indicative of *syn* addition of the alkylcopper complexes to acetylenes, a result which is observed for several types of carbometallation (or insertion) reactions.[10]

In summary, the procedure described in this chapter is representative of a very general, highly stereoselective approach to trisubstituted alkenes. The usefulness of this methodology has already been demonstrated in total synthesis.[8, 13e, f, h, k, m, n, o, 14]

1. Department of Chemistry, State University of New York, Stony Brook, NY 11794. Present address: Department of Chemistry, University of Notre Dame, Notre Dame, IN 46556.
2. House, H. O.; Chu, C.-Y.; Wilkins, J. M.; Umen, M. J. *J. Org. Chem.* **1975**, *40*, 1460.
3. (a) Wuts, P. G. M. *Synth. Commun.*, **1981**, *11*, 139; (b) Theis, A. B.; Townsend, C. A. *Synth. Commun.* **1981**, *11*, 157.
4. Watson, S. C.; Eastham, J. F. *J. Organomet. Chem.* **1967**, *9*, 165.
5. Mukhopadhyay, T.; Seebach, D. *Helv. Chim. Acta* **1982**, *65*, 385.
6. See reference 8 for an extensive discussion of the occurrence, synthetic applications, and methods of preparation of trisubstituted alkenes.
7. (a) Reucroft, J.; Sammes, P. G. *Quart. Rev., Chem. Soc.* **1971**, *25*, 135; (b) Faulkner, D. J. *Synthesis* **1971**, 175; (c) Arora, A. S.; Ugi, I. K. *Methoden Org. Chem. (Houben-Weyl)* **1972**, *5/1b*, 728; (d) Corey, E. J.; Long, A. K. *J. Org. Chem.* **1978**, *43*, 2208.
8. Marfat, A.; McGuirk, P. R.; Helquist, P. *J. Org. Chem.* **1979**, *44*, 3888.
9. Normant, J. F.; Bourgain, M. *Tetrahedron Lett.* **1971**, 2583.
10. Normant, J. F.; Alexakis, A. *Synthesis* **1981**, 841.
11. Kleijn, H.; Westmijze, H.; Meijer, J.; Vermeer, P. *Recl. Trav. Chim. Pays-Bas* **1981**, *100*, 249.
12. LaLima, N. J., Jr.; Levy, A. B. *J. Org. Chem.* **1978**, *43*, 1279.
13. (a) Crandall, J. K.; Collonges, F. *J. Org. Chem.* **1976**, *41*, 4089; (b) Bouet, G.; Mornet, R.; Gouin, L. *J. Organomet. Chem.* **1977**, *135*, 151; (c) Corey, E. J.; Arai, Y.; Mioskowski, C. *J. Am. Chem. Soc.* **1979**, *101*, 6748; (d) Duboudin, J.-G.; Jousseaume, B. *Synth. Commun.* **1979**, *9*, 53; (e) Sato, T.; Kawara, T.; Sakata, K.; Fujisawa, T. *Bull. Chem. Soc. Jpn.* **1981**, *54*, 505; (f) Parr, W. J. E. *J. Chem. Res. (S)* **1981**, 354; (g) De Chirico, G.; Fiandanese, V.; Marchese, G.; Naso, F.; Sciacovelli, O. *J. Chem. Soc., Chem. Commun.* **1981**, 523; (h) Baker, R.; Billington, D. C.; Ekanayake, N. *J. Chem. Soc., Chem. Commun.* **1981**, 1234; (i) Ashby, E. C.; Smith, R. S.; Goel, A. B. *J. Org. Chem.* **1981**, *46*, 5133; (j) Knight, D. W.; Ojhara, B. *Tetrahedron Lett.* **1981**, *22*, 5101; (k) Furber, M.; Taylor, R. J. K.; Burford, S. C. *J. Chem. Soc., Perkin Trans. 1* **1986**, 1809; (l) Negishi, E.; Zhang, Y.; Cederbaum, F. E.; Webb, M. B. *J. Org. Chem.* **1986**, *51*, 4082; (m) O'Connor, B.; Just, G. *J. Org. Chem.* **1987**, *52*, 1801; (n) McMurry, J. E.; Bosch, G. K. *J. Org. Chem.* **1987**, *52*, 4885; (o) Stang, P. J.; Kitamura, T. *J. Am. Chem. Soc.* **1987**, *109*, 7561.
14. Marfat, A.; McGuirk, P. R.; Helquist, P. *J. Org. Chem.* **1979**, *44*, 1345.

DICHLOROVINYLATION OF AN ENOLATE: 8-ETHYNYL-8-METHYL-1,4-DIOXASPIRO[4.5]DEC-6-ENE

(1,4-Dioxaspiro[4.5]dec-6-ene, 8-ethynyl-8-methyl-)

A.

1) LDA, (Me₂N)₃PO, THF, −78°C
2) Cl₂C=CHCl, warm to room temperature, remove solvents and HN(CHMe₂)₂
3) (i-Bu)₂Al-H, toluene, 0°C
4) Acidic workup

B.

1) HO~~OH , PhH, reflux, catalytic p-TsOH, remove solvents and p-TsOH
2) 2 equiv. BuLi, THF, −78°C

Submitted by ANDREW S. KENDE and PAWEL FLUDZINSKI[1]
Checked by P. WOVKULICH, F. BARCELOS, and GABRIEL SAUCY

1. Procedure

Caution! Hexamethylphosphoric triamide and trichloroethylene are suspected carcinogens. All operations with either one should be performed in an efficient hood. The use of disposable gloves is highly recommended. Glassware should be rinsed with copious amounts of water into separate waste containers before removal from the hood.

A. *4-[(E)-1,2-Dichlorovinyl]-4-methyl-2-cyclohexen-1-one* (**1**). A dry, 3-L, one-necked, round-bottomed flask is equipped with a magnetic stirrer and a 500-mL pressure-equalizing dropping funnel. The dropping funnel is fitted with a rubber septum and the air in the system is replaced with dry nitrogen (Note 1). The flask is charged with 1500 mL of anhydrous tetrahydrofuran (Note 2) and 38.9 g (54 mL, 0.38 mol) of diisopropylamine (Note 3). The flask is cooled to 0°C with an ice bath. A 1.51 M hexane solution of butyllithium (255 mL, 0.38 mol) is added dropwise with stirring over a 30-min period. The resulting lithium diisopropylamide is cooled to −78°C with a dry ice-acetone bath (Note 4). A solution of 57.8 g (0.38 mol) of 3-ethoxy-6-methyl-2-cyclo-hexen-1-one (Note 5) in 400 mL of anhydrous tetrahydrofuran is added dropwise with stirring at −78°C over a 90-min period, followed immediately by the addition of 68 g (66 mL, 0.38 mol) of neat hexamethylphosphoric triamide (Note 6) over a 5-min period. The solution is stirred at −78°C for 45 min, followed by the dropwise addition of 52.6 g (36 mL, 0.40 mol) of neat trichloroethylene (Note 7). The solution is allowed to warm to room temperature slowly over a 6-hr period. As the solution warms, the color changes from pale yellow to olive green, to pale red, and finally to black. After 6 hr (Note 8) the solution is quenched with 1000 mL of water and the organic phase is separated. The aqueous phase is extracted four times with 250 mL of diethyl ether. The organic phases are combined and washed four times with 750 mL of water and twice with 750 mL of brine and dried over magnesium sulfate. The solvent is removed on a rotary evaporator and recovered starting material is removed by fractional distillation at 91–93°C (1 mm) through a 15-cm Vigreux column. The residual crude 6-[(E)-1,2-dichlorovinyl]-3-ethoxy-6-methyl-2-cyclohexen-1-one (Note 9) is dissolved in 400 mL of toluene and placed in a dry, 2-L, one-necked, round-bottomed flask, equipped with a mechanical stirrer; a 500-mL pressure-equalizing dropping funnel is fitted with a rubber septum and the air in the system is replaced with dry nitrogen (Note 1). The solution is cooled to 0°C with an ice bath. A 1 M hexane solution of diisobutylaluminum hydride (400 mL, 0.40 mol) (Note 10) is added dropwise with stirring at 0°C over a 1-hr period. The solution is stirred for 2 additional hr at 0°C. To quench the reaction 200 mL of methanol is carefully added to the stirred reaction mixture, followed slowly at first then more rapidly with 400 mL of water and then 300 mL of 10% sulfuric acid solution. After the mixture is stirred for 10 min, it is transferred to a separatory funnel and 500 mL of 10% sulfuric acid solution is added. The separatory funnel is shaken vigorously for 5 min and the organic phase is separated. The aqueous phase is extracted four times with 300 mL of diethyl ether. The organic phases are combined and washed twice with 300 mL of saturated sodium bicarbonate solution, twice with 300 mL of water, twice with 300 mL of brine, and dried over magnesium sulfate. Solvent removal on a rotary evaporator followed by short-path distillation at reduced pressure affords 31–34 g (40–44%, based on 3-ethoxy-6-methyl-2-cyclohexen-1-one) of 4-[(E)-1,2-dichlorovinyl]-4-methyl-2-cyclohexen-1-one (1) as a colorless oil, bp 75–78°C (0.1 mm) (Note 11).

B. *8-Ethynyl-8-methyl-1,4-dioxaspiro[4.5]dec-6-ene* (2). A dry, 1-L, one-necked, round-bottomed flask is equipped with a magnetic stirrer, Dean–Stark trap, and reflux condenser. The flask is charged with 500 mL of benzene, 12.0 g (0.059 mol) of 4-[(E)-1,2-dichlorovinyl]-4-methyl-2-cyclohexen-1-one, 12.2 g (11 mL, 0.20 mol) of ethylene glycol and 40 mg (a catalytic amount) of p-toluenesulfonic acid. After the solution is refluxed for 24 hr, it is poured into 200 mL of saturated sodium bicarbonate solution. The organic phase is separated and the aqueous phase is extracted four times with 50 mL of diethyl ether. The organic phases are combined and washed twice with 100 mL of water and once with 100 mL of brine and are dried over magnesium sulfate. The solvent is removed on a rotary evaporator, and the resulting crude 8-[(E)-1,2-dichloro-vinyl]-8-methyl-1,4-dioxaspiro[4.5]dec-6-ene (Note 12) is dissolved in 200 mL of anhydrous tetrahydrofuran and placed in a dry, 1-L, one-necked, round-bottomed flask equipped with a magnetic stirrer and a 500-mL pressure-equalizing dropping funnel. The dropping funnel is fitted with a rubber septum and the air in the system is replaced with dry nitrogen (Note 1). The solution is cooled to −78°C with a dry ice–acetone bath (Note 4). A 1.51 M hexane solution of butyllithium (76 mL, 0.12 mol) is added dropwise with stirring at −78°C over a 30-min period. The solution is stirred at −78°C for 2 hr, the cold bath is removed, and stirring is continued for 90 min. The solution is poured into 100 mL of water and the organic phase is separated. The aqueous phase is extracted four times with 25 mL of diethyl ether. The organic phases are combined and washed twice with 75 mL of water, twice with 75 mL of brine, and dried over magnesium sulfate. Solvent removal on a rotary evaporator followed by short-path distillation at reduced pressure yields 5.5-6.3 g (52–60%) of 8-ethynyl-8-methyl-1,4-dioxaspiro[4.5]dec-6-ene as a colorless oil, bp 88–90°C (1 mm) (Note 13).

2. Notes

1. This procedure is accomplished by alternatively evacuating and filling the funnel with dry nitrogen two times; an oil bubbler is used to maintain a slight positive pressure throughout the reaction.

2. Tetrahydrofuran is freshly distilled from sodium and benzophenone, as is all tetrahydrofuran used in this procedure.

3. Diisopropylamine is distilled from calcium hydride.

4. The flask is cooled with the dry-ice–acetone bath for 1 hr before the next addition to insure complete cooling of the solution.

5. See *Org. Synth. Coll. Vol. VII* **1990**, 208.

6. Hexamethylphosphoric triamide (HMPA) is freshly distilled from calcium hydride. Because of the suspected carcinogenicity of HMPA, the editors of *Organic Synthesis* rechecked all procedures in which it has been used. In the case of the present procedure, it was found that the decrease in yield in Step A was very large (55% → 16%) and that the product was a mixture of **1** and the corresponding chloroalkyne. Thus DMPU, the suggested substitute solvent,[2] is unsatisfactory in the present case.

7. Trichloroethylene is freshly distilled from phosphorus pentoxide.

8. On a smaller scale, the reaction warms to room temperature more quickly and can be worked up after 4 hr. Extended reaction times (e.g., overnight) lead to the formation of by-products.

9. Distillation is not necessary at this point. Spectroscopic data for 6-[(*E*)-1,2-dichlorovinyl]-3-ethoxy-6-methyl-2-cyclohexen-1-one is as follows: ^{1}H NMR (CDCl$_3$) δ: 1.38 (t, 3 H, $J = 6$), 1.48 (s, 3 H), 1.8–2.7 (m, 4 H), 3.96 (q, 2 H, $J = 6$), 5.44 (s, 1 H), 6.36 (s, 1 H). A purified sample (bp 140–142°C, 1 mm) gave satisfactory analyses. Anal. calcd. for C$_{11}$H$_{14}$Cl$_2$O$_2$: C, 53.02; H, 5.68. Found: C, 53.20; H, 5.43.

10. Diisobutylaluminum hydride was purchased from Aldrich Chemical Company, Inc. Since the reagent is not titrated, excess is used to ensure complete reduction.

11. Spectroscopic data for 4-[(*E*)-1,2-dichlorovinyl]-4-methyl-2-cyclohexen-1-one are as follows: ^{1}H NMR (CDCl$_3$) δ: 1.50 (s, 3 H), 1.8–2.8 (m, 4 H), 5.92 (d, 1 H, $J = 10$), 6.34 (s, 1 H), 7.04 (d, 1 H, $J = 10$); ms (mass spectrum) (75 eV) m/e 204; IR (CHCl$_3$) cm^{-1}: 1680. Anal. calcd. C$_9$H$_{10}$Cl$_2$O: C, 52.71; H, 4.91. Found: C, 53.08, H, 5.03.

12. Spectroscopic data for 8-(*E*-1,2-dichlorovinyl)-8-methyl-1,4-dioxa-spiro[4.5]dec-6-ene is as follows: ^{1}H NMR (CDCl$_3$) δ: 1.36 (s, 3 H), 1.6–2.6 (m, 4 H), 3.88–4.08 (m, 4 H), 5.56 (d, 1 H, $J = 10$), 6.08 (d, 1 H, $J = 10$), 6.28 (s, 1 H).

13. Spectroscopic data for 8-ethynyl-8-methyl-1,4-dioxaspiro[4.5]dec-6-ene are as follows: ^{1}H NMR (CDCl$_3$) δ: 1.32 (s, 3 H), 1.6–2.2 (m, 4 H), 2.12 (s, 1 H), 3.88–4.04 (m, 4 H), 5.64 (AB q, 2 H, $J = 10$); ms (75 eV) m/e 178; IR (neat) cm^{-1}: 3290, 2100. Anal. calcd. for C$_{11}$H$_{14}$O$_2$: (at C-11) C, 74.13; H, 7.92.

Found: C, 73.96; H, 7.78. ^{13}C NMR (CDCl$_3$) δ: 28.1 (at C-11), 30.5 (at C-9 or C-10), 31.4 (at C-8), 34.6 (at C-9 or C-10), 64.0 (at C-2 or C-3), 64.3 (at C-2 or C-3), 68.4 (at C-13), 88.3 (at C-12), 104.5 (at C-5), 126.3 (at C-6), 137.0 (at C-7).

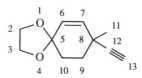

3. Discussion

Trichloroethylene serves as an effective reagent for the dichlorovinylation of lithium enolates of several conjugated ketones. Under similar reaction conditions, 2,6-dimethyl-cyclo-2-hexen-1-one and 2-ethyl-5-methoxy-1-tetralone give the analogous dichloro-vinyl adduct in comparable yield.[3] This procedure represents an heretofore unknown, uncatalyzed[4] carbon–carbon bond forming reaction between enolates and a polychlo-roolefin that can subsequently provide access to α- and γ-acetylenic ketones.[5]

1. Department of Chemistry, University of Rochester, Rochester, NY 14627.
2. Seebach, D. *Chimia* **1985**, *39*, 147.
3. Kende, A. S.; Benechie, M.; Curran, D. P.; Fludzinski, P.; Swenson, W.; Clardy, J. *Tetrahedron Lett.* **1978**, 4513.
4. For examples of Ni-catalyzed vinylation and arylation of enolates by bromides and iodides, see Millard, A. A.; Rathke, M. W. *J. Am. Chem. Soc.* **1977**, *99*, 4833.
5. The trichloroethylene condensation has been shown to proceed by way of dichloroacetylene as an obligatory intermediate in a carbanion chain mechanism. See Kende, A. S.; Fludzinski, P. *Tetrahedron Lett.* **1982**, *23*, 2369, 2373; Kende, A. S.; Fludzinski, P.; Hill, J. M.; Swenson, W.; Clardy, J. *J. Am. Chem. Soc.* **1984**, *106*, 3551.

PALLADIUM-CATALYZED SYNTHESIS OF 1,4-DIENES BY ALLYLATION OF ALKENYLALANES: α-FARNESENE

(1,3,6,10-Dodecatetraene, 3,7,11-trimethyl-)

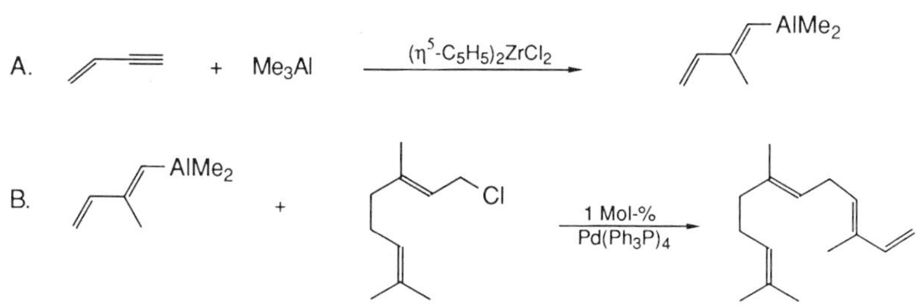

Submitted by Ei-ichi Negishi[1] and Hajime Matsushita[2]
Checked by Pauline J. Sanfilippo and Andrew S. Kende

1. Procedure

Caution! Trimethylalane (Note 1) is highly pyrophoric. It must be kept and used under a nitrogen atmosphere.

A. *(E)-(2-Methyl-1,3-butadienyl)dimethylalane.* An oven-dried, 1-L, two-necked, round-bottomed flask equipped with a magnetic stirring bar, a rubber septum, and an outlet connected to a mercury bubbler is charged with 7.01 g (24 mmol) of dichlorobis (η^5-cyclopentadienyl)zirconium (Note 2) and flushed with nitrogen. To this are added sequentially at 0°C 100 mL of 1,2-dichloroethane (Note 3), 12.48 g (120 mmol) of a 50% solution of 1-buten-3-yne in xylene (Note 4), and 120 mL (240 mmol) of a 2 M solution of trimethylalane in toluene (Note 1). The reaction mixture is stirred for 12 hr at room temperature and used in the next step without further treatment (Note 5).

B. *(3E, 6E)-3,7,11-Trimethyl-1,3,6,10-dodecatetraene (α-farnesene).* To the solution of (E)-(2-methyl-1,3-butadienyl)dimethylalane prepared above are added 17.25 g (100 mmol) of geranyl chloride (Note 6) and 1.15 g (1 mmol) of tetrakis(triphenylphosphine)palladium (Note 7) dissolved in 100 mL of dry tetrahydrofuran (Note 8), while the reaction temperature is controlled below 25–30°C with a water bath. After the reaction mixture is stirred for 6 hr at room temperature, 250 mL of 3 N hydrochloric acid is slowly added at 0°C. The organic layer is separated and the aqueous layer is extracted twice with pentane. The combined organic layer is washed with water, saturated aqueous sodium bicarbonate, and water again. After the organic extract is dried over anhydrous magnesium sulfate, the solvent is removed thoroughly using a rotary evaporator (15–20 mm), and the crude product is passed through a short (15–20-cm) silica gel column (60–200 mesh) using hexane as an eluent (Note 9). After the hexane is evaporated using a rotary evaporator, the residue is distilled using a 12-cm Vigreux column to give 16.70 g (83% based on geranyl chloride) of α-farnesene as a colorless liquid, bp 63–65° (0.05 mm) (Note 10).

2. Notes

1. The submitters used trimethylalane available in a cylinder from Ethyl Corporation. Both neat trimethylalane and its 2 M solution in toluene gave comparable results. The toluene solution of trimethylalane is also available from Aldrich Chemical Company.

2. The submitters used dichlorobis (η^5-cyclopentadienyl)zirconium available from Aldrich Chemical Company. This chemical is sufficiently air-stable to be handled in air.

3. The 1,2-dichloroethane available from Aldrich Chemical Company was distilled from phosphorus pentoxide before use. Although less effective, dichloromethane may also be used in the carbometallation step.

4. The submitters used a 50% solution of 1-buten-3-yne in xylene, available from Chemical Samples Company. For transferring this solution, the following procedure may be recommended. An ampule containing the solution is cooled with an ice–salt bath, opened, and capped with a rubber septum. A weighed measuring flask capped with a rubber septum is cooled with the ice–salt bath. To this is introduced the cooled solution by means of a double-tipped needle, and the weighed solution is then introduced to the reaction flask by means of a double-tipped needle.

5. The reaction mixture containing (E)-(2-methyl-1,3-butadienyl)dimethylalane may be stored at room temperature for at least a few days. Although it appears to be stable at room temperature for a much longer period of time, its thermal stability has not been carefully determined. The cross-coupling reaction in Section B should require only one equivalent of the alkenylalane, and its yield by gas chromatographic examination is 90–100%. It is practical, however, to use ca. 20% excess of 1-buten-3-yne for preparing the alkenylalane so as to achieve a high-yield conversion of geranyl chloride into α-farnesene.

6. Geranyl chloride was prepared by treating geraniol, available from Aldrich Chemical Company, with carbon tetrachloride and triphenylphosphine according to an *Organic Syntheses* procedure (Calzada, J. G.; Hooz, J. *Org. Synth., Coll. Vol. VI* **1988,** 634).

7. Tetrakis(triphenylphosphine)palladium was prepared by treating palladium chloride, available from Matthey Bishop, Inc., with hydrazine hydrate in the presence of triphenylphosphine according to an *Inorganic Syntheses* procedure.[3] The submitters used a freshly prepared, shiny yellow, crystalline sample of the palladium complex. On standing for an extended period of time (more than a few weeks), its color gradually darkens. Even such samples are effective in many palladium-catalyzed cross-coupling reactions,[4] but have not been tested in this reaction. Tetrakis(triphenylphosphine)palladium is also available from Aldrich Chemical Company.

8. Tetrahydrofuran available from Aldrich Chemical Company was distilled from sodium and benzophenone.

9. The main purpose of this filtration is to remove traces, if any, of palladium-containing compounds that might induce undesirable transformations, such as isomerization and polymerization, during the subsequent distillative workup.

10. The submitters reported bp 73–75°C (0.05 mm). Gas chromatographic examination of the reaction mixture with a hydrocarbon internal standard indicates that
α-farnesene is formed in 98% yield, based on geranyl chloride, essentially as a
single product (>98%). The product obtained by this procedure shows the
following properties: n_D^{23} 1.4977; IR (neat) cm^{-1}: 3080(w), 2960(s), 2900(s),
1664(w), 1635(m), 1601(m), 981(m), 883(s); ^{1}H NMR [CDCl$_3$, (CH$_3$)$_4$Si] δ:
1.59 (s, 3 H), 1.63 (s, 3 H), 1.66 (s, 3 H), 1.74 (s, 3 H), 2.03 (m, 4 H), 2.82
(t, $J = 6$, 2 H); ^{13}C NMR [CDCl$_3$, (CH$_3$)$_4$Si] δ: 11.62, 16.07, 17.63, 25.69,
26.89, 27.35, 39.88, 110.37, 122.36, 124.50, 131.10, 131.74, 133.79, 135.55,
141.69.

3. Discussion

This procedure for the synthesis of α-farnesene[5] is representative of the palladium-
catalyzed stereo- and regiospecific coupling of allylic derivatives with alkenyl- and
arylmetals.[6] The use of neryl chloride in place of geranyl chloride gives the 6-Z isomer
of α-farnesene in 77% yield (>98% isomeric purity).[6] The high stereo- and regiospecificity (>98%) has been observed only with γ,γ-disubstituted allylic electrophiles.
With γ-monosubstituted allylic derivatives, varying amounts of stereo- and regioisomers
have been observed.[7]

Various allyl derivatives, such as those containing acyloxy, dialkylaluminoxy, and
trialkylsilyloxy groups, also react with alkenylalanes in the presence of a palladium–
phosphine catalyst,[7] and the synthesis of α-farnesene has been achieved by using geranyl
acetate. Although the observed yields are ca. 20% lower than those observed with geranyl
chloride, a careful comparison of the two derivatives has not been performed. In general,
the order of reactivity of various leaving groups is: $-Cl > -OAc > -OPO(OR)_2 >$
$-OSiR_3$.

In addition to alkenylalanes, readily obtainable by either hydroalumination[8] or
carboalumination[9] of alkynes, alkenylzirconium derivatives,[6,10] obtainable by
hydrozirconation[11] of alkynes, undergo a related alkenyl–allyl coupling reaction. In a
related aryl–allyl coupling reaction catalyzed by palladium complexes, arylmetals
containing magnesium, zinc, and cadmium, in addition to those containing aluminum
and zirconium, give the expected cross-coupled products. The yields with zinc or
cadmium tend to be higher than those with aluminum or zirconium, whereas magnesium,
in this respect, is inferior to aluminum or zirconium.[7] Related reactions of
alkenylboranes[12] and alkenylmercury compounds[13] are also known, but their applicability to the selective synthesis of 1,4-dienes of terpenoid origin, such as α-farnesene,
is unknown.

The synthesis of 1,4-dienes via cross-coupling can, in principle, be achieved by either
the reaction of allylmetals with alkenyl electrophiles or by the reaction of alkenylmetals
with allyl electrophiles. The reaction of π-allylnickel derivatives with alkenyl halides[14]
represents the former approach and can be highly regioselective. Stereo- and regiode-
fined alkenylmetals containing aluminum,[15] boron,[16] silicon[17] and copper[18] have been
reported to react with allylic electrophiles producing 1,4-dienes. With the possible
exception of the organocopper reaction, the scope of these uncatalyzed reactions is
practically limited to γ-unsubstituted allylic halides. Finally, the nickel-catalyzed reaction
of Grignard reagents with allylic electrophiles[19] is also known, but the reaction is gener-

ally nonselective. Nor does it appear that the reaction has been applied to the synthesis of 1,4-dienes.

1. Department of Chemistry, Purdue University, West Lafayette, IN 47907.
2. On leave from the Japan Tobacco & Salt Public Corporation.
3. Coulson, D. R. *Inorg. Synth.* **1972,** *13,* 121.
4. For reviews, see (a) Negishi, E. In "Aspects of Mechanism and Organometallic Chemistry," Brewster, J. H., Ed.: Plenum: New York, 1978, p. 285; (b) Negishi, E. *Acc. Chem. Res.* **1982,** *15,* 340.
5. Murray, K. E. *Aust. J. Chem.* **1969,** *22,* 197.
6. Matsushita, H.; Negishi, E. *J. Am. Chem. Soc.* **1981,** *103,* 2882.
7. Negishi, E.; Chatterjee, S.; Matsushita, H. *Tetrahedron Lett.* **1981,** *22,* 3737.
8. For a review, see Mole, T.; Jeffery, E. A. "Organoaluminum Compounds"; Elsevier: Amsterdam, 1972.
9. For a review, see Negishi, E. *Pure Appl. Chem.* **1981,** *53,* 2333.
10. Hayashi, Y.; Riediker, M.; Temple, J. S.; Schwartz, J. *Tetrahedron Lett.* **1981,** *22,* 2629. For a related reaction that is stoichiometric in palladium, see Temple, J. S.; Schwartz, J. *J. Am. Chem. Soc.* **1980,** *102,* 7381.
11. For a review, see Schwartz, J. *J. Organomet. Chem. Library* **1976,** *1,* 461.
12. Miyaura, N.; Yano, T.; Suzuki, A. *Tetrahedron Lett.* **1980,** 2865.
13. Larock, R. C.; Bernhardt, J. C.; Driggs, R. J. *J. Organomet. Chem.* **1978,** *156,* 45.
14. Semmelhack, M. F. *Org. React.* **1972,** *19,* 115.
15. (a) Lynd, R. A.; Zweifel, G. *Synthesis* **1974,** 658; (b) Baba, S.; Van Horn, D. E.; Negishi, E. *Tetrahedron Lett.* **1976,** 1927; (c) Eisch, J. J.; Damasevitz, G. A. *J. Org. Chem.* **1976,** *41,* 2214; (d) Uchida, K.; Utimoto, K.; Nozaki, H. *J. Org. Chem.* **1976,** *41,* 2215.
16. Yamamoto, Y.; Yatagai, H.; Sonoda, A.; Murahashi, S. I. *J. Chem. Soc., Chem. Commun.* **1976,** 452.
17. Yoshida, J.; Tamao, K.; Takahashi, M.; Kumada, M. *Tetrahedron Lett.* **1978,** 2161.
18. (a) Normant, J. F.; Bourgain, M. *Tetrahedron Lett.* **1971,** 2583; (b) Corey, E. J.; Cane, D. E.; Libit, L. *J. Am. Chem. Soc.* **1971,** *93,* 7016; (c) Raynolds, P. W.; Manning, M. J.; Swenton, J. S. *J. Chem. Soc., Chem. Commun.* **1977,** 499; (d) Alexakis, A.; Cahiez, G.; Normant, J. F. *Synthesis* **1979,** 826.
19. For a review, see Felkin, H.; Swierczewski, G. *Tetrahedron* **1975,** *31,* 2735.

ALKYLATION OF THE ANION FROM BIRCH REDUCTION OF *o*-ANISIC ACID: 2-HEPTYL-2-CYCLOHEXENONE

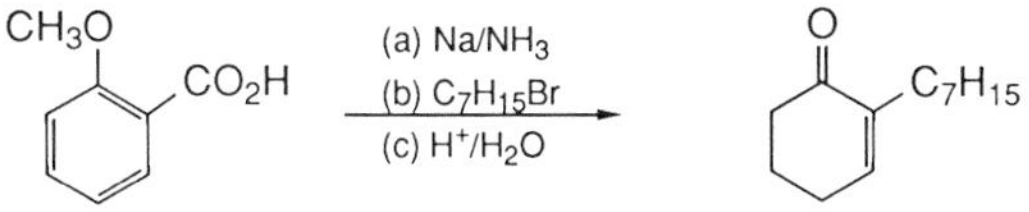

Submitted by D. F. TABER, B. P. GUNN, and I-CHING CHIU[1]
Checked by M. F. SEMMELHACK and E. STELTER

1. Procedure

Caution! Liquid ammonia should be used only in a well-ventilated hood.

2-Heptyl-2-cyclohexenone. A 1-L, three-necked, round-bottomed flask is charged with 15.2 g (0.1 mol) of *o*-anisic acid (Note 1) and 100 mL of tetrahydrofuran (Note 2). An acetone–dry ice condenser and mechanical stirrer are put in place, the flask is immersed in an acetone–dry ice bath, and 400 mL of ammonia is distilled in (Notes 3, 4). The resulting thick white suspension (the ammonium salt of the acid) is stirred mechanically. Sodium, washed sequentially with xylenes and ether, is added in small pieces. The suspension dissolves to give a pale yellow solution which, upon introduction of more sodium, changes to the characteristic blue color of excess sodium. When the deep blue color persists, a mixture of 1-bromoheptane (21.49 g, 0.12 mol) and 1.0 mL (2.4 mmol) of 1,2-dibromoethane is added in one portion. The blue color is discharged immediately, leaving a yellow solution. The acetone–dry ice bath and the condenser are removed, and the ammonia is allowed to evaporate under a gentle stream of nitrogen.

The residue is diluted with 700 mL of water, and the resulting aqueous solution is washed with three 40-mL portions of dichloromethane, acidified with cold concentrated HCl, and extracted with five 40-mL portions of 1,2-dichloroethane. The combined 1,2-dichloroethane extracts are placed in a 500-mL, one-necked, round-bottomed flask bearing a reflux condenser; water (50 mL), concentrated HCl (50 mL), and hydroquinone (300 mg) are added; and the mixture is heated at reflux under a positive pressure of nitrogen for 30 min. The mixture is cooled to 25°C, the layers are separated, and the organic layer is washed with 60 mL of 0.5 *M* aqueous sodium bicarbonate solution. The organic phase is dried over anhydrous potassium carbonate, concentrated by rotary evaporation at aspirator vacuum, and distilled through a 10-cm Vigreux column to yield a center cut, bp 100–104°C (0.02 mm), 9.0–11.5 g (46–59%) (Notes 5–7).

2. Notes

1. *o*-Anisic acid was obtained from Aldrich Chemical Company, Inc.
2. Tetrahydrofuran was dried and made oxygen-free by boiling over sodium/benzophenone ketyl under argon, and distilling just before use.
3. Reduction in refluxing liquid ammonia (-33°C) led to substantial cleavage of the methoxyl group with resultant formation of alkylated dihydrobenzoic acid.

4. Arrangements for cooling or condensing the liquid ammonia over sodium in a preliminary drying operation could be made, but were not necessary. The results reported here were achieved by simply passing ammonia gas from a cylinder into the cold reaction system through heavy Tygon tubing.

5. The spectral properties of 2-heptyl-2-cyclohexenone are as follows: IR (CCl_4) cm^{-1}: 2920, 2860, 1670, 1455, 1435, 1370, 1170, 1120, 1095, 905; ^1H NMR($CDCl_3$) δ: 0.85 (br t, 3 H, $J = 7$), 1.28 (br s, 10 H), 1.8–2.6 (m, 8 H), 6.70 (br s, 1 H); n_D^{26} 1.4738.

6. Before distillation, the crude enone contained substantial amounts of the β,γ-isomer. As an alternative to equilibrium on distillation, this mixture could be converted to the α,β-isomer by stirring with 0.1 M sodium methoxide in methyl alcohol under nitrogen at 0°C for 2 hr.

7. Professor L. N. Mander has advised us that addition of 1.0 equivalent of potassium t-butoxide prior to addition of sodium metal significantly improved the yield of this procedure: Hoole, J. M.; Mander, L. N.; Woolias, M. *Tetrahedron Lett.* **1982,** *23,* 1095.

3. Discussion

Cyclohexenones with 2-alkyl substituents are usually prepared by alkylation of dihydroresorcinol followed by enol ether formation, reduction, and hydrolysis.[2b] A variety of other approaches have been employed.[2] The procedure outlined here is simple, occurring in essentially one pot, using commercially available starting materials. The alkylating agent can equally well be an alkyl iodide or p-toluenesulfonate ester. A variety of other alkylating agents have been employed using an earlier, unoptimized version of this procedure.[3]

1. Department of Pharmacology, School of Medicine, Vanderbilt University, Nashville, TN 37232.
2. (a) Alkylation of cyclohexenone: Conia, J.-M.; Le Craz, A. *Bull. Soc. Chim. Fr.* **1960,** 1934. (b) Alkylation of cyclohexane-1,3-dione followed by enol ether formation, reduction, and hydrolysis: Angell, M. F.; Kafka, T. M. *Tetrahedron* **1969,** *25,* 6025. (c) Bromination-dehydrobromination of a 2-alkylcyclohexanone: Warnhoff, E. W.; Martin, D. G.; Johnson, W. S. *Org. Synth. Coll. Vol. IV* **1963,** 162. (d) Sulfenylation of a 2-alkylcyclohexanone followed by oxidation and elimination: Trost, B. M.; Salzmann, T. N. *J. Am. Chem. Soc.* **1973,** *95,* 6840. (e) Selenation of a 2-alkylcyclohexanone followed by oxidation and elimination: Reich, H. J.; Renga, J. M.; Reich, I. L. *J. Am. Chem. Soc.* **1975,** *97,* 5434. (f) Reduction of a 2-alkylpyridine followed by hydrolysis and aldol condensation: Danishefsky, S.; Cain, P. *J. Org. Chem.* **1975,** *40,* 3607. (g) Oxidation of a 1-alkylcyclohexene: Belyaev, V. F. *Zhidko-faznoe Okislenie Nepredel'nykh Organ. Soedin., Sb.* **1961,** *No. 1,* 97–104; *Chem. Abstr.* **1963,** *58,* 4435d.
3. Taber, D. F. *J. Org. Chem.* **1976,** *41,* 2649.

HEXAFLUOROACETONE

(2-Propanone, 1,1,1,3,3,3-hexafluoro-)

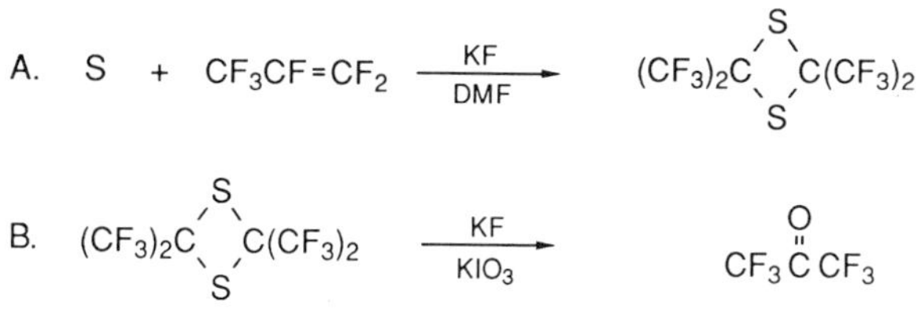

A. S + $CF_3CF=CF_2$ $\xrightarrow[\text{DMF}]{\text{KF}}$ $(CF_3)_2C\underset{S}{\overset{S}{\diagdown\diagup}}C(CF_3)_2$

B. $(CF_3)_2C\underset{S}{\overset{S}{\diagdown\diagup}}C(CF_3)_2$ $\xrightarrow[\text{KIO}_3]{\text{KF}}$ $CF_3\overset{O}{\overset{\|}{C}}CF_3$

Submitted by MICHAEL VAN DER PUY and LOUIS G. ANELLO[1]
Checked by EVAN D. LAGANIS and BRUCE E. SMART

1. Procedure

Caution! Hexafluoroacetone and its precursor are toxic. Both procedures should be conducted in an efficient hood.

A. *2,2,4,4-Tetrakis(trifluoromethyl)-1,3-dithietane.* A 500-mL, three-necked flask is fitted with a good magnetic stirring bar, thermometer, water-cooled condenser, and a fritted gas inlet tube (Note 1). The outlet of the condenser is attached to a tared $-78°C$ cold trap and the inlet tube is connected via flexible tubing to a graduated $-78°C$ cold trap into which 60 mL (96 g, 0.64 mol) of hexafluoropropene has been condensed under nitrogen. The flask is charged with 3 g of potassium fluoride and is flamed gently under vacuum. The apparatus is cooled while purging with nitrogen. Sulfur (23 g, 0.72 mol) and 200 mL of dry dimethylformamide are then added (Note 2). The reaction mixture is heated to 40–45°C with stirring. The heat source is removed, the stopcock on the trap containing the hexafluoropropene is opened, and the trap is gently thawed. The rate of hexafluoropropene bubbling into the reaction mixture is adjusted to about 0.6 mL (1 g)/min by cooling or warming the trap containing the hexafluoropropene (Notes 3–5). When all the hexafluoropropene has been added, the reaction mixture is cooled to $-20°C$ to $-30°C$ and quickly filtered under suction (Note 6). The filtercake is transferred to an Erlenmeyer flask and is allowed to melt. Water (50 mL) is added, and the mixture is filtered. The lower liquid phase is separated, washed with 50 mL of water, and distilled through a 20-cm Vigreux column at atmospheric pressure to give 93.0–99.4 g (80–85%) of product, bp 106–108°C (Note 7).

B. *Hexafluoroacetone.* A 1-L, three-necked flask is fitted with a sealed mechanical stirrer, thermometer, and condenser. A $-78°C$ glass trap is attached to the condenser via flexible tubing. While the system is purged with nitrogen, 3 g of potassium fluoride is added and the flask and potassium fluoride are flame-dried (Note 8). After the flask has cooled, 300 mL of dry dimethylformamide, 80 g (0.374 mol) of powdered potassium iodate, and 60 g (0.165 mol) of 2,2,4,4-tetrakis(trifluoromethyl)-1,3-dithietane are added (Note 9). The stirrer and water condenser are started and the reaction mixture is heated over a 45-min period to 149°C and is kept at 149°C for an additional 15 min. The heat source is then removed, and a slow stream of nitrogen is used to flush any remaining product gas into the cold trap (Note 8). The condensate is transferred under vacuum to a tared, evacuated-gas cylinder (Note 10). The cylinder contains 37.0–39.9 g (68–73%) of material (Note 11). This material is distilled to give 35.0–37.6 g (64–69%) of pure product, bp $-28°C$ [lit.[2] bp $-27°C$] (Note 12).

2. Notes

1. The checkers dried the glassware overnight at $150°C$ in an oven and assembled it hot under a nitrogen purge.

2. The checkers obtained potassium fluoride, potassium iodate, dimethylformamide (reagent grades), and sulfur (sublimed) from Fisher Scientific Co. The submitters purchased hexafluoropropene from PCR Research Chemicals, Inc.; the checkers used hexafluoropropene from E. I. du Pont de Nemours & Company, Inc. The potassium fluoride was predried overnight in a vacuum oven at $110°C$. The sulfur was dried in a vacuum desiccator and the dimethylformamide was distilled from P_2O_5 prior to use.

3. The mixture of dimethylformamide, sulfur, and potassium fluoride turns brown prior to the addition of hexafluoropropene, which quickly brings the color back to bright yellow. The submitters report that the reaction mixture will turn blue or green prior to the addition of hexafluoropropene, if the dimethylformamide is dry (less than ca. 0.05% water).

4. The reaction is moderately exothermic. The temperature rises to about $55°C$ and remains there as the reaction proceeds.

5. With good stirring, the reaction proceeds as fast as the hexafluoropropene is added. The dry ice trap attached to the condenser should be checked periodically, however. When the required amount of hexafluoropropene is added, little or no undissolved sulfur remains.

6. 2,2,4,4-Tetrakis(trifluoromethyl)-1,3-dithietane melts at $24°C$. Thus, this operation must be done quickly to minimize product loss.

7. The product is more than 99% pure by GLPC (6 ft $\times \frac{1}{8}$ in. 20% FS-1265 on 60/80 Gaschrome R, 50–200°C) and by ^{19}F NMR (CDCl$_3$) ϕ: -73.3 (s). The submitters report that they obtained 78–90 g of 98% pure product, bp $110°C$.

8. The nitrogen initially should come from the cold trap, itself cooled under a nitrogen flush. At the end of the reaction, the flow of nitrogen should be reversed. This can be done by replacing the thermometer with a gas inlet tube.

9. The submitters report that a ratio of KIO$_3$ to 2,2,4,4-tetrakis(trifluoromethyl)-1,3-dithietane of 2.26 is near the optimum since a ratio of 2.5 did not increase the yield, whereas a ratio of 2.0 gave 5–10% lower yields.

10. This transfer is best done on a vacuum manifold system equipped with a manometer. The trap and a stainless-steel cylinder of 100–300-mL capacity are attached via vacuum tubing to the manifold system, cooled in liquid nitrogen baths, and evacuated to 0.5–1 mm. The system is closed and the trap is removed from its cold bath and is slowly thawed. The volatile material in the trap is

transferred to and condensed in the cylinder at such a rate that no positive pressure builds up in the closed system.

11. The submitters report collecting 45–50 g of product (98% pure or better by GLPC on a 10-ft $\times \frac{1}{8}$-in. Porapak P column) in the cold trap attached to reaction vessel. The checkers found that the trap contained relatively nonvolatile material, principally dimethylformamide, in addition to the desired product.

12. The checkers used a 30-cm jacketed, low-temperature spinning band column for this distillation. The IR spectrum of the distilled product is identical to that of an authentic sample; IR (vapor) cm^{-1}: 1806 (C=O).

3. Discussion

Earlier methods of preparing 2,2,4,4-tetrakis(trifluoromethyl)-1,3-dithietane (hexafluorothioacetone dimer, HFTA dimer) include the reaction of hexafluoropropene (HFP) and sulfur over a carbon bed at 425°C[3] and the reaction of HFP and sulfur in tetramethylene sulfone at 120°C in the presence of potassium fluoride (autoclave).[4] Dimethylformamide appears to be a far superior solvent for this reaction, permitting the use of atmospheric pressure and modest temperatures, as well as affording a cleaner product.

The generation of hexafluoroacetone (HFA) from HFTA dimer has been accomplished by the hot-tube oxidation with nitric oxide at 650°C (high temperature converts dimer into monomer).[5] The present method uses the more convenient interconversion of dimer to monomer effected by potassium fluoride in dimethylformamide. This permits many reactions to be conducted on the very reactive monomer without actually isolating it.

For occasional laboratory synthesis of HFA, the present method offers distinct advantages of convenience (cost, workup, standard equipment) over other known methods. These include the epoxidation of HFP followed by isomerization of the epoxide to HFA,[6] the high-temperature halogen exchange of hexachloroacetone with Cr^{3+}/HF,[7] and permanganate oxidation of the extraordinarily toxic perfluoroisobutylene.[8]

Hexafluoroacetone is a reactive electrophile. It reacts with activated aromatic compounds (e.g., phenol), and can be condensed with olefins, dienes, ketenes, and acetylenes. It forms adducts with many compounds containing active hydrogen (e.g., H_2O or HCN). Reduction of HFA with $NaBH_4$ or $LiAlH_4$ affords the useful solvent hexafluoroisopropyl alcohol. The industrial importance of HFA arises largely from its use in polymers and as an intermediate in monomer synthesis.[9]

1. Allied Chemical Co., Buffalo Research Laboratory, Buffalo, NY 14210.
2. Middleton, W. J. "Hexafluoroacetone and Derivatives," in "Kirk-Othmer: Encyclopedia of Chemical Technology," 3rd ed.; Grayson, M., Ed.; Wiley, New York, 1980; Vol. 10, pp. 881–890.
3. Martin, K. V. J. Chem. Soc. 1964, 2944.
4. Dyatkin, B. L.; Sterlin, S. R.; Zhuravkova, L. G.; Martynov, B. I.; Mysov, E. I.; Knunyants, I. L. Tetrahedron 1973, 29, 2759.
5. Middleton, W. J.; Sharkey, W. H. J. Org. Chem. 1965, 30, 1384.

6. (a) Atkins, Jr., G. M. U.S. Patent 3775439, 1973; *Chem. Abstr.*, **1974**, *80*, 59845; (b) Moore, E. P.; Milian, A. S. U.S. Patent 3321515, 1967; *Chem. Abstr.* **1967**, *67*, 116581.
7. Anello, L. G.; Nychka, H. R.; Woolf, C. French Patent 1369784, 1964; *Chem. Abstr.* **1965**, *62*, 1570.
8. Brice, T. J.; LaZerte, J. D.; Hals, L. J.; Pearlson, W. H. *J. Am. Chem. Soc.* **1953**, *75*, 2698.
9. References can be found in reference 2 and in the review; Krespan, C. G.; Middleton, W. J. *Fluorine Chem. Rev.* **1967**, *1*, 145.

HEXAHYDRO-2-(1*H*)-AZOCINONE

[2(1*H*)-Azocinone, hexahydro-]

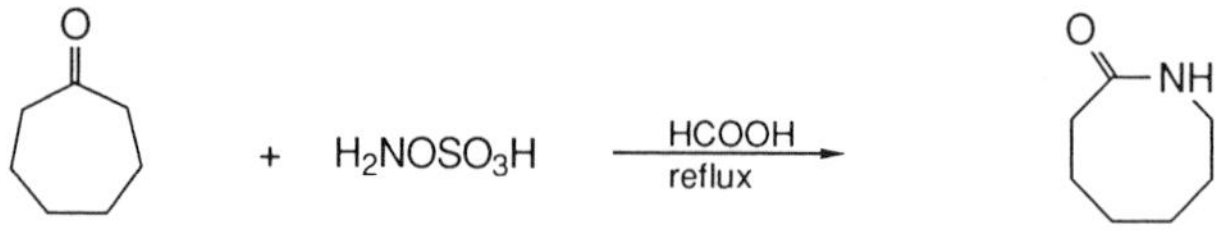

$$\text{(cycloheptanone)} + H_2NOSO_3H \xrightarrow[\text{reflux}]{\text{HCOOH}} \text{(hexahydro-2-(1}H\text{)-azocinone)}$$

Submitted by GEORGE A. OLAH and ALEXANDER P. FUNG[1]
Checked by DAVID VARIE and EDWIN VEDEJS
Rechecked by SCOTT THOMPSON and CLAYTON H. HEATHCOCK

1. Procedure

A 100-mL, three-necked flask is equipped with a magnetic stirring bar, a pressure-equalizing dropping funnel, and a reflux condenser connected to a nitrogen flow line. The system is dried with a heat gun while it is flushed with dry nitrogen. The reaction vessel is then cooled in a water bath while a light positive pressure of nitrogen is maintained. The flask is charged with hydroxylamine-*O*-sulfonic acid[2] (8.48 g, 0.075 mol) (Note 1) and 95–97% formic acid (45 mL) (Note 2). A solution of cycloheptanone (5.61 g, 0.05 mol) (Note 3) in 15 mL of 95–97% formic acid is added with stirring over a 3-min period. After addition is complete, the reaction mixture is heated under reflux for 5 hr and then cooled to room temperature. The reaction mixture is quenched with 75 mL of ice–water. The aqueous solution is slowly neutralized to pH ~7 with 6 *N* sodium hydroxide (Note 4) and extracted with three 100-mL portions of chloroform. The combined organic layers are dried with anhydrous magnesium sulfate. After removal of the solvent on a rotary evaporator, the product hexahydroazocinone is purified by distillation to give 4.6 g (72%), bp 94–96°C/0.2 mm, (short-path apparatus), lit[4] bp 133–135°C/4 mm (Note 5).

2. Notes

1. The hydroxylamine-*O*-sulfonic acid used by the submitters was purchased from Ventron Corporation and used directly. However, it can be readily prepared in the laboratory.[3,4]
2. Formic acid 95–97% was obtained from the Aldrich Chemical Company.

3. Commercial cycloheptanone (bp 179°C) obtained from MCB, Inc. was used directly.

4. An external ice–salt bath is used.

5. The product exhibits the following spectra: ^{1}H NMR (CDCl$_3$) δ: 1.6–1.8 (m, 6 H, CH$_2$), 2.40 (3 H, m), 2.57 (m, 2 H, CH$_2$CO), 3.31 (m, 2 H, CH$_2$−N), 7.16 (br, 1 H, NH); IR (cm^{-1}): 3270, 3200, 1650; GLC analysis: 20% SE-30, 60/80 on Chrom-W, $\frac{1}{8}$-in × 20-ft column, 180°C: one peak.

3. Discussion

The procedure described here is a one-step conversion of cycloheptanone into hexa-hydro-2(1H)-azocinone. The method is general and is characterized by good yields, mild conditions, and easy preparation of the product in pure form from readily available starting materials. Several methods are described in the patent literature for simultaneous oximation of ketones and rearrangement of the corresponding oxime, including the use of hydroxylamine and sulfuric acid,[6,7] or by employing primary nitroparaffins as a source of hydroxylamine.[8,9] The present method has been shown[10] to be applicable to a wide variety of lactams (C$_5$ ~ C$_{12}$). In the specific case of hexahydroazocinone, the yield from cycloheptanone (60–63%) appears lowers than for the conventional two-step method,[11,12] but the latter requires isolation of the intermediate oxime.

1. Institute of Hydrocarbon Chemistry and Department of Chemistry, University of Southern California, University Park, Los Angeles, CA 90007.

2. For a recent review on hydroxylamine-O-sulfonic acid, see Wallace, R. G., *Aldrichimica Acta* **1980**, *13*, 3.

3. Rathke, M. W.; Millard, A. A. *Org. Synth. Coll. Vol. VI* **1988**, 943.

4. Coffman, D. D.; Cox, N. L.; Martin, E. L.; Mochel, W. E.; Van Natta, F. J. *J. Polym. Sci.* **1948**, *3*, 85.

5. Donaruma, L. G.; Heldt, W. Z. *Org. React.* **1960**, *11*, 1.

6. Novotny, A. U.S. Patent 2579851, 1951; *Chem. Abstr.* **1952**, *46*, 6668.

7. Barnett, C.; Cohn, I. M.; Lincoln, J. U.S. Patent 2754298, 1956; *Chem. Abstr.* **1957**, *51*, 2853.

8. Hass, H. B.; Riley, E. F. *Chem. Rev.* **1943**, *32*, 373.

9. Novotny, A. U.S. Patent 2569114, 1951; *Chem. Abstr.* **1952**, *46*, 5078.

10. Olah, G. A.; Fung, A. P. *Synthesis* **1979**, 537.

11. Yields of 97 and 88% are reported for the oximation and Beckmann rearrangement steps, respectively, but no experimental details are given.[12] An earlier publication reports <50% yield in the second step.[4]

12. McKay, A. F.; Tarlton, E. J.; Petri, S. I.; Steyermark, P. R.; Mosely, M. A. *J. Am. Chem. Soc.* **1958**, *80*, 1510.

HEXAMETHYL DEWAR BENZENE

(Bicyclo[2.2.0]hexa-2,5-diene, 1,2,3,4,5,6-hexamethyl-)

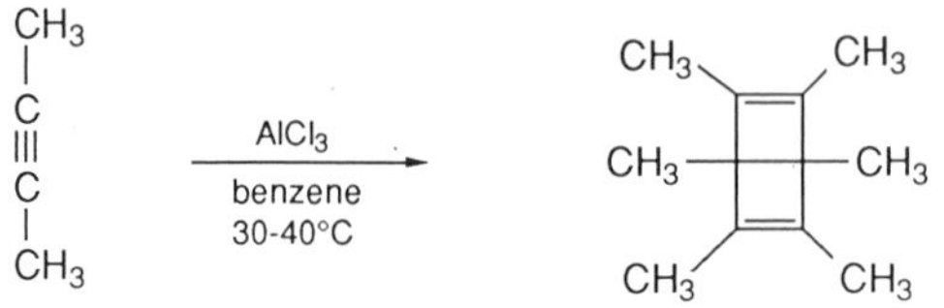

Submitted by SAMI A. SHAMA and CARL C. WAMSER[1]
Checked by RETO NAEF, DIETER SEEBACH, and BEAT WEIDMANN

1. Procedure

Caution! Benzene has been identified as a carcinogen; OSHA has issued emergency standards on its use. All procedures involving benzene should be carried out in a well-ventilated hood, and glove protection is required.

A 250-mL, three-necked, round-bottomed flask containing a 2.5-cm magnetic stirring bar is equipped with a Dewar-type reflux condenser containing ice, a dropping funnel, and a gas inlet tube. A calcium chloride drying tube is attached to the condenser and the apparatus is flushed with dry deoxygenated nitrogen (Note 1). The gas inlet tube is then replaced by a thermometer, and a suspension of 5.0 g of aluminum trichloride in 50 mL of benzene is introduced into the flask (Note 2). A solution of 100 g (1.85 mol) of 2-butyne (Notes 3 and 4) in 50 mL of cold dry benzene is added, with vigorous stirring, through the dropping funnel, over a period of 1 hr. During the addition, the temperature of the reaction mixture is kept between 30 and 40°C through the use of a water bath. Stirring is continued for 5 hr at 30–40°C after the addition has been completed. The catalyst is then decomposed by pouring the mixture onto 50 g of crushed ice in a 500-mL separatory funnel, whereupon the dark brown color turns pale yellow. When the ice has melted completely, the organic layer is separated, washed with two 25-mL portions of cold water, dried over anhydrous potassium carbonate, and filtered. Benzene and unreacted butyne (Note 5) are removed in a rotary evaporator using a water bath at 40°C and a water aspirator vacuum. The residual liquid is distilled through a short-path distillation head under reduced pressure using a capillary. The yield is 38–50 g (38–50%) of hexamethyl Dewar benzene, bp 43°C/10 mm, mp 7–8°C, n_D^{20} 1.4480 (Notes 6 and 7).

2. Notes

1. Commercial nitrogen is deoxygenated by bubbling it through a trap containing an alkaline pyrogallol solution.[2] The gas is then dried by passing it through a potassium hydroxide tower. The checkers used argon as an inert atmosphere.

2. Aluminum trichloride is purified by sublimation under reduced pressure and the benzene is dried over sodium wire before use. The checkers used sublimed $AlCl_3$ as supplied by Merck (Darmstadt).

3. 2-Butyne was purchased from Chemical Samples Company or from Fluka AG.

4. The bottle containing 2-butyne (bp 27°C) should be chilled thoroughly before opening.

5. About 20 g of 2-butyne may be collected in an ice-cooled receiver if the dried solution is concentrated by distillation through a 25-cm Vigreux column rather than by evaporation. The checkers do not recommend this mode of workup, nor did they use a column for distilling the Dewar benzene, to avoid prolonged heating of the bicyclic system.

6. The spectral properties of hexamethyl Dewar benzene are as follows: ^{1}H NMR (CDCl$_3$) δ: 1.07 (s, 6 H), 1.58 (s, 12 H).

7. Hexamethyl Dewar benzene undergoes thermal isomerization[3,4] and reacts with acids[5] and transition-metal ions.[6] It should be stored in a freezer in a tightly sealed bottle. Hexamethyl Dewar benzene is reportedly a carcinogen,[7] and care must be taken to avoid contact with the skin or inhalation of its vapor.

3. Discussion

The present procedure is that of Schäfer[8,9] and is the first method available for large-scale preparation of a Dewar benzene. Other syntheses of compounds containing the Dewar benzene skeleton have generally involved photochemical isomerization of the corresponding benzene isomer.[10]

The present procedure represents a novel reaction, bicyclotrimerization. The intermediate dimeric complex of AlCl$_3$ with tetramethylcyclobutadiene has been isolated, and addition of different alkynes to this complex provides a synthetic route to a variety of substituted Dewar benzenes.[11]

1. Department of Chemistry, California State University, Fullerton, CA 92634. Present address: Department of Chemistry, Portland State University, Portland, OR 97207.

2. Gordon, A. J.; Ford, R. A. "The Chemist's Companion"; Wiley-Interscience: New York, 1972; p. 438.

3. Oth, J. F. M. *Angew. Chem., Int. Ed. Engl.* **1968**, *7*, 646.

4. Adam, W.; Chang, J. C. *Int. J. Chem. Kinet.* **1969**, *1*, 487.

5. Hogeveen, H.; Volger, H. C. *Recl. Trav. Chim. Pays-Bas* **1968, 87**, 385.

6. Bishop, K. C., III. *Chem. Rev.* **1976, 76**, 461.

7. Dannenberg, H.; Brachmann, I.; Thomas, C. Z. *Krebsforsch.* **1970, 74**, 100; *Chem. Abstr.* **1970, 73**, 1933a.

8. Schäfer, W. *Angew. Chem., Int. Ed. Engl.* **1966, 5**, 669.

9. Schäfer, W.; Hellman, H. *Angew. Chem., Int. Ed. Engl.* **1967, 6**, 518.

10. van Tamelen, E. E. *Acc. Chem. Res.* **1972, 5**, 186.

11. Rantwijk, S. Van; Timmermans, G. J.; Van Bekkum, H. *Recl. Trav. Chim. Pays-Bas* **1976, 95**(2), 39.

SELECTIVE HYDROBORATION OF A 1,3,7-TRIENE: HOMOGERANIOL

[3,7-Nonadien-1-ol, 4,8-dimethyl-, (*E*)-]

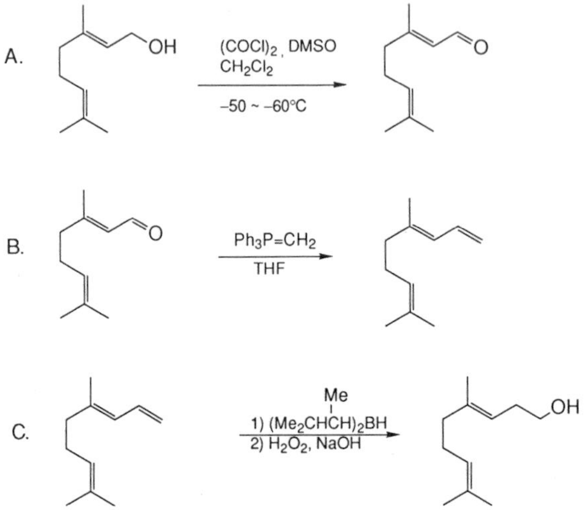

A.
$$\text{(structure)} \xrightarrow[\text{$-50 \sim -60°C$}]{\substack{\text{(COCl)}_2\text{, DMSO} \\ \text{CH}_2\text{Cl}_2}} \text{(structure)}$$

B.
$$\text{(structure)} \xrightarrow[\text{THF}]{\text{Ph}_3\text{P=CH}_2} \text{(structure)}$$

C.
$$\text{(structure)} \xrightarrow[\text{2) H}_2\text{O}_2\text{, NaOH}]{\substack{\text{Me} \\ | \\ \text{1) (Me}_2\text{CHCH)}_2\text{BH}}} \text{(structure)}$$

Submitted by ERIC J. LEOPOLD[1]
Checked by SHRIDHAR HEGDE and ROBERT M. COATES

1. Procedure

A. *Geranial.* A 2-L, three-necked, round-bottomed flask is dried in an oven and equipped with a mechanical stirrer, a thermometer, a Claisen adapter, and two pressure-equalizing dropping funnels. The flask is charged with 500 mL of dichloromethane (Note 1) and 20 mL (29.2 g, 0.23 mol) of oxalyl chloride (Note 2). The solution is stirred and cooled at -50 to $-60°C$ as 34 mL (37.5 g, 0.48 mol) of dimethyl sulfoxide (Note 3) in 100 mL of dichloromethane is added dropwise at a rapid rate. After 5 min 30.8 g (0.2 mol) of geraniol (Note 4) is added dropwise over 10 min maintaining the temperature at -50 to $-60°C$. After another 15 min, 140 mL of triethylamine is added dropwise while keeping the temperature at or below $-50°C$. Stirring is continued for 5 min, after which time the mixture if allowed to warm to room temperature and 700 mL of water is added. The aqueous layer is separated and extracted with two 300-mL portions of dichloromethane. The organic layers are combined, washed with two 100-mL portions of saturated sodium chloride, and dried over anhydrous magnesium sulfate. The filtered solution is concentrated to 500 mL by rotary evaporation and washed successively with 1% hydrochloric acid until it is no longer basic. The dichloromethane solution is washed with water, 5% sodium carbonate, water, and saturated sodium chloride before drying over anhydrous magnesium sulfate. Rotary evaporation of the solvent gives ca. 30 g of crude product. Distillation in a Kugelrohr apparatus (Note 5) with an oven temperature of 80–85°C (1 mm) affords 27.3–28.5 g (90–94%) of geranial, n_D^{24} 1.4870 (Note 6).

B. *(E)-4,8-Dimethyl-1,3,7-nonatriene.* A 1-L, three-necked, round-bottomed flask equipped with a pressure-equalizing dropping funnel, thermometer, magnetic stirring bar, and serum caps (Note 7) is charged with 50 g (0.12 mol) of methyltriphenylphosphonium iodide (Note 8) and 320 mL of tetrahydrofuran (Note 9) and is flushed with argon. The flask is cooled in an ice bath and the suspension is stirred under a positive pressure of argon, while about 0.2–0.6 mL of 2.05 M phenyllithium in 30 : 70 ether : cyclohexane (Notes 10 and 11) is added dropwise until the suspension develops a permanent yellow color (Note 12). Then 56 mL (0.115 mol) of 2.05 M phenyllithium is added dropwise over 10 min. The ice bath is removed, and the orange suspension containing excess phosphonium salt is stirred at room temperature for 30 min. The reaction mixture is stirred and cooled at 0–5°C, and 17.2 g (0.11 mol) of geranial in 50 mL of tetrahydrofuran is added dropwise over 10 min. The dropping funnel is rinsed with a small amount of tetrahydrofuran. The mixture is stirred at room temperature for 2 hr. The light-orange mixture is hydrolyzed by adding 2 mL of methanol, and most of the solvent is removed on a rotary evaporator until a slurry results (Note 13). The slurry is diluted with 200 mL of petroleum ether (bp 60–68°C), and the supernatant solution is decanted and filtered through 150 g of Celite on a Büchner funnel. The solids remaining in the flask are heated with three 100-mL portions of hot petroleum ether, and the supernatant solutions are also filtered through Celite. The filtrate is concentrated by rotary evaporation to a yellowish liquid that is filtered through 150 g of Florisil on a Büchner funnel, and the Florisil is washed with 300 mL of petroleum ether. Rotary evaporation of the eluate provides ca. 15 g of clear liquid, which on distillation in a Kugelrohr apparatus with an oven temperature of 60–70°C (2 mm) gives 13.1–13.5 g (77–80%) of the triene, n_D^{22} 1.4871 (Notes 14 and 15).

C. *Homogeraniol.* A 250-mL, three-necked, round-bottomed flask is equipped with a magnetic stirring bar, thermometer, pressure-equalizing dropping funnel, and a gas inlet tube to maintain a positive argon pressure within the apparatus (Note 7). The flask is charged with 102 mL (94.8 mmol) of 0.93 M diborane in tetrahydrofuran (Note 16), and the contents are cooled to −30°C. The diborane solution is stirred as 22.1 mL (0.21 mol) of 2-methyl-2-butene (Note 17) is added rapidly. Stirring is continued for 2 hr while maintaining the temperature at 0–2°C. A 500-mL, three-necked, round-bottomed flask equipped with a magnetic stirring bar, thermometer, pressure-equalizing dropping funnel, and a gas inlet tube to keep a positive pressure of argon (Note 7) is charged with 13.0 g (86.7 mmol) of *(E)*-4,8-dimethyl-1,3,7-nonatriene and 35 mL of tetrahydrofuran (Note 9). The contents are stirred and cooled at 0°C as the solution of disiamylborane in the first flask is transferred via a cannula to the pressure-equalizing dropping funnel attached to the second flask. After approximately 20 mL of disiamylborane is transferred to the dropping funnel via a cannula, the dropwise addition of the disiamylborane is started while the transfer continues. The remainder of the disiamylborane solution in the first flask is kept at 0°C. After the 1-hr addition is completed, stirring is continued for 1 hr at 0°C and overnight at room temperature (15 hr). Excess disiamylborane is destroyed by adding 2 mL of ethanol, the mixture is cooled to 0°C, and 33 mL of 3 M sodium hydroxide is added rapidly. Stirring and cooling at −10°C are continued as 33 mL of chilled 30% hydrogen peroxide is slowly added (Note 18). The reaction mixture is stirred at room temperature for 3 hr, the layers are separated, and the aqueous layer is extracted with two 75-mL portions of ether (Note 19). The combined organic layers are washed with two 25-mL portions of saturated sodium chloride and dried over anhydrous magnesium sulfate. Evaporation of the solvent gives ca. 21 g of crude product that is purified by chromatography on 400 g of silica gel packed in a 7.5-cm × 20-cm column. The column is eluted with dichloromethane and 100-mL fractions are collected, the first two of which are discarded. Elution is continued by collecting the 100-mL fractions in a weighed flask and evaporating the solvent under reduced pressure until a constant weight of product is obtained (nine 100-mL fractions). Distillation of the residue in a Kugelrohr apparatus with an oven temperature of 150°C (0.02 mm) gives 12.6–13.2 g (88–91%) of homogeraniol, n_D^{21} 1.4740 (Note 20).

2. Notes

1. Dichloromethane was distilled from calcium hydride and stored over Linde 4A molecular sieves.

2. Oxalyl chloride was distilled immediately before use.

3. Dimethyl sulfoxide was distilled from calcium hydride and stored over Linde 3A molecular sieves.

4. Geraniol was obtained from Aldrich Chemical Company, Inc. (Gold Label) and used without purification.

5. Kugelrohr ovens are available from Rinco Instrument Co., Inc., 5035 Prairie St., P.O. Box 167, Greenville, IL 62246.

6. Thin-layer chromatographic analysis of the product by the submitter on silica gel with 20% ethyl acetate in hexane as developing solvent showed one spot, R_f 0.5. Gas chromatographic analysis showed the presence of 1.5% of the cis isomer by coinjection with 40 : 60 cis : trans citral mixture available from Aldrich Chemical Company, Inc. The ^{1}H NMR spectral data for the product are as follows δ: 1.61 (s, 3 H, CH$_3$), 1.69 (s, 3 H, CH$_3$), 2.17 (s, 3 H, CH$_3$), 2.19–2.23 (m, 4 H, CH$_2$CH$_2$), 5.06 (br s, 1 H, vinyl H at C-6), 5.88 (d, 1 H, J = 8, vinyl H at C-2), 9.99 (d, J = 8, CHO).

7. The glassware was dried in an oven at 150°C, assembled while still hot, and alternately evacuated and flushed with argon.

8. Methyltriphenylphosphonium iodide was prepared by the following procedure. Triphenylphosphine was recrystallized from ethanol and dried over phosphorus pentoxide under reduced pressure for 12 hr. A solution of 39 g (0.15 mol) of triphenylphosphine and 10.0 mL (22.8 g, 0.16 mol) of iodomethane in 105 mL of benzene was allowed to stir at room temperature for 12 hr. The precipitate was filtered, washed with benzene, and dried over phosphorus pentoxide under reduced pressure for 12 hr. The yield was 57 g (94%), mp 189°C (lit.[2] mp 182°C). The reagent is also available from Aldrich Chemical Company, Inc.

9. Tetrahydrofuran was distilled from sodium–benzophenone ketyl.

10. The phenyllithium solution was purchased from Aldrich Chemical Company, Inc. The checkers used 64 mL (0.115 mol) of 1.8 M phenyllithium in 75 : 25 benzene : ether, which was purchased from Alfa Products, Morton Thiokol, Inc.

11. The submitter states that the slight excesses of phenyllithium (5%) and methyltriphenylphosphonium iodide (10%) specified ensure complete conversion of the aldehyde and simplify the purification of the product since the excess phosphonium salt is readily removed during filtration through Florosil.

12. The addition of 0.2–0.6 mL of the phenyllithium solution presumably destroys small amounts of moisture or other impurities.

13. The submitter cautions against evaporating all the solvent; the triphenylphosphine oxide will tenaciously occlude the product, and the yield will be reduced.

14. A gas chromatographic analysis of the product by the submitter on a 15 M capillary column coated with silicone oil SE-54 at 70°C exhibited one peak (98%).

15. An index of refraction of 1.4826 at 20°C is reported[3] for the product. The spectral properties of the product are as follows: IR (neat) cm^{-1}: 3080, 1645, 1600, 1345, 990, 900; ^{1}H NMR (CDCl$_3$) δ: 1.61 (3 H, CH$_3$), 1.68 (s, 3 H, CH$_3$), 1.76 (s, 3 H, CH$_3$ at C-4), 1.95–2.12 (broad, 4 H, CH$_2$CH$_2$), 4.80–5.15 (broad, 3 H, vinyl H), 5.85 (d, 1 H, $J = 10$, vinyl H at C-3), 6.55 (3 d, $J = 10, 10, 17$, vinyl H at C-2).

16. The diborane solution was obtained from Aldrich Chemical Company, Inc. It was titrated[4] before use, although the submitter states that this is not necessary. The solution was transferred from the stock solution to the reaction flask via a cannula. The checkers first transferred the diborane solution via a cannula into a graduated cylinder that was capped with a rubber septum and purged with nitrogen. The specified volume was then transferred into the reaction vessel.

17. 2-Methyl-2-butene was obtained from Aldrich Chemical Company, Inc. and was distilled from calcium hydride.

18. The oxidation of organoboranes is exothermic, and efficient cooling and slow addition are necessary to keep the temperature near 0°C.[5]

19. The checkers observed the separation of a heavy, white precipitate presumed to be a borate salt during the addition of hydrogen peroxide. After the three-phase mixture had been stirred at room temperature for 3 hr, the liquid layers were decanted into a separatory funnel. The solid remaining in the flask was washed with two 75-mL portions of ether and these washings were used to extract the aqueous layer.

20. Indices of refraction of 1.4722 at 22°C and 1.4718 at 26°C are reported for homogeraniol.[6,7] The spectral properties of the product are as follows: IR (neat) cm^{-1}: 3330, 2960 (sh), 2920, 1448, 1435 (sh, m), 1374 (m), 1108 (w), 1045 (s), 875 (w); ^{1}H NMR (CDCl$_3$) δ: 1.60 (s, 3 H, CH$_3$ at C-4), 1.64 (s, 3 H, CH$_3$), 1.68 (s, 3 H, CH$_3$ at C-4), 1.95–2.15 (s, 4 H, CH$_2$CH$_2$), 2.30 (m, 2 H, CH$_2$CH$_2$OH), 3.60 (t, 2 H, $J = 7$, CH$_2$OH), 4.95–5.25 (m, 2 H, vinyl H).

3. Discussion

Homogeraniol is an important intermediate in syntheses of squalene,[6] aplysistatin,[8] dendrolasin,[9] and juvenile hormone analogs.[10] The present procedure affords an efficient, stereoselective method for preparing (*E*)-homogeraniol, contaminated by at most 1–2% of the *Z* isomer.

In Step A geraniol is oxidized to geranial (citral) by Swern's modification of the Moffat oxidation.[11] The stereoisomeric purity of the product is at least 98%. This procedure is readily conducted on a large-scale and requires only a 4-hr time period, including distillation of oxalyl chloride. The oxidation of geraniol to pure (*E*)-geranial may also be accomplished by Collin's oxidation with chromium trioxide–dipyridine complex,[12] or by use of activated manganese dioxide.[13] However, these methods require large amounts of reagents and solvents for 0.2-mol-scale preparations.

The Wittig methylenation of geranial to (*E*)-4,8-dimethyl-1,3,7-nonatriene is best carried out with phenyllithium in tetrahydrofuran as described in Section B. The use of butyllithium in tetrahydrofuran or ether–hexane[3] affords the triene in only 50–60% yield. When the ylide was generated with sodium hydride or potassium *tert*-butoxide in dimethyl sulfoxide by the submitter, the Wittig reaction gave triene containing 10–20% of the *Z* isomer. Step C illustrates the selective hydroboration of a diene with disiamylborane.[14] The reaction is best carried out by adding preformed disiamylborane to the triene. Lower yields of homogeraniol were obtained by the submitter when the triene was added to the borane reagent.

Homogeraniol has been prepared by reduction of homogeranic acid with lithium aluminum hydride,[6] by cyclopropylcarbinol rearrangement to homogeranyl bromide and subsequent displacement of the bromide,[15] by zirconium-catalyzed syn addition of trimethylaluminum to an acetylene precursor followed by reaction with ethylene oxide,[7] and by hydroxymethylation of geranyl chloride with diisopropoxymethylsilylmethyl Grignard reagent.[16] Homogeranic acid has been prepared by base-catalyzed hydrolysis of the nitrile,[6,9,17] copper-catalyzed S_N2-type alkylation of β-isopropenyl-β-propiolactone with dimethylallyl Grignard reagent,[18] alkylation of methoxy(phenylthio)methyllithium with geranyl chloride and subsequent chromic acid oxidation,[19] and carboxylation of geranyl phenyl sulfone followed by reductive desulfonation.[20] Although homogeranic acid prepared by nitrile hydrolysis and by β-isopropenyl-β-propiolactone alkylation[18] is a 70 : 30 mixture of *E* and *Z* isomers,[9,21a] the *E* form may be isolated by crystallization at $-10°C$[6] or by preparative gas chromatography of their *tert*-butyl esters.[21b] Homogeraniol prepared by acid-catalyzed cyclopropylcarbinyl to homoallyl rearrangement[15] is also a mixture of *E* and *Z* isomers.[22]

1. Department of Chemistry, Stanford University, Stanford, CA 94305. Present address: SmithKline Beckman, Palo Alto, CA 94304.
2. Bestmann, H. J. *Chem. Ber.* **1962**, *95*, 58–63.
3. Pattenden, G.; Weedon, B. C. L. *J. Chem. Soc. (C)* **1968**, 1984–1997.
4. Brown, H. C. "Organic Syntheses via Boranes"; Wiley: New York, 1975; p. 241.
5. Zweifel, G.; Brown, H. C. "Organic Reactions"; Wiley: New York, 1963; Vol. 13, pp. 1–54.
6. Cornforth, J. W.; Cornforth, R. H.; Mathew, K. K. *J. Chem. Soc.* **1959**, 2539–2547.
7. Kobayashi, M.; Valente, L. F.; Negishi, E.-i. *Synthesis* **1980**, 1034–1035.
8. Hoye, T. R.; Kurth, M. J. *J. Am. Chem. Soc.* **1979**, *101*, 5065–5067.
9. Kojima, Y.; Wakita, S.; Kato, N. *Tetrahedron Lett.* **1979**, 4577–4580.
10. (a) Prestwich, G. D.; Eng, W.-S.; Roe, R. M.; Hammock, B. D. *Arch. Biochem. Biophys.* **1984**, *228*, 639–645; (b) Vig, O. P.; Trehan, I. R.; Kad, G. L.; Ghose, J. *Indian J. Chem., Sect. B* **1983**, *22B*, 515–516.

11. Mancuso, A. J.; Huang, S.-L.; Swern, D. *J. Org. Chem.* **1978**, *43*, 2480–2482.
12. Ratcliffe, R. W. *Org. Synth., Coll. Vol. VI* **1988**, 373.
13. Corey, E. J.; Gilman, N. W.; Ganem, B. E. *J. Am. Chem. Soc.* **1968**, *90*, 5616–5617.
14. (a) Brown, H. C.; Zweifel, G. *J. Am. Chem. Soc.* **1961**, *83*, 1241–1246; (b) Zweifel, G.; Nagase, K.; Brown, H. C. *J. Am. Chem. Soc.* **1962**, *84*, 190–195.
15. Julia, M.; Julia, S.; Guégan, R. *Bull. Soc. Chem. Fr.* **1960**, 1072–1079.
16. Tamao, K.; Ishida, N.; Kumada, M. *J. Org. Chem.* **1983**, *48*, 2120–2122.
17. (a) Barnard, D.; Bateman, L. *J. Chem. Soc.* **1950**, 926–932; (b) King, F. E.; Grundon, M. F. *J. Chem. Soc.* **1950**, 3547–3552; (c) Hoye, T. R.; Kurth, M. J. *J. Org. Chem.* **1978**, *43*, 3693–3697.
18. Fujisawa, T.; Sato, T.; Kawashima, M.; Nakagawa, M. *Chem. Lett.* **1981**, 1307–1310.
19. Mandai, T.; Hara, K.; Nakajima, T.; Kawada, M.; Otera, J. *Tetrahedron Lett.* **1983**, *24*, 4993–4996.
20. Gosselin, P.; Maignan, C.; Rouessac, F. *Synthesis* **1984**, 876–881.
21. (a) Gosselin, P.; Rouessac, F. *Tetrahedron Lett.* **1982**, *23*, 5145–5146; (b) Gosselin, P.; Rouessac, F. *C. R. Seances Acad. Sci., Ser. 2* **1982**, *295*, 469–471.
22. Brady, S. F.; Ilton, M. A.; Johnson, W. S. *J. Am. Chem. Soc.* **1968**, *90*, 2882–2889.

α-HYDROXYLATION OF A KETONE USING *o*-IODOSYLBENZOIC ACID: α-HYDROXYACETOPHENONE VIA THE α-HYDROXY DIMETHYLACETAL

(Ethanone, 2-hydroxy-l-phenyl-)

Submitted by ROBERT M. MORIARTY, KWANG-CHUNG HOU, INDRA PRAKASH, and S. K. ARORA[1]

Checked by JANICE KLUNDER and K. BARRY SHARPLESS

1. Procedure

A 250-mL, two-necked, round-bottomed flask is equipped with a magnetic stirring bar, 100-mL pressure-equalized addition funnel to which is attached a drying tube, and a stopper. Anhydrous methanol (80 mL) (Note 1) is added to the flask, which is cooled to 5–10°C. Stirring is begun and 8.4 g (0.15 mol) of powdered potassium hydroxide is added. Acetophenone (6.0 g; 0.05 mol) (Note 2) dissolved in 20 mL of methanol is added dropwise over a period of 10 min. After the solution is stirred for 15 min, 14.52 g (0.055 mol) of *o*-iodosylbenzoic acid (Note 3) is added during 30 min. The ice bath is removed and the resultant yellow-colored slurry is stirred overnight at room temperature to give a clear red solution (Note 4). The mixture is concentrated under reduced pressure in a rotary evaporator until one-half of the methanol is removed, and then 30 mL of water is added followed by extraction with four 50-mL portions of dichloromethane. The combined dichloromethane extracts are washed with two 10-mL portions of water, and the combined organic extracts are dried over anhydrous magnesium sulfate for 1 hr. After filtration, the methylene chloride is removed under reduced pressure in a rotary evaporator, and the crude acetal is distilled to give a fraction at 73–76°C (0.4 mm) weighing 6.0 g (65%) (Note 5). The acetal is of high purity, as shown by spectral analysis (Note 6).

α-Hydroxyacetophenone. In a 500-mL, round-bottomed flask equipped with a magnetic stirrer are placed 6.0 g (0.33 mol) of α-hydroxy dimethylacetal and 100 mL of dichloromethane. Stirring is begun and the flask is cooled to about 10°C with ice water. Aqueous 5% sulfuric acid (100 mL) is added dropwise from a pressure-equalized addition funnel and the mixture is stirred for another 30 min. The dichloromethane layer is separated and the aqueous layer is extracted twice with 25-mL portions of dichloromethane. The combined extracts are washed with two 10-mL portions of water and dried over anhydrous magnesium sulfate, and the solvent is removed under reduced pressure using a rotary evaporator. The resulting yellow crystalline solid is recrystallized from carbon tetrachloride to give a white crystalline material, mp 86–87.5°C (lit.[2] mp 86–87°C), yield 3.7 g (83%) (Note 7).

2. Notes

1. Anhydrous methanol is obtained by treatment with magnesium methoxide, obtained by refluxing 50 mL of methanol, 5 g of magnesium turnings, and 0.5 g of sublimed iodine together until the iodine color disappears. The 1 L of methanol is added and the system is kept at reflux for 1 hr and distilled to yield purified methanol (bp 64.5°C).
2. Acetophenone was used as purchased from Fisher Scientific Company.
3. *o*-Iodosylbenzoic acid was used as purchased from Sigma Chemical Company.
4. TLC (ethyl acetate:hexane) shows residual starting material.
5. The α-hydroxy dimethylacetal obtained must be used immediately in the next step because at room temperature it undergoes a dimerization reaction by loss of two molecules of methanol.
6. The spectral properties of the product are as follows: IR (neat) cm^{-1}: 3470 (—OH): ^{1}H NMR (CDCl$_3$) δ: 1.83 (s, 1 H, OH), 3.23 (s, 6 H, (OCH$_3$)$_2$), 3.73 (s, 2 H, CH$_2$), 7.27–7.67 (m, 5 H, Ar H); ^{13}C NMR (CDCl$_3$) δ: 139.3 (s), 128.4 (d), 127.4, (d) 102.4 (s), 65.3 (t) 49.1 (q); mass spectrum: m/e 151 (M$^+$—OCH$_3$ 100%), 105 (29.7%), 91 (31.7%), 77 (7.0%).
7. The product has the following spectral properties: ^{1}H NMR (CDCl$_3$) δ: 3.63 (s, 1 H, OH), 4.86 (s, 2 H, CH$_2$), 7.25–7.90 (m, 5 H, Ar H); ^{13}C NMR (CDCl$_3$) δ: 198.6 (s), 134.4 (s), 129.1 (d), 127.8 (d), 65.6 (t).

3. Discussion

The procedure reported here provides a convenient method for the α-hydroxylation of ketones that form enolates under the reaction conditions. The reaction has been applied successfully to a series of para-substituted acetophenones, 1-phenyl-1-propanone, 3-pentanone, cyclopentanone, cyclohexanone, cycloheptanone, cyclododecanone, 2-methylcyclohexanone, 2-norbornanone, and benzalacetone.[3] In the case of a steroidal example it was shown that a carbon–carbon double bond and a secondary hydroxyl group are not oxidized.[4] A primary amino function, as in the case of *p*-aminoacetophenone, is not affected.[5] Similarly, a tertiary amino ketone such as tropinone undergoes the α-hydroxylation reaction.[5]

The present procedure using *o*-iodosylbenzoic acid is an improvement over our original method, which uses either iodosylbenzene or diacetoxyphenyliodine(III).[6,7,8] The advantage of the present method is the solubility of the product iodobenzoic acid under the basic reaction conditions. Thus the α-hydroxy dimethylacetal may be isolated by direct extraction. Using the original procedure, both carboxylic acids and esters underwent high yield α-hydroxylation.[8]

The pathway by which the reactions are considered to occur involves attack of the enolate anion at the I=O bond of *o*-iodosylbenzoic acid followed by reductive elimination of *o*-iodobenzoic acid upon addition of methoxide to the carbonyl group. Ring opening of the epoxide thus formed yields the hydroxy dimethylacetal:

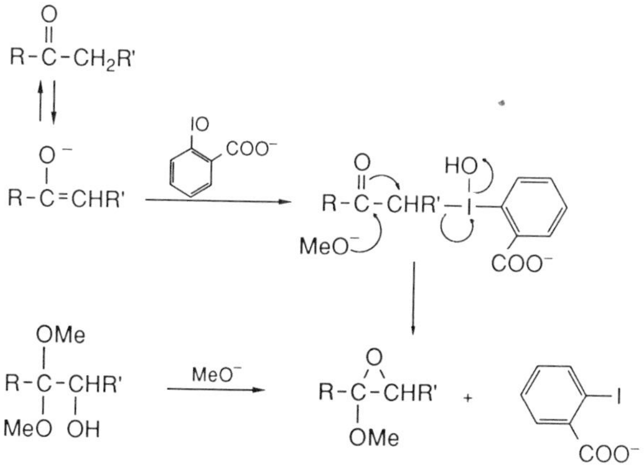

Other methods for α-hydroxy ketone synthesis are as follows: addition of $^3O_2$ to an enolate followed by reduction of the α-hydroperoxy ketone using triethyl phosphite;[9] the molybdenum peroxide–pyridine/HMPA oxidation of enolates;[10] photooxygenation of enol ethers followed by triphenylphosphine reduction;[11] the epoxidation of trimethylsilyl enol ethers by peracid;[12] the oxidation of trimethylsilyl enol ethers by osmium tetroxide in *N*-methylmorpholine *N*-oxide;[13] and, finally, the classical method of hydrolysis of an α-bromo ketone.[14]

1. Department of Chemistry, University of Illinois at Chicago, Chicago, IL 60680. This work was supported by the donors of the Petroleum Research Fund, administered by the ACS under grant PRF 14773-AC1.
2. Linnell, W. H.; Roushdi, I. M. *Quart. J. Pharm. Pharmacol.* **1941**, *14*, 270; *Chem. Abstr.* **1942**, *36*, 2545.

3. Moriarty, R. M.; Hou, K.-C. *Tetrahedron Lett.* **1984**, *25*, 691.
4. In this case iodosobenzene instead of *o*-iodosobenzoic acid was used (Moriarty, R. M.; John, L. S.; Du, P. C. *J. Chem. Soc., Chem. Commun.* **1981**, 641).
5. Moriarty, R. M.; Prakash, O.; Karalis, P.; Prakash, I. *Tetrahedron Lett.* **1984**, *25*, 4745.
6. Moriarty, R. M.; Hu, H.; Gupta, S. C. *Tetrahedron Lett.* **1981**, *22*, 1283.
7. Moriarty, R. M.; Gupta, S. C.; Hu, H.; Berenschot, D. R.; White, K. B. *J. Am. Chem. Soc.* **1981**, *103*, 686.
8. Moriarty, R. M.; Hu, H. *Tetrahedron Lett.* **1981**, *22*, 2747.
9. Gardner, J. N.; Carlon, F. E.; Gnoj, O. *J. Org. Chem.* **1968**, *33*, 3294.
10. Vedejs, E.; Engler, D. A.; Telschow, J. E. *J. Org. Chem.* **1978**, *43*, 188.
11. Friedrich, E.; Lutz, W. *Chem. Ber.* **1980**, *113*, 1245.
12. Rubottom, G. M.; Vazquez, M. A.; Pelegrina, D. R. *Tetrahedron Lett.* **1974**, 4319.
13. McCormick, J. P.; Tomasik, W.; Johnson, M. W. *Tetrahedron Lett.* **1981**, *22*, 607.
14. Catsoulacos, P.; Hassner, A. *J. Org. Chem.* **1967**, *32*, 3723.

SILYLATION OF 2-METHYL-2-PROPEN-1-OL DIANION: 2-(HYDROXYMETHYL)ALLYLTRIMETHYLSILANE

A.

B.

Submitted by Barry M. Trost, Dominic M. T. Chan, and Thomas N. Nanninga[1]
Checked by Paul R. Jenkins and Ian Fleming

1. Procedure

Caution! Part A should be carried out in an efficient hood, since the reagents are noxious.

A. *2-(Trimethylsiloxymethyl)allyltrimethylsilane.* An oven-dried (Note 1) 2-L, three-necked, round-bottomed flask is equipped with an airtight mechanical stirrer (Note 2), a 500-mL pressure-equalizing dropping funnel (Note 3), and a reflux condenser. The top of the condenser is connected to a three-way stopcock with one branch connected to a nitrogen source and the other to a variable pressure oil pump with a dry ice trap (Note 4). The apparatus is flamed dry under a steady stream of nitrogen. The flask is charged with 836 mL (1.07 mol) of a 1.28 M solution of butyllithium in hexane (Note 5). The bulk of the hexane is removed at reduced pressure with stirring until a thick oil is obtained (Note 6a). The system is carefully recharged with nitrogen. Alternatively, the use of 107 mL of 10 M butyllithium in hexanes is more convenient and gives similar yields (Note 6b). The butyllithium is then cooled in an ice bath and 500 mL of anhydrous ether is added (Note 7), followed by 160 mL of tetramethylethylenediamine (Note 8). The mixture is stirred for a few minutes and 34 mL, 29.14 g (0.404 mol) of 2-methyl-2-

propen-1-ol (Note 9) is added dropwise via a syringe over 22 min (Note 10). An immediate, vigorous reaction occurs and the lithium alkoxide precipitates as a white solid. Approximately 350 mL of tetrahydrofuran (Note 11) is added and the resultant slightly cloudy yellow solution is allowed to warm to room temperature over ca. 4 hr (Note 12). The reaction is stirred for 39 hr (Note 13), at which time the dianion separates as a dark-red gummy material from the deep orange solution. The mixture is cooled to ca. −30°C (Note 14) and 230 mL (1.81 mol) of chlorotrimethylsilane (Note 15) is added all at once over ca. 20 sec. The reaction turns milky white (Note 16). After 5 min, the dry ice bath is removed and the mixture is stirred for a further 15 min at room temperature. The reaction mixture is added in two portions with swirling to 1.5 L of ether in two 2-L conical flasks, after which 1 L of saturated aqueous sodium bicarbonate is added very carefully to destroy excess chlorotrimethylsilane (Note 17). The two layers are separated and the aqueous phase is extracted with a further 1.5 L of ether. The combined organic layer is then washed with 1 L of water, two 1-L portions of saturated aqueous copper sulfate solution, and 400 mL of water. The solution is dried over anhydrous potassium carbonate and the solvent is removed by atmospheric distillation (Note 18). Careful distillation of the residual oil through a 27-cm Vigreux column at reduced pressure gives a forerun of 4.25 g, bp 29–57°C (4 mm), and 45.8 g (52%) of 2-(trimethylsiloxy)allyltrimethylsilane as a colorless liquid, bp 57–59°C (4 mm) (Note 19).

B. *2-(Hydroxymethyl)allyltrimethylsilane*. A 500-mL round-bottomed flask equipped with a magnetic stirring bar is charged with 21.10 g (0.0975 mol) of 2-(trimethylsiloxymethyl)allyltrimethylsilane in 170 mL of tetrahydrofuran (Note 11) and 44 mL of ca. 1 N aqueous sulfuric acid (Note 20). The resultant two-phase mixture is then stirred vigorously for 1.5 hr at room temperature. Solid anhydrous potassium carbonate is added carefully until bubbling subsides. The layers are separated and the aqueous layer is extracted with 100 mL of ether. The combined organic layers are dried over anhydrous potassium carbonate and distilled at atmospheric pressure to remove the solvents (Note 18). The remaining liquid is distilled at reduced pressure to give a forerun, 0.4 g, bp 22–54°C (4 mm), and 10.95 g (78%) of 2-(hydroxymethyl)allyltrimethylsilane as a colorless liquid, bp 54–56°C (2 mm) (Note 21).

2. Notes

1. All glassware was dried in an oven at over 100°C overnight.
2. The use of a magnetic stirrer is not advisable since the formation of the gum-like dianion prevents efficient stirring. A mechanical stirrer with a ground-glass shaft bearing lubricated with mineral oil is recommended.
3. The funnel is capped with a rubber septum. For ease of operation, volume markings, corresponding to the amounts of reagents to be added, are put on the addition funnel.
4. The function of the trap is to condense the hexane from the butyllithium solution. The checkers used a 1-L three-necked flask fitted with a short delivery tube (a quick-fit air-bleed tube was used), stopper, and rubber tubing connection. The submitters used a water aspirator and a 1-L filter flask with a drying tower between.

5. Butyllithium in hexane was purchased by the checkers from Pfizer Chemicals Ltd., UK, and manufactured by the Lithium Corporation of America. It was titrated using the double titration method with dibromoethane and transferred to the addition funnel using a cannula. The submitters used a 1.58 M solution from the Foote Mineral Company; they found that the yield of product was reduced to ca. 42% when only 2 equiv of the lithium reagent was used.

6. (a) One should try to remove as much hexane as possible from the butyllithium solution (i.e. greater than 90%) because the purity of the product depends on the polarity of the reaction medium. A warm-water bath was used to facilitate solvent removal. The checkers used a variable pressure oil pump with the vacuum adjusted to ca. 10–20 mm. (b) Butyllithium as a 10 M solution in hexanes is supplied by Aldrich Chemical Company, Inc. The procedure for its use in place of solutions of lower concentration is identical to that in the text except for deletion of the solvent evaporation step. As 10 M butyllithium is quite viscous and pyrophoric it should be handled with large-bore needles.

7. Ether was distilled from sodium ketyl of benzophenone. The dissolution of butyllithium in ether was slightly exothermic.

8. Tetramethylethylenediamine was obtained from Aldrich Chemical Company and distilled from calcium hydride before use.

9. 2-Methyl-2-propen-1-ol, purchased from Aldrich Chemical Company, was distilled from anhydrous potassium carbonate. It was added directly to the butyllithium solution using a long needle. The checkers quickly replaced the pressure-equalizing dropping funnel with a serum cap to carry out this addition. The funnel was fitted to a small dry flask to prevent the introduction of moisture during the addition period and replaced on the reaction flask immediately afterward.

10. The reaction of the alcohol with butyllithium is quite vigorous with evolution of butane.

11. Tetrahydrofuran was distilled from sodium ketyl of benzophenone.

12. The checkers renewed the ice bath when additions were complete and allowed the flask to remain in the ice bath without addition of fresh ice.

13. Dianion formation appears to be essentially complete within 24 hr. However, a reaction time of 36 hr is recommended by the submitters to ensure complete reaction.

14. An extremely violent reaction is observed if the dianion is quenched above 0°C, with ether boiling off at an uncontrollable rate. The submitters observed that if the chlorotrimethylsilane addition is performed at a lower temperature, the reaction temperature will remain below that of the boiling point of ether. A dry-ice bath made up of 80 : 20 (v/v) ethanol : water was used; the checkers measured a bath temperature of −55°C and kept the reaction in the bath for 15 min before adding chlorotrimethylsilane.

15. Chlorotrimethylsilane was distilled from tributylamine before use. Both of these reagents were obtained from the Aldrich Chemical Company.

16. The submitters observed the appearance of a brown color at this point. The checkers obtained a brown color only after the reaction mixture was added to

ether. In a run at half scale the reaction mixture remained milky white for 35 min and turned brown only when ether (500 mL) was added to it.

17. The submitters observed more precipitate on dilution with ether and recommended that the aqueous workup be performed in a hood.

18. The submitters distilled most of the solvent using a bath temperature increasing up to 100°C. The checkers used a rotary evaporator with a hot water bath.

19. A variable pressure oil pump was used in this distillation. Approximately 10 g of a volatile component, consisting mostly of hexamethyldisiloxane, was obtained at room temperature (15 mm) before the forerun. The forerun contained the desired product and mineral oil from the butyllithium solution. The pot residue was about 5 g. The submitters find the disilyl compound thus obtained is contaminated with a trace amount of mineral oil and 4–6% of a vinylsilane, probably 2-methyl-1-trimethylsiloxy-3-trimethylsilyl-2-propene. This impurity becomes quite significant if the reaction medium is less polar than the one described (e.g., too much hexane from butyllithium is allowed to remain behind). The spectral properties of the desired product determined by the checkers are as follows: IR (neat) cm^{-1}: 2955, 1643, 1636, 1250, 1085, 885–830; ^{1}H NMR (chloroform-d, 90 MHz) δ: 0.03 [s, 9 H, $CH_2Si(CH_3)_3$], 0.14 [s, 9 H, $OSi(CH_3)_3$], 1.50 [broad s, 2 H, $CH_2-Si(CH_3)_3$], 3.93 [broad s, 2 H, $CH_2-OSi(CH_3)_3$], 4.62 (m, 1 H, vinyl H), 4.92 (m, 1 H, vinyl H). The checkers observed small NMR peaks assigned to mineral oil at δ 0.9 and 1.28 and peaks assigned to 2-methyl-1-trimethylsiloxy-3-trimethylsilyl-2-propene at δ 1.87 and 4.1. When the reaction was carried out at half scale the quantity of the latter impurity was not measurable from the NMR integral; however, a run at full scale gave about 10% of the impurity as estimated from the NMR integral. The product from the run at half scale had bp 56–57°C (2 mm); the submitters, bp 65°C (5.5 mm).

20. The acid solution was prepared by adding 13.5 mL of concentrated sulfuric acid to 500 mL of distilled water.

21. A variable-pressure pump is used for the distillation. The forerun consisted of mineral oil contaminant and product. The allylic alcohol is not very stable at room temperature but can be maintained indefinitely in the refrigerator at 0 to −6°C. The spectral properties of the alcohol were determined by the checkers as follows: IR (neat) cm^{-1}: 3600–3100, 2950, 1643, 1637, 1247, 1050, 885–830; ^{1}H NMR (chloroform-d, 90 MHz) δ: 0.02 [s, 9 H, $CH_2Si(CH_3)_3$], 1.51 [s, 2 H, $CH_2-Si(CH_3)_3$], 2.16 (broad s, 1 H, OH), 3.92 (broad s, 2 H, CH_2-OH), 4.62 (m, 1 H, vinyl H), 4.98 (m, 1 H, vinyl H).

The checkers observed small NMR peaks assigned to mineral oil at δ 0.82 and 1.51 and peaks assigned to 2-methyl-1-trimethylsiloxy-3-trimethylsilyl-2-propene at δ 0.10, 1.87, and 4.1. When the reaction was carried out at half scale the quantity of the latter impurity was reduced; the product from the run at half scale had bp 54–55°C (3 mm); submitters, bp 53–54°C (1.6 mm).

3. Discussion

Compound **1**, 2-(hydroxymethyl)allyltrimethylsilane, represents a conjunctive reagent that can be considered as the equivalent of zwitterion **2**, possessing a nucleophilic allyl anion synthon and an electrophilic allyl cation synthon in the same molecule. It has been

employed in a three-carbon condensative ring-expansion reaction,[2] a [3 + 2] annulation with cyclic enones,[3,4] and the total synthesis of coriolin.[5] Acetylation of the allylic alcohol gives 2-(acetoxymethyl)allyltrimethylsilane, which undergoes palladium(0) catalyzed annulation with electron-deficient olefins to produce methylenecyclopentanes via the trimethylenemethane–palladium complex.[4] This cycloaddition has served as a key step in synthetic approaches directed toward natural products such as brefeldin A[6] and albene,[7] and the polyquinanes.[8] Additions to heteroatom unsaturation can also be induced to occur to form oxygen and nitrogen heterocycles[9] as well as carbocycles.[10] It has also been employed to generate the TMM complex of transition metals other than palladium.[11]

The present procedure provides a convenient two-step route to 2-(hydroxymethyl)allyltrimethylsilane using relatively inexpensive reagents. Other approaches require additional steps and expensive chloromethyltrimethylsilane.[3,12]

Dianion formation from 2-methyl-2-propen-1-ol seems to be highly dependent on reaction conditions. Silylation of the dianion generated using a previously reported method[13] was unsuccessful in our hands. The procedure described here for the metalation of the allylic alcohol is a modification of the one reported for formation of the dianion of 3-methyl-3-buten-1-ol.[14] The critical variant appears to be the polarity of the reaction medium. In solvents such as ether and hexane, substantial amounts (15–50%) of the

vinylsilane **3** are observed. Very poor yields of the desired product were obtained in dimethoxyethane and hexamethylphosphoric triamide, presumably because of the decomposition of these solvents under these conditions. Empirically, the optimal solvent seems to be a mixture of ether and tetrahydrofuran in a ratio (v/v) varying from 1.4 to 2.2; in this case **3** becomes a very minor component.

A similar procedure has been employed to silylate the dianion of 3-methyl-3-buten-2-ol (67% yield).[15] In systems where such internal activation is not possible (e.g., 2-methyl-2-cyclohexen-1-ol), dianion formation can be performed in hexane to give a 75% yield of the corresponding disilyl compound.[16]

1. The Chemistry Department, University of Wisconsin, 1101 University Avenue, Madison, WI 53706. Present address of B.M.T.: Department of Chemistry, Stanford University, Stanford, CA 94305.
2. Trost, B. M.; Vincent, J. E. *J. Am. Chem. Soc.* **1980,** *102,* 5680. Also see Molander, G. A.; Andrews, S. W. *Tetrahedron Lett.* **1986,** *27,* 3115.
3. Knapp, S.; O'Connor, U.; Mobilio, D. *Tetrahedron Lett.* **1980,** *21,* 4557.
4. Trost, B. M.; Chan, D. M. T. *J. Am. Chem. Soc.* **1979,** *101,* 6429, 6432. For a review, see Trost, B. M. *Angew. Chem. Int. Ed. Engl.* **1986,** *25,* 1.

5. Trost, B. M.; Curran, D. P. *J. Am. Chem. Soc.* **1981,** *103*, 7380.

6. Chan, D. M. T. Ph.D. Thesis, University of Wisconsin, Madison, WI, 1982.

7. Trost, B. M.; Renaut, P. *J. Am. Chem. Soc.* **1982,** *104*, 6668.

8. Cossy, J.; Belotti, D.; Pete, J. R. *Tetrahedron Lett.* **1987,** *28*, 4547; Baker, R.; Keen, R. B. *J. Organomet. Chem.* **1985,** 285, 419.

9. Trost, B. M.; King, S. A. *Tetrahedron Lett.* **1986,** *27*, 5971; Trost, B. M.; King, S.; Nanninga, T. N. *Chem. Lett.* **1987,** 15; Jones, M. D.; Kemmitt, R. D. W. *J. Chem. Soc., Chem. Commun.* **1986,** 1201.

10. Heathcock, C. H.; Smith, K. M.; Blumenkopf, T. A. *J. Am. Chem. Soc.* **1986,** *108*, 5022, Molander, G. A.; Shubert, D. C. *J. Am. Chem. Soc.* **1987,** *109*, 576, 6877.

11. Jones, M. D.; Kemmitt, R. D. W.; Platt, A. W. G. *J. Chem. Soc., Dalton Trans.* **1986,** 1411.

12. Reduction of α-trimethylsilylmethylacrylic acid by lithium aluminum hydride prepared according to Hosomi, A.; Hashimoto, H.; Sakurai, H. *Tetrahedron Lett.* **1980,** *21*, 951. Trost, B. M.; Curran, D. P., unpublished results.

13. Carlson, R. M. *Tetrahedron Lett.* **1978,** 111.

14. Cardillo, G.; Contento, M.; Sandri, S. *Tetrahedron Lett.* **1974,** 2215.

15. Trost, B. M.; Nanninga, T. N., unpublished results. Also see Trost, B. M.; Chan, D. M. T. *J. Am. Chem. Soc.* **1981,** *103*, 5972.

16. Trost, B. M.; Hiemstra, H. *J. Am. Chem. Soc.* **1982,** *104*, 886.

2-HYDROXYMETHYL-2-CYCLOPENTENONE

Submitted by Amos B. Smith, III, Stephen J. Branca, Michael A. Guaciaro, Peter M. Wovkulich, and Abner Korn[1]
Checked by F. Pigott and G. Saucy

1. Procedure

A. *2-Bromo-2-cyclopentenone.* In a well-ventilated hood, a solution of 18.98 g (231.2 mmol) of 2-cyclopentenone (Note 1) in 150 mL of carbon tetrachloride is added to a 1-L, three-necked, round-bottomed flask fitted with a mechanical stirrer, thermometer, and an addition funnel. The solution is chilled to 0°C with an ice bath and a solution of 40.5 g (253.4 mmol, 13.0 mL) of bromine in 150 mL of carbon tetrachloride is added dropwise during 1 hr. Then a solution of 35.1 g (346.8 mmol, 48.3 mL) of triethylamine in 150 mL of carbon tetrachloride is added dropwise over 1 hr with vigorous stirring while the reaction is held at 0°C. Stirring is continued for an additional 2 hr at room temperature; the resulting dark suspension is filtered with suction and the filtercake washed with carbon tetrachloride. The filtrate and washings are combined and washed with two 100-mL portions of 2 N hydrochloric acid, one 100-mL portion of saturated sodium bicarbonate solution, one 100-mL portion of water, and one 100-mL portion of saturated sodium chloride solution. The resultant solution is dried over anhydrous magnesium sulfate, filtered, and the solvent removed under reduced pressure. Distillation of the resultant oil (69–78°C, 1.0 mm) afforded 23.7 g (147.2 mmol, 64%) (Note 2) of a white crystalline solid, mp 36–37°C, (lit.[2] mp 39–39.5°C) (Note 3).

B. *2-Bromo-2-cyclopentenone ethylene ketal.* A solution of 22.00 g (136.7 mmol) of freshly distilled 2-bromo-2-cyclopentenone, 21.80 g (351.2 mmol) of ethylene glycol, 1.5 L of benzene (Note 4), and 60 mg of *p*-toluenesulfonic acid monohydrate is heated at reflux for 64 hr (Note 5), with azeotropic removal of water, in a 3-L, round-bottomed flask, equipped with a Dean-Stark trap, condenser, and Drierite drying tube. The solution is cooled to room temperature, dried with potassium carbonate, and filtered by vacuum through 15 g of Celite. The filtercake is washed with 150 mL of benzene. Removal of the solvent under reduced pressure yields a mobile yellow oil. Distillation (65–67°C, 0.7 mm) affords 22.4 g (109.0 mmol, 80%) (Note 6) of the ketal (Note 7).

C. *2-Hydroxymethyl-2-cyclopentenone.* The apparatus, as illustrated in Figure 1 (Note 8), is flame-dried while dry nitrogen is passed through. Paraformaldehyde (13 g, 433 mmol) (Note 9) is then added to the 250-mL flask (the generator) and 19.4 g (94.6 mmol) of freshly distilled 2-bromo-2-cyclopentenone ethylene ketal in 300 mL of dry tetrahydrofuran containing 2 mg of 2,2'-bipyridyl (Note 10) is added to the 500-mL flask (the reaction flask). The reaction flask is chilled to −78°C with an acetone/dry ice

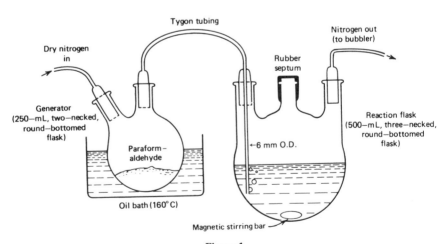

Figure 1

bath and 46.0 mL (105.8 mmol) of butyllithium (2.3 M in hexane) (Note 11) is added dropwise by syringe through the rubber septum during 1 hr. The resultant red solution is stirred for 1 hr and then warmed to $-30°C$ using a methanol–water $(2 : 3)$/dry ice bath. An oil bath previously heated to 160°C is applied to the generator and the monomeric formaldehyde thus generated is bubbled into the reaction mixture via a steady stream of dry nitrogen until the red color of the indicator is discharged (approximately 45 min). The reaction is quenched by the addition of 10 mL of saturated ammonium chloride solution and the resultant mixture is poured into 100 mL of a saturated sodium chloride solution. This mixture is extracted four times with 100-mL portions of methylene chloride; the methylene chloride extracts are combined and dried over magnesium sulfate. After filtration, evaporation of the solvent under reduced pressure affords 8.1–10.7 g (Note 12) of crude 2-hydroxymethyl-2-cyclopentenone ethylene ketal as a viscous liquid (Note 13).

Without purification, 8.1 g of the above crude ketal is added to a solution consisting of 1.0 g of oxalic acid, 5 mL of water, and 40 mL of methylene chloride. The resultant mixture is stirred for 5 hr at room temperature. At the end of this period the solution is filtered through 50 g of magnesium sulfate impregnated with 1.0 g of potassium carbonate (Note 14). Evaporation of the solvent from the filtrate affords a solid which, after purification by short-path distillation (70–80°C, 0.1 mm) (Note 15), gives 4.9 g (46%, based on bromoketal) (Note 16) of pure 2-hydroxymethyl-2-cyclopentenone, mp 68–69°C (off-white crystals) (Notes 17 and 18).

2. Notes

1. Cyclopentenone is commercially available from the Aldrich Chemical Company, Inc., or may be prepared according to the procedure of DePuy; see *Org. Synth., Coll. Vol. V* **1973**, 326.

2. The checkers obtained somewhat higher yields (i.e., 66 and 77%).

3. Pure 2-bromo-2-cyclopentenone obtained by recrystallization from diethyl ether-hexane[2] displayed the following spectroscopic properties: IR (CCl_4) cm^{-1}: 1720 (s), 1595 (m); ^{1}H NMR (CCl_4, 60 MHz) δ: 2.35–2.60 (m, 2H), 2.60–2.91 (m, 2 H), 7.40 (t, 1 H, $J = 2$).

4. *Caution: Benzene is a potential carcinogen!*

5. The reaction progress was monitored by TLC analysis (silica gel) using hexane-ethyl acetate $(4 : 1$, v/v) with 3.5% methanolic phosphomolybdic acid as indicator: bromoketal, R_f 0.37, bromoketone, R_f 0.15.

6. The bromoketal appears to be somewhat unstable and should be used as soon as possible after preparation. Some decomposition was observed during distillation.

7. Pure 2-bromo-2-cyclopentenone ethylene ketal displayed the following spectroscopic properties: IR (CCl_4) cm^{-1}: 2975 (s), 2950 (s), 2880 (s), 1615 (w); ^{1}H NMR (CCl_4, 60 MHz) δ: 1.95–2.55 (m, 4 H), 3.71–4.01 (m, 2 H), 4.01–4.33 (m, 2 H), 6.05 (t, 1 H, $J = 2$).

8. The checkers found that it is important to employ tubing of wide bore (6-mm o.d.) to conduct the gaseous formaldehyde from the generation flask into the reaction flask to avoid the possibility of the tube becoming plugged.

9. Prior to use paraformaldehyde was dried overnight in high vacuum (0.1 mm) over phosphorus pentoxide.

10. The reagent 2,2′-bipyridyl, available from the Aldrich Chemical Company, Inc., appears red in solutions containing organolithium and organomagnesium reagents[3] and is thereby an excellent indicator. Its use here allows addition of the precise amount of gaseous formaldehyde.

11. Butyllithium is available commercially from Alfa Products, Morton Thiokol, Inc.

12. The checkers found this crude product to contain 84.5% of the desired ketal, based on GC analysis.

13. Although 2-hydroxymethyl-2-cyclopentenone ethylene ketal could be purified by Kugelrohr distillation (88–100°C, 0.10 mm) this was not necessary for successful completion of the subsequent hydrolysis step. Pure 2-hydroxymethyl-2-cyclopentenone ethylene ketal possesses the following spectroscopic properties: IR (CCl_4) cm^{-1}: 3470–3500 (s), 1616 (w); ^{1}H NMR (CCl_4, 60 MHz) δ: 1.68–2.17 (m, 2 H), 2.17–2.58 (m, 2 H), 2.58–2.93 (br s, 1 H), 3.87 (s, 4 H), 3.98–4.16 (m, 2 H), 5.81–6.03 (m, 1 H).

14. The function of the potassium carbonate is to neutralize the oxalic acid as the solution passes through.

15. The short-path distillation of 2-hydroxymethyl-2-cyclopentenone is carried out without a water condenser. Furthermore, to prevent solidification of the distillate in the condenser, gentle warming of the condenser with a heat gun may be necessary.

16. The submitters had obtained a 70% yield for this two-step sequence, the crucial step being the reaction with formaldehyde.

17. Pure 2-hydroxymethyl-2-cyclopentenone displayed the following spectroscopic properties: IR ($CHCl_3$) cm^{-1}: 3400–3450 (s), 1680 (s), 1630 (m); ^{1}H NMR ($CDCl_3$, 60 MHz) δ: 2.27–2.84 (m, 4 H), 3.00 (br s, 1 H), 4.33 (d, 2 H, $J =$ 1), 7.60 (m, 1 H).

18. The overall yield from cyclopentenone to 2-hydroxymethyl-2-cyclopentenone over a number of runs was found to be in the range of 23–28%. The submitters had obtained 34.5%.

3. Discussion

The procedure reported here provides an efficient method for the construction of a wide variety of α,β-unsaturated ketones directly from the parent enone (i.e., $1 \rightarrow 2$), which does not require intervention of the thermodynamic dienolate. To our knowledge, a *general* solution for this recurring synthetic problem is unavailable, although Corey et al.,[4] Fuchs,[5] and Stork and Panaras[6] have independently developed a reverse polarity (umpolung) strategy for α-arylation of α,β-unsaturated ketones. Central to their approach was the generation of an effective latent equivalent for α-ketovinyl cation **3**. Such a strategy, however, is limited in that it depends critically upon the availability of the requisite alkyl or aryl organocuprate or magnesium reagent.

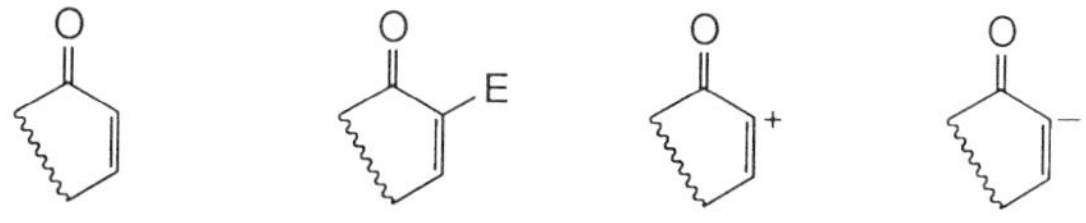

1	2	3	4
		α-Ketovinyl cation	α-Ketovinyl anion

A more versatile, as well as a more direct approach for the conversion of **1** to **2** employs the ethylene ketal of α-bromo-α,β-enones (e.g., **5**) as a latent equivalent of α-ketovinyl anion **4**.[7] Indeed, independent studies by Ficini and Depezay,[8] House and McDaniel,[9] and Manning et al.[10] as well as our own[11] suggested that such a general strategy would be viable. To illustrate this approach, we record here the preparation of the very useful synthon α-hydroxymethyl-2-cyclopentenone:

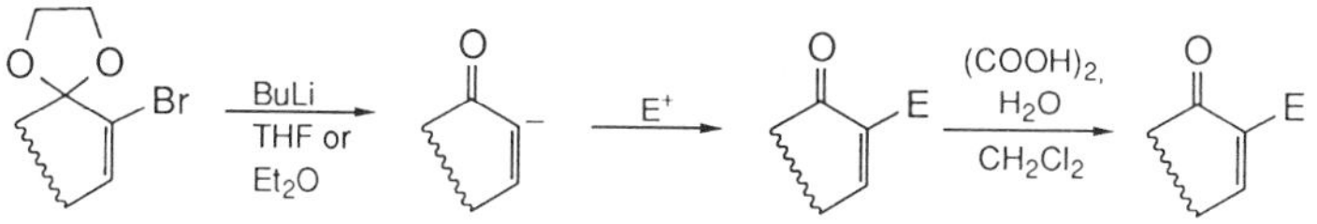

The overall efficiency of this sequence demonstrates, we believe, the considerable promise that α-bromoketals of α,β-enones hold as latent α-ketovinyl anion equivalents. In particular, we note the feasibility of introducing the very useful trimethylsilyl, tri-*n*-butyltin, and phenylselenenyl substituents.

1. Department of Chemistry, University of Pennsylvania, PA 19104.
2. DiPasquo, V. J.; Dunn, G. L.; Hoover, R. E. *J. Org. Chem.* **1968**, *33*, 1454.
3. Watson, S. C.; Eastham, J. F. *J. Organometal. Chem.* **1967**, *9*, 165; see also House, H. O.; Gall, M.; Olmstead, H. D. *J. Org. Chem.* **1971**, *36*, 2361.
4. Corey, E. J.; Melvin, Jr., L. S.; Haslanger, M. F. *Tetrahedron Lett.* **1975**, 3117.
5. Fuchs, P. L. *J. Org. Chem.* **1976**, *41*, 2935.
6. Stork, G.; Panaras, A. A. *J. Org. Chem.* **1976**, *41*, 2937.
7. A latent equivalent of a β-ketovinyl anion was recently reported by Okamura; see Hammond, M. L.; Mourino, A.; Okamura, W. H. *J. Am. Chem. Soc.* **1978**, *100*, 4907.
8. Ficini, J.; Depezay, J.-C. *Tetrahedron Lett.* **1969**, 4794.
9. House, H. O.; McDaniel, W. C. *J. Org. Chem.* **1977**, *42*, 2155.
10. Manning, M. J.; Raynolds, P. W.; Swenton, J. S. *J. Am. Chem. Soc.* **1976**, *98*, 5008; sell also Raynolds, P. W.; Manning, M. J.; Swenton, J. S. *J. Chem. Soc., Chem. Commun.* **1977**, 499.
11. Guaciaro, M. A.; Wovkulich, P. M.; Smith, III, A. B. *Tetrahedron Lett.* **1978**, 4661; see also Branca, S. J.; Smith, III, A. B. *J. Am. Chem. Soc.* **1978**, *100*, 7767.

ONE-STEP HOMOLOGATION OF ACETYLENES TO ALLENES: 4-HYDROXYNONA-1,2-DIENE

(1,2-Nonadien-4-ol)

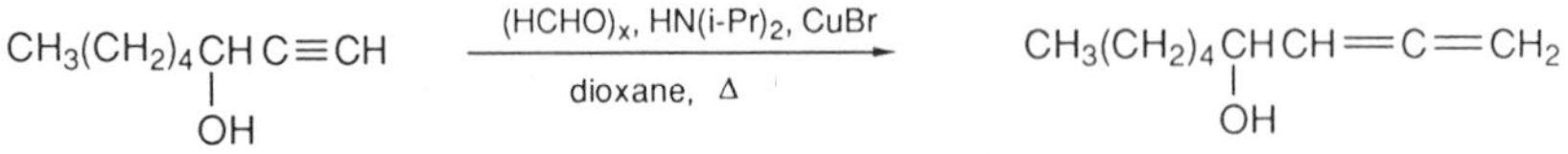

$$CH_3(CH_2)_4 \underset{\underset{OH}{|}}{CH} C\equiv CH \xrightarrow[\text{dioxane, }\Delta]{(HCHO)_x,\ HN(i\text{-}Pr)_2,\ CuBr} CH_3(CH_2)_4 \underset{\underset{OH}{|}}{CH} CH = C = CH_2$$

Submitted by Pierre Crabbé, Bahman Nassim, and Maria-Teresa Robert-Lopes[1]
Checked by Jeffrey S. Stults and Edwin Vedejs

1. Procedure

In a 500-mL, three-necked flask, equipped with a thermometer, stirrer, and a reflux condenser with drying tube, are placed 12.6 g (0.1 mol) of 1-octyn-3-ol, 154 mL of dioxane, 7.24 g (0.0504 mol) of cuprous bromide, 7.4 g of paraformaldehyde, and 18.54 g (0.183 mol) of diisopropylamine (Note 1). The resulting mixture is gently refluxed and stirred for 2 hr and then cooled to room temperature and filtered through a Celite plug. The dark-brown filtrate is concentrated under vacuum (Rotavapor) to a gummy residue and then diluted with 50 mL of water followed by 100 mL of ether and acidified with 6 N hydrochloric acid to pH 2. The ether–water layers are decanted from any residue, the ether layer is separated, and the aqueous solution is extracted with ether (5 × 50 mL). The ether extracts are combined and washed with small portions of water until pH 6.5 is reached. The organic layer is then washed with saturated sodium chloride solution and dried over anhydrous MgSO$_4$. After removal of ether by distillation through a 20-cm Vigreux column (water aspirator vacuum) while heating on a water bath, ≤40°C, the residual liquid is fractionated under reduced pressure through a 10-cm Vigreux column. The main fraction is collected at 41–42.5°C(0.15 mm) to give 8.65 g of pure allene (Note 2), with additional fractions of a less pure material.

2. Notes

1. Cuprous bromide and 1-octyn-3-ol were used as supplied by the Aldrich Chemical Company, Inc. Dioxane was dried over sodium–benzophenone and distilled, and diisopropylamine was distilled from barium oxide.

2. The spectral properties of 4-hydroxynona-1,2-diene are as follows: IR (neat) cm^{-1}: 3500 (OH), 1960 (C=C=C), 850 (=CH), 2900–2850 (CH). ^{1}H NMR (CDCl$_3$) δ: 0.65–1.7 (m); 4.15 (1 H, m); 4.8 (2 H, d of d, J = 2.6 Hz); 5.22, (1 H, q, J = 6 Hz).

3. Discussion

Although allenes were characterized long ago as a distinct class of organic substances, they have only recently received proper attention from chemists, in particular for their potential in organic synthesis.[2] A number of methods are known for the transformation of acetylenes into allenes,[3] but few are known to allow the homologation of an acetylenic group into a propadiene functionality.

A general procedure for the homologation of acetylenic compounds into allenes is described. The reaction conditions are mild and appear to be general, so that they can be applied to plain acetylenic substances as well as to acetylenic alcohols, ethers, and esters. This procedure is essentially a one-step reaction. As such, it is simpler and faster than the previously reported technique that involves the conversion of an acetylenic compound into the Mannich base, the formation of its quaternary ammonium salt and the reduction of this salt with lithium aluminum hydride.[4] Of great advantage over previously available methodology are the mild conditions, as well as the clean and fast procedure, which make this a method of choice for an efficient conversion of acetylenes to allenes.[5]

1. Department of Chemistry, University of Missouri, Columbia, MO 65211.
2. (a) Taylor, D. R. *Chem. Rev.* **1967,** *67,* 317 and references cited therein. (b) Griesbaum, K. *Angew. Chem. Int. Ed. Engl.* **1966,** *5,* 933; (c) Mavrov, M. V.; Kucherov, V. F. *Russ. Chem. Rev.* **1967,** *36,* 233.
3. See: Fischer, H. In "The Chemistry of Alkenes," Patai, S., Ed.; Wiley-Interscience: New York, 1964; Chapter 13, p. 1025.
4. Galantay, E.; Basco, I.; Coombs, R. V. *Synthesis* **1974,** 344.
5. (a) Crabbé, P.; André, D.; Fillion, H. *Tetrahedron Lett.* **1979,** 893; (b) Crabbé, P.; Fillion, H.; André, D.; Luche, J. L. *J. Chem. Soc., Chem. Commun.* **1979,** 859; (c) Searles, S.; Li, Y.; Nassim, B.; Lopes, M. T.; Tran, P. T.; Crabbé, P. *J. Chem. Soc., Perkin Trans. 1* **1984,** 747.

HYDROXYLATION OF ENOLATES WITH OXODIPEROXYMOLYBDENUM(PYRIDINE)(HEXAMETHYLPHOSPHORIC TRIAMIDE), $MoO_5 \cdot Py \cdot HMPA$(MoOPH): 3-HYDROXY-1,7,7-TRIMETHYLBICYCLO[2.2.1]HEPTAN-2-ONE

(Bicyclo[2.2.1]heptan-2-one, 3-hydroxy-1,7,7-trimethyl-)

A. MoO_3 $\xrightarrow[\text{2) (Me}_2\text{N)}_3\text{PO}]{\text{1) H}_2\text{O}_2}$ $MoO_5 \cdot H_2O \cdot (Me_2N)_3PO$

B. $MoO_5 \cdot H_2O \cdot (Me_2N)_3PO$ $\xrightarrow{\text{0.2 mm}}$ $MoO_5 \cdot (Me_2N)_3PO$ $\xrightarrow{\text{pyridine}}$

$MoO_5 \cdot Py \cdot (Me_2N)_3PO$

C. $\xrightarrow[\text{2) MoO}_5 \cdot \text{Py} \cdot \text{(Me}_2\text{N)}_3\text{PO}]{\text{1) LDA}}$

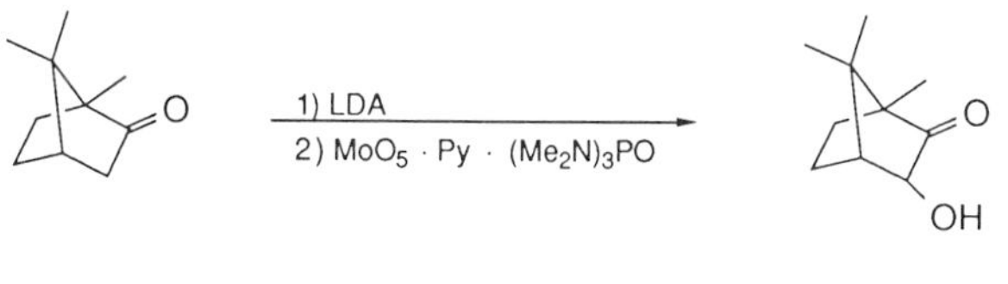

Submitted by EDWIN VEDEJS and S. LARSEN[1]
Checked by GORDON HILL and K. BARRY SHARPLESS

1. Procedure

Caution! Reactions using peroxides should be performed behind a safety shield to minimize explosion hazards (Note 1). Hexamethylphosphoric triamide (HMPA) and methanol are toxic and must be handled in a hood (Note 2).

A. *Oxodiperoxymolybdenum(aqua)(hexamethylphosphoric   triamide),   MoO$_5$· H$_2$O·HMPA.*[2] A 500-mL, three-necked flask is fitted with an internal thermometer and a mechanical paddle stirrer. The flask is charged, with stirring, with 30 g (0.2 mol) of molybdenum oxide (MoO$_3$) (Note 3) and 150 mL of 30% hydrogen peroxide (H$_2$O$_2$) (Note 4). An oil bath equilibrated at 40°C is placed under the rection mixture and heating is continued until the internal temperature reaches 35°C. The heating bath is removed and replaced by a water bath to control the mildly exothermic reaction so that an internal temperature of 35–40°C is maintained. After the initial exothermic period (approximately 30 min), the reaction flask is placed in the 40°C oil bath and stirred a total of 3.5 hr to form a yellow solution with a small amount of suspended white solid (Note 5).

After cooling to 20°C, the solution is filtered through a 1-cm mat of Celite pressed into a coarse-porosity sintered-glass filter. The yellow filtrate is cooled to 10°C (with an ice bath and magnetic stirring) and 37.3 g (0.21 mol) of hexamethylphosphoric triamide (HMPA) (Note 2) is added dropwise over 5 min, resulting in the formation of a yellow crystalline precipitate. Stirring is continued for a total of 15 min at 10°C, and the product is filtered using a Büchner funnel and pressed dry with a spatula. After 30 min in the funnel (aspirator vacuum), the filtercake is transferred to a 1-L Erlenmeyer flask. Methanol (20 mL) is added and the mixture is stirred in the 40°C bath. More methanol is slowly added until the crystals have dissolved. Cooling the saturated solution in the refrigerator gives yellow needles. The crystal mass is broken up with a spatula, the product is filtered and washed with 20–30 mL of cold methanol to give MoO$_5$·H$_2$O·HMPA, 46–50 g (59–64%) (Notes 5 and 6).

B. *Oxodiperoxymolybdenum(pyridine)(hexamethylphosphoric   triamide)   MoO$_5$· Py·HMPA=MoOPH.*[2] The recrystallized product from above is dried over phosphorus oxide (P$_2$O$_5$) in a vacuum desiccator, shielded from the light, for 24 hr at 0.2 mm to give a somewhat hygroscopic yellow solid, MoO$_5$·HMPA. A 36.0-g (0.101 mol) portion of MoO$_5$·HMPA is dissolved in 150 mL of dry tetrahydrofuran (THF) (Note 7) and the solution is filtered through a Celite mat, if needed, to remove a small amount of amorphous precipitate. The filtrate is then stirred magnetically in a 20°C water bath while 8.0 g (0.101 mol) of dry pyridine (Note 8) is added over 10 min. The crystalline, yellow product is collected on a Büchner funnel, washed with dry tetrahydrofuran (25 mL) and anhydrous ether (200 mL) and dried in a vacuum desiccator (1 hr, 0.2 mm) to yield 36–38 g (51–53% overall from MoO$_3$) of finely divided yellow crystalline MoO$_5$·Py·HMPA (Note 9).

The product is stored in a dark glass jar inside a second container partly filled with Drierite, and the container is kept in the refrigerator. Before opening the jar, the container is allowed to warm to room temperature to avoid condensation of moisture inside. Properly stored MoOPH is a freely flowing crystalline powder and can be used over a period of several months (Note 10).

C. *Hydroxylation of camphor: 1,7,7-trimethyl-3-hydroxybicyclo[2.2.1]-heptan-2-one.* A solution of lithium diisopropylamide (LDA) is prepared as follows: A 250-mL, three-necked flask and magnetic stirrer are flame-dried under a slow stream of nitrogen. After cooling, the flask is charged with 40 mL of approximately 1.5 M butyllithium in hexane (Note 11) under nitrogen flow using a syringe. The flask is cooled in a dry ice–acetone bath and 9.2 mL (66 mmol) of diisopropylamine (Note 8) is added by syringe, followed by 40 mL of dry THF (Note 7). The resulting LDA solution is allowed to reach room temperature under a slow flow of nitrogen. For titration, 0.312 g (2 mmol) of menthol is dissolved in 5 mL of dry THF with a few crystals of phenanthroline (Note 12) under nitrogen flow at $-70°C$. The LDA solution is added dropwise (using a nitrogen-purged syringe) to the stirred menthol solution until the yellow color of menthoxide–phenanthroline turns to the rust color of LDA–phenanthroline (2.67 mL of LDA solution is needed, 0.75 M).

An aliquot of 47.1 mL (35.3 mmol) of LDA solution is transferred by nitrogen-filled syringe into a nitrogen-swept, 500-mL, three-necked flask equipped with a magnetic stirrer and a device for addition of solid MoOPH. The latter is an L-shaped glass tube with male joints at each end. A round-bottomed flask containing 20.9 g (48.1 mmol) of MoOPH is wired to the L-tube, which is wired to the reaction vessel at such an angle that rotation of the L-tube causes addition of MoOPH to the enolate. The MoOPH container is temporarily suspended using clamps, and the entire apparatus is maintained under a slow flow of nitrogen. After the LDA solution is cooled in a dry ice–acetone bath, 4.88 g (32.1 mmol) of camphor (Note 13) in 200 mL of dry THF is added dropwise with stirring over 0.5 hr. Then 10 min later the reaction is placed in a dry ice–carbon tetrachloride bath and after 15 min the MoOPH is added over 1–2 min by rotating the L-tube and gently tapping it to dislodge the powder. The reaction immediately turns orange and eventually fades to a pale tan (Note 14). Stirring is continued at approximately $-23°C$ for 20 min, and the reaction is quenched by adding 100 mL of saturated aqueous sodium sulfite (Na_2SO_3). Vigorous stirring is maintained and the mixture is allowed to warm to room temperature. After 10 min at $20°C$, the mixture is shaken with 100 mL of saturated sodium chloride solution, and the aqueous layer is extracted twice with 70 mL of ether. The combined organic layers are washed once with a mixture of 50 mL of 10% aqueous hydrochloric acid and 50 mL of saturated sodium chloride solution. The hydrochloric acid–sodium chloride aqueous layer is back-extracted with 50 mL of ether, and the combined organic layers are dried over $MgSO_4$, filtered, and evaporated under an aspirator to yield a blue–green oil. Residual molybdenum salts are removed by filtration over 100 g of silica gel (Note 15) in a 2.5-cm column wet-packed and eluted with 1 : 1 ether–hexane. The product is eluted with approximately 750 mL of ether–hexane. Evaporation (aspirator) yields a white semisolid, which is crystallized from 15 mL of hexane at $-20°C$ and collected by washing with hexane cooled to $-70°C$. The mother liquors are crystallized in a similar manner from 2-4 mL of hexane to give a total of five crops, 4.14 g (77%) of colorless needles, mp 170–183°C, a 5 : 1 mixture of endo : exo isomers (Note 16). Recrystallization did not affect the isomer ratio (literature mp; endo, 192–195°C, exo, 210–211°C).[3]

2. Notes

1. There are no reports that $MoO_5 \cdot H_2O \cdot HMPA$, $MoO_5 \cdot HMPA$, or $MoO_5 \cdot Py \cdot HMPA$ are shock-sensitive. On heating on a hot plate, the crystalline solids ignite and burn, but do not detonate. These compounds can be stored in a refrigerator with precautions to exclude light. Prolonged storage at room

temperature in the light may cause decomposition with gas evolution and an exotherm sufficient to crack a glass container.

2. Hexamethylphosphoric triamide is toxic and a suspected carcinogen.

3. Molybdenum oxide was obtained from Mallinckrodt, Inc.

4. Hydrogen peroxide was obtained from Mallinckrodt, Inc.

5. Failure to maintain the internal temperature below 40°C results in formation of amorphous, insoluble products.

6. The purity of this material is decisive because the quality of subsequent products cannot be improved by recrystallization because of some decomposition.

7. Tetrahydrofuran was distilled from sodium–benzophenone and stored under nitrogen.

8. Pyridine was distilled from barium oxide.

9. The product melts with vigorous evolution of gas at 103–105°C.

10. Prolonged exposure to light, or failure to control exothermic reactions in prior steps, results in a sticky product that smells of pyridine. No method for purifying partly decomposed MoOPH has been found, and "sticky" product should not be used for enolate hydroxylation. Suspect material can be decomposed by stirring with aqueous sodium sulfite (Na₂SO₃) solution.

11. Butyllithium was obtained from the Foote Mineral Company.

12. Menthol and phenanthroline were obtained from the Aldrich Chemical Company, Inc.

13. Camphor was obtained from Eastman Organic Chemicals.

14. The colors are somewhat substrate-dependent. Some enolate hydroxylations acquire a green–blue color.

15. Silica gel, 60–200 mesh, was obtained from Davison Chemical Division.

16. The endo : exo ratio is determined by comparing the NMR CHOH signal areas of the endo (4.21 ppm, d, $J = 4.8$ Hz) and exo (3.75 ppm, br s) isomers.

3. Discussion

Enolate hydroxylation is a problem of long standing. Direct oxygenation succeeds with the fully substituted enolates of certain α,α-disubstituted ketones[4] and a variety of carboxylic acid derivatives (ester anions, acid dianions, amide anions),[5] but the reaction of enolates, $RCH=C(O^-)R'$ or $CH_2=C(O^-)R'$, with oxygen results in complex products of overoxidation. The stable molybdenum peroxide reagent $MoO_5 \cdot Py \cdot HMPA$ (MoOPH),[2] first prepared by Mimoun, allows the conversion of $RCH=C(OLi)R'$ into $RCH(OH)COR'$ in generally good yields (Table I).[6] In some cases, the α-diketone is formed as a by-product. The MoOPH reagent also hydroxylates branched or unbranched ester, amide, and nitrile anions.[6,7] For unknown reasons, MoOPH hydroxylations seldom give complete conversion of enolates into products, and recovery of 5–15% of the starting carbonyl substrate is to be expected.

Methyl ketone enolates are hydroxylated by MoOPH, but the products tend to undergo condensation with the starting enolate, resulting in poor yields.[6] Methyl ketone hydroxylation has been described by Moriarty, using $C_6H_5I=O/CH_3OH/OH^-$.[8]

TABLE I
CONVERSION OF RCH=C(OLi)R' INTO RCH(OH)COR'

Ketone	Oxidation Temperature (°C)	α-Hydroxy Ketone (%)	α-Diketone (%)
Valerophenone	−22	60	13
	−44	62	<2
Deoxybenzoin	−44	34	26
Isobutyrophenone	−22	65	
α-Tetralone	−22	48	
Camphor	−22	77	<2
	−22 → 60, 16 hr	44	11
4,4-Diphenylcyclo- hexanone	−22	46	
2-Phenylcyclohexanone	−44	70	<5
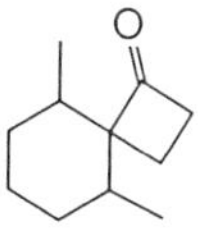 (structure: bicyclic ketone with methyl groups)	−22	81	
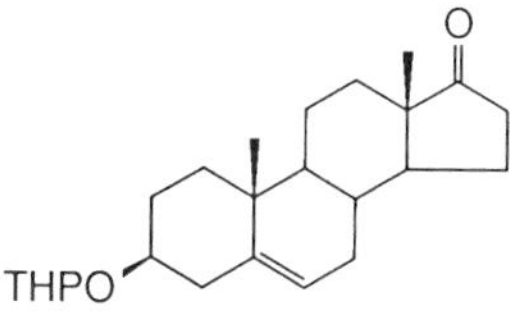 (steroid structure, THPO)	−44	75 (16 α-OH)	

Several indirect methods for conversion of enolates into α-hydroxycarbonyl compounds are known. The most versatile is the reaction of enol silanes with *meta*-chloroperbenzoic acid developed by Rubottom.[9] This technique is often successful with substrates that are oxidized inefficiently by the MoOPH technique. Another alternative is to use the oxaziridine reagents developed by Davis et al.[10]

The method described for MoOPH hydroxylation of the camphor enolate is representative for ketone enolate hydroxylations, but optimization in each individual case to determine the best temperature and concentration is recommended. Large-scale oxidations may benefit from addition of reagent in several portions over time, and enolates that are sensitive to self-condensation may give higher yields if enolate is added slowly to excess MoOPH.

1. Department of Chemistry, University of Wisconsin, Madison, WI 53706.
2. Mimoun, H.; Seree de Roche, I.; Sajus, L. *Bull. Soc. Chim. Fr.* **1969**, 1481.

3. Thoren, S. *Acta Chem. Scand.* **1970,** *24,* 93.
4. Bailey, E. J.; Barton, D. H. R.; Elks, J.; Templeton, J. F. *J. Chem. Soc.* **1962,** 1578; Gardner, J. N.; Carlon, F. E.; Gnoj, O. *J. Org. Chem.* **1968,** *33,* 3294; Gardner, J. N.; Popper, T. L.; Carlon, F. E.; Gnoj, O.; Herzog, H. L. *J. Org. Chem.* **1968,** *33,* 3695.
5. Konen, D. A.; Silbert, L. S.; Pfeffer, P. E. *J. Org. Chem.* **1975,** *40,* 3253; Moersch, G. W.; Zwiesler, M. L. *Synthesis* **1971,** 647; Adam, W.; Cueto, O.; Ehrig, V. *J. Org. Chem.* **1976,** *41,* 370; Wasserman, H. H.; Lipshutz, B. H. *Tetrahedron Lett.* **1975,** 1731; Cuvigny, T.; Hullot, P.; Larchevêque, M.; Normant, H. *C. R. Hebd. Seances Acad. Sci., Ser. C.* **1975,** *218,* 251; Corey, E. J.; Ensley, H. E. *J. Am. Chem. Soc.* **1975,** *97,* 6908.
6. Vedejs, E.; Engler, D. A.; Telschow, J. E. *J. Org. Chem.* **1978,** *43,* 188.
7. Vedejs, E.; Telschow, J. E. *J. Org. Chem.* **1976,** *41,* 740.
8. Moriarty, R. M.; Hu, H.; Gupta, S. C. *Tetrahedron Lett.* **1981,** *22,* 1283.
9. Rubottom, G. M.; Vazquez, M. A.; Pelegrina, D. R. *Tetrahedron Lett.* **1974,** 4319; Hassner, A.; Reuss, R. H.; Pinnick, H. W. *J. Org. Chem.* **1975,** *40,* 3427; Rubottom, G. M.; Marrero, R. *J. Org. Chem.* **1975,** *40,* 3783.
10. Davis, F. A.; Haque, M. S.; Ulatowski, T. G.; Towson, J. C. *J. Org. Chem.* **1986,** *51,* 2402; Davis, F. A.; Vishwakarma, L. C.; Billmers, J. M.; Finn, J. *J. Org. Chem.* **1984,** *49,* 3241.

α-HYDROXY KETONES FROM THE OXIDATION OF ENOL SILYL ETHERS WITH *m*-CHLOROPERBENZOIC ACID: 6-HYDROXY-3,5,5-TRIMETHYL-2-CYCLOHEXEN-1-ONE

(2-Cyclohexen-1-one, 6-hydroxy-3,5,5-trimethyl-)

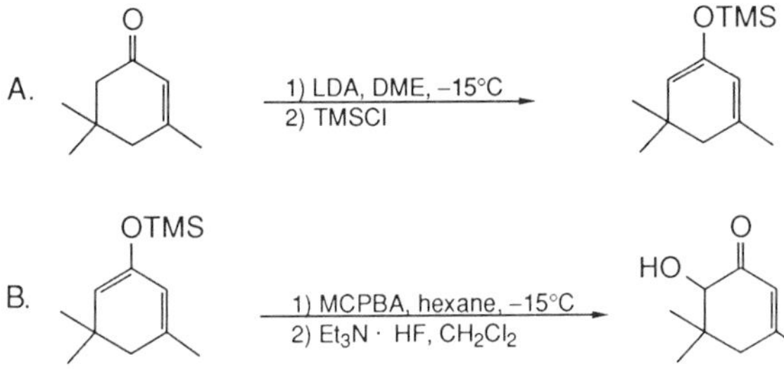

Submitted by GEORGE M. RUBOTTOM, JOHN M. GRUBER, HENRIK D. JUVE, JR., and DAN A. CHARLESON[1]
Checked by JUDY BOLTON and IAN FLEMING

1. Procedure

A. *4,6,6-Trimethyl-2-trimethylsilyloxycyclohexa-1,3-diene.* A 500-mL, three-necked, round-bottomed flask is fitted with a reflux condenser (center neck), Teflon-covered magnetic stirring bar, ground-glass stopper, and rubber septum. The apparatus is connected, through the reflux condenser, to a nitrogen source and a bubbler (Note 1). After the flask is flushed with nitrogen, it is charged with 150 mL of dry dimethoxy-ethane (DME) (Note 2) and 11.25 mL (80.4 mmol) of freshly distilled diisopropylamine (Note 3). The flask is immersed in a methanol–ice bath and cooled to an external temperature of −15°C. Over a period of about 5 min, butyllithium, 49.8 mL (79.6 mmol) (Note 4), is added, with continuous stirring, with a syringe through the septum. After an additional 15 min of stirring, 10.0 g (72.4 mmol) of freshly distilled isophorone (Note 5) is added neat over a 10-min period. The bright-yellow solution is stirred for an additional 10 min at −15°C. At this point, 17.5 mL (137.6 mmol) of freshly distilled chlorotrimethylsilane (TMSCl) (Note 6) is rapidly introduced through the septum. After the addition is complete (ca. 20 sec), the white slurry is stirred for an additional 2 hr at room temperature. The apparatus is then dismantled, the two outside necks of the flask are stoppered with ground-glass stoppers, and the center neck is attached to a rotary evaporator. Solvent is removed under reduced pressure and the residue is treated with 100 mL of pentane. The slurry is filtered through a sintered-glass filter and the filtrate is concentrated on a rotary evaporator. The residue is distilled at reduced pressure to give 13.5–13.9 g (88–91%) of pure 4,6,6-trimethyl-2-trimethylsiloxycyclohexa-1,3-diene, bp 54–57°C (1.5 mm), 37–39°C (0.01 mm) [lit.[2] bp 45–49°C (0.05 mm)] (Note 7).

B. *6-Hydroxy-3,5,5-trimethyl-2-cyclohexen-1-one.* A 500-mL, three-necked, round-bottomed flask is fitted with an adapter with a stopcock connected to a nitrogen source and a bubbler (center neck), two ground-glass stoppers, and a Teflon-covered magnetic stirring bar (Note 1). After the system is flushed with nitrogen, the flask is charged with 300 mL of dry hexane (Note 8) and 10.0 g (47.5 mmol) of 4,6,6-trimethyl-2-trimethylsiloxycyclohexa-1,3-diene. The flask is immersed in a methanol–ice bath and cooled to an external temperature of −15°C and then, with stirring, the solution is treated with a slurry which contains 10.6 g (52.3 mmol) of *m*-chloroperbenzoic acid (MCPBA) (Note 9) and 50 mL of dry hexane (Note 10). When the addition is complete (ca. 1.5 min), the resulting slurry is stirred at −15°C for 20 min and then at 30°C (water bath) for 2 hr. The mixture is filtered through a sintered-glass filter into a 500-mL, round-bottomed flask and the solvent is removed under reduced pressure using a rotary evaporator. If solid remains in the residue, 10–15 mL of pentane is added, filtration is repeated, and solvent is again removed under reduced pressure. The flask is fitted with a Teflon-covered stirring bar and the residue is treated with 150 mL of dry methylene chloride (Note 11) and 11.5 g (95.0 mmol) of triethylammonium fluoride (Et₃NHF) (Note 12). After the solution is stirred for 2 hr at room temperature, it is transferred to a separatory funnel and extracted with saturated aqueous sodium bicarbonate solution (2 × 100 mL), 100 mL of 1.5 N hydrochloric acid, and saturated aqueous sodium bicarbonate solution (2 × 50 mL). The organic layer is dried with anhydrous magnesium sulfate, and filtered, and solvent is removed from the filtrate using a rotary evaporator. The residue is then freed of the last traces of solvent by pumping, with stirring, at reduced pressure (ca. 2.0 mm) (Note 13); the residue solidifies. The round-bottomed flask is attached to a short-path distillation apparatus and the residue is distilled at reduced pressure. After a small forerun, the main fraction, bp 73–75°C (1.3 mm), is collected (Note 14). This fraction solidifies and is triturated with 3–5 mL of petroleum ether (bp 30–60°C) at −15°C (ice–methanol) to remove traces of isophorone. When the crystalline residue is dried in a stream of nitrogen, pure 6-hydroxy-3,5,5-trimethyl-2-cyclohexen-1-one is obtained: 4.8–5.1 g (66–70%), mp 44.5–45°C [lit.[3] mp 45–46°C]. The forerun and the material left in the still head after distillation are combined (Note 15) and treated with the petrolum ether that was used to triturate the main fraction. Crystallization gives an additional 0.2–0.3 g (3–4%) of the hydroxy ketone, mp 44.5–45°C. Thus the total weight of the 6-hydroxy-3,5,5-trimethyl-2-cyclohexen-1-one is 5.1–5.4 g (70–73%) (Note 16).

2. Notes

1. All glassware was dried in an oven for 2 hr at 110°C before use. All reactions were carried out under an atmosphere of nitrogen. The checkers used a balloon filled with nitrogen rather than a bubbler.

2. Dimethoxyethane (DME) (Aldrich Chemical Company, Inc.) was dried over lithium aluminum hydride and distilled just before use. The submitters have found that DME is the solvent of choice in this reaction and is preferred over the more commonly used tetrahydrofuran (THF).

3. Diisopropylamine, bp 80–80.5°C (699 mm), (Aldrich Chemical Company, Inc.) was distilled under a static atmosphere of nitrogen just prior to use.

4. Butyllithium (Aldrich Chemical Company, Inc.) was a 1.6 M solution in hexane. The submitters used the method of Ronald[4] to check titer. It is essential to the success of the reaction that this value be checked with accuracy.

5. Isophorone, bp 85–87°C (10 mm), (Aldrich Chemical Company, Inc.) was distilled immediately before use.

6. Chlorotrimethylsilane (TMSCl), bp 54–55°C (699 mm), (Aldrich Chemical Company, Inc.) was distilled under a static atmosphere of nitrogen just prior to use.

7. The product has the following spectroscopic properties: n_D^{25} 1.4509; IR (neat) cm^{-1}: 3040 (vinyl CH), 1660 (C=COTMS), 1610 (C=C); ^{1}H NMR (CDCl$_3$) δ: 0.21 [s, 9 H, Si(CH$_3$)$_3$], 0.98 (s, 6 H, 2 CH$_3$), 1.75 (broad s, 3 H, vinyl CH$_3$), 1.92 (broad s, 2 H, CH$_2$), 4.52 (broad s, 1 H, vinyl H on carbon 1), 5.40 (multiplet, 1 H, vinyl H on C-3); mass spectrum, m/z (relative abundance using 15 eV): 210 (M$^+$, 28), 196 (17), 195 (100), 179 (9); metastable (m*): 164.3 (195 → 179). Anal. calcd. for C$_{12}$H$_{22}$OSi: C, 68.50; H, 10.54. Found: C, 68.50; H, 10.52. A gas chromatographic analysis using a 0.25-in. × 6.0-ft column packed with 12.5% SE-52 at a column temperature of 130°C and gas flow rate of 90 mL/min showed the purity of the product to be greater than 95%. The impurities present were a small amount of unreacted isophorone and a trace of an unidentified material.

8. Hexane was purified in 1.5-L batches by sequential washing with concentrated sulfuric acid (5 × 50 mL) and water (3 × 100 mL), drying (CaCl$_2$), and distillation. The pure hexane is stored over Linde 4A molecular sieves.

9. m-Chloroperbenzoic acid (MCPBA) (Aldrich Chemical Company, Inc.) containing 15% m-chlorobenzoic acid was obtained commercially and used without purification.

10. It is convenient to stir the 85% MCPBA in hexane while the flask is being charged with the diene. Addition of the slurry with a pipette is the method used by the submitters. The checkers poured it in directly from a beaker, and washed the beaker with 10 mL of hexane.

11. Methylene chloride is dried by distillation from calcium chloride.

12. Triethylammonium fluoride is prepared by the method of Hünig.[5] The purity of this reagent seems to determine the amount of color that results in the crude hydroxy ketone. Stirring for a period of time greater than 2 hr results in lower yields and is to be avoided.

13. Stirring is crucial to prevent serious bumping when the crude hydroxy ketone solidifies. The checkers simply swirled the flask continuously without incident.

14. Taking a small forerun serves to concentrate residual isophorone in this fraction. Care must also be taken not to overcool the distillation head which may cause crystallization of the hydroxy ketone throughout the system.

15. A small amount of methylene chloride is used to wash the still head. This solvent is then removed (rotary evaporator) prior to the addition of the petroleum ether. Petroleum ether is used if recrystallization is needed.

16. The product has the following spectroscopic properties: IR (Nujol mull) cm^{-1}: 3360 (OH), 3040 (vinyl CH), 1670, 1635 (C=C=C=O); ^{1}H NMR (CDCl$_3$) δ: 0.79 (s, 3 H, CH$_3$ on C-5 trans to OH), 1.14 (s, 3 H, CH$_3$ on C-5 cis to OH), 1.88 (s, 3 H, vinyl CH$_3$), 2.13 (d, 1 H, J = 18, AB doublet for H on C-4), 2.35 (d, 1 H, J = 18, AB doublet for H on C-4), 3.52 (d, 1 H, J = 2, OH), 3.78 (d, 1 H, J = 2, H on C-6), 5.70 (broad s, 1 H, vinyl H); mass spectrum, m/z (relative abundance using 15 eV): 154 (M$^+$, 24), 125 (10), 111 (14), 83 (100), 82 (96), 72 (44); metastables (m*): 101.5 (154 → 125), 80.0 (154 → 111). Anal. calcd. for C$_9$H$_{14}$O$_2$: C, 70.10; H, 9.15. Found: C, 69.81; H, 9.50. A gas chromatographic analysis using a 0.25-in. × 6.0-ft column packed with 12.5% SE-52 at a column temperature of 158°C and a gas flow rate of 90 mL per minute showed the purity of the product to be greater than 98%. In some runs, a trace of isophorone could be detected (ca. 2%).

3. Discussion

The preparation of α-hydroxy carbonyl compounds has been accomplished by the oxidation of enolates using both oxygen[6] and MoO$_5$ · Py · HMPA · (MoOPH).[7] Acyl anion equivalents offer another route to this useful class of compounds.[8] The procedure presented here for the synthesis of 6-hydroxy-3,5,5-trimethyl-2-cyclohexen-1-one illustrates the use of MCPBA oxidation of an enol silyl ether as a method for obtaining an α-hydroxy enone. The procedure is a scale-up of a published synthesis.[9]

4,6,6-Trimethyl-2-trimethylsiloxycyclohexa-1,3-diene has been reported by Conia,[2] who used the standard "kinetic method" of House[10] for synthesis of the compound. The current method adapts this synthesis by employing DME as solvent and by using a nonaqueous workup which was previously noted by Ainsworth for the preparation of silyl ketene acetals.[11] These changes lead to higher yields of pure enol silyl ethers in general, and are recommended as a standard method.

6-Hydroxy-3,5,5-trimethyl-2-cyclohexen-1-one has been prepared in 22% yield by lead(IV) acetate oxidation of isophorone followed by hydrolysis of the resulting acetate.[3] The MCPBA method gives high yields of both α-hydroxy enones[12] and ketones[13] and is extremely general in scope. The method is also viable for the synthesis of α-hydroxy acids[14] and α-hydroxy esters.[15] The method fails with the enol silyl ethers of both lactones[15] and aldehydes.[16]

Since the double-bond placement in enol silyl ethers is predictable and controllable,[13] the method allows the regiospecific introduction of α-hydroxy groups. Omission of the fluoride treatment permits isolation of α-trimethylsiloxy carbonyl compounds,[17] while treatment of enol silyl ethers, first with MCPBA, then with triethylammonium fluoride-acetic anhydride gives the corresponding α-acetoxy carbonyl compounds.[9] The probable mechanism of the MCPBA oxidation of enol silyl ethers has also been discussed.[18]

1. Department of Chemistry, University of Idaho, Moscow, ID 83843.
2. Girard, C.; Conia, J. M. *J. Chem. Res. (M)* **1978**, 2351–2385; *J. Chem. Res. (S)* **1978**, 182–183.
3. Fort, A. W. *J. Org. Chem.* **1961**, *26*, 332–334.
4. Winkle, M. R.; Lansinger, J. M.; Ronald, R. C. *J. Chem. Soc., Chem. Commun.* **1980**, 87–88.
5. Hünig, S.; Wehner, G. *Synthesis* **1975**, 180–182.
6. Wasserman, H. H.; Lipshutz, B. H. *Tetrahedron Lett.* **1975**, 1731–1734.
7. Vedejs, E.; Engler, D. A.; Telschow, J. E. *J. Org. Chem.* **1978**, *43*, 188–196.
8. Hase, T. A.; Koskimies, J. K. *Aldrichimica Acta* **1981**, *14*, 73–77.
9. Rubottom, G. M.; Gruber, J. M. *J. Org. Chem.* **1978**, *43*, 1599–1602.
10. House, H. O.; Czuba, L. J.; Gall, M.; Olmstead, H. D. *J. Org. Chem.* **1969**, *34*, 2324–2336.
11. Ainsworth, C.; Chen, F.; Kuo, Y.-N. *J. Organomet. Chem.* **1972**, *46*, 59–71.
12. (a) Reference 9; (b) Anderson, R. C.; Gunn, D. M.; Murray-Rust, J.; Murray-Rust, P.; Roberts, J. S. *J. Chem. Soc., Chem. Commun.* **1977**, 27–28.
13. For numerous examples, see: (a) Rasmussen, J. K. *Synthesis* **1977**, 91–110; (b) Rubottom, G. M. *J. Organometal. Chem. Lib.* **1979**, *8*, 263–377; **1980**, *10*, 277–424; **1981**, *11*, 267–414; **1982**, *13*, 127–269.
14. Rubottom, G. M.; Marrero, R. *J. Org. Chem.* **1975**, *40*, 3783–3784.
15. Rubottom, G. M.; Marrero, R. *Synth. Commun.* **1981**, *11*, 505–511.
16. Hassner, A.; Reuss, R. H.; Pinnick, H. W. *J. Org. Chem.* **1975**, *40*, 3427–3429.
17. (a) Schlessinger, R. H.; Nugent, R. A. *J. Am. Chem. Soc.* **1982**, *104*, 1116–1118; (b) Danishefsky, S.; Kerwin, J. F., Jr.; Kobayashi, S. *J. Am. Chem. Soc.* **1982**, *104*, 358–360; (c) Musser, A. K.; Fuchs, P. L. *J. Org. Chem.* **1982**, *47*, 3121–3131.
18. Rubottom, G. M.; Gruber, J. M.; Boeckman, R. K., Jr.; Ramaiah, M.; Medwid, J. B. *Tetrahedron Lett.* **1978**, 4603–4606.

IMIDAZOLE-2-CARBOXALDEHYDE

(1*H*-Imidazole-2-carboxaldehyde)

A.

B.

C.

D.

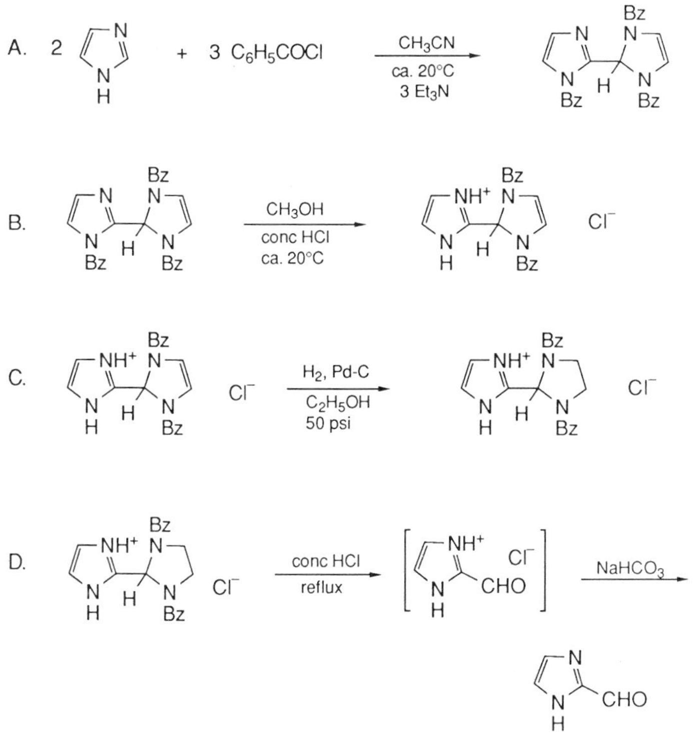

Submitted by LEONARD A. M. BASTIAANSEN, PIETER M. VAN LIER, and ERIK F. GODEFROI[1]
Checked by NANCY ACTON and ARNOLD BROSSI

1. Procedure

A. *1-Benzoyl-2-(1,3-dibenzoyl-4-imidazolin-2-yl)imidazole.*[2] A 12-L, wide-mouthed, round-bottomed vessel fitted with an efficient air-driven stirrer and thermometer is charged with 68 g (1.0 mol) of imidazole (Note 1), 202 g (2 mol) of triethylamine (Note 2), and 1000 mL of acetonitrile (Note 3). To the mixture is added dropwise over a 1-hr period and with external cooling 281 g (2.0 mol) of benzoyl chloride (Note 2); the temperature is maintained at 15–25°C. After addition is complete, stirring is continued for another hour at ambient temperature. With continued stirring 1 L of ether and 5 L of water are introduced, whereupon the temperature is brought to 5°C. The crystalline product is removed by filtration and is sucked dry with the aid of a rubber dam. Rinsing of the filtercake with, successively, water, acetone, and ether gives, on air-drying, 181–190 g (80–85%) of product, mp 197–198°C (Note 4). This product (5 g), taken up in 50 mL of boiling 90% dimethylformamide diluted while hot with water to the cloud point, gives, on cooling, 4.5 g of analytically pure stout prisms, mp 202–203°C (Note 5).

B. *2-(1,3-Dibenzoyl-4-imidazolin-2-yl)imidazole hydrochloride.*[3] Into a 3-L beaker equipped with an air-driven stirrer are successively introduced 150 g (0.335 mol) of dry, unrecrystallized 1-benzoyl-2-(1,3-dibenzoyl-4-imidazolin-2-yl) imidazole, 500 mL of technical-grade methyl alcohol, and 30 mL of concentrated hydrochloric acid. The mixture is stirred as the solids gradually dissolve. After 1 hr a clear yellow solution results, which is allowed to stand for another 5 hr, during which time white solid product begins to precipitate. Technical diethyl ether (1500 mL) is next added, and the mixture is allowed to stand overnight. Filtration and rinsing of the crystals with fresh ether and ultimate air drying furnish 114–119 g (89–93%) of product, mp 238–239°C (Note 6). Analytically pure material, obtained on recrystallizing a small sample from methyl alcohol-ether, has mp 240–241°C

C. *2-(1,3-Dibenzoylimidazolidin-2-yl)imidazole hydrochloride.*[3] A 1000-mL Parr hydrogenation bottle is charged with 38.0 g (0.10 mol) of dry, unrecrystallized 2-(1,3-dibenzoyl-4-imidazolin-2-yl)imidazole hydrochloride suspended in 300 mL of 95% reagent-grade ethyl alcohol. Then 2 g of 10% palladium on carbon (Note 7) is cautiously added (Note 8). The reaction vessel is now attached to the Parr hydrogenator and, after alternate evacuation and flushing with hydrogen gas, is shaken under a 50-psi atmosphere of hydrogen. Gas uptake ceases after absorption of 1 mol-equiv per mole of substrate; this requires ca. 2 hr. The catalyst is removed by vacuum filtration through Hyflow, the filtercake is rinsed with three portions of 95% ethyl alcohol (Note 9), and the filtrate is stripped to leave solid, impure product. This is triturated with 200 mL of ice-cold acetone; filtration and rinsing of the solids with fresh acetone and ultimately with ether yield 33.2–35.8 g (87–94%) of air-dried material, mp 225–226°C (Note 10). An analytically pure sample from isopropyl alcohol-ether melts at 225–226°C.

D. *Imidazole-2-carboxaldehyde.*[3] A solution of 19.1 g (0.05 mol) of dry, unrecrystallized 2-(1,3-dibenzoylimidazolidin-2-yl) imidazole hydrochloride in 200 mL of concentrated hydrochloric acid is refluxed for 22 hr (Note 11). The mixture is then chilled on ice, causing deposition of benzoic acid, which is removed by filtration (Note 11). The filtrate, on evaporation, leaves a residue that is first digested with 100 mL of 95% ethyl alcohol and then cooled on ice. The remaining solids are essentially pure ethylenediamine dihydrochloride and are filtered off (Note 12). Filtrate solvent is again removed under reduced pressure to leave solid residue. This is dissolved in 40 mL of water. Addition of solid sodium bicarbonate until foaming ceases causes imidazole-2-carboxaldehyde to crystallize. The mixture is chilled on ice, and the product is filtered and washed with ice-water to give, after thorough drying, 3.2–3.7 g (67–77%) of beige crystals, mp 206–207°C. Analytical material, prepared from water, has mp 206–207°C (Note 13).

2. Notes

1. Imidazole is a bulk chemical available from the Badische Anilin- & Sodafabrik AG, 6700 Ludwigshafen/Rhein, West-Germany. The checkers used Aldrich imidazole, 99%, from Aldrich Chemical Company, Inc.

2. Triethylamine and benzoyl chloride, both 99.5% pure, were purchased from Fluka. The checkers used material from Aldrich Chemical Company, Inc.

3. Acetonitrile, 99%, was obtained from Aldrich-Europe, B-2340 Beerse, Belgium.

4. Observing the theoretical stoichiometry, i.e., 2 equiv of imidazole and 3 equiv each of benzoyl chloride and triethylamine, resulted in significantly lower product yields.

5. The ^{1}H NMR spectrum (CDCl$_3$) corresponded to that in the literature[2]: δ 6.43 (s, 2, vinyl protons), 7.07 (t, 2, imidazole protons), 8.05 (s, 1, methine proton). Analysis calculated for C$_{27}$H$_{20}$N$_4$O$_3$: C, 72.31; H, 4.49; N, 12.49. Found: C, 72.27; H, 4.54; N, 12.53.

6. The checkers obtained variable melting points that were accompanied by decomposition and depended on the rate of heating. ^{1}H NMR (CD$_3$OD) δ: 6.57 (s, 2), 7.50 (m, 13, aromatic protons). Analysis calculated for C$_{20}$H$_{16}$N$_4$O$_2$ · HCl: C, 63.07; H, 4.56; N, 14.71. Found: C, 63.02; H, 4.60; N, 14.79.

7. Merck-Schuchardt "Hydrierkatalisator," purchased from E. Merck, Darmstadt. The checkers used Pd/C from Alfa Products, Morton Thiokol Inc.

8. Direct introduction of a dry hydrogenation catalyst into an alcoholic system has been known to bring about spontaneous ignition. This risk may be obviated by addition of a slurry of 2.0 g of catalyst in 15 mL of water to the substrate in 285 mL of absolute ethyl alcohol.

9. Filtercakes of fresh, spent hydrogenation catalysts are known to be pyrophoric and should not be sucked completely dry.

10. The checkers obtained an oily foam that remained oily on adding cold acetone. The oily material became solid on adding and evaporating benzene (2 × 100 mL). The checkers obtained variable melting points accompanied by decomposition. ^{1}H NMR (CD$_3$OD) included an *AA'BB'* system centered around δ 4.17 (4, CH$_2$CH$_2$). Analysis calculated for C$_{20}$H$_{18}$N$_4$O$_2$·HCl: C, 62.74; H, 5.00; N, 14.64. Found: C, 62.96; H, 4.98, N, 14.42.

11. Benzoic acid sublimes into the condenser and will plug a small-bore condenser.

12. The benzoic acid and ethylenediamine dihydrochloride isolated after the cited reaction time amount to ca. 90%.

13. Imidazole-2-carboxaldehyde has been reported to melt at 204°C,[4] 195–205°C,[5] 195°C,[6] 202°C,[8] and 190–196°C.[7] The material prepared has an ^{1}H NMR spectrum corresponding to that of the literature[8]: δ 7.43 (s, 2, imidazole protons), 9.67 (s, 1, CHO). Analysis calculated for C$_4$H$_4$N$_2$O: C, 49.99; H, 4.19; N, 29.16. Found: C, 50.05; H, 4.26; N, 28.96.

3. Discussion

Synthesis of imidazole-2-carboxaldehyde has previously been reported by manganese dioxide oxidation of the corresponding carbinol,[4] by acid-promoted cyclization of *N*-(2,2-diethoxyethyl)-2,2-diethoxyacetamidine,[5] and by methods centering around formylation of appropriately protected 2-imidazolelithium reagents.[6,7] The present method constitutes an optimization of the route recently reported from our laboratories.[3] Inexpensive, commercially available bulk chemicals are utilized to give imidazole-2-carboxaldehyde via high-yield processes mostly in open vessels and at ambient temperatures. All products are isolated directly from the reaction mixtures in a high state of purity without resorting to extractions, distillations, or recrystallizations. The by-products, benzoic acid and ethylenediamine dihydrochloride, are recovered in nearly quantitative yields. Waste and environmental pollution are kept to a minimum.

1. Department of Organic Chemistry, Eindhoven University of Technology, The Netherlands.
2. Regel, E. *Justus Liebigs Ann. Chem.* **1977,** 159–168.
3. Bastiaansen, L. A. M.; Godefroi, E. F. *J. Org. Chem.* **1978,** *43,* 1603–1604.
4. Schubert, H.; Rudorf, W. D. *Angew. Chem.* **1966,** *78,* 715.
5. English, J. P.; Berkelhammer, G. U.S. Patent 3 812 189, 1974; *Chem. Abstr.* **1974,** *81,* 49293j.
6. Iversen, P. E.; Lund, H. *Acta Chem. Scand.* **1966,** *20,* 2649–2657.
7. Kirk, K. L. *J. Org. Chem.* **1978,** *43,* 4381–4383.
8. Gebert, U.; von Kerékjártó, B. *Justus Liebigs Ann. Chem.* **1968,** *718,* 249–259.

Z-1-IODOHEXENE

(1-Hexene, 1-iodo-, (Z)-)

A. C_4H_9Li + $CuBr \cdot Me_2S$ ⟶ $LiCu(C_4H_9)_2$

B. $LiCu(C_4H_9)_2$ + $HC \equiv CH$ ⟶ $\left[\begin{array}{c} H \quad H \\ C_4H_9 \end{array} \right]_2 CuLi$

C. $\left[\begin{array}{c} H \quad H \\ C_4H_9 \end{array} \right]_2 CuLi$ + I_2 ⟶ $\begin{array}{c} H \quad H \\ C_4H_9 \quad I \end{array}$

Submitted by A. ALEXAKIS, G. CAHIEZ, and J. F. NORMANT[1]
Checked by J. GABRIEL, P. KNOCHEL, and DIETER SEEBACH

1. Procedure

A. *Preparation of an ether solution of lithium dibutylcuprate.* A 500-mL flask (Figure 1) with a side arm is equipped with a magnetic stirring bar, rubber septum, and three-way stopcock, on top of which is attached a rubber balloon, D. A Pt-100-thermometer, E, is inserted into the flask through the septum (Note 1). The air in the flask is replaced by dry nitrogen (Note 2). The flask is charged with 10.8 g (0.0525 mol) of cuprous bromide–dimethyl sulfide complex (Note 3) and 100 mL of ether, then immersed in a bath at $-50°C$; 0.10 mol of butyllithium, ca. 1.6 M solution in hexane (Note 4), is added dropwise, with stirring, via a syringe inserted through the rubber septum, at such a rate that the temperature of the reaction mixture does not exceed $-20°C$.

After the addition is complete, stirring is continued at $-30°C$ for 10 min to produce a gray–blue or dark-blue solution of the cuprate (Note 5).

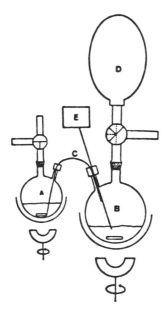

Figure 1

B. *Preparation of a solution of lithium di(Z-hexenyl)cuprate.* A needle connected to an acetylene supply (Note 6) is introduced through the rubber septum of the flask, with its end at least 1 cm below the surface of the cuprate solution. The stopcock is fully opened toward the balloon, the solution is cooled to −50°C, and 2.64 L (0.11 mol) of acetylene (Note 6) is bubbled into the stirred cuprate solution, the temperature of which should not rise above −25°C. The gas inlet is removed and the greenish solution is stirred at −25°C for 30 min.

C. *Preparation of Z-iodohexene.* A dry, 100-mL flask with a sidearm is charged with 26.7 g (0.105 mol) of iodine, equipped with a stirring bar, three-way stopcock, and rubber septum, and flushed with argon as described above (Section A). The iodine is dissolved by introducing, with stirring, 30 mL of tetrahydrofuran through the septum with a syringe. Flask A, which contains the iodine solution, is connected to flask B, which contains the vinyl cuprate solution as shown in Figure 1. The cuprate solution is kept between −60° and −50°C while the iodine solution is pushed through the Teflon tubing, C. Then the cooling bath is removed and the temperature is allowed to rise to −10°C, whereupon a precipitate of copper(I) iodide is formed and the mixture turns yellow. After 10 min at −10°C, a mixture of 100 mL of saturated aqueous ammonium chloride and 10 mL of saturated sodium bisulfite is added with vigorous stirring. The mixture is filtered by suction through 10 g of Celite on a sintered-glass funnel (No. 3), the contents of the funnel are washed twice with 50 mL of ether, and the filtrate is separated into two layers (Note 7). The inorganic layer is washed twice with 50 mL of pentane, and the combined organic layers are washed with aqueous sodium bisulfite

(Note 8) and saturated ammonium chloride solution and dried over anhydrous $MgSO_4$. The solvents are removed by distillation through a 20-cm Vigreux column at atmospheric pressure. A spatula of copper powder is added to the residue, and the stirred mixture is distilled under reduced pressure through a 10-cm Vigreux colum to give 13.5–15.5 g (65–75%) of Z-1-iodohexene, bp 47°C/(15 mm) (Notes 9 and 10).

2. Notes

1. The technique used here has been described previously by the checkers.[2] Instead, the submitters used a dry 500-mL, three-necked flask equipped with a variable-speed mechanical stirrer, a 100-mL pressure-equalizing dropping funnel topped by a gas inlet and a Claisen head containing a low-temperature thermometer (−70°C to +35°C), and a bubbler. A stream of nitrogen followed from the gas inlet.

2. This manipulation is described in detail in *Org. Synth., Coll. Vol. VI* **1988,** 869.

3. This complex[3] should be used when the organolithium is in solution in a hydrocarbon solvent. For organolithium reagents prepared in ether (see Note 4), the same complex may be used or, more conveniently, copper iodide (CuI) can be used. The CuI purchased from Prolabo or Merck & Company, Inc. may be used directly. Other commercial sources of CuI (Fluka, Aldrich Chemical Company, Inc., Alfa Products, Morton Thiokol, Inc.) furnish a salt that affords better results when purified. First 1 mol of CuI is stirred for 12 hr with 500 mL of anhydrous tetrahydrofuran, then filtered on a sintered-glass funnel (No. 3), washed twice with 50 mL of anhydrous tetrahydrofuran, once with 50 mL of anhydrous ether, and finally dried under reduced pressure (0.1 mm) for 4 hr.

4. Butyllithium was used as purchased from Aldrich Chemical Company, Inc., Fluka, or Metallgesellschaft (Frankfurt). Ethereal solutions of butyllithium may also be used. Other organolithium compounds are easily prepared in ether; the following is representative.

   Under an atmosphere of argon, a solution of butyl bromide (137 g, 1 mol) in anhydrous ether (500 mL) is added with stirring to small chips of lithium containing 1–2% of sodium (15.5 g, 2.2 g-atom) in ether (150 mL). The reaction starts after the addition of about 40 mL of butyl bromide solution at room temperature. The temperature rises and the lithium metal becomes bright. If the reaction does not start, the addition of a small amount of 1,2-dibromoethane (1 mL) is often effective. Then the reaction mixture is cooled (−5°C to −10°C) and addition of the butyl bromide solution is continued slowly (about 4 hr). At the end of the addition, the solution is stirred for 2 hr at −5 to −10°C; then the reaction mixture is allowed to warm to room temperature. After 2 hr, excess lithium metal is removed. For many purposes, the use of a clear solution, obtained after the reaction mixture has stood overnight at 0 to −5°C, is preferable. Butyllithium in ether can be stored under an argon atmosphere without decomposition for 15 days at 0°C or for 2 months at −15°C.

5. During all of the operations, the rate of stirring is adjusted to avoid splashing the wall of the flask; above −10°C, thermal decomposition of the cuprate occurs. This is indicated by the presence of a black suspension, which is also formed if

a copper(I) salt of insufficient purity is used, or when oxygen gets into the reaction flask.

6. The proper volume of acetylene is measured with a water gasometer as described in *Org. Synth., Coll. Vol. I* **1941**, 230, with two modifications: (a) Two traps immersed in an acetone–dry ice bath at $-65\,°C$ are placed between the acetylene tank and the gasometer in order to remove acetone; (b) the washing bottles between the gasometer and the reaction flask are replaced by a drying tube (2-cm $\times$ 30-cm column packed with anhydrous $CaCl_2$). The apparatus must be flushed with acetylene in order to remove all traces of oxygen. Acetylene dissolved in acetone is most appropriate. Acetylene obtained from tanks that contain solvents such as dimethylformamide (or other solvents) gave lower yields of carbocupration.

7. If a precipitate appears in the filtrate, filtration is repeated until two layers can be clearly distinguished.

8. A mixture of 10 mL of saturated sodium bisulfite and 50 mL of water is used. One or more washings with sodium bisulfite solution are necessary if iodine is present.

9. The sample thus obtained is $>99\%$ pure by GC analysis (3% OV 101 in a 2-m $\times$ 4-mm glass column, on Chromosorb G, with an injection temperature of $175\,°C$, raised $100\,°C$ in 5 min, then $5\,°C/min$).

10. The ^1H NMR spectrum of Z-1-iodohexene (in CCl_4) is as follows: δ: 0.94 (m, 3 H), 1.42 (m, 4 H, $-CH_2-$), 2.12 (m, 2 H, $-CH_2-C=$), 6.12 (m, 2 H).

3. Discussion

1-Iodoalkenes of the Z configuration are usually prepared by hydroboration of 1-iodoalkynes. The present method affords a product of higher configurational purity and constitutes an easier way to obtain such compounds in high yield, starting from less expensive reagents. In addition, the reaction can be performed easily on a larger scale (the submitters have prepared up to 1.8 mol of dialkenyl cuprate). The Z-1-iodo-1-alkenes shown in Table I have been prepared by the submitters.

TABLE I

EXAMPLES OF ALKENYL IODIDE PREPARATION FROM CARBOCUPRATION

Entry	Organolithium	Product[a]	Yield (%)
1	EtLi	EtCH=CHI	72
2	n-C_5H_{11}Li	$(n$-$C_5H_{11})$CH=CHI	89
3	n-C_7H_{15}Li	$(n$-$C_7H_{15})$CH=CHI	90
4	EtCH=CHCH$_2$CH$_2$Li	EtCH=CHCH$_2$CH$_2$CH=CHI	79
5	RO(CH$_2$)$_3$Li[b]	HO(CH$_2$)$_3$CH=CHI[c]	58
6	RO(CH$_2$)$_8$Li[b]	HO(CH$_2$)$_8$CH=CHI[c]	70

[a]All alkenes, reactants and products, are Z.
[b]R = CHMeOEt.
[c]After acid hydrolysis.

This reaction illustrates a stereoselective preparation of (Z)-vinylic cuprates,[4,5] which are very useful synthetic intermediates. They react with a variety of electrophiles such as carbon dioxide,[5,6] epoxides,[5,6] aldehydes,[6] allylic halides,[7] alkyl halides,[7] and acetylenic halides;[7] they undergo conjugate addition to α,β-unsaturated esters,[5,6] ketones,[6] aldehydes,[6] and sulfones.[8] Finally, they add smoothly to activated triple bonds[6] such as $HC{\equiv}C-OEt$, $HC{\equiv}C-SEt$, and $HC{\equiv}C-CH(OEt)_2$. In most cases these cuprates transfer both alkenyl groups. The uses and applications of the carbocupration reaction have been reviewed recently.[9] The configurational purity in the final product is at least 99.9% Z in the preceding transformations.

1. Laboratoire de Chimie des Organoéléments, Université Pierre et Marie Curie, 4, place Jussieu, F-75230 Paris Cédex 05, France.
2. Seebach, D.; Weller, T.; Protschuk, G.; Beck, A. K.; Hoekstra, M. S. *Helv. Chim. Acta* **1981**, *64*, 716–735.
3. House, H. O.; Chu, C.-Y.; Wilkins, J. M.; Umen, M. J. *J. Org. Chem.* **1975**, *40*, 1460–1469.
4. Alexakis, A.; Cahiez, G.; Normant, J. F. *J. Organomet. Chem.* **1979**, *177*, 293–298.
5. Alexakis, A.; Normant, J.; Villiéras, J. *Tetrahedron Lett.* **1976**, 3461–3462.
6. Alexakis, A.; Cahiez, G.; Normant, J. F. *Tetrahedron* **1980**, *36*, 1961–1969.
7. Alexakis, A.; Cahiez, G.; Normant, J. F. *Synthesis* **1979**, 826–830.
8. De Chirico, G.; Fiandanese, V.; Marchese, G.; Naso, F.; Sciacovelli, O. *J. Chem. Soc., Chem. Commun.* **1981**, 523–524.
9. Normant, J. F.; Alexakis, A. *Synthesis* **1981**, 841–870.

CONVERSION OF EPOXIDES TO β-HYDROXY ISOCYANIDES:
trans-2-ISOCYANOCYCLOHEXANOL

(Cyclohexanol, 2-isocyanato-, trans-)

Submitted by PAUL G. GASSMAN and THOMAS L. GUGGENHEIM[1]
Checked by CURTIS E. ADAMS and K. BARRY SHARPLESS

1. Procedure

Caution! Trimethylsilyl cyanide is very toxic. All reactions in this sequence should be carried out in a hood.

A. *[(trans-2-Isocyanocyclohexyl)oxy]trimethylsilane.* A 100-mL, three-necked flask equipped with a reflux condenser, constant-pressure dropping funnel, magnetic stirring bar, and drying tube is charged with 20.2 g (204 mmol) of trimethylsilyl cyanide (Note 1), 60 mg (0.19 mmol) of anhydrous zinc iodide (Note 2), and 5 mL of dry methylene chloride (Note 3). The constant-pressure dropping funnel is charged with 10.0 g (102 mmol) of cyclohexene oxide (Note 4) and 5 mL of dry methylene chloride. The reaction mixture is heated to reflux and the cyclohexene oxide–methylene chloride solution is added dropwise to the refluxing reaction mixture over a 30-min period. After the addition is complete, the reaction mixture is refluxed for 4 hr and then allowed to cool to room temperature. The reaction mixture is transferred to a one-necked flask and the solvent and the excess trimethylsilyl cyanide are removed under reduced pressure on a rotary evaporator (Note 5). The residue is vacuum-distilled through a 3-in. Vigreux distillation column to yield 15.74 g (78%) of [(*trans*-2-isocyanocyclohexyl)oxy]trimethylsilane, bp 69–70°C (1.5 mm) (Note 6).

B. *trans*-2-*Isocyanocyclohexanol.* A 250-mL, one-necked, round-bottomed flask is charged with 13.72 g (70 mmol) of [(*trans*-2-isocyanocyclohexyl)oxy]trimethylsilane, 12.12 g (210 mmol) of potassium fluoride (Note 7), and 100 mL of methanol. The reaction mixture is stirred magnetically for 5 hr at room temperature (23°C). The methanol is removed under reduced pressure on a rotary evaporator to yield a white slurry. This slurry is added to the top of a 250-g, 60–200-mesh silica gel chromatography column and the column is eluted with 20% ethyl acetate–80% hexane solvent mixture (Note 8). The solvent is removed from those fractions containing the product under reduced pressure on a rotary evaporator to afford an oil that is redissolved in methylene chloride, and the solution is filtered. The methylene chloride is removed from the filtrate under reduced pressure on a rotary evaporator to yield 8.46 g (68 mmol, 97%) of white, crystalline *trans*-2-isocyanocyclohexanol, mp 57.0–59.5°C (Note 9).

2. Notes

1. Trimethylsilyl cyanide was prepared shortly before use according to the procedure of Livinghouse, T. *Org. Synth., Coll. Vol. VII,* **1990,** 517. The checkers used trimethylsilyl cyanide as supplied from Aldrich Chemical Company, Inc.

2. Anhydrous zinc iodide was purchased from Alfa Products, Morton Thiokol, Inc., and used without further purification. In one run the checkers used 0.25 mmol of ZnI_2 and obtained a better yield than when they used 0.19 mmol of ZnI_2 (84% yield instead of 73%).

3. Commercial methylene chloride is dried by distillation from calcium hydride prior to use.

4. Cyclohexene oxide was purchased from Aldrich Chemical Company, Inc., and was used without purification.

5. The checkers also carried out this process in a fume hood. All glassware was rinsed afterward with 10% KOH solution or rinsed with acetone and the rinses mixed with 10% KOH. The resulting KOH solutions were treated with Chlorox overnight before being discarded.

6. This pure, colorless liquid showed the following physical properties: IR (neat) cm^{-1}: 2950, 2870, 2145, 1454, 1267, 1255, 1144, 1114, 1065, 1028, 931, 894, 884, 844, and 758; ^{1}H NMR (60 MHz, $CDCl_3$/TMS) δ: 0.17 (s, 9 H), 0.95–

2.30 (br m, 8 H), 3.00–3.73 (br m, 2 H); ^{1}H NMR (250 MHz, CDCl$_3$) δ: 0.15 (s, 9 H), 1.25 (m, 3 H), 1.56 (m, 1 H), 1.67 (m, 2 H), 1.86 (m, 1 H), 2.13 (m, 1 H), 3.28 (m, 1 H), 3.56 (m, 1 H); density 0.882 g/mL.

7. Potassium fluoride was purchased from the Fisher Scientific Company.

8. Approximately 100-mL fractions are collected. The progress of the chromatography is followed by analysis of the eluting fractions with thin-layer chromatography developed with iodine vapor. The checkers achieved equal success using 120 g of 70–230-mesh silica in a 30-mm × 250-mm column.

9. The product showed the following physical properties: IR (KBr) cm^{-1}: 3470, 3400, 2965, 2945, 2870, 2175, 1450, 1376, 1328, 1302, 1240, 1123, 1090, 1081, 1007, 919, 856, and 851; ^{1}H NMR (60 MHz, CDCl$_3$/TMS) δ: 0.70–2.40 (br m, 8 H), 2.85 (d, 1 H, J = 5), 3.00–3.90 (br m, 2 H); ^{1}H NMR (250 MHz, CDCl$_3$) δ: 1.27 (m, 3 H), 1.56 (m, 1 H), 1.71 (m, 2 H), 2.02 (m, 1 H), 2.16 (m, 1 H), 2.35 (d, 1 H, J = 4), 3.30 (m, 1 H), 3.60 (m, 1 H).

3. Discussion

This method of preparation of *trans*-isocyanocyclohexanol is a version of our literature procedure.[2] It represents a general procedure that gives comparable yields with a wide variety of epoxides.[2] The method described is a new approach to the synthesis of isocyanides. Traditionally, isocyanides have been prepared by dehydration of formamides, the reaction of dihalocarbenes with primary amines, and the reaction of active halides and olefins with cyanides.[3–5]

Isocyanides are useful intermediates because of their diverse reactivity.[4] The β-hydroxy isocyanides, which are prepared readily by our general procedure, are particularly useful because of their straightforward conversion to β-amino alcohols in acids and their catalyzed cyclization to oxazolines.[2]

1. Department of Chemistry, University of Minnesota, Minneapolis, MN 55455.
2. Gassman, P. G.; Guggenheim, T. L. *J. Am. Chem. Soc.* **1982**, *104*, 5849–5850.
3. Sandler, S. R.; Karo, W. "Organic Functional Group Preparations"; Academic Press: New York, 1972; Vol. III, pp. 179–204.
4. Ugi, I, "Isonitrile Chemistry"; Academic Press: New York, 1971.
5. Ugi, I.; Fetzer, U.; Eholzer, U.; Knupfer, H.; Offermann, K. *Angew. Chem. Int. Ed. Engl.* **1965**, *4*, 472–484.

2,3-*O*-ISOPROPYLIDENE-D-ERYTHRONOLACTONE

(Furo[3,4-*d*]-1,3-dioxol-4(3a*H*)-one, dihydro-2,2-dimethyl-(3a*R-cis*)-)

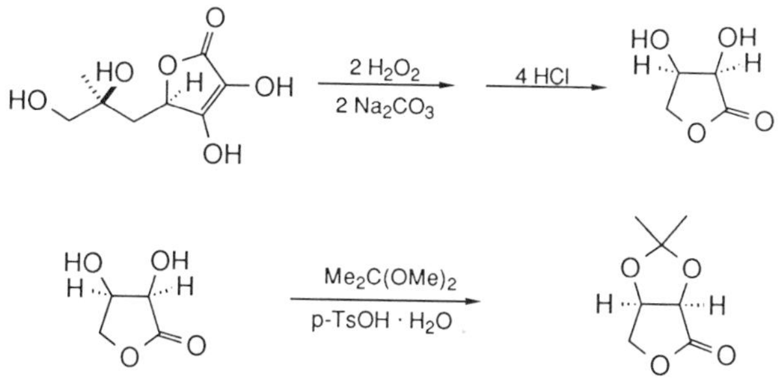

Submitted by NOAL COHEN, BRUCE L. BANNER, ANTHONY J. LAURENZANO, and LOUIS CAROZZA[1]
Checked by LEE A. FLIPPIN and CLAYTON H. HEATHCOCK

1. Procedure

A 1-L, three-necked, round-bottomed flask fitted with a thermometer, addition funnel, and an air motor-driven paddle stirrer is charged with 35.2 g (0.20 mol) of erythorbic acid (Note 1) and 500 mL of deionized water. The solution is stirred with ice bath cooling (Note 2), and 42.4 g (0.40 mol) of anhydrous, powdered sodium carbonate (Note 3) is added in small portions (Note 4). The resulting yellow solution (Note 5) is stirred with ice-bath cooling while 44 mL (0.45 mmol) of 31.3% by weight aqueous hydrogen peroxide (Note 6) is added dropwise over a 10-min period. The internal temperature rises from 6 to 19°C (Note 7). The solution, containing a few solid particles, is stirred for 5 min with ice bath cooling, during which time the internal temperature continues to rise to 27°C. The flask is now immersed in a water bath that is heated to 42°C. The solution is stirred for 30 min, during which time the internal temperature reaches a maximum of 42°C (Note 8). Norit A (8 g) is added in portions over 10 min to decompose the excess peroxide and the mixture is heated on a steam bath with continued stirring for 30 min, at which point gas evolution has essentially ceased and a negative starch–iodide test is observed. The internal temperature reaches and is kept at 75–78°C. The hot mixture is filtered with suction on a Celite pad into a 2-L, three-necked, round-bottomed flask and the filtercake is washed, in several small portions, with a total of 100 mL of deionized water. The combined filtrate and washes are acidified to pH 1 by the *cautious* (Note 9) addition of 150 mL (0.90 mol) of 6 *N* aqueous hydrochloric acid, in portions, with swirling. The acidic solution is concentrated with a rotary evaporator at 50°C/water aspirator pressure. The residue is dried at 50°C/0.2 mm to give 84.6 g of a pale-yellow solid residue containing D-erythronolactone, oxalic acid, and sodium chloride (Notes 10 and 11). To this material is added 175 mL of acetone (Note 13) and the mixture is swirled to loosen the solids caked on the sides of the flask. A 50-g portion of anhydrous, powdered magnesium sulfate (Note 14) is now added and the mixture is stirred by means of an air motor-driven paddle stirrer as 350 mL (2.85 mol) of 2,2-dimethoxypropane (Note 15) is added in one portion. To the stirred mixture is added 0.42 g (0.0022 mol) of *p*-toluenesulfonic acid monohydrate at room temperature. The slurry is blanketed with nitrogen and stirred at room temperature for 18 hr. In a 2-L, three-necked, round-bottomed flask fitted with a thermometer and an air motor-driven paddle stirrer, a mixture of 500 mL of anhydrous ether and 61.3 mL (0.44 mol) of triethylamine (Note 16) is cooled in an ice bath to 5°C. The reaction mixture is decanted into this solution. The residual solids are rinsed with 60 mL of ether, which is also decanted into the triethylamine solution. After being stirred for a few minutes (Note 17), the mixture is filtered with suction on a 600-mL, coarse, sintered-glass funnel. The solids are washed thoroughly with a total of 300 mL of anhydrous ether by slurrying three times on the funnel with the vacuum turned off; the vacuum is then applied to draw the wash ether through the funnel. The filtrate and washes are combined and concentrated with a rotary evaporator at water aspirator pressure, and the residue is dried at 45°C/0.5 mm to give 34.3 g of a pale-yellow solid (Note 18). This material is dissolved in approximately 150 mL of 1 : 1 hexanes : ethyl acetate and the solution (Note 19) is adsorbed on a column of 200 g of silica gel (Note 20) packed in 1 : 1 hexanes : ethyl acetate. The column is eluted with a total volume of 2 L of 1 : 1 hexanes : ethyl acetate (Note 21). The eluate is concentrated with a rotary evaporator at aspirator pressure and the solid residue is dried under high vacuum to afford 27.3 g of a colorless solid. This material, contained in a 1-L, one-necked, round-bottomed flask, is treated with 150 mL of anhydrous ether and the mixture is refluxed on a steam bath for 5 min to dissolve all the solid. The solution is removed from the steam bath and treated with 225 mL of hexanes. An immediate precipitate results. The mixture is refrigerated (0°C) for 3.5 hr and then filtered with suction. The solid is washed with a total of 100 mL of hexanes, in small portions, and then dried under high vacuum at 20°C. There is obtained 23.6 g (74.7%) of 2,3-*O*-isopropylidene-D-erythronolactone as a white solid, mp 65.5–66°C, $[\alpha]_D^{25}$ −113.8° (*c* 1.11 H_2O) (Notes 22–25).

2. Notes

1. Erythorbic acid is the same compound as D-isoascorbic acid, available from the Aldrich Chemical Company, Inc. This substance is also known as *araboascorbic acid*.

2. The internal temperature is 6°C initially.

3. Sodium carbonate was obtained from the Fisher Scientific Company.

4. Vigorous evolution of carbon dioxide is observed. The internal temperature rises to 8°C.

5. A few particles of undissolved sodium carbonate may remain.

6. Aqueous hydrogen peroxide was obtained from the Fisher Scientific Company. The lot analysis given on the bottle is used to calculate the volume of hydrogen peroxide solution required. Approximately 10% molar excess of peroxide appears to be required to provide a clean product.

7. The oxidation is quite exothermic. Attempts to increase the concentrations of the reactants led to an exotherm that was difficult to control and was complicated by the precipitation of solids, which hampered stirring.

8. A small amount of gas evolution is noted during this period.

9. Evolution of carbon dioxide is vigorous.

10. It is essential that all the water be removed at this point and that a constant weight of approximately 84 g be obtained.

11. If desired, D-erythronolactone can be isolated at this point by treatment of the residue with boiling ethyl acetate. On this scale, the solid is triturated at reflux with 325 mL of ethyl acetate for 5 min. The solution is decanted and the trituration is repeated with 130 mL of ethyl acetate. The combined solutions are cooled to 5°C and filtered. The solid is washed in portions with a total of 400 mL of cold ethyl acetate. After air drying, there is obtained 15.4 g (77.0%) of D-erythronolactone as a white solid, mp 97.5–99.5°C, $[\alpha]_D^{25}$ − 72.8° (H_2O, c 0.498) (Note 12).

12. The physical properties of D-erythronolactone are as follows: Lit.[2] mp 104–105°C, $[\alpha]_D^{20}$ −73.2° (H_2O, c 0.533).

13. Acetone was obtained from Fisher Scientific Company.

14. The drying agent is added to remove any residual moisture and to facilitate the subsequent filtration.

15. 2,2-Dimethoxypropane was obtained from the Aldrich Chemical Company, Inc.

16. Triethylamine was obtained from Eastman Chemical Products, Inc.

17. The mixture is alkaline to pH paper.

18. TLC analysis of the crude product (1 : 3 hexane : ethyl acetate, EM Silica Gel 60 F-254 plates) reveals the desired acetonide lactone to be the major component (R_f 0.6) with one minor, less polar impurity and several minor, more polar impurities. The ^{1}H NMR and IR spectra of a pure sample of the less polar impurity (an oil) were compatible with the following structure:

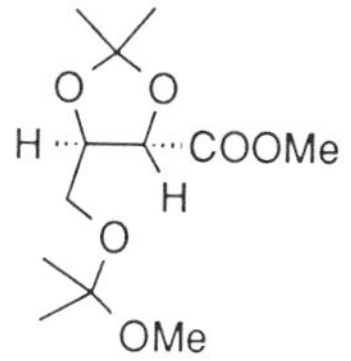

^{1}H NMR (100 MHz, CDCl$_3$) δ: 1.31 (2 s, 6 H, (CH$_3$)$_2$C), 1.39 (s, 3 H, C$_2$—CH$_3$), 1.59 (s, 3 H, C$_2$—CH$_3$), 3.20 (s, 3 H, OCH$_3$), 3.41 (dd, 1 H, J = 6, 10.5, CH$_2$O), 3.57 (dd, 1 H, J = 4.5, 10.5, CH$_2$O), 3.76 (s, 3 H, CO$_2$CH$_3$), 4.49 (m, 1 H, H$_5$), 4.67 (d, 1 H, $J_{4,5}$ = 7, H$_4$); IR (CHCl$_3$) cm^{-1}: 1760, 1735 (ester C=O).

19. A small amount of insoluble material is present.

20. EM Silica Gel 60, 0.063–0.2 mm was used. The column dimensions are approximately 1.75 in. × 14 in.

21. TLC is utilized to ensure that all of the desired product is eluted from the column. This procedure removes the minor, polar impurities present in the crude product that appear at or near the origin of the TLC plate.

22. This material is homogeneous on TLC analysis; ^{1}H NMR (100 MHz, CDCl$_3$) δ: 1.37 (s, 3 H, C$_2$—CH$_3$), 1.46 (s, 3 H, C$_2$—CH$_3$), 4.42 (d, 2 H, $J_{6,6a}$ = 2, H$_6$), 4.75 (d, 1 H, $J_{3a,6a}$ = 6, H$_{3a}$), 4.89 (dt, 1 H, $J_{3a,6a}$ = 6, $J_{6,6a}$ = 2, H$_{6a}$); IR (CHCl$_3$) cm^{-1}: 1786 (γ-lactone C=O).

23. The physical properties are as follows: lit.[3] mp 68–68.5°C, $[\alpha]_D^{20}$ −112° (H$_2$O, c 1.5).

24. The reaction sequence has been run on a 176-g (1.0-mol) scale with no loss in yield.

25. The checkers obtained 22.5 g (71.1%) of product as a white solid, mp 68.0–68.5°C, $[\alpha]_D^{20}$ −123.4° (H$_2$O, c 0.96). It is important that crystallization from the ether–hexane mixture be carried out at 0°C. In one run in which crystallization was carried out at 8°C, the checkers obtained only 15.3 g (48.4%) of product, mp 65.5–66.0°C.

3. Discussion

2,3-O-Isopropylidene-D-erythronolactone and the corresponding lactol, 2,3-O-isopropylidene-D-erythrose, are useful chiral synthons in the total synthesis of certain natural products such as the leukotrienes[4a] and pyrrolizidine alkaloids.[4b] The lactol is readily available from the lactone, in excellent yield, by reduction with diisobutylaluminum hydride.[4,5] 2,3-O-Isopropylidene-L-erythrose has been employed as the starting material in an enantioselective synthesis of (+)-15S-prostaglandin A$_2$.[6] Optically pure, selectively protected, polyfunctional C$_4$-units such as these have great potential in synthesis if readily available, in substantial quantity, from inexpensive members of the pool of chiral starting materials.[7]

D-Erythronolactone and/or its isopropylidene derivative have been prepared starting from L-rhamnose,[8] D-ribose,[9] D-ribonolactone,[3] potassium D-glucuronate,[10] D-glucose,[11] and erythorbic acid,[2] by optical resolution of racemic erythronolactone,[12] and asymmetric total synthesis.[13] 2,3-O-Isopropylidene-D-erythrose has been obtained from D-arabinose by a route that does not involve the intermediacy of the lactone.[14] All of these processes suffer from either relatively low overall yields or the requirement of a large number of individual stages. The procedure described here, which is based on a similar oxidative degradation of L-ascorbic acid (vitamin C) to L-threonic acid,[15] is undoubtedly the most expeditious route to the acetonide of D-erythronolactone available. In addition, the starting material, erythorbic acid, is an inexpensive and readily available substance, commonly used as a food preservative. It is pertinent to note that recently L-ascorbic acid has itself found synthetic utility as a precursor to (R)-glycerol acetonide, an important C$_3$ chiral synthon.[16]

1. Research and Development Division, Hoffmann-La Roche, Inc., Nutley, NJ 07110.
2. Weidenhagen, R.; Wegner, H. *Chem. Ber.* **1939**, *72*, 2010–2020. This process involves treatment of erythorbic acid with p-tolyldiazonium bisulfate followed by aqueous hydrolysis of the resulting oxalyl hydrazide intermediate, giving D-erythronolactone.
3. Mitchell, D. L. *Can. J. Chem.* **1963**, *41*, 214–217.
4. (a) Cohen, N.; Banner, B. L.; Lopresti, R. J.; Wong, F.; Rosenberger, M.; Liu, Y.-Y.; Thom, E.; Liebman, A. A. *J. Am. Chem. Soc.* **1983**, *105*, 3661–3672; (b) Buchanan, J. G.;

Jigajinni, V. B.; Singh, G.; Wightman, R. H. *J. Chem. Soc., Perkin Trans 1* **1987**, 2377–2384.

5. Ireland, R. E.; Wilcox, C. S.; Thaisrivongs, S. *J. Org. Chem.* **1978**, *43*, 786–787.
6. Stork, G.; Raucher, S. *J. Am. Chem. Soc.* **1976**, *98*, 1583–1584.
7. Seebach, D.; Hungerbuhler, E. In "Modern Synthetic Methods 1980," Scheffold, R., Ed.; Salle + Sauerlander; Frankfurt/Aarau, 1980; pp. 91–171; Vasella, A. In "Modern Synthetic Methods 1980," Scheffold, R., Ed.; Salle + Sauerlander; Frankfurt/Aarau, 1980; pp. 173–267; Fischli, A. In "Modern Synthetic Methods 1980," Scheffold, R., Ed.; Salle + Sauerlander; Frankfurt/Aarau, 1980; pp. 269–350; Fraser-Reid, B.; Anderson, R. C. In "Progress in the Chemistry of Organic Natural Products," Herz, W.; Griesbach, H.; Kirby, G. W., Eds.; Springer-Verlag: New York; 1980; Vol. 39, pp. 1–61; Hanessian, S. "Total Synthesis of Natural Products: The Chiron Approach"; Pergamon Press: Oxford; 1983; Scott J. W. In "Asymmetric Synthesis," Morrison, J. D.; Scott, J. W. Eds.; Academic Press: Orlando; 1984; Vol. 4, pp. 1–226.
8. Baxter, J. N.; Perlin, A. S. *Can. J. Chem.* **1960**, *38*, 2217–2225.
9. Hardegger, E.; Kreis, K.; El Khadem, H. E. *Helv. Chim. Acta* **1951**, *34*, 2343–2348.
10. Gorin, P. A. J.; Perlin, A. S. *Can. J. Chem.* **1956**, *34*, 693–700.
11. Perlin, A. S.; Brice, C. *Can. J. Chem.* **1955**, *33*, 1216–1221; MacDonald, D. L.; Crum, J. D.; Barker, R. *J. Am. Chem. Soc.* **1958**, *80*, 3379–3381; Barker, R.; MacDonald, D. L. *J. Am. Chem. Soc.* **1960**, *82*, 2301–2303.
12. Glattfeld, J. W. E.; Forbrich, L. R. *J. Am. Chem. Soc.* **1934**, *56*, 1209–1210; Jelinek, V. C.; Upson, F. W., *J. Am. Chem. Soc.* **1938**, *60*, 355–357.
13. Mukaiyama, T.; Yamaguchi, M.; Kato, J. *Chem. Lett.* **1981**, 1505–1508.
14. Ballou, C. E. *J. Am. Chem. Soc.* **1957**, *79*, 165–166.
15. Isbell, H. S.; Frush, H. L. *Carbohydr. Res.* **1979**, *72*, 301–304.
16. Jung, M. E.; Shaw, T. J. *J. Am. Chem. Soc.* **1980**, *102*, 6304–6311; Takano, S.; Hirotoshi, N.; Ogasawara, K. *Heterocycles* **1982**, *19*, 327–328; Abushanab, E.; Bessodes, M.; Antonakis, K. *Tetrahedron Lett.* **1984**, *25*, 3841–3844; Nicolaou, K. C.; Zipkin, R. E.; Dolle, R. E.; Harris, B. D. *J. Am. Chem. Soc.* **1984**, *106*, 3548–3551; Wei, C. C; DeBernardo, S.; Tengi, J. P.; Borgese, J.; Weigele, M. *J. Org. Chem.* **1985**, *50*, 3462–3467; Irie, H.; Igarishi, J.; Matsumoto, K.; Yanagawa, Y.; Nakashima, T.; Ueno, T.; Fukami, H. *Chem. Pharm. Bull.* **1985**, *33*, 1313–1315; Sletzinger, M.; Verhoeven, T. R.; Volante, R. P.; McNamara, J. M.; Corley, E. G.; Liu, T. M. H. *Tetrahedron Lett.* **1985**, *26*, 2951–2954; Huberschwerlen, C. *Synthesis,* **1986**, 962–964; Tanaka, A.; Yamashita, K. *Synthesis* **1987**, 570–573.

o-ISOTHIOCYANATO-(E)-CINNAMALDEHYDE

[2-Propenal, 3-(2-isothiocyanatophenyl)-, (E)-]

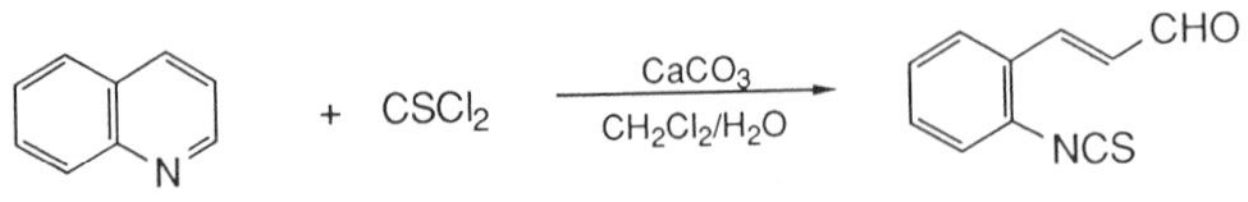

Submitted by R. FARRAND and R. HULL[1]
Checked by K. E. FAHRENHOLTZ and G. SAUCY

1. Procedure

Caution! This reaction should be carried out in a good hood because of the toxicity of thiophosgene.

A 1000-mL (Note 1), multinecked flask is provided with an efficient stirrer, vented outlet, thermometer, and 250-mL dropping funnel. The flask is surrounded by an ice-water bath and charged with 62.5 mL (68.4 g, 0.53 mol) of quinoline, 250 mL of dichloromethane, 55 g (0.55 mol) of finely powdered calcium carbonate, and 250 mL of water. The mixture is stirred vigorously, cooled to 10°C, and maintained at 10–15°C as a solution of 37.5 mL (56.5 g, 0.49 mol) of thiophosgene (Note 2) in 120 mL of dichloromethane is added over 15 min. There is very little exotherm or foaming. The cooling bath is removed and the reaction mixture is stirred vigorously at ambient temperature overnight. The reaction is then filtered through a bed of filter aid. The layers are separated and the aqueous layer is extracted with 50 mL of dichloromethane. The combined organic layers are washed twice with 150 mL of 2 N hydrochloric acid (Note 3), then with 150 mL of water, and dried over anhydrous magnesium sulfate. Concentration under reduced pressure gives 95–103 g of crude material (Note 4). This is dissolved with heating in 400 mL of cyclohexane, decolorizing carbon is added, and the mixture is filtered through a bed of filter aid. The filtrate is heated under reflux for 2 hr (Note 5) and allowed to cool with stirring (Note 6). The resulting solid is isolated by filtration, washed with cyclohexane, and dried in a vacuum oven at 40°C to give 78–83 g (84–89%) of o-isothiocyanato-(E)-cinnamaldehyde as cream crystals, mp 77–79°C (Note 7).

2. Notes

1. The reaction has been carried out on 10 times these quantities with no difficulty.
2. The checkers used an older bottle of thiophosgene and obtained an 84% yield (based on thiophosgene). A subsequent run was carried out with Aldrich "85% in CCl₄" thiophosgene found by analysis to contain 63% thiophosgene (therefore 89.4 g was used) and an 89% yield was obtained. A subsequent run on an unanalyzed bottle of the same lot number using 89.4 g gave a 100% yield (92% based on quinoline). It is suggested that thiophosgene be analyzed before use (Note 8).
3. These two washes remove unreacted quinoline.

4. The crude material consists of a mixture of *Z* and *E* isomers, with *Z* predominating. If workup of the reaction is delayed, more of the less soluble *E* isomer is formed, complicating subsequent filtration.

5. This additional heating completes the isomerization of the *Z* to the *E* isomer.

6. Subsequent breakup and filtration of the solid are facilitated if this solution is transferred and allowed to cool with stirring in a large-mouth container such as a beaker.

7. Melting points were taken in open capillaries on a Thomas-Hoover melting point apparatus. The crude material can be purified by dissolving it in dichloromethane, passing the solution over a plug of silica gel, and concentrating the solution with the addition of ether. The recrystallized material has essentially the same melting point and is colorless. The spectral properties of *o*-isothiocyanato-(*E*)-cinnamaldehyde are as follows: IR (Nujol) cm^{-1}: 2075 (NCS) and 1670 (conjugated CHO); ^{1}H NMR (CDCl$_3$) δ: 6.75 (d of d, 1 H, J = 16 and 7.5, CH=CH*CHO*), 7.4 (m, 4 H, aromatic H), 7.8 (d, 1 H, J = 16, ArC*H*=CH), 9.78 (d, 1 H, J = 7.5, C*HO*).

8. Thiophosgene mixed with CCl$_4$ can be analyzed as follows: a 0.5-mL aliquot of the reagent is mixed with a warm mixture of 15 mL of 30% hydrogen peroxide and 15 mL of 1 *N* sodium hydroxide. The mixture is shaken occasionally during 20 min (overnight gives the same titer) and diluted to 200 mL with water. Liberated Cl$^-$ is then titrated with mercuric nitrate.

3. Discussion

This procedure is an example of a simple fission reaction of *N*-heterocyclic compounds by thiophosgene and base[2] wherein the dihydro intermediate **1** undergoes ring fission to yield the *Z*-isothiocyanate **2**, which isomerizes in situ to the *E*-isomer **3**. The reaction may be applied to certain substituted quinolines,[3,4] isoquinoline,[2] pyridine,[5] benzoxazole,[6] benzimidazole,[6,7] and oxazole[8] derivatives, but not to benzothiazole.[6]

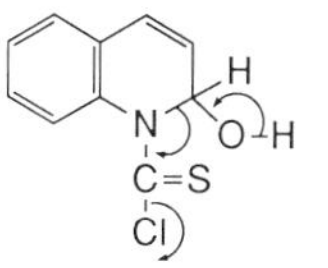

1 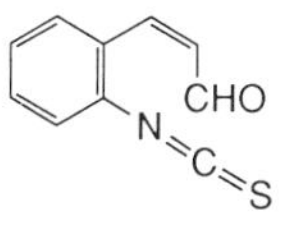   2 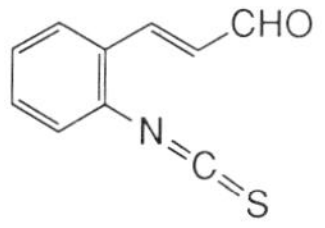   3

The ortho-substituted isothiocyanates are valuable intermediates for the preparation of a variety of heterocyclic compounds; for example, *o*-isothiocyanato-(*E*)-cinnamaldehyde with sodio diethyl malonate undergoes facile cyclization to 3-formylquinoline-2(1*H*)-thione,[9] which in turn may be used for the preparation of tricyclic[9,10] and large ring heterocyclic compounds.[11]

1. Imperial Chemical Industries Limited, Pharmaceutical Division, Alderley Park, Macclesfield, Cheshire SK10 4TG, England.

2. Hull, R. *J. Chem. Soc. (C)* **1968,** 1777–1780.
3. Hull, R.; van den Broek, P. J.; Swain, M. L. *J. Chem. Soc., Perkin Trans. 1* **1975,** 922–925.
4. Hull, R.; Swain, M. L.; *J. Chem. Soc., Perkin Trans. 1,* **1976,** 653–660.
5. Boyle, F. T.; Hull, R. *J. Chem. Soc., Perkin Trans. 1,* **1974,** 1541–1546.
6. Faull, A. W.; Hull, R. *J. Chem. Res., Synop.* **1979,** 148.
7. Hull, R. *Synth. Commun.* **1979,** *9,* 477–481.
8. Faull, A. W.; Hull, R. *J. Chem. Res., Synop.* **1979,** 240–241.
9. Hull, R. *J. Chem. Soc., Perkin Trans. 1* **1973,** 2911–2914.
10. Brown, K. J.; Meth-Cohn, O. *Tetrahedron Lett.* **1974,** 4069–4072.
11. Griffiths, D.; Hull, R. *J. Heterocycl. Chem.* **1977,** *14,* 1097–1098.

PHOTOPROTONATION OF CYCLOALKENES: LIMONENE TO p-MENTH-8-EN-1-YL METHYL ETHER

[Cyclohexane, 1-methoxy-1-methyl-4-(1-ethenyl-1-methyl-)]

Submitted by F. P. Tise and P. J. Kropp[1]
Checked by R. L. Amey and R. E. Benson

1. Procedure

A 250-mL photochemical reactor (see Figure 1) is fitted with a cylindrical Vycor filter sleeve, a 450-W Hanovia mercury lamp, and a watercooled condenser which is connected to a mineral oil bubbler. Tubing attachments are made so that water is circulated through the condenser and then through the Vycor filter sleeve. The tube leading from the bottom of the reaction vessel and containing the glass frit is connected in series to a trap fitted with a fritted filter stick and then to a trap that is connected to a nitrogen source. The system is flushed with nitrogen, and sufficient anhydrous methanol is placed in the trap containing the fritted stick to provide for a methanol-saturated gas stream during the course of the reaction (Note 1).

The nitrogen-flushed reactor is charged with a solution of 20.0 g (147 mmol) of (+)-limonene (Note 2), 5.0 g (53 mmol) (Note 3) of phenol, and 5 drops of concentrated sulfuric acid in 210 mL (167 g, 5.2 mol) of anhydrous methanol (Note 4). Water flow through the condenser is started (Note 5), and the nitrogen flow is adjusted to provide good agitation of the contents of the vessel. After 15 min, irradiation is started and the reaction followed by GLC (Note 6), with 48 hr being the approximate time needed for essentially complete conversion (Note 7).

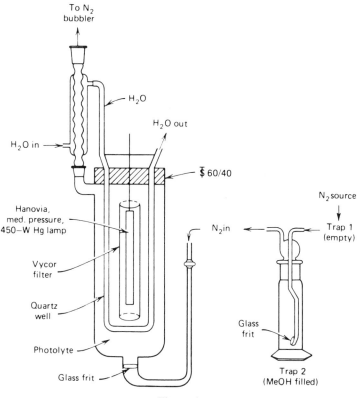

Figure 1

The solution is poured into 900 mL of 5% aqueous sodium hydroxide solution containing 125 g of sodium chloride, and the mixture is extracted with two 100-mL portions of ether. The ether layers are combined, washed with 50 mL of saturated sodium chloride solution, and dried over anhydrous sodium sulfate. The drying agent is removed by filtration and the filtrate is concentrated with a Büchi rotary evaporator. After a preliminary distillation to separate the product from a small amount of nonvolatile material, the liquid is distilled at reduced pressure through a Teflon spinning band column (47 cm × 7 mm). The material that distills at 90–95°C (10 mm) is collected to give 12.8–13.2 g (52–53%) of a mixture of *cis*- and *trans*-p-menth-8-en-1-yl methyl ether (Notes 8–10).

2. Notes

1. The submitters used a reactor with a joint that was capped with a rubber septum fitted with two syringe needles, which were attached by means of a Y-tube to a single nitrogen line. To one of these needles is attached a piece of 1.70-mm-o.d. polyethylene tubing of sufficient length to reach to the bottom of the reaction

vessel. By use of pinchcocks, nitrogen can be passed through either of the two needles. The solution was stirred with a magnetic stirring bar.

2. (+)-Limonene was obtained from Aldrich Chemical Company, Inc. and distilled before use.

3. The checkers used reagent available from Fisher Scientific Company.

4. The checkers used fresh, acetone-free, absolute methanol available from Fisher Scientific Company.

5. For best results the cooling water should pass through the condenser first and then through the immersion well. This arrangement lessens evaporation of methanol.

6. The submitters used a 3-m × 3.2-mm stainless steel column packed with 20% SF-96 on Chromasorb W (60–80 mesh) and a He flow rate of 60 mL/min. With a temperature program of 4 min at 50°C followed by an increase of 10°C/min to a maximum of 200°C, the retention times were 17.9 and 18.7 min.

7. The checkers found that the reaction was impeded by the formation of a yellow film on the immersion well with very little further conversion occurring after 30 hr of irradiation.

8. The checkers used a 2.4-m × 3.2-mm column packed with 7% SE-30 and 3% Silar on Chromasorb W (60–80 mesh) at 160°C. The retention time was 1.56 min for the trans isomer and 1.81 min for the cis isomer at a He flow rate of 55 mL/min.

9. The spectral properties of the product (approximately 60% cis:40% trans isomers) are as follows: IR (neat) cm^{-1}: 3080 (=C—H); 2964, 2939, 2860, 2825, (C—H); 1645 (C=C); 1464, 1453, and 1442 (overlapping peaks); 1370, 1124, and 1082 (C—OC); 885 (=CH). ^{1}H NMR (CDCl$_3$) δ: 1.10 [s, 3 H, CH$_3$ (trans)], 1.19 [s, 3 H, CH$_3$ (cis)], 1.30–2.00 (8 H, —CH$_2$-), 1.71 [s, 3 H, CH$_3$ (cis/trans)], 3.14 [s, 3 H, OCH$_3$ (trans)], 3.21 [s, 3 H, OCH$_3$ (cis)], 4.69 [s, 2 H, =CH$_2$ (cis/trans)].

10. The submitters state that similar irradiation of 20.0 g of cyclohexene, 5.0 g of phenol, and 1.5 mL of concentrated sulfuric acid for 24 hr afforded cyclohexyl methyl ether in 70% yield.

3. Discussion

Acid-catalyzed, ground-state additions to limonene generally afford a mixture of products resulting from competing protonation of both double bonds.[2] In one case in which selective reaction was observed, attack occurred at the acyclic C$_8$—C$_9$ double bond.[3]

The photoprotonation of cycloalkenes, described in this procedure, is believed to proceed via initial light-induced cis → trans isomerization of the alkene.[4] The resulting highly strained trans isomer undergoes facile protonation. This procedure permits the protonation of cyclohexenes and cycloheptenes under neutral or mildly acidic conditions.[5] Since the process is irreversible, high levels of conversion to addition products can be achieved.

Photoprotonation is generally specific for cyclohexenes and cycloheptenes. Smaller-ring cycloalkenes are incapable of undergoing cis → trans isomerization, and the trans isomers of larger-ring or acyclic analogues have insufficient strain to undergo ready protonation. Thus, in addition to facilitating protonation of cycloalkenes, the procedure affords a means of selectively protonating a double bond contained in a six- or seven-membered ring in the presence of another double bond contained in an acyclic, exocyclic, or larger-ring cyclic environment.[6] When conducted in non-nucleophilic media, the photoprotonation procedure is also useful for effecting the isomerization of 1-alkylcyclohexenes and -heptenes to their exocyclic isomers.[4]

1. Department of Chemistry, University of North Carolina, Chapel Hill, NC 27599-3290.
2. For a review of the chemistry of limonene, see Verghese, J. *Perfum. Essent. Oil Rec.* **1968**, *59*, 439–454.
3. Kuczynski, L.; Kuczynski, H. *Rocz. Chem.* **1951**, *25*, 432–453.
4. For reviews of the photochemistry of alkenes see Kropp, P. J. *Mol. Photochem.* **1978**, *9*, 39–65 and *Organic Photochemistry* **1979**, *4*, 1–142.
5. There is a fine balance between the acidity of the alcohol and the basicity of the trans olefin. For example, 1-methylcyclohexenes undergo photoprotonation in methanol whereas cyclohexenes require the addition of small amounts of acid. In the present example, the addition of a small quantity of acid reduces the competing formation of the exocyclic isomer, *p*-mentha-1(7),8-diene.
6. For an earlier report on the photoprotonation of (+)-limonene, see Kropp, P. J. *J. Org. Chem.* **1970**, *35*, 2435–2436.

ANODIC OXIDATION OF *N*-CARBOMETHOXYPYRROLIDINE: 2-METHOXY-*N*-CARBOMETHOXYPYRROLIDINE

(1-Pyrrolidinecarboxylic acid, 2-methoxy-, methyl ester)

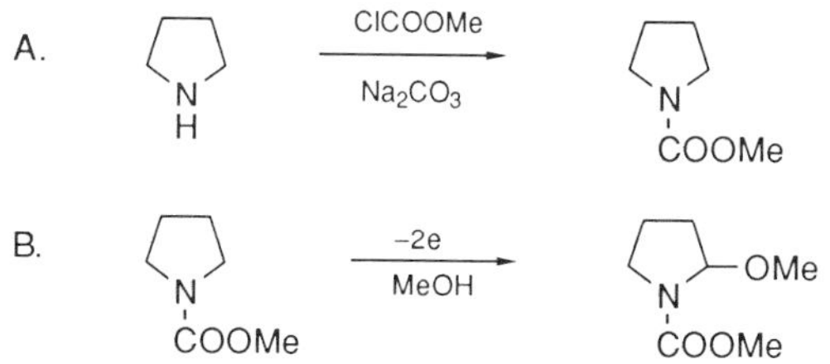

Submitted by T. Shono, Y. Matsumura, and K. Tsubata[1]
Checked by B. Schaer, G. Reymond, V. Toome, and Gabriel Saucy

1. Procedure

A. *N-Carbomethoxypyrrolidine.* A 1-L, three-necked, round-bottomed flask is equipped with a 200-mL pressure-equalizing dropping funnel, a Graham condenser protected by a calcium chloride tube, and a mechanical stirrer. The flask is charged with

200 g (1.89 mol) of sodium carbonate (Note 1), 400 mL of methylene chloride (Note 2), and 71 g (1 mol) of pyrrolidine (Note 3). The dropping funnel is charged with 103 g (1.1 mol) of methyl chlorocarbonate (Note 3), which is added with stirring over a 2-hr period at a rate that sustains a gentle reflux. After the addition of methyl chlorocarbonate is completed, the reaction mixture is stirred overnight at room temperature. The white precipitate is filtered with suction through a coarse Büchner funnel and washed three times with 100 mL of methylene chloride. The filtrate is concentrated on a vacuum rotary evaporator at a bath temperature of 30°C. The crude oily product is distilled under reduced pressure through a Claisen flask to yield 119–121 g (92–94%) of N-carbomethoxypyrrolidine, bp 64°C/1.3 mm.

B. *2-Methoxy-N-carbomethoxypyrrolidine.* A solution of N-carbomethoxypyrrolidine (12.7 g, 0.098 mol) and tetraethylammonium p-toluenesulfonate (0.83 g, 0,0027 mol) (Note 3) in 83 mL of methanol (Note 2) is added into an undivided jacketed cell (Note 4) equipped with two graphite-rod anodes and two graphite-rod cathodes, (Note 5) a thermometer, an exit tube for venting purposes, and a magnetic stirring bar. The carbon rods (0.6 cm in diameter, immersed 5.5 cm into the solution, resulting in a working electrode surface of 21.3 cm^2 and a current density of 46.9 mA/cm^2) are spaced 4.5 mm apart. The anode rods and the two cathode rods are connected with No. 22 copper wire as shown in Figure 1. During the electrolysis (Note 6), the temperature of the reaction mixture is maintained at 10–15°C (Note 7) by cooling with tap water. After 2.34F/mol of electricity (1 A, 6 hr; the voltage between the anode and cathode was 19–24 V for the example in Figure 1) has been passed through, (Note 8) the current is stopped and the solvent is removed under reduced pressure. The residue is dissolved in 120 mL of methylene chloride and washed with aqueous NaCl (20 mL). The aqueous NaCl wash is reextracted with methylene chloride (2 × 30 mL). The methylene chloride phases are combined and dried over magnesium sulfate. The solvent is evaporated and the residue is distilled, employing a 5 cm Vigreux column and an oil bath at 80–90°C (Note 9). The yield is 12.3–13.0 g (78–83%), bp 48–55°C/0.2–0.5 mm (Note 10).

2. Notes

1. Sodium carbonate, anhydrous powder, supplied by J. T. Baker Chemical Company, is used directly.

2. Methylene chloride was purchased from Fisher Scientific Co.

3. Pyrrolidine and methyl chlorocarbonate were purchased from the Aldrich Chemical Company, and used without further purification.

4. The cell is shown in Figure 1.

5. The electrodes were purchased from Princeton Applied Research (PAR); spectroscopic grade, Lot #174/78. This grade is not necessarily the best type of graphite for electrochemical purposes, but it was the only one immediately available. The submitters used graphite plates, purchased from Tokai Carbon Company, Inc., as electrodes, but they note that these are not the only electrode material usable in this reaction.

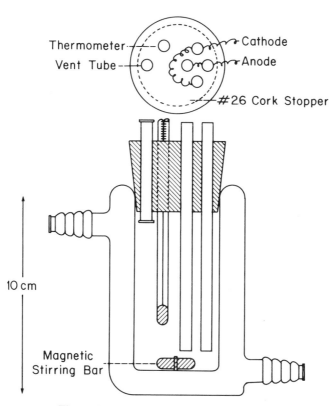

Figure 1. Electrolysis cell for methoxylation.

6. Princeton Applied Research (PAR) Potentiostat-Galvanostat, Model 173/179 was used.

7. According to the submitters, the temperature should be kept below 50°C; otherwise lower yields are observed.

8. According to the submitters, if more than 2.2–2.5F/mol of electricity is passed *N*-carbomethoxy-2,5-dimethoxypyrrolidine forms. The by-product can be separated by distillation (bp 64–65°C/1.0 mm).

9. An oil bath temperature higher than 100°C results in the formation of unsaturated carbamate formed by the elimination of methanol from the methoxylated carbamate. Accordingly, in the anodic methoxylation of carbamates having high boiling points, the product must be purified by column chromatography in order to avoid formation of the unsaturated carbamates.

10. The product has the following spectral properties: IR (liquid film) cm^{-1}: 2940, 2880, 1685, 1440, 1370, 1185, 1080, 950, 825, 770; ^{1}H NMR (CCl$_4$) δ: 1.48–2.21 (m, 4 H, CH$_2$ at C$_3$ and C$_4$ of pyrrolidine ring), 3.25 (s, 3 H, methoxy

CH$_3$), 3.08–3.52 (m, 2 H, CH$_2$ at C$_5$ of pyrrolidine ring), 3.64 (s, 3 H, ester CH$_3$), 5.06 (m, 1 H, CH at C$_2$ of pyrrolidine ring).

3. Discussion

This procedure describes anodic α-methoxylation of carbamates (2), which are derived from primary and secondary amines (1).[2,3]

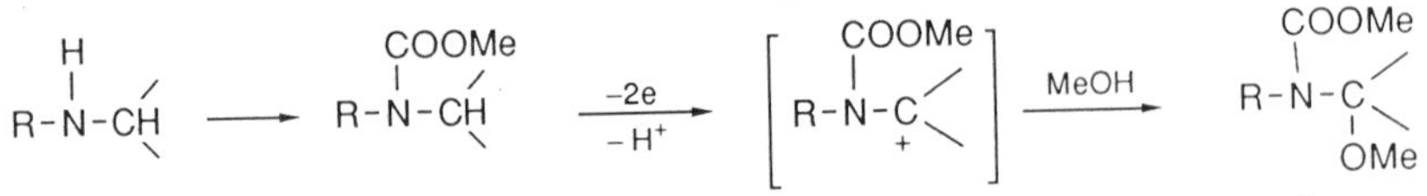

The intermediate cations (3) are trapped with methanol to yield α-methoxycarbamates, 4, which are sufficiently stable to be stored for a long period. Table I shows other examples of anodic synthesis of 4.

The high regioselectivity in the methoxylation of unsymmetrical carbamates is remarkable (see 2-pipecoline carbamate and N-carbomethoxyproline methyl ester in Table I). The methoxylation always takes place in the order of CH$_3$— > $\diagdown$CH$_2$ > —CH.

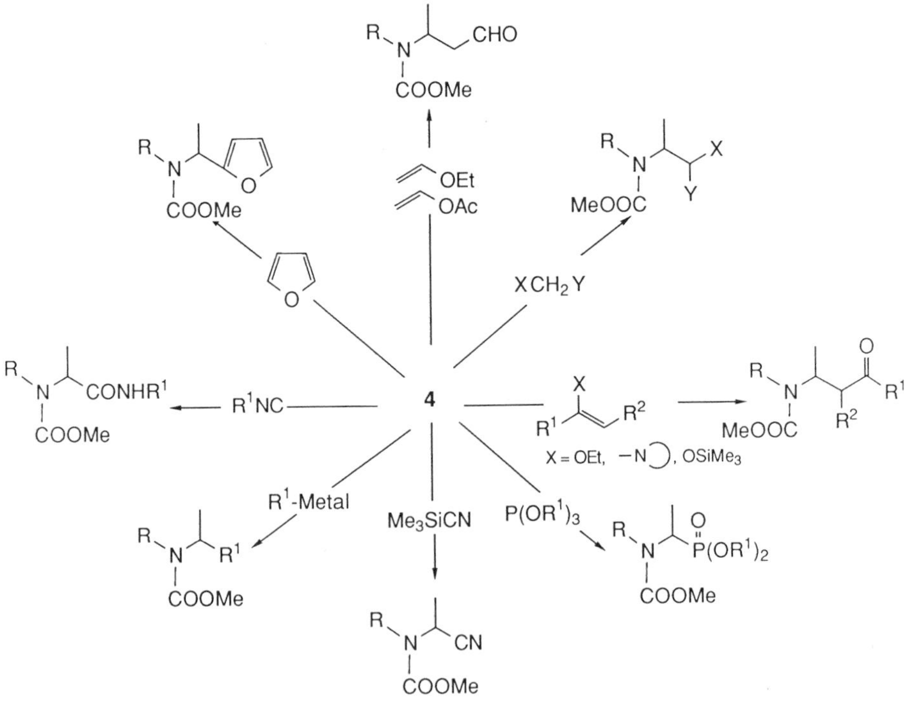

Figure 2

TABLE I
ANODIC SYNTHESES OF α-METHOXYCARBAMATES

Carbamate	Electricity Passed (F/mol)	α-Methoxycarbamate	Yield (%)
(piperidine, N-COOMe)	2.7	(piperidine, α-OMe, N-COOMe)	86
(2-methylpiperidine, N-COOMe)	2.6	(methylpiperidine, α-OMe, N-COOMe)	69
(pyrrolidine-2-COOMe, N-COOMe)	2.5	(MeO-pyrrolidine-2-COOMe, N-COOMe)	87
(oxazolidinone, NH)	3.0	(oxazolidinone, OMe, NH)	89
(morpholine, N-COOMe)	2.7	(morpholine, OMe, N-COOMe)	55
CH₃CH₂-NHCOOMe	4.85	CH₃CH(OMe)-NHCOOMe	83
CH₃CH₂CH₂CH₂-NHCOOMe	10.2	CH₃CH₂CH₂CH(OMe)-NHCOOMe	88
(CH₃)₂CHCH₂-NHCOOMe	6.0	(CH₃)₂CHCH(OMe)-NHCOOMe	70
(CH₃)₂CHCH₂CH₂-NHCOOMe	7.1	(CH₃)₂CHCH₂CH(OMe)-NHCOOMe	77
(CH₃)₂N-COOMe	2.1	CH₃N(CH₂OMe)-COOMe	72
CH₃N(CH₂COOMe)-COOMe	3.2	MeOCH₂N(CH₂COOMe)-COOMe	94

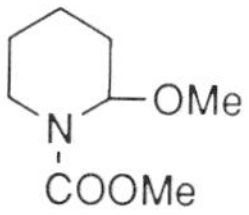

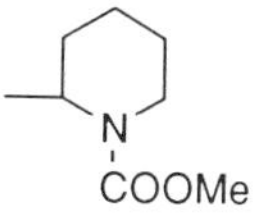

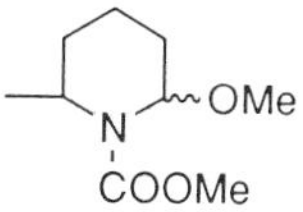

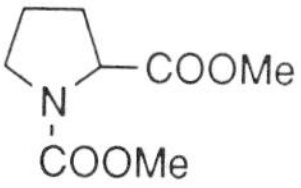

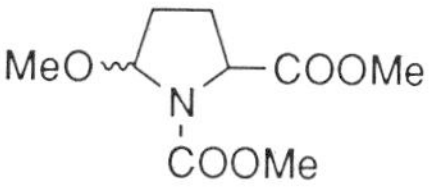

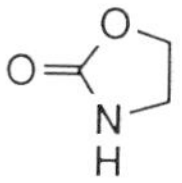

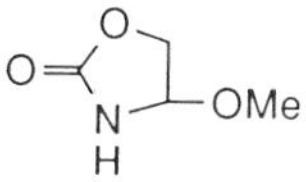

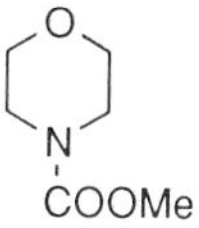

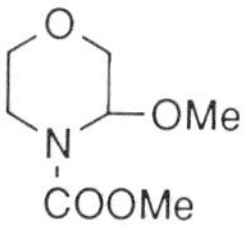

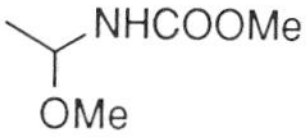

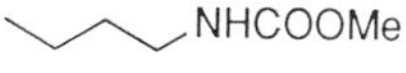

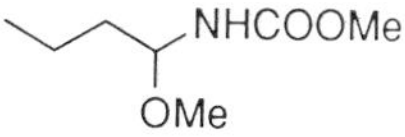

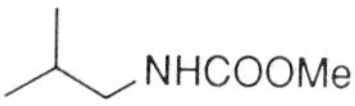

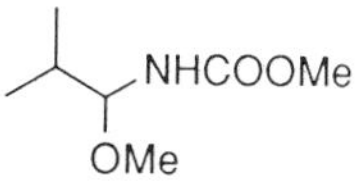

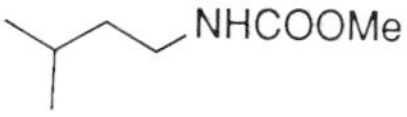

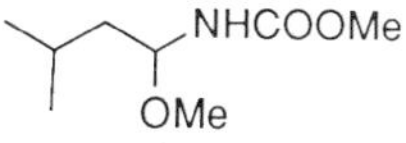

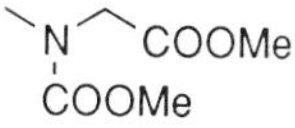

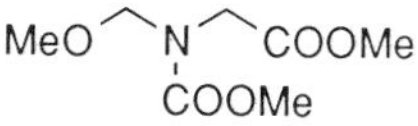

α-Methoxycarbamates (**4**) are useful intermediates in organic syntheses, since treatment of **4** with Lewis acids or Brønsted acids regenerates **3** which can be trapped with a variety of nucleophiles. Thus, physiologically active compounds such as alkaloids,[3] amino acids,[4] nitrogen-containing phosphorus compounds,[5] and pyridoxine[6] can be synthesized using **4** as key starting compounds. Figure 2 summarizes the transformations.[3-7]

1. Department of Synthetic Chemistry, Faculty of Engineering, Kyoto University, Yoshida, Sakyo, Kyoto 606, Japan.
2. Shono, T.; Hamaguchi, H.; Matsumura, Y. *J. Am. Chem. Soc.* **1975,** *97,* 4264.
3. Shono, T.; Matsumura, Y.; Tsubata, K. *J. Am. Chem. Soc.* **1981,** *103,* 1172.
4. Shono, T.; Matsumura, Y.; Tsubata, K. *Tetrahedron Lett.* **1981,** *22,* 2411.
5. Shono, T.; Matsumura, Y.; Tsubata, K. *Tetrahedron Lett.* **1981,** *22,* 3249.
6. Shono, T.; Matsumura, Y.; Tsubata, K.; Takata, J. *Chem. Lett.* **1981,** 1121.
7. Shono, T.; Matsumura, Y.; Tsubata, K.; Sugihara, Y.; Yamane, S.; Kanazawa, T.; Aoki, T. *J. Am. Chem. Soc.* **1982,** *104,* 6697.

PREPARATION AND DIELS–ALDER REACTION OF A HIGHLY NUCLEOPHILIC DIENE: *trans*-1-METHOXY-3-TRIMETHYLSILOXY-1,3-BUTADIENE AND 5β-METHOXYCYCLOHEXAN-1-ONE-3β,4β-DICARBOXYLIC ACID ANHYDRIDE

(Silane, [(3-methoxy-1-methylene-2-propenyl)oxy]trimethyl-)

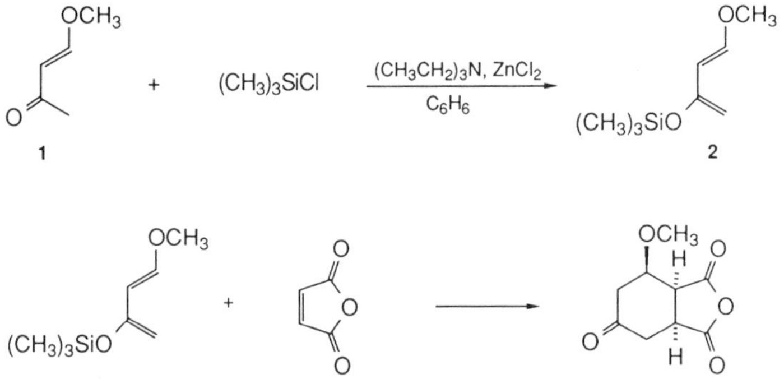

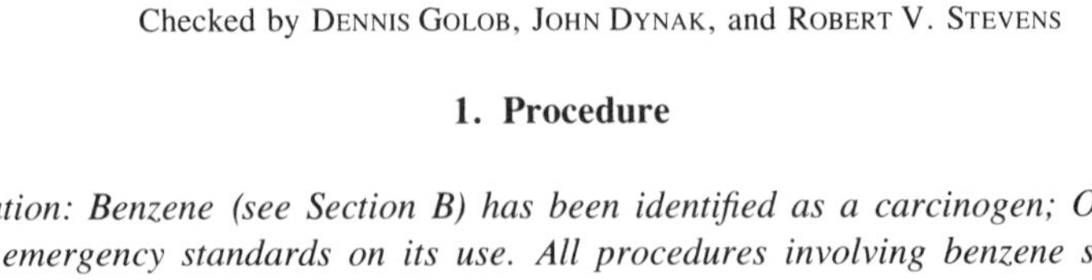

Submitted by SAMUEL DANISHEFSKY, TAKESHI KITAHARA, and PAUL F. SCHUDA[1]
Checked by DENNIS GOLOB, JOHN DYNAK, and ROBERT V. STEVENS

1. Procedure

Caution: Benzene (see Section B) has been identified as a carcinogen; OSHA has issued emergency standards on its use. All procedures involving benzene should be carried out in a well-ventilated hood, and glove protection is required.

A. *Preparation of the zinc chloride.* Reagent-grade zinc chloride (50 g) is placed in an evaporating dish and heated in a fume hood with a Fisher burner until no more water vapor is driven off. The hot dish is rapidly transferred to a glove bag that has been maintained under nitrogen. After the zinc chloride has cooled to a transparent glassy solid, it is ground to a fine powder with a mortar and pestle. The solid is transferred to a tightly stoppered bottle and stored in a desiccator over Drierite.

B. *Preparation of 1-methoxy-3-trimethylsiloxy-1,3-butadiene.* Triethylamine (575 g, 5.7 mol) is stirred mechanically in a three-necked flask (Note 1). To this is added 10.0 g (0.07 mol) of zinc chloride prepared as described above. The mixture is stirred at room temperature under nitrogen for 1 hr. A solution of 250 g (2.50 mol) of 4-methoxy-3-buten-2-one (from Aldrich Chemical Company, Inc.) in 750 mL of benzene is added all at once. Mechanical stirring is continued for 5 min. Chlorotrimethylsilane (542 g, 5.0 mol) is added rapidly. The reaction mixture first turns pink, then red, and finally brown. Heat is evolved and the reaction is kept below 45°C by cooling in an ice bath. After 30 min, the mechanically stirred solution is heated by a heating mantle to 43°C (Note 2). This temperature is maintained for 12 hr. The reaction mixture becomes very thick during this time. After the mixture cools to ambient temperature, it is poured, with mixing, into 5 L of ether. The solid material is filtered through Celite. The Celite and solid material are removed and stirred with 4 L more of ether and refiltered through Celite. The combined ether washings are evaporated under reduced pressure (rotary evaporator) to a brown, sweet-smelling oil. The oil is transferred to a 1-L, single-necked flask equipped with an 18-in. Vigreux column (Note 3). Careful fractional distillation under water vacuum affords a forerun of approximately 16 g that boils at 70–78°C (22 mm). This fraction consists of impure diene that contains 4-methoxy-3-buten-2-one. The main fraction boils at 78–81°C (23 mm) and consists of 245 g of diene (Note 4) with approximately 5–10% of 4-methoxy-3-buten-2-one (Note 5). This material is suitable for most purposes. If higher purity is desired, the second fraction may be redistilled under reduced pressure through an 18-in. Vigreux column to afford 200 g (46%) of *trans*-1-methoxy-3-trimethylsiloxy-1,3-butadiene (Note 6).

C. *5β-Methoxycyclohexan-1-one-3β,4β-dicarboxylic acid anhydride.* To 3.00 g (0.174 mol) of 1-methoxy-3-trimethylsiloxy-1,3-butadiene at 0°C (ice bath) is added a total of 980 mg (0.01 mol) of freshly sublimed maleic anhydride in portions of 70–80 mg each over a period of 25 min. When the addition is complete, the ice bath is removed and the clear solution is stirred for 15 min at room temperature (Note 7). Three 5-mL portions of a solution of tetrahydrofuran (35 mL) and 0.1 N hydrochloric acid (15 mL) are added and the solution is stirred for 1 min. The remaining acid solution (35 mL) is added all at once and the resulting solution is poured into 100 mL of chloroform and treated with 25 mL of water. The organic layer is separated and the aqueous layer is extracted four times with 100-mL portions of chloroform. The extracts are combined and dried over anhydrous magnesium sulfate. The solvent is then removed under reduced pressure (Note 8) to provide 2.0 g of an oil which solidifies. Pentane (10 mL) is added to the oily solid and small portions of ether (total of 6 mL) are added; trituration is continued until the crystals become free flowing. The crystals are isolated by filtration and washed with 10 mL of 2 : 1 pentane/ether to afford 1.75 g (90%) of the anhydride, mp 87–89°C. Further recrystallization affords an analytically pure sample, mp 97–98°C.

2. Notes

1. The checkers dried all reagents by allowing them to stand over molecular sieves (Type 4A), with the exception of triethylamine, which was dried over potassium hydroxide pellets. The reaction flask was flame-dried. Because of evolution of triethylamine hydrochloride that was encountered during addition of the chloro-trimethylsilane and in the workup, the reaction should be carried out in a hood.

2. The checkers did not cool the reaction, which allowed the temperature to rise to 55°C. After 30 min, the solution was heated overnight with a heating mantle. After 12 hr, the reaction temperature was 67°C.

3. A 16-in. Widmer column packed with 3-mm glass helices may also be used for the distillation.

4. *Caution! When the temperature begins to drop, heating must be stopped. Otherwise, on occasion, a violent reaction may occur with formation of a gas and rapid expansion of the residual tars.*

5. The checkers performed this distillation at a lower pressure (1–10 mm) through a similar Vigreux column to yield 225 g of clear liquid containing fluffy white material (triethylamine hydrochloride) that could not be removed by filtration. The purity of this distillate, determined by NMR, was 90 : 10 (diene : ketone). No forerun was obtained which contained more than 15% ketone.

6. The checkers carefully redistilled the impure distillate through the same previously mentioned distillation apparatus under water vacuum. Six fractions of various amounts were collected and combined to yield (1) a forerun of 64 g, bp 70–78°C (23–25 mm), purity 77 : 23 (diene : ketone); and (2) 145 g of pure diene, bp 78–81°C (23–25 mm). This second distillation seemed to remove the triethylamine hydrochloride from the product.

7. The reaction mixture is initially yellow, but turns colorless when the solution is warmed to room temperature.

8. When the chloroform extract is concentrated, care must be exercised to avoid overheating. The temperature should be no greater than 40°C.

3. Discussion

The procedure described here is a scale-up of the published method[2] for the preparation of 1-methoxy-3-trimethylsiloxy-1,3-butadiene (**2**) from readily available reagents. The preparation of this diene has recently been complemented by a report of the preparation of 1,3-bis(trimethylsiloxy)-1,3-butadiene,[3] and earlier by a reported synthesis of a 1,3-dialkoxy-1,3-butadiene.[4]

The electron-donating nature of this diene confers high reactivity and orientational specificity in its reaction with unsymmetrical dienophiles.[5] This fact, coupled with the readily available conversion to the α,β-unsaturated ketone from the imparted functionality, makes 1-methoxy-3-trimethylsiloxy-1,3-butadiene (2) a potentially very valuable reagent in organic synthesis. The general reaction scheme is illustrated below:

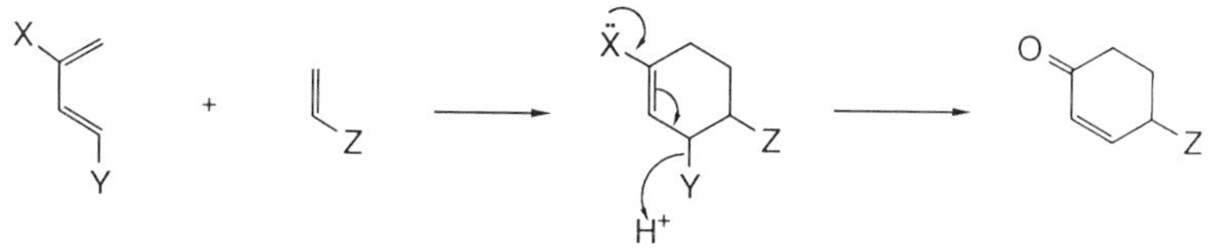

The high reactivity of the diene is shown by reaction with notoriously unreactive dienophiles such as 1-carbomethoxycyclohexene, 2,5-dihydrobenzoic acid methyl ester,[6] and 2-methylcyclohex-2-en-1-one to give, after mild work-up, the corresponding α,β-unsaturated ketones in quite respectable yields.[5]

The Diels–Alder reaction with maleic anhydride is illustrative of the high reactivity and potential utility of this diene.

1. Department of Chemistry, University of Pittsburgh, Pittsburgh, PA 15260. Current address of S.D.: Department of Chemistry, Yale University, New Haven, CT 06520.
2. Danishefsky, S.; Kitahara, T. *J. Am. Chem. Soc.* **1974**, *96*, 7807.
3. Ibuka, T.; Mori, Y.; Inubushi, Y. *Tetrahedron Lett.* **1976**, 3169.
4. Wolinsky, J.; Lozin, R. B. *J. Org. Chem.* **1970**, *35*, 1986.
5. Danishefsky, S.; Kitahara, T. *J. Org. Chem.* **1975**, *40*, 538.
6. Danishefsky, S.; Kitahara, T.; Schuda, P. F.; Etheredge, S. J. *J. Am. Chem. Soc.* **1976**, *98*, 3028.

PHOTOCYCLIZATION OF AN ENONE TO AN ALKENE: 6-METHYLBICYCLO[4.2.0]OCTAN-2-ONE

(Bicyclo[4.2.0]octan-2-one, 6-methyl-)

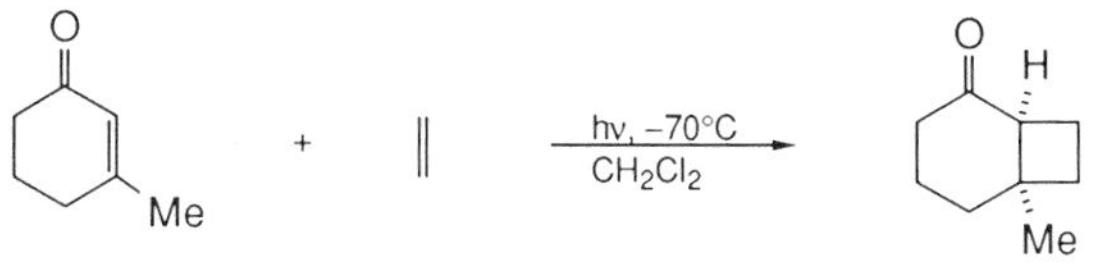

Submitted by R. L. CARGILL,[1a] J. R. DALTON, G. H. MORTON, and W. E. CALDWELL[1]
Checked by BARRY A. WEXLER, AMOS B. SMITH, III, and CARL R. JOHNSON

1. Procedure

The irradiation apparatus (Note 1) is charged with a solution of 25.0 g (0.277 mol) of 3-methyl-2-cyclohexenone (Note 2) in reagent-grade dichloromethane (Note 3). A gas outlet tube to an efficient hood is placed in one 14/20 standard taper joint; in the other, there is a stopper that can be removed for periodic sampling. The cooling water is turned on (Note 4) and the apparatus is immersed in a dry-ice/2-propanol bath while the chilled solution is saturated with ethylene (Note 5). The lamp is inserted into the well and turned on (Note 6). Progress of the irradiation is conveniently followed by gas chromatography (Note 7). After about 8 hr, most of the starting material has reacted. At this time, the lamp is turned off and the apparatus removed from the cooling bath. The reaction mixture is degassed with a slow stream of nitrogen while it warms to room temperature, dried over magnesium sulfate, and concentrated with a rotary evaporator at a temperature below 30°C (Note 8). The product is isolated by distillation to afford 27–28 g (86–90%) of 6-methylbicyclo[4.2.0]octan-2-one, bp 62–65°C (3.5 mm) (Notes 9 and 10).

2. Notes

1. The apparatus is similar to one described earlier.[2] A triple-walled Dewar is constructed of Pyrex according to Figure 1. For further discussion concerning this immersion well, contact Joel M. Babbitt, Glassblower, Department of Chemistry, University of South Carolina, Columbia, SC 29208. The evacuated jacket permits the safe use of circulating tap water as a lamp coolant even when irradiations are conducted in a dry ice bath. A further advantage is that three layers of Pyrex constitute an effective filter for light in the 280–300-nm region so that secondary photolysis of cycloadducts is not usually observed. The irradiation flask is a cylindrical vessel of suitable volume fitted with a coarse, fritted disc for gas dispersion and a flanged lip. The light source is either a GE H1000-A36-15, Westinghouse H-36GV-1000, or equivalent lamp with the outer globe removed, used in conjunction with a GE 35-9627-6009 ballast. These lamps are available from the General Electric Company, Lamp Division, Charlotte, North Carolina.

2. This material can be purchased from Aldrich Chemical Company or prepared from Hagemann's ester.[3]

3. The volume of solution will vary depending on the exact volume of the apparatus, the temperature, and the miscibility of gaseous reactant in the solvent. The solution should completely surround the lamp, but should not overflow the vessel. The submitters used a volume of 1100 mL, and the checkers used 200 mL.

4. If the flow of cooling water is stopped while the apparatus is cold, the water may freeze and crack the immersion well. The vacuum jacket provides greater insurance against this problem than is available in the commercially available wells used with the usual 450 watt lamps.[2]

5. CP-grade ethylene (Matheson) was used without purification. A flow of ca. 100 mL/min of ethylene for 2–3 hr is adequate for saturation. Gas flow is continued throughout the irradiation in order to maintain a high concentration of ethylene and for stirring.

6. The lamp will not start if it is too cold or too hot. The practice of blowing nitrogen over the lamp to remove ozone is not recommended as this cools the lamp and decreases its output significantly, resulting in an unnecessarily long irradiation period.

7. The submitters used a Varian 1200 FID chromatograph with a 7% Carbowax 20 M on Chromosorb Q, 8-ft × 0.125-in. column, a carrier gas (N_2) flow rate of

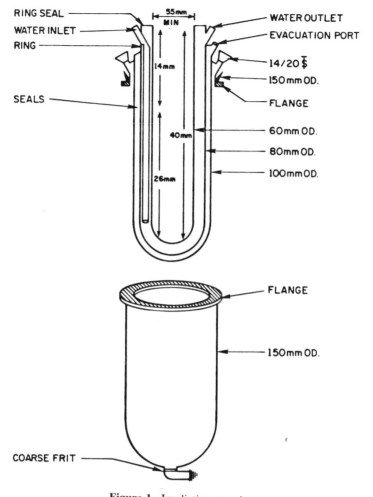

Figure 1. Irradiation vessel.

40 mL/min; column, 160°C; injector, 220°C; and detector, 215°C. Retention times were 3-methyl-2-cyclohexenone, 4.2 min, and 6-methylbicyclo[4.2.0] octan-2-one, 3.9 min, respectively.

8. If the solvent is removed without care, a considerable amount of volatile product may be lost.

9. This material is contaminated with ~10% of 3-methylcyclohexenone. Material of greater purity can be obtained by extending the time of irradiation, carrying out an efficient distillation of product, or decomposing starting material with potassium permanganate prior to distillation.

10. The product has the following spectral properties: IR (CCl$_4$) cm^{-1}: 1700; ^{1}H NMR (CCl$_4$) δ: 1.21 (s, 3 H, methyl), 1.9 (m, 11 H, all other protons); ^{13}C

NMR (C_6D_6) δ (based on δ C_6D_6 128.00): 211.63, 51.34, 40.86, 39.45, 35.26, 31.20, 28.84; 21.45, 20.35; mass spectrum (m/e) 138.1041 (parent ion).

3. Discussion

Although photochemical cycloadditions have gained acceptance in synthetic chemistry,[4] most such reactions are limited to a relatively small scale. The use of a 1000-W street lamp permits the irradiation of up to 1 mol of substrate in less time than 0.2 mol can be irradiated with the conventional 450-W lamps. Thus, under optimum conditions, the submitters were able to add ethylene to 3-methylcyclohexenone on a 20-g scale in 48 hr (80%) with a 450-W lamp; with the apparatus described here 94 g of this enone was condensed with ethylene in 8 hr (91%).

Some general points regarding photochemical cycloadditions deserve mention: (1) since the reaction is first-order in olefin, the concentration of olefin (especially gaseous olefins) is of critical importance; therefore, the cycloadditions are carried out at low temperature (in some cases, however, low temperature can be detrimental[5]); (2) since lamp output deteriorates with lamp age, the rates of otherwise identical cycloadditions

TABLE I
PREPARATIVE-SCALE CYCLOADDITIONS

Entry	Enone	Weight (g)	Olefin	Time (hr)	Product(s)	Yield (%)	Ref.
1.	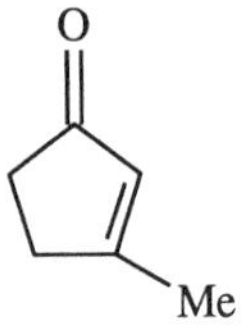	25	C_2H_4	12	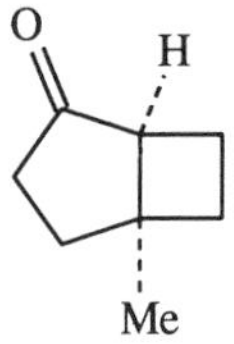	90	6
2.	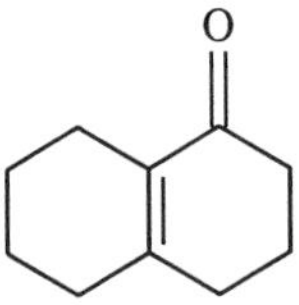	10	C_2H_4	6	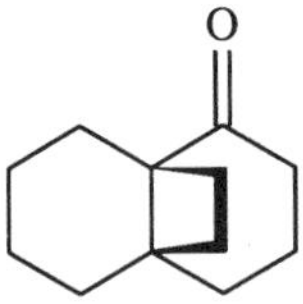	71	7
3.	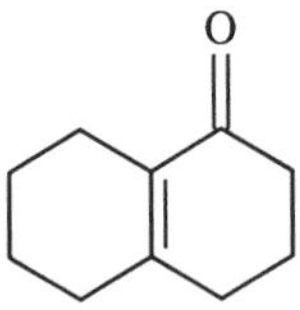	20	ClHC=CHCl	10	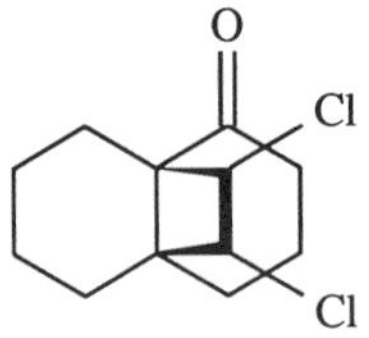	93[a]	8

[a] A mixture of cis and trans olefins was used; a mixture of diastereomeric products was obtained. Both olefins give similar mixtures.

are unlikely to be the same; therefore, it is of critical importance that the progress of each photochemical reaction be followed by some suitable means (GLC, IR, UV, NMR, etc); and (3) as long as all the incident light of appropriate wavelength is absorbed by the enone, the reaction proceeds at a rate independent of enone concentration; thus, the highest concentration of enone at which dimerization can be avoided is optimal.

Several examples of preparative cycloadditions are listed in Table I.

1. Department of Chemistry, University of South Carolina, Columbia, SC 29208. Present address: Cargill Interests, Ltd., P.O. Box 992, Longview, TX 75606.
2. Bloomfield, J. J.; Owsley, D. C. *Org. Photochem. Synth.* **1976**, *2*, 36.
3. Smith, L. I.; Rouault, G. F. *J. Am. Chem. Soc.* **1943**, *65*, 631.
4. For reviews, see Eaton, P. E. *Acc. Chem. Res.* **1968**, *1*, 50; de Mayo, P. *Acc. Chem. Res.* **1971**, *4*, 41; Bauslaugh, P. G. *Synthesis* **1970**, 287; Sammes, P. G. *Synthesis* **1970**, 636; Dilling, W. L. *Chem. Rev.* **1966**, *66*, 373; **1969**, *69*, 845; Wiesner, K. *Tetrahedron* **1975**, *31*, 1655; Dilling, W. L. *Photochem. Photobiol.* **1977**, *25*, 605; **1977**, *26*, 557.
5. Loutfy, R. O.; de Mayo, P. *J. Am. Chem. Soc.* **1977**, *99*, 3559.
6. Cargill, R. L.; Wright, B. W. *J. Org. Chem.* **1975**, *40*, 120.
7. Peet, N. P.; Cargill, R. L.; Bushey, D. F. *J. Org. Chem.* **1973**, *38*, 1218.
8. Rae, I. D.; Umbrasas, B. N. *Aust. J. Chem.* **1975**, *28*, 2669.

METHYL α-(BROMOMETHYL)ACRYLATE

(2-Propenoic acid, 2-(bromomethyl)-, methyl ester)

A. $(HOCH_2)_2C(CO_2C_2H_5)_2$ + HBr $\xrightarrow[\text{6 hr}]{\text{reflux}}$ $(BrCH_2)_2CHCO_2H$

B. $(BrCH_2)_2CHCO_2H$ + CH_3OH $\xrightarrow[\text{ClCH}_2\text{CH}_2\text{Cl}]{\text{CH}_3\text{SO}_3\text{H}}$ $(BrCH_2)_2CHCO_2CH_3$

C. $(BrCH_2)_2CHCO_2CH_3$ + $(CH_3CH_2)_3N$ $\longrightarrow$ $CH_2{=}\underset{\overset{|}{CH_2Br}}{C}CO_2CH_3$

Submitted by John M. Cassady, Gary A. Howie, J. Michael Robinson, and Ioannis K. Stamos[1]
Checked by Paul R. West and Orville L. Chapman

1. Procedure

Caution! Methyl α-(bromomethyl)acrylate is a potent vesicant and lachrymator and should be handled with care. All operations should be carried out in an efficient hood in order to avoid contact.

A. *β,β'-Dibromoisobutyric acid.* To a 5-L, single-necked flask, equipped with a heating mantle, 22-cm Vigreux distillation head, thermometer, 30-cm water-cooled condenser with adapter, and 1-L, ice-cooled receiving vessel (Note 1) is added 440 g (2.0 mol) of diethyl bis(hydroxymethyl)malonate (Notes 2[2] and 3) and 3540 mL of concentrated aqueous hydrobromic acid (Note 4). Heating for 6 hr at vigorous reflux gives 2400 mL of aqueous distillate (Note 5). The undistilled concentrate is poured into a 3-L beaker, cooled overnight at $-15°C$ (Note 6), and filtered through a 500-mL fritted-glass Büchner funnel using aspirator vacuum. After suction air drying for 6 hr, drying is continued for 6 days in a vacuum desiccator containing active Drierite and under 10 mm of initial vacuum (Note 7) to give 332 g (67.5%) of β,β'-dibromoisobutyric acid as a brown solid. Distillation of the filtrate to remove an additional 850 mL of aqueous hydrobromic acid (Note 8), followed by cooling and filtration, gives an additional 34.0 g (6.9%) of solid. Crude product, obtained in 74–85% yield (Note 9[3]), is suitable for use without further purification (Note 10).

B. *Methyl β,β'-dibromoisobutyrate*. In a 200-mL, round-bottomed flask fitted with a reflux condenser are placed 61.5 g (0.25 mol) of β,β'-dibromoisobutyric acid, 25 g (0.78 mol) of commercial methanol, 75 mL of ethylene dichloride, and 0.2 mL of methanesulfonic acid (Notes 11[4] and 12). The reaction mixture is heated under reflux for 24 hr. The solution is cooled to room temperature, diluted with about 200 mL of methylene chloride, and neutralized with dilute, cold sodium bicarbonate solution (Note 13). The organic layer is dried over anhydrous sodium sulfate and concentrated on a rotary evaporator to remove most of the methylene chloride. Fractional distillation of this residue under reduced pressure (the receiver is cooled with an ice–salt mixture) yields 48.8 g (75%) of product, bp 64–65°C (0.3 mm).[3,5,6]

C. *Methyl α-(bromomethyl)acrylate*. In a dry, 250-mL three-necked flask, equipped with a mechanical stirrer, reflux condenser, and an addition funnel, 20 g (0.077 mol) of methyl β,β'-dibromoisobutyrate (Note 14[5,6]) in 50 mL of anhydrous benzene (Note 15[7]) is stirred vigorously. Triethylamine (Notes 16[8] and 17) (7.7 g, 0.076 mol) in 50 mL of benzene is introduced dropwise at a rate of about 3 mL per min. After the addition is complete the mixture is stirred for an additional 1 hr at room temperature, refluxed for 1 hr, and then cooled to 20°C. The reaction mixture is filtered with suction and the amine salt washed twice with 20 mL of benzene. The filtrate and washings are combined in a round-bottomed flask and concentrated on a rotary evaporator at 30–35°C to remove most of the benzene. The residue is transferred to a small distillation apparatus and fractionally distilled at reduced pressure using an oil bath at 50–55°C. The yield of ester collected at bp 35–37°C (1.3 mm) is 11.0 g (80%) (Notes 18–21).

2. Notes

1. Cooling the receiving vessel greatly reduces loss of ethyl bromide.

2. Diethyl bis(hydroxymethyl)malonate was prepared up to an 8.0-mol scale by the method of Block.[2] After suction filtration to remove the drying agent, the dried diethyl ether extracts were concentrated directly on a Büchi rotary evaporator at aspirator vacuum using a bath temperature of 50°C; concentration was continued for ca. 2 hr after removal of the ether. The crude, oily diethyl bis(hydroxymethyl)malonate, obtained in 94–96% yield, solidified on standing and was suitable for use without further purification. The malonate can be stored at room temperature with no special precautions.

3. The checkers ran this reaction on a 20% scale [starting with 88 g (0.4 mol) of diethyl bis(hydroxymethyl)malonate]. At this scale, yields between 63 and 75% were realized.

4. Initial experiments used commercial 48% aqueous hydrobromic acid. In subsequent runs no decrease in yields was apparent when recovered distillate boiling at or above 110°C was substituted for the commercial acid.

5. Approximately 45 additional min of heating was required to reach distillation temperature. The first 780 mL (excluding ethyl bromide) of aqueous distillate boiled below 110°C and was discarded. The remaining distillate was recycled as described in Note 4.

6. Cooling in a refrigerator freezing compartment is satisfactory. The beaker should be sealed (e.g., using Saran Wrap) to prevent escape of corrosive fumes.

7. After the solid was dried in the desiccator, weight reductions of up to 10% were observed.

8. Special care must be used toward the end of the distillation to avoid overheating caused by removal of too much solvent. Overheating can result in an intractable gummy residue.

9. Failure to distill the maximum amount of concentrated hydrobromic acid, higher crystallization temperatures, and/or washing with water may account for the lower (66%) reported[3] yield.

10. Storage at room temperature (under nitrogen or in a filled, sealed container) for periods in excess of 1 year resulted in no significant deterioration of the crude acid as judged by its suitability for use in step B. Preparation of acid was done on a 0.5–3.4-mol scale with no significant variation in yield.

11. These conditions are patterned after a general procedure for esterification reported by Clinton and Laskowski.[4]

12. The checkers ran this reaction on a 50% scale [starting with 30.75 g (0.125 mol) of β,β'-dibromoisobutyric acid] and obtained yields ranging from 66 to 67%.

13. A brown, emulsified layer, which separates on long standing, is formed between the organic and aqueous layers. This layer can also be taken up with an additional 200 mL of methylene chloride and dried with a sufficient amount of anhydrous sodium sulfate to recover the organic layer.

14. It is recommended that methyl β,β'-dibromoisobutyrate[5,6] that has been purified by fractional distillation be used, since the presence of acidic compounds reduces the yield and the presence of any hydroxyl function gives a product mixture that cannot be purified by simple distillation.

15. The preparation of anhydrous benzene has been described.[7]

16. Commercial triethylamine is conveniently purified by two distillations from a 2% solution of phenyl isocyanate.[8]

17. In a parallel experiment, ethyldiisopropylamine (9.82 g, purified as in Note 16) was mixed with a solution of 20 g of methyl β,β'-dibromoisobutyrate in 100 mL of dry benzene. The reaction mixture was stored at room temperature for 10 hr

and gently refluxed for 1 hr under nitrogen in the dark. After workup and distillation the yield of the product was 80%.

18. Distillation at higher temperatures results in viscous residues with considerably reduced yields of the product. The receiver should be immersed in an acetone-dry ice bath in order to prevent loss of the product to the trap of the vacuum line.

19. The product is stable for long periods of time if kept under an inert atmosphere in the absence of light and in the refrigerator.

20. Ethyl α-(bromomethyl)acrylate is prepared similarly, bp 38–42°C (0.8 mm).

21. The checkers obtained a 76% yield.

3. Discussion

Although methyl and ethyl α-(bromomethyl)acrylate are used extensively as synthetic intermediates in the preparation of a variety of organic compounds,[9-16] many of biological importance, they are not commercially available and their preparation in good yield on a large scale is therefore of interest. The procedures outlined above represent useful modifications of published literature routes to these compounds.

The procedure for the elimination of HBr from the dibromo ester is a modification of the method of Lawton and co-workers for sui generis generation of the methyl[5,9] or ethyl ester[10] during a reaction. Methyl α-(bromomethyl)acrylate has also been prepared by bromination of methyl methacrylate in 700°C steam[17] and by dehydrohalogenation with sodium acetate in acetic acid.[6] Ethyl α-(bromomethyl)acrylate has been prepared by dehydrohalogenation with the monosodium salt of ethylene glycol[3 18] and ethyl diisopropylamine.[11] The latter reaction was reported by Öhler et al. with no experimental details for the elimination reaction. The use of triethylamine as reported in this procedure appears to be the most efficient and convenient method for dehydrobromination to these acrylate esters.

1. Department of Medicinal Chemistry and Pharmacognosy, School of Pharmacy and Pharmacal Sciences, Purdue University, West Lafayette, IN 47907.
2. Block, Jr., P. *Org. Synth., Coll. Vol. V* **1973**, 381.
3. Ferris, A. F. *J. Org. Chem.* **1955**, *20*, 780.
4. Clinton, R. O.; Laskowski, S. C. *J. Am. Chem. Soc.* **1948**, *70*, 3135.
5. McEwen, J. M.; Nelson, R. P.; Lawton, R. G. *J. Org. Chem.* **1970**, *35*, 690.
6. Tanaka, A.; Nakata, T.; Yamashita, K. *Agric. Biol. Chem.* **1973**, *37*, 2365.
7. Johnson, G. D. *Org. Synth., Coll. Vol. IV* **1963**, 900.
8. Breslow, R.; Posner, J. *Org. Synth., Coll. Vol. V* **1973**, 514.
9. Dunham, D. J.; Lawton, R. G. *J. Am. Chem. Soc.* **1971**, *93*, 2074.
10. Nelson, R. P.; McEwen, J. M.; Lawton, R. G. *J. Org. Chem.* **1969**, *34*, 1225.
11. Öhler, E.; Reininger, K.; Schmidt, U. *Angew. Chem. Int. Ed. Engl.* **1970**, *9*, 457.
12. Rosowsky, A.; Papathanasopoulos, N.; Lazarus, H.; Foley, G. E.; Modest, E. J. *J. Med. Chem.* **1974**, *17*, 672.
13. Marschall, H.; Vogel, F.; Weyerstahl, P. *Chem. Ber.* **1974**, *107*, 2852.
14. Cassady, J. M.; Li, G. S.; Spitzner, E. B.; Floss, H. G.; Clemens, J. A. *J. Med. Chem.* **1974**, *17*, 300.
15. Howie, G. A.; Stamos, I. K.; Cassady, J. M. *J. Med. Chem.* **1976**, *19*, 309.

16. Cousse, H.; Bonnaud, B. *J. Labelled Compd. Radiopharm.* **1976,** *12,* 491.

17. Hoffenberg, D. S.; Zaccardo, L. M. U.S. Patent 3213072; *Chem. Abstr.* **1965,** *63,* 17908h.

18. Dickstein, J.; Bodnar, M.; Hoegerle, R. M. U.S. Patent 3094554; *Chem. Abstr.* **1963,** *59,* 12647h.

METHYL DIFORMYLACETATE

(2-Propenoic acid, 2-formyl-3-hydroxy-, methyl ester)

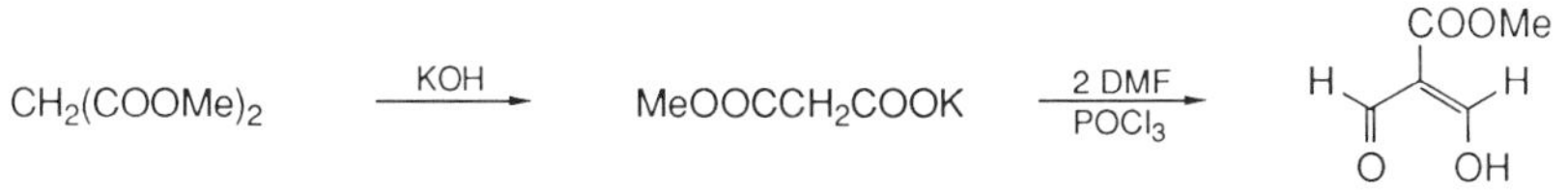

$$CH_2(COOMe)_2 \xrightarrow{KOH} MeOOCCH_2COOK \xrightarrow[POCl_3]{2\ DMF} \underset{O\quad OH}{\overset{COOMe}{H\diagup\diagdown H}}$$

Submitted by C. R. Hutchinson, M. Nakane, H. Gollman, and P. L. Knutson[1]
Checked by David J. Wustrow and Andrew S. Kende

1. Procedure

A. *Potassium monomethyl malonate.* Dimethyl malonate (Note 1, 264.2 g, 2.0 mol) is dissolved in anhydrous methanol (Note 2, 1150 mL) contained in a dry, 3-L, one-necked flask containing a large magnetic stirring bar and protected from atmospheric moisture with a calcium sulfate-filled drying tube. The solution is stirred magnetically and cooled to ice-water bath temperature. Potassium hydroxide pellets (112.2 g, 2.0 mol) are added rapidly to the cold solution and the reaction mixture is allowed to warm to room temperature with stirring overnight. The colorless crystals of potassium monomethyl malonate that form are recovered by suction filtration through a Büchner funnel and washed with anhydrous diethyl ether. The combined filtrate and diethyl ether wash are concentrated at 30°C to a volume of ca. 750 mL on a rotary evaporator. The resulting crystalline precipitate is recovered as before by filtration and washing and combined with the first crop of crystals to give 220 g (71%) of potassium monomethyl malonate as fine colorless needles, mp 204–207°C. These crystals are dried under vacuum (0.1 mm) before use in the following reaction.

B. *Methyl diformylacetate.* Freshly distilled phosphorus oxychloride (612 g, 4 mol) is added dropwise with constant stirring at ambient temperature (Note 3) to dimethyl-formamide (1460 g) contained in a 3-L, three-necked flask equipped with a mechanical paddle stirrer, immersion thermometer, and a 500-mL pressure-equalizing addition funnel fitted with a calcium chloride-filled drying tube. The reaction mixture warms up and turns to a dark reddish-brown color during addition of the phosphorus oxychloride and formation of the Vilsmeier reagent $[(CH_3)_2N^+=CHCl\ Cl^-]$. The addition funnel is replaced with a 10-in. long West condenser (Note 4), and then the reaction mixture is cooled to 0°C by immersing the reaction flask in an ice-salt water bath. The cooling bath is removed and potassium monomethyl malonate (206 g, 1.32 mol) is added to the stirred reaction mixture in ten equal portions over a 30 min period (Notes 3 and 5), keeping the temperature of the mixture below 90°C. The dark-brown mixture then is stirred and heated on a water bath at 90°C for 4 hr. Gas (CO_2 plus HCl) evolves initially from the reaction on heating (Note 6). The thermometer is replaced with a glass stopper, the condenser is fixed for distillation by addition of a distilling head and vacuum distillation receiver, and the reaction solvent is removed from the flask by distillation at ca. 2 mm on a steam bath (Note 7). The resulting dark-brown liquid is poured onto ice (4 kg, Note 3). A saturated aqueous solution of potassium carbonate (1.3 kg) is added slowly to the ice-cold crude reaction product with constant stirring until the pH of the mixture stabilizes at ca. 11. Considerable foaming and gas evolution (CO_2) occur during the addition of the base. The resulting basic solution is stirred magnetically at ambient temperature for 48 hr and then extracted with ethyl acetate in four 1-L portions. The organic phases are discarded, and the aqueous phase is saturated with potassium chloride (500 g) by stirring at ambient temperature until no more salt dissolves. This mixture is mixed with ice (1 kg), slowly acidified to pH 1 with ice-cold 12 N hydrochloric acid, and then thoroughly extracted with four 2-L portions of diethyl ether (Note 8). The combined cold ether extracts are washed with a saturated aqueous solution of potassium chloride (4 L) and dried over anhydrous sodium sulfate (500 g) for 1 hr. The solution is decanted from the desiccant, combined with a 500-mL diethyl ether wash of the desiccant, concentrated under reduced pressure to ca. 500 mL, and redried over anhydrous sodium sulfate. After removal of the desiccant by gravity filtration, the diethyl ether is removed by rotary evaporation at water aspirator pressure and 25°C. Fractional distillation of the resulting liquid residue at 2 mm with a N_2 bleed capillary through a Claisen head first gives a little dimethylformamide. When dimethylformamide ceases to distill (Note 9), the receiver is cooled in a dry ice–ethanol bath and the methyl diformylacetate distilled at 58–61°C to give 86–94 g (50–55%) of a colorless, solid distillate, which melts at about 10°C (Notes 10 and 11). Methyl diformylacetate prepared in this way is stable for at least 6 months if stored at −20°C.

2. Notes

1. All reagents are used as received from commercial suppliers unless stated otherwise.
2. Reagent grade methanol is made anhydrous by refluxing over $Mg(OCH_3)_2$ according to the method of Vogel.[2]
3. All the following operations must be done in a hood.
4. The desiccant in the drying tube should be replaced with fresh calcium chloride and the drying tube fitted onto the top of the condenser.
5. The salt is added by replacing the condenser with a glass powder funnel, quickly pouring the crystalline solid through the funnel into the flask, and then replacing the powder funnel with the condenser.
6. The condenser can be cooled to prevent loss of solvent that is carried out of the reaction mixture by the escaping gases.
7. The volume of dimethylformamide distillate is ca. 1000 mL, and distillation is stopped when no more liquid distills from the reaction mixture.
8. The solvent extractions and washings should be done as rapidly as is possible since the crude methyl diformylacetate is not stable to small amounts of acid or base over long periods.

9. Dimethylformamide ceases to distill at a temperature less than 30°C.

10. Considerable product is lost if the receiver is not chilled to a low temperature.

11. The submitters have obtained the same yield when this procedure was done on scales from 0.5 to 1.3 mol.

3. Discussion

Methyl diformylacetate can be prepared from ketene and trimethyl orthoformate,[3] or methyl propiolate and methanol,[4] via formylation of the methyl 3,3-dimethoxypropanoate intermediate (equation 1). The present procedure is better because it avoids the tedious preparation of ketene,[3] affords a superior yield,[3] or is much cheaper[4] than the other two methods. A fourth method[5] for its preparation (equation 2) should permit the preparation of any ester of diformylacetic acid that is stable to Birch reduction and ozonolysis conditions. However, this method is not convenient for use above a 0.1-mol scale, nor recommended for reasons of safety because of the amount of an O_2/O_3, mixture needed at larger scales.

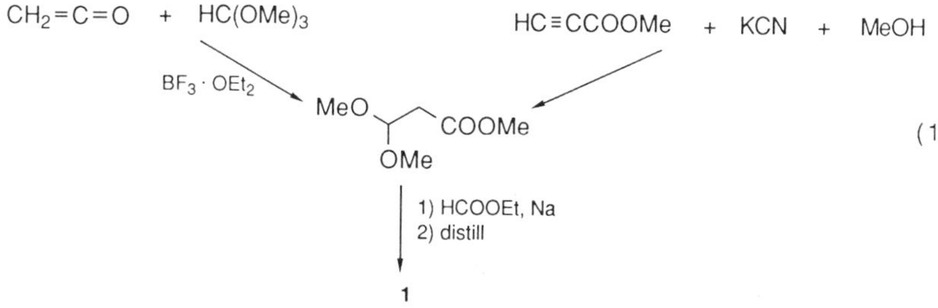

1. School of Pharmacy, University of Wisconsin, Madison, WI 53706.
2. Vogel, A. I. "A Textbook of Practical Organic Chemistry," 3rd ed.; Wiley: New York, 1956, p. 167.
3. Büchi, G.; Carlson, J. A.; Powell, J. E., Jr.; Tietze, L.-F. *J. Am. Chem. Soc.* **1973**, *95*, 540.
4. Baldwin, S. W. personal communication, 1979; Walia, J. S.; Walia, A. S. *J. Org. Chem.* **1976**, *41*, 3765.
5. Nakane, M.; Gollman, H.; Hutchinson, C. R.; Knutson, P. L. *J. Org. Chem.* **1980**, *45*, 2536.

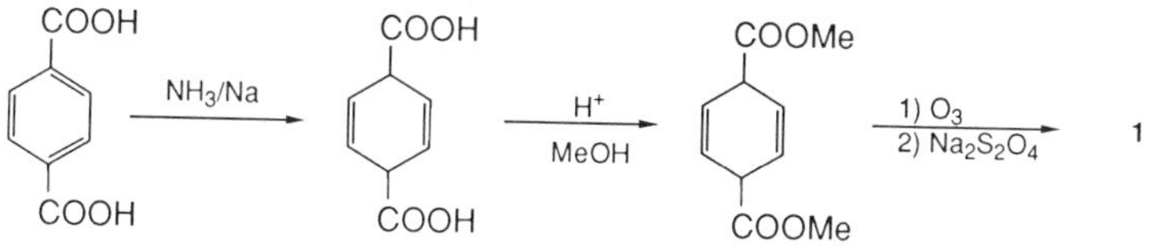

SELECTIVE HALOGEN–LITHIUM EXCHANGE REACTIONS OF 2-(2'-HALOPHENYL)ETHYL HALIDES: SYNTHESIS OF 4,5-METHYLENEDIOXYBENZOCYCLOBUTENE AND 1-PHENYL-3,4-DIHYDRO-6,7-METHYLENEDIOXYISOQUINOLINE

(Cyclobuta[*f*]-1,3-benzodioxole, 5,6-dihydro- and 1,3-dioxolo[4,5-*g*]isoquinoline, 7,8-dihydro-5-phenyl-)

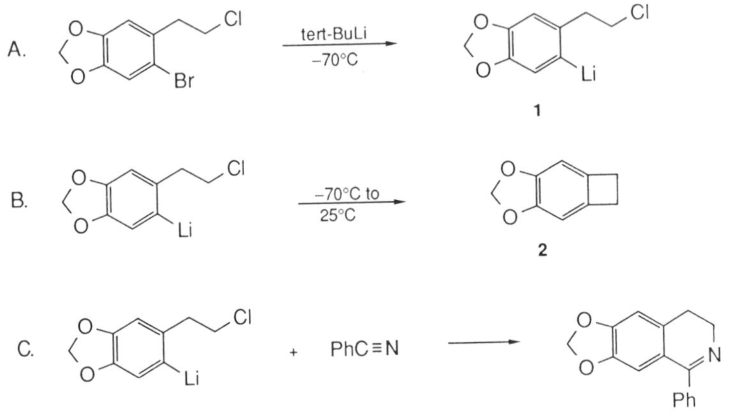

Submitted by DENNIS J. JAKIELA, PAUL HELQUIST,[1] and LAWRENCE D. JONES[2]
Checked by NEVILLE D. EMSLIE and IAN FLEMING

1. Procedure

A. *2-(2'-Lithio-4',5'-methylenedioxyphenyl)ethyl chloride.* A 500-mL, three-necked, round-bottomed flask, equipped with a magnetic stirring bar, 50-mL pressure-equalizing addition funnel (Note 1), low-temperature thermometer, and a three-way stopcock having a vertically oriented tube capped with a rubber septum and a horizontal tube connected to a source of dry nitrogen and vacuum, is charged with 10.0 g (37.9 mmol) of 2-(2'-bromo-4',5'-methylenedioxyphenyl)ethyl chloride (Note 2). The assembled apparatus is evacuated and refilled with nitrogen three times. Freshly distilled diethyl ether (200 mL) (Note 3) is added to the flask by means of a double-ended needle (0.5 m in length) inserted through the vertical tube of the stopcock while a slight vacuum is applied to the apparatus. A slightly positive pressure of nitrogen is then maintained in the apparatus throughout the course of the reaction. The solution is cooled in a dry ice–acetone bath (Note 4). The glass jacket (or styrofoam cup) (Note 1), which surrounds the addition funnel, is filled with powdered dry ice, and 33 mL of a 2.3 *M* solution of *tert*-butyllithium (76 mmol) in pentane (Note 5) is added to the addition funnel by means of a syringe. After 10 min the lithium reagent is added dropwise to the flask over a period of 1 hr, while the temperature of the reaction mixture is maintained below −60°C. The solution of the resulting aryllithium reagent 1 is then used in either of the two reactions described below.

B. *4,5-Methylenedioxybenzocyclobutene.* The reaction mixture from Step A is simply allowed to warm to room temperature over a period of several hours, during which time a white precipitate forms. After 18 hr, 100 mL of water is slowly added and the mixture is transferred to a 500-mL separatory funnel. As the mixture is shaken, the solid dissolves in the aqueous phase, which becomes light brown. The aqueous layer is extracted with two 75-mL portions of diethyl ether, and the combined organic layers are reduced in volume to 150 mL by rotary evaporation, washed with 75 mL of water and then 75 mL of saturated aqueous sodium chloride solution, dried over magnesium sulfate, filtered, and concentrated to dryness by rotary evaporation to give 5.6 g of pale-yellow solid (Note 6). This crude product is transferred to a large, dry ice-cooled sublimation apparatus (Note 7) and sublimed over a 6-hr period at 35°C (0.07 mm), at which time a dark-brown oil remains in the bottom of the apparatus. The vacuum is released by filling the apparatus with nitrogen, and the cooled portion of the apparatus is allowed to warm to room temperature. Pure 4,5-methylenedioxybenzocyclobutene, **2** (5.1–5.2 g, 91–93%) is obtained as colorless crystals, mp 60–62°C (Note 8).

C. *1-Phenyl-3,4-dihydro-6,7-methylenedioxyisoquinoline.* The reaction mixture containing the aryllithium intermediate is stirred for 15 min (internal temperature −65 to −68°C), and then 4.3 mL (42 mmol) of distilled benzonitrile is added quickly. The mixture is allowed to warm gradually to room temperature and the stirring is continued overnight. The yellow solution (Note 9) is diluted with 25 mL of ether, the mixture is poured into a 1-L separatory funnel, and the reaction flask is rinsed with an additional 75 mL of ether. The combined ether solutions are washed with 150 mL of water and then extracted with three 75-mL portions of 10% (w/w) hydrochloric acid. The combined acid extracts are rendered basic by the addition of 100 mL of 20% (w/w) aqueous sodium hydroxide solution, and the resulting milky white mixture is extracted with three 75-mL portions of dichloromethane. The combined organic extracts are washed with 50 mL of water and 50 mL of saturated aqueous sodium chloride solution, dried over magnesium sulfate, and concentrated to dryness by rotary evaporation, to give 8.96 g (94% crude yield) of orange–tan solid. This material is purified by recrystallization from ethyl acetate : acetone 2 : 1 (v : v) to give a first crop (6.8 g), and by flash chromatography[3] of the residue from the mother liquor, using 150 g of 230–400-mesh silica gel (Merck), a 40-mm diameter column, and elution with 10 : 1 (v : v) ethyl acetate : methanol. A fast-moving orange band and a slower moving lemon–yellow band can be seen clearly on the column. The lemon-yellow band is collected from the column and evaporation gives a second crop (1.4 g) of comparably pure material. The total yield of the pale-yellow isoquinoline is 8.2 g (86%), mp 135–137°C (Note 10).

2. Notes

1. The checkers used a home-made, glass-jacketed funnel sealed with a rubber septum. The submitters cut one side and part of the bottom of a styrofoam cup and held this in place with tape around the lower part of the addition funnel.

2. This starting material is prepared in three steps from commercially available (from Research Organic/Inorganic Chemical Corp., Belleville, NJ) 3,4-methylenedioxyphenylacetic acid according to well-established procedures that have

been applied to similar compounds.[4] As an alternative starting material that could be used in a closely related fashion, 3,4-methylenedioxyphenylacetonitrile is available from Aldrich Chemical Company, Inc. First, 16.0 g (88.8 mmol) of the acid, recrystallized from chloroform, is dissolved in 50 mL of tetrahydrofuran, and the solution is added to a suspension of 5.98 g (158 mmol) of lithium aluminum hydride powder in 225 mL of distilled diethyl ether (Note 3) at 0°C. *(Caution: Lithium aluminum hydride is very sensitive to mechanical shock and very reactive toward moisture and other protic substances; its dust is very irritating to skin and mucous membranes. It should not be allowed to come into contact with metallic species of apparatus, including metal spatulas, because of the potential danger of metal ion-promoted detonation.)* The mixture is stirred at 25°C for 16 hr and is then quenched[5] by the careful, dropwise addition of 6 mL of 15% aqueous sodium hydroxide, and finally 18 mL of water. *(Caution: The reaction of excess lithium aluminum hydride with water is very exothermic and produces a large volume of hydrogen gas.)* The resulting mixture is stirred for 1 hr and is then subjected to vacuum filtration. The white solid that is retained is washed with three 50-mL portions of diethyl ether, and the combined filtrates are concentrated by rotary evaporation to give 13.1 g (89%) of 2-(3',4'-methylenedioxyphenyl)ethanol as a clear yellow oil, bp 136–140°C (0.003 mm). Next, 12.7 g (76.5 mmol) of this compound and 7.4 mL (91.5 mmol) of pyridine are dissolved in 200 mL of dichloromethane at 0°C, and 4.3 mL (83.9 mmol) of neat bromine is added to the solution over a 4-min period. After the solution has been stirred at 25°C for 16 hr, it is washed with three 50-mL portions of 2 N hydrochloric acid, two 50-mL portions of saturated aqueous sodium sulfite, two 50-mL portions of water, and 50 mL of saturated aqueous sodium chloride. The organic layer is then dried over anhydrous magnesium sulfate and concentrated by rotary evaporation to give 18.6 g (99.5%) of yellow solid. Recrystallization from a mixture of 160 mL of hexane and 60 mL of ethyl acetate gives 14.6 g (78%) of 2-(2'-bromo-4',5'-methylenedioxyphenyl)ethanol as light yellow needles, mp 93–94°C. Finally, 9.95 mL (123 mmol) of distilled pyridine and 8.75 mL (120 mmol) of distilled thionyl chloride are added separately to a solution of 14.4 g (58.8 mmol) of the preceding product and 180 mL of chloroform at 25°C. The mixture is heated at reflux for 18 hr, cooled to 25°C, washed with 40 mL of 1 N hydrochloric acid, 40 mL of 5% aqueous sodium carbonate, two 40-mL portions of water, and 40 mL of saturated aqueous sodium chloride, dried over anhydrous magnesium sulfate, and concentrated by rotary evaporation to give 14.0 g (90%) of brown crystals. Distillation gives 13.0 g (84%) of an oil (bp 130–134°C at 0.006 mm), which solidifies to give the final product as colorless crystals: ^{1}H-NMR (CDCl$_3$) δ: 3.08 (t, 2 H, $J = 6.8$), 3.67 (t, 2 H, $J = 6.8$), 5.95 (s, 2 H), 6.74 (s, 1 H), and 6.98 (s, 1 H); mp 47.0–47.5°C (corrected).

3. Commercially available anhydrous diethyl ether is distilled under nitrogen from a solution of the sodium benzophenone radical anion generated by treating a solution of 10 g of benzophenone and 1 L of ether with 10 g of sodium ribbon until a dark-blue or purple color persists.

4. Although the dry ice–acetone bath itself attains a temperature of $-78°C$, the lowest temperature achieved by the solution within the flask is only $-68°C$.

5. *Caution: tert-Butyllithium is pyrophoric in air; excess quantities of the reagent in the syringe should be discarded very carefully.* The checkers used the reagent available from Aldrich Chemical Company Ltd., England and standardized it by double titration with ethylene dibromide and hydrochloric acid.[6]

6. The submitters also ran the reaction on smaller scales using from 0.5 to 5.0 g of starting material and regularly obtained a crude yield of 98–105% at this stage.

7. The sublimation apparatus should have a least a 1-cm separation between the upper surface of the crude solid to be sublimed and the bottom of the cooling surface in order to avoid splattering of the oily residue onto the purified product near the end of the sublimation procedure.

8. The product showed the following spectral properties: ^1H NMR ($CHCl_3$) δ: 3.00 (s, 4 H), 5.75 (s, 2 H), and 6.50 (s, 2 H).

9. At this stage, the submitters had a brick-red reaction mixture that became yellow on dilution with ether.

10. The product showed the following spectral properties: ^1H-NMR ($CDCl_3$) δ: 2.67 (t, 2 H, $J = 7.5$), 3.73 (t, 2 H, $J = 7.5$), 5.83 (s, 2 H), 6.63 (s, 2 H), and 7.37 (m, 5 H).

3. Discussion

The halogen–metal exchange reaction was pioneered by Gilman and coworkers,[7] who established that substituted aryl bromides would exchange efficiently with butyllithium and that the reaction was of synthetic value provided the substituent was not reactive toward alkyl- or aryllithium reagents. More recently, Parham[4e, 8] and others[4f, 9] further defined the scope and limitations of this reaction by demonstrating that haloarenes substituted with electron-withdrawing (CO_2H, CN, CO_2R) or electron-donating [OR, OCH_2O, $(-CH_2-)_nX$, where X = Br, Cl] functional groups would selectively exchange with alkyllithium reagents at low temperature. While a detailed mechanistic evaluation is not within the scope of this discussion, the halogen–metal exchange reaction has been shown to be reversible and rapid at $-75°C$, and, in the exchange of alkyllithium with a haloarene, the equilibrium reaction favors formation of the lithioarene.[7, 10, 11]

As exemplified in the present procedure, the reaction has been optimized and extended in scope; it affords functionalized benzocyclobutenes as well as substituted isoquinolines in high yields. Benzocyclobutenes have been used as intermediates in the synthesis of many naturally occurring alkaloids,[12] steroids,[13] polycyclic terpenoids,[14] and anthracycline antibiotics.[15] The traditional routes leading to the preparation of benzocyclobutenes have been reviewed[16] and have involved (1) Cava's cyclization of *o*-quinodimethane intermediates (via reaction of sodium iodide with α,α,α′,α′-tetrabromo-*o*-xylene) (2) thermal extrusion of sulfur dioxide from 1,3-dihydroisothianaphthene 2,2-dioxide, (3) dehydrogenation of the Diels–Alder adducts of 1,4-butadienes and cyclobutenes, and (4) Wolff rearrangement of α-diazoindanones. More recent methods include (1) thermal rearrangement of p-tolylcarbene,[17] (2) thermal decomposition of 3-isochromanones,[12] and (3) cobalt-catalyzed cyclizations of acetylenic compounds.[18] Many of these methods for synthesizing functionalized benzocyclobutenes involve (a) multistep routes, (b) unusual or relatively unavailable starting materials, (c) low overall yields, or (d) special apparatus. The method of halogen–metal exchange demonstrates a high degree of selectivity for formulation of the lithioarene intermediate, is broad in scope without loss of procedural simplicity, and provides a high-yield route to benzocyclobutenes of general synthetic utility by direct cyclization of readily available 2-(2′-lithiophenyl)ethyl chlorides.[4e, f, 9b]

The lithioarene intermediate has also been shown to be of use in the synthesis of the isoquinoline ring system. This ring system is common to a variety of natural products that possess useful physiological activity. Several methods have been developed for the synthesis of isoquinolines; the most commonly used routes are the Bischler–Napieralski and the Pictet–Spengler reactions.[19, 20] These methods involve electrophilic, aromatic substitution in the key ring-forming steps with the limitation that best results are obtained only when the aromatic ring bears electron-donating substituents. The present method permits use of substrates either with or without electron-donating groups on the aromatic nucleus since generation of the lithioarene has been shown to be relatively independent of the nature of the substituents.[8a, d]

1. Department of Chemistry, State University of New York, Stony Brook, NY 11794. Partial support was provided by the National Institutes of Health (Grant No. CA22741). Present address of Paul Helquist: Department of Chemistry, University of Notre Dame, Notre Dame, IN 46556.
2. Johnson Matthey Inc., Wayne, PA 19087; formerly of FMC Corporation, Princeton, NJ 08540, which we acknowledge for partial support of this work.
3. Still, W. C.; Kahn, M.; Mitra, A. *J. Org. Chem.* **1978,** *43,* 2923.
4. (a) Bickelhaupt, F.; Stach, K.; Thiel, M. *Chem. Ber.* **1965,** *98,* 685; (b) Gilman, H.; Marrs, O. L. *J. Org. Chem.* **1965,** *30,* 325; (c) Lane, C. F.; Myatt, H. L.; Daniels, J.; Hopps, H. B. *J. Org. Chem.* **1974,** *39,* 3052 (d) Semmelhack, M. F.; Chong, B. P.; Stauffer, R. D.; Rogerson, T. D.; Chong, A.; Jones, L. D. *J. Am. Chem. Soc.* **1975** *97,* 2507–2516; (e) Parham, W. E.; Jones, L. D.; Sayed, Y. A. *J. Org. Chem.* **1976,** *41,* 1184; (f) Bradsher, C. K.; Hunt, D. A. *Org. Prep. Proc. Int.* **1978,** *10,* 267.
5. Fieser, L. F.; Fieser, M. "Reagents for Organic Synthesis"; Wiley: New York, 1967; Vol. 1, p. 584.
6. Whitesides, G. M.; Casey, C. P.; Kreiger, J. K. *J. Am. Chem. Soc.* **1971,** *93,* 1379.
7. Jones, R. G.; Gilman, H. *Org. React.* **1951,** *6,* 339.
8. (a) Parham, W. E.; Sayed, Y. A. *J. Org. Chem.* **1974,** *39,* 2053; (b) Parham, W. E.; Jones, L. D.; Sayed, Y. *J. Org. Chem.* **1975,** *40,* 2394; (c) Parham, W. E.; Jones, L. D. *J. Org. Chem.* **1976,** *41,* 2704.
9. (a) Hergrueter, C. A.; Brewer, P. D.; Tagat, J.; Helquist, P. *Tetrahedron Lett.* **1977,** 4145; (b) Brewer, P. D.; Tagat, J.; Hergrueter, C. A.; Helquist, P. *Tetrahedron Lett.* **1977,** 4573; (c) Ponton, J.; Helquist, P.; Conrad, P. C.; Fuchs, P. L. *J. Org. Chem.* **1981,** *46,* 118; (d) Parham, W. E.; Bradsher, C. K.; Hunt, D. A. *J. Org. Chem.* **1978,** *43,* 1606; Bradsher, C. K.; Reames, D. C. *J. Org. Chem.* **1978,** *43,* 3800; (f) Bradsher, C. K.; Hunt, D. A. *J. Org. Chem.* **1980,** *45,* 4248; (g) Reames, D. C.; Hunt, D. A.; Bradsher, C. K. *Synthesis* **1980,** 454; (h) Toth, J. E.; Fuchs, P. L. *J. Org. Chem.* **1987,** *52,* 473.
10. Gilman, H.; Jones, R. G. *J. Am. Chem. Soc.* **1941,** *63,* 1441.
11. The use of two equivalents of alkyllithium reagent to effect lithium–halogen exchange reactions

most efficiently was developed by Seebach and Neuman; see Neuman, H.; Seebach, D. *Chem. Ber.* **1978,** *111*, 2785.

12. Spangler, R. J.; Bechmann, B. G.; Kim, J. H. *J. Org. Chem.* **1977,** *42*, 2989 and references cited therein.

13. (a) Kametani, T.; Matsumoto, H.; Nemoto, H.; Fukumoto, K. *J. Am. Chem. Soc.* **1978,** *100*, 6218; (b) Nicolaou, K. C.; Barnette, W. E. *J. Chem. Soc., Chem. Commun.* **1979,** 1119; (c) Oppolzer, W.; Roberts, D. A. *Helv. Chim Acta* **1980,** *63*, 1703; (d) Grieco, P. A.; Takigawa, T.; Schillinger, W. J. *J. Org. Chem.* **1980,** *45*, 2247; (e) Kametani, T.; Suzuki, K.; Nemoto, H. *J. Org. Chem.* **1980,** *45*, 2204.

14. Kametani, T.; Hirai, Y.; Shiratori, Y.; Fukumoto, K.; Satoh, F. *J. Am. Chem. Soc.* **1978,** *100*, 554.

15. Wiseman, J. R.; French, N. I.; Hallmark, R. K.; Chiong, K. G.; *Tetrahedron Lett.* **1978,** 3765.

16. Klundt, I. L. *Chem. Rev.* **1970,** *70*, 471; see also Cava, M. P.; Deana, A. A.; Muth, K. *J. Am. Chem. Soc.* **1960,** *82*, 2524 and references cited therein; Thummel, R. P.; Nutakul, W. *J. Org. Chem.* **1977,** *42*, 300; Radlick, P.; Brown, L. R. *J. Org. Chem.* **1973,** *38*, 3412.

17. Hedaya, E.; Kent, M. E. *J. Am. Chem. Soc.* **1971,** *93*, 3283.

18. Vollhardt, K. P. C. *Acc. Chem. Res.* **1977,** *10*, 1.

19. For extensive discussions concerning the synthesis, pharmacology, and other properties of isoquinolines, see: (a) McCorkindale, N. J. In "The Alkaloids," Grundon, M. F., Senior Reporter; The Chemical Society: London, 1976; Vol. 6, Chapter 8, and the earlier volumes in this series; (b) "The Alkaloids: Chemistry and Physiology"; Manske, R. H. F., Ed.; Academic Press: New York, 1975; Vol. XV, Chapters 3 and 5, and the earlier volumes of this series; (c) Kametani, T. "The Chemistry of the Isoquinoline Alkaloids"; Hirokawa Publishing Co, Inc.; Tokyo, 1969; Vol. 1, Kinkodo Publishing Co.: Sendai 1974; Vol. 2; (d) Shamma, M. "The Isoquinoline Alkaloids"; Academic Press: New York, 1972; (e) Kametani, T. In "The Total Synthesis of Natural Products"; ApSimon, J., Ed.; Wiley-Interscience: New York, 1977; Vol. 3, pp. 1–272.

20. For other more recent methods for the synthesis of isoquinolines, see references 19a and 19c and other works cited therein, in addition to: (a) Clive, D. L. J.; Wong, C. K.; Kiel, W. A.; Menchen, S. M. *J. Chem. Soc., Chem. Commun.* **1978,** 379; (b) Barrett, A. G. M.; Barton, D. H. R.; Falck, J. R.; Papaioannou, D.; Widdowson, D. A. *J. Chem. Soc. Perkin Trans. 1* **1979,** 652; (c) Kozikowski, A. P.; Ames, A. *J. Org. Chem.* **1980,** *45*, 2548; (d) Mendelson, W. L.; Spainhour, C. B.; Jones, S. S.; Lam, B. L.; Wert, K. L. *Tetrahedron Lett.* **1980,** *21* 1393.

METHYLENE KETONES AND ALDEHYDES BY SIMPLE, DIRECT METHYLENE TRANSFER: 2-METHYLENE-1-OXO-1,2,3,4-TETRAHYDRONAPHTHALENE

[1(2H)-Naphthalenone, 3,4-dihydro-2-methylene-]

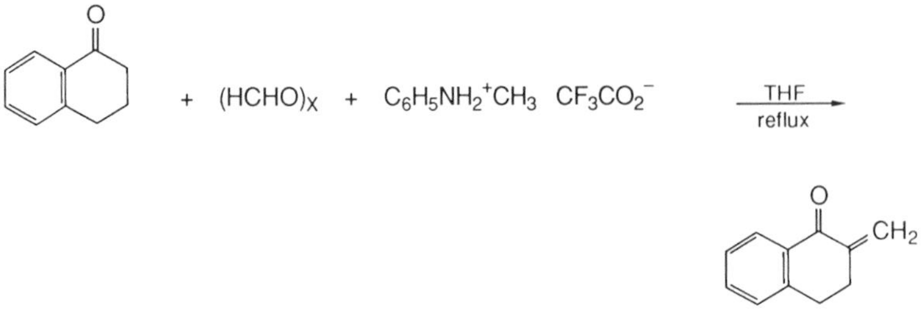

Submitted by JEAN-LOUIS GRAS[1]
Checked by KERRY J. GOMBATZ and GEORGE BÜCHI

1. Procedure

A 250-mL flask equipped with a reflux condenser is charged with 6.75 g (0.225 mol) of paraformaldehyde (Note 1) and 16.57 g (0.075 mol) of N-methylanilinium trifluoroacetate (Note 2). A solution of 7.30 g (0.05 mol) of α-tetralone (Note 3) in 50 mL of dry tetrahydrofuran (Note 4) is added at room temperature. The N-methylanilinium trifluoroacetate dissolves, and the magnetically stirred mixture is refluxed for 4 hr under a nitrogen atmosphere (Note 5). During this time a red color develops and the paraformaldehyde dissolves after 2 hr. After 4 hr the heating oil bath is removed and the red solution allowed to cool for 10 min. Diethyl ether (100 mL) is gradually added under efficient magnetic stirring, which induces the separation of a red gum. The ethereal solution is decanted from the red gum into a separatory funnel and washed with 50 mL of half-saturated sodium bicarbonate solution. The red gum is triturated with 50 mL of diethyl ether, and the resulting ethereal solution is then used to extract the washing water (Note 6). The combined organic layers are dried over magnesium sulfate. Filtration and concentration of the extract, first on a rotary evaporator then under high vacuum, afford 8.05–8.6 g of a heavy red oil (Note 7). Trituration of this oil with 70 mL of diethyl ether precipitates impurities and causes some polymerization. Filtration through Celite and concentration under high vacuum give 6.8–7.2 g (86–91%) of material that solidifies in a freezer (Note 8). Further purification by column chromatography over silica gel affords analytically pure material (mp 46–46.5°C) but lowers the yield to 70–82%.

2. Notes

1. Paraformaldehyde is sometimes sold commercially under the label "polyoxymethylene," and commercial polyoxymethylene (Prolabo—France) was used.
2. This crystalline white salt can be obtained by adding dropwise 1 mol of commercial trifluoroacetic acid (Fluka AG) to a stirred solution of 1 mol of commercial N-methylaniline (Fluka AG) in 1 L of dry diethyl ether in a nitrogen atmosphere

with cooling in an ice bath. After addition the solution is stirred magnetically for 1 hr. The white precipitate that forms is filtered, washed with 100 mL of pentane, and dried overnight in a desiccator under high vacuum. The salt (195 g, 88%) thus obtained had mp 66.5°C.

3. Commercial α-tetralone, 95% pure, was purchased from Fluka AG and used without purification.

4. Tetrahydrofuran was distilled from the ketyl prepared from benzophenone and sodium, but the reaction does not suffer from moisture. Dioxane can also be used, but the iminium salt polymerizes rapidly at the reflux temperature of this solvent (101°C). To avoid polymerization the N-methylanilinium trifluoroacetate should be added in portions to the reaction mixture.

5. The reaction can be monitored by TLC. The α-methylene ketones exhibit higher R_f values than starting material when eluted in a diethyl ether-pentane (1 : 1) solvent system.

6. Workup and isolation should be completed in minimum time to avoid polymerization of the product. The heavy red gum thus obtained is soluble in methylene chloride and contains some β-methylene-α-tetralone.

7. A TLC analysis reveals a major component accompanied by two minor, more polar impurities. Because of the relative instability of α-methylene carbonyl compounds, isolation of these substances is associated with dimerization or polymerization. This crude material exhibits satisfactory NMR and IR data and can be used as such for many synthetic purposes.

8. The checkers found that storage of this material at room temperature results in the total conversion to polymer in less than 12 hr. The stability of the product is greatly increased if it is stored at temperatures below −5°C. The spectral properties are as follows: IR (CCl$_4$) cm^{-1}: 3065, 3030, 1680, 1620, 1604, 918; ^{1}H NMR (CCl$_4$) δ: 2.9 (singlet, 4), 5.37 (thin multiplet, 1), 6.17 (thin multiplet, 1), 7.3 (multiplet, 3), 8.07 (multiplet, 1).

3. Discussion

The procedure described herein demonstrates a general synthetic method to form α-methylene ketones by direct methylene transfer. A number of methods have been previously described and reviewed.[2,3] The advantages of direct methylene transfer for the formation of α-methylene ketones are the aprotic, nearly neutral conditions utilized. Although the reaction is not regiospecific, it is highly sensitive to steric hindrance, and transfer occurs at the less hindered site of unsymmetrical ketones. The reaction has been applied to cyclic and acyclic ketones[4] and extended to the synthesis of vinyl ketones[5] and α-methylenealdehydes. It is not applicable to γ- or δ-lactones, or strained cyclic ketones such as norcamphor or cyclobutanone. With cyclohexanone, cyclopentanone,[6] or aldehydes as substrates, preformation of the iminium intermediate

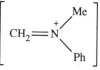 is recommended prior to the addition of the carbonyl compound.

This can be achieved by heating the reagents to reflux in tetrahydrofuran for 20 min, followed by the addition of the carbonyl compound at reflux temperature or lower, if necessary. When higher reflux temperatures are required, dioxane can be used as a solvent. Addition in portions of N-methylanilinium trifluoroacetate to the reaction mixture minimizes polymerization of the iminium intermediate. In some cases, large-scale experiments may suffer from polymerization; it is recommended that the reaction be quenched before completion.

1. Laboratoire de Synthèse Organique, Université d'Aix-Marseille III, Rue Henri Poincaré, 13397 Marseille Cedex 13.
2. For a review see Tramontini, M. *Synthesis* **1973**, 703–775.
3. Stork, G.; D'Angelo, J. *J. Am. Chem. Soc.* **1974**, *96*, 7114–7116; Holy, N. C.; Wang, Y. F. *J. Am. Chem. Soc.* **1977**, *99*, 944–946; Ksander, G. M.; McMurry, J. E.; Johnson, M. *J. Org. Chem.* **1977**, *42*, 1180–1185; Paterson, I.; Fleming, I. *Tetrahedron Lett.* **1979**, 995–998; Desolms, S. J. *J. Org. Chem.* **1976**, *41*, 2650–2651; Danishefsky, S.; Kitahara, T.; McKee, R.; Schuda, P. F. *J. Am. Chem. Soc.* **1976**, *98*, 6715–6717; Hayashi M.; Mukaiyama, T. *Chem. Lett.*, **1987**, 1293.
4. Gras, J. L. *Tetrahedron Lett.* **1978**, 2111–2114.
5. Gras, J. L. *Tetrahedron Lett.* **1978**, 2955–2958.
6. Disanayaka, B. W.; Weedon, A. C. *Synthesis*, **1983**, 952.

METHYL 4-HYDROXY-2-BUTYNOATE

(2-Butynoic acid, 4-hydroxy-, methyl ester)

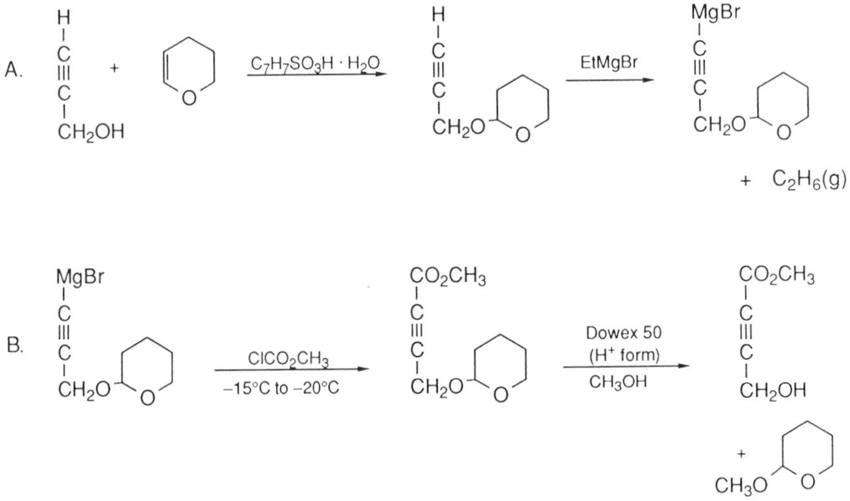

Submitted by R. A. EARL[1] and L. B. TOWNSEND[2]
Checked by G. SAUCY and G. WEBER

1. Procedure

Caution! Acetylenic compounds are potentially explosive and methyl 4-hydroxy-2-butynoate is a potent vesicant (Note 1).

A. *Tetrahydropyranyl derivative of propargyl alcohol* [*tetrahydro*-2-(*2-propynyl-oxy*)-*2H-pyran*]. Two crystals (ca. 10 mg) of *p*-toluenesulfonic acid monohydrate are added to 268 g (291.3 mL, 3.2 mol) of warm (60°C) dihydropyran (Note 2) in a 1-L, three-necked, round-bottomed flask equipped with a magnetic stirring bar, a thermometer, a dropping funnel containing 168 g (174.5 mL, 3.0 mol) of propargyl alcohol (Note 2) and a reflux condenser fitted with a drying tube. Stirring is started, and the propargyl alcohol is added (Note 3) as a thin stream during a period of ca. 30 min. The reaction is mildly exothermic, and the temperature is maintained at 60–65°C by controlling the rate of addition of the propargyl alcohol and by occasional external cooling with an ice bath. After the addition is completed, the temperature is monitored for another 30 min; slight cooling is sometimes necessary to keep the temperature in the range of 60–65°C. The reaction mixture is stirred for a total of 1.5 hr after the addition is completed, and then 0.5 g of powdered sodium bicarbonate is added and the mixture stirred for another hour. The mixture is then gravity-filtered into a 1-L, round-bottomed flask. The reaction mixture is distilled through a 45-cm Vigreux column under reduced pressure (Note 15). A small forerun (ca. 40 mL) with a bp of 45°C (15–20 mm) is followed by the product, bp 47–50°C (3.5–5 mm), 330–355 g (78–92%) (Note 15); n_D^{22}1.4559 (Note 16); ^{1}H NMR (90 MHz, neat) δ: 1.18–1.93 (br m, 6, $H_{2',2''}$, $H_{3',3''}$, $H_{4',4''}$); 2.47 (t, 1, $J = 2$, $C{\equiv}C{-}H$), 3.17–3.84 (m, 2, $H_{5',5''}$), 4.03 (d, 2, $J = 2$, $C{\equiv}C{-}CH_2O$), 4.63 (s, 1, H_1); IR (neat) cm^{-1}: 3300 ($C{\equiv}C{-}H$), 2117 ($C{\equiv}C$ stretch).

B. *Methyl 4-hydroxy-2-butynoate.* One mole of ethylmagnesium bromide (Note 4) in diethyl ether is poured into a dry (Note 5) 2-L, three-necked, round-bottomed flask fitted with a mechanical stirrer and a glass stirrer bearing (Note 6), a dropping funnel fitted with a nitrogen-inlet tube, and an efficient condenser fitted with a drying tube. Stirring is started, and a solution of 140 g (1.0 mol, 141 mL) of the tetrahydropyranyl derivative of propargyl alcohol in 1-L of dry (Note 7) tetrahydrofuran is added during ca. 30 min (Note 8). Stirring is continued for an additional 1.5 hr, during which time a dry 3-L, three-necked flask is fitted with a mechanical stirrer, immersion thermometer, and dropping funnel. The 3-L flask is charged with a solution of 104 g (1.10 mol, 85.4 mL) of methyl chloroformate (Note 2) in 250 mL of tetrahydrofuran and the contents stirred and cooled to −20°C with a dry ice-acetone bath. Under gentle nitrogen pressure the acetylenic Grignard reagent is transferred portionwise through a ¼-in. polypropylene tube to the dropping funnel attached to the 3-L, three-necked flask (Note 9). The acetylenic Grignard reagent is then added dropwise during 1.5 hr to the well-stirred solution of methyl chloroformate in tetrahydrofuran while the temperature is maintained at −15 to −20°C by external cooling. After the addition is completed, the light-brown reaction mixture is stirred another 30 min at −15°C, followed by another 1.5 hr at ice temperature. The reaction mixture is then stored without stirring for 12 hr at +3°C, during which time the remaining magnesium salts separate from solution. The salts are removed by filtration (Note 10) and washed with three 150-mL portions of cold (0°C), dry toluene. The supernatant and washings are combined and concentrated (Note 11) to ca. 500-mL volume. The dark-brown solution is then washed five times with 100-mL portions of saturated brine followed by drying over anhydrous sodium sulfate. The solution is concentrated to remove the toluene and then dissolved (Note 12) in 1 L of anhydrous methanol; 25 mL of Dowex 50-X4 cation resin (H^+ form, prewashed with anhydrous methanol) is then added and the mixture stirred for 1.5 hr at 25°C. The ion-exchange resin is removed by filtration through a sintered-glass filter and is then washed with two 50-mL portions of anhydrous methanol. Solvent and 2-methoxytetrahydropyran are removed by concentration using a water aspirator and then an oil pump at 0.5-mm pressure. The residue from the concentration is then treated a second time (Note 13) with 1 L of anhydrous methanol and 25 mL of Dowex 50, followed by concentration as before. The residue is then distilled through a Claisen head to give methyl 4-hydroxy-2-butynoate (Note 14), 69–74 g (60–65%), by 66–69°C/0.2 mm, n_D^{22} 1.4684 (Note 17); ^{1}H NMR [$(CD_3)_2SO$] δ: 3.79 (s, 3, OCH_3), 4.31 (d, 2, CH_2, $J = 6$), 5.57 (t, 1, OH); IR (neat) cm^{-1}: 3410 (OH), 2240 ($-C\equiv C-$), 1715 (ester).

2. Notes

1. Acetylenic compounds are potentially explosive, and all concentrations and distillations should be carried out behind a safety shield. Methyl 4-hydroxy-2-butynoate is a potent vesicant that causes painful burns on contact with skin. All operations should be carried out in an efficient fume hood and gloves should be worn at all times.

2. As supplied by Aldrich Chemical Company, Inc. (97% purity).

3. The general method of Robertson,[3] whereby toluenesulfonic acid monohydrate is added to a mixture of an alcohol and dihydropyran, is not recommended for this preparation since the reaction is rather exothermic. Reaction temperatures below 60°C are to be avoided for the same reason since unreacted reagents accumulate and the reaction may suddenly get out of hand with resulting boiling and colorization of the reaction mixture.

4. Ethylmagnesium bromide was obtained from Aldrich Chemical Company, Inc. in the form of a 3 M solution in diethyl ether containing 133.3 g of ethylmagnesium bromide. Alternately, the ethylmagnesium bromide could be prepared by a standard procedure.[4]

5. Glassware was dried in an oven at 110°C, assembled while still hot, and flushed with dry nitrogen as the assembly cooled to room temperature. All reactions involving the Grignard reagents were carried out under an atmosphere of dry nitrogen and in a fume hood.

6. The bearing was lubricated with mineral oil.

7. Anhydrous tetrahydrofuran from MCB, Inc. was used for the reactions. Freshly opened bottles gave no effervescence when mixed with powdered calcium hydride. If smaller amounts of tetrahydrofuran are used, the acetylenic Grignard reagent often crystallizes out of the reaction mixture.

8. Vigorous gas evolution (highly flammable ethane gas) and boiling take place during the addition.

9. This type of transfer technique[5] is preferred to open-air transfer to minimize losses due to hydrolysis by atmospheric moisture.

10. Filtration and subsequent washing of the hygroscopic salts are best carried out by replacing the dropping funnel in the reaction vessel with a sintered-glass filter stick. The reaction mixture is kept under a slightly positive nitrogen pressure while the supernatant is led from the filter stick through a polypropylene tube to a suction flask that is kept under a slightly negative pressure with the help of an aspirator.

11. Concentrations were carried out using a rotary evaporator and at a pressure of 12–15 mm and a temperature not exceeding 35°C unless otherwise noted.

12. The crude product at this stage shows the following ^{1}H NMR ($CDCl_3$) δ: 1.22–1.97 (m, 6 H, $H_{2',2''}$, $H_{3',3''}$, $H_{4',4''}$), 3.26–4.08 (m, 2, $H_{5',5''}$), 3.76 (s, 3, OCH_3), 4.36 (s, 2, $C \equiv C - CH_2$), 4.81 (bs, 1, H_1).

13. Distillation of the residue after the first treatment with Dowex 50 and methanol gives a product containing 7–10% of the tetrahydropyranyl derivative of methyl 4-hydroxy-2-butynoate. Removal of the by-product 2-methoxytetrahydropyran and retreatment with Dowex 50 and methanol gives a product containing only 0.5–1.5% of the undeblocked alcohol.

14. The distillate often turns a light pink or yellow color in the receiver flask.

15. The checkers found that distillation at a pressure of 15–20 mm (submitters) gave a somewhat lower yield (78–84%). A slight yield improvement (78–94%) was obtained by the checkers by using lower pressure in the distillation.

16. The submitters reported n_D^{22} 1.4595.

17. The submitters reported n_D^{22} 1.4720.

3. Discussion

The preparation of the tetrahydropyranyl derivative of propargyl alcohol is a modification of a published[3] general procedure that is simple and useful for large-scale preparations.

Methyl 4-hydroxy-2-butynoate has been prepared[6] in 83% yield by treatment of 4-hydroxy-2-butynoic acid with 2% sulfuric acid in methanol and in 65% yield by carboxylation in an autoclave of the Grignard reagent of 1-(tetrahydropyran-2'-yloxy)prop-2-yne followed by treatment with 10% sulfuric acid in methanol. 4-Hydroxy-2-butynoic acid has been prepared[6] in 65% yield by treating the Grignard reagent of propargyl alcohol with carbon dioxide in an autoclave for 24 hr followed by acidic hydrolysis with aqueous 10% sulfuric acid. 4-Hydroxy-2-butynoic acid has also been prepared[7] in an unspecified yield by bubbling carbon dioxide for 14 days through a suspension of the Grignard derivative of propargyl alcohol in ether.

The first part of the procedure illustrates a method for the preparation of the tetrahydropyranyl derivative of an alcohol which requires no extraction or wash procedures during the workup of the product.

The second part of the preparation illustrates a very efficient, mild method for the preparation of a highly reactive α,β-acetylenic ester via the carbomethoxylation of the

Grignard reagent of a terminal acetylenic compound with methyl chloroformate. This preparation of methyl 4-hydroxy-2-butynoate obviates the necessity[6] of carrying out the carboxylation of an acetylenic Grignard reagent in an autoclave. This procedure also eliminates the necessity[6] of carrying out the continuous ether extraction of 4-hydroxy-2-butynoic acid from an aqueous phase.

The use of a mixture of a strongly acidic cation-exchange resin and methanol to remove a tetrahydropyranyl protecting group offers a very mild method of deblocking that does not require the use of a base during the workup.

Methyl 4-hydroxy-2-butynoate has been used[6] as a starting material for the preparation of a δ-hydroxy-α,β-acetylenic ester. It has also been employed[8] as a dipolarophile in a 1,3-dipolar cycloaddition reaction that resulted in the first synthesis of 8-aza-3-deazaguanosine.

1. Hercules, Inc., Bacchus Works, Magna, UT 84064.
2. Department of Medicinal Chemistry, College of Pharmacy, University of Michigan, Ann Arbor, MI 48109.
3. Robertson, D. N. *J. Org. Chem.* **1960,** *25,* 931–932.
4. Dreger, E. E. *Org. Synth., Coll. Vol. I* **1941,** 306–308.
5. Kramer, G. W.; Levy, A. B.; Midland, M. M. in H. C. Brown, "Organic Synthesis *via* Boranes"; Wiley: New York, 1975.
6. Henbest, H. B.; Jones, E. R. H.; Walls, I. M. S. *J. Chem. Soc.* **1950,** 3646–3650.
7. Lespiéau, R. *Ann. Chim.* **1912,** *27,* 178.
8. Earl, R. A.; Townsend, L. B., *Can. J. Chem.* **1980,** *58,* 2550–2561.

ASYMMETRIC HYDROBORATION OF 5-SUBSTITUTED CYCLOPENTADIENES: SYNTHESIS OF METHYL (1R,5R)-5-HYDROXY-2-CYCLOPENTENE-1-ACETATE

[2-Cyclopentene-1-acetic acid, 5-hydroxy-, methyl ester, (1R-trans)-]

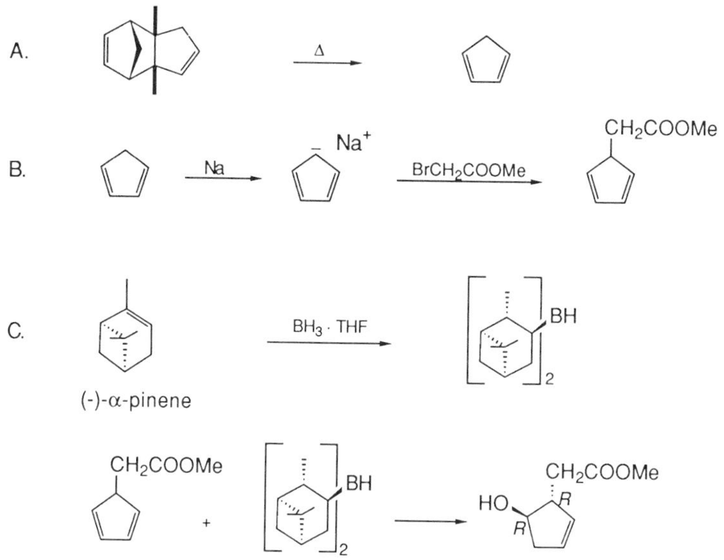

Submitted by JOHN J. PARTRIDGE, NARESH K. CHADHA, and MILAN R. USKOKOVIC[1]
Checked by BAI DONG-LU and CLAYTON H. HEATHCOCK

1. Procedure

Caution! Methyl bromoacetate, used in Step B, is intensely irritating to eyes and skin. The preparation of the ester should be carried out in an efficient hood.

A. *Pyrolysis of dicyclopentadiene to form cyclopentadiene.* Cyclopentadiene is prepared from its dimeric form by distillation according to the method of Moffett.[2] The apparatus for the distillation is assembled as shown in Figure 1. The equipment consists of a 250-mL flask, a Friedrichs condenser fitted with a Haake Model FE hot water circulator, a Claisen head, a thermometer, a gas inlet tube, and a collection receiver that is cooled to −78°C in a dry ice–acetone bath.

In the 250-mL flask is placed 100 mL of dicyclopentadiene (Note 1). The material is heated at reflux (bath temperature 200–210°C) under a nitrogen atmosphere (Note 2). After a 5-mL forerun is collected and discarded, the collection receiver is cooled to −78°C and 25 mL (0.30 mol) of cyclopentadiene is rapidly distilled at bp 36–42°C. A slight positive pressure of nitrogen is maintained throughout the distillation to prevent moisture from entering the system.

The distilled cyclopentadiene is stored at −78°C until it is used (Note 3). Residual dicyclopentadiene can be reused until it solidifies on cooling.

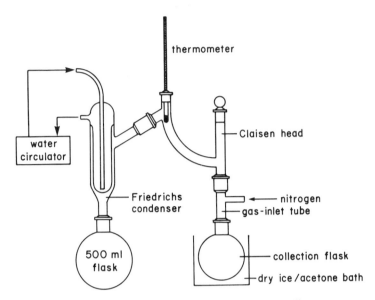

Figure 1. Apparatus for producing cyclopentadiene.

B. *Preparation in situ of methyl 2,4-cyclopentadiene-1-acetate (Note 4).* Cyclopentadienylsodium is prepared by modification of the methods of King[3] and Hafner[4] (Note 5).

In a 500-mL, three-necked Morton flask fitted with a condenser, mechanical stirrer, and gas inlet tube is placed 8.6 g (0.375 g-atom) of sodium and 75 mL of dry xylene (Note 6); the unstirred mixture is heated at reflux under a nitrogen atmosphere. After the xylene has reached its boiling point and the sodium has melted, the solution is rapidly stirred to produce a very fine-grained sodium sand. Quickly the heating mantle is removed and stirring stopped (Note 7). After cooling, the xylene is pipetted or siphoned away from the sodium sand and stored for future use.

The sand is washed with 3 × 25 mL of dry tetrahydrofuran (Note 8) and then is layered with 100 mL of dry tetrahydrofuran, and the mixture is cooled to −10°C (Note 9) under a nitrogen atmosphere. A solution of 25 mL (0.30 mol) of cyclopentadiene in 25 mL of tetrahydrofuran is added dropwise using a dropping funnel. After the addition is complete, the mixture is stirred overnight at room temperature, by which time hydrogen evolution has ceased. In the absence of air, the solution ranges from near colorless to bright pink (Note 10).

In a 1-L, three-necked flask fitted with a 200-mL pressure-equalizing dropping funnel, mechanical stirrer, and a gas inlet tube is placed 45.9 g (0.30 mol) of methyl bromoacetate (Note 11) and 75 mL of tetrahydrofuran and the mixture is cooled to −78°C in an inert atmosphere.

The solution of ca. 0.30 mol of cyclopentadienylsodium is decanted from residual sodium sand with a U-tube into the dropping funnel (Note 12) and is added dropwise over a 2-hr period (Note 13). A white precipitate of sodium bromide forms during the addition. The heterogeneous solution is stirred overnight at −78°C to ensure complete formation of methyl 2,4-cyclopentadiene-1-acetate.

C. *Asymmetric hydroboration with (+)-di-3-pinanylborane to form methyl (1R,5R)-5-hydroxy-2-cyclopentene-1-acetate (Note 4)*. The (+)-di-3-pinanylborane is prepared from (−)-α-pinene by a modification[5,6] of the method of Brown[7] (Note 14).

In a 2-L, three-necked flask fitted with a condenser, mechanical stirrer, and a gas inlet tube is placed 90.0 g (0.66 mol) of (−)-α-pinene (Note 15). The flask is cooled to 0°C and under an inert atmosphere a total of 300 mL (0.30 mol) of 1 *M* borane in tetrahydrofuran (Note 16) is added dropwise over a 1-hr period. The solution is stirred for 18 hr at 0°C, during which time a white precipitate of (+)-di-3-pinanylborane forms. This solution is then cooled to −78°C. The ca. 0.30 mol solution of methyl 2,4-cyclopentadiene-1-acetate (Section B) is transferred at −78°C to a 500-mL pressure-equalizing dropping funnel through a U-tube in an inert atmosphere and is added rapidly, in one portion, to the stirring solution of di-3-pinanylborane at −78°C. After this mixture is stirred for 6 hr at −78°C, the bath temperature is allowed to rise to 0°C and the mixture is stirred for 16 hr at 0°C to complete the hydroboration reaction.

To the reaction mixture is added dropwise 90 mL of 3 *N* aqueous sodium hydroxide, followed by 90 mL of 30% hydrogen peroxide (Note 17). The mixture is stirred for 30 min to complete the oxidation process. A total of 3 g of sodium bisulfite, 5 g of sodium chloride, and 125 mL of ether are added and the mixture is stirred for 10 min (Note 18). On standing, the reaction mixture separates into two layers, which are separated with a 1-L separatory funnel. The organic layer is washed with brine (2 × 50 mL). The water layer and the brine washes are combined and extracted with ether (3 × 125 mL). All the organic layers are then combined and dried over anhydrous magnesium sulfate. Filtration and removal of solvent under reduced pressure yield 110 g of a pale-yellow oil containing the desired product as well as (+)-isopinocampheol, and (−)-α-pinene (Note 19). The product mixture is dissolved in 250 mL of ether and is extracted with 1 *M* aqueous silver nitrate solution (3 × 100 mL). The aqueous layers are combined and back-extracted once with 50 mL of ether. The ether layers containing (+)-isopinocampheol are discarded.

The aqueous layers containing the silver(I) complex of methyl (1R,5R)-5-hydroxy-2-cyclopentene-1-acetate are then treated with an excess of saturated brine to precipitate silver chloride and free the desired product. After precipitation is complete, the water layer is decanted from the solid silver chloride. The solids are washed with ether (4 × 100 mL) and each ether layer is used to extract the water layer (Note 20). The combined ether layers are washed with 50 mL of brine and dried over anhydrous magnesium sulfate. Filtration and removal of solvent under reduced pressure yield 16–19 g of crude product. The product is distilled through a 4-in.-Vigreux column at 0.1 mm pressure to yield 12.8–14.7 g (27–31%) of methyl (1R,5R)-5-hydroxy-2-cyclopentene-1-acetate, bp 74–78°C at 0.1 mm, $[\alpha]_D^{25}$ −132° (CH$_3$OH, *c* 1.06) (Notes 21 and 22).

2. Notes

1. Dicyclopentadiene was obtained from Ace Scientific (TX 315), practical grade, 95%.

2. The Haake water circulator was employed with the circulating water temperature at 50°C. This allows only cyclopentadiene to distill.

3. Cyclopentadiene is stable at −78°C but dimerizes readily at room temperature.

4. Steps B and C must be run concurrently.

5. The efficient formation of cyclopentadienylsodium is of paramount importance for the entire reaction sequence. Variations in the yield of methyl 2,4-cyclopentadiene-1-acetate have been traced to the degree of efficiency in producing a fine sodium sand that is used to produce cyclopentadienylsodium. In the alkylation reaction of cyclopentadienylsodium with methyl bromoacetate, the entire process must be carried out in an inert dry atmosphere at −78°C. At higher temperatures, the desired product can undergo undesired dimerization and double-bond migration side reactions. Methyl 2,4-cyclopentadiene-1-acetate, once formed, is used immediately.

6. Xylene was obtained from Fisher Scientific Company. The xylene is initially dried over sodium and is saved and reused in making additional batches of sodium sand.

7. If stirring continues while the xylene cools, the sodium sand coagulates into a large lump.

8. Tetrahydrofuran was obtained from Fisher Scientific Company. The tetrahydrofuran employed was freshly distilled from lithium aluminum hydride. Care should be exercised in drying tetrahydrofuran; cf. *Org. Synth., Coll. Vol. V* **1973,** 976. The checkers also examined the use of tetrahydrofuran that had been dried by distillation from sodium–benzophenone ketyl. When material that has been purified in this manner is used, the fine sodium sand coagulates, giving small porous lumps. No such coagulation occurs when tetrahydrofuran that has been dried by distillation from lithium aluminum hydride is used. However, the method of drying had no effect on overall yield of final product.

9. A bath of carbon tetrachloride containing a little dry ice is used for cooling.

10. The efficient formation of cyclopentadienylsodium was found to be the product-limiting step for the reaction sequence. *If air is present or if the sodium sand is not fine-grained, quantitative formation of cyclopentadienylsodium cannot be assumed.* Residual sodium sand may be washed with tetrahydrofuran, dried in a nitrogen atmosphere, and weighed to determine approximately the extent of cyclopentadienylsodium formation.

11. Methyl bromoacetate was obtained from Ace Scientific (MX 755).

12. Care must be taken during this transfer to minimize exposure of the cyclopentadienylsodium to air. Trace amounts of oxygen cause the formation of a dark brown color and brown solid in the solution.

13. The drip rate should be adjusted so that the dropping funnel is not plugged by crystalline cyclopentadienylsodium.

14. After the asymmetric hydroboration–oxidation sequence is completed, the desired product is separated via its silver(I) complex from (+)-isopinocampheol. The desired product can also be isolated by column chromatography.

15. (−)-α-Pinene was obtained from Chemical Samples Company. The (−)-α-pinene was distilled from sodium metal: bp 155–156°C; $[\alpha]_D^{25}$ −47° (neat).

16. Borane–tetrahydrofuran was obtained from Alfa Products, Morton Thiokol Inc.

17. The hydrogen peroxide oxidation is a very exothermic process and efficient cooling and stirring are necessary.

18. After the addition of ether, some inorganic salts precipitate. The checkers found it advantageous to remove this solid by suction filtration. The solid was washed with ether, which was combined with the organic solution.

19. Vacuum distillation does not effectively purify the desired product from the other impurities.

20. Methyl (1R,5R)-5-hydroxy-2-cyclopentene-1-acetate is found in both the aqueous layer and occluded with the solid silver chloride.

21. In like manner and employing (+)-α-pinene [bp 155–156°C; $[\alpha]_D^{25}$ +47° (neat)], the sequence affords the methyl (1S,5S)-5-hydroxy-2-cyclopentene-1-acetate, bp 74–77°C (0.1 mm); $[\alpha]_D^{25}$ +131° (CH$_3$OH, c 1.03).

22. The checkers used (−)-α-pinene (bp 155°C, $[\alpha]_D^{22}$ −42° (neat)) from Aldrich Chemical Company, Inc. and obtained a product having bp 75–80°C (0.15 mm) and $[\alpha]_D^{21}$ −126° (CH$_3$OH, c 0.039).

3. Discussion

Several highly enantioselective asymmetric hydroboration reactions with prochiral olefins have been reported[8] with the di-3-pinanylborane reagents (diisopinocamphenylboranes) discovered by Brown and Zweifel.[9] Recently, alternative reagents such as the mono-3-pinanylboranes (monoisopinocamphenyl boranes)[10,11] and (+)-dilongifolylborane[12] have been used in effecting asymmetric hydroborations on prochiral olefins. With the di-3-pinanylborane reagents, the cis-disubstituted olefins[8,9] and 5-substituted cyclopentadienes[5,6] yield alcohols of high optical purity (80–95% e.e.). Lower asymmetric inductions (20–40% e.e.) occur when 1,1-disubstituted, trans-disubstituted, or trisubstituted olefins are employed as substrates. However, significantly higher enantioselective hydroborations occur when these olefins are treated with the mono-3-pinanylboranes[10,11] and (+)-longifolylborane.[12] Tetrasubstituted olefins have not successfully been asymmetrically hydroborated with any of these reagents.

Several racemic cis- or trans-2-alkyl-3-cyclopenten-1-ols have been prepared by multistep sequences from cyclopentadiene[13-16] or from substituted 1,3-dienes.[17] However, optically active cis- and trans-2-alkyl-3-cyclopenten-1-ols have been prepared directly by asymmetric hydroboration reactions using prochiral 5-substituted cyclopentadienes as substrates.[5,6] This asymmetric hydroboration method, described above, gives moderate yields of highly optically active trans-2-alkyl-3-cyclopenten-1-ols (94–96% e.e.), which are readily converted into the corresponding cis isomers.[5,6] Several of these substances are intermediates in the synthesis of such natural products as the monoterpene glycoside loganin,[5] the carbohydrate daunosamine,[18] and the prostaglandins such as PGF$_{2\alpha}$.[6]

TABLE I

ASYMMETRIC HYDROBORATION OF 5-SUBSTITUTED CYCLOPENTADIENES

Substituent	Alkylating Agent	Hydroborating Agent	Yield (%)	Absolute Stereochemistry	Enantiomeric Excess (%)[a]	Reference
R = CH₃	CH₃I	(+)-Di-3-pinanylborane	40–50	R,R	94–96	5,19
	CH₃I	(−)-Di-3-pinanylborane	40–50	S,S	94–96	5,19
	(CH₃)₂SO₄	(+)-Di-3-pinanylborane	2	R,R	—	19
R = CH₂CO₂CH₃	BrCH₂CO₂CH₃	(+)-Di-3-pinanylborane	40–50	R,R	94–96	6,19
	BrCH₂CO₂CH₃	(−)-Di-3-pinanylborane	40–50	S,S	94–96	6,19
	ClCH₂CO₂CH₃	(+)-Di-3-pinanylborane	Trace	R,R	—	19
R = CH₂CO₂-t-Bu	BrCH₂CO₂-t-Bu	(+)-Di-3-pinanylborane	Trace	R,R	—	19

[a]The percent enantiomeric excess was determined by HPLC analysis of products esterified with pure (S)-α-methoxy-α-trifluoromethyl-phenylacetyl chloride (Mosher Reagent).[20,21]

1. Department of Natural Products Chemistry, Hoffmann-La Roche Inc., Nutley, NJ 07110.
2. Moffett, R. B. *Org. Synth., Coll. Vol. IV* **1963**, 238. For a slightly different procedure, see Korach, M.; Nielsen, D. R.; Rideout, W. H. *Org. Synth., Coll. Vol. V* **1973**, 414.
3. King, R. B.; Stone, F. G. A. *Inorg. Synth.* **1963**, *7*, 99.
4. Hafner, K.; Kaiser, H. *Org. Synth., Coll. Vol. V* **1973**, 1088.
5. Partridge, J. J.; Chadha, N. K.; Uskokovic, M. R. *J. Am. Chem. Soc.* **1973**, *95*, 532.
6. Partridge, J. J.; Chadha, N. K.; Uskokovic, M. R. *J. Am. Chem. Soc.* **1973**, *95*, 7171.
7. For a detailed description of the original procedure for preparing (+)-di-3-pinanylborane, see Zweifel, G.; Brown, H. C. *Org. Synth.* **1972**, *52*, 59. For an improved procedure, applied to the preparation of the levorotatory enantiomer, see Lane, C. F.; Daniels, J. J. *Org. Synth., Coll. Vol. VI* **1988**, 719.
8. For a review of asymmetric hydroboration reactions, see Brown, H. C.; Jadhav, P. K.; Mandal, A. K. *Tetrahedron* **1981**, *37*, 3547.
9. Brown, H. C.; Zweifel, G. *J. Am. Chem. Soc.* **1961**, *83*, 2544.
10. Brown, H. C.; Yoon, N. M. *J. Am. Chem. Soc.* **1977**, *99*, 5514.
11. Mandal, A. K.; Jadhav, P. K.; Brown, H. C. *J. Org. Chem.* **1980**, *45*, 3543 and references therein.
12. Jadhav, P. K.; Brown, H. C. *J. Org. Chem.* **1981**, *46*, 2988.
13. Fried, J.; Sih, J. C.; Lin, C. H.; Dalven, P. *J. Am. Chem. Soc.* **1972**, *94*, 4343.
14. Fried, J.; Lin, C. H. *J. Med. Chem.* **1973**, *16*, 429.
15. Evans, D. A.; Crawford, T. C.; Fujimoto, T. T.; Thomas, R. C. *J. Org. Chem.* **1974**, *39*, 3176.
16. Evans, D. A.; Crawford, T. C.; Thomas R. C.; Walker, J. A. *J. Org. Chem.* **1976**, *41*, 3947.
17. Danheiser, R. L.; Martinez-Davila, C.; Auchus, R. J.; Kadonaga, J. T. *J. Am. Chem. Soc.* **1981**, *103*, 2443.
18. Grethe, G.; Sereno, J.; Williams, T. H.; Uskokovic, M. R. *J. Org. Chem.* **1983**, *48*, 5315.
19. Partridge, J. J.; Chadha, N. K.; Uskokovic, M. R., unpublished results.
20. Dale, J. A.; Dull, D. L.; Mosher, H. S. *J. Org. Chem.* **1969**, *34*, 2543.
21. Uskokovic, M. R.; Lewis, R. L.; Partridge, J. J.; Despreaux, C. W.; Pruess, D. L. *J. Am. Chem. Soc.* **1979**, *101*, 6742.

PREPARATION OF LOW-HALIDE METHYLLITHIUM

(Lithium, methyl-)

$$CH_3Li \quad + \quad Li(1\% \, Na) \quad \xrightarrow[25°C]{ether} \quad CH_3Li \quad + \quad LiCl$$

Submitted by Michael J. Lusch, William V. Phillips, Ronald F. Sieloff,
Glenn S. Nomura, and Herbert O. House[1]
Checked by Gregory S. Bisacchi and Robert V. Stevens

1. Procedure

Caution! The fine lithium dispersion used in this preparation, once washed to remove the mineral oil coating, will ignite spontaneously if exposed to air. Also, the methyl chloride and ether used are very volatile and highly flammable. The entire preparation, including the disposal of any residual lithium, should be performed in an efficient hood with a safety shield in front of the apparatus. A suitable dry-powder fire extinguisher should be kept at hand to extinguish any fires resulting from the accidental spillage of the washed lithium dispersion or of the methyllithium solution.

A dry, 1-L, three-necked, round-bottomed flask equipped with a large Teflon-covered magnetic stirring bar, a thermometer, and a dry ice condenser (Note 1) is flushed with argon (Note 2), then capped with a serum stopper, and subsequently maintained under a positive pressure of argon (Note 3). A 30% dispersion of lithium metal (in mineral oil) containing 1% sodium (13.9 g, 2.00 g-atom of lithium) (Note 4) is rapidly weighed and transferred to the flask.

The lithium is washed three times by transferring approximately 150-mL portions of anhydrous ethyl ether (Note 5) into the flask through the serum stopper by forced siphon through a stainless-steel cannula, stirring the resulting suspension of lithium briefly, allowing the lithium to rise to the surface, and finally withdrawing the major part of the underlying ether by forced siphon through a cannula. Anhydrous ethyl ether (500 mL) is added to the resultant oil-free lithium. Methyl chloride gas (bp $-24°C$, $d^{-24°C}$ 0.99 g/mL) from a compressed gas cylinder is passed through a flask containing Linde 4A molecular sieves and into a dry, 100-mL Pyrex graduated cylinder equipped with a 24/40 standard taper joint attached to a Claisen adapter and dry ice condenser, and cooled to $-24°C$ with a bath of dry ice–acetone (Figure 1). When 52.7 mL (52.5 g, 1.04 mol) of liquid methyl chloride has been collected, the adapter and condenser are removed, several boiling chips are added to the cold ($-24°C$) graduated cylinder, and the cylinder is stoppered with a rubber septum through which is inserted a stainless-steel cannula. The other end of this cannula is inserted through the rubber septum of the flask so that its tip is just above the liquid surface of the reaction flask. Dry ice–acetone is then added to the condenser attached to the reaction flask. Vigorous stirring of the ethereal lithium dispersion is begun and the methyl chloride is added over approximately a 1.5-hr period. The rate at which methyl chloride is distilled into the reaction vessel is controlled by slight cooling or warming of the graduated cylinder that contains the liquid methyl chloride. During addition, the initial grey suspension changes to a brown to

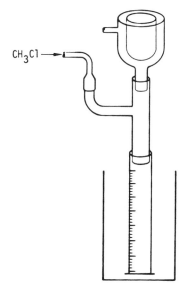

CH₃Cl —▶

Figure 1. Condensing the methyl chloride.

purple suspension; by the end of the addition, little, if any, lithium metal should be seen floating on the surface of the ether solution when stirring is interrupted. After the addition of methyl chloride is complete, the reaction mixture is stirred at 25°C for an additional 0.5–1 hr and then allowed to stand overnight or longer (Note 6) at 25°C under a static argon atmosphere, whereupon the precipitated lithium chloride settles to the bottom of the flask. The dry ice condenser and thermometer are removed from the flask and replaced with rubber septa. The supernatant methyllithium solution is transferred by forced siphon using a large-gauge cannula through a glass wool pad (Note 7) into a receiving flask previously flushed with an inert gas (Figure 2). The receiving flask, which contains the

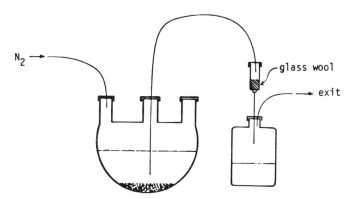

N₂ —▶

glass wool

exit

Figure 2. Decanting the methyllithium solution.

filtrate, a pale-yellow solution of methyllithium, is removed (Note 8) and stored in a refrigerator for 12–24 hr during, which time an additional small quantity of lithium chloride separates as fine crystals. The resulting supernatant solution is transferred with a stainless-steel cannula and a slight positive pressure of argon or nitrogen into one or more suitable oven-dried, nitrogen-filled storage bottles capped with rubber septa. Two 1-mL aliquots of the solution are removed with a hypodermic syringe for a modified Gilman titration (Note 9) and a 5-mL aliquot is removed with a hypodermic syringe to determine the halide concentration (Note 10). The solution contains 1.40–1.77 M methyllithium accompanied by 0.07–0.09 M lithium chloride corresponding to a 70–89% yield of methyllithium. If this solution is protected from oxygen and moisture, it may be stored at 0–25°C for several months (and remain active).

2. Notes

1. The dry ice condenser used with the apparatus should have sufficient condensing capacity to prevent the loss of significant amounts of methyl chloride; a condenser 38 cm long and 3.8 cm in diameter was suitable.

    Since finely divided lithium floats on the surface of the solvent and will be in contact with the atmosphere in the reaction vessel, an argon atmosphere, rather than a nitrogen atmosphere, should be used to avoid formation of the insoluble reddish-brown lithium nitride.

3. A slight positive pressure of argon was maintained in the vessel throughout the reaction by using an argon line connected to both a bubbler containing Nujol and the inlet on the dry ice condenser.

4. A dispersion in mineral oil of 30% (by weight) of lithium containing 1% by weight of sodium is marketed by Alfa Products, Morton Thiokol, Inc. This oil-coated dispersion may be exposed to the air during transfer and weighing and is conveniently transferred from its container by pouring through a wide-mouth funnel. Small quantities of the dispersion that adhere to the apparatus may be disposed of by rinsing in a stream of warm water to lower the viscosity of the oil and allow the suspended lithium to react with water at a controlled rate. To dispose of large quantities of this dispersion (or any quantity of lithium powder no longer coated with oil), the material should be suspended in anhydrous ether under an argon atmosphere and *tert*-butyl alcohol should be added dropwise to the suspension until all of the lithium metal has been consumed. Since hydrogen is liberated during these disposal procedures, they should be performed in an efficient hood.

5. Anhydrous ethyl ether was distilled from lithium aluminum hydride immediately before use.

6. Although most of the lithium chloride separates from the ether solution as a finely divided solid during the reaction, additional small quantities of lithium chloride continue to separate for 12–14 hr. After standing overnight, a typical reaction contains a precipitate of finely divided brownish-pink solid below a clear, pale-yellow solution.

7. A convenient filter was constructed by packing glass wool, previously dried in an oven, into a 20-mL Luer-Lok syringe barrel fitted with a 15-gauge needle.

The syringe barrel was capped with a serum stopper. A large-diameter cannula (at least 15 gauge) should be used to transfer the methyllithium solution from the flask to the filter since smaller-gauge cannulas are frequently plugged by solid particles.

8. As soon as the receiver containing the methyllithium solution has been removed and stoppered, the residual solids in the reaction flask and the filtration apparatus should be rinsed into another receiver with anhydrous ether under an atmosphere of argon or nitrogen. The ether slurry of solids, which may contain some unchanged lithium metal, should be treated cautiously in a hood with *tert*-butyl alcohol to consume any residual lithium metal before the mixture is discarded.

9. One 1-ml aliquot is added to 1.0 mL of freshly distilled 1,2-dibromoethane (bp 132°C) in an oven-dried flask that contains a static atmosphere of nitrogen or argon. After the resulting solution has been allowed to stand at 25°C for 5 min, it is diluted with 10 mL of water and titrated for base content (residual base) to a phenolphthalein endpoint with standard 0.100 M hydrochloric acid. The second 1-mL aliquot is added cautiously to 10 mL of water and then titrated for base content (total base) to a phenolphthalein endpoint with standard aqueous 0.100 M hydrochloric acid. The methyllithium concentration is the difference between the total base and residual base concentrations.[2] Alternatively, the methyllithium concentration may be determined by titration with a standard solution of *sec*-butyl alcohol employing 2,2′-bipyridyl as an indicator.[3a,b]

10. To determine the concentration of chloride ion,[3c,d] a 5-mL aliquot of the methyllithium solution is cautiously added to 25 mL of water and the resulting solution is acidified with concentrated sulfuric acid and then treated with 2–3 mL of ferric ammonium sulfate [Fe(NH$_4$)(SO$_4$)$_2$ 12 · H$_2$O] indicator solution and 2–3 mL of benzyl alcohol. The resulting mixture is treated with 10.0 mL of standard aqueous 0.100 M silver nitrate solution and then titrated with standard aqueous 0.100 M potassium thiocyanate solution to a brownish-red endpoint.

3. Discussion

Although ethereal solutions of methyllithium may be prepared by the reaction of lithium wire with either methyl iodide[4] or methyl bromide[5] in ether solution, the molar equivalent of lithium iodide or lithium bromide formed in these reactions remains in solution and forms, in part, a complex with the methyllithium.[6] Certain of the ethereal solutions of methyllithium currently marketed by several suppliers including Alfa Products, Morton Thiokol, Inc., Aldrich Chemical Company, and Lithium Corporation of America, Inc., have been prepared from methyl bromide and contain a full molar equivalent of lithium bromide. In several applications such as the use of methyllithium to prepare lithium dimethylcuprate[7] or the use of methyllithium in 1,2-dimethyoxyethane to prepare lithium enolates from enol acetates or trimethylsilyl enol ethers,[3b] the presence of this lithium salt interferes with the titration and use of methyllithium. There is also evidence indicating that the stereochemistry observed during addition of methyllithium to carbonyl compounds may be influenced significantly by the presence of a lithium salt in the reaction solution.[8] For these reasons it is often desirable to use ethereal solutions of methyllithium that do not contain an equivalent amount of lithium iodide or lithium bromide.

The reaction of lithium with methyl chloride in ether solution produces a solution of methyllithium from which most of the relatively insoluble lithium chloride precipitates. Ethereal solutions of "halide-free" methyllithium, containing 2–5 mol % of lithium chloride, were formerly marketed by Foote Mineral Company and by Lithium Corporation of America, Inc., but this product has been discontinued by both companies. Comparable solutions are also marketed by Alfa Products and Aldrich Chemical Company; these solutions have a limited shelf life and older solutions have often deteriorated badly even before the container is opened. Since an ether solution of methyl chloride reacts very slowly with lithium wire used in reactions with methyl bromide or methyl iodide, the present procedure[9] uses a finely divided suspension of lithium metal containing 1% (by weight) of sodium[6,10] to achieve a rapid reaction with methyl chloride. The finely divided lithium containing 1% sodium is marketed as a 30% (by weight) dispersion in mineral oil and must be washed free of this protective hydrocarbon diluent before use in order to avoid contamination of the final methyllithium reagent with a substantial amount of a mixture of high molecular weight hydrocarbons. Since lithium is less dense than common organic solvents such as diethyl ether or pentane, the washing procedure must be done with special care to avoid starting a fire with the pyrophoric, finely divided lithium.[2b] Finely divided lithium with somewhat higher or lower percentages of sodium are expected to work equally well.

1. School of Chemistry, Georgia Institute of Technology, Atlanta, GA 30332.
2. (a) Whitesides, G. M.; Casey, C. P.; Krieger, J. K. *J. Am. Chem. Soc.* **1971,** *93,* 1379; (b) Linstrumello, G.; Krieger, J. K.; Whitesides, G. M. *Org. Synth.* **1976,** *55,* 103. (Lithium wire containing 1% sodium is no longer available commercially.)
3. (a) Watson, S. C.; Eastham, J. F. *J. Organomet, Chem.* **1967,** *9,* 165; (b) Gall, M.; House, H. O. *Org. Synth., Coll. Vol. VI* **1988,** 121; (c) Kolthoff, I. M.; Sandell, E. B. "Textbook of Quantitative Inorganic Analysis," 2nd ed.; Macmillan: New York, 1947; pp. 572–574; (d) Skoog, D. A.; West, D. M. "Fundamentals of Analytical Chemistry," 2nd ed.; Holt, Rinehart, and Winston: New York, 1969; pp. 233–234.
4. Schollkopf, U.; Paust, J.; Patsch, M. R. *Org. Synth., Coll. Vol. V* **1973,** 859.
5. Wittig, G.; Hesse, A. *Org. Synth., Coll. Vol. VI* **1988,** 901.
6. Wakefield, B. J. "The Chemistry of Organolithium Compounds"; Pergamon: New York, 1974; pp. 8–11, 21–25.
7. (a) Eliel, E. L; Hutchins, R. O.; Knoeber, Sr. M. *Org. Synth., Coll. Vol. VI* **1988,** 442; (b) Muchmore, D. C. *Org. Synth., Coll. VI* **1972,** 762.
8. Ashby, E. C.; Noding, S. A. *J. Org. Chem.* **1979,** *44,* 4371.
9. We are indebted to Dr. W. Novis Smith, formerly with Foote Mineral Company, and now with Stauffer Chemical Company, Dobbs Ferry, New York, for supplying the general preparative procedure that we have adapted to the laboratory-scale preparation described here.
10. Kamienski, C. W.; Esmay, D. L. *J. Org. Chem.* **1960,** *25,* 1807.

β-ALKYL-α,β-UNSATURATED ESTERS FROM ENOL PHOSPHATES OF β-KETO ESTERS: METHYL 2-METHYL-1-CYCLOHEXENE-1-CARBOXYLATE

(1-Cyclohexene-1-carboxylic acid, 2-methyl, methyl ester)

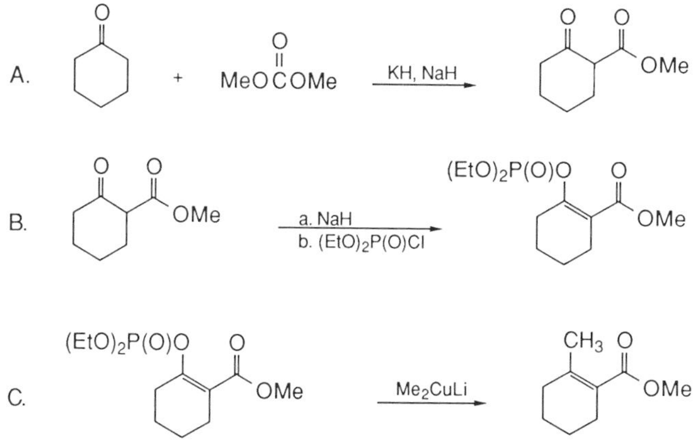

Submitted by MARGOT ALDERDICE, F. W. SUM, and LARRY WEILER[1]
Checked by STEPHEN P. ASHBURN, CLARK H. CUMMINS, and ROBERT M. COATES

1. Procedure

A. *Methyl 2-oxocyclohexanecarboxylate.* A 500-mL, three-necked, round-bottomed flask is equipped with a mechanical stirrer, a reflux condenser, and a pressure-equalizing dropping funnel bearing a nitrogen inlet (Note 1). The flask is flushed with nitrogen and charged with 18.02 g (0.20 mol) of dimethyl carbonate, 50 mL of anhydrous tetrahydrofuran, and 6.12 g (0.25 mol) of sodium hydride (Note 2). The suspension is stirred and heated to reflux temperature, at which time the slow, dropwise addition of 7.80 g (0.080 mol) of cyclohexanone in 20 mL of dry tetrahydrofuran is begun. After 2 min, 0.306 g (0.0076 mol) of powdered potassium hydride *(Caution! Dry potassium hydride is pyrophoric.)* (Note 3) is added to initiate the reaction. The addition of cyclohexanone is continued over a 1-hr period. The mixture is stirred and heated at reflux for another 30 min, cooled in an ice bath for 15–20 min, and hydrolyzed by slowly adding 75 mL of 3 M aqueous acetic acid. The contents of the flask are poured into 100 mL of aqueous sodium chloride, and the aqueous mixture is extracted with four 150-mL portions of chloroform. The organic layers are combined, dried with anhydrous sodium sulfate, and concentrated at room temperature with a rotary evaporator. Distillation of the residual liquid under reduced pressure gives 9.8–10.8 g (79–87%) of methyl 2-oxocyclohexanecarboxylate as a colorless liquid, bp 53–55°C (0.35 mm) (Note 4).

B. *Methyl 2-(diethylphosphoryloxy)-1-cyclohexene-1- carboxylate.* A 250-mL, two-necked, round-bottomed flask is equipped with a magnetic stirring bar, a rubber septum, and a gas inlet tube connected to a nitrogen source and a mineral oil bubbler (Note 1).

The flask is flushed with nitrogen and charged with 1.58 g (0.0329 mol) of a 50% dispersion of sodium hydride in mineral oil (Note 5). The sodium hydride is freed from the mineral oil by washing with four 40-mL portions of anhydrous diethyl ether (Note 6) and withdrawing the supernatant solvent with a syringe, after which 120 mL of anhydrous ether is added. The mixture is stirred and cooled in an ice bath as 4.68 g (0.0300 mol) of methyl 2-oxocyclohexanecarboxylate (Note 7) in 10 mL of ether is added at a moderately rapid rate such that vigorous but controlled evolution of hydrogen occurs (Note 8). The resulting creamy suspension is stirred at 0°C for another 30 min, after which time 4.5 mL (5.37 g, 0.031 mol) of diethyl chlorophosphate (Note 9) is injected through the septum with a syringe. The ice bath is removed, the mixture is stirred at room temperature for an additional 3 hr, and 0.6 g of solid ammonium chloride is added. Stirring is continued for 30 min, and the salts are then separated by suction filtration through a medium porosity fritted-glass funnel. Concentration of the filtrate under reduced pressure affords 8.18–8.63 g of the enol phosphate that is used in Step C without purification (Note 10).

C. *Methyl 2-methyl-1-cyclohexene-1-carboxylate.* A 250-mL, three-necked, round-bottomed flask is equipped with a magnetic stirring bar, a rubber septum, a pressure-equalizing addition funnel, and an inlet tube connected to a nitrogen source and a mineral oil bubbler (Note 1). The flask is charged with 8.03 g (0.042 mol) of copper(I) iodide (Note 11) and 50 mL of dry diethyl ether (Note 6), flushed with nitrogen, and cooled in an ice bath. The mixture is stirred and cooled as 92.7 mL (0.084 mol) of 1.1 M methyl-lithium in diethyl ether (Note 12) is added quickly through the septum by means of a syringe. The resulting clear and colorless, or light tan, solution of lithium dimethylcu-prate is then cooled in a carbon tetrachloride–dry ice slush bath maintained at $-23°C$ (Note 13). A solution of 8.18–8.63 g (ca. 0.028–0.030 mol) of the enol phosphate in 35 mL of dry ether is added from the addition funnel over 5–10 min. Stirring and cooling are continued for 3 hr, after which time the dark-purple solution is poured into a 1-L Erlenmeyer flask containing 75 mL of ice-cold 5% hydrochloric acid saturated with sodium chloride (Note 14). The mixture is stirred, or shaken vigorously, and cooled in an ice bath for 5–10 min to complete the hydrolysis. A 150-mL portion of 15% aqueous ammonia is added to the gray suspension and the mixture is swirled vigorously for a few minutes until the organic layer becomes clear and the aqueous layer turns bright blue. The mixture is transferred to a separatory funnel, the aqueous layer is withdrawn, and the organic phase is washed with 50 mL of 15% aqueous ammonia. The aqueous layers are combined and extracted with one 100-mL portion of ether. The combined ethereal layers are washed with two 50-mL portions of saturated sodium chloride, dried with anhydrous magnesium sulfate, and concentrated by rotary evaporation. Distillation of the remaining 4.25–5.47 g of liquid in a short-path distillation apparatus affords 3.99–4.17 g (86–90% based on β-keto ester) of methyl 2-methyl-1-cyclohexene-1-carbox-ylate, bp 96–97°C (27 mm) (Notes 15–17).

2. Notes

1. The glassware was dried in an oven at 125°C and assembled while warm.
2. Dimethyl carbonate is available from Aldrich Chemical Company, Inc. The checkers dried the tetrahydrofuran immediately before use by distillation from

the sodium ketyl of benzophenone under a nitrogen atmosphere. The submitters purchased sodium hydride (50% oil dispersion) from Alfa Products, Morton Thiokol, Inc. The checkers used 12.24 g of a 50% dispersion of sodium hydride in mineral oil obtained from the same supplier. The dispersion was washed with three portions of pentane to remove the mineral oil and the remaining sodium hydride was allowed to dry under nitrogen.

3. The submitters used a 35% dispersion of potassium hydride in mineral oil supplied by Alfa Products, Morton Thiokol, Inc.; the mineral oil was separated by washing the dispersion with five portions of dry hexane. The checkers used a 25% dispersion of potassium hydride in mineral oil obtained from the same source but without removing the mineral oil. The oil remained in the distillation pot when the product was distilled.

4. A boiling point of 68°C (0.8 mm) and a melting point of 25°C have been reported for methyl 2-oxocyclohexanecarboxylate.[2] The spectral properties of the product are as follows: IR (liquid film) cm^{-1}: 1745, 1715, 1615; ^{1}H NMR (CDCl$_3$) δ: 1.62 (m, 4 H, 2 CH$_2$), 2.22 (m, 4 H, 2 CH$_2$), 3.37 (t, 0.25 H, J = 7 Hz, CH at C-2 in keto form), 3.74 (s, 3 H, CH$_3$), 12.10 (s, 0.75 H, enol OH).

5. The sodium hydride-mineral oil dispersion was purchased from Alfa Products, Morton Thiokol, Inc.

6. Diethyl ether was dried by the submitters by refluxing over lithium aluminum hydride and was distilled immediately before use. The checkers distilled diethyl ether from the sodium ketyl of benzophenone before use.

7. A mixture of methyl and ethyl 2-oxocyclohexanecarboxylate, available from Aldrich Chemical Company, Inc., may also be used. The product obtained is a mixture of methyl and ethyl 2-methylcyclohexene-1-carboxylates.

8. No gas evolution was observed by the checkers in some runs in which an older lot of sodium hydride was used. In this case, the cooling bath was removed and the mixture was allowed to stir at room temperature until the bubbling ceased.

9. Diethyl chlorophosphate, supplied by Aldrich Chemical Company, Inc., was used by the submitters without purification and was handled in a glove bag under an atmosphere of dry nitrogen in a well-ventilated hood. The reagent was distilled and stored under nitrogen by the checkers. Aliquots were withdrawn with a syringe as needed.

10. The spectral properties of the enol phosphate are as follows: IR (CHCl$_3$) cm^{-1}: 1715, 1660, 1290, 1030; 90 MHz ^{1}H NMR (CDCl$_3$) δ: 1.3–1.9 (m, 4 H, CH$_2$CH$_2$), 1.35 (t, 6 H, J = 7 Hz, OCH$_2$CH_3), 2.3 (m, 4 H, allylic CH$_2$), 3.68 (s, 3 H, OCH$_3$), 4.15 (quintet, 4 H, J = 7 Hz, OCH_2CH$_3$).

11. Copper(I) iodide, supplied by either MC and B Manufacturing Chemists or Fisher Scientific Company, was purified by recrystallization from water saturated with potassium iodide.[3,4] The wet powder was washed successively with ethanol, acetone, and ether, and dried by heating overnight at 100°C in an evacuated drying pistol containing phosphorus pentoxide.[4a,5] The submitters advise that the compound should not be dried by heating in air.[5] When oven-dried copper(I) iodide was used in this procedure, the yield of product was somewhat lower (77–88%) and as much as 10–20% of 1-acetyl-2-methylcyclohexene was formed.

It is probable that the presence of small amounts of copper(II) impurities is responsible for the increased proportion of this by-product.[4b,6] Purified copper(I) iodide may be stored under nitrogen without change for several months.[4a]

12. Ethereal methyllithium (as the lithium bromide complex) was obtained by the submitters from Aldrich Chemical Company, Inc. The checkers used 1.19 M methyllithium–lithium bromide complex in ether supplied by Alfa Products, Morton Thiokol, Inc. The concentration of the methyllithium was determined by titration with 1.0 M tert-butyl alcohol in benzene using 1,10-phenanthroline as indicator.[7] The submitters report that ethereal methyllithium of low halide content purchased from Alfa Products, Morton Thiokol, Inc., gave similar results.

13. The coupling reaction between lithium dimethylcuprate and acyclic enol phosphates must be carried out between -47 and $-98°C$ for stereoselective formation of β-methyl-α,β-unsaturated esters.

14. The submitters have found that the reaction may also be hydrolyzed with a solution of 60 mL of saturated ammonium chloride and 15 mL of concentrated aqueous ammonia. The ethereal layer is then washed with 15% aqueous ammonia until the aqueous layer is no longer blue. When lithium di-n-butylcuprate is used, the yields are often improved by adding 1-bromobutane to the reaction mixture before hydrolysis with aqueous ammonium chloride.

15. The product exhibits the following spectral properties: IR (CHCl$_3$) 1720, 1640, 1080 cm^{-1}; ^{1}H NMR (CDCl$_3$) δ: 1.3–1.7 (m, 4 H, CH$_2$CH$_2$), 1.8–2.4 (m, 4 H, allylic CH$_2$), 1.97 (s, 3 H, CH$_3$), 3.69 (s, 3 H, OCH$_3$); MS (70 eV) m/e (assignment, relative intensity): 154 (M$^+$, 50%), 95 ($-$CO$_2$CH$_3$, 100%).

16. The purity of the product was determined by the checkers by GLC analysis using the following column and conditions: 3-mm × 1.8-m column, 5% free fatty acid phase (FFAP) on acid-washed Chromasorb W (60–80 mesh) treated with dimethyldichlorosilane, 90°C (1 min) then 90° to 200°C (15°C/min). The chromatogram showed a major peak for methyl 2-methyl-1-cyclohexene-1-carboxylate preceded by two minor peaks for methyl 1-cyclohexene-1-carboxylate and 1-acetyl-2-methylcyclohexene. The areas of the two impurity peaks were 5–6% and 0.5–2% that of the major peak. The purity of the product seems to depend on careful temperature control during the reaction. The total amount of the two impurities was 14–21% in runs conducted at about -15 to $-20°C$ or at temperatures below $-23°C$.

17. The submitters purified the product by distillation in a Kugelrohr apparatus with an oven temperature of 85–88°C (20 mm) and obtained 3.80–3.85 g (88–89%). The purity of the product was 93–96% according to GLC analysis. The major impurity (2–6%) was 1-acetyl-2-methylcyclohexene.

The product may also be purified by flash chromatography[8] using 19/1 (v/v) petroleum ether, (bp 30–60°C)/ethyl acetate as eluant. A column of 2-cm diameter was packed to a height of 25 cm with Kieselgel 60 (230–400 mesh) supplied by BDH Chemicals Ltd. In one run chromatography of 4.19 g of crude product afforded 3.70 g (88%) of the α,β-unsaturated ester that was completely free of the more polar by-product, 1-acetyl-2-methylcyclohexene. However, the checkers found that the other by-product, methyl 1-cyclohexene-1-carboxylate, is not readily separated by flash chromatography.

3. Discussion

This procedure illustrates a new method for the preparation of β-alkyl-α,β-unsaturated esters by coupling lithium dialkylcuprates with enol phosphates of β-keto esters.[9] The procedure for the preparation of methyl 2-oxocyclohexanecarboxylate described in Section A is based on one reported by Ruest, Blouin, and Deslongchamps.[2] Methyl 2-methyl-1-cyclohexene-1-carboxylate has been prepared by esterification of the corresponding acid with diazomethane[10] and by reaction of methyl 2-chloro-1-cyclohexene-1-carboxylate with lithium dimethylcuprate.[11]

The formation of β-alkyl-α,β-unsaturated esters by reaction of lithium dialkylcuprates or Grignard reagents in the presence of copper(I) iodide, with β-phenylthio-,[12,13] β-acetoxy-,[14,15] β-chloro-,[11,16] and β-phosphoryloxy-α,β-unsaturated esters[9] has been reported. The principal advantage of the enol phosphate method is the ease and efficiency with which these compounds may be prepared from β-keto esters. A wide variety of cyclic and acyclic β-alkyl-α,β-unsaturated esters has been synthesized from the corresponding β-keto esters.[9] However, the method is limited to primary dialkylcuprates. Acyclic β-keto esters afford (Z)-enol phosphates that undergo stereoselective substitution with lithium dialkylcuprates with predominant retention of stereochemistry (usually >85–98%). It is essential that the cuprate coupling reaction of the acyclic enol phosphates be carried out at lower temperatures (-47 to $-98°C$) to achieve high stereoselectivity. When combined with the γ-alkylation of methyl acetoacetate dianion,[17] this method provides a facile means of isoprenoid chain extension.[18] The procedures have been employed to advantage in syntheses of (E,E)-10-hydroxy-3,7-dimethyldeca-2,6-dienoic acid,[18a] latia luciferin,[18b] and mokupalide.[18c] β-Diketones may be converted to β-alkyl-α,β-unsaturated ketones in a similar manner.[9]

1. Department of Chemistry, University of British Columbia, Vancouver, British Columbia, Canada, V6T 1Y6.
2. Ruest, L.; Blouin, G.; Deslongchamps, P. *Synth. Commun.* **1976**, *6*, 169–174.
3. Kauffman, G. B.; Teter, L. A. *Inorg. Synth.* **1963**, *7*, 9–12.
4. (a) Linstrumelle, G.; Krieger, J. K.; Whitesides, G. M. *Org. Synth.* **1976**, *55*, 103–113; (b) Smith, J. G.; Wikman, R. T. *Synth. React. Inorg. Metal-Org. Chem.* **1974**, *4*, 239–248.
5. Kauffmann, G. B.; Pinnell, R. P. *Inorg. Synth.* **1960**, *6*, 3–6.
6. House, H. O.; Chu, C.-Y.; Wilkins, J. M.; Umen, M. J. *J. Org. Chem.* **1975**, *40*, 1460–1469.
7. Gall, M.; House, H. O. *Org. Synth., Coll. Vol. VI* **1988**, 121.
8. Still, W. C.; Kahn, M.; Mitra, A. *J. Org. Chem.* **1978**, *43*, 2923–2925.
9. Sum, F. W.; Weiler, L. *Can. J. Chem.* **1979**, *57*, 1431–1441.
10. Rhoads, S. J.; Chattopadhyay, J. K.; Waali, E. E. *J. Org. Chem.* **1970**, *35*, 3352–3358.
11. Heathcock, C. H.; Leong, J. H., unpublished results cited by Clark, R. D.; Heathcock, C. H. *J. Org. Chem.* **1976**, *41*, 636–643 (see footnote 24).
12. Posner, G. H.; Brunelle, D. J. *J. Chem. Soc., Chem. Commun.* **1973**, 907–908.
13. (a) Kobayashi, S.; Takei, H.; Mukaiyama, T. *Chem. Lett.* **1973**, 1097–1100; (b) Kobayashi, S.; Mukaiyama, T. *Chem. Lett.* **1974**, 1425–1428.

14. (a) Casey, C. P.; Marten, D. F. *Synth. Commun,* **1973**, *3*, 321–324; (b) Casey, C. P.; Marten, D. F.; Boggs, R. A. *Tetrahedron Lett.* **1973**, 2071–2074; (c) Casey, C. P.; Marten, D. F. *Tetrahedron Lett.* **1974**, 925–928.
15. Quannès, C.; Langlois, Y. *Tetrahedron Lett.* **1975**, 3461–3464.
16. Harding, K. E.; Tseng, C.-Y. *J. Org. Chem.* **1978**, *43*, 3974–3977.
17. (a) Huckin, S. N.; Weiler, L. *J. Am. Chem. Soc.* **1974**, *96*, 1082–1087; (b) Huckin, S. N.; Weiler, L. *Can. J. Chem.* **1974**, *52*, 2157–2164.
18. (a) Sum, F. W.; Weiler, L. *J. Chem. Soc., Chem. Commun.* **1978**, 985–986; (b) Sum, F. W.; Weiler, L. *Tetrahedron Lett.* **1979**, 707–708; (c) Sum, F. W.; Weiler, L. *J. Am. Chem. Soc.* **1979**, *101*, 4401–4403.

OPTICALLY ACTIVE EPOXIDES FROM VICINAL DIOLS VIA VICINAL ACETOXY BROMIDES: (S)-(−)- and (R)-(+)-METHYLOXIRANE

Submitted by Martin K. Ellis and Bernard T. Golding[1]
Checked by Stephen H. Montgomery and Clayton H. Heathcock

1. Procedure

A. *(S)-(+)-Propane-1,2-diol.* Into a three-necked, 500-mL, round-bottomed flask fitted with a mechanical stirrer, dropping funnel and reflux condenser are placed 10.8 g (0.284 mol) of lithium aluminum hydride and 200 mL of dry ethyl ether. To this slurry is added, from the dropping funnel, 33 g (0.28 mol) of ethyl L-(−)-lactate (Note 1) in 150 mL of dry ethyl ether at a rate that maintains a steady reflux. The heterogeneous mixture is stirred for 3 hr. Then 25 mL (1.39 mol) of water is carefully added and stirring is continued for a further 1.5 hr. The mixture is filtered and the white solid (LiOH) is washed well with ether and dichloromethane. The organic phases are combined, dried over magnesium sulfate, and concentrated at reduced pressure with a rotary evaporator to give a portion of the crude product (3 g). Aqueous 1 M sulfuric acid is added to the solid until the milky suspension is just acidic (pH 6–6.5). The suspension is subjected to continuous extraction with twice its volume of dichloromethane (about 500 mL) for 168 hr. The dichloromethane layer is dried over magnesium sulfate and concentrated at reduced pressure with a rotary evaporator. The crude products are combined and distilled at reduced pressure to obtain 14.4–15.6 g (68–73%) of (S)-(+)-propane-1,2-diol, bp 52–56°C (0.5 mm), as a colorless liquid (Note 2).

B. *(S)-(−)-2-Acetoxy-1-bromopropane.* A three-necked, 100-mL, round-bottomed flask fitted with a magnetic stirring bar, dropping funnel, and reflux condenser is charged with 7.6 g (0.1 mol) of (S)-(+)-propane-1,2-diol. A solution of 45% w/v hydrogen bromide–acetic acid (71 g, 0.3 mol) (Note 3) is added from the dropping funnel with cooling over ca. 5 min. The homogeneous solution is stirred at room temperature for 45 min, after which time it is added to 200 mL of water and the mixture neutralized immediately with solid sodium carbonate (Note 4). The neutral solution is extracted three times with 150 mL of ethyl ether, the organic phases are combined, dried over magnesium sulfate, and concentrated at reduced pressure with a rotary evaporator. Distillation of the crude product at reduced pressure affords 14.1–15.4 g (78–85%) of (S)-$(-)$-2-acetoxy-1-bromopropane, bp 54–57°C (7 mm), as a colorless liquid (Note 5).

C. *(S)-$(-)$-Methyloxirane.* To a three-necked, 100-mL, round-bottomed flask equipped with a magnetic stirring bar, pressure-equalizing dropping funnel, and 10-cm Vigreux column connected to an efficiently cooled condenser and receiver are added 9.05 g (50 mmol) of the acetoxybromopropane and 20 mL of dry 1-pentanol. The solution is stirred at room temperature and 41.66 mL (50 mmol) of 1.2 M potassium pentoxide in 1-pentanol (Note 6) is added from the dropping funnel over ca. 20 min. A white precipitate of potassium bromide is observed. After addition is complete, the flask is warmed in an oil bath at ca. 130–145°C to attain distillation (Note 7). The product, (S)-$(-)$-methyloxirane, 2.0–2.35 g (69–81%), is collected as a colorless liquid, bp 34–35°C (Note 8).

2. Notes

1. Ethyl L-$(-)$-lactate was purchased from Fluka AG, Buchs, Switzerland and was used directly. Checkers found that fresh ethyl lactate purchased from Fluka is only 97–98% e.e. by ^{19}F NMR spectroscopy on the Mosher ester.

2. An optical rotation of $[\alpha]_D^{16}$ +20.3 (H_2O, c 7.5), [lit.[2] $[\alpha]_D^{20}$ +20.7° (H_2O, c 7.5))] was observed for this product. It had the following spectral properties: IR (liquid film, polystyrene reference) cm^{-1}: 3350 (s), 2970 (m), 2930 (m), 2870 (m), 1455 (m), 1375 (m); ^{1}H NMR (CDCl$_3$) δ: 1.15 (d, 3 H, —CH$_3$), 3.40 [q, 1 H, H$_2$C(OH)—] and 3.59 [q, 1 H, H$_2$C(OH)—], 3.89 [m, 1 H, —CH(OH)CH$_3$], —OH resonances variable.

3. Hydrogen bromide–acetic acid, 45%, was purchased from BDH Chemicals Ltd., Poole, England. The checkers used hydrobromic acid (30–32% in acetic acid, 4.1 M) from Fisher Scientific, 711 Forbes Ave., Pittsburgh, PA 51219.

4. Approximately 80 g of sodium carbonate is required. On addition of solid sodium carbonate a considerable amount of frothing occurs. To prevent the loss of product, the addition of the reaction mixture to the water and subsequent neutralization with solid sodium carbonate is performed in a 2-L beaker.

5. An optical rotation of $[\alpha]_D^{20}$ −13.7° (CHCl$_3$, c 5.8), [lit.[2] $[\alpha]_D^{23}$ −13.55 (CHCl$_3$, c 5.8)] was observed for (S)-$(-)$-2-acetoxy-1-bromopropane. (R)-$(+)$-2-Acetoxy-1-bromopropane, obtained from (R)-$(-)$-propane-1,2-diol,[3,4] gave an optical rotation of $[\alpha]_D^{18}$ +14.1° (CHCl$_3$, c 5.8). Both enantiomers of acetoxybromopropane had the following spectral properties: IR (liquid film, polystyrene ref.) cm^{-1}: 2980 (w), 2937 (w), 1735 (s), 1450 (w), 1425 (w), and 1370 (s); ^{1}H NMR (CCl$_4$) δ: 1.34 (d, 3 H, CH$_3$), 2.10 (s, 3 H, —OCOCH$_3$), 3.38 (d, 2 H, —CH$_2$Br), and 4.97 [m, 1 H, —CH(OCOCH$_3$)CH$_3$] due to 2-acetoxy-1-bromopropane (94% by integration) and 1.70 (3 H) and 4.16 (3 H) due to 1-acetoxy-2-bromopropane (6%).

6. Potassium pentoxide in 1-pentanol is prepared by dissolving freshly cut potassium in dry, freshly distilled 1-pentanol under nitrogen. The molarity of this solution may be determined by titration against standard aqueous acid.

7. The oil bath is preheated to 120–130°C. It is then transferred to a prewarmed heater with stirrer on a lab jack below the reaction flask. The oil bath can then be moved into position with the aid of the lab jack.

8. An optical rotation of $[\alpha]_{D}^{20}$ −18.7° (CCl$_4$, c 5.83), [lit.[2] $[\alpha]_{D}^{22}$ −18.55° (CCl$_4$, c 5.84))] was observed for (S)-(−)-methyloxirane. (R)-(+)-Methyloxirane, obtained from (R)-(+)-acetoxybromopropane (Note 5), gave an optical rotation of $[\alpha]_{D}^{18}$ +19.13° (CCl$_4$, c 5.66), [lit.[2] $[\alpha]_{D}^{18}$ +18.7° (CCl$_4$, c 5.83)], bp 34–35°C, and a range of yields within the limits of those obtained for (S)-(−)-methyloxirane. Both enantiomers of methyloxirane had the following spectral properties; ^{1}H NMR (CCl$_4$) δ: 1.27 (d, 3 H, −CH$_3$), 2.27 (q, 1 H, −CH(O)CH$_2$), 2.59 [t, 1 H, −CH(O)CH$_2$] and 2.83 [m, 1 H, H$_3$C−CH(O)CH$_3$] ppm.

3. Discussion

This procedure illustrates the stereospecific conversion of 1,2-diols into vicinal acetoxy bromides by hydrogen bromide in acetic acid.[2] The acetoxy bromides that are formed are easily transformed into epoxides by base treatment. In the examples presented, the base is used in a high-boiling solvent to facilitate isolation of epoxide by direct distillation from the reaction mixture (see also references 5–8). For other examples, a solvent may be used that is either more volatile than the epoxide (e.g., methanol[2]) or easily removed by aqueous workup and solvent extraction of the epoxide (e.g., ethane-1,2-diol[9]). The hydrogen bromide–acetic acid method is superior to the preparation of epoxides from 1,2-diols via 1-O-sulfonate esters, because any contaminating 2-O-sulfonate ester will detract from the optical purity of the epoxide.[10] The optical purities of the samples of (R)- and (S)-methyloxirane prepared as described were better than 98%, according to complexation chromatography and ^{1}H NMR analysis with chiral shift reagent.[11,12] Other procedures for preparing (R)-[13,14] and (S)-methyloxirane have been described.[15–17] These compounds are valuable starting materials for preparing a variety of optically active natural products (nonactin,[18] sulcatol,[19] recifeiolide,[20] methyl-1,6-dioxaspiro[4.5]decanes[21]), drugs (e.g., N-2-hydroxypropyl-6,7-benzomorphans[22]) and for studies of stereoregular polymerizations.[23]

1. Department of Chemistry and Molecular Sciences, University of Warwick, Coventry CV4 7AL, UK. Present address: Department of Organic Chemistry, The University, Newcastle upon Tyne, NE1 7RU, UK.
2. Golding, B. T.; Hall, D. R.; Sakrikar, S. *J. Chem. Soc., Perkin Trans.* 1 **1973**, 1214–1220.
3. Levene, P. A.; Walti, A. *Org. Synth., Coll. Vol. II* **1943**, 5–6.
4. Levene, P. A.; Walti, A. *Org. Synth., Coll. Vol. II* **1943**, 545–547.
5. Howes, D. A.; Brookes, M. H.; Coates, D.; Golding, B. T.; Hudson, A. T. *J. Chem. Res.* (S) **1983**, 9; (M) **1983**, 217–228.
6. Mori, K.; Tamada, S. *Tetrahedron* **1979**, 35, 1279–1284; Mori, K.; Sasaki, M.; Tamada, S.; Suguro, T.; Masuda, S. *Tetrahedron* **1979**, 35, 1601–1605.
7. Schurig, V.; Koppenhoefer, B.; Buerkle, W. *J. Org. Chem.* **1980**, 45, 538–541.
8. Schmidt, U,; Talbiersky, J.; Bartkowiak, F.; Wild, J. *Angew. Chem. Int. Ed. Engl.* **1980**, 19, 198–199.

9. Mori, K. *Tetrahedron* **1977**, *33*, 289–294.
10. Mori, K. *Tetrahedron* **1976**, *32*, 1101–1106.
11. Schurig, V.; Koppenhoefer, B.; Buerkle, W. *Angew. Chem. Int. Ed. Engl.* **1978**, *17*, 937.
12. Golding, B. T.; Sellars, P. J.; Wong, A. K. *J. Chem. Soc., Chem. Commun.* **1977**, 570–571.
13. Hillis, L. R.; Ronald, R. C. *J. Org. Chem.* **1981**, *46*, 3348–3349.
14. Pirkle, W. H.; Rinaldi, P. L. *J. Org. Chem.* **1978**, *43*, 3803–3807.
15. Ghirardelli, R. G. *J. Am. Chem. Soc.* **1973**, *95*, 4987–4990.
16. Gombos, J.; Haslinger, E.; Schmidt, U. *Chem. Ber.* **1976**, *109*, 2645–2647.
17. Seuring, B.; Seebach, D. *Helv. Chim. Acta* **1977**, *60*, 1175–1181.
18. Schmidt, U.; Gombos, J.; Haslinger, E.; Zak, H. *Chem. Ber.* **1976**, *109*, 2628–2644.
19. Johnston, B. D.; Slessor, K. N. *Can. J. Chem.* **1979**, *57*, 233–235.
20. Utimoto, K.; Uchida, K.; Yamaya, M.; Nozaki, H. *Tetrahedron Lett.* **1977**, 3641–3642.
21. Hintzer, K.; Weber, R.; Schurig, V. *Tetrahedron Lett.* **1981**, *22*, 55–58.
22. Rahtz, D.; Paschelke, G.; Schroeder, E. *Eur. J. Med. Chem.-Chim. Ther.* **1977**, *12*, 271–278; *Chem. Abstr.* **1977**, *87*, 201833u.
23. Spassky, N.; Dumas, P.; Sepulchre, M.; Sigwalt, P. *J. Polym. Sci. Polym. Symp.* **1975**, *52*, 327–349.

METHYL PHENYLACETYLACETATE FROM PHENYLACETYL CHLORIDE AND MELDRUM'S ACID

(Benzenebutanoic acid, β-oxo-, methyl ester)

Submitted by Y. Oikawa, T. Yoshioka, K. Sugano, and Osamu Yonemitsu[1]
Checked by Michael J. Taschner, Hans P. Märki, and Clayton H. Heathcock

1. Procedure

Into a 300-mL, round-bottomed flask equipped with a dropping funnel and a magnetic stirrer is placed a solution of 23.75 g (0.165 mol) of recrystallized Meldrum's acid (Note 1) in 65 mL of anhydrous dichloromethane. The flask and its contents are cooled in an ice bath, and 32.5 mL (0.40 mol) of anhydrous pyridine (Note 2) is added with stirring under an argon atmosphere over a 10-min period. To the resulting colorless clear solution is added a solution of 25.0 g (0.16 mol) of freshly distilled phenylacetyl chloride (Note 3) in 50 mL of anhydrous dichloromethane over a 2-hr period. After the addition is complete, the resulting orange, cloudy reaction mixture is stirred for 1 hr at 0°C, then for an additional 1 hr at room temperature. The reaction mixture is diluted with 35 mL of dichloromethane, and then poured into 100 mL of 2 *N* hydrochloric acid containing crushed ice. The organic phase is separated and the aqueous layer extracted twice with 25-mL portions of dichloromethane. The organic phase and the extracts are combined, washed twice with 25-mL portions of 2 *N* hydrochloric acid and 30 mL of saturated sodium chloride solution, and dried over anhydrous sodium sulfate. The solvent is removed with a rotary evaporator to yield an acyl Meldrum's acid (Note 4) as a pale-yellow solid.

The solid acyl Meldrum's acid, without purification, is refluxed in 250 mL of anhydrous methanol for 2.5 hr. The solvent is removed with a rotary evaporator, and the residual oil is distilled under reduced pressure to give 25.2 g (82%) of methyl phenylacetylacetate as a colorless liquid, bp 126–128°C/(0.6 mm).

2. Notes

1. Meldrum's acid, 2,2-dimethyl-1,3-dioxane-4,6-dione, is available from the Aldrich Chemical Company, Inc. It may also be prepared from malonic acid and acetone.[2] It is used in this preparation after recrystallization from acetone or from acetone–hexane. The checkers found that a final product of significantly lower purity is obtained if the Meldrum's acid is not recrystallized.

2. The checkers used pyridine that had been distilled from calcium hydride.

3. Phenylacetyl chloride is supplied by Wako Pure Industries, Ltd. (Japan) and the Aldrich Chemical Company, Inc. It is distilled before use, bp 95–96°C/(12 mm). The checkers found the distilled commercial material to be slightly pink. However, material of this quality gave a good yield of pure product.

4. The product, 2,2-dimethyl-5-phenylacetyl-1,3-dioxane-4,6-dione, is isolated in its enol form in 97% yield. If desired, it may be further purified by recrystallization from ether–hexane to give pale-yellow prisms, mp 96–97°C (dec). The checkers recrystallized the material from dichloromethanehexane and obtained 65% yield of material, mp 94–96°C (dec) and 7%, mp 84–90°C. The ^{1}H NMR spectrum of this compound has absorptions at δ 1.65 (s, 6 H), 4.30 (s, 2 H), 7.20 (s, 5 H), and 15.0 (br s, 1 H).

3. Discussion

Because β-keto esters are among the most important intermediates in organic synthesis, many methods have been developed for their synthesis.[3] However, it is still desirable to have a general and practical method for preparation of β-keto esters of the general type $RCOCH_2CO_2R'$, and thence by alkylation with alkyl halides compounds of the type $RCOCHR''CO_2R'$.[4] The available synthetic methods can be classified broadly in three categories: those involving acetoacetic esters,[5] those involving mixed malonic esters,[6] and those involving malonic acid half esters.[7] The procedure described herein[8] may be classified as one of the malonic ester methods. The procedure consists of two simple steps and utilizes readily accessible starting materials. When the carboxylic acid chloride is not available, the carboxylic acid may be condensed with Meldrum's acid in the presence of a condensing agent such as ethyl phosphorocyanidate.[9]

Methanolysis or ethanolysis of an acyl Meldrum's acid is performed simply by refluxing in methanol or ethanol solution. The products are methyl or ethyl β-keto esters, and they can usually be purified by distillation. When a higher ester (such as benzyl, tert-butyl, or trichloroethyl) is required, it is easily prepared by refluxing the acyl Meldrum's acid in benzene containing about 3 equiv of the appropriate alcohol.

Recently, Melillo and co-workers applied this Meldrum's acid method with some modifications to the synthesis of thienamycin. A carboxylic acid was treated with carbonyldiimidazole, followed by treatment with Meldrum's acid to give an acyl Meldrum's acid, which was converted to a β-keto p-nitrobenzyl ester by refluxing in acetonitrile containing p-nitrobenzyl alcohol.[10]

1. Faculty of Pharmaceutical Sciences, Hokkaido University, Sapporo 060, Japan.
2. Davidson, D.; Bernhard, S. A. *J. Am. Chem. Soc.* **1948**, *70*, 3426.
3. For reviews, see Hauser, C. R.; Hudson, B. E. *Org. React.* **1942**, *1*, 266; House, H. O. "Modern Synthetic Reactions," 2nd ed.; W. A. Benjamin: Menlo Park, CA, 1972, p. 734.
4. Durst, H. D.; Liebeskind, L. *J. Org. Chem.* **1974**, *39*, 3271; Brändström, A.; Junggren, U. *Acta Chem. Scand.* **1969**, *23*, 2204.
5. Shriner, R. L.; Schmidt, A. G.; Roll, L. J. *Org. Synth.*, *Coll. Vol. II* **1943**, 266; Guha, M.; Nasipuri, D. *Org. Synth.*, *Coll. Vol. V* **1973**, 384; Viscontini, M.; Merckling, N. *Helv. Chim. Acta* **1952**, *35*, 2280.
6. Breslow, D. S.; Baumgarten, E.; Hauser, C. R. *J. Am. Chem. Soc.* **1944**, *66*, 1286; Taylor, E. C.; McKillop, A. *Tetrahedron* **1967**, *23*, 897; Bowman, R. E.; Fordham, W. D. *J. Chem. Soc.* **1951**, 2758; Pichat, L.; Beaucourt, J.-P. *Synthesis* **1973**, 537.
7. Ireland, R. E.; Marshall, J. A. *J. Am. Chem. Soc.* **1959**, *81*, 2907; Bram, G.; Vilkas, M. *Bull. Soc. Chim. Fr.* **1964**, 945; Pollet, P.; Gelin, S. *Synthesis* **1978**, 142; Brooks, D. W.; Lu, L. D.-L.; Masamune, S. *Angew. Chem. Int. Ed. Engl.* **1979**, *18*, 72.
8. Oikawa, Y.: Sugano, K.; Yonemitsu, O. *J. Org. Chem.* **1978**, *43*, 2087.
9. Shioiri, T.; Yokoyama, Y.; Kasai, Y.; Yamada, S. *Tetrahedron* **1976**, *32*, 2211; Takamizawa, A.; Sato, Y.; Tanaka, S. *Yakugaku Zasshi* **1965**, *85*, 298; *Chem. Abstr.* **1965**, *63*, 9940h.
10. Melillo, D. G.; Shinkai, I.; Liu, T.; Ryan, K.; Sletzinger, M. *Tetrahedron Lett.* **1980**, *21*, 2783.

2-METHYL-3-PHENYLPROPANAL

(Benzenepropanal, α-methyl-)

$$C_6H_5I \ + \ CH_2{=}CCH_2OH \ + \ Et_3N \ \xrightarrow{\ Pd(OCOCH_3)_2\ } \ C_6H_5CH_2CHCHO$$

Submitted by S. A. Buntin and R. F. Heck[1]
Checked by C. M. Tice and C. H. Heathcock

1. Procedure

A 250-mL, three-necked, round-bottomed flask, equipped with a mechanical stirrer and a reflux condenser, is charged with 0.49 g (2.2 mmol) of palladium acetate (Note 1), 20.4 g (100 mmol) of iodobenzene, 9.0 g (125 mmol) of 2-methyl-2-propen-1-ol, 12.6 g (125 mmol) of triethylamine, and 32.5 mL of acetonitrile (Note 2). The reaction vessel is placed in an oil bath at 100°C and the solution is heated to reflux for 11 hr under a nitrogen atmosphere. The reaction mixture is allowed to cool to room temperature and transferred to a 500-mL separatory funnel with the aid of 100 mL of ether and 100 mL of water. The organic layer is washed five times with 100 mL portions of water. The combined aqueous layers are reextracted with 100 mL of ether. The organic layers are combined, dried over anhydrous sodium carbonate, and filtered. The organic layer is concentrated and distilled under reduced pressure. The product, 2-methyl-3-phenyl-propanal, 12.05 g (82%), has a boiling range of 52–58°C at 0.40 mm (Note 3).

2. Notes

1. Palladium acetate was prepared by the method of Stephenson et al.[2] A suitable material is also available from the Strem Chemical Company or Alfa Inorganics.

2. Iodobenzene, 2-methyl-2-propen-1-ol, and triethylamine were obtained from the Aldrich Chemical Company, Inc. Acetonitrile was obtained from the J. T. Baker Chemical Company. All these reagents were used as received.

3. The 2-methyl-3-phenylpropanal is 90% pure by GLC. The product mixture contains 6% of another isomer, 2-methyl-2-phenylpropanal, and a small amount of 2-phenyl-2-propen-1-ol. A completely pure sample of the aldehyde is readily obtained by stirring the crude aldehyde with excess saturated aqueous sodium bisulfite solution for several hours, filtering the solid bisulfite adduct, washing with ether, and liberating the aldehyde with excess aqueous sodium bicarbonate. Redistillation gives the completely pure aldehyde in about 60% yield.

3. Discussion

The reaction of allylic alcohols and aryl halides in the presence of a palladium catalyst has been used in the past to prepare various β-arylaldehydes. The procedure described here is essentially that of Heck and Melpolder.[3] A similar reaction has been carried out with bromobenzene and 2-methyl-2-propen-1-ol in hexamethylphosphoric triamide (HMPT) as solvent with sodium bicarbonate as base. A variety of other bases have also been used.[4] 2-Methyl-3-phenylpropanal has been prepared by reacting palladium acetate and phenylmercuric acetate with 2-methyl-2-propen-1-ol.[5]

The aldehyde is also obtained by the hydroformylation of allylbenzene.[6] An alternative method involves benzylation of 2-ethylthiazoline followed by reduction with aluminum amalgam and cleavage with mercuric chloride.[7] A sixth method of preparation is the phenylation of 2-vinyl-5,6-dihydro-1,3-oxazine with phenylmagnesium bromide followed by methylation and hydrolysis.[8] Finally, arylation of 2-methyl-2-propen-1-ol with phenyldiazonium salts catalyzed by zero-valent palladium complexes gives the title aldehyde.[9]

1. Department of Chemistry, University of Delaware, Newark, DE 19711.
2. Stephenson, T. A.; Morehouse, S. M.; Powell, A. R.; Heffer, J. P.; Wilkinson, G. *J. Chem. Soc.* **1965,** 3632–3640.
3. Heck, R. F.; Melpolder, J. B. *J. Org. Chem.* **1976,** *41,* 265.
4. Chalk, A. J.; Magennis, S. A. *J. Org. Chem.* **1976,** *41,* 273.
5. Heck, R. F. *Org. Synth., Coll. Vol. VI* **1988,** 815.
6. Lai, R.; Ucciani, E. *C. R. Hebd. Seances Sci., Ser. C* **1972,** *275,* 1033; *Chem. Abstr.* **1973,** *78,* 42963g.
7. Durandetta, J. L.; Meyers, A. I. *J. Org. Chem.* **1975,** *40,* 2021.
8. Adickes, H. W.; Kovelesky, A. C.; Malone, G. R.; Meyers, A. I.; Nabeya, A.; Nolen, R. L.; Politzer, I. R.; Portnoy, R. C. *J. Org. Chem.* **1973,** *38,* 36.
9. Kikukawa, K.; Matsuda, T. *Chem. Lett.* **1977,** 159.

(+)-(7aS)-7a-METHYL-2,3,7,7a-TETRAHYDRO-1H-INDENE-1,5-(6H)-DIONE

[1H-Indene-1,5(6H)-dione, 2,3,7,7a-tetrahydro-7a-methyl-, (S)-]

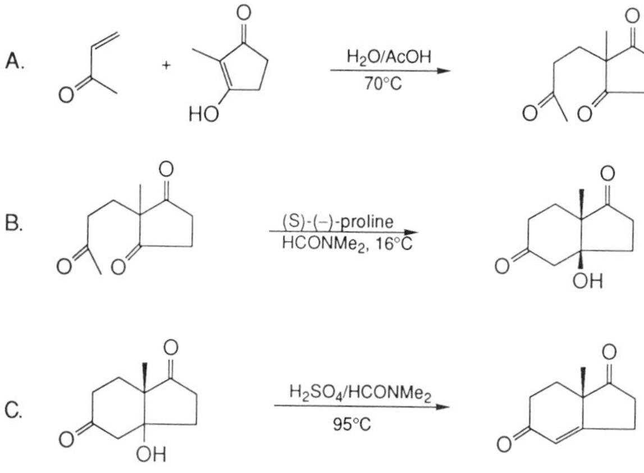

Submitted by ZOLTAN G. HAJOS[1] and DAVID R. PARRISH[2]
Checked by STUART REMINGTON, DAVID LUST, and GABRIEL SAUCY

1. Procedure

Caution! Part A should be performed in a well-ventilated hood because methyl vinyl ketone is a lachrymator.

A. *2-Methyl-2-(3-oxobutyl)-1,3-cyclopentanedione.* A 1.0-L, three-necked, round-bottomed flask equipped with a condenser, magnetic stirring bar, and thermometer is charged with 112.1 g (1.0 mol) of 2-methyl-1,3-cyclopentanedione (Note 1), 230 mL of deionized water, 3.0 mL of glacial acetic acid, and 140 mL (120.96 g, 1.72 mol) of methyl vinyl ketone (Note 2). The system is shielded from light with aluminum foil and placed under a slight positive pressure of nitrogen. The flask is placed in an oil bath and the temperature is raised to 70°C. The reaction is monitored by gas chromatography (GLC, Note 3) until complete (1–2 hr). The mixture is cooled, transferred to a separatory funnel, and extracted with 500 mL and then two 100-ml portions of dichloromethane. The combined extracts are washed with 500 and 100 mL of saturated brine. The combined brine wash is extracted with a further two 100-mL portions of dichloromethane. The total dichloromethane extract is dried over sodium sulfate and filtered. The solvent is removed on a rotary evaporator at 45°C (70 mm). Drying on the rotary evaporator at 40–45°C (0.03 mm) for 16 hr gives 181.8 g (100%) of the desired triketone as an orange oil (Notes 4 and 5).

B. *(+)-(3aS,7aS)-2,3,3a,4,7,7a-Hexahydro-3a-hydroxy-7a-methyl-1H-indene-1,5(6H)-dione.* A 500-mL, three-necked, round-bottomed flask equipped with a magnetic stirring bar and a nitrogen inlet is charged with 188 mL of N,N-dimethylformamide (Note 6) and 863 mg (7.5 mmol) of S-(−)-proline (Notes 7 and 8). The mixture is degassed four times by alternate evacuation and refilling with nitrogen. The system is shielded from light with aluminum foil and the contents of the flask are stirred in a 15–16°C bath (Note 9) for 1.0 hr. To the resultant suspension is added 45.5 g (0.25 mol) of the 2-methyl-2-(3-oxobutyl)-1,3-cyclopentanedione prepared in Step A. A total of 62.5 mL of N,N-dimethylformamide is used to ensure complete transfer. The degassing procedure is repeated four times and stirring at 15–16°C (Note 10) is continued for 40–120 hr (Note 11) as the mixture becomes yellow and then brown. The reaction is monitored for completeness by TLC (Note 12). The solution of the desired ketol (Note 13) is used directly in Step C.

C. (+)-(7aS)-7a-Methyl-2,3,7,7a-tetrahydro-1H-indene-1,5(6H)-dione.   A 100-mL, three-necked, round-bottomed flask equipped with a magnetic stirring bar, pressure-equalizing dropping funnel, and nitrogen inlet is charged with 50 mL of N,N-dimethylformamide (Note 6). The contents of the flask are cooled to −20°C with a dry ice–acetone bath and 2.70 mL (4.97 g, 48.6 mmol) of concentrated sulfuric acid is added over 5–10 min at a rate to maintain a temperature of −15 to −20°C (Note 14).

The flask containing the solution of the (+)-(3aS,7aS)-2,3,3a,4,7,7a-hexahydro-3a-hydroxy-7a-methyl-1H-indene-1,5(6H)-dione in N,N-dimethylformamide is placed in an oil bath and heated to 95°C. When the temperature reaches 70–75°C, an 18.8-mL aliquot of the concd sulfuric acid in N,N-dimethylformamide solution is added in one portion. The reaction mixture is heated to 95°C for 3.0 hr. After 1.0 hr, an additional 7.5-mL aliquot of the concd sulfuric acid in N,N-dimethylformamide solution is added in one portion. The reaction is monitored for completeness by GLC (Note 15) and cooled. The solvent is removed on a rotary evaporator at 45°C (0.3–0.5 mm) to give a brown oil. The material is taken up in 375 mL of dichloromethane. The solution is washed with two 190-mL portions of 2.0 N sulfuric acid solution that have been saturated with sodium chloride, two 190-mL portions of saturated sodium bicarbonate solution that have been saturated with sodium chloride, and 190 mL of saturated brine. Each aqueous wash is extracted, in turn, with the same two 190-mL portions of dichloromethane. The combined dichloromethane solutions are dried over sodium sulfate and filtered, and the solvent is removed on a rotary evaporator at 40°C (70 mm) to give 38.8–39.6 g of oily, brown semisolid. This material is taken up in 78 mL of ethyl acetate and the solution is applied to a dry column of 78 g of silica gel (Note 16). The column is eluted with 600 mL of ethyl acetate, and the total eluate is stripped of solvent on a rotary evaporator at 40°C (70 mm) to give 37.2–38.8 g of tan crystalline solid. The solid is subjected to bulb-to-bulb distillation[3] (Note 17) at 120–135°C (0.1 mm) to give 35.9–36.9 g of a slightly yellowish (cream white) solid, mp 56–61°C, $[\alpha]_D^{25}$ +324–329° (toluene, c 1.0) (Notes 18–20). This material is taken up in 74 mL of ether at reflux. The solution is brought at reflux to the point of turbidity with 19 mL of hexanes. The mixture is seeded, allowed to stand at ambient temperature for 2 hr, and then chilled in a 17°C water bath for 30 min (Note 21). The solid is collected by filtration on medium porosity sintered glass, washed with two 12-mL portions of cold (3°C) 1 : 1 v/v ether : hexanes and dried at 20°C (70 mm) to give 28.7–31.3 g (70–76%) of white crystalline solid (Note 22), mp 64–66°C, $[\alpha]_D^{25}$ +347.5–349° (toluene, c 1.0) (Note 23), purity by GLC 99.4–99.5% (Notes 24–26).

2. Notes

1. 2-Methyl-1,3-cyclopentanedione, 98%, purchased from the Aldrich Chemical Company, Inc., was used. Material prepared according to Hengartner, U.; Chu, V. *Org. Synth., Coll. Vol. VI* **1988**, 774 was determined by the checkers to be equally satisfactory.

2. Methyl vinyl ketone, technical grade, purchased from the Aldrich Chemical Company, Inc., was fractionally distilled into ca. 1% w/v hydroquinone shortly before use. The fraction boiling at 33–36°C (120 mm) was used.

3. Analyses were carried out on a Hewlett Packard HP 5840 A gas chromatograph operated isothermally at 150°C. A 25-m capillary column packed with cross-linked phenylmethylsilicone was employed. 2-Methyl-1,3-cyclopentanedione and 2-methyl-2-(3-oxobutyl)-1,3-cyclopentanedione had retention times of ca. 7 min 12.5 min, respectively.

4. If desired, pure triketone can be isolated by distillation of the crude triketone through a Vigreux column. The yield of light yellow oil, bp 115–120°C (0.2–0.3 mm), is 80–89%.

5. The triketone has the following spectral properties: IR (neat) cm^{-1}: 1770, 1725; ^{1}H NMR (CDCl$_3$) δ: 1.12 (s, 3 H, CH$_3$), 2.22 (s, 3 H, CH$_3$CO), 2.8 (m, 4 H, COCH$_2$CH$_2$CO).

6. *N,N*-Dimethylformamide, purchased from the Fisher Scientific Co., was mixed with 10% v/v toluene and distilled at atmospheric pressure. After all of the toluene had been distilled (head temperature to 148°C), vacuum was cautiously applied. The fraction of *N,N*-dimethylformamide that distilled at 78–82°C (56–65 mm) was collected and stored under nitrogen prior to use.

7. L-(–)-Proline [(*S*)-configuration], 99$^+$%, purchased from the Aldrich Chemical Company, Inc., was employed. The material was finely ground in a mortar and pestle immediately before use.

8. The L-(–)-proline was established by the checkers to be of >99.8% (estimated level of detection) enantiomeric purity by conversion to *N*-pentafluoropropionyl-L-(–)-proline isopropyl ester and GLC analysis on a 50-m glass capillary column containing the chiral phase, Chirasil-Val (Quadrex, Inc.). Analyses were performed on a Hewlett-Packard HP 5710 A instrument operated isothermally at 140°C. Racemic proline was used as a control.

9. The checkers used a flask with a built-in jacket. Water at 15–16°C was continuously circulated through the jacket.

10. Temperature control in this reaction is critical. At higher temperatures, the enantioselectivity of the reaction drops off significantly, while at lower temperatures, the reaction time becomes unacceptably long.

11. The reaction time varied substantially from run to run, but generally complete conversion was observed in 48–72 hr.

12. E. Merck silica gel F-254 plates were used, with 20 : 1 v/v dichloromethane : methanol as eluent. The plates were developed by drying, spraying with 9 : 1 v/v deionized water : concentrated sulfuric acid, light drying with a hot air gun, spraying with 3% w/v vanillin solution in ethanol, and strong heating

with the hot air gun. The approximate R_f values observed were 0.67 (starting triketone) and 0.37 (product ketol). In addition, a minor spot at R_f 0.59 (enone arising from dehydration of the ketol) was seen.

13. If desired, the ketol can be isolated as follows. The reaction mixture from 18.0 g of distilled 2-methyl-2-(3-oxobutyl)-1,3-cyclopentanedione is evaporated on a rotary evaporator at 45°C (0.3 mm) to give 22.0 g of brown oil. A solution of this material in 200 mL of ethyl acetate is filtered through 80 g of J. T. Baker silica gel. Elution with ca. 1.3 L of ethyl acetate in 200-mL fractions is monitored by TLC (Note 12). The fractions containing the desired product are combined and stripped of solvent on a rotary evaporator at 45°C (70 mm). Final drying on the rotary evaporator at 45°C (0.3 mm) gives 18.0 g (100%) of crude ketol as a slightly oily, brown solid having the following spectral properties: IR (CHCl$_3$) cm^{-1}: 3600, 3500–3300, 1742, 1722; ^{1}H NMR (CDCl$_3$) δ: 1.26 (s, 3 H, CH$_3$), 2.63 (s, 2 H, COCH$_2$COH). Further purification of the compound by crystallization from ether (ca. 50% recovery) gives material of mp 118–119°C, $[\alpha]_D^{25}$ +59.8° (lit.[5] mp 119–119.5°C, $[\alpha]_D^{25}$ +60.4°).

14. The solution is prepared immediately before use and kept at −20°C.

15. The GLC system described in Note 3 was employed. The intermediate ketol and product enone had retention times of ca. 23 and 16.5 min, respectively. A trace of ketol (<1%) is observed at the end of the reaction.

16. E. Merck silica gel 60 (70–230 mesh) was used. The column dimensions were 3.2 cm × 60 cm.

17. A Kugelrohr apparatus purchased from the Aldrich Chemical Company, Inc. was used. The receiving bulb was cooled with an ice–water bath. The temperature indicated is that of the oven air bath.

18. The ratio of rotations obtained in toluene and benzene has been determined to be 1.00 : 1.03. The rotation of enantiomerically pure material in toluene, based on the accepted[4] value of +362° in benzene, is 351°. The enantiomeric purity at this stage is thus 92–94% (Note 19).

19. Attempts by both the submitters and checkers to find a method other than optical rotation to determine the enantiomeric purity have been unsuccessful.

20. Material of this purity is satisfactory for many synthetic purposes; see reference 3.

21. Further cooling results in a higher recovery of material. However, the melting point and rotation of the samples thus obtained are lower.

22. The compound is somewhat unstable. It is best stored in an amber bottle under nitrogen at 3°C.

23. The enantiomeric purity of the purified material is thus 99.0–99.4% (see Note 18).

24. GLC analysis was carried out on a Hewlett Packard HP 5710 A gas chromatograph operated isothermally at 155°C. A 50-m capillary column of OV-17 on fused silica was employed. The enone had a retention time of ca. 14.5 min.

25. The material has the following spectral properties: UV (CH$_3$OH) λ 235 nm (ε = 11,200); IR (CHCl$_3$) cm^{-1}: 1746, 1665; ^{1}NMR (CDCl$_3$) δ: 1.31 (s, 3 H), 7a-CH$_3$), 5.97 (broad, S, 1 H, vinylic-H).

26. Steps B and C have been scaled up to the 2.0-mol level with no loss in yield or enantiomeric purity.

3. Discussion

The (S)-(−)-proline-catalyzed asymmetric aldol cyclization of the triketone to the optically active bicyclic aldol product, followed by dehydration to the optically active enedione, (+)-7a-methyl-(7aS)-2,3,7,7a-tetrahydro-1H-indene-1,5(6H)-dione, has been described, and two alternative reaction mechanisms have been suggested by the submitters.[5] The exact mechanism of the extremely high asymmetric induction in the crucial conversion of the prochiral triketone to the optically active ketol still needs to be clarified.[6a-c]

The synthesis of the triketone has been included (Step A of the procedure), since identification of the crystalline compound originally claimed[7] to be the triketone has been shown to be in error.[8] After completion of our work, the triketone was correctly characterized by another research group.[9]

Asymmetric aldol cyclization of the triketone with (S)-(−)-proline can also be effected in solvents other than N,N-dimethylformamide; acetonitrile is outstanding.[5]

Of the asymmetric amino acid reagents investigated, (S)-(−)-proline gave the highest optical yield (93.4%); (−)-trans-4-hydroxyproline gave 73.1%, and (S)-(−)-azetidinecarboxylic acid gave 63.9% optical yields in the asymmetric synthesis of the optically active bicyclic ketol.

The use of (R)-(+)-proline in acetonitrile induced the asymmetric aldol cyclization of the triketone to the enantiomeric ketol, (−)-(3aR,7aR)-2,3,3a,4,7,7a-hexahydro-3a-hydroxy-7a-methyl-1H-indene-1,5(6H)-dione.[10]

The ethyl homolog of the triketone, 2-ethyl-2-(3-oxobutyl)-1,3-cyclopentanedione, has been converted with (S)-(−)-proline in N,N-dimethylformamide to (+)-(3aS, 7aS)-7a-ethyl-2,3,3a,4,7,7a-hexahydro-3a-hydroxy-1H-indene-1,5(6H)-dione in good yield.[5] This, in turn, could be dehydrated to the homologous bicyclic enedione, (+)-(7aS)-7a-ethyl-2,3,7,7a-tetrahydro-1H-indene-1,5(6H)-dione.[5]

Circular dichroism studies of the 7a-methyl bicyclic ketol suggested, and a single-crystal X-ray diffraction study of the racemic compound confirmed, the cis conformation with an axial 7a-methyl and an equatorial 3a-hydroxy group in the six-membered ring of the bicyclic system. On the other hand, similar measurements of the 7a-ethyl bicyclic keto established the alternate possible cis conformation to avoid the 1,3-diaxial interactions between the angular ethyl group and the C-4 and C-6 axial hydrogens.

Dehydration of the optically active bicyclic ketols in refluxing benzene with a little p-toluenesulfonic acid could readily be effected without loss of optical purity.[5] It has been shown by a research group at Schering A. G., Berlin, Germany that the triketone can be converted directly to the optically active enedione with 0.5 eq. of (S)-(−)-proline and 0.25 equiv. of 1 N aqueous $HClO_4$ in refluxing acetonitrile.[11]

The optically active bicyclic enedione, (+)-7a-methyl-(7aS)-2,3,7,7a-tetrahydro-1H-indene-1,5(6H)-dione, was prepared first by microbiological means,[4] and its absolute stereochemistry has been established.[12] The compound was later prepared by optical resolution.[13]

The products of this highly efficient asymmetric synthesis are important intermediates in natural product chemistry, such as the total synthesis of steroids and prostaglandins.

1. Formerly with Hoffmann-La Roche Inc., Nutley, NJ 07110. Present address: 36 Bainbridge St., Princeton, NJ 08540.
2. Roche Research Center, Hoffmann-La Roche Inc., Nutley, NJ 07110.
3. Micheli, R. A.; Hajos, Z. G.: Cohen, N.; Parrish, D. R.; Portland, L. A.; Sciamanna, W.; Scott, M. A.; Wehrli, P. A. *J. Org. Chem.* **1975,** *40,* 675–681.
4. Acklin, W.; Prelog, V.: Prieto, A. P. *Helv. Chim. Acta* **1958,** *41,* 1416–1424.
5. Hajos, Z. G.; Parrish, D. R. *J. Org. Chem.* **1974,** *39,* 1615–1621.
6. (a) Buchschacher, P.; Cassal, J.-M.; Fürst, A.; Meier, W. *Helv. Chim. Acta* **1977,** *60,* 2747– 2755; (b) Brown, K. L.; Damm, L.; Dunitz, J. D.; Eschenmoser, A.; Hobi, R.; Kratky, C. *Helv. Chim Acta* **1978,** *61,* 3108–3135; (c) Agami, C.; Meynier, F.; Puchot, C.; Guilhem, J.; Pascard, C. *Tetrahedron* **1984,** *40,* 1031–1038.
7. Boyce, C. B. D.; Whitehurst, J. S. *J. Chem. Soc.* **1959,** 2022–2024.
8. Hajos, Z. G.; Parrish, D. R. *J. Org. Chem.* **1974,** *39,* 1612–1615.
9. Crispin, D. J.; Vanstone, A. E.; Whitehurst, J. C. *J. Chem. Soc. C* **1970,** 10–18.
10. Hajos, Z. G.; Parrish, D. R. U.S. Patent 3975442, August 17, 1976; *Chem. Abstr.* **1977,** *86,* P43296u.
11. Eder, U.; Sauer, G.; Wiechert, R. *Angew. Chem. Int. Ed. Engl.* **1971,** *10,* 496–497.
12. Acklin, W.; Prelog, V. *Helv. Chim. Acta* **1959,** *42,* 1239, 1247.
13. Hajos, Z. G.; Parrish, D. R.; Oliveto, E. P. *Tetrahedron* **1968,** *24,* 2039–2046.

(*S*)-8a-METHYL-3,4,8,8a-TETRAHYDRO-1,6(2*H*,7*H*)-NAPHTHALENEDIONE

[1,6(2*H*,7*H*)-Naphthalenedione, 3,4,8,8a-tetrahydro-8a-methyl-, (*S*)-]

Submitted by PAUL BUCHSCHACHER,[1] A. FÜRST,[1] and J. GUTZWILLER[1]
Checked by P. S. BELICA, P. S. MANCHAND, and GABRIEL SAUCY

1. Procedure

Caution! Part A should be performed in a well-ventilated hood because methyl vinyl ketone is a lachrymator.

A. *2-Methyl-2-(3-oxobutyl)-1,3-cyclohexanedione.* A 1-L, round-bottomed flask equipped with a thermometer and a reflux condenser capped with an argon-inlet tube is charged with 126.1 g (1 mol) of 2-methyl-1,3-cyclohexanedione (Note 1) and 300 mL of distilled water. To the well-stirred suspension are added 3 mL of acetic acid, 1.1 g of hydroquinone, and 142 g (167 mL, 2 mol) of freshly distilled methyl vinyl ketone (Note 2). The reaction mixture is stirred under argon at 72–75°C for 1 hr, cooled to room temperature, treated with sodium chloride (103 g), and poured into a separatory funnel containing ethyl acetate (400 mL). The organic phase is collected and the aqueous phase is reextracted twice with ethyl acetate (150 mL each time). The combined extracts are washed with two 200-mL portions of saturated brine, dried over anhydrous magnesium sulfate, and filtered, and the filtrate is evaporated at 40°C under reduced pressure (water aspirator) on a rotary evaporator. The residue is kept under high vacuum (1.0 mm) at 40°C for 30 min to give 210.8 g of crude 2-methyl-2-(3-oxobutyl)-1,3-cyclo-hexanedione (**1**, "trione") as a pale-yellow oil, homogeneous by thin-layer chromatography (Note 3). This crude material is used in Step B.

B. *(S)-8a-Methyl-3,4,8,8a-tetrahydro-1,6(2H,7H)-naphthalenedione (3-S).* A 3-L, one-necked, round-bottomed flask, equipped with an argon-inlet tube and containing a magnetic stirrer, is charged with 5.75 g (0.05 mol) of finely ground L-proline (Note 4) and a solution of 210.8 g of crude trione **1** (from Step A) in 1 L of anhydrous dimethyl sulfoxide (Note 5). The mixture is stirred at room temperature (ca. 25°C) under argon for 120 hr, the magnetic bar is removed, and the solvent is removed under high vacuum (1.0 mm) at 65°C (Note 6) on a rotary evaporator to give 206.9 g of a dark reddish–violet oil. The oil is dissolved in toluene (100 mL) and is absorbed on a column (9 cm × 60 cm) of silica gel (1.5 kg, 70–230 mesh) (Note 7), which was previously packed in hexane. Elution is carried out under a slight positive pressure of argon (ca. 1 atm) (Note 8) initially with 1 L of hexane : ethyl acetate (5 : 1) and then with a 3 : 2 mixture of hexane : ethyl acetate taking 300-mL fractions. The progress of the purification is monitored by thin-layer chromatography (Note 9): no product is observed until ca. 5 L of eluant is collected. Fractions containing the product are combined, and the solvents are removed under reduced pressure (water aspirator) at 45–50°C. The residue is then kept under high vacuum (0.1 mm) at 40°C for 30 min to give 154.2 g of an orange-colored oil, which became glassy and sometimes crystalline on standing at room temperature, $[\alpha]_D^{25}$ +68° (toluene, *c* 1.5) (Note 10). This material is dissolved in 535 mL of ether and is filtered through a fluted filter paper to remove small particles. The flask is rinsed with 500 mL of ether, and this is passed through the filter paper. After cooling to 3°C, the combined filtrates are seeded with a few crystals of pure 3-S (Note 11), and the mixture is left undisturbed at −20°C for 18 hr. Most of the supernatant liquid is carefully decanted without agitation, and the crystals are collected by filtration. The flask is rinsed with cold (0°C) 50% ether in hexane and the rinse is used to wash the crystals. The crystals are dried for 16 hr under high vacuum at room temperature to yield 85.9 g of (S)-enedione (first crop), mp 49–50°C, $[\alpha]_D^{25}$ +96.9° (toluene, *c* 1.2). The combined filtrate and washings are evaporated to give 67.1 g of an orange-colored oil, which is dissolved in 604 mL of ether, cooled to 3°C, and seeded with (R,S)-enedione (Note 12). The mixture is left at -20°C for 18 hr, and the supernatant liquid is carefully decanted (no agitation). The wet crystals are then collected by filtration, washed with cold (0°C) 50% ether in hexane, and dried under reduced pressure at room temperature to give 36.3 g of racemic material ($3R + 3S$). The filtrate and washing are evaporated to give 30.6 g of an oil, which is dissolved in 100 mL of ether and filtered through a fluted filter paper. The flask is rinsed with 114 mL of ether, and filtered through the fluted filter paper, and the combined filtrates are cooled to 3°C and seeded with crystals of the pure 3-S. The mixture is left at -20°C overnight, the supernatant liquid is carefully decanted without much agitation and the wet crystals are collected by filtration and washed with cold (0°C) 50% hexane in ether. After drying 15.3 g of light amber-colored crystals (second crop), mp 49–50°C, $[\alpha]_D^{25}$ +97.3° (toluene, c 1.0). The total yield of (S)-enedione is 101.2 g (56.8%) (Note 13).

2. Notes

1. 2-Methyl-1,3-cyclohexanedione[2] was obtained from Aldrich Chemical Company, Inc. or Fluka and had mp 208–210°C.

2. Methyl vinyl ketone, bp 34°C/120 mm, was obtained from Aldrich Chemical Company, Inc. or Fluka.

3. Thin-layer chromatography was performed on silica gel with ethyl acetate : hexane (3 : 2). Visualization of the spots was achieved by spraying the plates with 10% ceric sulfate in 10% sulfuric acid, heating the plates to ca. 120°C, and spraying again with 10% phosphomolybdic acid in isopropyl alcohol. The product has R_f 0.50; 2-methyl-1,3-cyclohexanedione has R_f 0.30.

4. S-($-$) Proline was obtained from Aldrich Chemical Company, Inc. It is also available from Ajinomoto GmbH or Degussa.

5. Dimethyl sulfoxide was dried over Linde 4A molecular sieves. Anhydrous, deaerated N,N-dimethylformamide was preferred by the submitters (Note 14).[3]

6. The temperature should be kept below 70°C.

7. Silica gel was purchased from EM Reagents, E. Merck, Darmstadt, Germany. The submitters preferred to do the preliminary purification by fractional vacuum distillation using a Hickman-type short-path distillation head. The main fraction distills as a light orange oil, bp 126–130°C (0.02 mm) (Note 14).

8. The procedure of W. C. Still[4] is used.

9. Silica gel and 60% ethyl acetate in hexane were used. The product, R_f 0.40, is visible under short-wavelength UV light, whereas the starting trione, also R_f 0.40, is not. Visualization is achieved as described in Note 3.

10. The material was melted at 55°C under reduced pressure (12 mm) on a rotary evaporator prior to sampling in order to measure the optical rotation on a homogeneous sample.

11. Compound 3-S was obtained from material having $[\alpha]_D^{25}$ +68° (toluene, c 1.5) that was prepared in another experiment. Thus, 28.2 g of this (S)-enedione is

dissolved in 90 mL of ether and the solution is left at −20°C for 18 hr. The crystals are collected by filtration without much agitation, washed with 30 mL of cold (0°C) 50% ether in hexane, and redissolved in 117 mL of ether. The solution is left at −20°C for 18 hr, and the crystals are collected by filtration, washed with 30 mL of cold (0°C) 50% ether in hexane, and dried under reduced pressure (1.0 mm) at room temperature to give 12.0 g of (S)-enedione, mp 50°C, $[\alpha]_D^{25}$ +100° (benzene, c 1.5); $[\alpha]_D^{25}$ +97° (toluene, c 1.0). It should be possible to prepare seed crystals from a small aliquot, but this was not attempted by the checkers.

12. Racemic Wieland–Miescher ketone was obtained from Aldrich Chemical Company, Inc. or prepared according to the procedure of Ramachandran and Newman.[5]

13. ^{1}H NMR studies (100 MHz, CDCl₃) using the shift reagent tris[3-(heptafluoropropylhydroxymethylene)-d-camphorato]europium(III) (purchased from Aldrich Chemical Company, Inc.) indicated that the two crops were enantiomerically pure. Under identical conditions (10 mg of reagent per 9.6 mg of substrate), absorption due to the vinyl proton at 5.86 in the racemate appeared as two peaks (1-Hz separation) of equal intensity.

14. The submitters' procedure using dimethylformamide in Step B and using distillation for isolation of the enantiomerically enriched ketone was checked by K. Job, A. K. Beck, and D. Seebach and proved equally satisfactory.

3. Discussion

Racemic 8a-methyl-3,4,8,8a-tetrahydro-1,6(2H,7H)-naphthalenedione (the Wieland–Miescher ketone)[5,6] is a versatile building block for the synthesis of steroids[7] and terpenoids.[8] The (S)-enantiomer, 3-S, was first obtained by microbiological means[9] and by classical resolution via a derived hemiphthalate.[10] The present synthesis[3] of 3-S is based[11] on the asymmetric intramolecular aldolization of the prochiral triketone 1 using S-(−)-proline catalytically. The product is obtained in 56% yield (from 2) and is enantiomerically pure on the basis of optical rotation and NMR spectroscopy determined in the presence of a chiral shift reagent. Despite numerous synthetic investigations and modifications of this asymmetric Robinson annulation,[12] the mechanism of enantiodifferentiation is still not fully understood;[13] see discussion in this volume, p. 367, relating to the asymmetric synthesis of the corresponding S-(+)-tetrahydro-7-methylindenedione.

1. F. Hoffmann-La Roche and Company, Ltd., Basle, Switzerland. Paul Buchschacher deceased, March 1982. Because of the death of one of the submitters and the retirement of another, the version of this procedure in *Org. Synth.* **1985**, *63*, 37–43 had not been reviewed by the submitters prior to publication. The modifications in the current procedure have been incorporated at their request.

2. Mekler, A. B.; Ramachandran, S.; Swaminanthan, S.; Newman, M. S. *Org. Synth., Coll. Vol. V* **1973**, 743–746.

3. Gutzwiller, J.; Buchschacher, P.; Fürst, A. *Synthesis* **1977**, 167–168.

4. Still, W. C.; Kahn, M.; Mitra, A. *J. Org. Chem.* **1978**, *43*, 2923–2925.

5. Ramachandran, S.; Newman, M. S. *Org. Synth., Coll. Vol. V* **1973**, 486–489.

6. Wieland, P.; Miescher, K. *Helv. Chim. Acta* **1950**, *33*, 2215–2228.

7. Danishefsky, S.; Cain, P.; Nagel, A. *J. Am. Chem. Soc.* **1975**, *97*, 380–387.

8. Spencer, T. A.; Weaver, T. D.; Villarica, R. M.; Friary, R. J.; Posler, J.; Schwartz, M. A. *J. Org. Chem.* **1968**, *33*, 712–719.

9. Prelog, V.; Acklin, W. *Helv. Chim. Acta* **1956**, *39*, 748–757.

10. Newkome, G. R.; Roach, L. C.; Montelaro, R. C.; Hill, R. K. *J. Org. Chem.* **1972**, *37*, 2098–2101.

11. (a) Eder, U.; Sauer, G.; Wiechert, R. *Angew. Chem. Int. Ed. Engl.* **1971**, *10*, 496–497; (b) Hajos, Z. G.; Parrish, D. R. *J. Org. Chem.* **1974**, *39*, 1615–1621; *Org. Synth., Coll. Vol. VII* **1990**, 363.

12. For reviews, see: (a) Cohen, N. *Acc. Chem. Res.* **1976**, *9*, 412–417; (b) Apsimon, J. W.; Sequin, R. P. *Tetrahedron* **1979**, *35*, 2797–2842; (c) Drauz, K.; Kleeman, A.; Martens, J. *Angew. Chem. Int. Ed. Engl.* **1982**, *21*, 584–608.

13. (a) Brown, K. L.; Damm, L.; Dunitz, J. D.; Eschenmoser, A.; Hobi, R.; Kratky, C. *Helv. Chim. Acta* **1978**, *61*, 3108–3135; (b) Agami, C.; Meynier, F.; Puchot, C.; Guilhem, J.; Pascard, C. *Tetrahedron* **1984**, *40*, 1031; (c) Agami, C.; Levisalles, J.; Sevestre, H. *J. Chem. Soc., Chem. Commun.* **1984**, 418–420.

THIATION WITH 2,4-BIS(4-METHOXYPHENYL)-1,3,2,4-DITHIADIPHOSPHETANE 2,4-DISULFIDE: N-METHYLTHIOPYRROLIDONE

(2-Pyrrolidinethione, 1-methyl-)

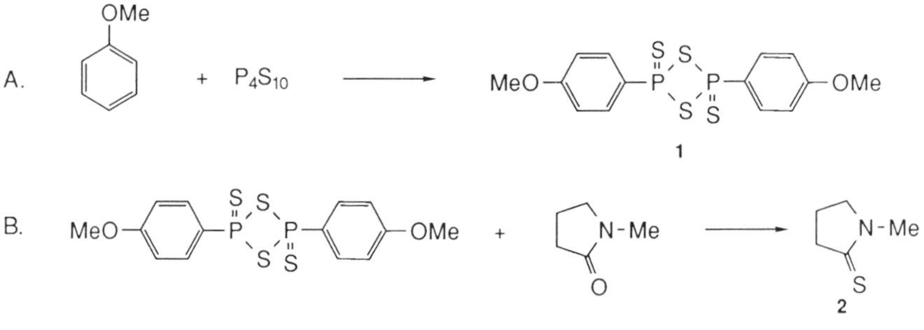

Submitted by I. Thomsen, K. Clausen, S. Scheibye, and S.-O. Lawesson[1]
Checked by Clayton H. Heathcock, Mark Sanner and Terry Rosen

1. Procedure

Caution! Preparation of 2,4-bis(4-methoxyphenyl)-1,3,2,4-dithiadiphosphetane 2,4-disulfide must be carried out in an efficient hood because hydrogen sulfide is evolved. Also, benzene has been identified as a carcinogen; OSHA has issued emergency standards on its use. All procedures involving benzene should be carried out in a well-ventilated hood, and glove protection is required.

A. *2,4-Bis(4-methoxyphenyl)-1,3,2,4-dithiadiphosphetane 2,4-disulfide (1).* A dry 1-L, three-necked, round-bottomed flask, fitted with a reflux condenser, mechanical stirrer, and ground-glass stopper, is charged with 111.0 g (0.25 mol) of phosphorus sulfide, P_4S_{10} (Note 1) and 270 g (2.5 mol) of anisole (Note 1). Stirring is commenced and the mixture is heated at reflux temperature by use of a heating mantle. After 1 hr, the solution is homogeneous and after a second hour 2,4-bis(4-methoxyphenyl)-1,3,2,4-dithiadiphosphetane 2,4-disulfide (1) begins to precipitate. The reaction mixture is allowed to cool to room temperature and the precipitate is filtered (Note 2) and washed with anhydrous ether (2×50 mL) and 50 mL of anhydrous chloroform (free of alcohols) to yield 160–165 g (79–82%) of pale-yellow crystals, mp 228°C (Notes 3 and 4).

B. *N-Methylthiopyrrolidone (2).* A 200-mL, three-necked, round-bottomed flask is fitted with a rubber septum, thermometer, magnetic stirring bar, and reflux condenser equipped with a nitrogen bubbler. The flask is charged with 19.8 g (19.3 mL, 0.20 mol) of *N*-methylpyrrolidone (Note 5) and 40.4 g (0.10 mol) of **1**, whereupon the temperature of the reaction mixture increases to 75–80°C. After 5 min, 35 mL of benzene (Note 6) is added by syringe and the mixture is stirred while being brought to reflux (Note 7). The mixture is heated at reflux for 2 hr (Note 8) and then cooled to room temperature, whereupon it again becomes heterogeneous. The benzene is removed with the aid of a rotary evaporator and the resulting yellow slurry is distilled under reduced pressure through a 5-cm Vigreux column to provide 23.0 g (100%) of *N*-methylthiopyrrolidone (**2**) as a yellow liquid, bp 94–97°C/0.03 mm (Note 9).

2. Notes

1. Commercial phosphorus sulfide, P_4S_{10}, is used without purification. Checkers used P_4S_{10} from Matheson, Coleman and Bell and from Alfa Products, Morton Thiokol, Inc. Best results (yield, melting point) were obtained with the Alfa sample, mp 291–295°C.

2. Excess anisole (137 g) can be recovered by distillation of the filtrate.

3. The product is somewhat hygroscopic and should be stored in an airtight container. It is also available as Lawesson's reagent from Aldrich, Fluka, and from Merck–Schuchard.

4. The checkers obtained 176 g (87%) of **1**, mp 228–231°C.

5. Commercial material from the Aldrich Chemical Company was stored over Linde 4A molecular sieves.

6. Benzene was distilled from and stored over sodium wire.

7. During this operation most of the yellow solid gradually dissolves, affording a clear yellow solution with small amounts of suspended solid. When reflux begins, the internal temperature of the reaction mixture is 95°C.

8. The reaction time can be decreased to 3 min by the use of toluene as solvent.

9. The purified product freezes when stored in a refrigerator. The spectral properties are as follows: ^{1}H NMR ($CDCl_3$) δ: 2.07 (quintet, 2 H, $J = 7$), 3.03 (t, 2 H, $J = 7$), 3.29 (s, 3 H), 3.77 (t, 2 H, $J = 7$). IR (neat): 1520 cm^{-1}.

3. Discussion

A variety of thiating reagents are known: H_2S,[2] H_2S/HCl,[3] H_2S_2/HCl,[4] $(Et_2Al)_2S$,[5] $(EtAlS)_n$,[6] SiS_2,[7] B_2S_3,[7] $PCl_5/Al_2S_3/Na_2SO_4$,[8] Na_2S/H_2SO_4,[9] P_2S_5[10] $P_2S_5/pyridine$,[11] P_2S_5/NEt_3,[12] $P_2S_5/NaHCO_3$,[13] $RPS(OR^1)_2$[14] $PSCl_x(NMe_2)_{3-x}$ ($X = 0$–3),[15] and SCNCOOEt.[16] The reagent described here, 2,4-bis(4-methoxyphenyl)-1,3,2,4-dithia-diphosphetane 2,4-disulfide (1),[17] offers a number of advantages as a thiating reagent. It is easily prepared in a simple one-step procedure employing commercially available starting materials. It has a satisfactory shelf life, provided it is protected from moisture. In contrast to commercial P_4S_{10}, compound 1 is a well-defined reagent that gives repro-ducible results, usually in high yield. Under defined conditions, certain selectivity has been observed.[18-20] Other methods for the preparation of analogs of 1 have been described.[21-23]

The thiation procedure described here[24] is an example of a general synthetic method for the conversion of carbonyl to thiocarbonyl groups. Similar transformations have been carried out with ketones,[25] carboxamides,[26-30] esters,[31-32] thioesters,[31] lactones,[18,33] thiolactones,[18] imides,[24] enaminones,[34] and protected peptides.[35]

1. Department of Chemistry, University of Aarhus, DK-8000 Aarhus, C, Denmark. S.-O. Lawesson deceased.
2. Paquer, D.; Smadja, S.; Vialle, J. *C. R. Hebd. Seances, Acad. Sci.*, *Ser. C* **1974**, *279*, 529.
3. Fournier, C.; Paquer, D.; Vazeux, M. *Bull. Soc. Chim. Fr.* **1975**, 2753.
4. Barillier, D.; Gy, C.; Rioult, P.; Vialle, J. *Bull. Soc. Chim. Fr.* **1973**, 277.
5. Ishii, Y.; Hirabayashi, T.; Imaeda, H.; Ito, K. Jpn. Patent 40441, 1974; *Chem. Abstr.* **1975**, *82*, 156074f.
6. Hirabayashi, T.; Inoue, K.; Tokota, K.; Ishii, Y. *J. Organomet. Chem.* **1975**, *92*, 139.
7. Dean, F. M; Goodchild, J.; Hill, A. W.; Murray, S.; Zahman, A. *J. Chem. Soc.*, *Perkin Trans. 1* **1975**, 1335.
8. Testa, E.; Fontanella, L.; Maffii, G. S. African Patent 6707088, 1968; *Chem. Abstr.* **1969**, *70*, 57602x.
9. Russell, G. A.; Tanikaga, R.; Talaty, E. R. *J. Am. Chem. Soc.* **1972**, *94*, 6125.
10. Wakabayashi, T.; Kato, Y.; Watanabe, K. *Jpn. Kokai Tokkyo Koho* **1978**, *78*, 56, 662; *Chem. Abstr.* **1979**, *90*, 22818h.
11. Barton, D. H. R. U.S. Patent 4011316, 1977; *Chem. Abstr.* **1977**, *87*, 53490n.
12. Machiguchi, T.; Hoshino, M.; Kitahara, Y. *Jpn. Kokai Tokkyo Koho* **1977**, *77*, 23, 066; *Chem. Abstr.* **1977**, *82*, 102153r.
13. Alper, H.; Currie, J. K.; Sachdeva, R. *Angew. Chem.* **1978**, *90*, 722; *Angew. Chem. Int. Ed. Engl.* **1978**, *17*, 689.
14. Sane, R. T.; Kamat, S. S. *Curr. Sci.* **1978**, *47*, 765; *Chem. Abstr.* **1979**, *90*, 6044x.
15. Pedersen, B. S.; Lawesson, S.-O. *Bull. Soc. Chim. Belg.* **1977**, *86*, 693.
16. Seitz, G.; Sutrisno, R. *Synthesis* **1978**, 831.
17. Lecher, H. Z.; Greenwood, R. A.; Whitehouse, K. C.; Chao, T. H. *J. Am. Chem. Soc.* **1956**, *78*, 5018.
18. Scheibye, S.; Kristensen, J.; Lawesson, S.-O. *Tetrahedron* **1979**, *35*, 1339.
19. Clausen, K.; Lawesson, S.-O. *Bull. Soc. Chim. Belg.* **1979**, *88*, 305.
20. Clausen, K.; Lawesson, S.-O. *Nouv. J. Chim.* **1980**, *4*, 43.
21. Grishina, O. N.; Andreev, N. A.; Babkina, E. I. U.S.S.R. Patent 475363, 1975; *Chem. Abstr.* **1975**, *83*, 164370k.
22. Maier, L. U.S. Patent 3336378, 1967; *Chem. Abstr.* **1968**, *68*, 22056d.

23. Baudler, M.; Valpertz, H. W. Z. *Naturforsch.* **1967**, *22b*, 222.
24. Shabana, R.; Scheibye, S.; Clausen, K.; Olesen S. O.; Lawesson, S.-O. *Nouv. J. Chim.* **1980**, *4*, 47.
25. Pedersen, B. S.; Scheibye, S.; Nilsson, N. H.; Lawesson, S.-O. *Bull. Soc. Chim. Belg.* **1978**, *87*, 223.
26. Scheibye, S.; Pedersen, B. S.; Lawesson, S.-O. *Bull. Soc. Chim. Belg.* **1978**, *87*, 229.
27. Scheibye, S.; Pedersen, B. S.; Lawesson, S.-O. *Bull. Soc. Chim. Belg.* **1978**, *87*, 299.
28. Fritz, H.; Hug, P.; Lawesson, S.-O. Logemann, E.; Pedersen, B. S.; Sauter, H.; Scheibye, S.; Winkler, T. *Bull. Soc. Chim. Belg.* **1978**, *87*, 525.
29. Clausen, K.; Pedersen, B. S.; Scheibye, S.; Lawesson, S.-O.; Bowie, J. H. *Org. Mass Spectrom.* **1979**, *14*, 101.
30. Clausen, K.; Pedersen, B. S.; Scheibye, S.; Lawesson, S.-O.; Bowie, J. H. *Int. J. Mass Spectrom. Ion Phys.* **1979**, *29*, 223.
31. Pedersen, B. S.; Scheibye, S.; Clausen, K.; Lawesson, S.-O. *Bull. Soc. Chim. Belg.* **1978**, *87*, 293.
32. Pedersen, B. S.; Lawesson, S.-O. *Tetrahedron* **1979**, *35*, 2433.
33. Baxter, S. L.; Bradshaw, J. S. *J. Org. Chem.* **1981**, *46*, 831.
34. Shabana, R.; Rasmussen, J. B.; Olesen, S. O.; Lawesson, S.-O. *Tetrahedron* **1980**, 36, 3047.
35. Clausen, K.; Thorsen, M.; Lawesson, S.-O. *Tetrahedron* **1981**, *37*, 3635.

OSMIUM-CATALYZED VICINAL OXYAMINATION OF OLEFINS BY CHLORAMINE-T: *cis*-2-(*p*-TOLUENESULFONAMIDO)CYCLOHEXANOL AND 2-METHYL-3-(*p*-TOLUENESULFONAMIDO)-2-PENTANOL

[Benzenesulfonamide, *N*-(2-hydroxycyclohexyl)-4-methyl-, *cis*] and Benzenesulfonamide, *N*-(1-ethyl-2-hydroxy-2-methylpropyl)-4-methyl-]

A. TsNClNa · 3 H₂O   +   [cyclohexene] $\xrightarrow[\text{phase-transfer conditions}]{\text{1\% OsO}_4\text{, CHCl}_3/\text{H}_2\text{O, 60°C}}$ [cyclohexanol with OH and NHTs]

B. TsNClNa · 3 H₂O   +   [2-methyl-2-pentene] $\xrightarrow{\text{1\% OsO}_4\text{, tert-BuOH, 60°C}}$ [product with NHTs and OH]

Submitted by EUGENIO HERRANZ and K. BARRY SHARPLESS[1]
Checked by RITA LOCHER, THOMAS WELLER, and DIETER SEEBACH

1. Procedure

Caution! Because of the volatility and toxic nature of OsO₄, these reactions should be carried out in a well-ventilated hood.

A. *cis-2-p(Toluenesulfonamido)cyclohexanol.* A 1-L, three-necked, round-bottomed flask is equipped with an efficient mechanical stirrer, thermometer, and reflux condenser. The flask is charged with 8.2 g (0.1 mol) of cyclohexene (Note 1), 250 mL of reagent grade chloroform (Note 2), and 10 mL (1 mmol) of osmium tetroxide catalyst solution (Note 3). To the resulting black solution is added a solution of 35.2 g (0.125 mol) of chloramine-T trihydrate (Note 4) and 1.1 g (5 mmol) of benzyltriethylammonium chloride (Note 5) in 250 mL of distilled water. Vigorous stirring is begun, and the reaction mixture is brought to 55–60°C by means of a heating mantle.

After 10 hr at 55–60°C, 14.2 g (0.1 mol) of sodium sulfite (Note 6) is added and the mixture is refluxed for 3 hr. The hot reaction mixture (Note 7) is transferred to a 1-L separatory funnel and allowed to stand for 10 min. The organic layer is collected in a 500-mL, round-bottomed flask. The aqueous layer is extracted once with 25 mL of CHCl₃ that is then combined with the original organic layer. Removal of solvent with a rotary evaporator provides a residue (Note 8) that is transferred to a 350-mL fritted-glass funnel and triturated successively with 200 mL and 100 mL of saturated sodium chloride solution containing 1% sodium hydroxide (Note 9) and finally with two 50-mL portions of distilled water.

The resulting solid is placed in a 500-mL Erlenmeyer flask and dissolved in a mixture of 250 mL of CHCl₃ and 25 mL of CH₃OH. Anhydrous magnesium sulfate (ca. 8–10 g) is added and the resulting suspension is stirred magnetically for 5 min. Filtration of this suspension through a Celite mat on a sintered-glass funnel (Note 10), followed by evaporation of the solvent, affords (after drying under reduced pressure) 20.3–22 g (75–81.2%) of almost pure *cis-2-(p*-toluenesulfonamido)cyclohexanol, mp 155–157°C (Note 11). The oxyaminated product may be purified further by washing with toluene to give 20–21.8 g (74.3–80.9%); mp 157–158°C (Note 12).

B. *2-Methyl-3-(p-toluenesulfonamido)-2-pentanol.* A 500-mL, three-necked, round-bottomed flask is equipped with an efficient mechanical stirrer, thermometer, and reflux condenser. The flask is charged with 8.4 g (0.1 mol) of 2-methyl-2-pentene (Note 1), 100 mL of reagent-grade *tert*-butyl alcohol (Note 2), 10 mL (1 mmol) of osmium tetroxide catalyst solution (Note 3), and 35.2 g (0.125 mol) of chloramine-T trihydrate (Note 4). Vigorous stirring is begun, and the reaction mixture is brought to 55–60°C by means of a heating mantle.

After ca. 20 hr at 55–60°C, the mixture is cooled to room temperature using a water bath, and then 1.1 g (0.03 mol) of sodium borohydride is added (Note 6). Stirring is continued at room temperature for about 1 hr. Removal of the solvent on a rotary evaporator gives an oil that is taken up in 100 mL of ethyl acetate and washed once with a solution that is prepared by mixing 100 mL of saturated sodium chloride solution containing 1% sodium hydroxide (Note 13) with 25 mL of distilled water. The organic layer is washed twice more with 200 mL of saturated sodium chloride solution containing 1% sodium hydroxide and finally with 100 mL of saturated sodium chloride solution (Notes 9 and 14). Addition of anhydrous magnesium sulfate, filtration through a column of 75 g of silica gel (Note 15), elution with ethyl acetate (Note 16), and evaporation of the solvent on a rotary evaporator provides 21.5 g of the crude oxyaminated product (Note 17). The solid is then washed twice with ether (Note 18) to give 13.8–14.9 g (51–

55%) of white, crystalline 2-methyl-3-(*p*-toluenesulfonamido)-2-pentanol, mp 96–97°C. Concentration of the ether yields an additional 4.0–5.0 g (15–18%) of oxyaminated product, mp 95–97°C (Note 19).

2. Notes

1. Cyclohexene and 2-methyl-2-pentene were used as commercially available.

2. The amount of solvent used is not critical. Several experiments have been performed at higher and lower concentrations and in all cases the yields were very much alike.

3. Osmium tetroxide was supplied commercially in 1-g amounts in sealed glass ampuls. The procedure we describe below should be followed to prepare the osmium tetroxide catalyst solution. Work in a well-ventilated hood. One ampul is scored in the middle, broken open, and the two halves are dropped into a clean brown bottle containing 39.8 mL of reagent grade *tert*-butyl alcohol and 0.20 mL of 70 or 90% *tert*-butyl hydroperoxide. The bottle is capped (use caps with Teflon liners) and then swirled to ensure dissolution of the OsO_4. Each milliliter of this stock solution contains 25 mg (ca. 0.1 mmol) of OsO_4. These solutions are stored in the hood at room temperature and seem to be very stable. We have also prepared five times more dilute solutions of OsO_4 in *tert*-butyl alcohol, which we use for small-scale experiments.[2]

4. Chloramine-T trihydrate (CT) was obtained commercially. Excess chloramine-T is used because we have observed traces of the α-ketosulfonamide in those cases where the oxyaminated product contains a secondary hydroxyl group. We have also observed that these α-ketosulfonamides are further oxidized under the reaction conditions in a process that consumes several moles of chloramine-T.

5. Benzyltriethylammonium chloride was used as purchased.

6. The rates of reduction of the osmate esters vary considerably. We found that, although the sulfite method (in the past we have also used sodium bisulfite) would reduce osmate esters from monosubstituted and 1,2-disubstituted olefins, osmate esters derived from trisubstituted and 1,1-disubstituted olefins were more inert to this treatment. Sodium borohydride reduces even these more hindered osmate esters rapidly at room temperature.

7. The oxyaminated product derived from cyclohexene is highly crystalline and begins to crystallize if the chloroform phase is allowed to cool.

8. The residue is dried under reduced pressure to remove the last traces of chloroform and *tert*-butyl alcohol, and then pulverized with a mortar and pestle.

9. In this way the *p*-toluenesulfonamide by-product along with some other impurities are removed from the oxyaminated product.

10. This treatment removes the suspended osmium particles from the solution.

11. GLC analysis revealed a purity of 99%.

12. The product obtained by this procedure is pure enough for most purposes. Its melt, however, is faintly cloudy. A product of higher purity, giving a clear melt, mp 158–159°C, can be obtained by recrystallization from about 10 mL of CHCl$_3$ per gram of oxyaminated product. The structural characterization of *cis*-2-(*p*-toluenesulfonamido)cyclohexanol, mp 158–159°C, is as follows. Anal. calcd. for C$_{13}$H$_{19}$NO$_3$S: C, 57.97; H, 7.11; N, 5.20. Found: C, 57.81; H, 6.98; N, 5.19. ^{1}H NMR (CDCl$_3$) δ: 1.2–1.9 (m, 8 H, C*H*), 2.26 (d, 1 H, *J* = 4, O*H*), 2.44 (s, 3 H, Ar C*H*$_3$), 3.30 (m, 1 H, NC*H*), 3.80 (m, 1 H, OC*H*), 5.30 (d 1 H, *J* = 7, N*H*), 7.55 (AA'BB' pattern, 4 H, *J* = 8, Ar*H*); IR (KBr pellet) cm^{-1}: 3420, 3150, 1305, 1285, 1145, 1085, 550 all (s); 2920, 2850, 1440, 970, 930, 890, 815, 660 all (m): 1590, 1370, 1250, 1195, 1185, 1060, 1000 all (w). GC analysis was carried out using the following conditions: 6-ft × 2-mm glass column, packed with 5% OV-17 on 80/100 GasChrom Q, at 70 → 250°C (32°C/min), retention time 9.50–9.60 min.

13. First a 1% solution of NaOH is prepared and then sodium chloride is added until saturation is reached. For the first washing, 25 mL of distilled water is added to 100 mL of the above solution in order to dissolve the inorganic salts present in the reaction mixture.

14. When ethyl acetate is used as the extracting solvent, rapid separation of the two phases was achieved. If a slight emulsion forms at the interphase during the last wash, the addition of celite and subsequent filtration improves the separation.

15. Silica Gel 60 (70–230 mesh ASTM) was used as obtained commercially. A column 50 cm long by 3.5 cm in diameter was used.

16. Approximately 700 mL of EtOAc was necessary to elute all the oxyaminated product from the silica gel column (monitoring by TLC elution with EtOAc is continued until a UV active spot does not appear on TLC). To speed up filtration, a slight pressure of 4 psi is applied.

17. The product is a yellowish-brown solid that usually crystallizes when the last traces of EtOAc are removed on the rotary evaporator. If problems are encountered in inducing crystallization, either high vacuum or addition of ether followed by concentration should yield the desired solid.

18. The solid was washed in a 60-mL, sintered-glass funnel, the first time with 30 mL of ether and the second time with 25 mL.

19. The product obtained by this procedure is relatively pure. However, a product of higher purity, giving a clear melt, mp 99–100°C, can be obtained by recrystallization from about 1 mL of toluene per gram of oxyaminated product.

3. Discussion

This osmium-catalyzed procedure provides the first practical and direct means for the cis addition of a hydroxyl group and an arylsulfonamido moiety (ArSO$_2$NH) to an olefinic bond. The resulting vicinal hydroxy arylsulfonamides may in some cases be useful in their own right, but they are easily transformed in a variety of selective and potentially useful ways. Some of the interesting transformations that we[3] have observed are shown below:

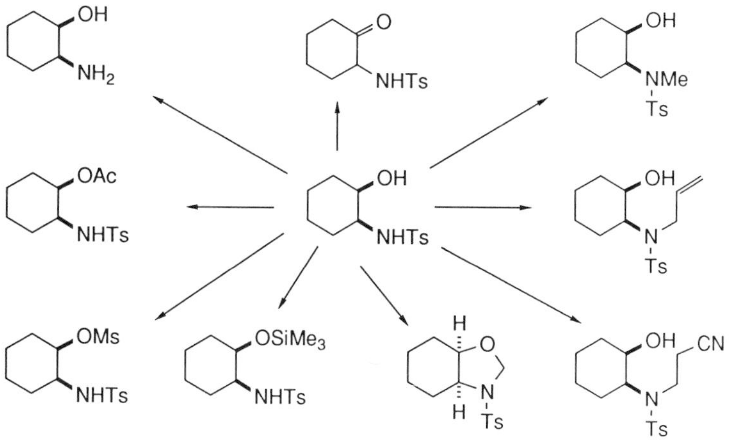

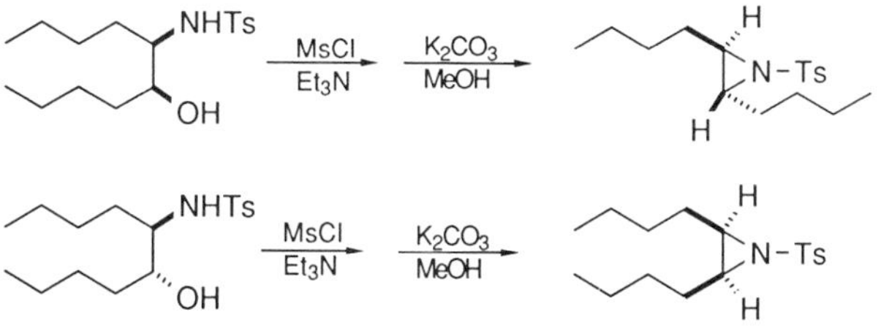

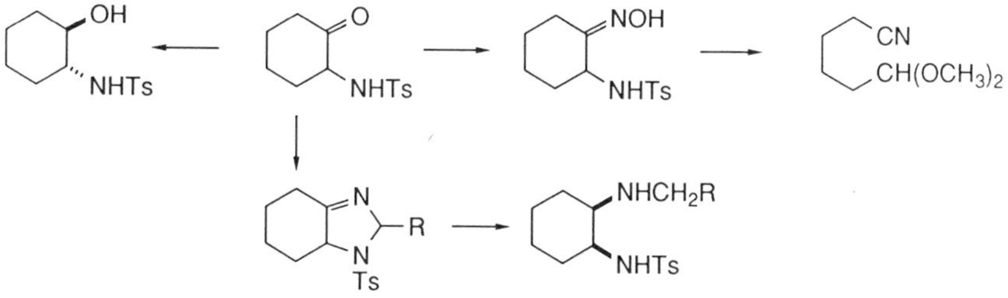

Procedure (Step) A is very effective for most monosubstituted and 1,2-disubstituted olefins. This method,[2] using phase-transfer conditions (PTC), has been developed recently in our laboratory and represents a substantial improvement over our former procedures.[4] Cyclooctene, (Z)-5-decene, stilbene, ethyl crotonate, and 1-decene are among the olefins that are readily oxyaminated under the conditions described in Procedure A.

It is important to point out that the work-up we have used in the case of cyclohexene is a peculiar one because of the exceptional crystallinity of the oxyamination product. Generally, removal of the p-toluenesulfonamide is accomplished by shaking the chloroform layer with a saturated sodium chloride solution containing 1% sodium hydroxide.

The chloramine derivatives ($ArSO_2NClNa$) of a variety of other arylsulfonamides (Ar = phenyl, o-tolyl, p-chlorophenyl, p-nitrophenyl, and o-carboalkoxyphenyl) have been used successfully in these catalytic oxyaminations. Since only chloramine-T (Ar = p-tolyl) and chloramine-B (Ar = phenyl) are commercially available, we have developed a convenient procedure for generating the chloramines in situ for use in the modification involving phase-transfer catalysis. One simply stirs a suspension of the arylsulfonamide with an equivalent of sodium hypochlorite (Clorox) until a homogeneous solution is obtained. When this solution is used in the PTC method (see 2 for experimental details), the yields of oxyaminated product are comparable with those obtained with isolated chloramine salts.

The PTC method gives poor results with trisubstituted and 1,1-disubstituted olefins. The oxyamination product may still form, but it is accompanied by a number of by-products. Fortunately, this class of olefins is successfully oxyaminated by the alternative procedure (B). Methylcyclohexene, α-methylstyrene, 2-methyl-2-hepten-6-one, and its ketal are examples of olefins that give oxyamination products in good yield following Procedure B.

Addition of a phase-transfer catalyst such as dicyclohexyl-18-crown-6 to the reaction mixture (in Procedure B) results in a faster reaction rate. However, there are no significant changes in the final yield of oxyamination product.

We have carried out experiments on a 1-mol scale in the case of cyclohexene and α-methylstyrene (in the cyclohexene 1-mol experiment, the reaction mixture was 2.5 times more concentrated than described here), and have realized 70–80% and 65–75% yields, respectively, of the oxyaminated products.

Procedure A does not succeed with diethyl fumarate and 2-cyclohexen-1-one. Both chloramine-T and part of the olefin are consumed, but the oxyamination product has not been detected in the reaction mixtures. It seems likely that it forms, but is unstable to the reaction conditions. Both of these olefins do form isolable oxyamination products under the milder conditions (room temperature) of a more recent oxyamination procedure.[5]

Procedure B does not succeed with tetramethylethylene and cholesterol, and it seems reasonable to anticipate negative results with most hindered tri- and tetrasubstituted olefins. No reaction occurs, and chloramine-T is not consumed.

The sulfonamide protecting group on the nitrogen may be undesirable in some cases. For this reason we have developed an analogous osmium-catalyzed procedure that effects cis addition of hydroxyl and carbamate (ROCONH) moieties across the olefinic linkage.[5] β-Amino alcohols with benzyloxycarbonyl (Z or CBZ) and tert-butoxycarbonyl (BOC) protecting groups on the nitrogen are accessible directly from the corresponding olefins by the new method.[5]

1. Department of Chemistry, Stanford University, Stanford, CA 94305. Present address: Department of Chemistry, Massachusetts Institute of Technology, Cambridge, Mass., 02139.
2. Herranz, E.; Sharpless, K. B. J. Org. Chem. 1978, 43, 2544.
3. Bäckvall, J.-E.; Oshima, K.; Sharpless, K. B. J. Org. Chem. 1979, 44, 1953.

4. (a) Sharpless, K. B.; Patrick, D. W.; Truesdale, L. K.; Biller, S. A. *J. Am. Chem. Soc.* **1975,** *97*, 2305; (b) Patrick, D. W.; Truesdale, L. K.; Biller, S. A.; Sharpless, K. B. *J. Org. Chem.* **1978,** *43*, 2628; (c) Sharpless, K. B.; Chong. A. O.; Oshima, K. *J. Org. Chem.* **1976,** *41*, 177.

5. (a) Herranz, E.; Biller, S. A; Sharpless, K. B. *J. Am. Chem. Soc.* **1978,** *100*, 3596; (b) Herranz, E.; Sharpless, K. B. *Org. Synth., Coll. Vol. VII* **1990,** 226.

2-METHYL-2-(TRIMETHYLSILOXY)PENTAN-3-ONE

(3-Pentanone, 2-methyl-2-[(trimethylsilyl)oxy]-)

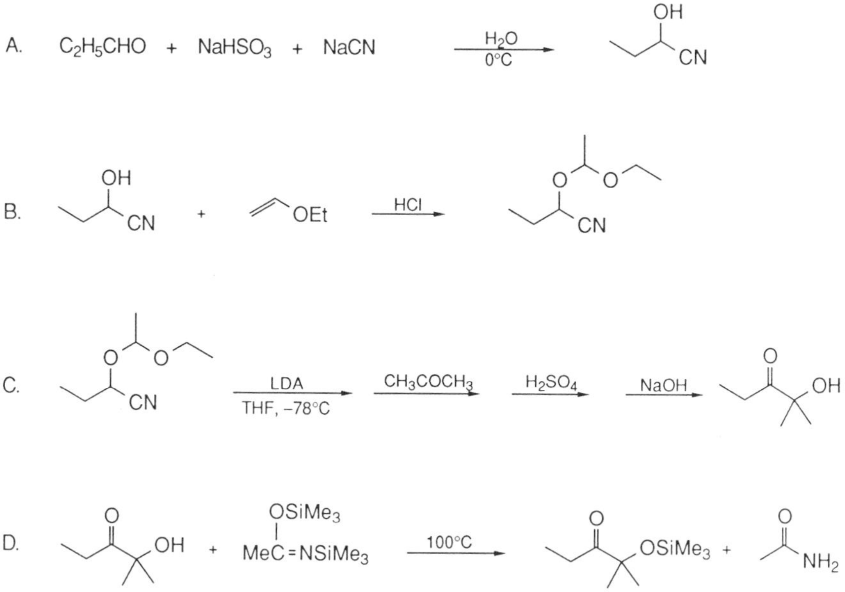

Submitted by STEVEN D. YOUNG, CHARLES T. BUSE, and CLAYTON H. HEATHCOCK[1]
Checked by JOSEPH R. FLISAK, SAMI FARAHAT, STAN S. HALL, HUGH W. THOMPSON, and GABRIEL SAUCY

1. Procedure

A. *2-Hydroxybutanenitrile.* A 3-L, three-necked, round-bottomed flask is fitted with a mechanical stirrer and thermometer and charged with 312 g (3.0 mol) of sodium bisulfite and 1050 mL of water. The stirrer is started and after the sodium bisulfite has dissolved, the flask is placed in an ice–salt bath. A solution of 147 g (3.0 mol) of sodium cyanide (Note 1) in 450 mL of water and 174 g (3.0 mol) of propanal (Note 2) are separately cooled to 0°C in ice–salt baths. When the temperature of the vigorously stirring sodium bisulfite solution has stabilized at 0°C the cold propanal is added in one portion.

The temperature of the reaction solution immediately increases to ca. 35°C and then returns to ca. 0°C. After 30 min the cold sodium cyanide solution is added in one portion. The reaction mixture again warms to ca. 15°C and then returns to ca. 0°C. This mixture is stirred for 2 hr at 0°C, during which time a thick white precipitate of sodium sulfite forms. The supernatant liquid is decanted into a 4-L separatory funnel and the precipitate is washed with 1 L of ice–water. The combined aqueous solution is extracted with three 1-L portions of ethyl ether. The combined ether extracts are washed with 1 L of saturated brine and dried by stirring (magnetic stirring bar) over magnesium sulfate for 2 hr. The solution is filtered through a coarse, sintered-glass funnel and the ether is removed with a rotary evaporator at water aspirator pressure. After the pH of the residue is adjusted to 5 with a few drops of concentrated hydrochloric acid (Note 3), the residue is distilled to give 154–192 g (60–75%) of 2-hydroxybutanenitrile, bp 108–114°C (30 mm), as a colorless liquid (Note 4).

B. *2-[(1'-Ethoxy)-1-ethoxy]butanenitrile.* A 1-L, three-necked, round-bottomed flask is equipped with a condenser topped with a calcium chloride drying tube, a magnetic stirring bar, a 500-mL pressure-equalizing addition funnel, and a thermometer. The flask is charged with 174 g (2.05 mol) of 2-hydroxybutanenitrile to which 0.5 mL of concentrated hydrochloric acid has been added. The addition funnel is charged with 221 g (3.07 mol) of ethyl vinyl ether (Note 5), which is then added dropwise to the stirring cyanohydrin at such a rate that the temperature is maintained at ca. 50°C. When the addition is complete, the mixture is heated to 90°C for 4 hr. The condenser is replaced with a distillation head and the dropping funnel and thermometer are replaced with stoppers. Direct distillation of the gold–yellow solution from the reaction flask yields 226–277 g (70–86%) of nearly pure 2-[(1'-ethoxy)-1-ethoxy]butanenitrile, bp 85–87°C (30 mm), as a colorless liquid (Note 6).

C. *2-Hydroxy-2-methylpentan-3-one.* A dry, 5-L, three-necked (including a thermometer well), round-bottomed flask is equipped with a mechanical stirrer, low-temperature thermometer, nitrogen inlet, rubber septum, and a 1-L, graduated, pressure-equalizing addition funnel that is sealed with a rubber septum. The flask is charged with 775 mL of dry tetrahydrofuran (Note 7) and 166 g (1.64 mol) of dry diisopropylamine (Note 8). The contents of the flask are cooled to −10°C (dry ice–acetone bath) and 1095 mL (1.6 mol) of 1.5 M butyllithium in hexane (Note 9), which has been transferred to the addition funnel by means of a 16-gauge cannula and argon pressure, is slowly added to the stirring solution at such a rate as to maintain a temperature of −10°C. After the addition is complete, 50 mL of dry THF is added to the addition funnel with a syringe to rinse the walls of the funnel; the rinse is added, and then the mixture is cooled to −75°C. The addition funnel is charged by syringe with 246 g (1.6 mol) of 2-[(1'-ethoxy)-1-ethoxy]butanenitrile, which is then added at such a rate that the temperature does not exceed −70°C. The mixture is stirred for 10 min and 104 g (1.8 mol) of dry acetone (Note 10) is added by syringe over a 30-min period at such a rate that the temperature of the reaction mixture does not exceed −70°C. When the addition is complete, the cooling bath is removed and the reaction mixture is allowed to warm to 0°C. The solution is poured into 1 L of water and the resulting mixture is concentrated at aspirator pressure with a rotary evaporator (30°C water bath) to remove the volatile organic compounds. The aqueous residue is extracted with three 1-L portions of methylene chloride. The organic extracts are combined and washed with two 500-mL portions of water and then concentrated with a rotary evaporator (25°C water bath) at aspirator pressure to obtain a yellow syrupy residue. This material is stirred with 680 mL of methanol and 340 mL of aqueous 5% sulfuric acid overnight at room temperature. The methanol is evaporated with a rotary evaporator (30°C water bath) at aspirator pressure and the yellow residue is extracted with three 1-L portions of ethyl ether. The combined ether extracts are shaken in a 4-L separatory funnel with 210 mL of 10 N aqueous sodium hydroxide for 15 min (Note 11). The layers are separated, and the ether layer is washed with 500 mL of brine and dried by stirring (magnetic stirring bar) over magnesium sulfate for 2 hr. The drying agent is removed by filtration through a coarse sintered-glass funnel and the ether is removed with a rotary evaporator (water bath below 40°C) at aspirator pressure. The yellow–orange liquid residue is distilled to obtain 82–115 g (45–63%) of 2-hydroxy-2-methylpentan-3-one, bp 57–65°C (15 mm), as a pale-yellow liquid (Notes 12 and 13).

D. *2-Methyl-2-(trimethylsiloxy)pentan-3-one.* A dry, 500 mL, three-necked, round-bottomed flask equipped with a mechanical stirrer, reflux condenser with a nitrogen inlet, and a thermometer is charged with 84 g (0.72 mol) of 2-hydroxy-2-methylpentan-3-one and 74 g (0.36 mol) of *N,O*-bis(trimethylsilyl)acetamide (Note 14). The mixture is heated at 100°C for 12 hr with stirring and then cooled to room temperature, at which point the mixture becomes a semisolid as the acetamide crystallizes. The semisolid mixture is diluted with 50 mL of water and stirred at room temperature for 1 hr (Note 15). After the stirring is stopped, 200 mL of hexane is added and the layers are separated. The aqueous layer is extracted with 100 mL of hexane. The combined hexane extracts are washed with four 100-mL portions of water and then dried over magnesium sulfate for 2 hr. After removal of the drying agent by filtration through a coarse sintered-glass funnel, the hexane is evaporated with a rotary evaporator (25°C water bath) at aspirator pressure. The crude, pale-yellow oil is distilled to afford 105–112 g (75–80%) of 2-methyl-2-(trimethylsiloxy)pentan-3-one, bp 71–75°C (15 mm), as a colorless liquid (Note 16).

2. Notes

1. *Caution! Sodium cyanide and 2-hydroxybutanenitrile are extremely toxic. Great care should be taken when using these materials. Reactions should be carried out in a well-ventilated hood and suitable protective clothing should be worn at all times.*

2. Propanal was obtained from Aldrich Chemical Company and was used without further purification.

3. If HCl is omitted, the cyanohydrin reverts to HCN and propanal on attempted distillation. The checkers found it necessary to ensure that the residue was acidic by adjusting the pH to 5 by testing the residue with wet pH paper.

4. The IR spectrum (neat) shows absorption at 3420, 2960, 2310, and 1460 cm^{-1}.

5. Ethyl vinyl ether was obtained from Aldrich Chemical Company and was used without further purification.

6. The IR spectrum (neat) shows absorption at 2970, 1425, and 1385 cm^{-1}. The C≡N absorption is not observed.

7. Tetrahydrofuran is distilled under a nitrogen atmosphere from sodium–benzophenone immediately prior to use.

8. Diisopropylamine is distilled under a nitrogen atmosphere from calcium hydride prior to use. It may be stored under nitrogen for 1 week without redistillation.

9. *Caution! Concentrated butyllithium may ignite spontaneously on exposure to air or moisture. Manipulations with this reagent should be performed with care.* The submitters used butyllithium, 1.5 M in hexane from Foote Mineral Company, and measured it by transferring the solution to a 2-L, graduated cylinder stoppered with a rubber septum with a 15-gauge cannula and argon. The solution was then transferred directly to the reaction vessel by the same procedure. The checkers used fresh butyllithium, 1.55 M in hexane under argon, from Aldrich Chemical Company. Stainless steel cannulas with deflected points (double-tip syringe needles) are available from Ace Glass, Inc. and Aldrich Chemical Co.

10. ACS Certified acetone was obtained from Fisher Chemical Company and distilled from Linde 3A molecular sieves immediately prior to use.

11. Periodic shaking (once every 3 min) is sufficient to effect cyanohydrin hydrolysis.

12. The ^{1}H NMR (200 MHz, CDCl$_3$) spectrum is as follows δ: 1.12 (t, 3 H, J = 7.2), 1.39 (s, 6 H), 2.59 (q, 2 H, J = 7.2), 3.85 (s, 1 H). The IR spectrum (neat) shows absorption at 3450, 2960, and 1705 cm^{-1}.

13. In one run, the checkers, at this point obtained 241 g, bp 116–120°C (12 mm) of the protected cyanohydrin (NMR), which had not been deprotected and hydrolyzed, rather than the expected product. In this case, the entire distillate was resubjected to the acid and base sequence, which afforded the desired product in a 61% overall isolated yield.

14. *N,O*-Bis(trimethylsilyl)acetamide was obtained from Aldrich Chemical Company and used without further purification.

15. This process is necessary to ensure hydrolysis of any unreacted *N,O*-bis(trimethylsilyl)acetamide, which inevitably contaminates the product if the step is omitted.

16. The ^{1}H NMR spectrum (200 MHz, CDCl$_3$) is as follows: δ: 0.15 (s, 9 H), 1.02 (t, 3 H, J = 7.2), 1.33 (s, 6 H), 2.67 (q, 2 H, J = 7.2). The IR spectrum (neat) shows absorptions at 2980 and 1720 cm^{-1}.

3. Discussion

2-Methyl-2-(trimethylsiloxy)pentan-3-one (**1**) is the prototype member of a series of α-trimethylsiloxy ketones that are useful for stereoselective aldol addition reactions (Equation 1).[2] β-Hydroxy ketones **2** may be converted into β-hydroxy acids,[2] β-hydroxy aldehydes,[3] and other β-hydroxy ketones.[4]

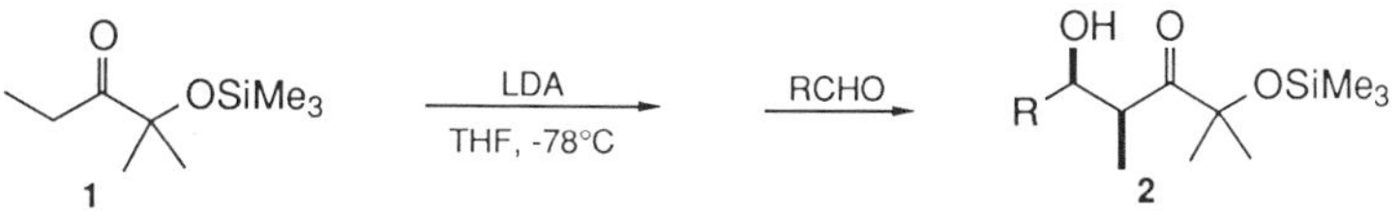

$$(1)$$

Compound **1** has also been prepared by the following methods. Addition of ethyl-magnesium bromide to the protected cyanohydrin of acetone, followed by hydrolysis and silylation provides **1** in 40% yield (Equation 2).[2] Metallation of 1-methoxypropene by butyllithium in pentane[5] gives 1-lithio-1-methoxypropene, which reacts with acetone to give, after hydrolysis and silylation, ketone **1**

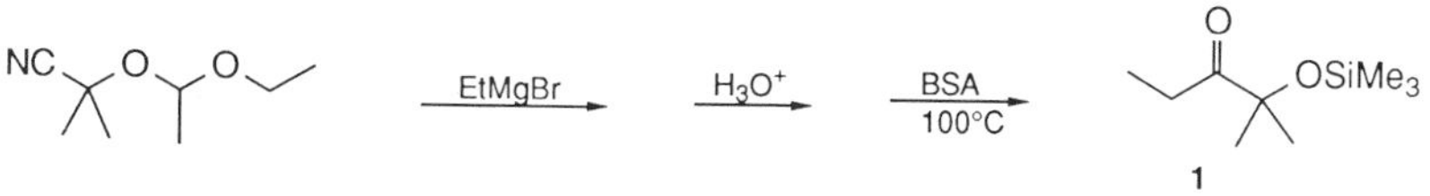

$$(2)$$

in 25–30% overall yield (Equation 3).[6] The trimethylsilyl ether of acrolein

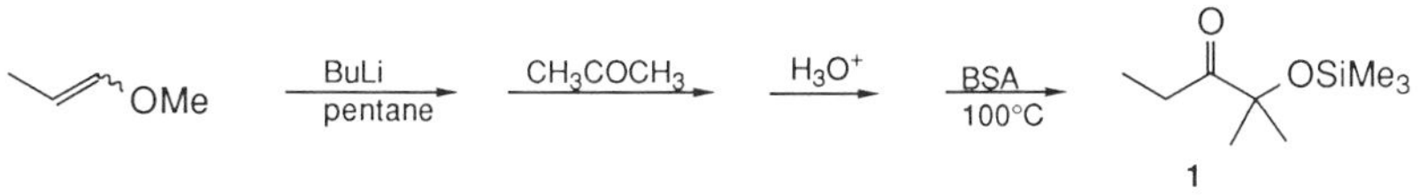

$$(3)$$

cyanohydrin, prepared by the method of Hünig,[7] may be metallated and added to acetone to provide an enone which is hydrogenated to **1** (Equation 4).[8] Although the

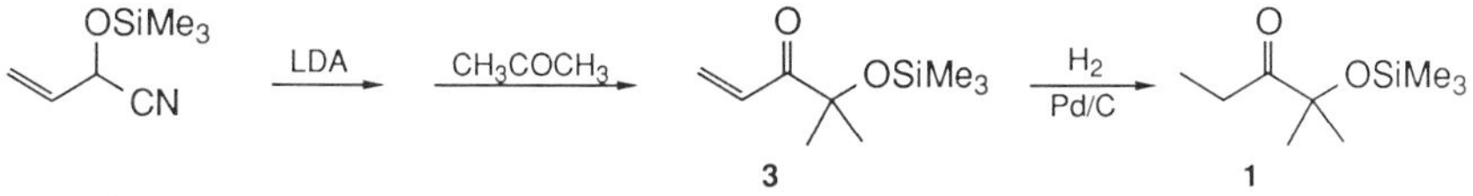

$$(4)$$

overall yield in this sequence can be quite high, the intermediate enone **3** polymerizes very readily, and the procedure is not reliable on a large scale. Compound **1** has also been prepared by methylation of the lithium enolate of the lower homolog, **4**, (Equation 5).[9] Although this alkylation provides

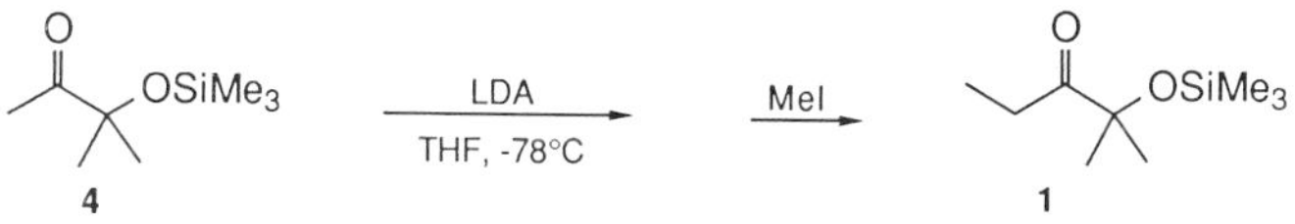

$$(5)$$

1 in 60% yield on a 2-mmol scale, the desired product is accompanied by a significant quantity of the dimethylated product, from which it is not easily separated.[8]

1. Department of Chemistry, University of California, Berkeley, CA 94720.
2. Heathcock, C. H.; Buse, C. T.; Kleschick, W. A.; Pirrung, M. C.; Sohn, J. E.; Lampe, J. *J. Org. Chem.* **1980**, *45*, 1066.
3. Heathcock, C. H.; Young, S. D.; Hagen, J. P.; Pirrung, M. C.; White, C. T.; VanDerveer, D. *J. Org. Chem.* **1980**, *45*, 3846.
4. White, C. T.; Heathcock, C. H. *J. Org. Chem.* **1981**, *46*, 191.
5. Baldwin, J. E.; Höfle, G. A.; Lever, O. W., Jr. *J. Am. Chem. Soc.* **1974**, *96*, 7125.
6. Young, S. D.; Heathcock, C. H., unpublished results.
7. Hünig, S.; Öller, M. *Chem. Ber.* **1981**, *114*, 959.
8. Montgomery, S. H.; Heathcock, C. H., unpublished results.
9. Smith, A. B., III; Levenberg, P. A.; Jerris, P. J.; Scarborough, R. M., Jr.; Wovkulich, P. M. *J. Am. Chem. Soc.* **1981**, *103*, 1501.

(*R,S*)-MEVALONOLACTONE-2-^{13}C

(2*H*-Pyran-2-one-^{13}C, tetrahydro-4-hydroxy-4-methyl-)

A.

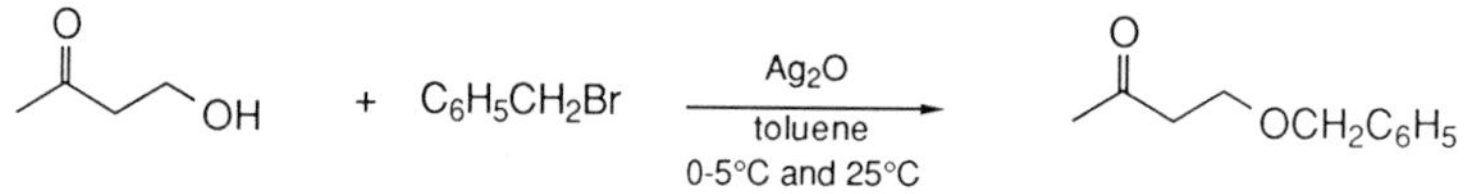

B.

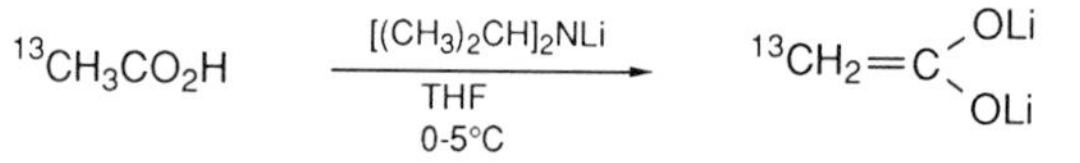

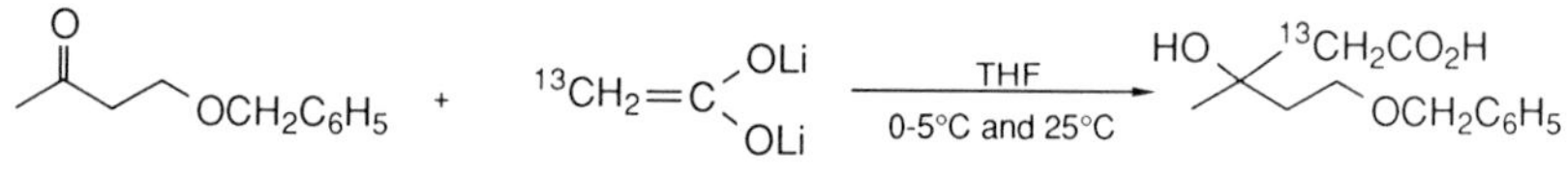

C.

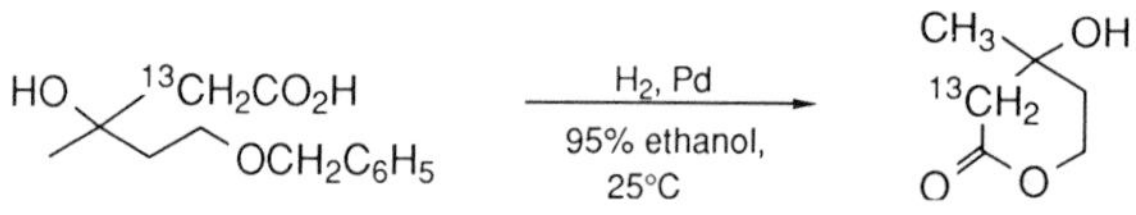

Submitted by MASATO TANABE and RICHARD H. PETERS[1]
Checked by PAULA M. ROACH, SUNG W. RHEE, and ROBERT M. COATES

1. Procedure

Caution! Benzyl bromide is a lachrymator. This procedure should be conducted in a ventilated hood.

A. *4-Benzyloxy-2-butanone.* A 100-mL, three-necked, round-bottomed flask is equipped with a magnetic stirring bar, a condenser mounted with a nitrogen inlet, and a pressure-equalizing dropping funnel (Note 1). The flask is charged with 4.40 g (0.050 mol) of 4-hydroxy-2-butanone (Note 2), 50 mL of dry toluene (Note 3), and 13.9 g (0.060 mol) of freshly prepared silver oxide (Note 4). The suspension is stirred and cooled in an ice bath while 12.0 g (0.070 mol) of benzyl bromide (Note 5) is added over ca. 5 min. The ice bath is removed, and the mixture is allowed to stir for 18 hr at room temperature (Note 6). The suspension is filtered through Celite, the filter cake is washed with two 50-mL portions of toluene, and the combined filtrates are evaporated under reduced pressure. The remaining liquid, which weighs 9.6–10.4 g, is dissolved in 15 mL of 5% tetrahydrofuran in hexane. The cloudy solution is applied to a 5-cm × 47.5-cm column prepared with 380–385 g of silica gel (Note 7) packed in 5% tetrahydrofuran in hexane. The column is eluted with 5% tetrahydrofuran in hexane, and 250-mL fractions are collected and analyzed by TLC (Note 8). A total of 12 or 13 fractions (3–3.25 L) is collected first to separate benzyl bromide, dibenzyl ether, and other minor by-products. The product is then eluted with 0.5–1.0 L of tetrahydrofuran, the solvent is evaporated, and the remaining 6.0–6.5 g of liquid is distilled under reduced pressure. After separation of a 0.7–1.0 g forerun, bp 30–68°C (0.2 mm), consisting mainly of benzyl alcohol, 3.87–4.33 g (43–49%) of 4-benzyloxy-2-butanone, bp 77–79°C (0.2 mm), n_D^{25} 1.5018 is collected (Note 9).

B. *5-Benzyloxy-3-hydroxy-3-methylpentanoic-2-^{13}C acid.* A 50-mL, three-necked, round-bottomed flask is equipped with a magnetic stirring bar, a rubber septum, a condenser connected to a nitrogen inlet, and a pressure-equalizing dropping funnel. The apparatus is purged with nitrogen and dried (Note 1), and the flask is charged with 1.79 g (2.4 mL, 0.0177 mol) of freshly distilled diisopropylamine and 6.5 mL of dry tetrahydrofuran (Note 10). The solution is stirred and cooled in an ice bath while 7.22 mL (0.0169 mol) of 2.34 M butyllithium in hexane (Note 11) is added from the dropping funnel over 30 min. After 30 min a solution of 0.439 g (0.00720 mol) of acetic acid-2-^{13}C (Note 12) in 3 mL of tetrahydrofuran is added by syringe over ca. 10 min. The solution is stirred and cooled in an ice bath for 3.5 hr, after which 1.30 g (0.0073 mol) of 4-benzyloxy-2-butanone in 4 mL of tetrahydrofuran is added by syringe over 15 min. Stirring is continued for 2 hr at 0°C and 18 hr at room temperature. The reaction mixture is cooled in an ice bath, hydrolyzed by adding 4.5 mL of water, and concentrated under reduced pressure to remove most of the tetrahydrofuran. The remaining aqueous suspension is basified by addition of 6 mL of aqueous 4% sodium hydroxide and extracted with 30 mL of diethyl ether. The ethereal layer is extracted with 40 mL of 4% sodium hydroxide, the combined alkaline extracts are cooled and acidified to pH 3 with ca. 10 mL of 18% hydrochloric acid, and the aqueous mixture is extracted with three 25-mL portions of ether. The combined ethereal extracts are dried over anhydrous sodium sulfate and evaporated. The remaining viscous, yellow liquid weighs 0.95–1.01 g (55–59%) and is used in part C without further purification (Note 13).

C. *Mevalonolactone-2-^{13}C.* In a 250-mL Parr hydrogenation bottle are placed 50 mL of 95% ethanol, 0.107 g (0.001 mol) of palladium black (Note 14), and 0.519 g (0.00217 mol) of 5-benzyloxy-3-hydroxy-3-methylpentanoic acid-2-^{13}C. The bottle is attached to a Parr hydrogenation apparatus (Note 15), charged to 50 psig with hydrogen and shaken at room temperature for 2 hr. The hydrogen is flushed from the bottle with nitrogen, and the suspension is filtered by gravity through a layer of Celite with a medium-porosity sintered-glass Büchner funnel to separate the catalyst. *Caution! The palladium is pyrophoric and must always be kept wet with ethanol during filtration to prevent contact with air* (Note 16). The bed of Celite and adhering catalyst is rinsed with three 5-mL portions of ethanol. The combined filtrates are returned to the Parr bottle, 0.107 g (0.001 mol) of fresh palladium black is added, and the hydrogenation is continued for another 8 hr. The catalyst is separated by filtration as previously described, and the combined filtrates are evaporated under reduced pressure. Distillation of the residual liquid with a Kugelrohr apparatus at 90–100°C and 0.01 mm affords 0.235–0.249 g (83–88%) of mevalonolactone-2-^{13}C as a slightly yellow oil (Note 17).

2. Notes

1. The apparatus was dried in an oven at 125°C and allowed to cool while a stream of nitrogen was passed through the condenser and out the dropping funnel. Alternatively the apparatus may be flushed with nitrogen and flamed dry. A nitrogen atmosphere was maintained within the apparatus during the subsequent operations.

2. 4-Hydroxy-2-butanone was purchased from Chemical Samples Company by the checkers and distilled, bp 56–58°C (5.0 mm). The submitters obtained the material from BASF Wyandotte Corporation, Parsippany, NJ 07054.

3. Toluene was dried over sodium wire for 36 hr.

4. The silver oxide was prepared by the following procedure. A solution of 6.9 g (0.172 mol) of sodium hydroxide in 200 mL of water was heated to 80–90°C and added to a solution of 30 g (0.177 mol) of silver nitrate in 200 mL of water, also heated to 80–90°C. The resulting hot suspension was quickly filtered, and the filter cake was washed with 200 mL of hot water, 200 mL of 95% ethanol, and 200 mL of absolute ethanol. The silver oxide was dried at 1 mm and weighed 17.8–18.3 g (87–89%).

5. Benzyl bromide was distilled before use, bp 89°C (14 mm).

6. The reaction is mildly exothermic, and the mixture becomes warm after the ice bath is removed. The checkers monitored the progress of the reaction by TLC on silica gel with 5% methanol in chloroform as developing solvent. After 2 hr the spot at R_f 0.42 for the starting alcohol has disappeared, and the formation of the spots at R_f 0.70 and 0.47 for the product and benzyl alcohol, respectively, appeared to be complete.

7. The checkers used silica gel 60 having particle sizes from 0.05 to 0.2 mm (70–270-mesh ASTM), supplied by Brinkmann Instruments, Inc., Westbury, NY. The submitters used 450 g of silica gel with 90–200 mesh purchased from Gallard-Schlesinger Chemical Manufacturing Corp., Carle Place, NY 11514.

8. Thin-layer chromatograms were performed by the checkers on plates coated with silica gel using chloroform as developing solvent. The R_f values for benzyl bromide, dibenzyl ether, 4-benzyloxy-2-butanone, and benzyl alcohol were 0.72, 0.61, 0.20, and 0.09, respectively. Chromatograms of the crude product showed

spots for these four components and in addition three minor spots at R_f 0.44, 0.40, and 0.36. The first six fractions (1.5 L) were combined and evaporated, affording 0.6–1.5 g of material judged to be mainly benzyl bromide. The following six or seven fractions (1.5–1.75 L) provided 1.6–3.2 g of material composed largely of dibenzyl ether.

9. The submitters obtained 4.5 g (51%) of product, bp 95°C (0.8 mm), n_D^{27} 1.5029. A boiling point of 88–91°C (0.5 mm) and a refractive index of n_D^{28} 1.5040 are reported for 4-benzyloxy-2-butanone.[2] The product was analyzed by the checkers. Anal. calcd. for $C_{11}H_{14}O_2$: C, 74.13; H, 7.92. Found: C, 73.87; H, 8.09. The product has the following spectral characteristics: IR (liquid film) cm^{-1}: 1725 and 1710 (split C=O), 1360, 1175, 1110, 1090, 740, 700; ^{1}H NMR (CDCl$_3$) δ: 2.17 (singlet, 3, CH_3), 2.70 (triplet, 2, $J = 6$, CH_2CH_2O), 3.78 (triplet, 2, $J = 6$, CH_2CH_2O), 4.50 (singlet, 2, $CH_2C_6H_5$), 7.32 (singlet, 5, C_6H_5).

10. Diisopropylamine was dried over potassium hydroxide pellets and distilled from barium oxide before use. The submitters purified tetrahydrofuran by distillation from lithium aluminum hydride. For a warning concerning the potential hazards of this procedure, see *Org. Synth., Coll. Vol. V*, **1973**, 976. The checkers distilled the solvent from the sodium ketyl of benzophenone.

11. Butyllithium in hexane is available from Alfa Products, Morton Thiokol, Inc. The submitters standardized the butyllithium solution by titration of diphenyl-acetic acid in tetrahydrofuran.[3] The concentration of the butyllithium solution was determined by the checkers by titration of a 1-mL aliquot in 10 mL of benzene with 1 M 2-butanol in xylene using 1,10-phenanthroline as indicator.[4]

12. Acetic acid-2-^{13}C of 90% isotopic purity was purchased by the checkers from Stohler Isotope Chemicals, Rutherford, NJ, and dried by distillation from phosphorus pentoxide in the following manner. A 0.5-g portion of the labeled acetic acid was transferred to a 5-mL flask containing 0.1 g of phosphorus pentoxide. The flask (A) was attached to a vacuum system (see Figure 1), chilled

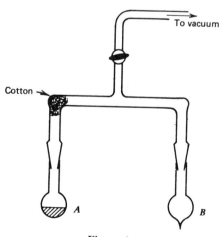

Figure 1

with a dry ice-acetone bath until the mixture solidified, and evacuated to 0.01 mm. The stopcock was closed, the cooling bath was moved to the receiver (flask B), and flask A was allowed to warm to room temperature. The distillation was completed by heating flask A to 60°C. Nitrogen was introduced into the system and flask B removed and stoppered. The recovery of acetic acid-2-^{13}C was 0.42–0.44 g (84–86%).

13. The submitters carried out the procedure in Section B at five times the scale described with 1.96 g (0.0321 mol) of acetic acid-2-^{13}C and obtained 4.01 g (56%) of product. The spectral properties of the product are as follows: IR (liquid film) cm^{-1}: 1710 (C=O); ^{1}H NMR (CDCl$_3$) δ: 1.31 (doublet, ca. 2.4, J = 4.5, ^{13}CH$_2$CCH$_3$), 1.31 (singlet, ca. 0.6, ^{12}CH$_2$CCH$_3$), 1.89 (multiplet, 2, CH$_2$CH$_2$O), 2.55 (doublet, ca. 1.6, J = 128, ^{13}CH$_2$CO$_2$H), 2.55 (singlet, ca. 0.4, ^{12}CH$_2$CO$_2$H), 3.70 (triplet, 2, J = 6, CH$_2$CH$_2$O), 4.52 (singlet, 2, C$_6$H$_5$CH$_2$), 7.35 (singlet, 5, C$_6$H$_5$). The product may be purified further by Kugelrohr distillation with an oven temperature of 100–110°C (0.015 mm).

14. The palladium black was purchased from Engelhard Industries Division, Engelhard Minerals and Chemicals Corporation, Iselin, NJ 08830. The checkers found that the hydrogenolysis may also be effected with 5% palladium on carbon, although 20 hr was required to achieve complete reaction.

15. The hydrogenation apparatus is available from Parr Instrument Company, Inc., Moline, IL 61265.

16. As a further precaution the checkers chilled the suspension in an ice bath prior to filtration.

17. The product was further purified by the checkers by recrystallization from ca. 1.5 mL of ether at 0°C. The recovery of white, crystalline mevalonolactone-2-^{13}C, mp 24–26°C, was 90–92%. The reported[2] melting point is 27–28°C. The spectral properties of the product are as follows: IR (liquid film) cm^{-1}: 3300 (OH), 1730 (C=O); 220–MHz ^{1}H NMR (CDCl$_3$) δ: 1.40 (doublet, ca. 2.4, J = 4.5, ^{13}CH$_2$CCH$_3$), 1.40 (singlet, ca. 0.6, ^{12}CH$_2$CCH$_3$), 1.90 (multiplet, 2, CH$_2$CH$_2$O), 2.54 and 2.67 (eight-line ABX multiplet, ca. 1.6, J_{AB} = 17, J_{AX} = 132, J_{BX} = 127, ^{13}CH$_A$H$_B$), 2.54 and 2.67 (AB doublet, 0.4, J = 17, ^{12}CH$_A$H$_B$), 4.47 (multiplet, 2, CH$_2$CH$_2$O).

3. Discussion

The important role of mevalonate is the biosynthesis of terpenes and sterols has been the impetus for the development of numerous syntheses of the parent mevalonolactone[5-9] and a host of labeled analogs.[7-9] Mevalonolactone has been prepared by reduction of dimethyl or diethyl 3-hydroxy-3-methylglutarate in two stages with lithium aluminum hydride and hydrogen[10] or sodium borohydride[11]; by reduction of monomethyl 3-hydroxy-3-methylglutarate with lithium borohydride[12,13] or sodium in liquid ammonia[13]; by reduction of mevaldic acid or its esters with borohydride,[11a,14,15] hydrogen,[14b] or NADPH in enzyme preparations[16]; by reduction of N-(diphenylmethyl)-3,4-epoxy-5-hydroxy-3-methylpentanamide with lithium borohydride followed by hydrolysis;[17] by oxidation of 3,5-dihydroxy-3-methylpentanal and its derivatives with hydrogen peroxide in acetic acid,[18] or formic acid,[8] or with aqueous bromine;[19] by oxidation of 3-methyl-1,3,5-pentanetriol with chromium trioxide[20a] or silver carbonate;[20b] by ozonolysis of 3-methyl-1-tetrahydropyranyloxy-5-hexen-3-ol;[6,21] by hydrolysis of 3,5-dihydroxy-3-methylpentanenitrile;[22] by degradation of linalool;[23] and by Reformatsky reactions of acetate with a variety of 4-substituted 2-butanones.

The Reformatsky reactions of methyl or ethyl bromoacetate with 4-acetoxy-,[2,24] 4-benzyloxy-,[2] 4-tetrahydropyranyloxy-,[2] 4-chloro-,[6] and 4,4-dimethoxy-2-butanone[14,18] have been carried out. The adducts were converted to mevalonolactone by hydrolysis and, in the case of the acetal reactant, by appropriate reduction and oxidation procedures. The same Reformatsky-type syntheses of mevalonolactone have also been performed using the lithium and magnesium carbanions of acetate esters[5,19,25,26] and the dianion of acetic acid[26,27] instead of the usual zinc reagent. The intramolecular Reformatsky reaction of 4-(bromoacetoxy)-2-butanone gives mevalonolactone directly.[28] A related route to mevalonolactone involves boron trifluoride-catalyzed cycloaddition of ketene to 4-acetoxy-2-butanone followed by hydrolysis.[18a]

Many of the procedures given above have been utilized for the preparation of mevalonolactone labeled with isotopes of carbon, hydrogen, and oxygen.[7-9] Mevalonolactone-^{14}C has been prepared with the label at all six positions: 1-,[7,29] 2-,[7,14a,24] 3-,[11b] 3'-,[20a,30] 4-,[7,18a] and 5-[14]C.[19] Preparations of singly and doubly labeled mevalonolactone-^{13}C have been reported recently: 2-,[26,31] 3-,[32] 4-,[33] 3',4-,[7,18a] 3,4-,[26,33b] and 4,5-[13]C.[34] The procedure described here[26] for the preparation of mevalonolactone-2-^{13}C is both convenient and economical compared to the usual Reformatsky methods since acetic acid-2-^{13}C is utilized directly in the condensation reaction, rather than methyl or ethyl bromoacetate. The overall yield of mevalonolactone-2-^{13}C is 46–52% based on acetic acid-2-[13]C.

1. Bio-Organic Chemistry Department, SRI International, Menlo Park, CA 94025.

2. Hoffman, C. H.; Wagner, A. F.; Wilson, A. N.; Walton, E.; Shunk, C. H.; Wolf, D. E.; Holly, F. W.; Folkers, K. *J. Am. Chem. Soc.* **1957**, *79*, 2316–2318.

3. Kofron, W. G.; Baclawski, L. C. *J. Org. Chem.* **1976**, *41*, 1879–1880.

4. Watson, S. C.; Eastham, J. F. *J. Organomet. Chem.* **1967**, *9*, 165–168; Gall, M.; House, H. O. *Org. Synth., Coll. Vol. VI*, **1988**, 121.

5. Dubois, J.-E.; Moulineau, C. *Bull. Soc. Chim. Fr.* **1967**, 1134–1140.

6. Gray, W. F.; Deets, G. L.; Cohen, T. *J. Org. Chem.* **1968**, *33*, 4532–4534.

7. Cornforth, R. H.; Popják, G. In "Methods in Enzymology," Clayton, R. B., Ed.; Academic Press: New York, 1969; Vol. 15, pp. 359–378.

8. Cornforth, J. W.; Cornforth, R. H. In "Natural Substances Formed Biologically from Mevalonic Acid," Goodwin, T. W., Ed.; Biochemical Society Symposium No. 29, 1969; Academic Press: London, 1970; pp. 5–15.

9. (a) Hanson, J. R. *Adv. Steroid Biochem. Pharmacol.* **1970**, *1*, 51; (b) Hanson, J. R. In "Biosynthesis," A Specialist Periodical Report; The Chemical Society: London, 1972; Vol. 1, pp. 41–43; (c) Hanson, J. R. In "Biosynthesis," A Specialist Periodical Report; The Chemical Society: London, 1977; Vol. 5, pp. 56–58.

10. Wolf, D. E.; Hoffman, C. H.; Aldrich, P. E.; Skeggs, H. R.; Wright, L. D.; Folkers, K. *J. Am. Chem. Soc.* **1957**, *79*, 1486–1487.

11. (a) Tschesche, R.; Machleidt, H. *Justus Liebigs Ann. Chem.* **1960**, *631*, 61–76; (b) v. Euw, J.; Reichstein, T. *Helv. Chim. Acta* **1964**, *47*, 711–724.

12. Cornforth, J. W.; Cornforth, R. H.; Popják, G.; Yengoyan, L. *J. Biol. Chem.* **1966**, *241*, 3970–3987.

13. Huang, F.-C.; Lee, L. F. H.; Mittal, R. S. D.; Ravikumar, P. R.; Chan, A. J.; Sih, C. J.; Caspi, E.; Eck, C. R. *J. Am. Chem. Soc.* **1975**, *97*, 4144–4145.
14. (a) Eggerer, H.; Lynen, F. *Justus Liebigs Ann. Chem.* **1957**, *608*, 71–81; (b) Shunk, C. H.; Linn, B. O.; Huff, J. W.; Gilfillan, J. L.; Skeggs, H. R.; Folkers, K. *J. Am. Chem. Soc.* **1957**, *79*, 3294–3295.
15. Blattmann, P.; Rétey, J. *J. Chem. Soc. Chem. Commun.* **1970**, 1393.
16. (a) Donninger, C.; Popják, G. *Biochem. J.* **1964**, *91*, 10p–11p; *Proc. Roy. Soc., Ser. B* **1965**, *163*, 465–491; (b) Blattman, P.; Rétey, J. *J. Chem. Soc., Chem. Commun.* **1970**, 1394; (c) Scott, A. I.; Phillips, G. T.; Reichardt, P. B.; Sweeny, J. G. *J. Chem. Soc., Chem. Commun.* **1970**, 1396–1397.
17. Cornforth, J. W.; Cornforth, R. H.; Donninger, C.; Popják, G. *Proc. Roy. Soc., Ser. B* **1965**, *163*, 492–514.
18. (a) Cornforth, J. W.; Cornforth, R. H.; Pelter, A.; Horning, M. G.; Popják, G. *Tetrahedron* **1959**, *5*, 311–339; (b) Popják, G.; Goodman, D. S.; Cornforth, J. W.; Cornforth, R. H.; Ryhage, R. *J. Biol. Chem.* **1961**, *236*, 1934–1947.
19. Pichat, L.; Blagoev, B.; Hardouin, J.-C. *Bull. Soc. Chim. Fr.* **1968**, 4489–4491.
20. (a) Escher, S.; Loew, P.; Arigoni, D. *J. Chem. Soc., Chem. Commun.* **1970**, 823–825; (b) Fétizon, M.; Golfier, M.; Louis, J.-M. Tetrahedron **1975**, *31*, 171–176.
21. Tamura, S.; Takai, M. *Bull. Agric. Chem. Soc. Jpn.* **1957**, *21*, 260.
22. Cornforth, J. W.; Ross, F. P.; Wakselman, C. *J. Chem. Soc., Perkin Trans. 1* **1974**, 429–432.
23. Cornforth, R. H.; Cornforth, J. W.; Popják, G. *Tetrahedron* **1962**, *18*, 1351–1354.
24. (a) Isler, O.; Rúegg, R.; Würsch, J.; Gey, K. F.; Pletscher, A. *Helv. Chim. Acta* **1957**, *40*, 2369–2373; (b) Cornforth, J. W.; Cornforth, R. H.; Popják, G.; Gore, I. Y. *Biochem. J* **1958**, *69*, 146–155.
25. Ellison, R. A.; Bhatnagar, P. K. *Synthesis* **1974**, 719.
26. Lawson, J. A.; Colwell, W. T.; DeGraw, J. I.; Peters, R. H.; Dehn, R. L.; Tanabe, M. *Synthesis* **1975**, 729–730.
27. Angelo, B. *C. R. Acad. Sci., Ser. C* **1970**, *271*, 865–867.
28. Hulcher, F. H.; Hosick, T. A. U.S. Patent 3 119 842; *Chem. Abstr.* **1964**, *60*, 10554g.
29. Tavormina, P. A.; Gibbs, M. H. *J. Am. Chem. Soc.* **1956**, *78*, 6210.
30. Phillips, G. T.; Clifford, K. H. *Eur. J. Biochem.* **1976**, *61*, 271–286.
31. Hanson, J. R.; Marten, T.; Siverns, M. *J. Chem. Soc., Perkin Trans. 1* **1974**, 1033–1036.
32. Banerji, A.; Jones, R. B.; Mellows, G.; Phillips, L.; Sim, K.-Y. *J. Chem. Soc., Perkin Trans. 1* **1976**, 2221–2228; Banerji, A.; Hunter, R.; Mellows, G.; Sim, K.-Y.; Barton, D. H. R. *J. Chem. Soc., Chem. Commun.* **1978**, 843–845.
33. (a) Seo, S.; Tomita, Y.; Tori, K. *J. Chem. Soc. Chem. Commun.* **1975**, 270–271; (b) Cane, D. E.; Levin, R. H. *J. Am. Chem. Soc.* **1976**, *98*, 1183–1188.
34. Evans, R.; Hanson, J. R.; Nyfeler, R. *J. Chem. Soc., Perkin Trans. 1* **1976**, 1214–1217.

REDUCTION OF KETONES TO HYDROCARBONS WITH TRIETHYLSILANE: m-NITROETHYLBENZENE

(Benzene, 1-ethyl-3-nitro-)

Submitted by James L. Fry, Steven B. Silverman, and Michael Orfanopoulos[1]
Checked by Jack W. Muskopf and Robert M. Coates

1. Procedure

Caution! Boron trifluoride gas is highly corrosive; this preparation should be conducted in a well-ventilated hood.

A dry, 250-mL, three-necked, round-bottomed flask is equipped with a magnetic stirring bar, a gas-inlet tube (Note 1), a pressure-equalizing dropping funnel, and a Dewar condenser cooled with ice-water and fitted with a drying tube containing anhydrous calcium sulfate (Drierite). A solution of 20.9 g (0.180 mol) of triethylsilane (Note 2) in 80 mL of dichloromethane (Note 3) is placed in the flask. The solution is stirred rapidly and cooled in an ice bath while boron trifluoride gas (Note 4) is introduced below the surface of the liquid at a moderate rate. The first appearance of white fumes at the exit of the drying tube (Note 5) indicates that the solution is saturated and the apparatus is filled with boron trifluoride. The flow of boron trifluoride is adjusted to a level sufficient to maintain a slight emission of white fumes from the drying tube, and a solution of 10.0 g (0.0606 mol) of m-nitroacetophenone (Note 6) in 30 mL of dichloromethane is added dropwise during 10 min. Stirring and cooling are continued for 30 min, after which the flow of boron trifluoride gas is stopped and the cooling bath is removed. The mixture is stirred at room temperature for 20 min, cooled again with the ice bath, and hydrolyzed by adding 20 mL of aqueous saturated sodium chloride. The upper organic layer is decanted from the salts, which are then dissolved in 50 mL of water. The aqueous solution is extracted with two 25-mL portions of pentane, which are combined with the original organic layer. The organic solution is washed with two 25-mL portions of saturated sodium chloride and dried with anhydrous sodium sulfate. Distillation at atmospheric pressure through a 10-cm Vigreux column serves to separate solvent and ca. 15 g of a mixture of primarily triethylsilane and triethylfluorosilane, bp 106–109°C (Note 7). The remaining liquid is distilled in a short-path distillation apparatus under reduced pressure, affording 8.33–8.38 g (91–92%) of pale yellow m-nitroethylbenzene, bp 120–121°C (15 mm), 134–135°C (22 mm), n_D^{25} 1.5330 (Note 8).

2. Notes

1. A Pasteur pipet clamped in an O-ring thermometer adapter served as a convenient, adjustable gas-inlet tube. The large end of the pipet was fitted with a small rubber

septum, through which a syringe needle was passed for introducing the boron trifluoride gas.

2. Triethylsilane was purchased from either Petrarch Systems, Inc., or PCR Research Chemicals, Inc. The reagent was dried with anhydrous sodium sulfate and distilled before use, bp 107–108°C.

3. Reagent-grade dichloromethane was extracted repeatedly with concentrated sulfuric acid, washed twice with water, dried with anhydrous calcium chloride, and distilled from phosphorus pentoxide before use.[2]

4. A cylinder of boron trifluoride gas was purchased from Linde Division, Union Carbide Chemical Corporation. The gas was passed through Teflon tubing to a 250-mL gas-washing bottle equipped with a fritted-glass inlet and containing a saturated solution of boric anhydride (ca. 16 g) in 150 mL of concentrated sulfuric acid. Another section of Teflon tubing was connected to the exit port of the gas-washing bottle and to the barrel of a 2-mL, gastight syringe that fitted into the syringe needle in the septum. Boric anhydride (boron oxide) is available from Fisher Scientific Company. The purpose of the boric anhydride-sulfuric acid scrubber is to remove hydrogen fluoride.[3] All ground-glass joints in the system were lined with Nalgene or Teflon standard-taper sleeves to prevent them from sticking together. Nalgene standard-taper sleeves are supplied by Nalge Sylron Corporation. Any stopcocks used should be Teflon. The checkers placed one empty 250-mL trap in the gas line before the gas-washing bottle and another after it.

5. The submitters recommend that the effluent gases be directed toward a water trap to reduce the amount of boron trifluoride and hydrogen fluoride released into the atmosphere.

6. m-Nitroacetophenone was supplied by Aldrich Chemical Company, Inc. and recrystallized from ethanol, mp 76–78°C. The compound may also be prepared by the method of Corson and Hazen.[4]

7. If desired, the mixture of triethylsilane and triethylfluorosilane can be reconverted into triethylsilane by reduction with lithium aluminum hydride.[5]

8. The product was analyzed by the checkers. Anal. calcd. for $C_8H_9NO_2$: C, 63.56; H, 6.00; N, 9.27. Found: C, 63.51; H, 6.17; N, 9.04. The spectral properties of the product are as follows: IR (CCl_4) cm^{-1}: 3100, 3075, 2975, 2940, 2880, 1532, 1348, 1095, 895; ^{1}H NMR $(CDCl_3)$ δ: 1.27 (triplet, 3, $J = 7.5$, CH_3), 2.74 (quartet, 2, $J = 7.5$, CH_2), 7.3–8.1 (multiplet, 4, aromatic H); ^{13}C NMR $(CDCl_3)$ δ: 14.7, 28.2, 120.4, 122.2, 128.8, 133.9, 145.9, 148.2.

3. Discussion

This procedure illustrates a method for the selective reduction of a carbonyl group to a methylene group in compounds having other potentially reducible functional groups. The method is applicable to the reduction of aryl aldehydes without electron-withdrawing ring substituents, aryl alkyl ketones, diaryl ketones, and dialkyl ketones.[6] Under the above conditions, aryl aldehydes having strongly electron-withdrawing ring substituents (viz., NO$_2$, CN) and alkyl aldehydes yield alcohols. Benzylic alcohols and secondary or tertiary aliphatic alcohols are also reduced to hydrocarbons under these reaction conditions,[7] as are some epoxides.[8] Functional groups that are not affected during ketone reductions include nitro, cyano, ether, ester, carboxylate, and ring halogens.

The effectiveness of this reduction procedure is related to the ability of free boron trifluoride to coordinate strongly to the carbonyl oxygen and to the strong driving force provided by the formation of the silicon-fluorine bond.[6] Similar carbonyl reductions using organosilicon hydrides in conjunction with Brønsted acids[9,10] or boron trifluoride etherate[11] are generally only successful with aryl ketones and aldehydes bearing electron-donating ring substituents; with other carbonyl substrates by-products other than hydrocarbons become the predominant products.

Other carbonyl reduction methods include the familiar Clemmensen[12] and Wolff-Kishner[13] reactions. These are usually conducted for extended periods of time at elevated temperatures under strongly acidic or basic conditions, respectively. Mixed metal hydride-Lewis acid reagents constitute strong reducing systems that are often effective in the deoxygenation of diaryl ketones and some aryl alkyl ketones. However, even the mixed lithium aluminum hydride-aluminum chloride[14] and sodium borohydride-boron trifluoride[15] reagents reduce dialkyl ketones only to the corresponding alcohols, often with the formation of significant amounts of olefinic by-products. m-Nitroethylbenzene has been prepared by reductive deamination of 4-amino-3-nitroethylbenzene[16] and by nitration of ethylbenzene and subsequent fractional distillation to separate the isomers.[17]

1. Department of Chemistry, The University of Toledo, Toledo, OH 43606.
2. Gordon, A. J.; Ford, R. A. "The Chemist's Companion: A Handbook of Practical Data, Techniques, and References"; Wiley: New York, 1972; p 434.
3. Lombard, R.; Stephan, J.-P. *Bull. Soc. Chim. Fr.* **1957,** 1369–1373.
4. Corson, B. B.; Hazen, R. K. *Org. Synth., Coll. Vol. II,* **1943,** 434.
5. Sommer, L. H.: Frye, C. L.; Parker, G. A.; Michael, K. W. *J. Am. Chem. Soc.* **1964,** *86,* 3271–3279.
6. Fry, J. L.; Orfanopoulos, M.; Adlington, M. G.; Dittman, W. R., Jr.; Silverman, S. B. *J. Org. Chem.* **1978,** *43,* 374–375.
7. Adlington, M. G.; Orfanopoulos, M.; Fry, J. L. *Tetrahedron Lett.* **1976,** 2955–2958.
8. Fry, J. L.; Marz, T. J. *Tetrahedron Lett.* **1979,** 849–852.
9. Kursanov, D. N.; Parnes, Z. N.; Bossova, G. I.; Loim, N. M.; Zdanovich, V. I. *Tetrahedron* **1967,** *23,* 2235–2242; Kursanov, D. N.; Parnes, Z. N.; Loim, N. M. *Synthesis* **1974,** 633–651.
10. West, C. T.; Donnelly, S. J.; Kooistra, D. A.; Doyle, M. P. *J. Org. Chem.* **1973,** *38,* 2675–2681.
11. Doyle, M. P.; West, C. T.; Donnelly, S. J.; McOsker, C. C. *J. Organomet. Chem.* **1976,** *117,* 129–140.
12. Vedejs, E. *Org. React.* **1975,** *22,* 401–422; Martin, E. L. *Org. React.* **1942,** *1,* 155–209.
13. Todd, D. *Org. React.* **1948,** *4,* 378–422; Szmant, H. H. *Angew. Chem. Int. Ed. Engl.* **1968,** *7,* 120–128.
14. Nystrom, R. F.; Berger, C. R. A. *J. Am. Chem. Soc.* **1958,** *80,* 2896–2898; Blackwell, J.; Hickinbottom, W. J. *J. Chem. Soc.* **1961,** 1405–1407; Brewster, J. H.; Osman, S. F.; Bayer, H. O.; Hopps, H. B. *J. Org. Chem.* **1964,** *29,* 121–123.
15. Thakar, G. P.; Subba Rao, B. C. *J. Sci. Ind. Res.* **1962,** *21B,* 583; *Chem. Abstr.* **1963,** *59,* 5117g; Pettit, G. R.; Green, B.; Hofer, P.; Ayres, D. C.; Pauwels, P. J. S. *Proc. Chem. Soc.* **1962,** 357.
16. Behal, A.; Choay, E. *Bull. Soc. Chim. Fr., Ser. 3* **1894,** *11,* 206–212; Beilstein, 5, H358.
17. For example, see Brown, H. C.; Bonner, W. H. *J. Am. Chem. Soc.* **1974,** *76,* 605–606.

2-NITROPROPENE

(1-Propene, 2-nitro)

$$HOCH_2CH(NO_2)CH_3 \xrightarrow{\ o\text{-}C_6H_4(CO)_2O\ } CH_2 = C(NO_2)CH_3$$

Submitted by Masaaki Miyashita, Tetsuji Yanami, and Akira Yoshikoshi[1]
Checked by D. Seebach, H. Siegel, and E. Wilka

1. Procedure

Caution! This procedure should be carried out in a hood since 2-nitropropene is a powerful lachrymator. Nitroolefins have a tendency to undergo "fume-offs" (which can be like explosions) near the end of a distillation, particularly if air is let in on a hot distillation residue (from a vacuum distillation).

A 250 mL, one-necked, round-bottomed flask equipped with a magnetic stirring bar is charged with 96.5 g (0.65 mol) of phthalic anhydride (Note 1) and 52.5 g (0.50 mol) of 2-nitro-1-propanol (Note 2). A 10-cm vacuum-insulated Vigreux column, a stillhead fitted with a thermometer, a condenser, a 50-mL, round-bottomed receiving flask, and a water aspirator are installed in due order, and the reaction vessel is placed in an oil bath and evacuated to 110 mm (Note 3). The bath temperature is raised to 150°C and maintained for 30 min while the phthalic anhydride melts to give a homogeneous solution. The receiving flask is immersed in an ice bath, stirring is started, and the bath temperature is raised to 180°C. As the reaction mixture darkens, green-colored 2-nitropropene is gradually distilled off with water, bp 50–65°C (110 mm). The bath temperature is held at 180–185°C until the distillation ceases (ca. 1 hr). The distillate is transferred into a 50-mL separatory funnel, and the lower layer is separated from water (Note 4) and dried over anhydrous magnesium sulfate. Redistillation under reduced pressure through a 10-cm vacuum-insulated Vigreux column (Note 5) gives 25.0–31.4 g (57–72%) of 2-nitropropene, which is collected in an ice-cooled receiving flask as a transparent green liquid, bp 56–57°C (86 mm), n_D^{20} 1.4348 [lit.[3] bp 58°C (90 mm), n_D^{19} 1.4292, d^{20}1.0492] (Note 5). The distilling flask is cooled to room temperature before the vacuum is released.

2. Notes

1. Commercial phthalic anhydride, purchased from Wako Pure Chemical Industries, Ltd. (Japan) or from Fluka AG, Buchs (Switzerland), was used without purification.

2. The checkers purchased 2-nitro-1-propanol (ca. 98% purity) from EGA-Aldrich and used it without further purification. The submitters prepared this reagent from nitroethane and formalin according to the procedure of Feuer,[2] yield 70–75%, bp 79–80°C (5 mm).

3. Lower pressure may cause a loss of the product because of its volatility.

4. The layers sometimes do not separate well. In this case a small amount of magnesium sulfate should be added.

5. It is important that the bath temperature be kept as low as possible to avoid fume-off decompositions. In the checked procedure the bath temperature was never allowed to exceed 80°C. Toward the end of the distillation the pressure was reduced to ca. 60 mm to achieve complete distillation. Although pure 2-nitropropene may be stored in a freezer as a low-melting solid for several weeks, it is recommended to prepare it immediately before use since it tends to polymerize and to darken slowly on storage. 2-Nitropropene polymerizes readily in the presence of a trace of alkali.

3. Discussion

The procedure described is essentially the same as that of Buckley and Scaife.[3] The yield has been increased from 55.5% up to 72% by using 1.3 mol eq of phthalic anhydride and by carefully controlling the pressure and cooling the receiving flask. Although 2-nitropropene has previously been prepared by pyrolysis of 2-nitro-1-propyl benzoate in 72% overall yield from 2-nitro-1-propanol,[4] the present method is preferred for its preparation since the procedure is much simpler and the product is directly obtainable from 2-nitro-1-propanol without first preparing its ester. It is also applicable to the preparation of 1-nitro-1-propene (58%),[5,6] 2-nitro-1-butene (82%),[7] and 2-nitro-2-butene (60%).[6,7] In general, aliphatic nitroolefins have the tendency to polymerize readily with alkali.

1. Chemical Research Institute of Non-Aqueous Solutions, Tohoku University, Sendai 980, Japan.
2. Feuer, H.; Miller, R. *J. Org. Chem.* **1961**, *26*, 1348–1357.
3. Buckley, G. D.; Scaife, C. W. *J. Chem. Soc.* **1947**, 1471–1472.
4. Blomquist, A. T.; Tapp, W. J.; Johnson, J. R. *J. Am. Chem. Soc.* **1945**, *67*, 1519–1524.
5. Miyashita, M.; Kumazawa, T.; Yoshikoshi, A. *J. Chem. Soc., Chem. Commun.* **1978**, 362–363.
6. Melton, J.; McMurry, J. E. *J. Org. Chem.* **1975**, *40*, 2138–2139.
7. Miyashita, M.; Yanami, T.; Yoshikoshi, A. *J. Am. Chem. Soc.* **1976**, *98*, 4679–4681.

CARBOXYLIC ACIDS FROM THE OXIDATION OF TERMINAL ALKENES BY PERMANGANATE: NONADECANOIC ACID

$$CH_3(CH_2)_{17}CH=CH_2 \xrightarrow[CH_2Cl_2/H_2O]{KMnO_4/Adogen\ 464} CH_3(CH_2)_{17}CO_2H$$

Submitted by DONALD G. LEE, SHANNON E. LAMB, and VICTOR S. CHANG[1]
Checked by DENNIS P. LORAH and ANDREW S. KENDE

1. Procedure

A 5-L, three-necked, round-bottomed flask fitted with a mechanical stirrer is placed in an ice bath and charged with 1000 mL of distilled water, 120 mL of 9 *M* sulfuric acid, 3.0 g of Adogen 464 (Note 1), 20 mL of glacial acetic acid, 1000 mL of methylene chloride, and 50 g of 1-eicosene (Note 2). The solution is rapidly stirred and 80 g (0.544

mol) of potassium permanganate is added in small portions over a 3-hr period (Note 3). Stirring is continued for an additional 18 hr at room temperature. The mixture is cooled in an ice bath, and 60 g of sodium bisulfite is added in small portions to reduce any precipitated manganese dioxide. The solution is acidified, if basic, with sulfuric acid and separated. The aqueous layer is extracted with two 400-mL portions of methylene chloride. The organic extracts are combined, washed with two 400-mL portions of water, washed once with brine, and concentrated to 400 mL on a rotary evaporator. The resulting mixture is heated to dissolve any precipitated product, a small amount of amorphous solid is removed by filtration, and the filtrate is cooled to 0°C. A first crop of white crystals (33-36 g, mp 67-68°C) is collected by suction filtration and washed with a minimum amount of ice-cold methylene chloride. Concentration of the mother liquor to 150 mL and cooling to 0°C yields a second crop of crystals (7-12 g). The combined products are dissolved, with heating, in 400 mL of methylene chloride and the pale-yellow solution is allowed to cool to room temperature, then *slowly* to −10°C. The white crystals (36-37 g) are collected, washed with a small amount of cold methylene chloride, and dried in vacuum overnight (mp 68-68.5°C, lit[2] 67-68°C). The yield is 75-77% (Note 4.)

2. Notes

1. Adogen 464, a methyl trialkylammonium (C_8–C_{10}) chloride, was obtained from Ashland Chemical Co.
2. Technical 1-eicosene was obtained from the Aldrich Chemical Company, Inc. and used without further purification. Analysis by quantitative catalytic hydrogenation over Pd-C and NMR spectroscopy indicated that it contained about 10% unreactive, saturated hydrocarbon material.
3. Potassium permanganate was "Baker Analyzed" reagent. About 7.5 g was added every 15 min.
4. The yield was calculated by assuming that the starting material contained 90% 1-eicosene (see Note 2).

3. Discussion

Potassium permanganate is the preferred reagent for the oxidative cleavage of carbon-carbon double bonds.[3] Because of its low solubility in nonpolar solvents, however, the reactions have traditionally been carried out in polar organic solvent systems. For example, the use of aqueous *tert*-butyl alcohol[4] and acetic anhydride[5] for this purpose has been described. An alternative approach involves the use of phase-transfer agents to solubilize permanganate ion in organic solvents and several examples of this approach have been reported.[6] Although benzene has often been used as the solvent,[2] it has been observed that methylene chloride is a superior solvent[7]; better yields are obtained and because of its greater volatility it is more easily removed at the conclusion of the reaction. In addition, methylene chloride is more resistant to oxidation by solubilized permanganate. Adogen 464 was used as the phase-transfer agent because the yields compared well with those obtained when other agents[8] were used and because it is both inexpensive and readily available. The solutions were maintained acidic to neutralize hydroxide ions

TABLE I

PHASE-TRANSFER-ASSISTED PERMANGANATE OXIDATIONS OF TERMINAL ALKENES

Alkene[a]	Product (Yield)[b]	Purification Method	mp or bp (°C)
1-Docosene	Heneicosanoic acid (84%)	Recrystallization (acetone)	72–74° (75°, 82°)[c]
1-Eicosene	Nonadecanoic acid (80%)	Recrystallization (methylene chloride)	68–68.5° (67–68°)[d]
1-Octadecene	Heptadecanoic acid (81%)	Recrystallization (pet. ether)	60–62° (60–61°)[e]
1-Hexadecene	Pentadecanoic acid (84%)	Recrystallization (ethanol-water)	52–53° (53°)[e]
1-Tetradecene	Tridecanoic acid (83%)	Recrystallization (acetone-water)	43–44° (44.5–45.5°)[e]
1-Decene	Nonanoic acid (92%)	Vacuum distillation	109–111°/2.4 mm (121°/4 mm)[e]
1-Octene	Heptanoic acid (70%)	Vacuum distillation	107–108°/7 mm (115–116°/11 mm)[e]
Styrene	Benzoic acid (96%)	Recrystallization (water)	121–122° (122°)[e]

[a]The alkenes were obtained from the Aldrich Chemical Company, Inc. Purity ranged from 99 to 87%.
[b]The purity of the starting material was taken into consideration in yield calculations.
[c]"Handbook of Chemistry and Physics," The Chemical Rubber Co., 52nd ed.
[d]Reference 2.
[e]"Dictionary of Organic Compounds," Eyre and Spottiswoode, 4th ed.

formed during the reduction of permanganate ($MnO_4^- + 2H_2O \rightarrow MnO_2 + 4OH^-$). In the absence of acetic acid the accumulation of base promotes certain side reactions and increases the stickiness of the manganese(IV) oxides which precipitate as the reaction proceeds.

The results obtained with a number of other representative terminal alkenes have been summarized in Table I.

1. Department of Chemistry, University of Regina, Regina, Canada S4S OA2.
2. Krapcho, A. P.; Larson, J. R.; Eldridge, J. M. *J. Org. Chem.* **1977**, *42*, 3749–3753.
3. Lee, D. G. In "Techniques and Applications in Organic Syntheses," Augustine, R. L., Ed.; Dekker: New York, 1969; Vol. 1, pp. 6–28.
4. ApSimon, J. W.; Chau, A. S. Y.; Craig, W. G.; Krehm, H. *Can. J. Chem.* **1967**, *45*, 1439–1445. In these reactions sodium metaperiodate was used as a cooxidant.
5. Sharpless, K. B.; Lauer, R. F.; Repic, O.; Teranishi, A. Y.; Williams, D. R. *J. Am. Chem. Soc.* **1971**, *93*, 3303–3304; Jensen, H. P.; Sharpless, K. B. *J. Org. Chem.* **1974**, *39*, 2314. These two references describe oxidation of olefins to α-diketones.
6. Starks, C. M. *J. Am. Chem. Soc.* **1971**, *93*, 195–199; Sam, D. J.; Simmons, H. E. *J. Am.*

Chem. Soc. **1972,** *94,* 4024–4025; Herriott, A. W.; Picker, D. *Tetrahedron Lett.* **1974,** 1511–1514.

7. Foglia, T. A.; Barr, P. A.; Malloy, A. J. *J. Am. Oil Chem. Soc.* **1977,** *54,* 858A; *Chem. Abstr.* **1978,** *88,* 37210m.

8. Lee, D. G.; Chang, V. S. *J. Org. Chem.* **1978,** *43,* 1532–1536.

SUBSTITUTED γ-BUTYROLACTONES FROM CARBOXYLIC ACIDS AND OLEFINS: γ-(*n*-OCTYL)-γ-BUTYROLACTONE

[2(3*H*)-Furanone, dihydro-5-octyl-]

$$4\ Mn(OAc)_2 \cdot 4\ H_2O \quad + \quad KMnO_4 \quad + \quad 8\ CH_3CO_2H \longrightarrow$$

$$5\ Mn(OAc)_3 \cdot 4\ H_2O \quad + \quad KOAc$$

$$Mn(OAc)_3 \cdot 4\ H_2O \quad + \quad 3\ (CH_3CO)_2O \longrightarrow Mn(OAc)_3 \cdot H_2O \quad + \quad 6\ CH_3CO_2H$$

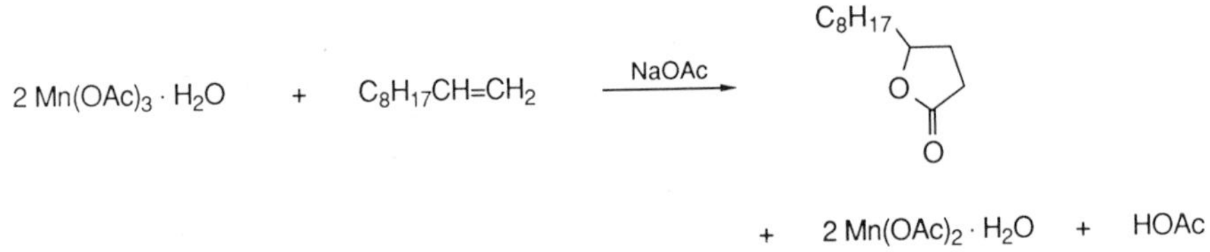

$$2\ Mn(OAc)_3 \cdot H_2O \quad + \quad C_8H_{17}CH{=}CH_2 \xrightarrow{\text{NaOAc}}$$

$$+ \quad 2\ Mn(OAc)_2 \cdot H_2O \quad + \quad HOAc$$

Submitted by E. I. Heiba, R. M. Dessau, A. L. Williams and P. G. Rodewald[1]

Checked by Gerald E. Lepone and Orville L. Chapman

1. Procedure

Benzene has been identified as a carcinogen; OSHA has issued emergency standards on its use. All procedures involving benzene should be carried out in a well-ventilated hood, and glove protection is required.

A 1-L, four-necked flask is fitted with a nitrogen inlet tube, stirrer, dropping funnel, and thermometer. Acetic acid (558 g) is introduced and 107.6 g (0.439 mol) of manganese acetate tetrahydrate (Note 1) is added with stirring and heating under nitrogen. When the temperature reaches 90°C, 16.5 g of solid potassium permanganate (0.104 mol) is added. After the temperature has again fallen to 90°C, 175 mL (189 g, 1.86 mol) of acetic anhydride (Note 2) is added. When the temperature rise has ceased, 44.0 g of 1-decene (0.312 mol) (Note 3) is introduced, followed at once by 250 g of anhydrous sodium acetate. The reaction mixture is then heated to reflux (134°C pot temperature). After 2 hr of reflux under nitrogen the reaction mixture, now clear yellow, is diluted with 1 L of water. The crude product is extracted into 200 mL of benzene, and the aqueous layer again washed with 100 mL of benzene. Benzene is distilled from the combined extracts to give 55.1 g of lactone and 1-decene. 1-Decene is removed by vacuum distillation, followed by the lactone, which distills at 98–99°C (0.05 mm) (Note 4). The yield of γ-(n-octyl)-γ-butyrolactone is 34.1 g (66% based on potassium permanganate. However, the lactone yield based on olefin consumed is greater than 95%.)

2. Notes

1. The checkers used manganous acetate tetrahydrate obtained from Fisher Scientific Company. This compound is more readily available than manganous acetate dihydrate used by the submitters and obtained from the Harshaw Chemical Company.

2. If the dihydrate is used, only 76.7 g (0.751 mol) of acetic anhydride is required.

3. 1-Decene was used as obtained from the Humphrey Chemical Company.

4. The checkers found the yield based upon olefin consumed to be 85%. This discrepancy could be accounted for by losses due to the high volatility of 1-decene at reduced pressure.

3. Discussion

This method has the advantage that it does not require the preparation and purification of solid manganic acetate dihydrate. Dehydration by various ratios of acetic anhydride to manganese shows that in this procedure the yield (35%) from the monohydrate is greater than that from the manganic acetate dihydrate. Further removal of all water from the manganic acetate by means of acetic anhydride does not improve the yield (66%).

This general procedure can be used to prepare a wide variety of substituted γ-butyr-olactones that depend on the structure of the olefin and the aliphatic acid used. The free-radical mechanism and scope of this reaction are described in detail in a paper by Heiba, Dessau, and Rodewald.[2]

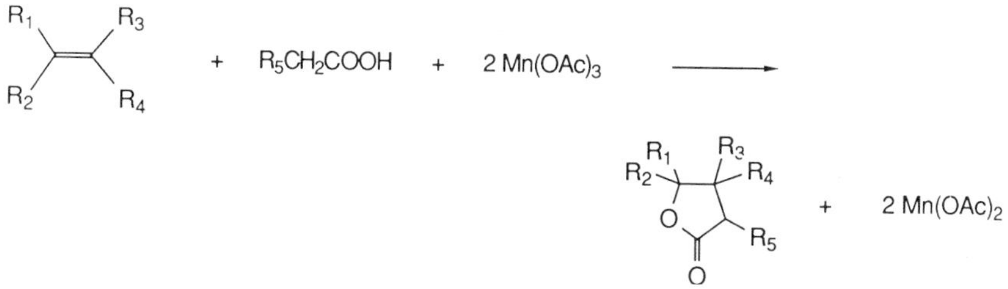

1. Mobil Research and Development Corporation, Central Research Division, P.O. Box 1025, Princeton, NJ 08540.

2. Heiba, E. I.; Dessau, R. M.; Rodewald, P. G. *J. Am. Chem. Soc.* **1974,** *96,* 7977–7981.

ASYMMETRIC REDUCTION OF α,β-ACETYLENIC KETONES WITH B-3-PINANYL-9-BORABICYCLO[3.3.1]NONANE: (R)-(+)-1-OCTYN-3-OL

[1-Octyn-3-ol, (R)-]

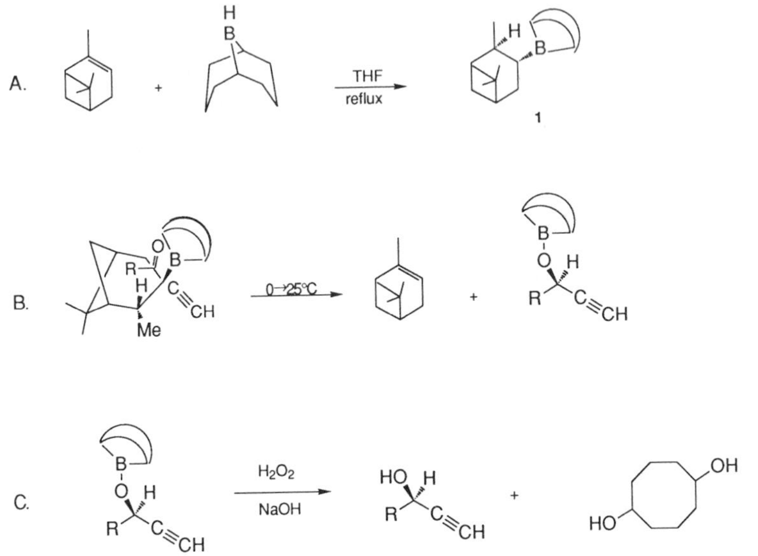

Submitted by M. MARK MIDLAND and RICHARD S. GRAHAM[1]
Checked by JOEL M. HAWKINS and K. BARRY SHARPLESS

1. Procedure

A. A 2-L, round-bottomed flask equipped with a septum-capped sidearm, magnetic stirring bar, reflux condenser, and stopcock adapter connected to a mercury bubbler is flame-dried while being flushed with nitrogen. A nitrogen atmosphere is maintained during the procedure through the oxidation step. After the apparatus is cooled, it is charged, via a double-ended needle,[2] with 800 mL of a 0.5 M tetrahydrofuran (THF) solution of 9-borabicyclo[3.3.1]nonane (9-BBN, 0.4 mol, Note 1). Then 61.3 g (71.5 mL, 0.45 mol) of (+)-α-pinene (Note 2) is added. After the solution is refluxed for 4 hr, the excess α-pinene and THF are removed by vacuum (Note 3) to provide a thick clear oil of neat B-3-pinanyl-9-borabicyclo[3.3.1]nonane, 1 (Note 4).

B. The flask is cooled to 0°C (ice bath) and 35.3 g (0.285 mol) of 1-octyn-3-one (Note 5) is added. After an initially exothermic reaction, the reaction is allowed to warm to room temperature. The reduction can be monitored by gas chromatography (Note 6), but generally 8 hr is required for completion. The color of the reaction mixture is initially light yellow and darkens to red at the end of the reduction.

C. Excess 1 is destroyed by adding 22 mL (0.3 mol) of freshly distilled propion-aldehyde and stirring for 1 hr at room temperature. Liberated α-pinene is then removed by vacuum (Note 7). Tetrahydrofuran, 200 mL, is added, followed by 150 mL of 3 M aqueous NaOH. Hydrogen peroxide (150 mL, 30%) is added dropwise (see *Caution* in Note 8). Oxidation is complete in 3 hr at 40°C. The reaction mixture is transferred to a separatory funnel and extracted with three 50-mL portions of ethyl ether. The ether layers are combined and dried with copious amounts of anhydrous magnesium sulfate, filtered, and concentrated by rotary evaporation to give an oil. Distillation at 60–65°C (3.0 mm) yields 31 g (0.245 mol) of 1-octyn-3-ol, 86% yield (Note 9). The distillation pot residue is a thick oil consisting for the most part of *cis*-1,5-cyclooctanediol. An NMR lanthanide shift study showed the alcohol to be 93% (R) and 7% (S), 86% e.e. (Notes 10 and 11).

2. Notes

1. A 0.5 M THF solution of 9-BBN is available from Aldrich Chemical Company in 800 mL bottles.

2. (+)-α-Pinene (90–92% e.e.) is available from Aldrich Chemical Company. The pinene was distilled from lithium aluminum hydride before use.

3. Most of the THF is removed by water aspirator vacuum. Excess pinene (0.05 mol, ~8 mL) is removed by applying a 0.05-mm vacuum for 2 hr while warming to 40°C with a water bath. The vacuum should be bled with nitrogen to maintain an inert atmosphere in the reaction flask. Recently Brown's group[3] has shown that reduction occurs at an enhanced rate with neat organoborane **1**. Excess **1**, 1.4 equiv per equivalent of 1-octyn-3-one, is used to provide a slight excess of reducing agent to increase the rate for this bimolecular process.

4. B-3-Pinanyl-9-borabicyclo[3.3.1]nonane, **1,** is also available from Aldrich Chemical Company under the tradename "R-Alpine-Borane."

5. 1-Octyn-3-one was obtained by standard Jones oxidation[4] of racemic 1-octyn-3-ol (in ~80% yield). Racemic 1-octyn-3-ol is available from Aldrich Chemical Company. It is essential to check the ketone for unreacted starting alcohol since racemic alcohol will contaminate the final, optically active product.

6. GLC can be used to monitor the disappearance of the acetylenic ketone. 1-Octyn-3-one is eluted just after α-pinene from a SE-30 6-ft column at 80°C. The checkers followed the disappearance of ketone by TLC (15% ethyl acetate in hexane).

7. This is the most convenient time for removal of α-pinene since α-pinene and 1-octyn-3-ol have similar boiling points, making separation by distillation difficult. Application of a 0.05-mm vacuum while the flask is warmed to 40°C for several hours will remove most of the α-pinene (0.4 mol, ~63.5 mL). Because of the volume of α-pinene, cold traps in the vacuum system may become plugged; therefore, the traps will have to be emptied several times. This provides a convenient method for recovery of liberated (+)-α-pinene.

8. Hydrogen peroxide oxidation of organoboranes is exothermic. Careful, dropwise addition of 30% hydrogen peroxide to the organoborane will provide sufficient heating to maintain a reaction temperature in the 40–50°C range.

9. 1-Octyn-3-ol has the following properties: bp 60–65°C (3.0 mm); IR (neat) cm^{-1}: 3315, 2950, 2860, 2120, 1475, 1380, 1120, 1060, 1025, 650; ^{1}H NMR (CDCl$_3$) δ: 0.86 (t, 3 H, J = 6.6, CH$_3$), 1.3–1.4 (m, 6 H), 1.65 (m, 2 H), 2.42 (d, 1 H, J = 2, C≡C—H), 3.0 variable (broad, 1 H, OH), 4.33 (m, 1 H); ^{13}C NMR (CDCl$_3$) δ: 72.6 (C-1), 85.1 (C-2), 62 (C-3), 37.4 (C-4), 31.3 (C-5), 24.6 (C-6), 22.4 (C-7), 13.9 (C-8); [α]$_D^{25}$ + 7.50° (neat, density 0.864 g/mL). It has been shown that optical rotation is an unreliable criterion of enantiomer purity of 1-octyn-3-ol.[5]

10. Commercially available Eu (hfc)$_3$, tris [3-(heptafluoropropylhydroxy-methylene)-d-camphorato]europium III, NMR shift reagent, was used as received from Aldrich Chemical Company. The proton on the chiral carbinol carbon was shifted downfield to ~11 ppm in CDCl$_3$. The R isomer was shifted ~0.5 ppm further downfield than the S isomer.

11. Optically pure (+)-1-octyn-3-ol may be obtained by recrystallization of the half-acid phthalate with (+)-α-methylbenzylamine (Aldrich Chemical Company). The half-acid phthalate salt is made by heating equal molar amounts of 1-octyn-3-ol and phthalic anhydride. This half acid phthalate derivative is a waxy solid that does not lend itself to recrystallization. Attempts to form crystalline salts of the phthalate derivative with achiral alkyl amines only lead to waxy solids or thick oils. The phthalic amine salt made with racemic 1-octyn-3-ol requires three to four recrystallizations from methylene chloride to resolve enantiomers.[6] The first recrystallization may take several days, with successive recrystallizations becoming easier. If the 86% e.e. 1-octyn-3-ol is used to make the phthalic amine salt only one facile recrystallization is needed to provide optically-pure alcohol. The pure amine salt melts at 132–134°C. The enantiomeric purity of the salt may be determined by NMR by observing the ethynyl hydrogen doublets at δ 2.48 (minor) and 2.52 (major) (CDCl$_3$ solvent).

3. Discussion

In this procedure, we describe a general method for the synthesis of alkynyl alcohols of high enantiomeric purity. The one-pot asymmetric reduction of 1-octyn-3-one with B-3-pinanyl-9-borabicyclo[3.3.1]nonane provides a mild and efficient method for the preparation of optically active 1-octyn-3-ol. The reduction occurs in good chemical yield and is virtually (>95%) stereospecific (correcting for the use of 90% e.e. α-pinene). The availability of optically pure α-pinene is a limiting factor in this method, but recently Brown's group has developed a process that provides enantiomerically pure α-pinene.[7] The reduction can be applied to prepare both enantiomers of 1-octyn-3-ol, since both enantiomers of α-pinene are commercially available; although commercial (−)-α-pinene is only 81.3% e.e.,[7] (−)-α-pinene of 92% e.e. is easily obtained by isomerizing commercial (−)-β-pinene (92% e.e.).[8] (Reducing agent 1, made with (−)-α-pinene, will provide (S)-(−)-1-octyn-3-ol). The α-pinene liberated (by β-hydride elimination) in the reduction may be recycled without loss of optical purity. Another attractive feature of this reduction is that organoborane 1 is a mild reagent and seldom affects other functional groups present within the acetylenic ketone. For base-sensitive systems that cannot tolerate the standard sodium hydroxide–hydrogen peroxide oxidation, an alter-

TABLE I
REDUCTIONS OF ALKYNYL KETONES WITH B-(3)-α-PINANYL-9-BBN

| Ketone RCOC≡CR' | | | Enantiomeric |
R	R'	Yield (%)^a	Excess (%)^b
Ph	Bu	72	89^c
Me	Ph	98	72(78)
Pr	C$_6$H$_{13}$	68	77^c
2-Pr	H	78	91(99)
(chromanyl structure: Ph–CH$_2$–O–substituted chromane)	Me	77	85 : 15^d
(chromanyl structure: Ph–CH$_2$–O–substituted chromane)	H	75	91 : 9^d
Me	COOEt	59	71(77)
C$_5$H$_{11}$	COOEt	72	85(92)
Ph	COOEt	64	92(100)
t-Bu	Me	0	
Me	t-Bu	62	73^c

^{a}Isolated yield based on starting ketone.
^{b}Determined by analysis of the Eu(dcm)$_3$ shifted NMR spectrum. The numbers in parentheses are corrected for 92% e.e. α-pinene.
^{c}100% optically pure (+)-α-pinene was used.
^{d}Diastereomeric ratio (R,R to R,S) determined by LC or NMR analysis of the mixture.

native workup using ethanolamine is available.[9] Table I illustrates the application of this reduction to other propargyl ketones.[9] In these cases, tetrahydrofuran was not removed prior to reduction. Removal of tetrahydrofuran provides a faster reaction and slightly higher optical purity.[3]

The most popular methods of preparing optically active 1-octyn-3-ol involve asymmetric reduction of 1-octyn-3-one with optically active alcohol complexes of lithium aluminum hydride or aluminum hydride.[10] These methods give optical purities and chemical yields similar to the method reported above. A disadvantage of these metal–hydride methods is that some require exotic chiral alcohols that are not readily available in both enantiomeric forms. Other methods include optical resolution of the racemic propargyl alcohol (100% e.e.)[6] (and Note 11) and microbial asymmetric hydrolysis of the propargyl acetates (∼15% e.e. for 1-heptyn-3-ol).[11]

1. Department of Chemistry, University of California, Riverside, CA 92521.
2. This book discusses techniques required to handle air-sensitive chemicals. Brown, H. C.; Kramer, G. W.; Levy, A. B.; Midland, M. M. "Organic Synthesis via Boranes"; Wiley: New York, 1975.
3. Brown, H. C.; Pai, G. G. J. Org. Chem. 1982, 47, 1606.

4. Eisenbraun, E. J. *Org. Synth., Coll. Vol. V.* **1973,** 310.

5. McClure, N. L.; Mosher, H. S., private communication.

6. Fried, J.; Lin, C.; Mehra, M.; Kao, W.; Dalven, P. *Ann. N.Y. Acad. Sci.* **1971,** *180,* 38.

7. Brown, H. C.; Jadhav, P. K.; Desai, M. C. *J. Org. Chem.* **1982,** *47,* 4583.

8. (a) Brown, C. A. *Synthesis* **1978,** 754; (b) Cocker, W.; Shannon, P. V. R.; Staniland, P. A. *J. Chem. Soc. C* **1966,** 41.

9. Midland, M. M.; McDowell, D. C.; Hatch, R. L.; Tramontano, A. *J. Am. Chem. Soc.* **1980,** *102,* 867.

10. (a) Cohen, N.; Lopresti, R. J.; Neukom, C.; Saucy, G. *J. Org. Chem.* **1980,** *45,* 582; (b) Brinkmeyer, R. S.; Kapoor, V. M. *J. Am. Chem. Soc.* **1977,** *99,* 8339; (c) Vigneron, J.-P.; Bloy, V. *Tetrahedron Lett.* **1979,** 2683; (d) Nishizawa, M.; Yamada, M.; Noyori, R. *Tetrahedron Lett.* **1981,** *22,* 247.

11. Mori, K.; Akao, H. *Tetrahedron* **1980,** *36,* 91.

PREPARATION OF CHIRAL, NONRACEMIC γ-LACTONES BY ENZYME-CATALYZED OXIDATION OF *meso*-DIOLS: (+)-(1*R*,6*S*)-8-OXABICYCLO[4.3.0]NONAN-7-ONE

[1(3*H*)-Isobenzofuranone, hexahydro-, (3a*S-cis*)-]

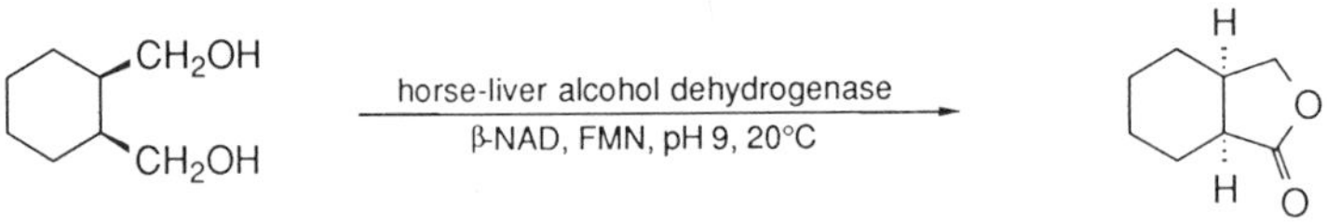

Submitted by J. BRYAN JONES and IGNAC J. JAKOVAC[1]

Checked by ROLAND H. WEBER, MAX F. ZÜGER, and DIETER SEEBACH

1. Procedure

In a 1-L Erlenmeyer flask are placed 475 mL of distilled water (Note 1) and 3.75 g (0.05 mol) of reagent-grade glycine, and the pH is adjusted to 9 by the careful addition of aqueous 10% sodium hydroxide. In the buffer solution thus obtained are dissolved 2.00 g (13.87 mmol) of *cis*-1,2-bis(hydroxymethyl) cyclohexane (Note 2), 0.58 g (0.852 mmol) of β-NAD (Note 3), and 7.8 g (16.2 mmol) of FMN (Note 4). To the clear orange solution obtained is added 80 units of horse-liver alcohol dehydrogenase (Note 5). After the solution is gently swirled for 1 min, the pH is readjusted to 9 and the mixture is kept at room temperature (Note 6) with the mouth of the flask loosely covered by a watchglass. After a few minutes the color of the solution begins to darken and after several hours becomes an opaque green–brown. The pH is readjusted to 9 after 6, 12, 24, 48, and 72 hr by the careful addition of aqueous 10% sodium hydroxide since the pH of the mixture drops progressively as the reaction proceeds. After 4 days (Note 7), the mixture is brought to a pH of ca. 13.3 by the addition of 20 mL of aqueous 50% sodium hydroxide solution. After 1 hr, the mixture is continuously extracted with chloroform for 10 hr (Note 8). The chloroform extract is discarded. The aqueous layer is acidified to pH 3 with concentrated hydrochloric acid and again extracted continuously for 15 hr with chloroform. To the green–orange solution are added charcoal (0.5 g), and magnesium sulfate. The dried and partially decolorized mixture is filtered through a bed of Celite, and the chloroform is removed under reduced pressure using a rotatory evaporator. The residual orange–green oil is distilled in a Kugelrohr apparatus to give 1.4–1.5 g (72–77% yield, Note 9) of (+)-(1R,6S)-8-oxabicyclo[4.3.0]nonan-7-one (>97% e.e., Note 10) as a colorless oil, bp 85–100°C (0.1–0.05 mm), mp 26–29°C, $[\alpha]_D^{22}$ +51.3° (CHCl$_3$, c 1.1) (Note 11).

2. Notes

1. It is not necessary to use doubly distilled or deionized water in this buffer preparation.

2. cis-1,2-Bis(hydroxymethyl)cyclohexane was purchased from Aldrich Chemical Company, Inc. (or EGA, D-Steinheim).

3. β-NAD is the standard biochemical abbreviation for the coenzyme β-nicotinamide adenine dinucleotide. The β-NAD used was of 95% purity and was purchased from Kyowa Hakko (USA), New York. It is also available from Sigma Chemical Company.

4. FMN is the standard biochemical abbreviation for flavin mononucleotide (or riboflavin phosphate). The sodium salt (95–97% pure) of FMN is used. This grade is inexpensive and is available from Sigma Chemical Company. Its purpose is to effect recycling[2] of the catalytic amount used of the much more costly NAD. A larger than stoichiometric amount of FMN is employed in order to ensure rapid recycling of the NAD.

5. Horse-liver alcohol dehydrogenase (HLADH or LADH, also called "equine-liver alcohol dehydrogenase") is the crystalline preparation (>98% protein) sold by Sigma Chemical Company. It is also available from Worthington and Boehringer. The amount added is quoted in units of activity since the activity of the enzyme from different sources can vary. For example, the Sigma enzyme is sold as having an activity of 1–2 units per milligram of protein. The enzyme used in this preparation had 1.5 units of activity per milligram. We have used Worthington–Boehringer enzyme with equal success. The activity of the enzyme diminishes slowly on prolonged storage, even at −20°C. For controlled results, the enzymatic activity may be determined prior to use and the requisite number of units used.

   The assay method of Dalziel[3] is convenient. In a recording UV spectrophotometer set at 340 nm is placed a 3-mL quartz cuvette containing 2.4 mL of 0.10 M glycine–sodium hydroxide buffer solution, pH 9, 500 μL of a 54 mM solution of ethanol in the same buffer, and 100 μL of a 15 mM solution of NAD, also in the same pH 9 buffer. The volume is made up to 3.0 mL, and the assay initiated by the addition of 10 μL of a 1 mg/mL solution of HLADH in 0.10 M "Tris-hydrochloric acid buffer," pH 7.4. The change in optical density at 340 nm is monitored at 25°C and the activity calculated from the following equation:

$$\text{Units of activity/mg protein} = \frac{\Delta \text{OD}_{340}/\text{min}}{6.23 \times \text{mg HLADH/mL of assay volume}}$$

If the preceding assay concentrations are followed exactly, this becomes:

$$\text{Units}/\text{mg protein} = \frac{\Delta OD_{340}/\text{min}}{20.75}$$

6. Ambient temperatures of up to 30°C can be employed but the reaction temperature should not be allowed to fall below 20°C.

7. The end of the reaction is checked by gas chromatography using 3% QF-1 or OV-101 on Chromosorb columns. The checkers used an OV-101, at 190°C oven temperature. A sample is extracted with ether. The organic layer is analyzed. At 20°C the reaction usually goes to completion within 4 days.

8. This removes residual starting material and other nonacidic impurities.

9. Scaling up the preparation is easily accomplished. It is best done by increasing the number of reaction vessels rather than by increasing the reaction volume. For example, 10 g of the *cis*-diol substrate can be oxidized simultaneously using 2.5 g in each of four 1-L Erlenmeyer flasks as described in the procedure described above. After 4 days, the reaction mixtures are combined prior to the chloroform extraction and the lactone is isolated.

10. The absolute configuration and optical purity of the lactone was established by its hydrolysis and epimerization to (1R,2R)-*trans*-2-hydroxymethylcyclo-hexanecarboxylic acid followed by lithium aluminum hydride reduction to (1R,2R)-*trans*-1,2-bis(hydroxymethyl)cyclohexane.[4] By ^{1}H NMR,[5] the e.e. was >97%.

11. The spectral properties of the product obtained were as follows: IR (thin film): C=O at 1770 cm^{-1}; ^{1}H NMR (CDCl$_3$) δ: 0.9–2.8 (m, 10 H, all cyclohexane H), 3.87–4.34 (m, 2 H, CH$_2$—O).

3. Discussion

Horse-liver alcohol degydrogenase is a well-documented enzyme capable of operating with high stereoselectivity on a broad structural range of alcohol and carbonyl substrates.[6] The present reaction proceeds via the pathway shown below, where NAD and NADH represent the oxidized and reduced forms, respectively, of the nicotinamide adenine dinucleotide coenzyme.

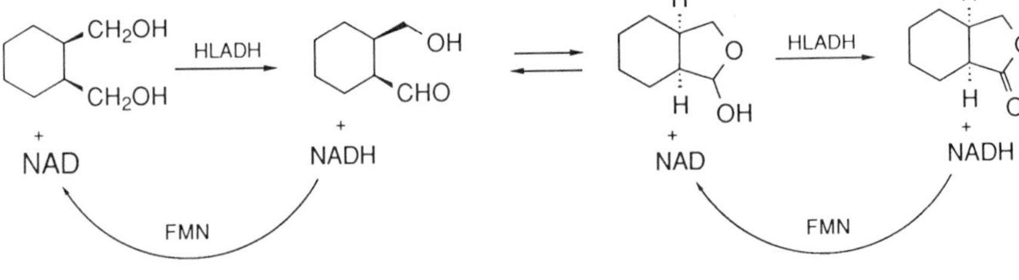

Chemical oxidations of diols to racemic lactones can be achieved by a broad spectrum of oxidizing agents.[7] At the present time, however, only the enzymatic route described can provide a versatile, one-step, access to such a wide range of highly enantiomerically enriched γ-lactones, useful as chiral building blocks for syntheses.

The lactones thus far obtained by this route have been assembled in Table I. Each

TABLE I
PREPARATION OF γ-LACTONES BY HLADH-CATALYZED OXIDATIONS OF *meso*-DIOLS
(YIELD[ref.])[a]

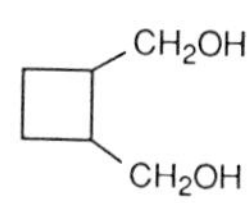

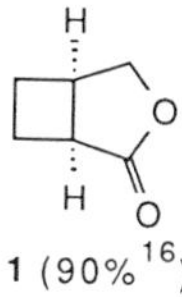

1 (90%[16])

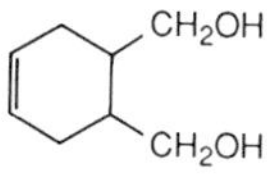

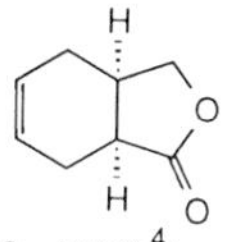

2 (80%[4])

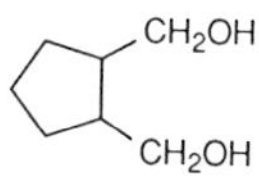

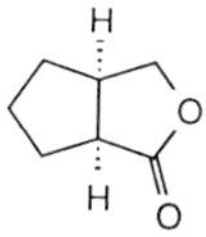

(72%[16])

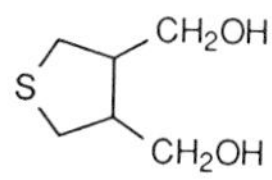

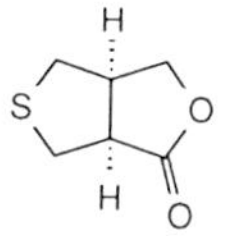

(81%[9])

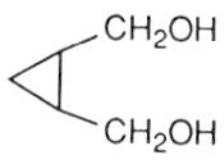

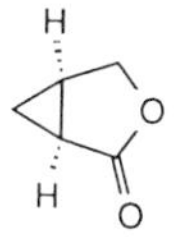

(68%[16])

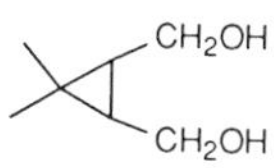

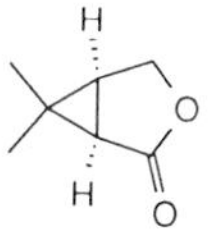

3 (71%[16])

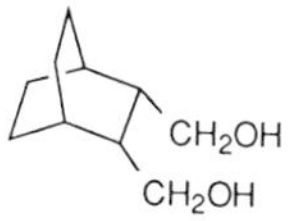

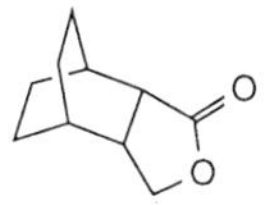

4 (87%[9])

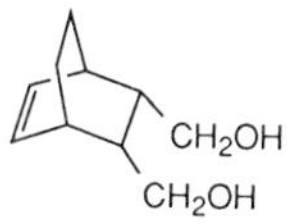

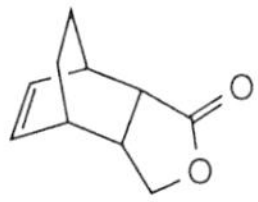

5 (64%[9])

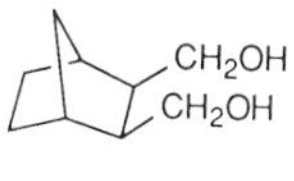

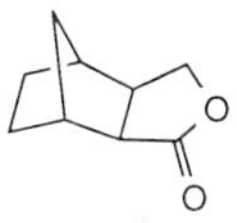

(73%[9])

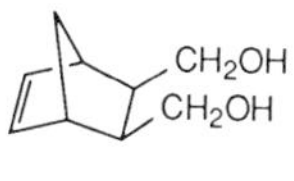

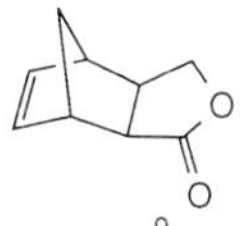

(74%[9])

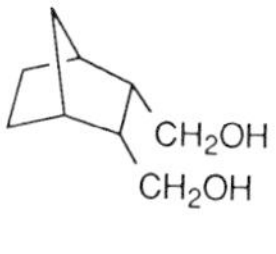

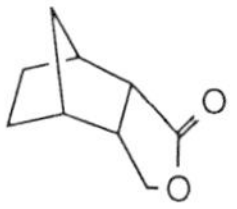

(86%[9])

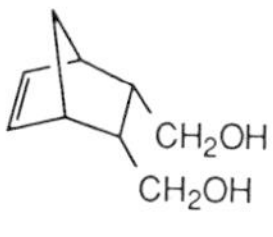

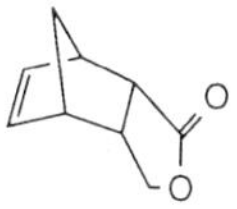

6 (64%[9])

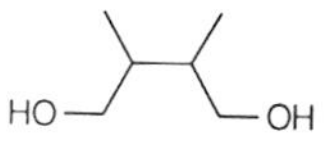

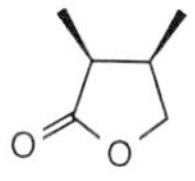

(65%[9])

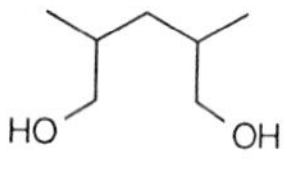

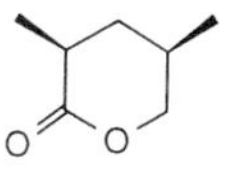

7 (65%[16])

[a]The optical purities and/or enantiomeric excesses were determined by ^{1}H NMR to be >97%;[5] 2 was obtained with 85% e.e.

oxidation proceeds in high chemical yield (65–90%) to give products of >97% enantiomeric excess.[5]

The maximum reaction time required for any one of the substrates shown in Table I is 7 days. In reaction mixtures that contain lactones **4** and **5**, minor amounts of the hemiacetal intermediates are present; they are removed during the extraction at pH 13. After chromatographic separation from any unreacted diols, they can be readily converted to the corresponding lactones by chemical oxidation with silver carbonate on Celite.[8]

The lactones shown in the Table include several representatives of recognized or potential value as starting materials in natural product synthesis. Lactone **1** is a precursor of grandisol,[9,10] lactone **3** of some pyrethroids,[9,11] lactone **6** of some prostaglandins,[9,12] and lactone **7** of multistriatin,[13] methynolide,[14] and monensin.[15]

1. Department of Chemistry, University of Toronto, Toronto, Ontario, Canada M5S 1A1.
2. Jones, J. B.; Taylor, K. E. *Can. J. Chem.* **1976**, *54*, 2969–2973.
3. Dalziel, K. *Acta Chem. Scand.* **1957**, *11*, 397–398.
4. Goodbrand, H. B.; Jones, J. B. *J. Chem. Soc., Chem. Commun.* **1977**, 469–470.
5. Jakovac, I. J.; Jones, J. B. *J. Org. Chem.* **1979**, *44*, 2165–2168.
6. Jones, J. B.; Beck, J. F. *Tech. Chem. (NY)* **1976**, *10*, 107–401.
7. Kano, S.; Shibuya, S.; Ebata, T. *Heterocycles* **1980**, *14*, 661–711.
8. Fétizon, M.; Golfier, M.; Louis, J.-M. *J. Chem. Soc., Chem. Commun.* **1969**, 1118–1119; Fétizon, M.; Golfier, M.; Louis, J.-M. *Tetrahedron* **1975**, *31*, 171–176.
9. Jakovac, I. J.; Ph.D. Thesis, University of Toronto, **1980**; Jones, J. B.; Finch, M. A. W.; Jakovac, I. J. *Can. J. Chem.* **1982**, *60*, 2007–2011.
10. Katzenellenbogen, J. A. *Science* **1976**, *194*, 139–148.
11. Elliott, M.; Janes, N. F. *Chem. Soc. Rev.* **1978**, *7*, 473–505.
12. Jones, G.; Raphael, R. A.; Wright, S. *J. Chem. Soc., Perkin Trans. I* **1974**, 1676–1683.
13. Bartlett, P. A.; Myerson, J. *J. Org. Chem.* **1979**, *44*, 1625–1627.
14. Nakano, A.; Takimoto, S.; Inanaga, J.; Katsuki, T.; Ouchida, S.; Inoue, K.; Aiga, M.; Okukado, N.; Yamaguchi, M. *Chem. Lett.* **1979**, 1019–1020.
15. Collum, D. B.; McDonald, III, J. H.; Still, W. C. *J. Am. Chem. Soc.* **1980**, *102*, 2118–2120.
16. Jakovac, I. J.; Ng, G.; Lok, K. P.; Jones, J. B. *J. Chem. Soc., Chem. Commun.* **1980**, 515–516; Jakovac, I. J.; Goodbrand, H. B.; Lok, K. P.; Jones, J. B. *J. Am. Chem. Soc.* **1982**, *104*, 4659–4665.

HOMOCONJUGATE ADDITION OF NUCLEOPHILES TO CYCLOPROPANE-1,1-DICARBOXYLATE DERIVATIVES: 2-OXO-1-PHENYL-3-PYRROLIDINECARBOXYLIC ACID

(3-Pyrrolidinecarboxylic acid, 2-oxo-1-phenyl-)

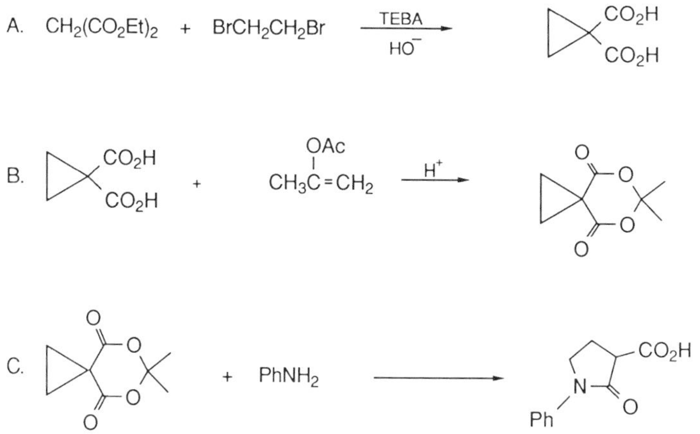

A. $CH_2(CO_2Et)_2$ + $BrCH_2CH_2Br$ $\xrightarrow[HO^-]{TEBA}$ (cyclopropane with CO_2H, CO_2H)

B. (cyclopropane with CO_2H, CO_2H) + $CH_3\overset{OAc}{\underset{|}{C}}=CH_2$ $\xrightarrow{H^+}$ (dioxaspiro dione structure)

C. (dioxaspiro dione structure) + $PhNH_2$ $\longrightarrow$ (pyrrolidinone with CO_2H, $N-Ph$, O)

Submitted by RAJENDRA K. SINGH and SAMUEL DANISHEFSKY[1]
Checked by M. R. CZARNY and M. F. SEMMELHACK

1. Procedure

A. *Preparation of cyclopropane 1,1-dicarboxylic acid (1)*. To a 1-L solution of aqueous 50% sodium hydroxide (Note 1), mechanically stirred in a 2-L, three-necked flask, was added, at 25°C, 114.0 g (0.5 mol) of triethylbenzylammonium chloride (Note 2). To this vigorously stirred suspension was added a mixture of 80.0 g (0.5 mol) of diethyl malonate and 141.0 g (0.75 mol) of 1,2-dibromoethane all at once. The reaction mixture was vigorously stirred for 2 hr (Note 3). The contents of the flask were transferred to a 4-L Erlenmeyer flask by rinsing the flask with three 75-mL portions of water. The mixture was magnetically stirred and cooled with an ice bath to 15°C, and then carefully acidified by dropwise addition of 1 L of concentrated hydrochloric acid. The temperature of the flask was maintained between 15 and 25°C during acidification. The aqueous layer was poured into a 4-L separatory funnel and extracted three times with 900 mL of ether. The aqueous layer was saturated with sodium chloride and extracted three times with 500 mL of ether. The ether layers were combined, washed with 1 L of brine, dried ($MgSO_4$), and decolorized with activated carbon. Removal of the solvent by rotary evaporation gave 55.2 g of a semisolid residue. The residue was triturated with 100 mL of benzene. Filtration of this mixture gave 43.1–47.9 g (66–73%) of **1** as white crystals, mp 137–140°C.

B. *6,6-Dimethyl-5,7-dioxaspiro[2.5]octane-4,8-dione (2)*. A suspension of 39.0 g (0.30 mol) of **1** and 33.0 g (0.33 mol) of freshly distilled isopropenyl acetate was stirred vigorously (magnetic stirrer). To this suspension was added dropwise over a period of 30 min, 0.5 mL of concentrated sulfuric acid. While being stirred for an additional 30 min, the solution became clear yellow, and then partly solidified after being kept at 5°C for 24 hr. After addition of 50 mL of cold water, the precipitated solid was filtered, washed with 10 mL of cold water, and air-dried to give 30.9 g of crude spiroacylal 2. The filtrate was extracted three times with 50-mL portions of ether. The combined organic layers were carefully washed with 50 mL of brine, dried (MgSO$_4$), and decolorized with activated carbon. Evaporation of the solvent gave an additional 7.8 g of spiroacylal 2 as a yellow solid. The combined samples of crude spiroacylal (38.7 g) were recrystallized from 110 mL of hexane and 25 mL of benzene to give 28.7–31.5 g (55–61%) of 2 as colorless needles, mp 65–67°C. Concentration of the above mother liquor to ca. 40 mL gave 0.80 g of a second crop of spiroacylal 2 as slightly yellow crystals, mp 58–60°C.

C. *2-Oxo-1-phenyl-3-pyrrolidinecarboxylic acid (3)*. To 1.70 g (10 mmol) of spiroacylal 2 was added 2.79 g (3 mmol) of aniline. The mixture became a homogeneous orange solution after 15 min and was allowed to stir at room temperature for 12 hr. The resulting crystalline mass was diluted with 150 mL of chloroform, washed three times with 10 mL of aqueous 10% hydrochloric acid, washed once with 20 mL of brine, dried (MgSO$_4$), and decolorized with a small amount of activated carbon. Concentration of the organic layer by rotary evaporation gave 5.27 g of a brown residue, which was recrystallized from chloroform-hexane to afford 4.86–5.07 g (79–82%) of the pyrrolidinone 3 as white crystals, mp 146–148°C (dec) (Note 4).

2. Notes

1. Aqueous 50% sodium hydroxide was prepared by dissolving 500 g of sodium hydroxide pellets in water and diluting to 1 L.

2. This compound is commercially available from Aldrich Chemical Company, Inc. Alternatively, it can be made very cheaply and simply by mixing benzyl chloride (1 equiv) with triethylamine (2.5 equiv). The mixture is allowed to stand for 4–7 days at room temperature. Filtration of the solid and drying in vacuum give triethylbenzylammonium chloride suitable for use in nearly quantitative yield.

3. Some exothermicity results on mixing, causing the temperature to rise to ca. 65°C.

4. At this temperature, after a few minutes, the lactam acid 3 suffers smooth decarboxylation to afford N-phenyl-2-pyrrolidinone. Alternatively, the acid can be esterified (methanol-hydrochloric acid), and the resulting 1-phenyl-3-carbomethoxypyrrolidin-2-one can be used for the introduction of other functionality at the 3-position.

3. Discussion

Previously cyclopropane-1,1-dicarboxylic acid had been prepared[2-4] by hydrolysis of the corresponding diester. The preparation of 1,1-dicarboalkoxycyclopropanes by a conventional double alkylation of diethyl malonate with 1,2-dibromoethane was severely complicated by the recovery of unreacted diethylmalonate. This required a rather difficult distillation to separate starting material and product. In fact, many commercially offered lots of cyclopropane diester contain extensive amounts of diethyl malonate. Furthermore, preparation of the diacid required a separate and relatively slow saponification of the diester.[5]

The procedure described here for compound **1** is a scale-up of a published method.[6] Phase-transfer catalysis[7] and concentrated alkali are used to effect a one-pot conversion of diethyl malonate to the cyclopropane diacid, which is easily obtained by crystallization. Apparently alkylation of the malonate system occurs either at the diester or monocarboxylate, monoester stage since the method fails when malonic acid itself is used as the starting material. This method of synthesizing doubly activated cyclopropanes has been extended to the preparation of 1-cyanocyclopropanecarboxylic acid (86%) by the use of ethyl cyanoacetate and 1-acetylcyclopropanecarboxylic acid (69%) by use of ethyl acetoacetate.[6]

The spiroacylal **2** is potentially a valuable agent in organic synthesis.[8] It is readily attacked by a variety of nucleophiles, including pyridine, to give ring-opened products bearing a stabilized carbanion. It is thus seen to be a synthetic equivalent of $\oplus CH_2-CH_2-CH(CO_2H)_2$ and $\oplus CH_2(CH_2)_2-CO_2H$, i.e., a homo-Michael acceptor. The general reaction is

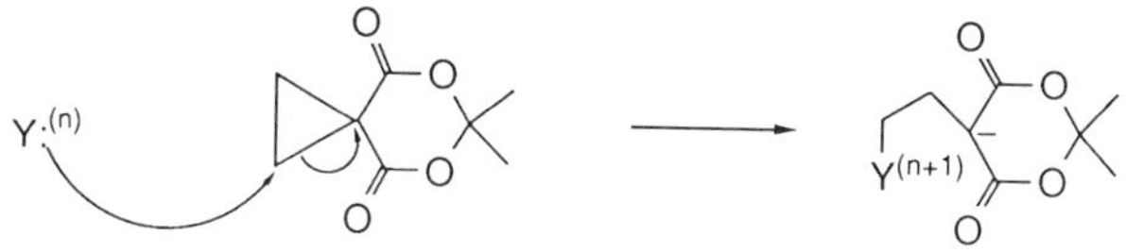

2

where Y = aniline, piperidine, pyridine, mercaptide, enolate, etc. Spiroacylal **2** was designed under the rationale that the constraint of the carbonyl groups into a conformation in which overlap of their π-orbitals with the "bent bonds" of the cyclopropane is assured should dramatically increase the vulnerability of the cyclopropane toward nucleophilic attack.[8] Experimental support for this notion is abundant.[8] Spiroacylal **2** is considerably more reactive than 1,1-dicarbethoxycyclopropane in such reactions. For instance, reaction of **2** with piperidine occurs at room temperature. The corresponding reaction in the case of the diester is conducted at 110°C.[5] Reactions with enolates also occur under mild conditions.[8] Compound **2** reacts with the weak nucleophile pyridine at room temperature to give a betaine.[8] An illustrative mechanism for the reaction of the acylal **2** with aniline to afford 2-oxo-1-phenyl-3-pyrrolidinecarboxylic acid (**3**) is

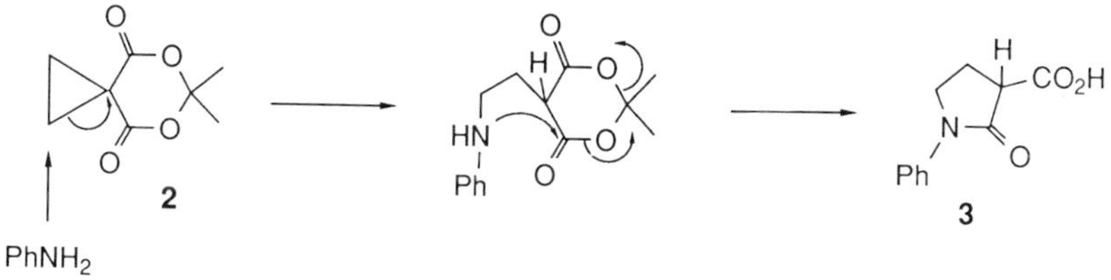

The synthesis of the spiroacylal **2** from the diacid **1** follows a procedure used by Scheuer in a different context.[9]

1. Department of Chemistry, Yale University, New Haven, CT 06520.
2. Bone, W. A.; Perkin, W. H. *J. Chem. Soc.* **1895,** *67,* 108.
3. Stewart, J. M.; Westbert, H. H. *J. Org. Chem.* **1965,** *30,* 1951–1955.
4. Dolfini, J. E.; Menich, K.; Corliss, P.; Cavanaugh, R.; Danishefsky, S.; Chakrabarty, S. *Tetrahedron Lett.* **1966,** 4421–4426.
5. Abell, P. I.; Tien, R. *J. Org. Chem.* **1965,** *30,* 4212–4215.
6. Singh, R. K.; Danishefsky, S. *J. Org. Chem.* **1975,** *40,* 2969–2970.
7. Dockx, J. *Synthesis* **1973,** 441–456.
8. Danishefsky, S.; Singh, R. K. *J. Am. Chem. Soc.* **1975,** *97,* 3239–3241.
9. Scheuer, P. J.; Cohen, S. G. *J. Am. Chem. Soc.* **1958,** *80,* 4933–4938.

SYNTHESIS OF 1,4-DIKETONES FROM SILYL ENOL ETHERS AND NITROOLEFINS: 2-(2-OXOPROPYL)CYCLOHEXANONE

[Cyclohexanone, 2-(2-oxopropyl)]

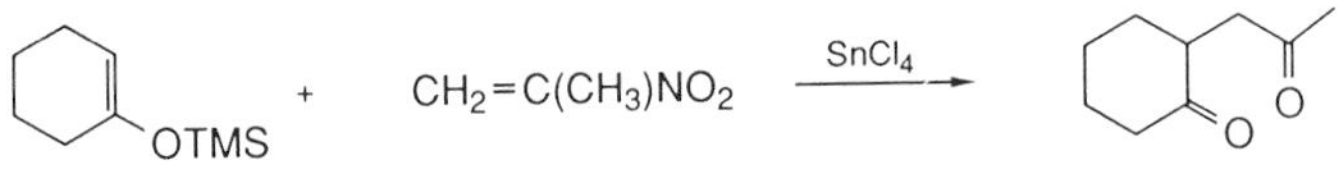

$$\text{(silyl enol ether)} + CH_2=C(CH_3)NO_2 \xrightarrow{\ SnCl_4\ } \text{(diketone product)}$$

Submitted by MASAAKI MIYASHITA, TETSUJI YANAMI, and AKIRA YOSHIKOSHI[1]
Checked by DONALD HILVERT, STEFAN KWIATKOWSKI, and DIETER SEEBACH

1. Procedure

Caution! This preparation should be carried out in a hood since 2-nitropropene is a powerful lachrymator and anhydrous stannic chloride is a skin irritant.

A 1-L, three necked, round-bottomed flask is fitted with a magnetic stirring bar and a pressure-equalizing dropping funnel to which is attached an oil bubbler, a rubber septum, and an argon or nitrogen inlet to maintain a static inert gas atmosphere in the reaction vessel throughout the reaction. The flask and dropping funnel are charged with 500 mL of dry methylene chloride and 40 mL (34 g, 0.20 mol) of 1-trimethylsiloxy-1-cyclohexene (Note 1), respectively. The flask is flushed with dry inert gas and immersed in a cooling bath at ca. −78°C (acetone or 2-propanol-dry ice). Stirring is started and 23 mL (52.1 g, 0.20 mol) of anhydrous stannic chloride (Note 2) is added rapidly through the rubber septum by means of a syringe. Then 20.0 mL (21.0 g, 0.23 mol) of 2-nitropropene (Note 3) is added through the rubber septum by a syringe over a period of 5–10 min, giving a green solution. The reaction mixture is further stirred at −78°C for 20 min, and then the silyl enol ether is added dropwise to the mixture over 1 hr, giving a faint yellow solution. After completion of the addition the resulting solution is stirred at ca. −78°C for an additional hour; then the bath temperature is gradually warmed to −5°C over a period of 3–3.5 hr while the stirring is continued (Note 4). The inert gas flow is stopped, the dropping funnel is replaced by a condenser, the magnetic stirrer is removed, and the flask is equipped with a heating mantle and an overhead stirring device. Then 280 mL of water are added, and the resulting heterogeneous mixture is vigorously stirred at reflux for 2 hr (Note 5). The mixture is subsequently cooled to room temperature and then poured into a 1-L separatory funnel and the methylene chloride layer is separated from the water. The aqueous layer is extracted once with 100 mL of methylene chloride, and the combined organic layers are washed twice with 160-mL portions of cold water (Note 6) and once with saturated brine, dried over anhydrous magnesium sulfate, and filtered. The solvent is removed on a rotary evaporator and the residual oil is distilled through a 10-cm Vigreux column under reduced pressure to yield 18.7–21.5 g (61–70%) of 2-(2-oxopropyl)cyclohexanone as a fragrant yellow liquid, bp 84–85°C (0.8 mm), n_D^{19} 1.4671 [lit.[5] bp 91–93°C (1.1 mm), n_D^{25} 1.4655] (Note 7).

2. Notes

1. This silyl enol ether was prepared according to the procedure of House,[2a] 80%, bp 75°C (21 mm) [lit.[2a] 74–75°C (20 mm)].

2. A fresh bottle of commercial anhydrous stannic chloride purchased from Wako Pure Chemical Industries, Ltd., Japan, or from Fluka AG, Buchs, Switzerland, was used without purification.

3. 2-Nitropropene[3] was freshly prepared before use.

4. The yellow solution becomes green on warming and finally turns yellow.

5. On addition of water the mixture turns purple, and after refluxing it becomes brown.

6. Although an insoluble white substance appears in the aqueous washings, it is discarded.

7. The checkers found refractive indices n_D^{19} 1.468 and 1.4665 or n_D^{25} 1.4657 and 1.4649.

3. Discussion

This procedure illustrates a recently published, simple, general method for the synthesis of 1,4-diketones from silyl enol ethers and nitroolefins.[4] 2-(2-Oxopropyl)cyclohexanone has been prepared by the reaction of the pyrrolidine enamine of cyclohexanone with bromoacetone (40%)[5] and by several other multistep processes.[6-8] However, the overall yields obtained by these routes have never exceeded 50% and some of the methods are laborious for large-scale preparations. The present method illustrates a mild and convenient one-pot reaction for the preparation of 1,4-diketones. In addition, the starting materials are readily accessible, the reaction proceeds regioselectively, and the yields of product are generally high. This process consists of the initial Michael addition of silyl enol ethers to nitroolefins, followed by a Nef reaction of the nitronate esters.[4] The scope of the reaction is shown in Table I. The 1,4-diketones thus obtained have been converted into corresponding cyclopentenones in high yields.[4]

TABLE I
1,4-Diketones Prepared from Silyl Enol Ethers and Nitroolefins

Silyl Enol Ether	Nitroolefin	Lewis Acid	1,4-Diketone	Yield (%)
⟨cyclohexene-OTMS⟩	2-Nitro-1-butene	TiCl₄	⟨diketone⟩	76
⟨methylcyclohexene-OTMS⟩	2-Nitropropene	TiCl₄	⟨diketone⟩	70
⟨methylcyclohexene-OTMS⟩	2-Nitro-1-butene	TiCl₄	⟨diketone⟩	82
⟨cyclopentene-OTMS⟩	2-Nitropropene	SnCl₄	⟨diketone⟩	70

1. Chemical Research Institute of Non-Aqueous Solutions, Tohuku University, Sendai 980, Japan.
2. (a) House, H. O.; Czuba, L. J.; Gall, M.; Olmsted, H. D. *J. Org. Chem.* **1969,** *34*, 2324–2336; (b) Ito, Y.; Fujii, S.; Nakatsuka, M.; Kawamoto, F.; Saegusa, T. *Org. Synth., Coll. Vol. VI* **1988,** 327.
3. Miyashita, M.; Yanami, T.; Yoshikoshi, A. *Org. Synth., Coll. Vol. VII* **1990,** 396.
4. Miyashita, M.; Yanami, T.; Yoshikoshi, A. *J. Am. Chem. Soc.* **1976,** *98*, 4679–4681.
5. Baumgarten, H. E.; Creger, P. L.; Villars, C. E. *J. Am. Chem. Soc.* **1958,** *80*, 6609–6612.
6. Grieco, P. A.; Pognowski, C. S. *J. Org. Chem.* **1974,** *39*, 732–734.
7. Risali, A.; Forchiassin, M.; Valentin, E. *Tetrahedron* **1968,** *24*, 1889–1898.
8. Brust, D. P.; Tarbell, D. S. *J. Org. Chem.* **1966,** *31*, 1251–1255.

ASYMMETRIC HYDROGENATION OF KETOPANTOYL LACTONE: D-(−)-PANTOYL LACTONE

[2(3H)-Furanone, dihydro-3-hydroxy-4,4-dimethyl-]

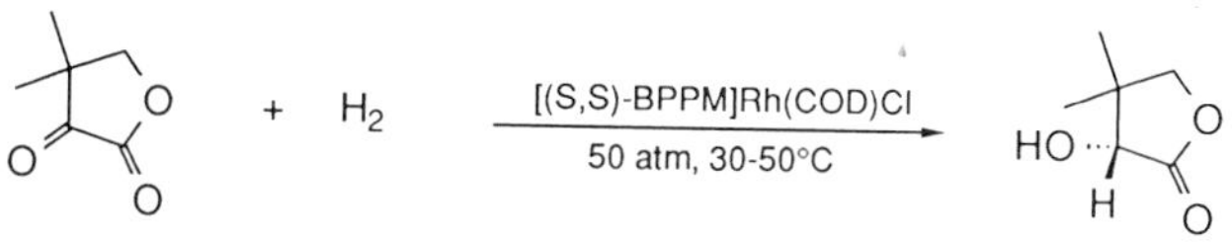

Submitted by I. OJIMA, T. KOGURE, and Y. YODA[1]
Checked by LARRY K. TRUESDALE, STANLEY D. HUTCHINGS, and GABRIEL SAUCY

1. Procedure

A. *Preparation of catalyst solution.* A 250-mL, round-bottomed flask fitted with a septum and magnetic stirring bar is charged with 486.9–488.2 mg (Note 1) (0.985–0.990 × 10^{-3} mol) of chloro(1,5-cyclooctadiene)rhodium(I) dimer (Note 2) and, under argon (Note 3), with 1.20 g (2.15 × 10^{-3} ml) of (2S, 4S)-N-*tert*-butyloxycarbonyl-4-diphenylphosphino-2-diphenylphosphinomethylpyrrolidine, (S,S)-BPPM (Note 4). The sealed flask is charged by cannula, under argon, with 150 mL of degassed benzene (Note 5) and stirred under argon for 15 min at room temperature. The catalyst is transferred by cannula, under argon, into the autoclave (see below).

B. *Asymmetric hydrogenation.* A stainless steel stirred autoclave with a total volume of 500 mL is charged with 25.6 g (0.2 mol) of ketopantoyl lactone (Notes 6–9). The autoclave is flushed with argon and the catalyst solution (see above) is added by cannula, under argon. The autoclave is sealed and hydrogenation is carried out at 40°C, 750-psig hydrogen and 950–1050 rpm for 48 hr (Note 10). Care should be taken to flush all the lines before connecting to the autoclave. After the autoclave is cooled to room temperature, it is vented and opened. The reaction mixture is then transferred to a 500-mL, round-bottomed flask and most of the solvent is removed by rotary evaporator. Distillation (Note 11) of this reddish solid affords 24–25.6 g (92–98%) (Note 1) of D-(−)-pantoyl lactone: bp 90–110°C (4 cm); $[\alpha]_D^{25}$ −39.3° to −42.4° (c 2, H_2O) (Note 12) (78 to 84% e.e.) (Notes 1 and 10).

The pantoyl lactone thus obtained (25.41 g), $[\alpha]_D^{25}$ −40.8° (80.5% e.e.) (Note 13) is refluxed with 75 mL benzene and 290 mL of UV-grade hexanes. The cloudy solution is stirred briskly overnight as solids form. Filtration of the solids and drying for 3 hr at 0.25 mm, 30°C in a vacuum oven affords 21.51 g of product; $[\alpha]_D^{25}$ −47.7° (94.27% e.e.). This material is again refluxed and crystallized (Note 14) from 30 mL of benzene and 116 mL of UV-grade hexanes to afford 19.97 g (77%) of product; $[\alpha]_D^{25}$ −49.87° (98.5% e.e.). Anal. calcd. for $C_6H_{10}O_3$: C, 55.37; H, 7.75. Found: C, 55.34; H, 7.57 (Note 15).

2. Notes

1. The reaction was done four times at this scale. The range represents the high and low amounts of catalyst precursor used over the four reactions.

2. Chloro(1,5-cyclooctadiene)rhodium (I) dimer is commercially available from Strem Chemicals, Inc., Newburyport, MA.

3. The addition and measurement of (S,S)-BPPM is most conveniently done in a dry box or glove bag under argon. A Schlenk tube apparatus can be used if these are not available.

4. (2S,4S)-N-tert-Butyloxycarbonyl-4-diphenylphosphino-2-diphenylphosphino-methylpyrrolidine, (S,S)-BPPM,[2,3] is commercially available from E. Merck, Darmstadt, West Germany and Kanto Chemical Company, Tokyo, Japan.

5. The submitters claim that tetrahydrofuran can also be used giving D-(−)-pantoyl lactone with 83.3–84.8% e.e. This was not checked.

6. (a) Ketopantoyl lactone is readily prepared by the oxidation of d,l-pantoyl lactone (Note 8) with bromine as follows.[4] Into a 500-mL, round-bottomed flask fitted with a mechanical stirrer, dropping funnel, condenser, and thermometer is charged 13.0 g (0.1 mol) of d,l-pantoyl lactone (Note 7) and 150 mL of carbon tetrachloride. The mixture is stirred and heated to reflux. Bromine (16.5 g, 0.103 mol) in 100 mL of carbon tetrachloride is slowly added from the dropping funnel over 3 hr. After 8 hr, generation of hydrogen bromide subsides and the red color of bromine almost disappears, indicating completion of the reaction. Dry air is bubbled through the solution to remove the remaining hydrogen bromide and the small quantity of bromine. The solvent is removed with a rotary evaporator and further evacuated with a vacuum pump to afford 12.8 g (100%) (Note 9) of almost pure ketopantoyl lactone. One recrystallization from 150 mL of carbon tetrachloride (heat to reflux and then cool to − 10°C) affords 11.6–12.2 g (90–95%) of pure ketopantoyl lactone, mp 66–67.5°C. (b) An alternative procedure preferred by the checkers to prepare highly pure ketopantoyl lactone follows. A 5-L, round-bottomed flask equipped with a mechanical stirrer, condenser, thermometer, and dropping funnel is charged with 700 g of Ca(OCl)$_2$ (analyzed as 20% active chlorine) and 1.5 L of acetonitrile dried overnight over Linde 4A molecular sieves. d,l-Pantoyl lactone (165 g) (Note 7) is dissolved in 500 mL of dried acetonitrile. The Ca(OCl)$_2$ slurry is stirred while ~ ¹/₇ of the pantoyl lactone solution is added. The temperature of the exothermic reaction is controlled with an ice bath to below 35°C. The remainder of the pantoyl lactone solution is added in ~ 75-mL aliquots over 25–30 min while taking care to control the temperature. The ice bath is removed and stirring is continued. After 3.5 hr, GLC analysis indicates 94% product. The reaction mixture is filtered and the solids are rinsed with acetonitrile. The crude product is dried on a rotary evaporator and further evacuated overnight to yield 105.6 g. The material is dissolved in methylene chloride, dried over Na$_2$SO$_4$, filtered through Celite, and concentrated under reduced pressure. The crude product (94.1 g) is then purified by refluxing and stirring overnight with 500 mL of ethyl ether. The slurry is allowed to stand at 5°C. The solids are filtered, washed with cold ether, and dried in a vacuum oven at room temperature for 6 hr to afford 80.8 g (86% recovery) of pure ketopantoyl lactone. Ketopantoyl lactone has also been reported to be easily prepared by the oxidation of d,l-pantoyl lactone with alkaline metal hypochlorite[5] or by reaction of sodium dimethylpyruvate with formaldehyde in the presence of potassium carbonate.[6]

7. *d,l*-Pantoyl lactone is very hygroscopic. Care must be taken during this oxidation that dry starting material is used and that water does not contaminate the reaction; the yield will fall drastically probably because of hydrolysis.

8. *d,l*-Pantoyl lactone is commercially available from Sigma Chemical Company, St. Louis, MO 63178.

9. GLC analysis indicates 97–98% yield. A simple GLC system to determine the relative completion of the reaction is a 3-ft × ⅛-in column packed with 10% Carbowax 20 M on Anakrom Q 90/100. With this column a program of 150 to 210°C at 8°/min and a 7-min hold gives baseline separation of ketopantoyl lactone at 2.75–3.2 min and pantoyl lactone at 3.7–3.95 min. The flow rate of the carrier gas is 20 mL/min.

10. When ketopantoyl lactone prepared by method 6b was used, the reaction was complete in 2 hr.

11. A bulb-to-bulb distillation using a Kugelrohr apparatus is most convenient.

12. The reported maximum rotation, $[\alpha]_D^{25}$, max, for pure D-(−)-pantoyl lactone is −50.7° (*c* 2.05, H_2O).[7]

13. The enantiomeric excess and the speed of reduction are both greatly influenced by impurities that are not detectable by GLC. Digestion in ether seems to remove these impurities better than recrystallization from CCl_4.

14. This recrystallization is very temperature-sensitive; for example, this purification was done at ambient temperature (28–30°C). The first recrystallization removes 3.7 g of *d,l*-pantoyl lactone and 0.2 g of D-(−)-pantoyl lactone. When the recrystallization was done at 5°C, twice as much solvent served to remove only 4.2 g of *d,l*-pantoyl lactone and none of the D-isomer.

15. The procedure described is a scaled-up version (20×) of the original submission worked out by the checkers.

3. Discussion

D-(−)-Pantoyl lactone is a key intermediate for the synthesis of pantothenic acid, which is a member of the vitamin B complex and is an important constituent of Coenzyme A. Although D-(−)-pantoyl lactone has been obtained by classical optical resolution using quinine, ephedrine, and other chiral amines, catalytic asymmetric synthesis appears to be more effective from a practical point of view.[8] One problem of the present approach was the availability of ketopantoyl lactone, but the recent method developed by Hoffmann–La Roche,[6] consisting in the condensation of sodium dimethylpyruvate with formaldehyde, may open a commercial route to ketopantoyl lactone. Thus, asymmetric reduction of ketopantoyl lactone now becomes an important route to D-(−)-pantoyl lactone. Asymmetric reduction of ketopantoyl lactone can also be achieved with microorganisms. For example, microbial reduction of ketopantoyl lactone using baker's yeast was reported to give ca. 72% e.e.,[9] and the specific strain of an ascomycete, *Byssochlamys fulva*, was reported to give D-(−)-pantoyl lactone with 95–100% e.e.[9] However, the isolation procedure from aqueous media in these microbial reductions—specifically: extraction, recovery of raw materials, and purification—is very troublesome because of the high solubility of the product in water. Consequently, the present method has considerable advantages from a synthetic point of view; for instance, (a) the yield of the reaction is virtually 100% and (b) isolation of the product is simple and convenient since the reaction is carried out in small amounts of nonaqueous media.

The present method has been successfully applied[10] to the asymmetric reduction of various α-keto carboxylates and α-keto lactones.

1. Sagami Chemical Research Center, Nishi-Ohnuma 4-4-1, Sagamihara, Kanagawa 229, Japan. Present address of I. Ojima: Department of Chemistry, SUNY Stony Brook, Stony Brook, NY 11794.
2. Achiwa, K. *J. Am. Chem. Soc.* **1976,** *98,* 8265.
3. Ojima, I.; Kogure, T.; Yoda, N. *J. Org. Chem.* **1980,** *45,* 4728.
4. Ojima, I. (Sagami Chemical Research Center), Japan Kokai Tokkyo Koho, JP 79 88257, 1979.
5. Schmid, M. (Hoffmann–La Roche), Swiss Patent Appl., 5883/78-3, 1978; Japan Kokai Tokkyo Koho, JP 79 160303, 1979.
6. Fizet, C. (Hoffmann–La Roche), Eur. Patent Appl. EP, 50, 721, 1982; *Chem. Abstr.* **1982,** *97,* 72241e.
7. (a) Stiller, E. T.; Harris, S. A.; Finkelstein, J.; Keresztesy, J. C.; Folkers, K. *J. Am. Chem. Soc.* **1940,** *62,* 1785; (b) Hill, R. K.; Chan, T. H. *Biochem. Biophys. Res. Commun.* **1970,** *38,* 181.
8. (a) Ojima, I.; Kogure, T.; Terasaki, T.; Achiwa, K. *J. Org. Chem.* **1978,** *43,* 3444; (b) Townsend, J. M.; Valentine, D., Jr. (Hoffmann–La Roche), U.S. Patent 4343741, 1982; *Chem. Abstr.* **1983,** *98,* 34761n.
9. Lanzilotta, R. P.; Bradley, D. G.; McDonald, K. M. *Appl. Microbiol.* **1974,** *27,* 130.
10. Ojima, I.; Kogure, T.; Achiwa, K. *J. Chem. Soc., Chem. Commun.* **1977,** 428.

PREPARATION OF 4-ALKYL-AND 4-HALOBENZOYL CHLORIDES: 4-PENTYLBENZOYL CHLORIDE

(Benzoyl chloride, 4-pentyl-)

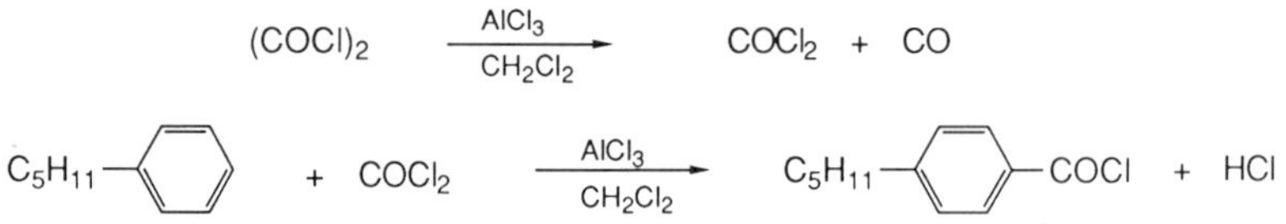

$$(COCl)_2 \xrightarrow[\text{CH}_2\text{Cl}_2]{\text{AlCl}_3} COCl_2 + CO$$

$$C_5H_{11}\text{—}\bigcirc + COCl_2 \xrightarrow[\text{CH}_2\text{Cl}_2]{\text{AlCl}_3} C_5H_{11}\text{—}\bigcirc\text{—}COCl + HCl$$

Submitted by MARY E. NEUBERT and D. L. FISHEL[1]
Checked by VINAY CHOWDHRY and R. E. BENSON

1. Procedure

Caution! Operations prior to vacuum distillation of the product should be done in a good hood since phosgene, carbon monoxide, and hydrogen chloride are present (Note 1). Rubber gloves should also be used to avoid contact with the reagents.

A 100-mL, three-necked, round-bottomed flask is fitted with a mechanical stirrer, 100-mL pressure-equalized addition funnel (Note 2) to which is attached a drying tube (Note 3), and a rubber septum. Dry methylene chloride (27 mL, Note 4) and 8.9 g

(0.067 mol) of aluminum chloride (Note 5) are added to the flask, stirring is begun, and 17.1 g (11.5 mL, 0.135 mol) of oxalyl chloride (Note 6) is added over 5 min by means of a syringe introduced through the septum (Note 7). The septum is replaced by a thermometer and a solution of 10 g (11.6 mL, 0.067 mol) of amylbenzene (Note 8) in 40 mL of dry methylene chloride is added dropwise over 1 hr with stirring while the temperature is maintained at 20–25°C. The reaction mixture is reduced to about half of the original volume by distillation of solvent and excess oxalyl chloride and/or phosgene (Note 9). Approximately 40 mL of fresh dry methylene chloride is added to the flask and the solution is cooled to 0°C in an ice-salt bath. The cold solution is slowly poured onto a stirred mixture of 170 g of crushed ice and 10 g of calcium chloride at a rate to maintain the temperature below 5°C (Note 10). The organic layer is rapidly separated from the aqueous layer and dried over anhydrous sodium sulfate. The mixture is filtered and the solvent is removed by distillation at reduced pressure. The residual liquid is dissolved in 50 mL of ether, and the resulting solution is cooled to 0°C, extracted with 5 mL of cold (0°C) 5% potassium hydroxide solution, and then washed twice with 15-mL portions of cold (0°C) water (Note 11). The ether solution is separated and dried over anhydrous sodium sulfate. The mixture is filtered and the solvent is removed by distillation at reduced pressure (Note 12). Distillation through a Vigreux column affords a small forerun and then 7.80–7.82 g (55%) of pure 4-pentylbenzoyl chloride, bp 95°C (0.20 mm) (Notes 13 and 14). The acid chloride is stable if kept in a sealed container to prevent hydrolysis.

2. Notes

1. Both phosgene and carbon monoxide were identified in IR spectra of gases generated from an equimolar mixture of oxalyl chloride and aluminum chloride at room temperature.

2. The submitters used a constant addition funnel.

3. Molecular sieves 4A available from Davison Chemical Co. were used.

4. The submitters state that the use of predried methylene chloride (stored overnight over 4A molecular sieves) gave the best results.

5. Use of either an excess of aluminum chloride or partially hydrolyzed aluminum chloride gives larger amounts of the by-product diaryl ketone at the expense of the acid chloride. The checkers used freshly opened containers of the anhydrous material available from Fisher Scientific.

6. Oxalyl chloride should be distilled if it is colored or contains solid. Studies by the submitters have shown that an excess of oxalyl chloride is needed for maximum conversion of the alkylbenzene to acid chloride. The checkers used oxalyl chloride available from Eastman Organic Chemicals.

7. The submitters added the oxalyl chloride through the funnel used to add amylbenzene.

8. The checkers used product available from Aldrich Chemical Company, Inc.

9. If excess oxalyl chloride (and/or phosgene) is not removed, the vigorous reaction with water during decomposition of the aluminum chloride complex contributes to hydrolysis of the product acid chloride by increasing the time needed to

complete this step. The more dilute solution achieved by additional solvent helps to prevent this hydrolysis as does maintenance of a low temperature during decomposition of the complex.

10. The calcium chloride-ice mixture helps to maintain a low temperature.

11. Changing the solvent to ether prior to the base extraction step (to remove carboxylic acid formed by hydrolysis) inhibits emulsion formation, particularly with the higher alkyl-substituted products.

12. The procedure may be interrupted at this point if the crude acid chloride is protected from moisture, although highest yields are obtained if distillation is done at once. Failure to remove water (even that associated with the sodium sulfate drying agent) before storage may result in anhydride formation during distillation because of the presence of free carboxylic acid.

13. Infrared analysis (neat, film) shows a carbonyl doublet at 1740, 1770 cm^{-1}, typical of 4-substituted benzoyl chlorides and thought to be due to Fermi resonance.[2,3] Contamination of the product with the anhydride can be detected by a doublet at 1720 and 1780 cm^{-1}, with the ketone by a singlet at 1650 cm^{-1}, and with the acid by a singlet at 1690 cm^{-1}.

14. The submitters obtained the product in 75% yield.

3. Discussion

This method is based on that of Fahim,[4] who isolated 4-alkylbenzoic acids in 40–60% yields by hydrolysis of the corresponding acid chlorides. The present improved procedure includes those conditions believed to be optimum for a one-step synthesis of 4-substituted benzoyl chlorides in good yields and apparently free of positional isomers, as indicated by gas chromatography/mass spectroscopy as well as ^{1}H and ^{13}C NMR analyses. The procedure has been used successfully for the synthesis of 4-halobenzoyl chlorides and several other aryl acid chlorides,[5,6] as well as for 4-alkylbenzoyl chlorides up through the decyl derivative. Some of these results are summarized in Table I. The reaction has been run on a 1-mol scale by the submitters with no difficulty.

The major by-product that can be isolated (3–6%) from the residue after distillation is the 4,4'-disubstituted benzophenone; formation of the ketone is minimized by using excess oxalyl chloride and by slow addition of a dilute solution of the alkylbenzene to the acylating agent. Ambient temperatures (20–25°C) appear to give optimum results; higher temperatures favor ketone formation and lower temperatures result in incomplete reaction for reasonable reactions times. Numerous additional reactions using 4-alkyl-benzenes indicate that the alkylbenzene solution can be added more rapidly as long as the temperature is maintained at 20–25°C. Maximum yields are obtained when the addition time is 30–60 min, apparently because longer times lead to loss of phosgene and reaction of the acid chloride with the alkylbenzene to give the ketone.

This method cannot be used to prepare acid chlorides of aromatic systems that contain substituents strongly activating for electrophilic substitution such as alkoxy groups (the major product is ketone), deactivating ring substituents (no reaction), or those that form stable acylium ions (major product is carboxylic acid). Mesitoic acid rather than the acid chloride was isolated from the acylation of mesitylene using these conditions, which confirms the results previously reported using similar conditions.[7]

TABLE I
4-Substituted Benzoyl Chlorides from Substituted Benzenes

Substituent	Yield (%)	bp (°C) (mm)
C_4H_9	66.5	113 (1.7)
i-C_4H_9	77.5	115 (1.6)
C_5H_{11}	75.3	136 (3.2)
C_6H_{13}	80.3	143 (1.3)
C_7H_{15}	79.3	160 (5)
C_9H_{19}	71.8	182 (2.6)
$C_{10}H_{21}$	68.0	169 (0.6)
F	84.4	50 (1.1)
Cl	77.7	86 (2.1)
Br	75.9	103 (2.5)
I	74.1	100 (0.7)

Previously, the most widely used method for preparation of 4-alkylbenzoyl chlorides on a laboratory scale has been from the benzoic acids obtained by oxidation of aromatic ketones, usually 4-alkylacetophenones.[8-14] The latter are usually prepared by acylating alkylbenzenes. Although this sequence gives high yields, it is lengthy (three completely separate steps) and the scale is restricted in the second step because of the large volumes required. The submitters state that they were unable to repeat the reported alkylation of toluic acid.[15] Methods that lead to formation of ortho and para isomeric intermediates are inconvenient since they require that the isomers be separated.[16-19]

This method provides easy access to 4-alkylbenzoyl chlorides, which are useful intermediates in the preparation of diaryl esters that have mesomorphic properties.[20] Benzoyl chlorides substituted in the 4-position also serve as starting materials for the preparation of aromatic aldehydes[21] and nitriles,[6,22] whereas the acids, derivable quantitatively from the acid chlorides, are good precursors via the Schmidt reaction to 4-substituted anilines.[23] This method has been used to prepare deuterated liquid crystalline anils from the anilines obtained by Schmidt rearrangement of 4-alkylbenzoic acids[24] and 4'-cyanobiphenyls[25] but was not successful in an attempt to prepare 2-alkyl-7-cyanofluorenes.[26]

1. Liquid Crystal Institute and Chemistry Department, Kent State University, Kent, OH 44242.
2. Flett, M. St. C. *Trans. Faraday Soc.* **1948**, *44*, 767.
3. Rao, C. N. R.; Venkataraghavan, R. *Spectrochim. Acta.* **1962**, *18*, 273.
4. Fahim, H. A. *Nature* **1948**, *162*, 526; *J. Chem. Soc.* **1949**, 520.
5. Neubert, M. E.; Carlino, L. T. *Mol Cryst. Liq. Cryst.* **1977**, *42*, 353.
6. Oh, C. S. "Abstracts of Papers," 174th National Meeting of the American Society, Chicago, IL, Aug. 1977; American Chemical Society: Washington, DC, 1977; Coll. 129.
7. Sokol, P. E. *Org. Synth., Coll. Vol. V* **1973**, 706.
8. Balle, G.; Wagner, H.; Nold, E. U.S. Patent 2 195 198, 1940; *Chem. Abstr.* **1940**, *34*, 5093⁹.
9. Zaki, A.; Fahim. H. *J. Chem. Soc.* **1942**, 307.
10. Martin, H.; Hirt, R.; Neracher, O. U.S. Patent 2 383 874, 1945; *Chem. Abstr.* **1945**, *39*, 5410².
11. Kirchner, F. K.; Bailey, J. H.; Cavallito, C. J. *J. Am. Chem. Soc.* **1949**, *71*, 1210.

12. Ogawa, M. *J. Chem. Soc. Jpn., Ind. Chem. Sect.* **1956**, *59*, 134; *Chem. Abstr.* **1957**, *51*, 1085h.
13. Price, C. C.; Cypher, G. A.; Krishnamurti, I. V. *J. Am. Chem. Soc.* **1952**, *74*, 2987.
14. Reinheimer, J. D.; Taylor, S. *J. Org. Chem.* **1954**, *19*, 802; Reinheimer, J. D.; List Jr., E. W. *Ohio J. Sci.* **1957**, *57*, 26; *Chem. Abstr.* **1957**, *51*, 6552c.
15. Greger, P. L. *J. Am. Chem. Soc.* **1970**, *92*, 1396.
16. Young, W. R.; Haller, I.; Green, D. C. *J. Org. Chem.* **1972**, *37*, 3707.
17. Fuson, R. C.; Larson, J. R. *J. Am. Chem. Soc.* **1959**, *81*, 2149.
18. Rinkes, I. J. *Recl. Trav. Chim. Pays-Bas* **1944**, *63*, 89.
19. Steinstrasser, R. *Z. Naturforsch.* **1972**, *27b*, 774.
20. Neubert, M. E.; Carlino, L. T.; D'Sidocky, R.; Fishel, D. L. in "Liquid Crystals and Ordered Fluids," Johnson, J. F.; Porter, R. S., Eds.; Plenum: New York, 1974; Vol. 2, p. 293.
21. Rachlin, A. I,; Gurien, H.; Wagner, D. P. *Org. Synth., Coll. Vol. VI* **1988**, 1007.
22. Direct conversion using phosphonitrilic chloride, Neubert, M. E.; Ferrato, J. P., unpublished results.
23. Fishel, D. L.; Neubert, M. E. unpublished results.
24. Neubert, M. E. *Mol. Cryst. Liq. Cryst.* **1985**, *129*, 327 (1985).
25. Oh, C. S. In "Liquid Crystals and Ordered Fluids," Vol. 3, Johnson, J. F.; Porter, R. S., Eds., Plenum Press: New York, 1978, p. 53.
26. Davison, I. R.; Hall, D. M.; Sage, I. *Mol. Cryst. Liq. Cryst.* **1985**, *129*, 17.

α-*tert*-ALKYLATION OF KETONES: 2-*tert*-PENTYLCYCLOPENTANONE

(Cyclopentanone, 2-*tert*-pentyl-)

Submitted by M. T. Reetz, I. Chatziiosifidis, F. Hübner, and H. Heimbach[1]
Checked by Kevin Kunnen and Carl R. Johnson

1. Procedure

A. *1-Trimethylsiloxycyclopentene.*[2] A 1-L, two-necked, round-bottomed flask is equipped with a mechanical stirrer and a reflux condenser having a drying tube (calcium chloride). The flask is charged with 200 mL of dimethylformamide (Note 1), 45 g (0.54 mol) of cyclopentanone (Note 2), 65.5 g (0.6 mol) of chlorotrimethylsilane (Note 2), and 185 mL (1.33 mol) of triethylamine (Note 1), and the mixture is refluxed for 17 hr (Note 3). The mixture is cooled, diluted with 350 mL of pentane, and washed four times with 200-mL portions of cold saturated aqueous sodium hydrogen carbonate. The aqueous phases are extracted twice with 100-mL portions of pentane and the combined organic phases are washed rapidly with 100 mL of ice-cold aqueous 2 N HCl and immediately thereafter with a cold saturated solution of sodium hydrogen carbonate. After the mixture has been dried over anhydrous magnesium sulfate, the pentane is removed by rotary evaporation. Distillation of the oily residue at 60°C (12 mm) using a 20-cm Vigreux column affords 50.1–51.6 g (60–62%) of 1-trimethylsiloxycyclopentene (1) as a colorless liquid (Note 4).

B. *2-tert-Pentylcyclopentanone.* A dry, 250-mL, three-necked, round-bottomed flask is fitted with a gas inlet, a gas bubbler, rubber septum, and magnetic stirrer. The apparatus is flushed with dry nitrogen or argon and charged with 120 mL of dry dichloromethane (Note 5), 15.6 g (0.10 mol) of 1-trimethylsiloxycyclopentene and 11.7 g (0.11 mol) of 2-chloro-2-methylbutane (Note 6). The mixture is cooled to −50°C (Note 7) and a cold (−50°C) solution of 11 mL (0.10 mol) of titanium tetrachloride (Note 8) in 20 mL of dichloromethane is added within 2 min through the rubber septum with the aid of a syringe. During this operation rapid stirring and cooling in maintained. Sunlight should be avoided. The reddish-brown mixture is stirred at the given temperature for an additional 2.5 hr and is then rapidly poured into 1 L of ice–water (Note 9). After the addition of 400 mL of dichloromethane, the mixture is vigorously shaken in a separatory funnel; the organic phase is separated and washed twice with 400-mL portions of water. The aqueous phase of the latter two washings is extracted with 200 mL of dichloromethane; the organic phases are combined and dried over anhydrous sodium sulfate. The mixture is concentrated using a rotary evaporator and the residue is distilled at 80°C (12 mm) (Note 10) to yield 9.2–9.5 g (60–62%) (Note 11) of 2-*tert*-pentylcyclopentanone as a colorless oil (Note 12).

2. Notes

1. Dimethylformamide and triethylamine were purchased from Baker (Baker Analyzed Reagent) and used without further purification.

2. Cyclopentanone and chlorotrimethylsilane were purchased from Aldrich Chemical Company and used without further purification.

3. According to the original procedure of House,[2] only 4 hr is needed, affording a 59% yield. However, the submitters found that an increase in reaction time raises the yield.

4. The spectral properties of the compound are as follows: ^{1}H NMR (CCl$_4$) δ: 0.2 (s, 9 H), 1.6–2.4 (m, 6 H), 4.4 (m, 1 H); IR (film) 1645 cm^{-1} (lit.[2] 1645 cm^{-1}).

5. Reagent-grade dichloromethane is dried by passing over a column of aluminum oxide (activity I).

6. The submitters purchased 2-chloro-2-methylbutane from Eastman Kodak Company. The checkers prepared the halide as follows. A separatory funnel was charged with 21.5 mL (0.2 mol) of 2-methyl-2-butanol and 100 mL of concentrated hydrochloric acid. The mixture was shaken vigorously with periodic venting for 10 min. The layers were separated and the 2-chloro-2-methylbutane

layer (upper) was washed several times with equal volumes of cold water. The product was dried over calcium chloride and distilled, bp 85°C.

7. The precise temperature is not critical. The checkers observed that the reaction proceeds in about the same time and yield at −78°C. However, at temperatures above −40°C a drop in yield may occur.

8. The titanium tetrachloride should be clean, colorless, and free of hydrogen chloride. The checkers used material freshly distilled in an argon atmosphere.

9. If sodium bicarbonate is used, large amounts of titanium oxide-containing emulsions tend to form that hamper the purification of the product.

10. The by-products consist of volatile cyclopentanone and an unknown high-boiling material, so that rapid vacuum transfer at room temperature and 0.02 mm is also possible. Extremely slow distillation at high temperatures should be avoided. The value of 72°C (2.2 mm) cited in the literature[3] seems to be in slight error.

11. The submitters ran the reaction on a 0.5 scale and reported yields of 63–68%.

12. The product is >96% pure as checked by gas chromatography (4% UCON LB 550X, Chromasorb G, AW-DMCS 80–100 mesh, 130°C). The spectral properties are as follows: IR (neat) cm^{-1}: 3050–2800, 1735, 1460, 1150; ^1H NMR (CCl_4) δ: 0.80 (J = 6 Hz, CH_3 of the ethyl group, which partially overlaps with the signals of the other two diastereotopic methyl groups), 0.82 (s), 0.92 (s), 1.15–2.25 (m); ^{13}C NMR ($CDCl_3$) δ: 7.78, 19.87, 23.72 (slightly broad), 25.57, 32.62, 34.70, 40.02, 55.39, 219.57.

3. Discussion

This procedure solves the long-pending problem of α-*tert*-alkylation of ketones. The generality is shown by the fact that a wide variety of structurally different ketones can be alkylated via the corresponding silyl enol ethers with good yields.[4] Variation of the alkylating agent is also possible, branched and cyclic tertiary alkyl halides reacting position specifically without signs of rearrangement.[4] Chemoselectivity studies reveal that esters, aromatic groups, and primary alkyl halide moieties are tolerated.[4] In the case of a sensitive enol ether such as that derived from acetone, titanium tetrachloride should be replaced by more mild Lewis acids such as zinc chloride, although the yields are lower.[5] Finally, it should be noted that any S_N1-reactive alkyl halide is likely to be a suitable alkylating agent in Lewis acid-promoted α-alkylation of carbonyl compounds. Indeed, aryl-activated secondary alkyl halides and acetates react in the same way.[6] Heteroatom substituted alkyl halides and acetates also react smoothly with enol silanes in the presence of ZnX_2.[4,6] Generally, such alkylating agents are unsuitable in classical enolate chemistry because of the ease of hydrogen halide elimination and/or the failure to react regiospecifically. The methods are thus complementary.

A related *tert*-butylation procedure in which the silyl enol ether is added to a mixture of titanium tetrachloride and *tert*-butyl chloride gives rise to distinctly lower yields.[7,8] This is also the case if the tertiary halide is added to a mixture of silyl enol ether and titanium tetrachloride.[5]

A number of alternative multistep procedures for the synthesis of α-*tert*-alkyl ketones are known, none of which possess wide generality. A previous synthesis of 2-*tert*-pentyl-cyclopentanone involved reaction of N-1-cyclopentenylpyrrolidine with 3-chloro-3-

methyl-1-butyne and reduction of the resulting acetylene (overall yield 46%).[3] However, all other enamines tested afford much lower yields.[3] Cuprate addition to unsaturated ketones may be useful in certain cases.[9] Other indirect methods have been briefly reviewed.[5]

1. Fachbereich Chemie der Universität, Hans-Meerwein-Strasse, 3550 Marburg, West Germany.
2. House, H. O.; Czuba, L. J.; Gall, M.; Olmstead, H. D. *J. Org. Chem.* **1969**, *34*, 2324.
3. Hennion, G. F.; Quinn, F. X. *J. Org. Chem.* **1970**, *35*, 3054.
4. Reetz, M. T.; Maier, W. F. *Angew. Chem., Int. Ed. Engl.* **1978**, *17*, 48; review: Reetz, M. T. *Angew. Chem. Int. Ed. Engl.* **1982**, *21*, 96,
5. Reetz, M. T.; Maier, W. F.; Heimbach, H.; Giannis, A.; Anastassiou, G. *Chem. Ber.* **1980**, *113*, 3734.
6. Reetz, M. T.; Hüttenhain, S.; Walz, P.; Löwe, U. *Tetrahedron Lett.* **1979**, 4971; Paterson, I. *Tetrahedron Lett.* **1979**, 1519; Reetz, M. T.; Schwellnus, K.; Hübner, F.; Massa, W.; Schmidt, R. E. *Chem. Ber.* **1983**, *116*, 3708.
7. Chan, T. H.; Paterson, I.; Pinsonnault, J. *Tetrahedron Lett.* **1977**, 4183.
8. Maier, W. F. Dissertation, Universität Marburg, **1978**.
9. Posner, G. H. *Org. React.* **1972**, *19*, 1; Corey, E. J.; Chen, R. H. K. *Tetrahedron Lett.* **1973**, 1611.

PERHYDRO-9b-BORAPHENALENE AND PERHYDRO-9b-PHENALENOL

(9b-Boraphenalene, dodecahydro-) and [Phenalen-9bα(2H)-ol, 3,3aα,4,5,6,6aα,7,8,9,9aβ-decahydro-]

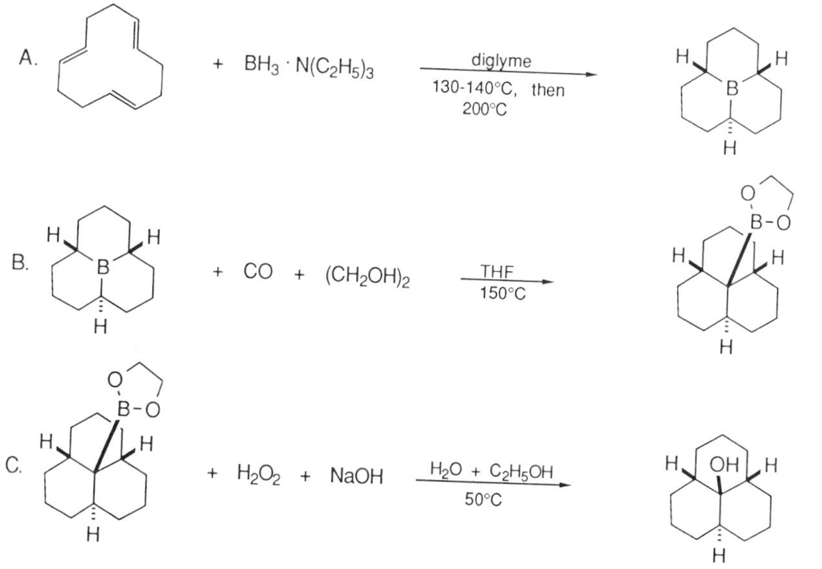

Submitted by EI-ICHI NEGISHI[1,3] and HERBERT C. BROWN[2,3]
Checked by A. J. COCUZZA and R. E. BENSON

1. Procedure

Caution! The products used and formed in Step A are extremely pyrophoric. Great care should be taken in conducting this step.

A. *cis,trans-Perhydro-9b-boraphenalene.* A 1-L, three-necked, round-bottomed flask is fitted with a septum, thermometer, magnetic stirring bar, and a 12-cm Vigreux column. A 2-L, two-necked receiving flask is attached to the Vigreux column and fitted with a nitrogen-inlet tube that is attached to a mercury bubbler device to permit a positive pressure on the system (Note 1). The entire system is flushed with nitrogen and, while the system is maintained under a static pressure of nitrogen, 500 mL (0.050 mol) of a 1.0 *M* solution of borane in tetrahydrofuran (THF, Note 2) is added to the reaction flask by means of a syringe. The flask is immersed in an ice–water bath, stirring is begun, and 50.6 g (0.50 mol) of triethylamine (Note 3) is added slowly over 15 min. After the addition is completed, the THF is removed by distillation at atmospheric pressure and 300 mL of dry diglyme (Note 4) is added. The resulting solution is heated to 130–140°C and a solution of 81 g (0.50 mol) of *trans,trans,trans*-1,5,9-cyclododecatriene (Note 5) in 100 mL of dry diglyme (Note 6) is added over 2 hr. At the end of this time, the diglyme is removed by distillation at atmospheric pressure and the residual oil is heated at 200°C for 6 hr (Note 7). After the reaction is cooled, the thermally treated product is used directly in Step B (Note 8). Product free of polymeric impurity can be obtained by distillation (Notes 9 and 10).

B. *cis,cis,trans-2-(Perhydro-9'b-phenalyl)-1,3,2-dioxaborole.* The 250-mL pressure vessel (Note 11) is fitted with a cap bearing a rubber septum. Two hypodermic needles are inserted into the vessel through the septum, one with the end close to the bottom of the vessel and the other with the end close to the top. The vessel is flushed with nitrogen, with the exit gas passing through a mercury bubbler device. One-fifth of the thermally treated, undistilled product obtained from Step A is dissolved in 50 mL of dry THF (Note 12) and added to the vessel by means of a syringe while a static pressure of nitrogen is maintained on the vessel. Ethylene glycol (18.6 g, 16.8 mL, 0.30 mol) (Note 13) is then added. The rubber septum is removed and the vessel is quickly connected to a cylinder of carbon monoxide (Note 14) and placed in a heating device capable of agitation. The vessel is agitated and the pressure increased with carbon monoxide to ca. 70 atm (ca. 1000 psi); the temperature is raised to 150°C. The vessel is maintained at this temperature for 2 hr and then cooled to room temperature and opened to the air. The contents of the vessel are transferred to a flask; the vessel is rinsed with two 50-mL portions of pentane, and the pentane is added to the product. The resulting solution is washed with 50 mL of water and dried over magnesium sulfate. The drying agent is removed by filtration, and the pentane removed by distillation to give 19.4 g of *cis,cis,trans*-2-(perhydro-9'b-phenalyl)-1,3,2-dioxaborole, a solid that can be further purified by recrystallization from pentane, mp 101–102°C (Note 15).

C. *cis,cis,trans-Perhydro-9b-phenalenol.* A 500-mL, three-necked, round-bottomed flask is fitted with a septum, thermometer, magnetic stirring bar, and a reflux condenser, which is connected to a nitrogen inlet and a mercury bubbler device. The system is flushed with nitrogen, and 50 mL of THF, 100 mL of 95% ethanol, and 19.4 g (0.0782 mol) of *cis,cis,trans*-2-(perhydro-9'b-phenalyl)-1,3,2-dioxaborole from Step B are added to the flask together with 37 mL (0.220 mol, 120% excess) of 6 *N* sodium hydroxide. The solution is stirred and, by means of a dropping funnel, 37 mL (~0.326 mol) of 30% hydrogen peroxide (Note 16) is added at such a rate that the temperature of the reaction mixture does not exceed 40°C. After the initial reaction has subsided, the reaction mixture is heated for 2 hr at 50°C to assure complete oxidation (Note 17). At the end of this time, 300 mL of pentane is added. The mixture is transferred to a separatory funnel, and the organic layer is separated and washed three times with 50-mL portions of water and then dried over magnesium sulfate. The mixture is filtered, and the solvent is removed by distillation to yield a solid (Note 18) which is recrystallized from cold pentane to give 10.9 g (71.7%) of *cis,cis,trans*-perhydro-9b-phenalenol, mp 75–76°C (Note 19).

2. Notes

1. All joints must be well greased and securely clamped. Even a minor leak is a fire hazard.

2. The checkers used a reagent available from Aldrich Chemical Company, Inc. Borane–THF was prepared by the submitters.[2] The direct use of borane–THF for hydroboration results in the formation of a polymeric, insoluble intermediate, which can be depolymerized by heating.

3. The checkers refluxed triethylamine, available from Eastman Organic Chemicals, with phenyl isocyanate and then isolated the amine by distillation. The submitters used a reagent available from Aldrich Chemical Company, Inc.

4. The checkers used a reagent available from Aldrich Chemical Company, Inc. The diglyme was distilled from sodium benzophenone ketyl prior to use. The submitters used a reagent available from the same source and distilled it from lithium aluminum hydride prior to use.

5. The checkers and the submitters used reagent available from Chemical Samples Co. The submitters state that other isomers such as *trans,trans,cis*-1,5,9-cyclo-dodecatriene or a mixture of isomers can be used.[3] In this case a slightly different isomer distribution is observed, and the yields of isolated product are somewhat lower. The checkers confirmed this observation, using trans,trans,cis reagent available from Aldrich Chemical Company, Inc.

6. A syringe pump was used with the syringe well greased with a polyhalo hydrocarbon lubricant. Alternatively, a pressure-equalizing dropping funnel can be used.

7. It is essential to heat the initially formed product to 200°C to achieve isomerization of the other isomers present to *cis,trans*-perhydro-9b-boraphenalene. When this thermal treatment is omitted, the desired product is contaminated with one major (30–40%) and several minor, unidentified, isomeric substances.

8. Perhydro-9b-boraphenalene is highly flammable. The transfer must be carried out with caution. The use of gloves is recommended to avoid direct contact with the organoborane. The transfer is most conveniently done under a slightly positive pressure of nitrogen using a broad-gauge (18-gauge), double-tipped needle.

9. The crude product is diluted with a small amount of dry THF and the solution transferred to a 100-mL distillation flask. Distillation through a 12-cm Vigreux column gives 58.0–60.1 g (66–68% yield) of a mixture of *cis,trans-* and *cis,cis-* perhydro-9b-boraphenalene, bp 113–114°C (9.5 mm), ^{1}H NMR (CDCl$_3$) δ: 0.7–2.2.

10. The submitters state that the composition of the distillate is 92 : 8 cis,trans : cis,cis isomer, based on GC analyses using an SE-30 column. The assigned stereochemistry is supported by the ^{1}H NMR spectrum of the pyridine complex.[3]

11. The checkers used a 250-mL Hastelloy pressure vessel. The submitters used a 250-mL autoclave available from American Instrument Co.

12. The checkers used a reagent available from Fisher Scientific Company. The submitters used a reagent available from Aldrich Chemical Company, Inc.

13. The checkers distilled the reagent available from E. I. du Pont de Nemours & Co. The submitters used a reagent available from Aldrich Chemical Company, Inc.

14. The checkers and submitters used a reagent available from Matheson Gas Products.

15. The checkers obtained the product in 87% crude yield using distilled boraphenalene. Recrystallization from pentane gave product in 66% yield, mp 101–102°C, with the following spectral characteristics: IR (KBr) cm^{-1}: 1185, 1200, 1250, 1310, 1385, 2860, and 2900–2950; ^{1}H NMR (CDCl$_3$) δ: 1.0–1.9 (m, 21 H), 4.11 (s, 4 H). The structure of the product has been confirmed by X-ray crystallography.[3]

16. The checkers and the submitters used reagent available from Fisher Scientific Company.

17. The submitters state that oxidation of the dioxaborole is unusually sluggish and urge the use of ethanol as a cosolvent and an excess of 6 N sodium hydroxide. They also urge monitoring of the reaction by GC. The checkers monitored the reaction by both GC and TLC analyses. GC analysis by the checkers was conducted using the following column and conditions: 3.2-mm × 2-m column, 7% SE 30/3% Silar on Gas Chrom Q (60–80 mesh), 170°C, 50 mL of nitrogen per min. The retention times for the perhydrophenalenol and dioxaborole are 6.4 and 15.2 min, respectively. For TLC analyses, Analtech silica gel plates bearing the material were eluted by 1 : 2 methylene chloride–petroleum ether, and visualized with phosphomolybdic acid: perhydrophenalenol, R_f 0.4; dioxaborole, R_f 0.9. The checkers found that the reaction was essentially complete after addition of the hydrogen peroxide. Additional heating did not lower the yield of product.

18. The submitters state that GC analysis of the solid indicates a 92 : 8 mixture of the cis,cis,trans and cis,cis,cis isomers.

19. Using recrystallized dioxaborole from Step B, the checkers obtained product, recrystallized from pentane, mp 75–76°C, in 85% yield, having the following spectral characteristics: IR (KBr) cm^{-1}: 1450 (s), 2860 (m), 2930 (s), and 3460 (m); ^{1}H NMR (CDCl$_3$) δ: 1.0–2.3; ^{13}C NMR (CDCl$_3$) δ: 21.25 (t), 26.52 (t),

27.42 (t), 29.79 (t), 29.12 (t), 33.86 (d), 44.32 (d), 73.24 (s) (undecoupled spectrum).

3. Discussion

Preparation of the two stereoisomers of perhydro-9b-boraphenalene was originally reported by Köster and Rotermund,[6] and the present procedure (Step A) is largely based on the procedure described by these authors. However, the original stereochemical assignment was incorrect and has been reversed.[3,7] Furthermore, these authors did not use the thermal treatment described above, which appears essential to achieve isomerization of other constitutional isomers into perhydro-9b-boraphenalene.[3] The original

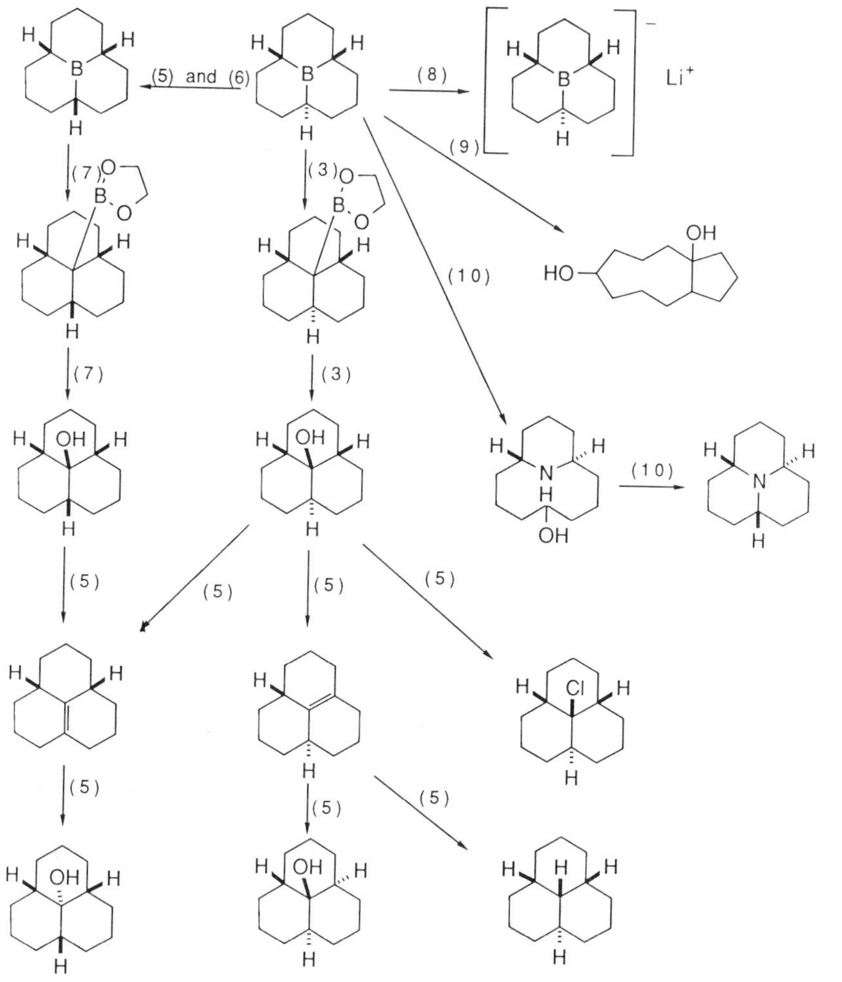

Scheme 1

procedure for isomerization of the cis,trans isomer to the all cis isomer has been satisfactory. Contrary to the claim made by these authors,[6] however, this isomerization does not lead quantitatively to the all cis isomer, but reaches an equilibrium, which consists of the all cis and cis,trans isomers in the ratio of 88 : 12; this ratio was also confirmed by reverse isomerization of the pure all cis isomer.[5]

cis,trans-Perhydro-9b-boraphenalene has been converted to lithium cis,cis,trans-perhydro-9b-boraphenalyl hydride by reaction with lithium hydride.[8] The tricyclic organoborane reported here has been converted to bicyclo[7.3.1]dodecane-1,5-diol[9] and trans-13-azabicyclo[7.3.1]tridecan-5-ol.[10] The latter has been converted to cis,trans-perhydro-9b-azaphenalene.[10]

The procedure reported here (Steps B and C) has been applied with minor modifications to the syntheses of the cis,cis,cis isomers of the 1,3,2-dioxaborole and perhydro-9b-phenalenol.[7] The two other stereoisomers, cis,trans,trans and trans,trans,trans, have been prepared from cis,cis,trans-perhydro-9b-phenalenol via the cis and trans isomers of $\Delta^{3a,9b}$-perhydrophenalene.[5] In addition, a few isomers of perhydrophenalenol and of perhydrophenalene and cis,cis,trans-9b-chloroperhydrophenalene have also been prepared from cis,cis,trans-perhydro-9b-phenalenol.[5] Some of the representative transformations are summarized in Scheme 1. (The numbers in parentheses refer to references.)

1. Department of Chemistry, Syracuse University, Syracuse, NY 13210. The current address is the same as Ref. 2.
2. Richard B. Wetherill Laboratory, Purdue University, West Lafayette, IN 47907.
3. The results described here were previously reported as a Communication: Brown, H. C.; Negishi, E. J. Am. Chem. Soc. 1967, 89, 5478–5480. A similar procedure has also been described elsewhere.[4]
4. Brown, H. C. "Organic Syntheses via Boranes"; Wiley: New York, 1975.
5. Dickason, W. C. Ph.D. Thesis, Purdue University, 1970.
6. Köster, R.; Rotermund, G. Angew Chem. 1960, 72, 563; Rotermund, G. W.; Köster, R. Liebigs Ann. Chem. 1965, 686, 153–166.
7. Brown, H. C.; Dickason, W. C. J. Am. Chem. Soc. 1969, 91, 1226–1228.
8. Brown, H. C.; Dickason, W. C. J. Am. Chem. Soc. 1970, 92, 709–710.
9. Yamamoto, Y.; Brown, H. C. J. Org. Chem. 1974, 39, 861–862.
10. Mueller, R. H. Tetrahedron Lett. 1976, 2925–2926.

2-PHENYL-2-ADAMANTANAMINE HYDROCHLORIDE

(Tricyclo[3.3.1.1^{3,7}]decan-2-amine, 2-phenyl, hydrochloride)

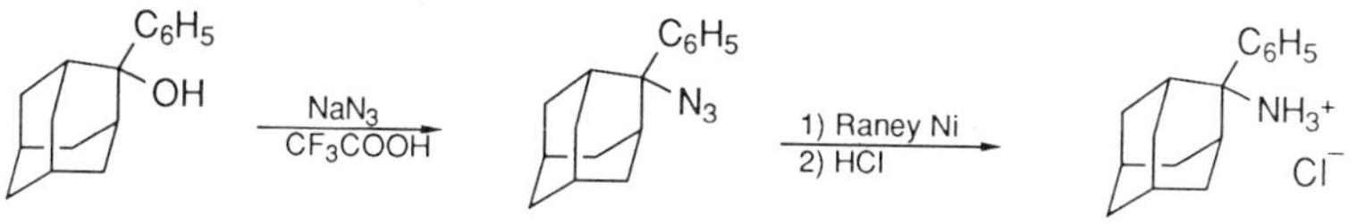

Submitted by ASHER KALIR and DAVID BALDERMAN[1]
Checked by CARL R. JOHNSON and DEBRA L. MONTICCIOLO

1. Procedure

Caution! The reaction should be carried out in a good hood.

A. *2-Azido-2-phenyladamantane.* A 500-mL, three-necked, round-bottomed flask equipped with a mechanical stirrer, a pressure-equalizing dropping funnel, and a thermometer is charged with 125 mL of chloroform and 13 g (0.2 mol) of sodium azide. The mixture is cooled with an ice–salt bath to −5°C to 0°C, and 37.5 mL (0.5 mol) of trifluoroacetic acid is added, followed after 5–10 min with 22.8 g (0.1 mol) of 2-phenyl-2-adamantanol (Note 1). The resulting slurry is stirred for 4 hr at 0°C and then allowed to reach room temperature overnight. The mixture is cautiously neutralized with a slight excess of 12–15% aqueous ammonia solution and transferred to a separatory funnel. The chloroform layer is separated, and the aqueous solution is extracted with 50 mL of chloroform. The combined organic extracts are washed with 50 mL of water, separated, and dried over magnesium sulfate. The solvent is removed in a rotary evaporator. The oily residue solidifies on cooling. The yield is 23.6–24.8 g (93–98%), mp 42–45°C. Recrystallization of a sample from 2-propanol raises the melting point to 47–48°C (Note 2).

B. *2-Phenyl-2-adamantanamine hydrochloride.* A solution of 24 g (0.095 mol) of the crude 2-azido-2-phenyladamantane in 75 mL of 2-propanol is placed in a 1-L beaker fitted with a mechanical stirrer, and heated in a water bath that can be removed quickly. Wet, active Raney nickel (Notes 3 and 4) is added in portions at 60–70°C with stirring until the evolution of nitrogen ceases (Note 5). The mixture is heated for an additional 10 min, filtered through a Büchner funnel, and washed with 75 mL of 2-propanol in such a manner that the catalyst is always covered with liquid (Note 6). The filtrate is concentrated in a rotary evaporator under reduced pressure. The crude residue is dissolved in 75 mL of toluene and treated with 22 mL of concentrated hydrochloric acid while stirring. The 2-phenyl-2-adamantanamine hydrochloride is collected, triturated with 50 mL of warm acetone, filtered again, and air-dried. The yield is 22.5–24.0 g (90–96%), and the product melts at 293–296°C (closed capillary) (Notes 7 and 8).

2. Notes

1. 2-Phenyl-2-adamantanol[2] is prepared by adding 25 g (0.167 mol) of 2-adamantanone (Note 3) in several portions to phenylmagnesium bromide, obtained from

40 g of bromobenzene and 6.5 g of magnesium turnings in 200 mL of diethyl ether. The solution is stirred for 1 hr and worked up with aqueous ammonium chloride. The organic layer is separated, dried over magnesium sulfate, concentrated, and the oily residue is crystallized from petroleum ether. The yield is 25.5 g (67%) of crystals melting at 77–78°C. The crude oily residue may be used in the next step without purification.

2. The product is characterized by IR (CCl$_4$) cm^{-1}: 2075; ^{1}H NMR (CCl$_4$) δ: 1.72 and 2.40 (s, 14 H), 7.20 (s, 5 H).

3. 2-Adamantanone was obtained from Aldrich Chemical Company, Inc. Active Raney nickel catalyst was obtained from W. R. Grace Company.

4. The amount of Raney nickel depends on its hydrogen content. Usually 25–35 g is sufficient.

5. A large vessel is required because of excessive frothing. The frothing may be controlled by adding a little cold 2-propanol, by removing the heating, or by stopping the stirrer.

6. *Caution! Dry catalyst is pyrophoric.*

7. The free 2-phenyl-2-adamantanamine may be liberated from the salt by adding a solution of ammonia or sodium hydroxide, extracting with toluene, concentrating, and distilling under reduced pressure; bp 120–122°C (0.15 mm); n_D^{17} 1.5850; ^{1}H NMR (CCl$_4$) δ: 1.30 (s, 2H, NH$_2$), 1.68 and 2.26 (br s, 14 H, adamantane protons), 7.1 (m, 5H, Ph).

8. Similarly, 2-butyl-2-adamantanamine hydrochloride, mp 300–305°C, is obtained from 2-butyl-2-adamantanol[3] in 30% yield.

TABLE I
AMINES FROM TERTIARY ALCOHOLS

Alcohol	Azide[a]		Amine		
	Bp (°C) (mm Hg)		Bp (°C) (mm Hg)	Yield (%)	Starting Material
2-Phenyl-2-propyl	106 (22)		100 (22)	66	α-Methylstyrene
1-Phenylcyclo-pentyl	139–140 (38)		128–130 (20)	40	Cyclopentanone
1-Phenylcyclo-hexyl			115–120 (5)	38	Cyclohexanone
2-Methyl-1-phenyl-cyclohexyl	90–91 (0.25)		150–153 (23)	66	2-Methyl-1-phenyl-cyclohexanol
1-Phenylcyclo-heptyl	153–155 (23)		163–165 (25)	45	Cycloheptanone
2-Phenyl-2-norbornyl	150–155 (25)		163–165 (28)	51	2-Norbornanone

[a]The azides contain up to 15–20% of the corresponding phenylalkenes.

3. Discussion

The present procedure is an example of preparation of tertiary phenylcarbinylamines, and is in many cases superior to methods based on the Ritter reaction,[4] and Hofmann[5] or Curtius degradation.[6] The availability of starting materials, fair yields of products, and the simplicity of operations (there is no need to isolate any intermediates or to use a hydrogenation apparatus) are the main advantages of this procedure. The azide synthesis is adapted from procedures described for the preparation of 1,1-diphenyl-2-azidoethane[7] and 1-phenyl-1-azidocyclohexane.[8] The azides are quite stable and could be distilled under reduced pressure. The amines and their substitution products are physiologically active agents.[4,9]

A number of compounds have been prepared by this method (the isolation of hydrochloride can be omitted), as listed in Table I.[10]

1. Israel Institute for Biological Research, Sackler School of Medicine, Tel Aviv University, Ness Ziona, 70 400, Israel.
2. Tanida, H.; Tsushima, T. *J. Am. Chem. Soc.* **1970**, *92*, 3397–3403.
3. Landa, S.; Vais, J.; Burkhard, J. *Collect. Czech. Chem. Commun.* **1967**, *32*, 570–575.
4. Maddox. H.; Godefroi, E. E.; Parcell, R. F. *J. Med. Chem.* **1965**, *8*, 230–235.
5. Kalir, A.; Pelah, Z. *Isr. J. Chem.* **1967**, *5*, 223–229.
6. Kaiser, C.; Weinstock, J. *Org. Synth.*, *Coll. Vol. VI* **1988**, 910.
7. Ege, S. N.; Sherk, K. W. *J. Am. Chem. Soc.* **1953**, *75*, 354–357.
8. Geneste, P.; Herrmann, P.; Kamenka, J. M.; Pons, A. *Bull. Soc. Chim. Fr.* **1975**, 1619–1626.
9. Kalir, A.; Edery, H.; Pelah, Z.; Balderman, D.; Porath, G. *J. Med. Chem.* **1969**, *12*, 473–477.
10. Balderman, D.; Kalir, A. *Synthesis* **1978**, 24–25.

CYANIC ACID ESTERS FROM PHENOLS: PHENYL CYANATE

(Cyanic acid, phenyl ester)

$$Br_2 \; + \; NaCN \; \xrightarrow[-5 \text{ to } 5°C]{H_2O} \; BrCN$$

$$BrCN \; + \; C_6H_5OH \; \xrightarrow[CCl_4, \, -5 \text{ to } 10°C]{(C_2H_5)_3N} \; C_6H_5OCN$$

Submitted by D. Martin[1] and M. Bauer
Checked by E. R. Holler, Jr. and R. E. Benson

1. Procedure

Caution! These operations, which involve toxic reagents, should be conducted in an efficient hood.

A 1-L, three-necked, round-bottomed flask equipped with a mechanical stirrer, thermometer, and a 200-mL pressure-equalizing dropping funnel with a stopper is charged with 160 g (50.9 mL, 1.0 mol) of bromine (Note 1) and 150 mL of water. The mixture is stirred rapidly while cooling in an ice–salt bath to $-5°C$, and a solution of 49.0 g (1.0 mol) of sodium cyanide in 150 mL of water is added dropwise over a 40–50 min period while maintaining the temperature of the reaction mixture at -5 to $5°C$. The resulting solution is stirred an additional 5–10 min (Note 2). A solution of 89.5 g (0.95 mol) of phenol in 300 mL of tetrachloromethane (Note 3) is added in one portion to the flask. The resulting mixture is stirred vigorously while 96.0 g (131 mL, 0.95 mol) of triethylamine is added dropwise over a 30–40-min period at such a rate that the temperature does not exceed 5–10°C. After an additional 15 min of stirring, the mixture is transferred to a separatory funnel, the organic phase is separated and the aqueous layer is extracted twice with 50-mL portions of tetrachloromethane. The organic phases are combined and washed three times with 50-mL portions of water and then dried over polyphosphoric anhydride (P_2O_5) (Note 4). The drying agent is removed by filtration and the solvent is removed by distillation under reduced pressure using a rotary evaporator at 20°C (25 mm). A few drops of polyphosphate ester (Note 5) are added to the remaining liquid and the product is distilled through a 20-cm Vigreux column to give 85–96 g (75–85%) of phenyl cyanate, bp 77–79° (13 mm), n_D^{20} 1.5094–1.5100, d_4^{20} 1.096. The product is a colorless liquid with a pungent odor (Note 6).

2. Notes

1. The chemicals used were commercially available products and were used without further purification. The checkers used sodium cyanide, phenol, and tetrachloromethane from Fischer Scientific Company, bromine from Matheson, Coleman and Bell, phosphoric anhydride from J. T. Baker Chemical Co., and triethylamine from Eastman Organic Chemicals.

2. The solution should develop a yellowish color.

3. The procedure can also be conducted using other water immiscible solvents such as ether, trichloromethane, and benzene.[2]

4. Other drying agents such as anhydrous calcium chloride can also be used. The desiccation must be done carefully since water is soluble in the product in the presence of phenol and may cause trimerization of the cyanate to a 1,3,5-triazine derivative.

5. A few drops of polyphosphate ester are a good drying agent and stabilizer.[3] The ester may be prepared by heating polyphosphoric anhydride in dry ether and trichloromethane for 40 hr followed by removal of the solvent.[4] The checkers found that the use of polyphosphate ester was essential to obtain the described yield.

6. The spectral properties of phenyl cyanate are as follows. IR(CCl_4) cm^{-1}: 2235 (m), 2261 (m), 2282 (S) ($\nu_{C\equiv N}$).[5] UV (cyclohexane) nm max (log ϵ): 216 (3.21), 256 (2.58), 262 (2.75), and 268 (2.67).[6] The product was further characterized by vapor-phase chromatography analysis using a 200-cm column containing 10% SE 52 on Chromosorb W/AW/DMCS at 140°C with a hydrogen flow rate of 70 mL/min and a retention time of 1.47 min.

3. Discussion

Although isocyanates have been known for some time, the isomeric cyanates were unknown until 1964. The latter were first prepared almost simultaneously by two different

TABLE I
CYANATES FROM HYDROXY COMPOUNDS

Hydroxy Compound	Cyanate	mp (°C) (bp, °C/mm)	Yield (%)
2-CH$_3$C$_6$H$_4$OH	2-CH$_3$C$_6$H$_4$OCN	(88–90/10)	81
4-CH$_3$C$_6$H$_4$OH	4-CH$_3$C$_6$H$_4$OCN	(90–91/10)	87
4-CH$_3$OC$_6$H$_4$OH	4-CH$_3$OC$_6$H$_4$OCN	22–26 (118–119/10)	91
2-ClC$_6$H$_4$OH	2-ClC$_6$H$_4$OCN	(112–113/13)	81
4-ClC$_6$H$_4$OH	4-ClC$_6$H$_4$OCN	38–39 (100–101/10)	87
2-CH$_3$OCOC$_6$H$_4$OH	2-CH$_3$OCOC$_6$H$_4$OCN	58–60	84
2-Naphthyl-OH	2-Naphthyl-OCN	(162–164/12)	95
4-NCOC$_6$H$_4$OH	4-NCOC$_6$H$_4$OCN	107–109	98
CCl$_3$CH$_2$OH	CCl$_3$CH$_2$OCN	(77–78/10)	75
CF$_3$CH$_2$OH	CF$_3$CH$_2$OCN	(29–30/13)	81

methods: (1) thermolysis of 5-aryl- or 5-alkyloxy-1,2,3,4-thiatriazoles[6,7] and (2) by reaction of phenols or alcohols with cyanogen halides.[8] Since their synthesis, cyanates have acquired considerable synthetic significance.[9-14] The simplified procedure described here for preparation of phenyl cyanate is a combination of the preparation of cyanogen bromide[15] and the cyanation of phenol in the presence of a base.[8] This procedure is also applicable to many other phenols, bisphenols, naphthols, and some acidic alcohols. Examples are given in Table I.

Aryl cyanates have activated cyano groups and undergo many reactions.[14] They are effective dehydrating and hydrogen sulfide-bonding agents in organic synthesis.[9-11,13,14] N-, O-, and S-nucleophiles (HX) add to the carbon atom of the cyano group to form the

$$X$$

corresponding carbonic acid imide esters (ArO—C=NH).[9-11,13,14] Transfer of the cyano group to a number of carbon nucleophiles also occurs.[9-11,13,14] Acyl halides (AcCl) add to the nitrogen atom of the cyano group to give N-acylated carbonic acid imide chlorides

$$Cl$$
$$|$$

(ArO—C=NH—Ac).[12-14] These compounds are useful starting materials for syntheses of heterocyclic compounds. The cyanates also undergo 1,3- and 1,4-dipolar cycloadditions involving the cyano group to give substituted azoles and azines.[9-11,13,14] Polycyclic trimerization of dicyanates to poly-s-triazines is of considerable importance.[16-18]

1. Academy of Sciences of GDR, Central Institute for Organic Chemistry, GDR-1199 Berlin.
2. Martin, D.; Bauer, M. GDR-Patent WP CO7c 211,614, 1979.
3. Martin, D.; Bauer, M.; Niclas, H.-J. GDR-Patent WP CO7c 207,625, 1978.
4. Kanaoka, Y.; Machida, M.; Yonemitsu, O.; Ban, Y. Chem. Pharm. Bull. 1965, 13, 1065.
5. Reich, P.; Martin, D. Chem. Ber. 1965, 98, 2063.
6. Martin, D. Chem. Ber. 1964, 97, 2689.
7. Jensen, K. A.; Holm, A. Acta Chem. Scand. 1964, 18, 826.
8. Grigat, E.; Pütter, R. Chem. Ber. 1964, 97, 3012.
9. Martin, D. Z. Chem. 1967, 7, 123.

10. Hedayatullah, M. *Bull. Soc. Chim. Fr.* **1967,** 416; **1968,** 1572.
11. Grigat, E.; Pütter, R. *Angew. Chem., Int. Ed. Engl.* **1967,** *6,* 206.
12. Grigat, E. *Angew. Chem., Int. Ed. Engl.* **1972,** *11,* 949.
13. Patai, S. (Ed.) "The Chemistry of Functional Groups. The Chemistry of Cyanates and their Thio Derivatives"; Wiley-Interscience: New York, 1977.
14. Martin, D.; Bacaloglu, R. "Organic Synthesis with Cyanic Acid Esters"; Akademie-Verlag: Berlin, GDR, 1980.
15. Hartman, W. W.; Dreger, E. E. *Org. Synth., Coll. Vol. II* **1943,** 150.
16. Kubens, R.; Schultheis, H.; Wolf, R.; Grigat, E. *Kunststoffe,* **1968,** *58,* 827; *Chem. Abstr.* **1969,** *70,* 88507z.
17. Pankratov, V. A.; Korshak, V. V.; Vinogradova, S. V.; Puchin, A. G. *Plaste Kaut.,* **1973,** *20,* 481; *Chem. Abstr.* **1973,** *79,* 53802a.
18. Weirauch, K. K.; Gemeinhardt, P. G.; Baron, A. L. *Soc. Plast. Eng., Tech. Pap.* **1976,** *22,* 317; *Chem. Abstr.* **1976,** *85,* 33838n.

TOSYLHYDRAZONE SALT PYROLYSES: PHENYLDIAZOMETHANES

(Benzenes, diazomethyl-)

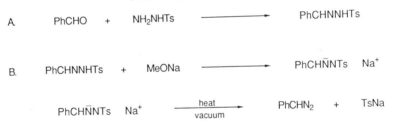

Submitted by XAVIER CREARY[1]
Checked by WEYTON W. TAM, KIM F. ALBIZATI, and ROBERT V. STEVENS

1. Procedure

Caution! Diazo compounds are presumed to be highly toxic and potentially explosive. All manipulations should be carried out in a hood. Although in numerous preparations we have never observed an explosion, all pyrolyses and distillations should routinely be carried out behind a safety shield.

A. *Benzaldehyde tosylhydrazone.* A 14.6-g sample (0.078 mol) of *p*-toluenesulfonylhydrazide (Note 1) was placed in a 125-mL Erlenmeyer flask and 25 mL of absolute methanol was added. The slurry was swirled as 7.50 g (0.071 mol) of freshly distilled benzaldehyde was added rapidly. A mildly exothermic reaction ensued and the *p*-toluenesulfonylhydrazide dissolved. Within a few minutes, the tosylhydrazone began to crystallize. After 15 min the mixture was cooled in an ice bath. The product was collected on a Büchner funnel, washed with a small amount of cold methanol, and dried under an aspirator vacuum. The dry benzaldehyde tosylhydrazone, mp 124–125°C, weighed 16.97–18.19 g (87–93%) and was not purified further.

B. *Phenyldiazomethane (Vacuum pyrolysis method).* In a 200-mL, single-necked, round-bottomed flask is placed 13.71 g (0.05 mol) of benzaldehyde tosylhydrazone. A 1.0 *M* solution (51 mL) of sodium methoxide in methanol (0.051 mol) (Note 2) is added via syringe and the mixture is swirled until dissolution is complete (Note 3). The methanol is then removed by a rotary evaporator. The last traces of methanol are removed by evacuation of the flask at 0.1 mm for 2 hr. The solid tosylhydrazone salt is broken up with a spatula and the flask is fitted with a vacuum take-off adapter and a 50-mL receiver flask. The system is evacuated at 0.1 mm and the receiver flask is cooled in a dry ice–acetone bath to about −50°C. The flask containing the salt is immersed in an oil bath and the temperature is raised to 90°C. (We recommend the use of a safety shield.) At this temperature, red phenyldiazomethane first begins to collect in the receiver flask. The temperature is raised to 220°C over a 1-hr period (Note 4). During this time red phenyldiazomethane collects in the receiver flask (Note 5). The pressure increases to 0.35 mm over the course of the pyrolysis. On completion of the pyrolysis the pressure drops to less than 0.1 mm.

The apparatus is disconnected and the 50-mL receiver flask that contains the crude phenyldiazomethane is fitted with a water-cooled short-path distillation head and a receiver flask cooled to about −50°C in a dry ice–acetone bath. The pressure is lowered to 1.5 mm and a trace of methanol collects in the receiver. A new receiver flask is connected and cooled to −50°C and the pressure is lowered to less than 0.2 mm. Red phenyldiazomethane distills below room temperature (Note 6). The yield of phenyldiazomethane, which is a liquid above −30°C, is 4.50–4.70 g (76–80%). The product should be used immediately or stored at a low temperature (−20 to −80°C) under nitrogen or argon (Notes 7–11); it is explosive at room temperature.

2. Notes

1. *p*-Toluenesulfonylhydrazide was obtained from Aldrich Chemical Company, Inc. and used without further purification.

2. The sodium methoxide solution was prepared by dissolving 2.30 g of sodium in absolute methanol and diluting it to 100 mL. If commercial sodium methoxide powder is used, it must be of high quality; otherwise the yield of phenyldiazomethane is lower.

3. Powdered sodium hydroxide can be used in place of sodium methoxide with no appreciable change in yield. Sodium hydroxide dissolves less readily in methanol.

4. When carried out on a small scale, pyrolysis is complete at lower temperatures (160–200°C).

5. Phenyldiazomethane solidifies at dry ice temperature. Care must be taken not to plug the vacuum take-off adapter; this occurs if the temperature of the receiver flask is too low. The receiver bath was maintained manually at about −50°C by addition of small pieces of dry ice to an acetone bath. We prefer to use this procedure rather than a chloroform–dry ice bath, which freezes at −63°C, because of the toxic nature of chloroform and the disposal problems associated with this solvent.

6. Slight warming with an oil bath at 30°C allows distillation to proceed at a reasonable rate. The bath should not be heated above this temperature. Gutsche and

Jason[2] report a boiling point of 37–41°C at 1.5 mm. Although we have never experienced any difficulty in numerous distillations, Gutsche and Jason[2] report that phenyldiazomethane "sometimes detonated violently during purification" by distillation. Therefore, we emphatically recommend that distillation be carried out below room temperature, behind a safety shield. On completion of the distillation, only a small amount of nonvolatile residue remained.

7. The checkers reported that a sample that was allowed to stand at room temperature for approximately 1 hr and then exposed to air decomposed violently after 5 min. In numerous preparations, when distilled phenyldiazomethane was immediately stored at −20°C or at −80°C under nitrogen, we never experienced any difficulty. We emphasize the need to keep phenyldiazomethane cold, and under nitrogen.

8. In runs on smaller scales, yields ranged from 84 to 91%.

9. The IR spectrum (CCl$_4$) shows an intense band at 4.83 μm (2060 cm^{-1}); ^{1}H NMR (CCl$_4$) δ: 4.79 (s, 1 H), 6.7–7.6 (m, 5 H).

10. Phenyldiazomethane shows no appreciable change on storage at −80°C for 3 months. Storage at −20°C led to significant decomposition after 2 weeks.

11. Traces of diazo compounds should be destroyed by addition to acetic acid.

3. Discussion

Diazo compounds have previously been prepared by a variety of methods. Some of these methods include hydrazone oxidations,[3] the reaction of diazomethane with acid chlorides,[4] the reaction of activated methylene compounds with tosyl azide,[5] decomposition of N-nitroso compounds,[6] diazotization of amines,[7] and pyrolysis of tosylhydrazone salts.[8-13] The present procedure for the preparation of phenyldiazomethane illustrates the vacuum pyrolysis method introduced by Shechter[12] for carrying out the Bamford–Stevens reaction.[9]

Phenyldiazomethane has been prepared by reaction of base with ethyl N-nitroso-N-benzylcarbamate,[13] N-nitroso-N-benzylurea,[14] and N-nitroso-N-benzyl-N'-nitroguanidine.[15] Staudinger's preparation[16] and that of Gutsche and Jason[2] employed mercuric oxide oxidation of benzaldehyde hydrazone. Yates and Shapiro[17] prepared phenyldiazomethane by basic cleavage of azibenzil. Bamford and Stevens[9] prepared phenyldiazomethane by solution pyrolysis of the salt of benzaldehyde tosylhydrazone. Closs and Moss[10] and Farnum[11] used variations of this solution pyrolysis method for the preparation of phenyldiazomethane. The vacuum pyrolysis method employed by Shechter[12] has also been used to prepare phenyldiazomethane.

The present procedure uses sodium methoxide in methanol for generation of the tosylhydrazone salt. This procedure gives the highest reported yield and, unlike other procedures, also gives pure diazo compounds free from solvents. This vacuum pyrolysis method appears applicable to the formation of relatively volatile aryldiazomethanes from aromatic aldehydes. Table I gives yields of diazo compounds produced by this vacuum pyrolysis method. The yields have not been optimized. The relatively volatile diazo esters, ethyl α-diazopropionate[18] and ethyl α-diazobutyrate, can also be prepared by this method.

TABLE I

FORMATION OF DIAZO COMPOUNDS BY VACUUM PYROLYSIS OF SODIUM SALTS OF TOSYLHYDRAZONES

Tosylhydrazone	Product	Yield (%)
p-MeC$_6$H$_4$CHNNHTs	p-MeC$_6$H$_4$CHN$_2$	52
m-MeC$_6$H$_4$CHNNHTs	m-MeC$_6$H$_4$CHN$_2$	55
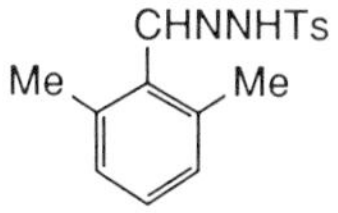 2,6-Me$_2$C$_6$H$_3$CHNNHTs (CHNNHTs, Me...Me)	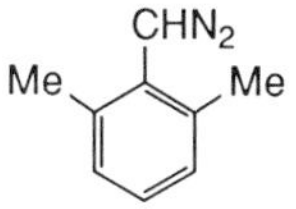 2,6-Me$_2$C$_6$H$_3$CHN$_2$ (CHN$_2$, Me...Me)	69
p-FC$_6$H$_4$CHNNHTs	p-FC$_6$H$_4$CHN$_2$	69
m-FC$_6$H$_4$CHNNHTs	m-FC$_6$H$_4$CHN$_2$	59
Me-C(=NNHTs)-COOEt	Me-C(=N$_2$)-COOEt	87
Et-C(=NNHTs)-COOEt	Et-C(=N$_2$)-COOEt	65

The major limitation of the vacuum pyrolysis method appears to be thermal decomposition of less volatile diazo compounds during the pyrolysis. The vacuum pyrolysis method was unsuccessful for the preparation of 1-naphthyldiazomethane and 3,5-dichlorophenyldiazomethane. However, such diazo compounds could be prepared from the corresponding tosylhydrazone salts by pyrolysis in ethylene glycol and extraction of the aryldiazomethane into hexane or ether. This procedure, as described by Goh,[19] permits the periodic extraction of the potentially labile diazo compound into an organic solvent while leaving the unreacted tosylhydrazone salt dissolved in the immiscible ethylene glycol phase. This solution pyrolysis method can also be used to prepare aryl diazo esters in high yields. This method is quite useful since the starting keto esters can be readily prepared in large quantities by reaction of the corresponding arylmagnesium bromides with diethyl oxalate.[20]

In a typical procedure, 0.14 g of sodium was dissolved in 10 mL of ethylene glycol by heating to 70°C and 0.0041 mol of tosylhydrazone was added. After heating with vigorous stirring for 5 min at 70–80°C, the mixture was cooled to about 35°C and 15 mL of hexane or ether was added with continued stirring. The organic extract was removed by pipette and the procedure was repeated a total of 5 times. The combined organic extracts were washed with 30 mL of 5% sodium hydroxide solution, with a saturated sodium chloride solution, and dried over magnesium sulfate. After filtration, the solvent was removed on a rotary evaporator to leave the diazo compound. Table II gives yields of diazo compounds prepared by this solution pyrolysis method.

TABLE II

FORMATION OF DIAZO COMPOUNDS BY PYROLYSIS OF SODIUM SALTS OF
TOSYLHYDRAZONES IN ETHYLENE GLYCOL

Tosylhydrazone	Temperature (°C)	Product	Yield (%)
CHNNHTs, Cl—ring—Cl (3,5-dichlorobenzaldehyde tosylhydrazone)	70[a]	CHN$_2$, Cl—ring—Cl	90
CHNNHTs (naphthalene)	80[a]	CHN$_2$ (naphthalene)	77
NNHTs, Ph—C(=)—COOEt	70[b]	N$_2$, Ph—C(=)—COOEt	86
NNHTs, p-MeC$_6$H$_4$—C(=)—COOEt	70[b]	N$_2$, p-MeC$_6$H$_4$—C(=)—COOEt	88
NNHTs, p-MeOC$_6$H$_4$—C(=)—COOEt	70[b]	N$_2$, p-MeOC$_6$H$_4$—C(=)—COOEt	76
NNHTs, p-CF$_3$C$_6$H$_4$—C(=)—COOEt	70[b]	N$_2$, p-CF$_3$C$_6$H$_4$—C(=)—COOEt	94

[a]The salt in ethylene glycol was heated at this temperature, cooled, and extracted periodically with hexane.
[b]Ether extraction.
[c]This product was further purified by distillation at less than 0.1 mm. The other products were *not* distilled.

1. Department of Chemistry, University of Notre Dame, Notre Dame, IN 46556.
2. Gutsche, C. D.; Jason, E. F. *J. Am. Chem. Soc.* **1956**, *78*, 1184–1187.
3. Smith L. I.; Howard, K. L. *Org. Synth., Coll. Vol. III* **1955**, 351–352; Murray, R. W.; Trozzolo, A. M. *J. Org. Chem.* **1961**, *26*, 3109–3112; Morrison, H.; Danishefsky, S.; Yates, P. *J. Org. Chem.* **1961**, *26*, 2617–2618; Allinger, N. L.; Freiberg, L. A.; Hermann, R. B.; Miller, M. A. *J. Am. Chem. Soc.* **1963**, *85*, 1171–1176; Ciganek, E. *J. Org. Chem.* **1965**, *30*, 4198–4204; Shepard, R. A.; Wentworth, S. E. *J. Org. Chem.* **1967**, *32*, 3197–3199; Creary, X. *J. Am. Chem. Soc.* **1980**, *102*, 1611–1618.
4. Bridson, J. N.; Hooz, J. *Org. Synth., Coll. Vol. VI* **1988**, 386. For leading references, see also Burke, S. D.; Grieco, P. A. *Org. React.* **1979**, *26*, 361–475.
5. Regitz, M.; Hocker, J.; Liedhegener, A. *Org. Synth., Coll. Vol. V* **1973**, 179–183 and references cited therein; Ledon, H. J. *Org. Synth., Coll. Vol. VI* **1988**, 414.
6. Moore, J. A.; Reed, D. E. *Org. Synth., Coll. Vol. V* **1973**, 351–355 and references cited therein.
7. Seale, N. E. *Org. Synth., Coll. Vol. IV* **1963**, 424–426 and references cited therein.
8. Blankley, C. J.; Sauter, F. J.; House, H. O. *Org. Synth., Coll. Vol. V* **1973**, 258–263 and references cited therein.

9. Bamford, W. R.; Stevens, T. S. *J. Chem. Soc.* **1952**, 4735–4740.

10. Closs, G. L.; Moss, R. A. *J. Am. Chem. Soc.* **1964**, *86*, 4042–4053.

11. Farnum, D. G. *J. Org. Chem.* **1963**, *28*, 870–872.

12. Kaufman, G. M.; Smith, J. A.; Vander Stouw, G. G.; Shechter, H. *J. Am. Chem. Soc.* **1965**, *87*, 935–937.

13. Hantzsch, A.; Lehmann, M. *Ber.* **1902**, *35*, 897–905.

14. Werner, E. A. *J. Chem. Soc.* **1919**, 1093–1102.

15. McKay, A. F.; Ott, W. L.; Taylor, G. W.; Buchanan, M. N.; Crooker, J. F. *Can. J. Res., Sect. B* **1950**, *28*, 683–688.

16. Staudinger, H.; Gaule, A. *Ber.* **1916**, *49*, 1897–1918.

17. Yates, P,; Shapiro, B. L. *J. Org. Chem.* **1958**, *23*, 759–760.

18. Sohn, M. B.; Jones, Jr., M.; Hendrick, M. E.; Rando, R. R.; Doering, W. V. E., *Tetrahedron Lett.* **1972**, 53–56.

19. Goh, S. H. *J. Chem. Soc. C* **1971**, 2275–2278.

20. Unpublished work from this laboratory. See also Nimitz, J. S.; Mosher, H. S. *J. Org. Chem.* **1981**, *46*, 211–213 for a synthesis of keto esters.

CONJUGATE ALLYLATION OF α,β-UNSATURATED KETONES WITH ALLYLSILANES: 4-PHENYL-6-HEPTEN-2-ONE

(6-Hepten-2-one, 4-phenyl-)

Submitted by HIDEKI SAKURAI, AKIRA HOSOMI, and JOSABRO HAYASHI[1]
Checked by TODD A. BLUMENKOPF and CLAYTON H. HEATHCOCK

1. Procedure

A 2-L, three-necked, round-bottomed flask is fitted with a dropping funnel (Note 1), mechanical stirrer, and reflux condenser attached to a nitrogen inlet. In the flask are placed 29.2 g (0.20 mol) of benzalacetone (Note 2) and 300 mL of dichloromethane (Note 3). The flask is immersed in a dry ice–methanol bath ($-40°C$) and 22 mL (0.20 mol) of titanium tetrachloride (Note 4) is slowly added by syringe to the stirred mixture. After 5 min, a solution of 30.2 g (0.26 mol) of allyltrimethylsilane (Notes 5 and 6) in 300 mL of dichloromethane is added dropwise with stirring over a 30-min period. The resulting red–violet reaction mixture is stirred for 30 min at $-40°C$ (Note 7), hydrolyzed by addition of 400 mL of H_2O, and, after the addition of 500 mL of ethyl ether with stirring, allowed to warm to room temperature. The nearly colorless organic layer is separated and the aqueous layer is extracted with three 500-mL portions of ethyl ether. The organic layer and ether extracts are combined and washed successively with 500 mL of saturated sodium bicarbonate and 500 mL of saturated sodium chloride, dried over anhydrous sodium sulfate, and evaporated at reduced pressure. The residue is distilled under reduced pressure through a 6-in. Vigreux column to give 29.2–30.0 g (78–80%) of 4-phenyl-6-hepten-2-one, bp 69–71°C (0.2 mm), n_D^{20} 1.5156, as a colorless liquid (Note 8).

2. Notes

1. A 500-mL dropping funnel, with pressure-equalizing arm, is used.
2. Benzalacetone is purchased from Wako Pure Chemical Ind., Ltd. or Aldrich Chemical Company, Inc.
3. Dichloromethane is dried over anhydrous calcium chloride, distilled, and stored over Linde 5A molecular sieves before use. The checkers distilled dichloromethane from calcium hydride immediately before use.
4. Titanium tetrachloride, purchased from Junsei Chemical Co., Ltd. is distilled before use. The checkers purchased titanium tetrachloride from the Fisher Scientific Company and distilled it from copper powder before use.
5. The starting allyltrimethylsilane can be prepared in satisfactory yield by the procedure of Sommer.[2] It can also be purchased from PCR, Inc.; Aldrich Chemical Company, Inc.; Fluka A. G., Petrarch Systems, Inc.; and Tokyo Kasei Kogyo Co., Ltd. The checkers employed material from Petrarch.
6. The use of more than 1.2 equiv of allyltrimethylsilane is essential for shortening the reaction time as well as to avoid contamination of the product by unreacted benzalacetone.
7. Disappearance of benzalacetone and appearance of product can be readily monitored by thin-layer or gas chromatographic analysis on a 1-m column packed with 20% Silicone SE-30 at 180°C. The reaction should be stopped as soon as disappearance of benzalacetone is confirmed.
8. Gas chromatographic analysis of the product on a 1-m column packed with 20% Silicone SE-30 at 180°C should give a single peak. The product has the following spectral properties: IR (film) cm^{-1}: 1710, 1630 (C=C); ^{1}H NMR (CDCl$_3$) δ: 1.97 (s, 3 H, CH$_3$CO), 2.35 (t, 2 H, J = 7.5, CH$_2$C=C), 2.72 (d, 2 H, J = 7.5, CH$_2$CO), 3.27 (quintet, 1 H, J = 7.5, PhCH), 4.8–5.1 (m, 2 H, CH$_2$=C), 5.4–5.9 (m, 1 H, CH=C), 7.0–7.4 (m, 5 H, aromatic).

3. Discussion

This procedure is general for the conjugate allylation of α,β-unsaturated ketones with allylsilanes.[3] Some representative examples are listed in Table I. The main advantages of the method are its wide generality and the ready availability of the necessary starting materials. The procedure is often useful for the preparation of δ,ε-unsaturated ketones that cannot be obtained in satisfactory yield by the use of allylcuprate (e.g., entry 13) reagents.[4] Another useful aspect of the reaction is the regiospecific coupling of the allyl group. Examples of this feature can be seen in entries 2 and 5. Although cyclic as well as acyclic α,β-unsaturated ketones give satisfactory results, the reaction is slower in sterically hindered systems (entries 13 and 14). However, even in these cases, good yields are obtained by using excess allylsilane and by conducting the reaction at higher

TABLE I

CONJUGATE ALLYLATION OF α,β-ENONES WITH ALLYLSILANES PROMOTED BY TITANIUM TETRACHLORIDE[a]

Entry	Allylsilane	α,β-Enone	Conditions Temp., °C, time	δ,ϵ-Enone	Yield (%)[b]
1	I[c]	$CH_2=CHCOCH_3$	-78, 1 min	$CH_2=CH(CH_2)_3COCH_3$	59
2	II[d]	$CH_2=CHCOCH_3$	-78, 3 hr	$CH_2=C(CH_3)_2CH_2CH_2COCH_3$	79
3	I	$(CH_3)_2C=CHCOCH_3$	25, 5 min	$CH_2=CHCH_2C(CH_3)_2CH_2COCH_3$	87
4	III[e]	$PhCH=CHCOCH_3$[f]	-78, 0.5 min	$CH_2=C(CH_3)CH_2CH(Ph)CH_2COCH_3$	69
5	IV[g]	$PhCH=CHCOCH_3$	-78, 5 hr	$CH_2=CHCH(CH_3)CH(Ph)CH_2COCH_3$	76
6	I	$PhCH=CHCOPh$	-78, 1 min	$CH_2=CHCH_2CH(Ph)CH_2COCH_3$	96
7	I	(cyclopent-2-enone)	-78, 2 hr	(3-allylcyclopentanone)	70
8	III	(cyclopent-2-enone)	-78, 10 min	(3-(2-methylallyl)cyclopentanone)	70
9	I	(2-heptyl-cyclopent-2-enone)	-78, 2 hr	(allyl heptyl cyclopentanone)	54
10	III	(2-butylidenecyclopentanone)	-78, 30 min	(structure)	82[h]
11	I	(cyclohex-2-enone)	-78, 1 hr	(3-allylcyclohexanone)	80[i]
12	III	(cyclohex-2-enone)	-78, 10 min	(3-(2-methylallyl)cyclohexanone)	99
13	I	(octahydronaphthalenone)	-78, 18 hr then -30, 5 hr	(allyl octahydronaphthalenone)	85[k]
14	I	(methyl octahydronaphthalenone)	-78, 2 hr then 0, 15 min	(allyl methyl octahydronaphthalenone)	88

[a]The reaction was carried out on a 1–20-mmol scale in dichloromethane.
[b]Yields after isolation by distillation or thin-layer chromatography.
[c]I = $Me_3SiCH_2CH=CH_2$.
[d]II = $Me_3SiCH_2CH=C(CH_3)_2$.
[e]III = $Me_3SiCH_2C(CH_3)=CH_2$.
[f]Three equivalents of the allylsilane were used.
[g]IV = $trans$-$Me_3SiCH_2CH=CHCH_3$.
[h]A [2 + 2] cycloadduct assigned the structure 1-methyl-1-trimethylsilylmethyl-3-n-propylspiro[3.4]octan-5-one was obtained in 19% yield.
[i]Bp 56–60°C (3 mm), n_D^{20} 1.4719.
[j]Two equivalents of the allylsilane were used.
[k]Bp 83–85°C (0.6 mm), n_D^{20} 1.5111. A diallylated product, assigned the structure 2,8a-diallyl-3,4,4a,5,6,7,8,8a-octahydronaphthalene, was obtained as a forerun in less than 5% yield.

temperature. Since the allyl group can be modified by the regioselective addition of various reagents to the double bond,[5,6] the method is applicable to the synthesis of a wider variety of compounds than are shown in the Table. By oxidation of the double-bond 1,5-diketones may be obtained.[7] Conjugate allylation with allylsilanes can be used in conjunction with a suitable electrophile to achieve "one-pot" double alkylation at the adjacent vinyl position of an α,β-unsaturated ketone.[8] The method has also been utilized in the synthesis of perhydroazulenones.[9] Allylsilanes also undergo regioselective, Lewis acid-catalyzed reaction with carbonyl compounds,[10] acetals,[11] α,β-unsaturated acetals,[12] acyl halides,[13] tertiary alkyl halides,[14] and oxiranes.[14] Such allylations can also be achieved by using allystannanes.[15]

1. Department of Chemistry, Faculty of Science, Tohoku University, Sendai 980, Japan.
2. Sommer, L. H.; Tyler, L. J.; Whitmore, F. C. *J. Am. Chem. Soc.* **1948**, *70*, 2872; Sakurai, H.; Hosomi, A.; Kumada, M. *J. Org. Chem.* **1969**, *34*, 1764; Abel, E. W.; Rowley, R. J. *J. Organomet. Chem.* **1975**, *84*, 199.
3. Hosomi, A.; Sakurai, H. *J. Am. Chem. Soc.* **1977**, *99*, 1673.
4. House, H. O.; Umen, M. J. *J. Org. Chem.* **1972**, *37*, 2841; House, H. O.; Fisher, Jr., W. F. *J. Org. Chem.* **1969**, *34*, 3615; House, H. O.; Wilkins, J. M. *J. Org. Chem.* **1978**, *43*, 2443.
5. Hosomi, A.; Saito, M.; Sakurai, H. *Tetrahedron Lett.* **1980**, *21*, 3783, **1979**, 429; Hosomi, A.; Sakurai, H. *Tetrahedron Lett.* **1978**, 2589; Hosomi, A.; Hashimoto, H.; Sakurai, H. *J. Org. Chem.* **1978**, *43*, 2551; *Tetrahedron Lett.* **1980**, *21*, 951; Hosomi, A.; Shirahata, A.; Sakurai, H. *Chem. Lett.* **1978**, 901.
6. For a review, see Sakurai, H. *Pure Appl. Chem.* **1982**, *54*, 1.
7. Hosomi, A.; Kobayashi, H.; Sakurai, H. *Tetrahedron Lett.* **1980**, *21*, 955; Yanami, T.; Miyashita, M.; Yoshikoshi, A. *J. Org. Chem.* **1980**, *45*, 607; Pardo, R.; Zahra, J.-P.; Santelli, M. *Tetrahedron Lett.* **1979**, 4557.
8. Hosomi, A.; Hashimoto, H.; Kobayashi, H.; Sakurai, H. *Chem. Lett.* **1979**, 245.
9. House, H. O.; Sayer, T. S. B.; Yau, C.-C. *J. Org. Chem.* **1978**, *43*, 2153.
10. Hosomi, A.; Sakurai H. *Tetrahedron Lett.* **1976**, 1295, **1977**, 4041; Calas, R.; Dunogues, J.; Deleris, G.; Pisciotti, F. *J. Organomet. Chem.* **1974**, *69*, C15.
11. Hosomi, A.; Endo, M.; Sakurai, H. *Chem. Lett.* **1976**, 941; Fleming, I.; Pearce, A.; Snowden, R. L. *J. Chem. Soc., Chem. Commun.* **1976**, 182.
12. Hosomi, A.; Endo, M.; Sakurai, H. *Chem. Lett.* **1978**, 499.
13. Pillot, J.-P.; Dunogues, J.; Calas, R. *Tetrahedron Lett.* **1976**, 1871.
14. Fleming, I.; Paterson, I. *Synthesis* **1979**, 446; Sasaki, T.; Usuki, A.; Ohno, M. *J. Org. Chem.* **1980**, *45*, 3559.
15. Hosomi, A.; Iguchi, H.; Endo, M.; Sakurai, H. *Chem. Lett.* **1979**, 977; Naruta, Y.; Ushida, S.; Maruyama, K. *Chem. Lett.* **1979**, 919; Maruyama, K.; Naruta, Y. *Chem. Lett.* **1978**, 431.

ENANTIOSELECTIVE ADDITION OF BUTYLLITHIUM IN THE PRESENCE OF THE CHIRAL COSOLVENT DDB: (R)-(+)-1-PHENYL-1-PENTANOL

[Benzenemethanol, α-butyl-, (R)-]

Submitted by DIETER SEEBACH and AUGUST HIDBER[1]
Checked by M. F. SEMMELHACK and CHARLES SHUEY

1. Procedure

As shown in Figure 1, a dry, 1-L, three-necked flask is equipped with an overhead stirrer bearing a four-bladed propeller of ca. 2.5-cm diameter driven by a strong, safely connected motor A (Note 1), a rubber septum, and a three-way stopcock. The air in the flask is replaced by dry argon or nitrogen, the pressure of which is maintained during the reaction at ca. 50 mm above atmospheric pressure with a mercury bubbler (Note 2). A second stirrer (motor B, Figure 1) to agitate the bath is attached next to the flask with the propeller just below the bottom of the flask. Finally, a 4.5-cm × 20-cm test tube is held next to the bath stirrer. The entire apparatus (Figure 1) is mounted well above the bench to allow for immersion of the flask, bath stirrer, and tube into cooling baths and for exchange of bulky bath containers with the aid of a lab jack. The flask is charged (Note 3) with 400 mL of 2-methylbutane (isopentane) (Note 4) and 24.6 g (27.5 mL, 0.12 mol) of (S,S)-(+)-N,N,N',N'-tetramethyl-1,4-diamino-2,3-dimethoxybutane (DDB) (Note 5). A methanol–dry ice bath is raised to immerse the flask and cool the contents to −78°C with slow stirring, whereupon 0.021 mol of butyllithium (13.5 mL of a 1.56 M solution in hexane) (Note 6) is added within a few minutes. A second cooling bath is prepared in a ca. 7-L Dewar cylinder (Note 7) by pouring liquid nitrogen into a stirred (glass rod) mixture of methylcyclohexane / isopentane (3 : 2) (Note 8) until about half of the liquid has solidified and a slush has been formed, the temperature of which is ca. −140°C (Note 9). The reaction flask is cooled to the lower temperature by exchanging baths and waiting for 15 min with bath stirring. The bath is temporarily lowered and cooled until again half frozen by pouring in liquid nitrogen and manual agitation (Note 10). From then on, cooling is kept constant by filling the tube in the stirred bath at intervals with liquid nitrogen (Note 10). A solution of 2.12 g (0.020 mol) of benzaldehyde (Note 11) in 20 mL of isopentane (Note 4) is added dropwise (Note 12) over 15 min to the vigorously stirred (ca. 1000 rpm) reaction mixture. After completion of the addition (ca. 0.5 hr), the bath is removed, the flask is warmed to ca. 0°C (Note 13), and the contents are poured into a 1-L separatory funnel containing 150 mL of ice-cold 2 N aqueous hydrochloric acid. The aqueous layer is extracted twice with 70 mL of hexane and saved for recovery of the chiral auxiliary agent DDB (Note 14). The

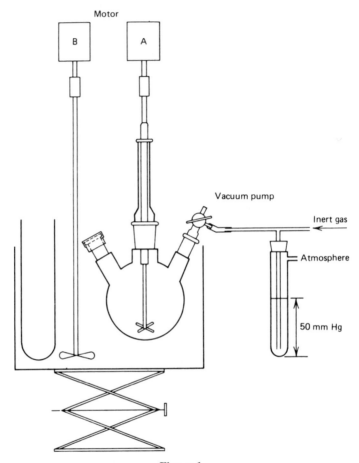

Figure 1

combined organic layers are sequentially washed with saturated aqueous bicarbonate and sodium chloride solutions and concentrated in a rotary evaporator to ca. 200 mL. The solution is then transferred to a 500-mL separatory funnel and vigorously shaken with 40 mL of a saturated aqueous sodium bisulfite solution to precipitate the bisulfite adduct of unreacted benzaldehyde (Note 15). After filtration (if necessary) the residue and the aqueous phase are washed with hexane. The combined organic solution is dried over anhydrous magnesium sulfate and concentrated by rotary evaporation. Simple distillation yields 2.60–2.95 g (80–90%) of 1-phenyl-1-pentanol, bp 54–56°C (0.02 mm), $[\alpha]_D = 6.13°$ (neat), (Note 16), optical yield 30% (Note 17).

2. Notes

1. The checkers used a conventional, flat, crescent-shaped Teflon blade, 8 cm long.
2. This is done as previously described in *Organic Syntheses* procedures: Seebach, D.; Beck, A. K. *Org. Synth., Coll. Vol. VI* **1988,** 869; Enders, D.; Pieter, R.;

Seebach, D. *Org. Synth., Coll. Vol. VI* **1988,** 542. All connections should be securely fastened.

3. All additions of solvents and reagents are carried out through the rubber septum with dry, appropriately sized, and argon-flushed syringes with hypodermic needles. Because of its low boiling point, it is advantageous to force isopentane into the 100-mL syringe by applying pressure to the storage flask.

4. Isopentane (bp 28°C, ~95% 2-methylbutane), was purchased from Fluka AG, freshly distilled from P_2O_5, and stored under inert gas pressure.

5. DDB is presently available from Aldrich Chemical Company, Inc. For its preparation, see p. 000 in this volume. DDB is hygroscopic and must be refluxed for some time and freshly distilled from lithium aluminum hydride (bp 38°C/0.01 mm) prior to use. The submitters used material with $[\alpha]_D$ 14.7°; the checkers' sample showed $[\alpha]_D$ 14.3°.

6. Butyllithium was purchased from Metallgesellschaft, Frankfurt, and titrated for active alkyllithium using diphenylacetic acid as an indicator: Kofron, W. G.; Baclawski, L. M. *J. Org. Chem.* **1976,** *41,* 1879.

7. If no such Dewar container is available, two appropriately sized plastic buckets with a layer of styrofoam particles between the inner and outer bucket can be used.

8. The mixture was used as purchased from Fluka AG. The submitters have occasionally used, as a bath liquid, petroleum ether (bp 40–60°C) of unknown composition or pure isopentane (mp −160°C). In such cases, temperature control is necessary; it was achieved with a platinum temperature sensor inside the reaction mixture.

9. The checkers used a thermocouple to verify the temperature of the cooling bath.

10. The coolant must not be poured directly into the bath, because local overcooling can cause partial freezing of the reaction mixture, which is clear and homogeneous before addition of the aldehyde. If freezing should occur, the flask is temporarily warmed slightly by removing the bath.

11. Benzaldehyde was obtained from Fluka AG or Aldrich Chemical Company, Inc., and freshly distilled under reduced pressure (40°C/3 mm).

12. Clear drops of the aldehyde solution must fall from the tip of the needle directly into the reaction mixture. If the needle is inserted too far, the aldehyde can freeze and clog the needle; it is thawed by extracting the needle tip into the upper, warmer part of the neck.

13. A slow method is to wait until the ice that has condensed on the walls of the flask has all melted. Alternatively, the flask may be immersed in a methanol bath.

14. The combined aqueous layers of several runs are saturated with potassium hydroxide by adding KOH pellets with cooling. DDB separates on top of the aqueous phase and is extracted with ether. Distillation leads to ~90% recovery (bp 42–43°C/0.05 mm).

15. The checkers observed no precipitate formation at this point.

16. In five runs carried out by the submitters at temperatures between −140 and −150°C, the specific rotations of phenylpentanol (d_4^{20} 0.967) ranged from

$[\alpha]_D$ 5.95 to 7.0° (29–34% optical yield; see Note 17). At dry ice temperature, the optical yields are only half as high.[2] The checkers obtained specific rotations of $[\alpha]_D$ 5.87° and 6.05° (28 and 29% optical yield).

17. For optically pure 1-phenyl-1-pentanol a specific rotation of $[\alpha]_D^{25}$ 20.7° (neat) is reported.[3]

3. Discussion

The optically active form of 1-phenyl-1-pentanol has been prepared by a variety of methods.[4,5] The present procedure is a modification and extended description of our previously published[2,6] chiral solvent method. DDB and other auxiliary agents from tartaric acid lead to a wide range of optically active products from achiral components with prochiral centers (enantioselective syntheses). A list of examples of DDB applications is found in the accompanying procedure describing its preparation from tartaric acid.

1. Laboratorium für Organische Chemie der Eidgenössischen Technischen Hochschule, ETH-Zentrum, Universitätstrasse 16, CH-8092 Zürich, Switzerland.
2. Seebach, D.; Kalinowski, H.-O.; Bastani, B.; Crass, G.; Daum, H.; Dörr, H.; DuPreez, N. P.; Ehrig, V.; Langer, W.; Nüssler, C.; Oei, H.-A.; Schmidt, M. *Helv. Chim. Acta* **1977**, *60*, 301; Langer, W.; Seebach, D. *Helv. Chim. Acta* **1979**, *62*, 1701, 1710; Seebach, D.; Crass, G.; Wilka, E.-M.; Hilvert, D.; Brunner, E. *Helv. Chim. Acta* **1979**, *62*, 2695.
3. Horeau, J.; Guetté, J. P.; Weidmann, R. *Bull. Soc. Chim. Fr.* **1966**, 3513.
4. For reviews of asymmetric syntheses, see: (a) Morrison, J. D.; Mosher, H. S. "Asymmetric Organic Reactions"; Prentice Hall: Englewood Cliffs, NJ, 1971; (b)Izumi, Y.; Tai, A. "Stereo-Differentiating Reactions"; Academic Press: New York, 1977; (c) Kagan, H. B.; Fiaud, J. C. *Top. Stereochem.* **1978**, *10*, 175.
5. Mukaiyama, T.; Soai, K.; Sato, T.; Shimizu, H.; Suzuki, K. *J. Am. Chem. Soc.* **1979**, *101*, 1455.
6. Seebach, D.; Hidber, A. *Chimia* **1983**, *37*, 449.

FORMYL TRANSFER TO GRIGNARD REAGENTS WITH
N-FORMYLPIPERIDINE: 3-PHENYLPROPIONALDEHYDE

$$PhCH_2CH_2Cl \; + \; Mg \; \longrightarrow \; PhCH_2CH_2MgCl$$

$$PhCH_2CH_2MgCl \; + \; \boxed{}N\text{-}CHO \; \longrightarrow \; PhCH_2CH_2CHO$$

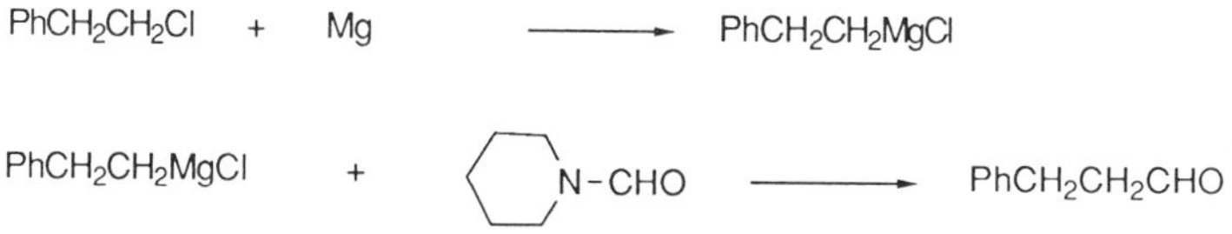

Submitted by George A. Olah and Massoud Arvanaghi[1]
Checked by David Heiler and Martin F. Semmelhack

1. Procedure

Magnesium (2.88 g, 0.12 mol), 300 mL of anhydrous tetrahydrofuran (Note 1), and 10 mg of iodine are placed in a 1-L, three-necked, round-bottomed flask fitted with a stirrer, dropping funnel with a pressure-equalizing tube, and a reflux condenser connected to a nitrogen flow line. Nitrogen is passed through the solvent for 15 min and a constant flow of nitrogen is maintained throughout the reaction. A solution of 14.06 g (0.1 mol) of (2-chloroethyl)benzene (Note 2) in 50 mL of tetrahydrofuran is placed in the dropping funnel. About 2 mL of this solution is added to the reaction mixture and the reaction is initiated by gently heating the flask (with a heat gun). Once the reaction has started, as evidenced by the disappearance of iodine color, the rest of the (2-chloroethyl)benzene solution is added dropwise at such a rate that a gentle reflux is maintained throughout the addition. The resulting solution is stirred for an additional 1 hr at 23°C, followed by heating at reflux for 8 hr. The reaction vessel is cooled to 0°C and a solution of 13.56 g (0.12 mol) of N-formylpiperidine (Note 3) in 50 mL of dry tetrahydrofuran is added dropwise (Note 4). The mixture is brought to 23°C and stirred for another 15 min.

The reaction mixture is quenched by the addition of 25 mL of ice water and slowly acidified to pH 2 with 75 mL of 3 N hydrochloric acid. The organic layer is separated and the aqueous layer is extracted with three 75-mL portions of ether. The extracts are combined with the original ether layer, washed successively with 50 mL of water, two 50-mL portions of aqueous 10% sodium bicarbonate, and 50 mL of saturated sodium chloride solution, and dried over anhydrous magnesium sulfate. After the magnesium sulfate is removed by filtration, the solvent is removed at aspirator vacuum on a rotary evaporator and the residue is distilled through a short column to give 8.8–10.2 g (66–76%) of 3-phenylpropionaldehyde, bp 87°C (1.0 mm) (Notes 5–7).

1. Technical-grade tetrahydrofuran was predried for a few days over sodium hydroxide. It was then heated under reflux over sodium wire with benzophenone until a permanent blue color developed and distilled with exclusion of atmospheric moisture. (*Caution: See p. 976 of Org. Synth., Coll. Vol. V for a warning regarding purification of tetrahydrofuran.*)

2. The (2-chloroethyl)benzene was purchased from Eastman Organic Chemicals and used without further purification.

3. *N*-Formylpiperidine was obtained from Reilly Tar and Chemicals or from Aldrich Chemical Company and used without further purification.

4. Too rapid addition of *N*-formylpiperidine should be avoided as it can result in a cake-like solid that hinders mixing of the reaction mixture. Efficient stirring is crucial to optimum yields.

5. The reported[2] boiling point for 3-phenylpropionaldehyde is 104–105°C (13 mm).

6. The spectral properties of the product are as follows: ^{13}C NMR (CDCl$_3$) δ: 27.9 (t, $-CH_2-CH_2-CHO$), 45.1 (t, $-CH_2-CHO$), 126.1 (d, *para*), 128.2 (d, *ortho*), 128.5 (d, *meta*), 140.2 (s, *ipso*), 201.4 (d, $-CHO$); ^{1}H NMR (CDCl$_3$) δ: 2.77 (m, $-CH_2-CHO$); 2.95 (m, $-CH_2-CH_2-CHO$), 7.16–7.33 (m, aromatic), 9.80 (t, $-CHO$); IR cm^{-1}: 2700, 1710.

7. (2-Bromoethyl)benzene can be used instead of (2-chloroethyl)benzene; anhydrous diethyl ether is used as the solvent instead of tetrahydrofuran.

3. Discussion

The procedure described here is a one-step conversion of (2-chloroethyl)benzene to 3-phenylpropionaldehyde. The method is general and characterized by good yields, mild conditions, and easy preparation of 3-phenylpropionaldehyde in pure form from readily available starting materials. Several methods are described in the literature for the preparation of 3-phenylpropionaldehyde, including dry distillation of calcium formate with calcium hydrocinnamate,[3] sodium amalgam reduction, and deprotection of cinnamaldehyde dimethyl acetal,[4] or formation from heterocyclic system.[5,6] The present method has been shown[7] to be applicable to a wide variety of organolithium and Grignard reagents.

1. Donald P. and Katherine B. Loker Hydrocarbon Research Institute and Department of Chemistry, University of Southern California, University Park, Los Angeles, CA 90089-1661.
2. "Dictionary of Organic Compounds"; Oxford University Press: New York, 1965.
3. Miller, W.; Rohde, G. *Ber.* **1890**, *23*, 1079.
4. Dollfus, W. *Ber.* **1893**, *26*, 1971.
5. Meyers, A. I.; Nabeya, A.; Adickes, H. W.; Politzer, I. R. *J. Am. Chem. Soc.* **1969**, *91*, 763.
6. Altman, L. J.; Richheimer, L. *Tetrahedron Lett.* **1971**, 4709.
7. Olah, G. A.; Arvanaghi, M. *Angew. Chem. Int. Ed. Engl.* **1981**, *20*, 878; Olah, G. A.; Arvanaghi, M. *Chem. Revs.* **1987**, *87*, 671.

PHENYL VINYL SULFONE AND SULFOXIDE

[Benzene, (ethenylsulfonyl)- and benzene, (ethenylsulfiny)-]

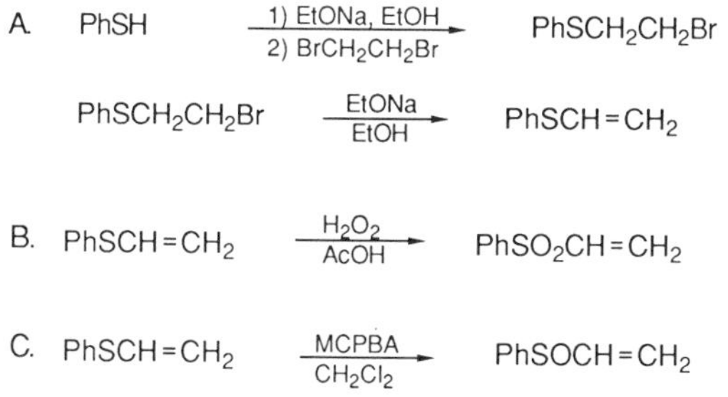

A. PhSH $\xrightarrow[\text{2) BrCH}_2\text{CH}_2\text{Br}]{\text{1) EtONa, EtOH}}$ PhSCH$_2$CH$_2$Br

PhSCH$_2$CH$_2$Br $\xrightarrow[\text{EtOH}]{\text{EtONa}}$ PhSCH=CH$_2$

B. PhSCH=CH$_2$ $\xrightarrow[\text{AcOH}]{\text{H}_2\text{O}_2}$ PhSO$_2$CH=CH$_2$

C. PhSCH=CH$_2$ $\xrightarrow[\text{CH}_2\text{Cl}_2]{\text{MCPBA}}$ PhSOCH=CH$_2$

Submitted by LEO A. PAQUETTE and RICHARD V. C. CARR[1]
Checked by WAYNE SCHNATTER and MARTIN F. SEMMELHACK

1. Procedure

Caution! 1-Phenylthio-2-bromoethane is a powerful alkylating agent that causes severe skin blistering. Although the present one-pot procedure eliminates the cumbersome handling of this intermediate, due care must be exercised to avoid exposure to this substance.

A. *Phenyl vinyl sulfide.* In a 1-L, three-necked, round-bottomed flask fitted with magnetic stirrer, condenser, addition funnel, and nitrogen-inlet tube is placed 400 mL of ethanol. Sodium metal (23 g, 1 g-atom), cut into small pieces, is added with stirring. When conversion to sodium ethoxide is complete (5–15 min), the stopper of the addition funnel is removed under a positive flow of nitrogen, and benzenethiol (110 g, 1 mol) is poured into the addition funnel. The stopper is put in place, and the benzenethiol is added over 15–20 min to the cloudy, gray sodium ethoxide solution. The reaction mixture warms spontaneously and becomes clear brown. At 25°C this solution is transferred by stainless-steel cannula (Note 1) over 45 min to a stirred solution of 1,2-dibromoethane (272 g, 1.45 mol) in ethanol (28 mL) contained in a 2-L, three-necked round-bottomed flask equipped with a mechanical stirrer, addition funnel, reflux condenser, nitrogen-inlet tube, and internal thermometer (Note 2). The reaction temperature is maintained at 25–30°C by cooling with an ice bath. The mixture is stirred under nitrogen for 30 min and treated for an additional 30 min with ethanolic sodium ethoxide prepared from 40 g (2.17 g-atom) of sodium and 800 mL of ethanol (Note 3). The resulting mixture is stirred at reflux for 8 hr (Note 4), cooled, and treated with 750 mL of benzene and 750 mL of water. The organic layer is separated, washed with water (2 × 50 mL) and brine (100 mL), and concentrated by rotary evaporation. The yellow oil that results is distilled to give 70–87 g (50–65%) of phenyl vinyl sulfide, bp 91–93°C/20 mm (Notes 5 and 6).

B. *Phenyl vinyl sulfone.* In a 250-mL, three-necked, round-bottomed flask fitted with a magnetic stirrer, condenser, addition funnel, and thermometer is placed 19.7 g (0.145

mol) of phenyl vinyl sulfide dissolved in 70 mL of glacial acetic acid. Hydrogen peroxide (30%, 56 mL, 0.5 mol) is added slowly at such a rate to maintain a reaction temperature of 70°C (Note 7). The reaction mixture is heated at reflux for 20 min, cooled, and treated with ether (150 mL) and water (200 mL). The organic phase is separated, washed with water (50 mL) and brine (50 mL), and concentrated at 70°C/0.3 mm for 3 hr to afford 18–19 g (74–78%) of phenyl vinyl sulfone as a colorless solid, mp 64–65°C. Although this material is sufficiently pure for most purposes, recrystallization from hexane affords colorless crystals, mp 66–67°C (Note 8).

C. *Phenyl vinyl sulfoxide.* A 500 mL, three-necked, round-bottomed flask equipped with a dropping funnel and magnetic stirrer is charged with 20 g (0.147 mol) of phenyl vinyl sulfide and 250 mL of dichloromethane. The solution is stirred and cooled to −78°C while a solution of *m*-chloroperbenzoic acid (25.4 g, 1.0 equiv) in 200 mL of dichloromethane is added dropwise during a 30-min period. The mixture is stirred and warmed to room temperature for 1 hr in a water bath at 30°C. The mixture is then poured into 300 mL of saturated sodium bicarbonate solution, and the mixture is extracted with three 250-mL portions of dichloromethane. The combined organic extracts are washed with three 250-mL portions of water and dried over anhydrous magnesium sulfate. The solvent is removed by rotary evaporation and the residual liquid is distilled to afford 15–16 g (68–70%) of phenyl vinyl sulfoxide as a colorless liquid, bp 98°C/0.6 mm (Notes 9 and 10).

2. Notes

1. The cannula is a stainless-steel tube, 16-gauge, sharpened to a needle at both ends, and 60 cm long. One end is placed through a rubber septum into the flask containing the 1,2-dibromoethane solution, while the other end is positioned under the surface of the benzenethiolate solution. Control of the nitrogen pressure allows slow transfer of the benzenethiolate solution.

2. The yield in the previously published method for the preparation of this sulfide is low, affording chiefly 1,2-bis(phenylthio)ethane.[2] The problem is overcome here by utilization of an inverse addition procedure.

3. Alternatively, dry powered sodium ethoxide may be substituted with a corresponding reduction of the reaction volume.

4. Thin-layer chromatographic analysis at this stage shows that 1-phenylthio-2-bromoethane is absent.

5. This product has the following spectral properties: IR (neat) cm^{-1}: 3040, 1585, 1475, 1435, 1085, 1020, 950, 735, and 680; ^{1}H NMR (chloroform-*d*) δ: 5.25 (superimposed doublets, 2 H, J = 12 and 18, terminal vinyl), 6.50 (dd, 1 H, J = 12 and 18, olefinic,), 7.32 (m, 5 H, aromatic).

6. When stored at room temperature, phenyl vinyl sulfide becomes yellow-colored within 1 day and a black syrup after 1 week. This decomposition can be substantially retarded by storage under a nitrogen or argon atmosphere in a freezer.

7. The submitter observed the temperature increase to 70°C during addition of the first 10 mL of hydrogen peroxide. The checkers noted that the mixture never rose in temperature to 70°C.

8. This product has the following spectral properties: IR (CHCl$_3$) cm^{-1}: 3020, 1445, 1380, 1315, 1145, 1080, and 965; ^{1}H NMR (chloroform-d) δ: 5.96 (d, 1 H, $J = 10$, olefinic), 6.33 (d, 1 H, $J = 17$, olefinic), 6.75 (dd, 1 H, $J = 10$ and 17, olefinic), 7.55 (m, 3 H, aromatic), 7.85 (m, 2 H, aromatic).

9. Earlier citations[3] report bp 105–110°C (1.5 mm) and 93–95°C (0.2 mm).

10. This product has the following spectral properties: IR (neat) cm^{-1}: 3025, 1720, 1680, 1480, 1440, 1045, 750, and 690; ^{1}H NMR (chloroform-d) δ: 5.63–6.17 (m, 2 H, olefinic H), 6.44–6.87 (m, 1 H, olefinic H), 7.10–7.55 (m, 5 H, aromatic H).

3. Discussion

The procedure for oxidation of the sulfide to the sulfone is based on that reported earlier by Bordwell and Pitt.[4] The synthetic utility of phenyl vinyl sulfone and sulfoxide derives not only from their ability to serve as excellent Michael acceptors toward such reagents as enolate anions and organometallics[5-12] but also as moderately reactive dienophiles in Diels–Alder reactions.[13-16] The resulting adducts, in turn, can be chemically modified so that these electron-deficient olefins serve as useful synthons for acetylene,[13] ethylene,[14] terminal olefins,[15] vinylsilanes,[17] and ketene[18] in [4 + 2] cycloadditions. Phenyl vinyl sulfone undergoes ready cycloaddition to Danishefsky's diene in the first step of a protocol for the regiospecific γ-alkylation of 2-cyclohexenones.[19] Furthermore, the ready lithiation of phenyl vinyl sulfones[20] and sulfoxides[21] represents a convenient route to α-(phenylsulfonyl) and α-(phenylsulfinyl)vinyllithium reagents.

The method described here for the preparation of phenyl vinyl sulfoxide is superior to that involving reaction of ethyl phenyl sulfinate with vinylmagnesium bromide.[13]

1. Department of Chemistry, The Ohio State University, Columbus, OH 43210.
2. Claisse, J. A.; Davies, D. I.; Alden, C. K. *J. Chem. Soc. (C)* **1966**, 1498.
3. Ford-Moore, A. H. *J. Chem. Soc.* **1949**, 2126; Barbierí, G.; Cinquini, M.; Colonna, S.; Montanari, F. *J. Chem. Soc. (C)* **1968**, 659.
4. Bordwell, F. G.; Pitt, B. M. *J. Am. Chem. Soc.* **1955**, 77, 572.
5. Kohler, E. P.; Potter, H. *J. Am. Chem. Soc.* **1935**, 57, 1316.
6. Posner, G. H.; Brunelle, D. J. *J. Org. Chem.* **1973**, 38, 2747; Posner, G. H.; Mallamo, J. P.; Miura, K. *J. Am. Chem. Soc.* **1981**, 103, 2886 and references cited therein.
7. Fiandanese, V.; Marchese, G.; Naso, F. *Tetrahedron Lett.* **1978**, 5131; De Chirico, G.; Fiandanese, V.; Marchese, G.; Naso, F.; Sciacovelli, O. *J. Chem. Soc., Chem. Commun.* **1981**, 523.
8. Cory, R. M.; Renneboog, R. M. *J. Chem. Soc., Chem. Commun.* **1980**, 1081.
9. Agawa, T.; Yoshida, Y.; Komatsu, M.; Ohshiró, Y. *J. Chem. Soc., Perkin Trans. 1* **1981**, 751.
10. Ponton, J.; Helquist, P.; Conrad, P. C.; Fuchs, P. L. *J. Org. Chem.* **1981**, 46, 118.
11. Koppel, G. A.; Kinnick, M. D. *J. Chem. Soc., Chem. Commun.* **1975**, 473.
12. Abbott, D. J.; Stirling, C. J. M. *J. Chem. Soc. (C)* **1969**, 818; Tsuchihashi, G.; Mitamura, S.; Inoue, S.; Ogura, K. *Tetrahedron Lett.* **1973**, 323; Tsuchihashi, G.; Mitamura, S.; Ogura, K. *Tetrahedron Lett.* **1973**, 2469; Barton, D. H. R.; Coates, I. H.; Sammes, P. G.; Cooper, C. M. *J. Chem. Soc., Perkin Trans. 1* **1974**, 1459; Tanikaga, R.; Sugihara, H.; Tanaka, K.; Kaji, A. *Synthesis* **1977**, 299; Sugihara, H.; Tanikaga, R.; Kaji, A. *Bull. Chem. Soc. Jpn.* **1978**, 151, 655; Hori, I; Oishi, T. *Tetrahedron Lett.* **1979**, 4087; Spry, D. O. *Tetrahedron Lett.* **1980**, 21, 1293.

13. Paquette, L. A.; Moerck, R. E.; Harirchian, B.; Magnus, P. D. *J. Am. Chem. Soc.* **1978,** *100*, 1597.
14. Carr, R. V. C.; Paquette, L. A. *J. Am. Chem. Soc.* **1980**, *102*, 853.
15. Little, R. D.; Brown, L. *Tetrahedron Lett.* **1980**, *21*, 2203.
16. Danishefsky, S.; Harayama, T.; Singh, R. K. *J. Am. Chem. Soc.* **1979**, *101*, 7008; Danishefsky, S.; Hirama, M.; Fritsch, N.; Clardy, J. *J. Am. Chem. Soc.* **1979**, *101*, 7013; Danishefsky, S.; Walker, F. *J. Am. Chem. Soc.* **1979**, *101*, 7018.
17. Daniels, R. G.; Paquette, L. A. *J. Org. Chem.* **1981**, *46*, 2901.
18. Little, R. D.; Myong, S. O. *Tetrahedron Lett.* **1980**, *21*, 3339.
19. Paquette, L. A.; Kinney, W. A.; *Tetrahedron Lett.* **1982**, *23*, 131; Kinney, W. A.; Crouse, G. D.; Paquette, L. A. *J. Org. Chem.* **1983**, *48*, 4986.
20. Eisch, J. J.; Galle, J. E. *J. Org. Chem.* **1979**, *44*, 3279.
21. Posner, G. H.; Tang, P. W.; Mallamo, J. P. *Tetrahedron Lett.* **1978**, 3995; Schmidt, R. R.; Speer, H.; Schmid, B. *Tetrahedron Lett.* **1979**, 4277.

2-ALKENYL CARBINOLS FROM 2-HALO KETONES: 2-*E*-PROPENYLCYCLOHEXANOL

[Cyclohexanol, 2-(1-propenyl)-, (*E*)-]

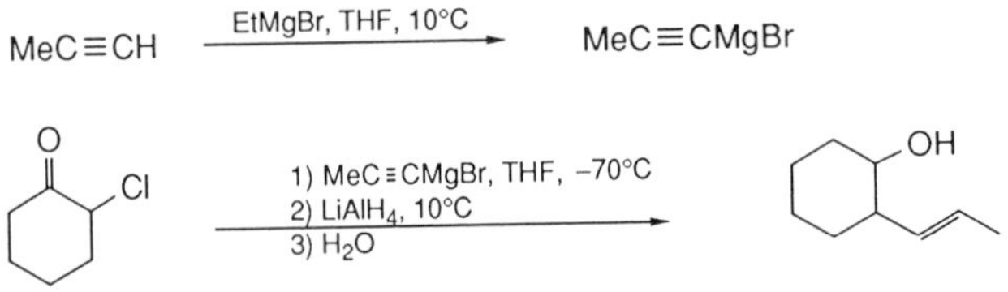

Submitted by P. A. WENDER, D. A. HOLT, and S. McN. SIEBURTH[1]
Checked by PEGGY A. RADEL and CLAYTON H. HEATHCOCK

1. Procedure

A dry, 5-L, four-necked, round-bottomed flask is equipped with an air-driven stirrer (Note 1), 250-mL pressure-equalizing dropping funnel, thermometer, rubber septum, and a nitrogen-inlet tube that, by means of a T-tube, is also connected to a gas bubbler. After being charged with 1200 mL of anhydrous tetrahydrofuran (Note 2), the flask is swept with dry nitrogen and maintained under an atmosphere of nitrogen throughout the remainder of the reaction. A solution of ethylmagnesium bromide in diethyl ether (1.1 mol, 380 mL, 2.9 *M*) is transferred to the flask and the flask is then cooled to below 10°C by means of an ice–water bath (Note 3). Propyne is bubbled through the cooled, stirred solution (Note 4) at such a rate that a small amount escapes through the nitrogen inlet–gas bubbler. Propyne addition is continued for 2.5 hr, at which time approximately 100 g (2.5 mol) of propyne has been used (Note 5) and the internal temperature has risen 5–10°C. The ice–water bath is then replaced with a dry ice–acetone bath and the mixture is cooled to ca. -70°C. A solution of 2-chlorocyclohexanone (1 mol, 132.6 g) (Note 6) in 50 mL of tetrahydrofuran is added dropwise from the addition funnel over

1.5 hr so as to maintain the temperature below −65°C (Note 7). After stirring for an additional 1.5 hr at −70°C (Note 8), the dry ice–acetone bath is replaced with an ambient-temperature water bath and the reaction mixture is allowed to warm slowly. When the temperature reaches 10°C, a solution of lithium aluminum hydride in tetrahydrofuran (1 mol, 1000 mL, 1 M) is added by cannula (Note 9). After addition of the lithium aluminum hydride, the mixture is stirred at ambient temperature for 3–5 hr, at which time the solids have dissolved and reaction is complete (Notes 10 and 11). The solution is then cooled to 5°C by means of an ice–water bath. The reaction is quenched by careful, dropwise addition of 38 mL of water over 2 hr to maintain the temperature below 20°C. The solution becomes somewhat cloudy at this point and 2000 mL of hexanes is added. The addition of 38 mL of aqueous 15% sodium hydroxide solution over 15 min is followed by the addition of 100 mL of water over 5 min. Some frothing occurs during the last addition of water and a large amount of white aluminum salts precipitates. After 5 min of stirring, 100 g of anhydrous sodium sulfate is added and stirring is continued for another 5 min. The thick mixture is then filtered by suction through Celite using a 200-mm-diameter Büchner funnel. The solids are removed from the funnel, thoroughly washed with 1500 mL of hot tetrahydrofuran (Note 12), and refiltered. This wash is repeated twice and the combined organic solutions are concentrated with a rotary evaporator (15 mm). The residual yellow liquid is distilled under reduced pressure to yield 118.8 g (85%) of 2-E-propenylcyclohexanol as a clear colorless liquid, bp 49–54°C (1 mm) (Note 13).

2. Notes

1. The use of a magnetic stirrer is more convenient than a mechanical stirrer for reactions conducted on small scale (<0.2 mol) or at low concentrations (<0.5 M). However, because of the difficulty encountered in stirring the sometimes thick suspensions associated with concentrated (e.g., 1 M) reaction mixtures, and because of the potentially disastrous results if a stir bar should fracture the flask wall during a large-scale preparation, the submitters strongly recommend the use of an air-driven overhead mechanical stirrer for such large-scale reactions. At a later stage in the reaction, when the propynylmagnesium bromide is cooled to −70°C, the THF solution becomes viscous and rather difficult to stir. The checkers recommend the use of a heavy-duty, air-driven stirrer such as the Fisher Scientific Model 14-508.5. An electrically driven overhead stirrer should not be used, as the hydrogen released during the quench represents a considerable explosion hazard.

2. The submitters used "Baker Analyzed" tetrahydrofuran (0.005% H_2O) without further purification or drying.

3. The submitters used ethylmagnesium bromide solution purchased from Aldrich Chemical Company, Inc. and found it most convenient to measure the required amount by transferring the solution by cannula to a nitrogen-flushed graduated cylinder fitted with a rubber septum. This measured amount of solution is then transferred to the flask by cannula. (See reference 2 for general techniques for handling air-sensitive reagents in this manner.) Some of the Grignard reagent precipitates at this temperature and concentration. There is a tendency for the

ethylmagnesium bromide to clog the cannula. The checkers found it convenient to use a cannula made from 2-mm stainless-steel tubing.

4. Propyne of 99.96% purity, purchased from Liquid Carbonic Company in a lecture bottle, was used without purification and was introduced to the flask by means of a Tygon tube that was attached to a 9-in., 18-gauge hypodermic needle.

5. The amount of propyne used is conveniently determined by weighing the lecture bottle before and after addition. In this case the submitters used an excess of alkyne to ensure complete consumption of the Grignard reagent. In cases where nongaseous, nonvolatile alkynes are to be used, stoichiometric amounts suffice.

6. The submitters used 2-chlorocyclohexanone purchased from Aldrich Chemical Company, Inc. without further purification. Alternatively, this compound can be easily prepared by chlorination of cyclohexanone.[3]

7. A flask should be used that is constructed in such a manner that the chloro ketone solution drips directly from the addition funnel into the reaction mixture. Any portion that flows along the sides of the flask will freeze.

8. Chloro alkoxide formation is essentially complete at this time and can be conveniently monitored by quenching a small aliquot and subjecting it to GLC analysis. Using a 50-m × 0.2-mm OV-1 capillary column at 110°C and a flow rate of 0.87 mL/min (H_2 carrier), the submitters found retention times of 3.2 min for 2-chlorocyclohexanone and 6.7 min and 7.2 for *trans-* and *cis-*1-propynyl-2-chlorocyclohexanols, respectively.

9. Lithium aluminum hydride in tetrahydrofuran was purchased from Aldrich Chemical Company, Inc. and was handled in the fashion described above for the Grignard solution (see Note 3). While solid lithium aluminum hydride can be used (with appropriate changes in the amount of solvent initially used), the hazards of handling this flammable and even explosive reagent (see references 4 and 5) can be reduced by using the preprepared solution. An excess of hydride reagent is necessary to facilitate complete reaction in a reasonable time. When only the stoichiomeric amounts of hydride reagent are used, the reaction is not complete even after several days at room temperature.

10. In reactions run at high concentration the reaction has a tendency to become slightly exothermic at some point, with the temperature increasing by as much as 30°C. Although the reaction is usually complete in less time, and with no reduction in yield of product, this exothermic reaction can be prevented by keeping the flask in a large ambient-temperature water bath, thus buffering temperature changes that apparently initiate the exothermic reaction. The progress of the reaction can be conveniently monitored by TLC or GLC analysis of a quenched aliquot. Using the same GLC conditions as described in Note 8, the retention times for *cis-* and *trans-*2-*E*-propenylcyclohexanols are 4.1 min and 3.9 min, respectively.

11. The checkers noted that a homogeneous solution occurs after about 2 hr. However, TLC and GLC analysis showed that reaction was not complete until 3.5–5 hr.

12. The tetrahydrofuran used for washing the filtercake should be tested for peroxides before use, since the final distillation is carried out almost to dryness.

13. GLC analysis indicated a purity greater than 98% and a cis : trans ratio of 1 : 2. These isomers can be separated by column chromatography and give the following ^{1}H NMR spectra: (CDCl$_3$) δ: cis: 1.1–2.0 (m, 12 H, CH$_2$, CH$_3$, OH), 2.2 (m, 1 H, allylic CH), 3.75 (m, 1 H, carbinol CH), 5.4–5.6 (m, 2 H, CH=CH); trans: 1.0–2.2 (m, 10 H, CH$_2$, OH, allylic CH), 1.70 (d, 3 H, J = 4.9, CH$_3$), 3.1 (m, 1 H, carbinol CH), 5.0–5.8 (m, 2 H, CH=CH).

3. Discussion

A variety of approaches have been employed to effect the preparation of α-alkenyl ketones and carbinols, including reactions of metallo alkenes with epoxides,[6] α-halo ketones,[7] or enolonium ion equivalents[8] and the reactions of ketone enolates with vinyl cation equivalents.[9] The procedure described here offers several advantages over existing methodology. Starting materials and reagents are all commercially available at low cost. Manipulations are simple, and the procedure can be carried out in a single operation, in a single flask, on a small or large (1-mol) scale and in high yield. In addition, as described in more detail elsewhere,[10] this method permits the use of cyclic as well as acyclic halo ketones, bromo ketones instead of chloro ketones, a variety of alkynes including acetylene, conjugated alkynes, 3-silyloxy functionalized alkynes, and other aluminum hydride reagents such as diisobutylaluminum hydride and lithium trimethoxyaluminum hydride. Furthermore, the method provides for complete control over alkene geometry and easy access to trisubstituted alkenes of defined stereochemistry.

TABLE I
PREPARATION OF ALKENYL CARBINOLS

Carbonyl	Alkynylide	Alkenyl Carbinol	Yield (%)
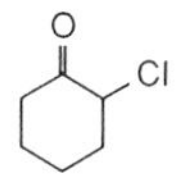	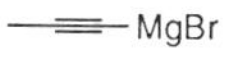 —≡—MgBr	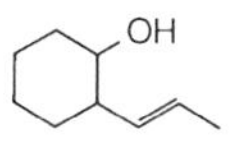	85
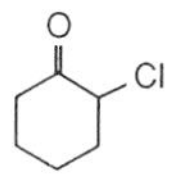	 ⌐≡—Li	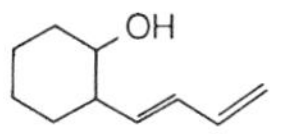	71
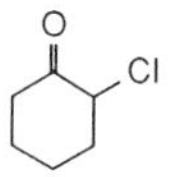	 ⌐≡—Li	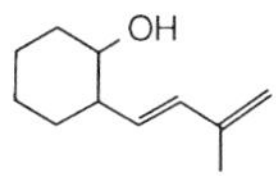	76
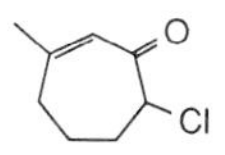	 ⌐≡—Li	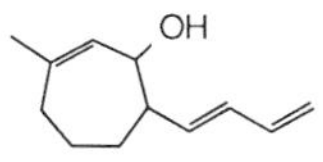	91
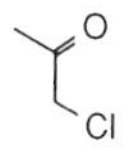	 ⌐≡—Li	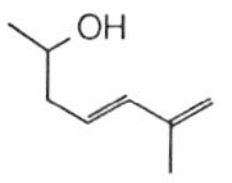	46

Mechanistically, the reaction proceeds through an alkynyl chloro alkoxide, which, when treated with the reducing agent, is hydroaluminated to yield the vinyl alanate, which subsequently undergoes a facile pinacol-like 1,2-rearrangement. Excess hydride reagent reduces the intermediate alkenyl ketone and the resulting 2-alkenyl carbinol is isolated on aqueous workup (Scheme 1). Table I contains representative examples.

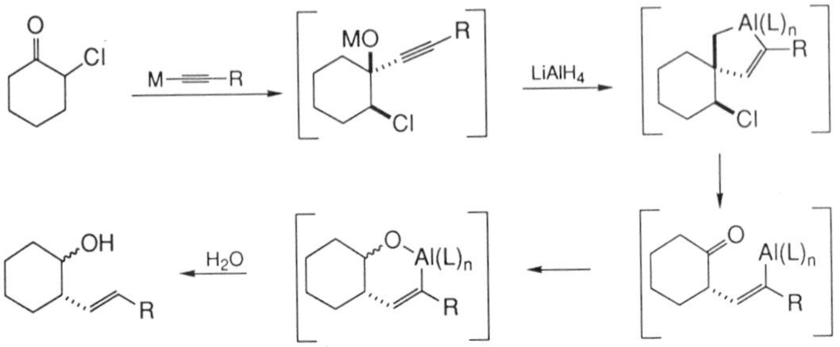

Scheme 1

1. Department of Chemistry, Harvard University, Cambridge, MA 02138 and Department of Chemistry, Stanford University, Stanford, CA 94305.
2. Aldrich Chemical Company, Inc., Bulletin No. A74, "Handling Air-Sensitive Solutions."
3. Newman, M. S.; Farbman, M. D.; Hipsher, H. *Org. Synth., Coll. Vol. III* **1955**, 188.
4. Pizey, S. S. "Synthetic Reagents"; Halstead Press: New York, 1974; Vol. I, Chapter 2.
5. Fieser, L. F.; Fieser, M. "Reagents for Organic Synthesis"; Wiley: New York, 1967; Vol. I, p. 583.
6. (a) Lipshutz, B. H.; Wilhelm, R. S.; Kozlowski, J. A.; Parker, D. *J. Org. Chem.* **1984**, *49*, 3928; (b) Crandall, J. K.; Arrington, J. P.; Hen, J. *J. Am. Chem. Soc.* **1967**, *89*, 6208.
7. (a) Marcou, A.; Normant, H. *C.R. Hebd. Seances Acad. Sci., Ser. C* **1960**, *250*, 359; (b) Nishino, M.; Kondo, H.; Miyake, A. *Chem. Lett.* **1973**, 667; (c) Kato, T.; Kondo, H.; Nishino, M.; Tanaka, M.; Hata, G.; Miyake, A. *Bull. Chem. Soc. Jpn.* **1980**, *53*, 2958; (d) Holt, D. A. *Tetrahedron Lett.* **1981**, *22*, 2243.
8. Wender, P. A.; Erhardt, J. M.; Letendre, L. J. *J. Am. Chem. Soc.* **1981**, *103*, 2114 and references cited therein; (b) Marino, J. P.; Jaen, J. C. *J. Am. Chem. Soc.* **1982**, *104*, 3165.
9. For example, see: (a) Koppel, G. A.; Kinnick, M. D. *J. Chem. Soc., Chem. Commun.* **1975**, 473; (b) van der Veen, R. H.; Cerfontain, H. *J. Chem. Soc., Perkin Trans.* 1 **1985**, 661; (c) Ohnuma, T.; Hata, N.; Fujiwara, H.; Ban, Y. *J. Org. Chem.* **1982**, *47*, 4713; (d) Kowalski, C. J.; Dung, J.-S. *J. Am. Chem. Soc.* **1980**, *102*, 7950; (e) Angoh, A. G.; Clive, D. L. J. *J. Chem. Soc., Chem. Commun.* **1984**, 534; (f) Clive, D. L. J.; Russell, C. G.; Suri, S. C. *J. Org. Chem.* **1982**, *47*, 1632; (g) Hudrlik, P. F.; Kulkarni, A. K. *J. Am. Chem. Soc.* **1981**, *103*, 6251; (h) Kende, A. S.; Fludzinski, P.; Hill, J. H.; Swenson, W.; Clardy, J. *J. Am. Chem. Soc.* **1984**, *106*, 3551; (i) Kosugi, M.; Hagiwara, I.; Sumiya, T.; Migita, T. *Bull. Chem. Soc. Jpn.* **1984**, *57*, 242; (j) Moloney, M. G.; Pinhey, J. T. *J. Chem. Soc., Chem. Commun.* **1984**, 965; (k) Chang, T. C. T.; Coolbaugh, T. S.; Foxman, B. M.; Rosenblum. M.; Simms, N.; Stockman, C. *Organometallics* **1987**, *6*, 2394; (l) Rosenblum, M.; Bucheister, A.; Chang, T. C. T.; Cohen, M.; Marsi, M.; Samuels, S. B.; Scheck, D.; Sofen, N.; Watkins, J. C. *Pure & Appl. Chem.* **1984**, *56*, 129; (m) Negishi, E.; Akiyoshi, K. *Chem. Lett.* **1987**, 1007; (n) Urabe, H.; Kuwajima, I. *Tetrahedron Lett.* **1983**, *24*, 4241.
10. Wender, P. A.; Holt, D. A.; Sieburth, S. M. *J. Am. Chem. Soc.* **1983**, *105*, 3348.

ENANTIOSELECTIVE EPOXIDATION OF ALLYLIC ALCOHOLS: (2S,3S)-3-PROPYLOXIRANEMETHANOL

[Oxiranemethanol, 3-Propyl-, (2S,3S)-]

$$
\text{HO}\diagdown\diagup\diagdown\diagup\diagdown \quad
\xrightarrow[\substack{\text{tert-BuOOH/PhMe} \\ \text{CH}_2\text{Cl}_2}]{\substack{\text{Ti[OCHMe}_2]_4 \\ \text{diethyl (2R,3R)-tartrate}}}
\quad \text{HO}\diagdown\diagup\diagdown\diagup\diagdown
$$

Submitted by J. Gordon Hill,[1] K. Barry Sharpless,[1] Christopher M. Exon,[2] and Ronald Regenye[2]

Checked by Mark H. Norman and Clayton H. Heathcock

1. Procedure

A 2-L, three-necked, round-bottomed flask equipped with a mechanical stirrer with Teflon blades, thermometer, and nitrogen inlet is charged with 1.00 L of methylene chloride (Note 1) and 39.9 mL (38.1 g, 0.134 mol) of titanium(IV) isopropoxide (Note 2). The flask content is stirred and cooled under nitrogen in a dry ice–ethanol bath to −70°C. To the flask is then added 33.1 g (27.5 mL, 0.161 mol) of diethyl (2R,3R)-tartrate (Note 3) and 25.0 g (0.25 mol) of E-2-hexen-1-ol (Note 4). A small volume of methylene chloride is used to ensure complete transfer of each material to the reaction flask. To the flask is then added 184.5 mL (0.50 mol) of 2.71 M anhydrous tert-butyl hydroperoxide in toluene (Note 5) that has been precooled to −20°C (Note 6). The addition causes a temperature increase to −60°C; the temperature of the reaction mixture is allowed to come to 0°C over a 2.0-hr period (Note 7).

A 4-L beaker equipped with a magnetic stirring bar and thermometer is charged with a solution of 125 g of ferrous sulfate and 50 g of tartaric acid in a total volume of 500 mL of deionized water. The solution is stirred and cooled by means of an ice–water bath to 10°C. When the epoxidation reaction mixture reaches 0°C, it is immediately (Note 8) poured into the stirred contents of the beaker. The resulting reaction is mildly exothermic, causing a temperature rise to ca. 20°C (Note 9). After the exothermic reaction has subsided and the temperature has begun to drop (ca. 5 min), the cooling bath is removed and the mixture is stirred at ambient temperature for 30 min. The contents of the beaker are transferred to a 2-L separatory funnel and the aqueous phase is separated and extracted with two 250-mL portions of ether. The combined organic layers are dried over sodium sulfate and filtered. The solvent is removed with a rotary evaporator at 35°C (70 mm) to give 85.9–89.9 g (Note 10) of pale-amber oil.

A 2-L, three-necked, round-bottomed flask equipped with a thermometer and a mechanical stirrer with Teflon blades is charged with a solution of the reaction product in 750 mL of ether. The contents of the flask are cooled in an ice–water bath to 3°C. To the flask is added a precooled (3°C) solution of 20 g (0.50 mol) of sodium hydroxide in 500 mL of brine (Note 11). The two-phase mixture is stirred vigorously for 1 hr with continued cooling (Note 12) and then is transferred to a separatory funnel. The aqueous phase is separated and extracted with two 150-mL portions of ether (Notes 13 and 14). The combined organic solution (Note 15) is dried over sodium sulfate and filtered.

Solvent removal with a rotary evaporator at 35°C (70 mm) followed by concentration with the rotary evaporator at 35°C (12 mm) for 1.0 hr gives 24.7–25.0 g of crude (2S,3S)-3-propyloxiranemethanol as a pale-amber oil (Note 16).

The crude product is distilled through a 10-cm Vigreux column (the receiving flask is cooled in an ice–water bath) to yield 22.45–22.84 g (80–81%) of (2S,3S)-3-pro-pyloxiranemethanol as a colorless liquid, bp 31–33°C (0.30–0.40 mm). Analysis by GC indicates a chemical purity of 89–93% (Note 17). The material is fractionally distilled through a 20-cm vacuum-jacketed Vigreux column to obtain 17.69–19.44 g (63–69%), $[\alpha]_D^{22}$ −38.1 to −38.6° (neat), $[\alpha]_D^{23}$ −46.2 to −48.6° (CHCl$_3$, c 1.0) of a colorless liquid. Analysis by GC indicates a chemical purity of 96–98% (Note 17). An enantio-meric purity of 96.4–97.5% is determined by ^{1}H NMR analysis of the derived acetate using Eu(hfc)$_3$, tris[3-(heptafluoropropylhydroxymethylene)-d-camphorato] europium(III), (Note 18) as the chiral shift reagent (Note 19). The enantiomeric purity may also be determined by GC analysis (Note 20) of the derived α-methoxy-α-(trifluo-romethyl)phenylacetic acid esters[3] (Notes 21 and 22). An alternative to the distillation method is purification by preparative HPLC (Notes 23 and 24) and bulb-to-bulb distil-lation (Note 25) to give 21.85 g (78%) of (2S,3S)-3-propyloxiranemethanol as a white solid, mp 19°C, $[\alpha]_D^{25}$ −46.6° (CHCl$_3$, c 1.0). Analysis by GC of material purified in this manner indicates a chemical purity of >99% (Note 14) and an enantiomeric purity of 96.8% (Notes 20–22).

2. Notes

1. Fisher Scientific Company methylene chloride, certified ACS grade containing 0.02% water was used; a fresh bottle was used for each run.

2. Titanium(IV) isopropoxide is available from the Aldrich Chemical Company, Inc.

3. (+)-Diethyl L-tartrate was obtained from the Aldrich Chemical Company, Inc.

4. trans-2-Hexen-1-ol was obtained from Alfa Products, Morton Thiokol, Inc. Analysis by GC (Hewlett-Packard HP 5710A; 50-m × 0.25-mm capillary column of bonded CPS-2 on fused silica; 120°C isothermal) with appropriate standards indicated an E-2-hexen-1-ol content of 96.1% with 0.9% Z-2-hexen-1-ol and 2.9% hexanol as impurities. trans-2-Hexen-1-ol from the Aldrich Chemical Company, Inc., could also be used. This material was of similar composition: 95.4% E-2-hexen-1-ol, 0.8% Z-2-hexen-1-ol and 3.2% hexanol.

5. Anhydrous tert-butyl hydroperoxide in toluene[4] was prepared starting with Aldrich Chemical Company, Inc. 70% aqueous tert-butyl hydroperoxide. A 500-g lot of this material was swirled in a separatory funnel with 1.0 L of toluene. The aqueous phase was removed and discarded. The organic solution was heated at reflux under nitrogen for 4 hr in a flask equipped with a Dean–Stark trap for water separation (for greater detail, see reference 4). The solution was cooled and stored under nitrogen at −20°C (Note 6). The content of tert-butyl hydro-peroxide was determined by ^{1}H NMR according to the equation

$$\text{Molarity} = \frac{X}{0.1X + 0.32Y}$$

where X = integration of *tert*-butyl resonance; Y = integration of methyl resonance.

6. Storage of the solution at $-20\,^{\circ}\text{C}$ is not necessary.[4] It does, however, provide a convenient method of precooling the material.

7. Addition of dry ice to the cooling bath was stopped. If the rate of temperature increase was too slow, the ethanol bath was lowered.

8. If the mixture is allowed to stand at $0\,^{\circ}\text{C}$ or to warm above this temperature, undesired by-products (TLC) are formed.

9. The temperature should be kept $\leq 20\,^{\circ}\text{C}$. In some cases, the addition of small amounts of ice to the reaction is necessary.

10. This weight can vary substantially, depending on the extent to which the solvents, especially toluene, have been stripped from the solution. Concentration need only be carried out until the weight is < 100 g.

11. This mixture is prepared by dissolving the sodium hydroxide in the brine at ambient temperature and then cooling the total to $3\,^{\circ}\text{C}$. The resultant cloudy, supersaturated suspension is used in toto. The use of this reagent ensures complete extraction (vide infra) of the somewhat water-soluble product. In addition, it minimizes contact of this material with the aqueous base, conditions which can lead to the Payne rearrangement.[5]

12. Saponification serves to remove the diethyl tartrate as well as to liberate any product which has been transesterified to form a tartrate ester.

13. GC analysis (Note 14) of a third extract showed no product.

14. GC analysis was performed on a column with the following properties: Hewlett-Packard HP 5702A, 2-m $\times$ 0.63-cm OV-101 column, programmed 70–$200\,^{\circ}\text{C}$ at $8\,^{\circ}\text{C}/\text{min}$.

15. GC analysis (Note 14) shows no diethyl $(2R,3R)$-tartrate, indicating that the saponification was complete.

16. Material of this quality is suitable for many synthetic purposes.

17. The properties of the column are as follows: Hewlett-Packard HP 5790A, 12-m $\times$ 0.2-mm cross-linked methyl silicone (fast analysis) column, programmed from 35 to $140\,^{\circ}\text{C}$ at $3\,^{\circ}\text{C}/\text{min}$.

18. The shift reagent was obtained from the Aldrich Chemical Company, Inc. Drying the Eu(hfc)$_3$ overnight with a drying pistol at $56\,^{\circ}\text{C}$ (refluxing acetone) under vacuum afforded optimum results.

19. The analytical sample of the acetate derivative is prepared as follows. Into a 5-mL, round-bottomed flask equipped with a magnetic stirring bar are placed 2 drops of the reaction product, 16 drops of acetic anhydride, and 32 drops of pyridine. The solution is stirred at ambient temperature for 2 hr, and the mixture is then transferred to a separatory funnel with the aid of 10 mL of methylene chloride. The methylene chloride solution is washed with two 10-mL portions of 1 M phosphoric acid, the organic layer is dried over $MgSO_4$, and the filtered solution is concentrated with a rotary evaporator to give approximately 20 mg of acetate as a colorless oil. A 5-μL sample of this crude acetate is dissolved in 0.5 mL of benzene-d_6 and transferred to an NMR tube. A solution of 75 mg of

Eu(hfc)$_3$ in 0.5 mL of benzene-d_6 is prepared. A 50-μL portion of the shift reagent solution is added to the acetate sample, the mixture is shaken well, and the ^{1}H NMR spectrum is recorded. Additional portions of shift reagent are added in 10-μL portions until the acetate methyl resonance (originally at δ = 1.65 ppm) shifts downfield to the region 2.3–3.1 ppm and shows baseline resolution of the resonances from the two enantiomers. A total of 50–90 μL of the shift reagent solution should be required to achieve the desired shift, at which point a chemical-shift difference of about 0.2 ppm should be obtained. The %ee is obtained by integration of the two acetate peaks.

20. The column had the following properties: Hewlett-Packard HP 5710A, 50-m × 0.25-mm capillary column of OV-17 (bonded) on fused silica; 175°C isothermal.

21. The analytical sample of α-methoxy-α-(trifluoromethyl)phenylacetic acid ester is prepared as follows. Into a 5-mL, capped, amber vial equipped with a magnetic stirring bar are placed 20 mg of the reaction product, 1.0 mL of methylene chloride, 87 mg of (+)-α-methoxy-α-(trifluoromethyl)phenylacetyl chloride (Note 22), 4 drops of triethylamine, and 1 crystal of 4-dimethylaminopyridine. The mixture is stirred at ambient temperature for 1.5 hr, at which point TLC (Note 26) indicates complete conversion to the ester. Addition of 4 drops of N,N-dimethyl-1,3-propanediamine and concentration on a rotary evaporator at 35°C (70 mm) affords a yellow oil. This material is filtered through 10 g of E. Merck silica gel 60 (70–230 mesh) with 9 : 1 hexanes–ethyl acetate until TLC analysis indicates no further product elution. The total eluate is concentrated on a rotary evaporator at 35°C (70 mm). The resultant colorless oil is subjected to GC analysis.

A sample of E-2-hexen-1-ol in methylene chloride was epoxidized at 20°C with m-chloroperoxybenzoic acid. The resultant racemic epoxy alcohol, on conversion to the diastereomeric (+)-α-methoxy-α-(trifluoromethyl) phenylacetic acid esters in the manner described above, provided a GC standard for determination of the enantiomeric excess obtained in the asymmetric epoxidation.

22. The acid chloride was prepared[3] from (+)-α-methoxy-α-(trifluoromethyl)phenylacetic acid, which was used as obtained from Aldrich Chemical Company, Inc.

23. Purification by preparative HPLC is accomplished as follows. The crude product is taken up in 60 mL of 4 : 1 hexanes : ethyl acetate. The solution is subjected to preparative HPLC (Note 24) using the same solvent system. Chromatography is monitored by TLC (Note 26) and the appropriate fractions are combined. Solvent removal with a rotary evaporator at 35°C (12 mm) gives a colorless oil. This material is subjected to bulb-to-bulb distillation at 75–90°C (8 mm).

24. A Waters Associates Prep LC/System 500 with two cartridges (1.0 kg) of PrepPak-500/Silica was used. The course of the chromatography was followed with a refractive index detector. Approximately 3.6 L of solvent was eluted prior to the product band.

25. A Kugelrohr apparatus purchased from the Aldrich Chemical Company, Inc. was used. The receiving bulb was cooled with an ice–water bath. The temperature indicated is the oven temperature.

26. E. Merck silica gel F-254 plates were used, with 2 : 1 hexanes : ethyl acetate as eluent. Visualization was effected by spraying with a 10% phosphomolybdic acid in ethanol solution followed by heating with a hot air gun. (2S,3S)-3-Propyloxiranemethanol had an R_f of ~0.3.

3. Discussion

Both the synthetic[6a] and mechanistic[6b] aspects of this asymmetric epoxidation process have been reviewed recently. While the process has great scope regarding the allylic alcohol substrate, two classes of substrates present difficulties. These limitations will be best appreciated by reference to the recent reviews;[6] however, the main problems are worth mentioning here. When difficulties arise, they are almost never due to the failure of the asymmetric epoxidation process itself but can be traced instead to the nature of the epoxy alcohol product.

Water-soluble products (e.g., 3- and 4-carbon epoxy alcohols) present obvious isolation problems that have been only partly solved. The other troublesome class of products includes those epoxy alcohols that are unstable under the epoxidation and/or isolation conditions. This latter class consists of the three main types shown below:

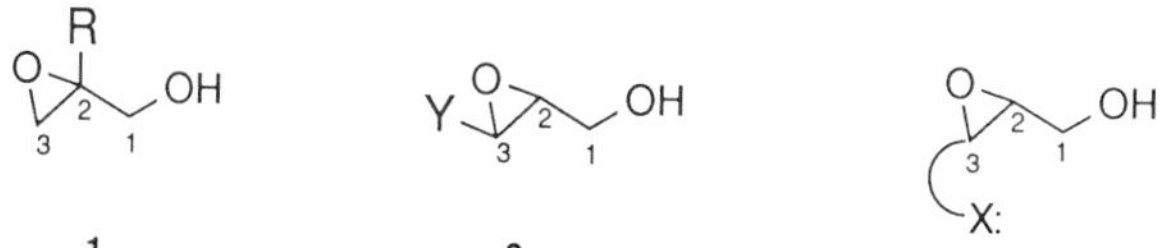

Type **1** is sensitive to nucleophilic opening at the primary epoxide carbon (C-3). Type **2** represents cases where the substituent Y facilitates opening at carbon-3 through resonance stabilization of the incipient carbonium ion. Finally, type **3** includes those cases in which the product bears a heteroatom substituent (X) placed so that a five- or six-membered ring results from anchimerically assisted opening of the epoxide at carbon-3. Not even these structural features (i.e., as in **1**, **2**, and **3**) are always fatal, for some representatives of types **1**, **2**, and **3** afford good yields of the desired epoxy alcohols.[6] Furthermore, when the structure of a given case is marginal, we have found that modification of the epoxidation and/or isolation procedures can lead to substantially improved yields. With such sensitive epoxy alcohols, milder isolation procedures are always employed (see references 6a and 9 for a discussion of these modified work up methods). However, certain substrates of types **1**, **2**, and **3** still fail completely with all current procedures. The best we have been able to do in these difficult cases is to use a strategy that actually takes advantage of the facile epoxide opening process.[6,7]

Questions are often asked concerning the catalytic nature of the asymmetric epoxidation. One notes that the present procedure calls for 50% catalyst. With very favorable substrates, one can realize complete conversion and >95% e.e. using as little as 2%

catalyst.[6] In the present case, the reaction stops at about 80% conversion using 2% catalyst and almost reaches completion with 10% catalysis.[8] The selection of 50% catalyst is a compromise aimed at rendering the procedure applicable to a wider range of substrates. In the literature, most applications of asymmetric epoxidation use 100% catalyst. This is rarely necessary, but ensures rapid and complete epoxidation in small-scale reactions where cost of the reagents is not an issue. The cases yielding epoxy alcohols that are sensitive to opening require the most catalyst, because the open-diol products are potent inhibitors of the epoxidation catalysis.[6a] If one wished to produce molar amounts of a given epoxy alcohol, it would be worthwhile to determine the optimum catalyst loading for the case at hand. In addition to the cost incentive, the isolation procedure becomes simpler as the amount of catalyst is decreased.

The aqueous tartaric acid workup procedure described here is the simplest method for removing the titanium species, but it should be used only with relatively stable epoxy alcohols that are not water-soluble. In this regard, the six-carbon epoxy alcohol made here probably represents the lower limit, as it is on the verge of water solubility. Of course, one cannot assume that this work up will succeed with all six-carbon or larger epoxy alcohols, for in addition to limited water solubility, the product must be fairly resistant to acid-catalyzed epoxide opening processes. *For water-soluble and/or acid-sensitive epoxy alcohols, the "sodium sulfate workup" is generally preferred.*[6a,9] When making a particularly sensitive epoxy alcohol, one should not only use this sodium sulfate workup but also modify the initial stage of the epoxidation process so that the reaction mixture only warms to $-20°C$ rather than to $0°C$.

Finally, two other practical points are worth mentioning. The early procedures for asymmetric epoxidation called for dilute solutions of sodium hydroxide to effect tartrate ester hydrolysis. For the reasons given in Note 11, one should always (unless one is certain that the epoxy alcohol is completely insoluble in water) use, instead, NaOH in brine. In this procedure, the excess *tert*-butyl hydroperoxide (TBHP) was destroyed early in the workup ($FeSO_4$), although this is not essential because dilute solutions of TBHP are not dangerous. Other methods for removing excess TBHP in these epoxidations have been reviewed.[6a] One of the simplest is to remove it as the azeotrope with toluene. We have removed up to 0.5 mol of TBHP by this means.[10]

1. Department of Chemistry, Massachusetts Institute of Technology, Cambridge, MA 02139.
2. Roche Research Center, Hoffmann–La Roche Inc., Nutley, NJ 07110.
3. Dale, J. A.; Dull, D. L.; Mosher, H. S. *J. Org. Chem.* **1969**, *34*, 2543–2549.
4. Hill, J. G.; Rossiter, B. E.; Sharpless, K. B. *J. Org. Chem.* **1983**, *48*, 3607–3608.
5. Payne, G. B. *J. Org. Chem.* **1962**, *27*, 3819–3822. For a discussion of this rearrangement and its synthetic implication, see Behrens, C. H.: Sharpless, K. B. *Aldrichimica Acta* **1983**, *16*, 67–79.
6. (a) Rossiter, B. E. In "Asymmetric Synthesis," Morrison, J. D., Ed.; Academic Press, Inc.: New York, 1985; Vol. 5, Chapter 7; (b) Finn, M. G.; Sharpless, K. B. In "Asymmetric Synthesis," Morrison, J. D., Ed.; Academic Press, Inc.: New York, 1985; Vol. 5, Chapter 8.
7. Lu, L. D.-L.; Johnson, R. A.; Finn, M. G.; Sharpless, K. B. *J. Org. Chem.* **1984**, *49*, 728–731.
8. Note, however, that in some cases, lower catalyst levels cause the enantiomeric excess to fall off. For example, the title compound is produced with 90% e.e. using 10% catalyst and with 97% e.e. using 50% catalyst.

9. The "sodium sulfate workup" should be used in place of the "tartaric acid workup," employed in the present procedure, whenever one is dealing with a water-soluble and/or acid-sensitive epoxy alcohol. The early stages of this alternate workup are as follows. The reaction mixture is removed from the freezer (ca. $-20°C$), and, while it is stirred (magnetic or mechanical depending on the scale), ether is added to the cold reaction mixture, followed immediately by a saturated sodium sulfate solution (no cooling bath is used at this stage and the ether is not precooled). We use 1 mL of saturated Na_2SO_4 solution per mmol of $Ti(O-i-Pr)_4$. (Note that this is about 3.3 times more than was recommended in an earlier[11] procedure.) The volume of ether added should be at least 1 mL per milliliter of saturated Na_2SO_4 solution used, and more ether is beneficial.

The heterogeneous mixture that results is stirred vigorously for about 2 hr at room temperature. It is then filtered through a Celite pad and the resulting orange–yellow paste is washed with several portions of anhydrous ether until the paste becomes somewhat granular. The orange–yellow layer is scraped off the Celite pad into an Erlenmeyer flask. Ethyl acetate is added along with a magnetic stirring bar and the resulting suspension is stirred vigorously for 5 min in boiling ethyl acetate. The slurry is then filtered through the same Celite pad, and the orange–yellow solid is washed once with hot ethyl acetate. Treatment of the filtrand in this manner is a key improvement that *usually increases the total isolated yield by 10–15%*. The combined filtrates are concentrated to afford crude product along with the tartrate diester and any excess TBHP. This material is ready for the next stage of the work-up, which involves removal of the tartrate ester. The present preparation describes (vide supra) hydrolysis of the ester with NaOH/brine. For alternate ways of separating the epoxy alcohols from the tartrate ester see reference 6a. (This reference also describes several ways for removing the TBHP.)

10. Hill, J. G., unpublished results.
11. Reed, L. A., III; Ito, Y.; Masamune, S.; Sharpless, K. B. *J. Am. Chem. Soc.* **1982**, *104*, 6468–6470.

ACID CHLORIDES FROM α-KETO ACIDS WITH α,α-DICHLOROMETHYL METHYL ETHER: PYRUVOYL CHLORIDE

(Propanoyl chloride, 2-oxo-)

$$CH_3\text{-}\underset{O}{\overset{O}{\|}}C\text{-}\underset{}{\overset{O}{\|}}C\text{-OH} \quad + \quad H\text{-}\underset{Cl}{\overset{Cl}{|}}C\text{-OCH}_3 \quad \xrightarrow{25\text{-}50°C} \quad CH_3\text{-}\underset{O}{\overset{O}{\|}}C\text{-}\underset{}{\overset{O}{\|}}C\text{-Cl} \quad + \quad H\text{-}\overset{O}{\overset{\|}{C}}\text{-OCH}_3$$

Submitted by Harry C. J. Ottenheijm and Marianne W. Tijhuis[1]
Checked by Larry A. Last and Robert M. Coates

1. Procedure

A 100-mL, two-necked, round-bottomed flask is equipped with a magnetic stirrer, pressure-equalizing dropping funnel, and a 1.2- × 24-cm vacuum-jacketed Vigreux column that is connected to a condenser, vacuum-take-off adapter, and fraction collector with three receiving flasks (Note 1). The vacuum-take-off adapter is attached to a calcium chloride drying tube which is connected to a water aspirator, and the flask is charged with 35.2 g (28.6 mL, 0.40 mol) of pyruvic acid (Note 2). The pyruvic acid is stirred at room temperature as 46.4 g (36.1 mL, 0.40 mol) of α,α-dichloromethyl methyl ether (Note 3) is added slowly over 30 min. Evolution of hydrogen chloride begins after a few minutes. When the addition is complete, the dropping funnel is removed and replaced by a glass stopper. The solution is stirred and heated at 50°C in an oil bath for 30 min (Note 4) while a few drops of methyl formate are collected as the first fraction (Note 5). The condenser is then cooled to -30°C (Note 6) and the receiving flasks are cooled to -50°C with chilled acetone. The aspirator is turned on and the pressure is adjusted to 190 mm. With the oil bath at 50°C, a second fraction, bp 25–35°C (190 mm), is collected. As soon as the head temperature begins to drop, the pressure is reduced to 120 mm and the temperature of the oil bath is raised slowly to 75°C. A third fraction, bp 35–40°C (120 mm), consisting mainly of pyruvoyl chloride is collected (Notes 7 and 8). The second and third fractions are combined to give 33–41 g of a mixture of pyruvoyl chloride and methyl formate, which is redistilled through a 1.4- by 18-cm vacuum-jacketed Vigreux column (Note 1). The condenser and the receiving flasks are cooled to -5°C with chilled acetone. The first fraction, weighing 2.3–11.4 g and consisting mainly of methyl formate, is collected at 25–30°C (190 mm) with an oil bath temperature of 60°C. When the pressure is reduced to 120 mm and the oil bath is maintained at 60°C, 18.6–21.2 g (44–50%) of pyruvoyl chloride, bp 43–45°C (120 mm), distills into the receiver as a light yellow liquid, n_D^{20} 1.4165 (Notes 9 and 10).

2. Notes

1. The glassware was dried for 16 hr in an oven at ca. 125°C and assembled while still warm. The checkers used a 27-cm Vigreux column insulated with glass wool instead of the vacuum-jacketed column.

2. Pyruvic acid, supplied by Aldrich Chemical Company, Inc., was freshly distilled: bp 59–62°C (14 mm).

3. α,α-Dichloromethyl methyl ether was purchased from Aldrich Chemical Company, Inc., and redistilled prior to use: bp 83–84°C. The reagent may also be prepared from methyl formate and phosphorus pentachloride.[2] Unlike chloromethyl methyl ether and bis(chloromethyl) ether, α,α-dichloromethyl methyl ether is reported to have no significant carcinogenic activity.[3] However, as a precaution, the compound should be handled with care in a well-ventilated hood.

4. At this temperature the intermediate, chloromethoxymethyl pyruvate, decomposes to pyruvoyl chloride and methyl formate.[4]

5. The submitters made no effort to collect methyl formate quantitatively. The checkers did not observe the formation of any condensate at this point.

6. This was accomplished by the checkers by passing acetone chilled with dry ice slowly through the condenser jacket. The coolant was contained in a 1-L separatory funnel which was connected to the condenser inlet with a section of Tygon tubing. The effluent was collected in a beaker and periodically returned to the separatory funnel reservoir.

7. Fractions 2 and 3 weighed 10.1–13.4 and 23.6–30.0 g, respectively. Proton NMR spectra of fraction 2 indicated a composition of 70–84% of methyl formate, 16–20% of pyruvoyl chloride, and 0–10% of unreacted starting materials. The composition of fraction 3 was 21–28% of methyl formate, 60–70% of pyruvoyl

chloride, and 1–20% of starting materials. The two fractions collected by the checkers boiled at 25–26°C (190 mm) and 40–46°C (120 mm).

8. For some reactions, such as simple esterification, it is not necessary to distill the acid chloride. The crude reaction mixture may be used provided the hydrogen chloride present is neutralized with an appropriate base.[5]

9. The submitters found that pyruvoyl chloride may be stored at −20°C in carbon tetrachloride solution or as the pure liquid in a sealed tube.

10. The product obtained by the checkers boiled at 48–51°C (120 mm) and was contaminated with ca. 5–10% of methyl formate and unreacted starting materials. The spectral properties of the product are as follows: IR (liquid film) cm^{-1}: 2900 (w), 1770 (s, broad), 1415 (m), 1355 (s), 1195 (s), 1130 (m), 1095 (m), 1005 (s), 875 (s); ^{1}H NMR (CDCl$_3$) δ: 2.51 (s, 3 H). The compound may be characterized as the p-nitroanilide derivative.[6]

3. Discussion

Most of the conventional reagents for the synthesis of acid chlorides from carboxylic acids are unsatisfactory for the preparation of α-keto acid chlorides. For example, the reaction of pyruvic acid with phosphorus halides does not give pyruvoyl chloride[7] whereas the use of phosgene[8] or oxalyl chloride[9,10] affords ether solutions of the acid chloride in low yield. Recently a useful preparation of pyruvoyl chloride from trimethylsilyl pyruvate and oxalyl chloride has been described.[11]

The use of α,α-dichloromethyl alkyl ethers for the conversion of carboxylic acids to acid chlorides was first reported by Heslinga et al. in 1957.[4] The submitters have found that the readily available α,α-dichloromethyl methyl ether[2] is the reagent of choice for the preparation of pyruvoyl chloride.[6] This simple and economical procedure has been used in other laboratories,[5,12,13] and the submitters have applied the method to the preparation of three other α-keto acid chlorides: 2-oxobutanoyl chloride (32%), 3-methyl-2-oxobutanoyl chloride (10%), and phenylglyoxylyl chloride (78%).[6]

1. Laboratory of Organic Chemistry, Department of Chemistry, University of Nijmegen, Toernooiveld, 6525 ED Nijmegen, The Netherlands.
2. Gross, H.; Rieche, A.; Höft, E.; Beyer, E. Org. Synth., Coll. Vol. V 1973 365–367.
3. Van Duuren, B. L.; Katz, C.; Goldschimidt, B. M.; Frenkel, K.; Sivak, A. J. Nat. Cancer Inst., USA 1972, 48, 1431–1439.
4. Heslinga, L.; Katerberg, G. J.; Arens, J. F. Recl. Trav. Chim. Pays-Bas 1957, 76, 969–981.
5. Binkley, R. W. J. Org. Chem. 1977, 42, 1216–1221.
6. Ottenheijm, H. C. J.; de Man, J. H. M. Synthesis 1975, 163–164.
7. Klimenko, E. Chem. Ber. 1980, 3, 465–468; Beckurts, H.; Otto, R. Chem. Ber. 1878, 11, 386–391; Bernton, A. Chem. Ber. 1925, 58, 661–663; Carré, P.; Jullien, P. C.R. Hebd. Seances Acad. Sci. 1936, 202, 1521–1523.
8. Wieland, T.; Köppe, H. Justus Liebigs Ann. Chem. 1954, 588, 15–23.
9. Kharasch, M. S.; Brown, H. C. J. Am. Chem. Soc. 1942, 64, 329–333.
10. Tanner, D. D.; Das, N. C. J. Org. Chem. 1970, 35, 3972–3974.
11. Häusler, J.; Schmidt, U. Chem. Ber. 1974, 107, 145–151.
12. Binkley, R. W. J. Org. Chem. 1976, 41, 3030–3031; Binkley, R. W.; Hehemann, D. G.; Binkley, W. W. J. Org. Chem. 1978, 43, 2573–2576.
13. Lopatin, W.; Sheppard, C.; Owen, T. C. J. Org. Chem. 1978, 43, 4678–4679.

RICINELAIDIC ACID LACTONE

(9-Octadecenoic acid, 12-hydroxy-, [(+)-(*R*)-*trans*]-, lactone)

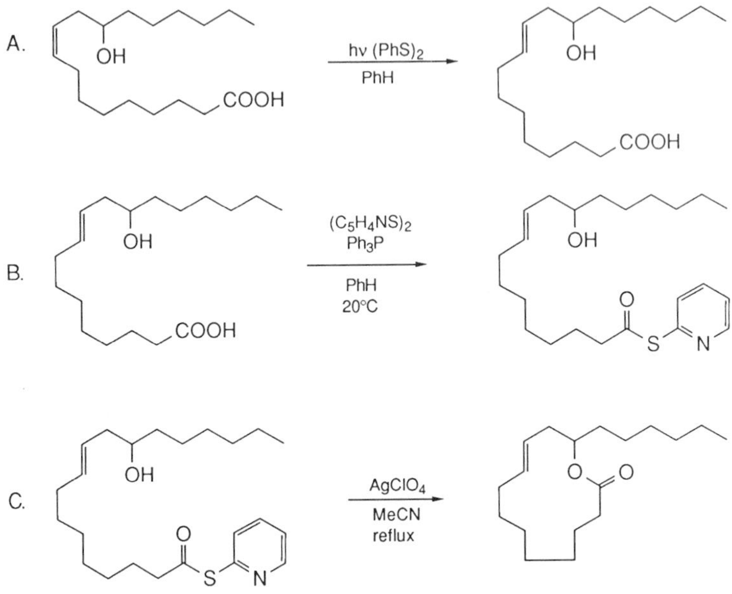

Submitted by ADOLF THALMANN, KONRAD OERTLE, and HANS GERLACH[1]
Checked by JAMES R. PRIBISH and EDWIN VEDEJS

1. Procedure

A. *Ricinelaidic acid.* Ricinoleic acid (Note 1) (39.75 g, 0.106 mol) and 586 mg (2 mol %) of diphenyl disulfide dissolved in 1000 mL of hexane are placed in a photochemical reactor (Note 2) and irradiated for 3 hr with a Philips HP(L) 250-W medium-pressure mercury lamp. After irradiation the solvent is removed under reduced pressure and the semisolid residue is recrystallized from 185 mL of hexane to yield 11.3 g of crude ricinelaidic acid, mp 39–43°C. The irradiation is repeated with the mother liquor under the same conditions to yield, after removal of the solvent and recrystallization of the residue from 135 mL of hexane, an additional 7.2 g, mp 38–42°C; total yield of crude ricinelaidic acid is 18.5 g (58%). The product after recrystallization from 220 mL of hexane weighs 15.6 g (49%), mp 43–45°C, and is suitable for the following step. Repeated recrystallization from hexane yields ricinelaidic acid with mp 51.0–51.5°C (Notes 3 and 4).

B. *Ricinelaidic acid S-(2-pyridyl)carbothioate.* In a dry, stoppered, 10-mL flask containing a magnetic stirring bar are placed 360 mg (1.2 mmol) of ricinelaidic acid (see above), 308 mg (1.4 mmol) of 2,2′-dipyridyl disulfide (Note 5), 1 mL of benzene, and 367 mg (1.4 mmol) of triphenylphosphine, and the mixture is stirred for 30 min. The resulting slurry is then dissolved in 55 mL of dry acetonitrile (Note 6).

C. *Ricinelaidic acid lactone.* Dry acetonitrile (100 mL), 3.5 mL of 1 M silver perchlorate in toluene (Notes 7 and 8), and a magnetic stirring bar are placed in a 500-mL flask equipped with a reflux condenser that carries a Hershberg dropping funnel. The solution is heated in an oil bath so that the boiling acetonitrile returns from the condenser at the rate of 5–10 drops per second (Note 9). Then the acetonitrile solution of the ricinelaidic acid S-(2-pyridyl)carbothioate is added dropwise during 1 hr through the condenser to the magnetically stirred refluxing silver perchlorate solution (Note 9). The slightly turbid mixture is boiled for an additional 15 min and the solvent is removed under reduced pressure in a rotary evaporator. The residue is diluted with 30 mL of 0.5 M potassium cyanide solution and the mixture containing suspended solids is extracted with three 50-mL portions of benzene. The benzene extracts are washed with 30 mL of water, dried with anhydrous magnesium sulfate, and filtered, and the solvent is removed under reduced pressure. Crude product is obtained as an oil (710 mg). It can be purified by chromatography on 40 g of silica gel (Note 10) with benzene as eluant. Fractions of 10 mL are collected at 30-min intervals. Fractions 7–19 contain 283–296 mg (84–88%) of ricinelaidic acid lactone (Note 11).

2. Notes

1. Technical-grade (80%) ricinoleic acid was obtained from Fluka AG Buchs, Switzerland or from Tridom Chemicals, Inc. Saponification of methyl ricinoleate[2] also gives suitable material.

2. The photochemical reactor used is quite similar to the one described in *Org. Synth., Coll. Vol. V* **1973,** 298.

3. The purity of the products has been checked by capillary gas liquid chromatography of the corresponding methyl ester obtained with ethereal diazomethane solution (Carlo Erba Fractovap 20-m glass capillary coated with UCON HB at 160°C). Ricinelaidic acid, mp 49–50°C, contains 4%, that with mp 51.0–51.5°C, less than 1% of ricinoleic acid. Submitters obtained higher yields (58%, mp 49–50°C), perhaps due to better quality starting material.

4. (+)-(R)-Ricinelaidic acid, mp 51.0–51.5°C, has an optical rotation of $[\alpha]_D$ +6.6° (C$_2$H$_5$OH, c 10).

5. 2,2'-Dipyridyl disulfide obtained from Fluka AG, Buchs, Switzerland, was recrystallized from hexane (30 mL/g) to yield a suitable product, mp 58–59°C.

6. Commercially available acetonitrile is distilled over phosphorus pentoxide.

7. Silver perchlorate monohydrate (9 g) (obtained from Fluka AG) is suspended in 110 mL of toluene together with a Teflon-coated magnetic stirring bar. The solution is magnetically stirred and heated in an oil bath until 70 mL of toluene has distilled.

8. The silver perchlorate solution may be substituted by 8.5 mL of 0.4 M silver trifluoromethanesulfonate (Fluka) in toluene.

9. This reflux rate is crucial for predilution of the carbothioate in the condenser. Lower reflux rates require an accordingly slower addition of the S-(2-pyridyl)carbothioate during 2–4 hr.

10. Silica gel 60 Merck in a 2.5-cm-diameter column was used.

11. The product distills at 110°C (0.01 mm) in a Kugelrohr distillation apparatus and has an optical rotation of $[\alpha]_D + 42°$ (CHCl$_3$, c 1).

3. Discussion

The silver ion-promoted lactonization of hydroxy-S-(2-pyridyl)carbothioates was introduced by the submitters[3] as a mild method for the synthesis of naturally occurring macrolides as, for example, nonactin[4] and recifeiolide[5] from the corresponding hydroxy acids. If the method of Mukaiyama et al.[6] is used for the formation of the S-(2-pyridyl)carbothioate, no protection of the hydroxyl group is needed in this step. The cited examples show that silver ion-promoted lactonization can be used to effect ring closure of base-sensitive and unsaturated acid-sensitive hydroxy acids in good yield.

Similar methods to effect lactonization have been proposed by Corey et al.[7] and Masamune et al.[8] The first consists of prolonged heating of hydroxy-S-(2-pyridyl)carbothioates in boiling xylene; the second is the mercury trifluoroacetate-promoted cyclization of a hydroxy-S-*tert*-butyl carbothioate.

Ricinelaidic acid was selected for the submitted procedure because it has a moderately complex structure and can be prepared easily from commercially available technical grade ricinoleic acid. This conversion represents an example of the facile cis–trans interconversion of olefins[9] caused by photochemically generated phenylthiyl radicals leading to the thermodynamic equilibrium.

1. Laboratorium für Organische Chemie, Universität Bayreuth, Postfach 3008, D-8580 Bayreuth, West Germany.
2. *Biochem. Prep.* **1952**, *2*, 104.
3. Gerlach, H.; Thalmann, A. *Helv. Chim. Acta* **1974**, *57*, 2661.
4. Gerlach, H.; Oertle, K.; Thalmann, A.; Servi, S. *Helv. Chim. Acta* **1975**, *58*, 2036.
5. Gerlach, H.; Oertle, K.; Thalmann, A. *Helv. Chim. Acta* **1976**, *59*, 755.
6. Araki, M.; Sakata, S.; Takei, H.; Mukaiyama, T. *Bull. Chem. Soc. Jpn.* **1974**, *47*, 1777.
7. Corey, E. J.; Nicolaou, K. C. *J. Am. Chem. Soc.* **1974**, *96*, 5614; Corey, E. J.; Nicolaou, K. C.; Melvin, L. S., Jr. *J. Am. Chem. Soc.* **1975**, *97*, 653, 654; Corey, E. J.; Nicolaou, K. C.; Toru, T. *J. Am. Chem. Soc.* **1975**, *97*, 2287; Corey, E. J.; Ulrich, P.; Fitzpatrick, J. M. *J. Am. Chem. Soc.* **1976**, *98*, 222.
8. Masamune, S.; Yamamoto, H.; Kamata, S.; Fukuzawa, A. *J. Am. Chem. Soc.* **1975**, *97*, 3513; Masamune, S.; Kamata, S.; Schilling, W. *J. Am. Chem. Soc.* **1975**, *97*, 3515.
9. Schulte-Elte, K. H.; Ohloff, G. *Helv. Chim. Acta* **1968**, *51*, 548 and references cited therein.

SPIRO[5.7]TRIDECA-1,4-DIEN-3-ONE

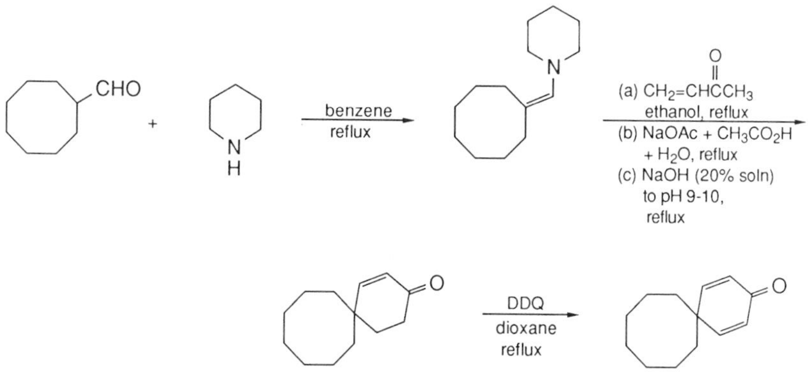

Submitted by VINAYAK V. KANE and MAITLAND JONES, JR.[1]
Checked by R. V. STEVENS and R. P. POLNIASZEK

1. Procedure

*Caution! The following reactions should be performed in an efficient hood to protect
the experimentalist from noxious vapors (piperidine and methyl vinyl ketone).*

A. *1-(Cyclooctylidenemethyl)piperidine.* Cyclooctanecarboxaldehyde (12.5 g, 0.089
mol) (Note 1) and piperidine (8.35 g, 0.098 mol) are dissolved in 115 mL of toluene
and placed in a 250-mL, one-necked flask equipped with a magnetic stirring bar and
Dean-Stark water separator, on top of which is a condenser fitted with a nitrogen-inlet
tube. The reaction mixture is placed under a nitrogen atmosphere, then brought to and
maintained at reflux with stirring for 6 hr, at which time the theoretical amount of water
(1.75 mL) has been collected. The reaction mixture is cooled and fractionally distilled
under reduced pressure (Note 2); toluene and excess piperidine are removed at 40°C
(0.5 mm), and the enamine product is distilled as a colorless liquid to yield 17.30 g
(0.084 mol, 93.6%) of 1-(cyclooctylidenemethyl)piperidine, bp 81–83°C (0.5 mm).

B. *Spiro[5.7]tridec-1-en-3-one.* A dry, 1-L, three-necked flask is equipped with a
Teflon stirring bar, condenser, pressure-equalizing dropping funnel, and nitrogen-inlet
tube. To this flask are introduced absolute ethanol (460 mL) (Note 3) and 1-(cyclooctylidenemethyl)piperidine (17.3 g, 0.084 mol). After the solution has been stirred for 5
min, methyl vinyl ketone (6.44 g, 0.092 mol) (Note 4) is added dropwise over a period
of 5 min. The solution is refluxed for 20 hr using a heating mantle. The mixture is
cooled and anhydrous sodium acetate (15.0 g), acetic acid (25.5 mL), and water (46
mL) are added. The mixture is brought to and maintained at reflux for 8 hr. The heat is
removed and the solution is cooled with ice water; aqueous sodium hydroxide (20%
solution, approximately 65 mL) is added until pH 9–10 is attained. The solution is
refluxed for another 15 hr; at the end of this period the reaction mixture is cooled. The

reaction mixture (600 mL) is divided equally into two 2-L separatory funnels and each portion is diluted with 600 mL of ice-cold water. Each separatory funnel is extracted with ether (3 × 125 mL). The ether extract is washed successively with aqueous 5% hydrochloric acid (125 mL) and saturated brine (3 × 170 mL), dried over anhydrous magnesium sulfate, and filtered. The solvent is removed on a rotary evaporator and the product is distilled under vacuum (Note 5) as a colorless liquid to yield 7.05–7.75 g (44–49%) of spiro[5.7]tridec-1-en-3-one, bp 95–125°C (0.5 mm) (Note 6).

C. *Spiro[5.7]trideca-1,4-dien-3-one.* Spiro[5.7]tridec-1-en-3-one (3.63 g, 0.0189 mol) and 2,3-dichloro-5,6-dicyano-1,4-benzoquinone (DDQ) (8.90 g, 0.0392 mol) (Note 7) are dissolved in 50 mL of dioxane (Note 8) in a 250-mL, one-necked flask equipped with a magnetic stirring bar and fitted with a condenser and drying tube. The reaction mixture is brought to and maintained at reflux with stirring for 6 hr. The mixture is cooled, filtered, and the dioxane removed in a rotary evaporator. The product is taken up in ether (125 mL), and the ether layer is washed with aqueous sodium hydroxide (15%, 4 × 60 mL). The combined aqueous layers are further extracted with ether (3 × 60 mL). The ether layers are combined and washed with saturated sodium chloride (4 × 60 mL), dried over anhydrous magnesium sulfate and filtered. The solvent is removed on the rotary evaporator to afford a crude yellow liquid. To this crude product are added silica gel (6.25 g) (Note 9) and enough ether to cover the silica gel. The ether is removed with a rotary evaporator so as to absorb the crude product on the silica gel. This silica gel dry powder is poured onto a column (12 in. long × 1.0 in. diameter) containing silica gel (50 g) in hexane. The column is eluted with hexane (70 mL) and then with an increasing amount of ethyl acetate/hexane (Note 10). The desired fractions are combined (Note 11) and solvent is removed under vacuum to afford spiro[5.7]trideca-1,4-dien-3-one (2.65 g, 73.7%), (Note 12).

2. Notes

1. Cyclooctanecarboxaldehyde was obtained from Aldrich Chemical Company, Inc. and used without purification.

2. 1-(Cyclooctylidenemethyl)piperidine is typical of most enamines in that it discolors rapidly when exposed to air and therefore must be handled under an inert atmosphere, preferably nitrogen.

3. Absolute ethanol, distilled and stored over molecular sieves, was used.

4. Methyl vinyl ketone (bp 35–36°C at 140 mm) was obtained from Aldrich Chemical Company, Inc. and distilled immediately before use.

5. A heating mantle was used for this distillation. A forerun of 25–95°C (0.5 mm) was discarded. The exact boiling point of spiro[5.7]tridec-1-en-3-one is 86°C (0.1 mm).

6. The product has the following spectral properties: ^{1}H NMR (CCl$_4$) δ: 1.62 (s, 14 H), 1.85 (br d, 2 H), 2.25 (m, 2 H), 5.68 (d, 1 H, J = 10 Hz), 6.75 (d, 1 H, J = 10 Hz).

7. 2,3-Dichloro-5,6-dicyano-1,4-benzoquinone, supplied by Aldrich Chemical Company, Inc. was used without further purification.

8. Dioxane was refluxed over potassium hydroxide pellets, distilled, and stored over molecular sieves.

9. Silica gel analytical reagent (60–200 mesh) was obtained from the J. T. Baker Chemical Co.

10. The silica gel column was eluted starting with hexane (70 mL), followed by 2% ethyl acetate–hexane (100 mL); 5% ethyl acetate–hexane (100 mL); 10% ethyl acetate–hexane (600 mL). The fractions were monitored with 20% ethyl acetate–hexane, using silica 7 GF plates (purchased from Analtech, Inc.), thickness 250 μm, 20 cm long × 5 cm wide. The plates were sprayed with 3% ceric sulfate and heated at 350°C to detect dienone and monoenone. Alternatively, silica gel 60 F-254 plates (purchased from EM Laboratories, Inc.), thickness 25 mm, 20 cm long × 5 cm wide may be used. Detection may be made with ultraviolet light. The ratio of 1 g of crude dienone to 15 g of silica gel is adequate for obtaining pure spiro[5.7]trideca-1,4-dien-3-one.

11. When 20% ethyl acetate/hexane is used, the monoenone, R_f 0.57, and the dienone, R_f 0.47 (Analtech Uniplate–Silica 7 GF), are obtained.

12. The product has the following spectral properties: ^{1}H NMR (CCl$_4$) δ: 1.65 (s, 14 H), 6.10 (d, 2 H), J = 10 Hz), 6.98 (d, 2 H, J = 10 Hz).

3. Discussion

This procedure illustrates a general method for preparing a wide range of spirocyclohexenones and hence spirocyclohexadienones. A number of intramolecular and intermolecular reactions are known to give spirodienones; however, these methods have limited synthetic application.[2] This procedure is superior[3] to that developed by Bordwell and Wellman[4] for side reactions such as aldol condensation of the aldehyde and polymerization of methyl vinyl ketone are avoided. These spirodienones are useful intermediates in the synthesis of paracyclophanes.[5,6]

Cyclopentanecarboxaldehyde (47%), cyclohexanecarboxaldehyde (41%), 1,2,5,6-tetrahydrobenzaldehyde (43%), cycloheptanecarboxaldehyde (41%), cyclooctanecarboxaldehyde (42%), cycloundecanecarboxaldehyde (36%), 5-norbornene-2-carboxaldehyde (32%), adamantanecarboxaldehyde (20%), and 1,2,3,4-tetrahydro-1-naphthylaldehyde (40%) gave corresponding spiroenones.[7] Spiroenones obtained from cyclohexanecarboxaldehyde, cycloheptanecarboxaldehyde and cyclooctanecarboxaldehyde were converted to the corresponding dienones using the dichlorodicyanobenzoquinone (DDQ). The yields for all three dienones are in the range of 56 to 58%.

1. Department of Chemistry, Princeton University, Princeton, NJ 08544.
2. Krapcho, A. P. *Synthesis* **1974**, 383.
3. Kane, V. V. *Synth. Commun.* **1976**, *6*, 237.
4. Bordwell, F. G.; Wellman, K. M. *J. Org. Chem.* **1963**, *28*, 1347, 2544.
5. Wolf, A. D.; Kane, V. V.; Levin, R. H.; Jones, Jr., M. *J. Am. Chem. Soc.* **1973**, *95*, 1680.
6. Kane, V. V.; Wolf, A. D.; Jones, Jr., M. *J. Am. Chem. Soc.* **1974**, *96*, 2643.
7. Yields are for the overall conversion.

2,2′ : 6′,2′ -TERPYRIDINE

A.

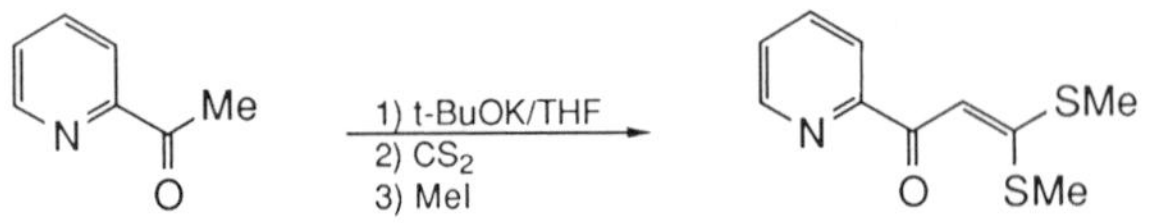

B.

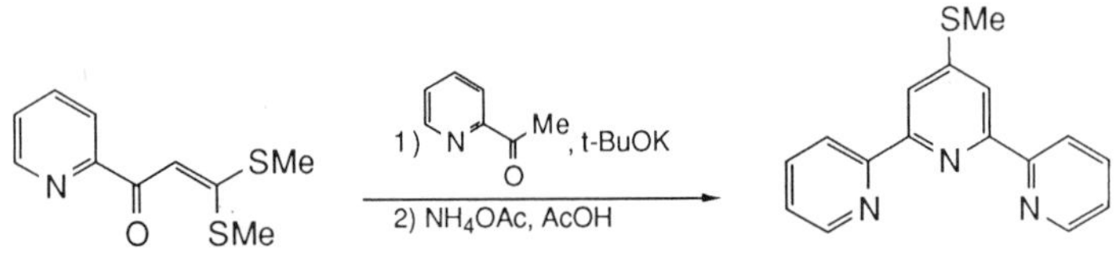

C.

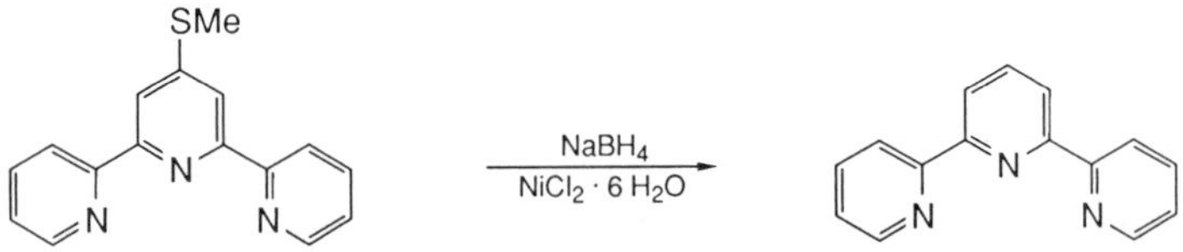

Submitted by KEVIN T. POTTS, PHILIP RALLI, GEORGE THEODORIDIS, and PAUL WINSLOW[1]
Checked by B. L. CHENARD and BRUCE E. SMART

1. Procedure

A. *3,3-Bis(methylthio)-1-(2-pyridinyl)-2-propen-1-one.* A 3-L, three-necked flask is equipped with an efficient mechanical stirrer, pressure-equalizing dropping funnel with needle valve, and a reflux condenser fitted with a nitrogen gas inlet tube that is attached to a mineral oil bubbler. The system is flushed with nitrogen, and while the system is maintained under a static pressure of nitrogen, the flask is charged with 1000 mL of dry tetrahydrofuran (Note 1) and 96.5 g (0.86 mol) of potassium *tert*-butoxide (Note 2). Freshly distilled 2-acetylpyridine (50.0 g, 0.41 mol) (Note 3) is then added dropwise over a period of 5–10 min (Note 4). To the resulting reaction mixture 32.7 g (0.43 mol) of carbon disulfide is added over a period of 30–35 min. After the addition is completed, 122.1 g (0.86 mol) of methyl iodide is added over 1 hr to the viscous, heterogeneous orange reaction mixture. After the tan reaction mixture is stirred for 12 hr at room temperature, it is poured into 2 L of iced water and allowed to stand for 4 hr. The solid that precipitates is collected by filtration and air-dried to give 56 g (61%) of yellow crystals, mp 106–107°C. The filtrate is diluted with water to a total volume of 4 L, and chilled to afford an additional 16.5 g (18%) of product, mp 104–107°C (Note 5).

B. *4'-(Methylthio)-2,2' : 6',2"-terpyridine (Note 6)*. A 1-L, three-necked round-bottomed flask fitted with a mechanical stirrer and a gas inlet tube is flushed with nitrogen and charged with 500 mL of anhydrous tetrahydrofuran and 22.4 g (0.20 mol) of potassium *tert*-butoxide. Freshly distilled 2-acetylpyridine (12.1 g, 0.10 mol) (Note 3) is added, the solution is stirred for 10 min, and 22.5 g (0.1 mol) of 3,3-bis(methylthio)-1-(2-pyridinyl)-2-propen-1-one is then added. The mixture is stirred for 12 hr at room temperature, during which time it turns bright red and a red solid precipitates (Note 7). The mixture is next treated with 77 g (1.0 mol) of ammonium acetate and 250 mL of glacial acetic acid. A distillation head fitted with a thermometer is attached to the flask and the tetrahydrofuran is removed by distillation over a 2-hr period. The residual brown solution is chilled to 15°C, treated with 400 g of ice, and allowed to stand for 3 hr. Water (400 mL) is added, the mixture is chilled to 15°C, and the gray material that precipitates is collected by filtration, washed with iced water (3 × 200 mL), and air-dried. The crude product is taken up in 250 mL of boiling ethanol and filtered. The filtercake is rinsed with 50 mL of hot ethanol, and the hot filtrates are combined, diluted with 150 mL of water, concentrated to a volume of 400 mL, and allowed to cool to room temperature. After the mixture is thoroughly chilled in an ice bath, the precipitate is collected by filtration, washed with 50% aqueous ethanol, and dried under reduced pressure (23°C, 0.1 mm) to give 20.6–21.4 g (74–77%) of 4'-(methylthio)-2,2' : 6',2"-terpyridine as gray needles, mp 118–119°C (Note 8). This material is sufficiently pure for use in the following step.

C. *2,2' : 6',2"-Terpyridine*. A 1-L, four-necked flask equipped with a mechanical stirrer, pressure-equalizing dropping funnel, thermometer, and a condenser fitted with a nitrogen gas inlet tube is flushed with nitrogen and charged with 300 mL of ethanol, 5.0 g (0.018 mol) of 4'-(methylthio)-2,2' : 6',2"-terpyridine and 42.8 g (0.180 mol) of finely ground nickel chloride hexahydrate (Note 9). The resultant green heterogeneous mixture is chilled in an ice bath while the system is maintained under a static pressure of nitrogen. To this chilled (0–5°C) mixture, a solution of 20.4 g (0.54 mol) of sodium borohydride in 128 mL of 40% aqueous sodium hydroxide is added dropwise over 4 hr (Note 10). After the addition is completed and the evolution of hydrogen subsides, the dark reaction mixture is refluxed for 12 hr. The hot mixture is then filtered through a Celite pad, and the pad is washed with hot ethanol (3 × 100 mL). The filtrates are combined and evaporated to dryness under reduced pressure to yield a gray solid residue (Note 11). This solid is suspended in 300–400 mL of water and chilled in an ice bath for 4 hr. The cold suspension is filtered and the gray solid is air-dried. The crude product is taken up in 100 mL of boiling hexane and filtered. The filtrate is concentrated to 50 mL, chilled in an ice bath, and filtered to give 2.48–2.53 g (59–60%) of 2,2' : 6',2"-terpyridine as cream-colored prisms, mp 84–86°C [lit.[2] mp 85–86°C] (Notes 12 and 13). The mother liquor is concentrated to 10 mL to give a second crop of 0.37–0.40 g (8.8–9.5%), mp 81–84°C.

2. Notes

1. The checkers used tetrahydrofuran that was distilled from lithium aluminum hydride (*Caution: See Org. Synth., Coll. Vol. V* **1973,** 976) and stored with a chip of sodium metal. Distillation from sodium/benzophenone is preferable.

2. Potassium *tert*-butoxide was obtained from the Aldrich Chemical Company, Inc.

3. The checkers obtained 2-acetylpyridine from the Aldrich Chemical Company, Inc. The submitters thank Reilly Tar & Chemical Corp. for a generous gift of 2-acetylpyridine used in their work.

4. A light yellow solid precipitates during this addition.

5. The product is pure by ^{1}NMR (CDCl$_3$) δ: 2.55 (s, 3 H), 2.65 (s, 3 H), 7.40 (d of d of d, 1 H, J = 1.5, 5.6, 7.5), 7.65 (s, 1 H), 7.85 (d of t, 1 H, J = 7.5, 2.0), 8.20 (d of t, 1 H, J = 7.5, 1.5), 8.65 (d of m, 1 H, J = 7.5); IR (KBr) cm^{-1}: 1484, 1471. Analytically pure material, mp 108–109°C, may be obtained by recrystallization from ethanol.

6. Recently, it has been shown that this product may be prepared without isolation of the precursors obtained in Step A.[3]

7. This solid is the potassium salt of the enedione intermediate.

8. The checkers also obtained material with mp 116–118°C. The submitters report product of unspecified purity with mp 120–122°C. The material obtained by the checkers shows the following ^{1}H NMR (CDCl$_3$) δ: 2.0 (s, impurity), 2.67 (s, 3 H), 7.30 (d of d of d, 2 H J = 1.8, 5.6, 8.0), 7.80 (d of t, 2 H, J = 1.8, 8.0), 8.35 (s, 2 H), 8.4–8.78 (m, 4 H). Mass spectrum m/e calculated: 279.0830. Found: 279.0815. IR (KBr) cm^{-1}: 1558, 1390. The combustion analyses for the products obtained by the checkers were within accepted limits for H, but off about 2% for C, and 0.6–0.8% for N.

9. Nickel chloride hexahydrate was obtained from the Fisher Scientific Company.

10. This reaction, which generates nickel boride,[4] is exothermic and evolves hydrogen. Frothing is prevented by keeping the reaction mixture at 0–5°C during addition of the sodium borohydride.

11. The submitters report obtaining tan material.

12. This material is analytically pure. Anal. calcd. for C$_{15}$H$_{11}$N$_3$: C, 77.23; H, 4.75; N, 18.01. Found C: 76.82; H, 4.69; N, 18.17. The product shows ^{1}H NMR (CDCl$_3$) δ: 7.33 (d of d of d, 2 H J = 1.5, 5.0, 8.0), 7.86 (d of t, 2 H, J = 2.0, 8.0), 7.96 (t, 1 H, J = 8.0 H), 8.45 (d, 2 H J = 8.0), 8.62 (d, 2 H, J = 8.0), 8.71 (d of m, 2 H).

13. The submitters report that 4'-(methylthio)-2,2' : 6',2"-terpyridine also can be conveniently reduced to 2',2" : 6',2"-terpyridine with Raney nickel in ethanol. The checkers found, however, that this procedure invariably gave product contaminated with 4'-ethoxy-2,2' : 6',2"-terpyridine. Raney nickel which was exhaustively washed with water to remove base still gave 15% of this by-product.

3. Discussion

The procedure described here is by far the most efficient synthesis of terpyridine.[5] Previous preparations include the dehydrogenation of pyridine with ferric chloride,[2] the Ullman reaction of 2-bromopyridine and 2,6-dibromopyridine,[6] the action of copper on 2-bromopyridine and 6-bromo-2,2'-dipyridyl,[6] the reaction of iodine or ferric chloride with 2,2'-bipyridyl,[6] and the reaction of 2,2'-bipyridyl with 2-lithiopyridine (40% yield).[7]

Terpyridine is a very effective chelating agent.

1. Department of Chemistry, Rensselaer Polytechnic Institute, Troy, NY 12181.
2. Morgan, G.; Burstall, F. H. *J. Chem. Soc.* **1937,** 1649.
3. Potts, K. T.; Usifer, D. A.; Guadalupe, A.; Abruna, H. D. *J. Am. Chem. Soc.* **1987,** *109,* 3961.
4. Truce, W. E.; Perry, F. M. *J. Org. Chem.* **1965,** *30,* 1316.
5. Potts, K. T.; Cipullo, M. J.; Ralli, P.; Theodoridis, G. *J. Am. Chem. Soc.* **1981,** *103,* 3584, 3583; *J. Org. Chem.* **1982,** *47,* 3027.
6. Burstall, F. H. *J. Chem. Soc.* **1938,** 1662.
7. Kauffmann, T.; König, J.; Woltermann, A. *Chem. Ber.* **1976,** *109,* 3864.

ELECTROHYDRODIMERIZATION OF AN ACTIVATED ALKENE: TETRAETHYL 1,2,3,4-BUTANETETRACARBOXYLATE

(1,2,3,4-Butanetetracarboxylic acid, tetraethyl ester)

$$2 \ CH_3CH_2OOCCH = CHCOOCH_2CH_3 \quad + \quad CH_3CH_2OH \quad \xrightarrow[\text{[(C}_4\text{H}_9)_4\text{N]ClO}_4]{\text{electricity}}$$

$$\underset{\displaystyle CH_3CH_2OOC \quad COOCH_2CH_3}{CH_3CH_2OOCCH_2CHCHCH_2COOCH_2CH_3}$$

$$+ \quad CH_3CHO$$

Submitted by D. A. WHITE[1]
Checked by CARL R. JOHNSON and DEBRA L. MONTICCIOLO

1. Procedure

The cell consists of a commercially available four-necked, 500-mL, round-bottomed flask equipped with a 34/45 standard-taper joint electrode assembly (Note 1), a 24/40 standard-taper joint purge and vent assembly, a mercury pool cathode (Note 2), a cathode contact (Note 3), a magnetic stirring bar (Note 4), and thermometer (inserted in a 10/18 standard-taper joint neck). The two platinum anodes of the electrode assembly (Note 1) are positioned in a horizontal plane ca. 1 cm above (Note 4) the mercury (cathode) surface.

To the cell are added diethyl fumarate (172 g, 1.0 mol) (Note 5), absolute ethanol (200 mL), and tetrabutylammonium perchlorate (3.41 g, 0.1 mol) (Note 6). The mixture is allowed to stand for 0.5 hr to allow complete dissolution of the tetrabutylammonium perchlorate. The cell is placed in a flowing-water bath in a hood.

The solution is electrolyzed with continuous magnetic stirring and nitrogen purging at a constant current (Note 7) until the theoretical quantity of electricity ($1.0F \equiv 1e^-$ per mole of diethyl fumarate) has been passed. The rate at which the cooling water in the bath flows is adjusted to maintain the electrolyte solution at 35°C during the first 2 hr of the electrolysis. It is then kept constant for the remainder of the electrolysis. After conditions have stabilized (ca. 1–2 hr of electrolysis), the reaction does not need constant attention, and may be allowed to run overnight.

The reaction mixture is transferred to a 2-L, round-bottomed flask with ethanol washing and the ethanol is removed by rotatory evaporation. Diethyl ether (1 L) is added to precipitate the electrolyte salt, which is collected by filtration and washed with ether. The crude electrolyte is obtained as a white solid (32–32.5 g, theory 34.1 g). The filtrate and washings were combined and evaporated to give a viscous brown oil, which was vacuum-distilled through a short Vigreux column (15 cm × 2.5 cm). After a forerun of 70 mL of material boiling below 150°C (0.15 mm), the product (92–96 g, 53–56%), bp 150–155°C (0.1 mm), is collected (Notes 8 and 9). The forerun contained diethyl maleate, diethyl fumarate, diethyl succinate, and diethyl ethoxysuccinate. The product is a mixture of diastereomers; on standing some meso isomer, mp 74–75°C, crystallizes.

2. Notes

1. The electrode assembly has been described (see synthesis of dimethyl decane-dioate, Note 1 and Figure 1, p. 182). In this case the electrodes have the same polarity and are electrically connected with a platinum wire dipping into the mercury contacts.

2. About 65 mL (860 g) of mercury was used, giving a pool with a surface diameter of ca. 6 cm.

3. A mercury-filled 6-mm o.d. glass tube with a platinum wire sealed through the lower end was used. The tube was bent to fit the contour of the flask. It was connected to the flask through a 24/40 standard-taper joint Teflon thermometer adapter (Ace Glass, Vineland, NJ). Contact to the mercury was made with a platinum wire as shown (Figure 1, p. 183).

4. A 20-cm × 0.5-cm Teflon coated stirring bar was used. This thickness (0.5 cm) is close to the maximum usable with an electrode gap of 1 cm. The rate of stirring was the maximum possible without breaking the mercury surface into droplets.

5. Diethyl fumarate, obtained from Aldrich Chemical Company, Inc., was used without prior purification. The submitters used diethyl maleate.

6. Tetrabutylammonium perchlorate, obtained from Eastman Organic Chemicals, was recrystallized from aqueous methanol (75%) and dried in vacuo.

7. The checkers used a Heath Schlumberger Model SP-2711 (30 V, 3 A) power supply at a current of 1.5 A. The cell voltage, initially 25 V, slowly rose to 30

V at the end of the electrolysis and the current dropped to ca. 1 A. The electrolysis required 17–24 hr. The submitters used a current of 1.0 A.

8. The submitters reported a yield of 135 g (78%). In part the reduced yields found by the checkers were caused by mechanical losses during distillation.

9. The product showed ^{1}H NMR (CDCl$_3$) δ: 1.25 (t, 12 H, CH$_3$), ca. 2.6 (m, 4 H,-COCH$_2$), ca. 3.3 (m, 2 H, CH), 4.2 (two overlapping q, 8 H, OCH$_2$). Analysis calculated for C$_{16}$H$_2$O$_8$: C, 55.5; H, 7.6%. Found: C, 55.5; H, 7.8. Molecular weight calculated: 346. Found (osmometrically in CHCl$_3$): 340, 338.

3. Discussion

This synthesis is an example of electrohydrodimerization of activated alkenes, the scope and mechanism of which have been recently reviewed.[2,3] The individual reactions combining to give the overall result.

$$2EtOOCCH=CHCOOEt + CH_3CH_2OH$$
$$\rightarrow EtOOCCH_2CH(COOEt)CH(COOEt)CH_2COOEt + CH_3CHO$$
$$(1)$$

are the cathodic reduction of the alkene to a dimer dianion (in the general case there are two major mechanisms by which the dianion may be formed and these are discussed in the references cited[2,3]),

$$2EtOOCCH=CHCOOEt + 2e^-$$
$$\rightarrow EtOOC\overline{C}HCH(COOEt)CH(COOEt)\overline{C}HCOOEt \quad (2)$$

the protonation of the dianion by ethanol,

$$EtOOC\overline{C}HCH(COOEt)CH(COOEt)\overline{C}HCOOEt + 2CH_3CH_2OH$$
$$\rightarrow EtOOCCH_2CH(COOEt)CH(COOEt)CH_2COOEt + CH_3CH_2O^-$$
$$(3)$$

and the anodic oxidation of ethanol.

$$CH_3CH_2OH \rightarrow CH_3CHO + 2H^+ + 2e^- \qquad (4)$$

In addition to providing an anode reaction (a suitable reaction at the "other" electrode is a necessity in any electrochemical reaction), reaction (4) also maintains the pH constant by producing protons to neutralize the ethoxide ions originating from reaction (3).

The present synthesis is an adaptation of a previously reported synthesis[4] in a divided cell (i.e., separate anode and cathode compartments). The overriding consideration in making this modification has been to simplify the operations involved and render the synthesis more attractive to chemists not well acquainted with electrochemical procedures. The main simplification achieved is that the pH is controlled internally via the anodic generation of protons as noted above (in the reported procedure,[4] this is achieved by periodic addition of acetic acid to the cathode compartment). A further simplification has been to run the reaction with a constant current rather than at controlled cathode potential. After the electrolysis has been initiated, the reaction requires no special attention. A small price is paid for the simplicity of the present synthesis in that the yield is somewhat lower than that obtained previously.[4] The major by-product formed is diethyl succinate, which results from a $2e^-$ reduction of diethyl fumarate or diethyl maleate:

$$EtOOCCH=CHCOOEt + 2e^- + 2CH_3CH_2OH$$
$$\rightarrow EtOOCCH_2CH_2COOEt + 2CH_3CH_2O^- \quad (5)$$

[cf. (2), which consumes $1e^-$ per mole of ester]. The occurrence of reaction (5) leads to incomplete consumption of ester after passage of the theoretical quantity of electricity (there may also be contributions from other sources).

The by-product, diethyl 2-ethoxybutanedioate, may be formed via base-catalyzed reaction in the vicinity of the cathode, where conditions may become quite basic.

1. Corporate Research Department, Monsanto Company, St. Louis, MO.
2. Baizer, M. M. In "Organic Electrochemistry," Baizer, M. M., Ed.; Marcel Dekker: New York, 1973.
3. Baizer, M. M.; Petrovich, J. P. *Progr. Phys. Org. Chem.* **1970,** *7,* 189–227.
4. Petrovich, M. M.; Baizer, M. R. *J. Electrochem. Soc.* **1969,** *116,* 749–756.

INDIRECT ELECTROLYSIS: TETRAMETHYL 1,1,2,2-ETHANETETRACARBOXYLATE

(1,1,2,2-Ethanetetracarboxylic acid, tetramethyl ester)

$$2\ CH_2(COOCH_3)_2 \xrightarrow[CH_3OH,\ NaI]{electricity} (CH_3OOC)_2CHCH(COOCH_3)_2 + H_2$$

Submitted by DONALD A. WHITE[1]
Checked by CARL R. JOHNSON and ROBERT C. ELLIOTT

1. Procedure

The preparation is carried out in a 500-mL, three-necked flask equipped with two graphite rod electrodes (Note 1). To the flask are added 132 g (1.0 mol) of dimethyl 1,3-propanedioate (Note 2), 15 g (0.10 mol) of sodium iodide, and 300 mL of methanol. A thermometer and reflux condenser are attached; the mixture is stirred and the solution formed is heated to 60°C. The heat source is removed. The solution is electrolyzed with a constant current of 2.0 A (Note 3) for 13.5 hr (Note 4) with gentle magnetic stirring. After a few minutes of electrolysis, the electrolyte begins to reflux gently and reflux is maintained throughout the electrolysis period by the heating effect of current passage (Note 5). Small granular crystals of the product begin to separate toward the end of the electrolysis period.

After electrolysis the reaction mixture is allowed to cool to room temperature and is filtered (Note 6). The crystalline residue is washed three times with 100-mL portions of methanol, dried by suction on the filter, and finally dried under vacuum. The product (88.4–91 g, 67–69%) is obtained as a white solid, mp 134–135°C.

2. Notes

1. The electrodes are $12 \times \frac{1}{4}$ in. graphite rods such as those used by glassblowers in shaping softened glass. They are attached as shown on page 183 via thermometer adaptors (Ace Glass Company, Vineland, NJ) and a specially made glass adaptor having two 10/18 and one 34/45 standard-taper joints. The electrodes should extend as far as possible into the electrolyte without interfering with the operation of the magnetic stirrer.

2. Dimethyl 1,3-propanedioate (dimethyl malonate) was obtained from Aldrich Chemical Company, Inc., and used as supplied.

3. A Heath/Schlumberger dc power supply, Model SP-2711, 30 V, 3 A, operating in its constant current mode, was used.

4. The current passed is $1.01F$ (1 faraday = 26.8 A-hr) and this is sufficient to convert 75–80% of the starting material to product. At higher conversions further oxidation occurs, leading to formation of tetramethyl ethenetetracarboxylate and hexamethyl 1,1,2,2,3,3-propanehexacarboxylate. The latter has solubility properties similar to those of the desired product. The product may be contaminated with the propanehexacarboxylate ester if the reaction is taken to higher conversions.

5. The cell voltage was initially 15 V and rose to 18 V at the end of the electrolysis. The cell voltage should be in the range 15–20 V so that the heat generated can be controlled by reflux. Since the cell voltage changes only slightly during the course of the electrolysis, a constant-voltage power supply could be used.

6. The filtrate contains only 2–5 g of the desired product; recovery is not worthwhile. The filtrate can, however, be reused as the electrolyte for conversion of further propanedioate ester.

7. The product may be recrystallized (from methanol), which gives material with mp 135–136°C.

8. The product gave an acceptable C,H analysis; the molecular weight by osmometry in chloroform was found to be 258 ($C_{10}H_{14}O_8$ in theory 262). The product showed ^1H NMR (CDCl$_3$) δ: 3.8, 4.2.

3. Discussion

The propanedioate (malonate) carbanion can be oxidized directly at an anode to give ethanetetracarboxylate esters, presumably via a radical intermediate.[2–4] Competing oxidation of solvent leads to a mixture of products[3,4] and for preparative purposes it is advantageous to carry out the reaction via indirect electrolysis as reported here. Indirect electrolysis refers to the continuous generation and regeneration of a reagent at an electrode, which interacts with substrate, as opposed to direct reaction of the substrate at the electrode. In the present case iodine is generated at the anode (1) and reacts with the cathodically generated (2) carbanion as shown (3) to give the desired overall reaction (4):

Anode reaction: $$2I^- \rightarrow I_2 + 2e^- \tag{1}$$

Cathode reaction: $$2CH_2(COOR)_2 + 2e^- \rightarrow 2^-CH(COOR)_2 + H_2 \tag{2}$$

Solution reaction: $$2^-CH(COOR)_2 + I_2 \rightarrow (ROOC)_2CHCH(COOR)_2 + 2I^- \tag{3}$$

Overall reaction: $$2CH_2(COOR)_2 \rightarrow (ROOC)_2CHCH(COOR)_2 + H_2 \tag{4}$$

The present procedure is based on literature reports[6,7] using indirect electrolysis involving electrogenerated halogens. Ethanetetracarboxylate esters have also been prepared by the chemical reaction of propanedioate carbanions with halogens.[8-10] The present procedure has the advantage of providing in situ generation of both the carbanion and the halogen from a small amount of added sodium halide. In other work it has been shown that the anodic formation of ethanetetracarboxylate ester can be paired with cathodic conversion of propenoate (acrylate) to hexanedioate (adipate) esters[11,12] In addition to these routes based on propanedioate esters, ethanetetracarboxylate esters have been obtained by electrocarboxylation of cis-butenedioate (maleate) esters.[13,14]

1. Corporate Research Department, Monsanto Company, St. Louis, MO 63166.
2. Okubo, T.; Tsutsumi, S. Technol. Rep. Osaka Univ. 1963, 13, 495; Chem. Abstr. 1964, 61, 6637e.
3. Brettle, R.; Parkin, J. G. J. Chem. Soc. C 1967, 1352–1357.
4. Brettle, R.; Seddon, D. J. Chem. Soc. C 1970, 1153–1154.
5. Mulliken, S. P. J. Am. Chem. Soc. 1893, 15, 526.
6. Okubo, T.; Tsutsumi, S. Bull. Chem. Soc. Jpn. 1964, 37, 1794–1797.
7. Osa, T.; Okhatsu, Y.; Tezuka, M. "Extended Abstracts"; 149th National Meeting of the Electrochemical Society, Washington, DC, 1976; p. 764.
8. Bischoff, C. A.; Rach, C. Chem. Ber. 1884, 17, 2781.
9. Bailey, W. J.; Anderson, W. J. J. Am. Chem. Soc. 1956, 78, 2287–2290.
10. Walker, J.; Appleyard, J. R. J. Chem. Soc. 1895, 67, 768.
11. Thomas, H. G.; Lux, E. Tetrahedron Lett. 1972, 965–968.
12. Baizer, M. M.; Hallcher, R. C. J. Electrochem. Soc. 1976, 123, 809–813.
13. Tyssee, D. A.; Wagenknecht, J. H.; Baizer, M. M.; Chruma, J. L. Tetrahedron Lett. 1972, 4809–4812.
14. Tyssee, D. A.; Baizer, M. M. J. Org. Chem. 1974, 39, 2819–2823.

DIELS-ALDER REACTION OF 1,2,4,5-HEXATETRAENE: TETRAMETHYL[2.2]PARACYCLOPHANE-4,5,12,13-TETRACARBOXYLATE

(Tricyclo[8.2.2.2^{4,7}]hexadeca-4,6,10,12,13,15-hexaene-5,6,11,12-tetracarboxylic acid, tetramethyl ester

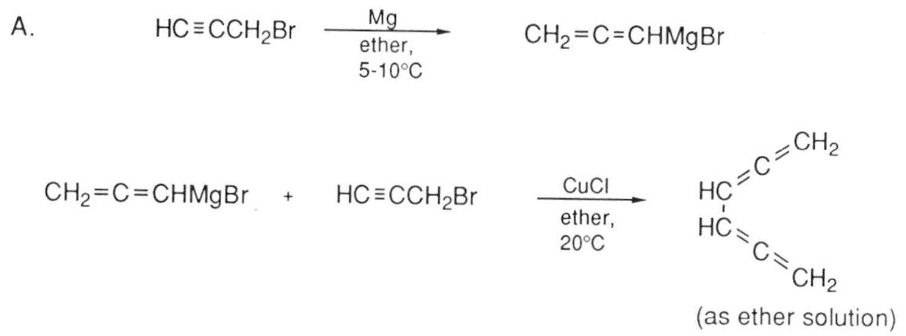

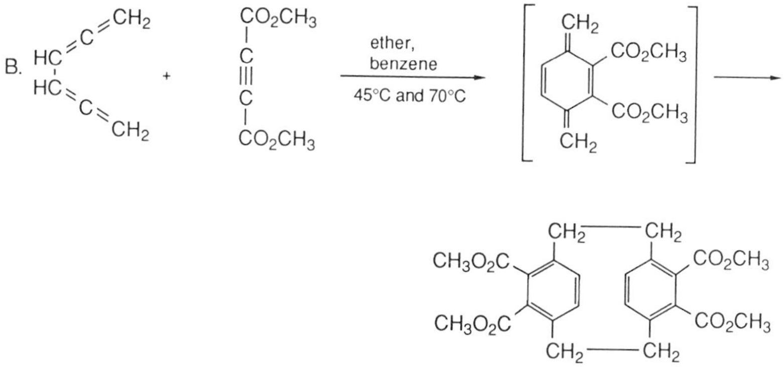

A. $HC \equiv CCH_2Br$ $\xrightarrow[\text{ether,}]{\text{Mg}}$ $CH_2 = C = CHMgBr$
5-10°C

$CH_2 = C = CHMgBr$ + $HC \equiv CCH_2Br$ $\xrightarrow[\text{ether,}]{\text{CuCl}}$ (structure)
20°C

(as ether solution)

B. (diagram)

Submitted by Henning Hopf, Ingrid Böhm, and Jürgen Kleinschroth[1]
Checked by Paul F. Sherwin and Robert M. Coates

1. Procedure

Caution! Propargyl bromide is poisonous and should be handled in a well-ventilated hood.

Benzene has been identified as a carcinogen; OSHA has issued emergency standards for its use. All procedures involving benzene should be carried out in a well-ventilated hood, and glove protection is required.

A. *1,2,4,5-Hexatetraene in ether solution.* A 2-L, four-necked, round-bottomed flask is equipped with a mechanical stirrer, a reflux condenser fitted with a drying tube containing anhydrous calcium sulfate (Drierite), a dropping funnel, and a thermometer

(Note 1). The flask is charged with 0.5 g (0.002 mol) of mercury(II) chloride and 29.2 g (1.2 mol) of magnesium turnings that have been crushed with a mortar and pestle, and the apparatus is flushed with nitrogen while being heated externally with a Bunsen burner to remove traces of moisture. In the cooled flask are placed 160 mL of anhydrous ethyl ether (Note 2) and 7.6 g (5.0 mL, 0.064 mol) of propargyl bromide (Note 3). The ether begins to reflux within 1 min, indicating that formation of the Grignard reagent has begun (Note 4). The mixture is cooled to 5° in an ice-salt bath and stirred vigorously as a solution of 135 g (89 mL, 1.13 mol) of propargyl bromide in 560 mL of anhydrous ether is added. The addition rate is adjusted so as to maintain the internal temperature between 5°C and 10°C (Note 5). The cooling bath is removed, and the dark green mixture is stirred for 45 min at room temperature (Note 6). A 2-g (0.02 mol) portion of finely pulverized, dry copper(I) chloride (Note 7) is added, and the mixture, which becomes a chocolate-brown color after 2–3 min, is stirred for 15 min at room temperature and cooled again to 5°C with either an ice-water bath or an ice-salt bath. Stirring is continued while a solution of 128 g (85 mL, 1.08 mol) of propargyl bromide in 100 mL of ether is added at a rate such that the internal temperature is kept at ca. 20°C (Note 8). The mixture becomes almost black, and two phases are discernible when the stirrer is stopped, especially toward the end of the addition. The cooling bath is removed and stirring is continued for 15 min at room temperature to complete the dimerization. The reaction mixture is cooled to 0°C with an ice-salt bath and stirred vigorously as 200 mL of 1 N aqueous hydrochloric acid is added (Note 9). The two-phase mixture is warmed to room temperature, and another 100 mL of 1 N hydrochloric acid is added. The ether layer is separated and washed with three 100-mL portions of water (Note 10). A few crystals (0.2–0.5 g) of hydroquinone are added to stabilize the reddish solution, which is then dried with anhydrous potassium carbonate. The drying agent is filtered, and the filtrate is concentrated to a volume of ca. 400 mL by distillation under nitrogen at atmospheric pressure with a 40-cm Vigreux column and a heating bath kept at 40–45°C. The concentrate is purified by vacuum transfer (Note 11), and the now colorless solution, which contains ca. 25–30 g (30–36%) of 1,2,4,5-hexatetraene, is stabilized by adding another 0.1–0.5 g of hydroquinone (Notes 12 and 13).

B. *Tetramethyl[2.2]paracyclophane-4,5,12,13-tetracarboxylate.* The 1-L flask from Step A containing ca. 25–30 g (0.32–0.38 mol) of 1,2,4,5-hexatetraene in ether solution is equipped with a magnetic stirring bar and a 40-cm Vigreux column for distillation at atmospheric pressure. A solution of 69.4 g (60 mL, 0.49 mol) of dimethyl acetylenedicarboxylate (Note 14) in 220 mL of benzene (Note 15) is added. The resulting solution is stirred and heated at 45°C for 5–7 hr and at 70°C for 20 hr (Note 16). The ether distils slowly at 45°C and rather rapidly at 70°C, the color of the solution changes from yellow to red–orange, and a white solid is gradually deposited. The mixture is cooled and filtered. Recrystallization of the solid from 600–800 mL of toluene affords 27–35 g of white crystalline product, mp 201.5–203°C. The original benzene filtrate and the toluene mother liquor, when evaporated to dryness and crystallized separately from 40–70 mL of toluene, provide 3–4 g and 2–3 g of product, respectively, having essentially the same melting point. The combined yield is 33–41 g (40–50% based on 1,2,4,5-hexatetraene) (Notes 17 and 18).

2. Notes

1. The checkers used a three-necked flask equipped with a Claisen adapter. The straight branch of the adapter was fitted with a thermometer, and the curved branch was mounted with a condenser bearing a nitrogen inlet. After the nitrogen-filled apparatus had been flamed dry and cooled, the dropping funnel was capped with a rubber septum. A nitrogen atmosphere was maintained in the apparatus at all times, and liquids were placed in the dropping funnel via syringe.

2. The checkers dried the ether by distillation from sodium benzophenone ketyl immediately before use.

3. Propargyl bromide may be purchased from Fluka. The reagent may also be prepared by the procedure of Gaudemar.[2] Propargyl bromide (97%) supplied by Tridom Chemical Inc. was dried by the checkers prior to use by stirring over Linde-type 4A molecular sieves under a nitrogen atmosphere for 2–3 days. The volumetric quantities given in the procedure are for 97% propargyl bromide, which has a density of 1.56 at 20°C according to a catalog from Tridom Chemical Inc.

4. The submitters initiated Grignard formation by adding a few milliliters of a solution of 142.8 g (1.2 mol) of propargyl bromide in 560 mL of anhydrous ether. If the reaction does not begin, they suggest that the flask be heated with a warm stream of air from a "heat gun."

5. The checkers found that the addition time varied from ca. 1 to 4 hr, depending on the temperature of the cooling bath and the extent to which it was stirred.

6. Some unreacted magnesium turnings remain in the flask at this time. However, the submitters recommend against heating the mixture to achieve further conversion, since the initially formed allenylmagnesium bromide will isomerize to 1-propynylmagnesium bromide.

7. The submitters used copper(I) chloride purchased from E. Merck, Darmstadt, which had a greenish tinge attributed to slight contamination by copper(II) salts. The reagent used by the checkers was supplied by J. T. Baker Chemical Company.

8. The checkers, using an ice-salt cooling bath, maintained the internal temperature between 7°C and 12°C. The addition time varied from 15 to 75 min depending on the efficiency of the cooling.

9. The internal temperature was maintained at 10–15°C by the submitters and 5–12°C by the checkers. The time required for the hydrolysis was reduced by the checkers by prior chilling of the hydrochloric acid.

10. Since the product is unstable to oxygen, the checkers tried to keep the ethereal solution under a nitrogen atmosphere during transfer and extractions.

11. The crude solution of 1,2,4,5-hexatetraene may also be employed in part B. However, yields are lower, and the purification of the product becomes tedious owing to the presence of insoluble by-products. The vacuum transfer may be accomplished with a simple distillation apparatus equipped with a magnetic stirring bar and a 1-L, round-bottomed flask as receiver in the following manner.

The crude ethereal solution is chilled to a glass with a liquid nitrogen bath, and the apparatus is evacuated to a pressure of 1–3 mm with a vacuum pump. The apparatus is isolated from the vacuum pump, the cooling bath is removed, and the ether glass is allowed to warm until it becomes mobile. The freeze–evacuate–thaw cycle is repeated two more times to complete the degassing process. The solution is again chilled with liquid nitrogen, the apparatus is evacuated to 1–3 mm, and the system is then isolated from the vacuum pump. The receiving flask is cooled with a liquid nitrogen or dry ice-isopropyl alcohol bath, the cooling bath is removed from the distilling flask, and the ether solution is stirred and allowed to warm to room temperature in the closed system. Once the ether solution becomes mobile, the flask may be warmed cautiously with a water bath at room temperature to speed up the vacuum transfer.

12. Pure 1,2,4,5-hexatetraene polymerizes readily when exposed to air at room temperature. However, solutions of the purified compound are stable for months at 0°C, especially if protected by an inert gas. Contact with air should be minimized to avoid inducing polymerization.

13. The amount of 1,2,4,5-hexatetraene in solution was estimated by the submitters from ^{1}H NMR spectra and GC analyses. The major product of the dimerization reaction is 1,2-hexadien-5-yne. The submitters have shown that this hydrocarbon does not interfere with the cycloaddition in Step B.

14. Dimethyl acetylenedicarboxylate is supplied by Aldrich Chemical Company, Inc.; E. Merck, Darmstadt; and Fluka. The reagent, bp 110–112°C (15 mm), was distilled before use.

15. The checkers carried out one run on one-tenth scale using toluene instead of benzene in part B. The yield of tetramethyl[2.2]paracyclophane-4,5,12,13-tetra-carboxylate, mp 202.5–204°C, was 2.91 g (35% based on 1,2,4,5-hexatetraene in part A).

16. The conditions and isolation procedure given are those used by the checkers. The submitters heated the solution first to 45°C, after which the temperature was gradually increased to 70°C over several hours. The solution was then heated at 70°C overnight, and the solvents were removed by rotary evaporation. The semisolid, yellow-orange residue was recrystallized from benzene or methanol. Concentration of the mother liquor afforded additional crops of crystalline product. The total yield of the tetraester was 25–30 g (30–36% based on 1,2,4,5-hexatetraene), mp 206–207°C.

17. The melting point of the product obtained by the checkers increased only slightly to 202–203.5°C on recrystallization from toluene, benzene, or methanol. The once-recrystallized product was analyzed by the checkers. Anal. calcd. for $C_{24}H_{24}O_8$: C, 65.45; H, 5.49. Found: C, 65.51; H, 5.49. The submitters obtained material of analytical purity by sublimation at 180°C (0.001 mm).

The spectral properties of the product are as follows: IR (potassium bromide) cm^{-1}: 1715, 1260, 1195, 1125, 1005, 870; ^{1}H NMR (CDCl$_3$) δ: 2.93–3.43 (nine-line $AA'BB'$ multiplet, 8, two CH_2CH_2), 3.83 (singlet, 12, four CO_2CH_3), 6.80 (singlet, 4, four aryl CH); proton-decoupled ^{13}C NMR (CDCl$_3$) δ (assignment): 33.27 (CH_2), 52.25 (OCH_3), 131.51 (CCO_2CH_3), 134.98 (CCH_2), 139.68 (CH), 168.40 (CO_2CH_3).

TABLE I
[2.2]PARACYCLOPHANES PREPARED BY DIELS-ALDER REACTION OF DISUBSTITUTED ACETYLENES WITH 1,2,4,5-HEXATETRAENE[13, 14]

R	mp (°C)	Yield (%)
CO_2CH_3	203	47
$CO_2C_2H_5$	133.5	30
$CO_2C(CH_3)_3$	205	20
CO_2H	365 (dec)	5
CN	320 (dec)	37
CF_3	174	21

18. The submitters have scaled up this procedure to prepare as much as 60 g of the paracyclophane in one run. However, since the volumes of solvents and flasks are quite large, the operations become rather cumbersome.

3. Discussion

[2.2]Paracyclophanes have been recognized for some time as interesting structures for stereochemical studies and for unusual intra- and intermolecular π-electron interactions.[3–6] The nonplanar, boatlike benzene rings[7] of these compounds have attracted the attention of numerous synthetic organic chemists[3–6] as well as theoreticians[8,9] and spectroscopists.[6,10]

The principal methods used previously for the preparation of [2.2]paracyclophanes have been reviewed several times[3–6] and include (1) intramolecular Wurtz coupling of appropriately substituted dihalides at high dilution; (2) ring contraction of cyclophanes having larger bridges by sulfone pyrolysis (the most versatile procedure currently known), Stevens rearrangement, and other extrusion reactions; and (3) the dimerization of transient p-quinodimethane intermediates (p-xylylenes), usually generated from p-xylene precursors by elimination reactions.[11] A p-quinodimethane is presumably also formed initially in Step B of this procedure. This cycloaddition route to p-quinodimethanes and [2.2]paracyclophanes, discovered for the first time in the submitters' laboratories, is attractive on account of the availability of the starting materials, the simplicity of the procedure, and the relatively large quantities of product that may be obtained. The approach appears to be fairly general, since both the bisallene and acetylene components may be varied (Table I).[12–14] However, methyl 2-butynoate, 2-butyne, diphenylacetylene, and bistrimethylsilylacetylene failed to react with 1,2,4,5-hexatetraene. Another limitation is the exclusive formation of [2.2]paracyclophanes with the anti configuration from disubstituted acetylenic dienophiles.

The substituted [2.2]paracyclophanes prepared by the present procedure have proven to be useful starting materials for the synthesis of cyclophanes with extended aromatic ring systems,[15] additional ethano bridges,[16,17] and chromium tricarbonyl complexes.[18]

1. Institut für Organische Chemie der Technischen Universität Braunschweig, Schleinitzstrasse, D-3300 Braunschweig, West Germany.
2. Gaudemar, M. Ann. Chim. (Paris) Ser. 13 1956, 1, 161–213.

3. For recent reviews on the chemical behavior of [2.2]paracyclophane, see: (a) Boekelheide, V. *Acc. Chem. Res.* **1980**, *13*, 65; (b) Hopf, H.; Kleinschroth, J. *Angew. Chem.* **1982**, *94*, 485: *Angew. Chem. Int. Ed. Engl.* **1982**, *21*, 469; (c) Boekelheide, V. *Top Curr. Chem.* **1983**, *113*, 87; (d) Hopf, H. In Keehn, P. M.; Rosenfeld, S. M. (Hrsg.); "The Cyclophanes," Academic Press, New York, 1983, S. 521 ff; (e) Heilbronner, E.; Yang, Z. *Top Curr. Chem.* **1984**, *115*, 1; (f) Gerson, F. ibid. **1984**, *115*, 57.

4. Cram, D. J.; Cram, J. M. *Acc. Chem. Res.* **1971**, *4*, 204–213.

5. Vögtle, F.; Neumann, P. *Synthesis* **1973**, 85–103.

6. Vögtle, F.; Neumann, P. P. *Top. Curr. Chem.* **1974**, *48*, 67–129.

7. Hopf, H. *Chem. Uns. Zeit* **1976**, *10*, 114; *Chem. Abstr.* **1976**, *85*, 191622s.

8. Lindner, H. J. *Tetrahedron* **1976**, *32*, 753–757 and references cited therein.

9. Misumi, S.; Iwamura, H.; Kihara, H.; Sakata, Y.; Umemoto, T. *Tetrahedron Lett.* **1976**, 615–618.

10. El-Sayed, M. A. *Nature* **1963**, *197*, 481–482.

11. Winberg, H. E. Fawcett, F. S. *Org. Synth., Coll. Vol. V* **1973**, 883–886.

12. Lenich, F. Th.; Hopf, H. *Chem. Ber.* **1974**, *107*, 1891–1902.

13. Böhm, I.; Herrmann, H.; Menke, K.; Hopf, H. *Chem. Ber.* **1978**, *111*, 523–527.

14. Blickle, P.; Hopf, H. *Tetrahedron Lett.* **1978**, 449–452.

15. Kleinschroth, J.; Hopf, H. *Tetrahedron Lett.* **1978**, 969–972.

16. Trampe, S.; Menke, K.; Hopf, H. *Chem. Ber.* **1977**, *110*, 371–372.

17. Gilb, W.; Menke, K.; Hopf, H. *Angew. Chem.* **1977**; *Angew. Chem. Int. Ed. Engl.* **1977**, *16*, 191.

18. Mourad, A. F.; Hopf, H. *Tetrahedron Lett.* **1979**, 1209–1212.

THIETE 1,1-DIOXIDE AND 3-CHLOROTHIETE 1,1-DIOXIDE

(2 H-Thiete 1,1-dioxide and 2 H-thiete, 3-chloro- 1,1-dioxide)

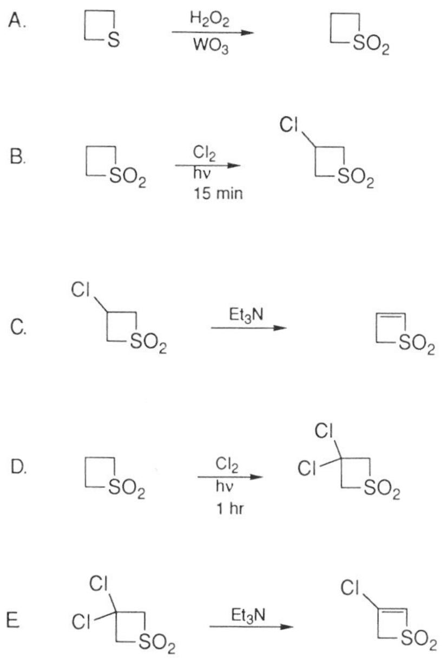

Submitted by Thomas C. Sedergran and Donald C. Dittmer[1]
Checked by M. F. Semmelhack, Elena M. Bingham, William A. Sheppard, and Joseph J. Bozell

1. Procedure

A. *Thietane 1,1-dioxide.* The pH of a solution of tungstic acid ($WO_3 \cdot H_2O$) (1.1 g, 0.044 mol) (Note 1) in 280 mL of distilled water is adjusted to 11.5 by addition of 10% aqueous sodium hydroxide; the white suspension of the tungstate catalyst is added to a 1-L, round-bottomed flask fitted with a mechanical stirrer and a pressure-equalizing addition funnel. The tungstic acid–water mixture is cooled to 0–10°C by means of an ice–salt bath; glacial acetic acid (50 mL) and trimethylene sulfide (thietane) (47.5 g, 0.641 mol, d 1.028) (Note 2) are added. The chilled mixture is stirred, and 30% hydrogen peroxide (189 mL) is added carefully by means of the addition funnel over a period of 2 hr (Note 3). The mixture is stirred at 0–10°C for an additional hour, transferred to an evaporating dish, and heated to near dryness on a steam bath. The resulting solid material is triturated five times with 100-mL portions of hot chloroform; any catalyst is removed by filtration. The chloroform solutions are combined and dried over anhydrous magnesium sulfate and the solvent is removed via a rotary evaporator to give a white solid (60.3–63.7 g, 0.57–0.60 mol, 88.7–93.7%), mp 74–76°C (lit.[2] mp 75.5–76°C).

B. *3-Chlorothietane 1,1-dioxide.* Thietane 1,1-dioxide (14.0 g, 0.132 mol) is placed in a three-necked, 500-mL, round-bottomed flask fitted with a magnetic stirrer, reflux condenser and a chlorine bubbler. *(Caution! Since chlorine is poisonous, the reaction involving it should be done in a good hood.)* Carbon tetrachloride (300 mL) is added to the flask (Note 4) and the suspension is irradiated by a 250-W sunlamp positioned as close as possible to the reaction flask without touching it (Note 5) while chlorine is bubbled through the solution for 15 min at a moderate rate (Note 6). A copious white precipitate forms and irradiation and addition of chlorine must be stopped at this point (or 10 min after the first appearance of a precipitate) to avoid dichlorination. The reaction mixture is cooled to room temperature and filtered to give a white, fluffy product (5.4–8.1 g, 30–44%) that is crystallized from chloroform, mp 136–137°C (lit.[3] mp 136.5–137.5°C).

C. *Thiete 1,1-dioxide.* A sample of 3-chlorothietane 1,1-dioxide (8.0 g, 0.057 mol) is dissolved in dry toluene (300 mL) (Note 7) in a 500-mL, two-necked, round-bottomed flask equipped with a reflux condenser, magnetic stirrer, heating mantle (or silicone oil bath), and thermometer. The reaction is heated to 60°C and triethylamine (28.7 g, 0.28 mol, 39.5 mL) is added through the condenser. The reaction mixture is stirred for 4 hr and triethylamine hydrochloride is removed by filtration and washed with toluene (100 mL). Toluene is removed on a rotary evaporator and the residue is recrystallized from diethyl ether–ethanol (Note 8) to give a white solid (4.5–4.8 g, 75–81%); mp 49–50°C (lit.[3] mp 52–54°C).

D. *3,3-Dichlorothietane 1,1-dioxide.* Thietane 1,1-dioxide (5.0 g, 0.047 mol) is placed in a 500-mL, three-necked, round-bottomed flask equipped with a reflux condenser, magnetic stirrer, and chlorine gas bubbler. Carbon tetrachloride (350 mL) is added and the solution is irradiated with a 250-W sunlamp (Note 5) while chlorine is bubbled through the stirred mixture for 1 hr (Note 9). Irradiation and chlorine addition are stopped and the reaction mixture is allowed to cool to room temperature. The product is collected by filtration as a white solid (4.0–4.4 g, 49–53%), mp 156–158°C[4] (Note 10). the product can be used without further purification or recrystallized from chloroform.

E. *3-Chlorothiete 1,1-dioxide.* A solution of 3,3-dichlorothietane 1,1-dioxide (4.0 g, 0.023 mol) in toluene (150 mL) is placed in a 250-mL, round-bottomed, two-necked flask equipped with a heating mantle (or silicone oil bath), magnetic stirrer, reflux condenser, and thermometer. The solution is heated to 60°C and triethylamine (2.54 g, 0.025 mol, 3.5 mL) is added dropwise through the condenser over a 10-min period. The solution is stirred for 2 hr at 60°C and cooled to room temperature. The triethylamine hydrochloride is collected by filtration and washed with hot toluene (50 mL). Removal of toluene on a rotary evaporator gives a white solid (2.7–3.0 g, 84–93%) that is recrystallized from chloroform-hexane, mp 118–120°C[4] (Note 11).

2. Notes

1. The tungstic acid was used as supplied by the Eastman Kodak Company.
2. The trimethylene sulfide was used as supplied by the Aldrich Chemical Company.

3. The addition rate of the hydrogen peroxide must be adjusted so that the temperature of the reaction mixture does not rise above 10°C. The yield is reduced if the temperature is allowed to rise above that point. The endpoint of the reaction, when excess peroxide is present, can be determined with potassium iodide–starch test paper. The yield also is reduced if more than a slight excess of hydrogen peroxide is used.

4. The sulfone is not completely dissolved at this point. The prescribed ratio of sulfone to carbon tetrachloride (0.0467 g mL) is important. If it is less (i.e., more carbon tetrachloride relative to sulfone), considerable 3,3-dichlorothietane 1,1-dioxide will be formed.

5. Any commercial sunlamp is satisfactory and should be used with eye protection. Carbon tetrachloride boils gently because of the heat from the lamp.

6. The submitters suggested adding the chlorine at such a rate that a constant yellow color is maintained in the solution or suspension. The checkers found that, depending on the rate of chlorine introduction, it took from 10 to 35 min for the appearance of the white precipitate. In each run, the monochlorinated product was contaminated with a small amount (5–10% by NMR integration) of either starting material or dichlorinated product. The checkers found that the optimum yield of monochlorinated product was obtained when the chlorine was bubbled into the solution through a ¼-in. glass tube at a rate estimated to be between 5–15 bubbles per second. The suspended sulfone dissolves as the reaction proceeds.

7. Toluene was dried over Linde 4A molecular sieves. Benzene may be used also.

8. The product is heated in about 25–30 mL of diethyl ether, and ethanol is added dropwise until a solution is obtained. The checkers found that the thiete sulfone could also be crystallized by gently heating the crude material in diethyl ether (~ 100 mL) until it dissolves, followed by cooling to $-15°C$.

9. If the reaction time is less than 1 hr, a mixture of monochloro- and dichlorosulfone is obtained.

10. The spectral properties of the product are as follows: IR (KBr disk) cm^{-1}: 2950 (m), 1370 (m, SO$_2$), 1310 (m), 1210 (m), 1140 (m, SO$_2$), 970 (m), 940 (m), 820 (w); ^{1}H NMR (chloroform-d) δ: 5.0 (s, 4 H, CH$_2$SO$_2$CH$_2$).

11. The spectral properties of the product are as follows: IR (KBr disk) cm^{-1}: 1540 (m, >C=C<), 1400 (w), 1300 (s, SO$_2$), 1210 (s), 1140 (s, SO$_2$), 1020 (m), 770 (m); ^{1}H NMR (chloroform-d) δ: 4.6 (s, 2 H, CH$_2$—SO$_2$), 6.8 (s, 1 H, CH=C).

3. Discussion

This preparation of thiete 1,1-dioxide is more direct and less tedious than previous methods.[3,5,6]

Oxidation of trimethylene sulfide catalyzed by tungstic acid[7] is preferred to the uncatalyzed reaction; yields are better and the reaction time is shortened by elimination of an induction period.

Selective chlorination of the 3-position of thietane 1,1-dioxide may be a consequence of hydrogen atom abstraction by a chlorine atom. Such reactions of chlorine atoms are believed to be influenced by polar effects, with preferential hydrogen abstraction occurring remotely from an electron withdrawing group.[8] The free-radical chain reaction may be propagated by attack of the 3-thietanyl 1,1-dioxide radical on molecular chlorine.

Conversion of 3-chlorothietane 1,1-dioxide to the 3-(N,N-dimethylamino) derivative followed by reduction, quaternization, and Hofmann elimination affords a convenient route to the highly reactive thiete (thiacyclobutene).[5,9]

The following compounds have been obtained from thiete 1,1-dioxide: substituted cycloheptatrienes,[10] benzyl α-toluenethiosulfinate,[11] pyrazoles,[12] naphthothiete 1,1-dioxides,[13] and 3-substituted thietane 1,1-dioxides.[14] It is a dienophile in Diels–Alder reactions[10,13,15] and undergoes cycloadditions with enamines, dienamines, and ynamines.[16] Thiete 1,1-dioxide is a source of the novel intermediate, vinylsulfene (CH_2=CHCH=SO_2), which undergoes cycloadditions to strained olefinic double bonds,[17] reacts with phenol to give allyl sulfonate derivatives[18] or cyclizes unimolecularly to give an unsaturated sultene.[18] Platinum[19] and iron[20] complexes of thiete 1,1-dioxide have been reported.

3-Chlorothiete 1,1-dioxide is a useful intermediate for the preparation of other 3-substituted thiete 1,1-dioxides via addition–elimination reactions.[4] It also undergoes Diels–Alder reactions with 1,3-butadiene and with 1,3-diphenylisobenzofuran.[4]

1. Department of Chemistry, Syracuse University, Syracuse, NY 13210.
2. Grishkevich-Trokhimovskii, E. *J. Russ. Phys. Chem. Soc.* **1916,** *48,* 880; *Chem. Abstr.* **1917,** *11,* 784; Grishkevich-Trokhimovskii, E. *Chem. Zentr.* **1923,** *III,* 773.
3. Dittmer, D. C.; Christy, M. E. *J. Org. Chem.* **1961,** *26,* 1324.
4. Sedergran, T. C.; Yokoyama, M.; Dittmer, D. C. *J. Org. Chem.* **1984,** *49,* 2408.
5. Chang, P. L.-F.; Dittmer, D. C. *J. Org. Chem.* **1969,** *34,* 2791.
6. Lamm, B.; Gustafsson, K. *Acta Chem. Scand.* **1974,** *B28,* 701.
7. Schultz, H. S.; Freyermuth, H. B.; Buc, S. R. *J. Org. Chem.* **1963,** *28,* 1140.
8. Kharasch, M. S.; Brown, H. C. *J. Am. Chem. Soc.* **1939,** *61,* 2142; Kharasch, M. S.; Brown, H. C. *J. Am. Chem. Soc.* **1940,** *62,* 925.
9. Dittmer, D. C.; Chang, P. L.-F.; Davis, F. A.; Iwanami, M.; Stamos, I. K.; Takahashi, K. *J. Org. Chem.* **1972,** *37,* 1111
10. Dittmer, D. C.; Ikura, K.; Balquist, J. M.; Takashina, N. *J. Org. Chem.* **1972,** *37,* 225.
11. McCaskie, J. E.; Nelsen, T. R.; Dittmer, D. C. *J. Org. Chem.* **1973,** *38,* 3048.
12. Dittmer, D. C.; Glassman, R. *J. Org. Chem.* **1970,** *35,* 999; DeBenedetti, P. G.; De Micheli, C.; Gandolfi, R.; Gariboldi, P.; Rastelli, A. *J. Org. Chem.* **1980,** *45,* 3646; DallaCroce, P.; DelButtero, P.; Maiorana, S.; Vistocca, R. *J. Heterocycl. Chem.* **1978,** *15,* 515.
13. Dittmer, D. C.; Takashina, N. *Tetrahedron Lett.* **1964,** 3809; Paquette, L. A. *J. Org. Chem.* **1965,** *30,* 629.
14. Dittmer, D. C.; Christy, M. E. *J. Am. Chem. Soc.* **1962,** *84,* 399.
15. Coxon, J. M.; Battiste, M. A. *Tetrahedron* **1976,** *32,* 2053.
16. Paquette, L. A.; Houser, R. W.; Rosen, M. *J. Org. Chem.* **1970,** *35,* 905.
17. Dittmer, D. C.; McCaskie, J. E.; Babiarz, J. E.; Ruggeri, M. V. *J. Org. Chem.* **1977,** *42,* 1910.
18. King, J. F.; deMayo, P.; McIntosh, C. L.; Piers, K.; Smith, D. J. H. *Can. J. Chem.* **1970,** *48,* 3704.
19. Reinhoudt, D. N.; Kouwenhoven, C. G.; Visser, J. P. *J. Organometal. Chem.* **1973,** *57,* 403.
20. McCaskie, J. E.; Chang, P. L.; Nelsen, T. R.; Dittmer, D. C. *J. Org. Chem.* **1973,** *38,* 3963.

(S)-$(+)$-2-$(p$-TOLUENESULFINYL)-2-CYCLOPENTENONE: PRECURSOR FOR ENANTIOSELECTIVE SYNTHESIS OF 3-SUBSTITUTED CYCLOPENTANONES

[2-Cyclopenten-1-one, 2-[(4-methylphenyl)sulfinyl]-, (S)-]

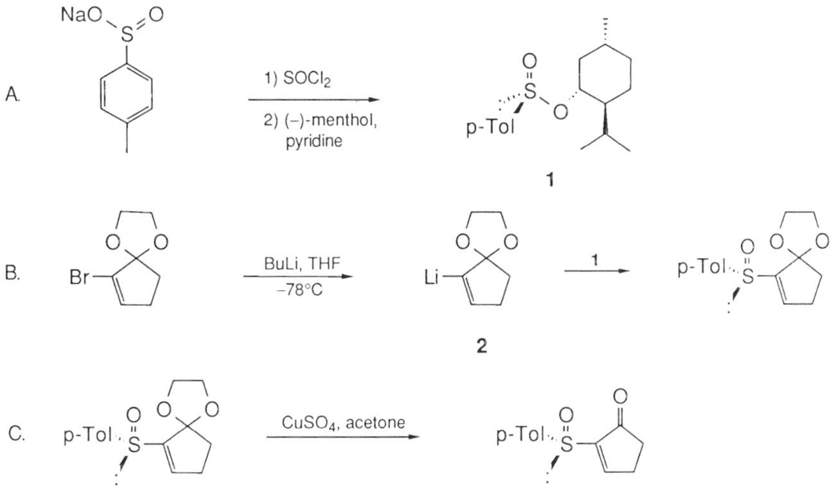

Submitted by MARTIN HULCE, JOHN P. MALLOMO, LEAH L. FRYE, TIMOTHY P. KOGAN, and GARY H. POSNER[1]
Checked by ERNEST B. CLARK, MICHEL CREVOISIER, HAN-YOUNG KANG, and ROBERT M. COATES.

1. Procedure

Caution! Part A should be conducted in an efficient fume hood to avoid exposure to sulfur dioxide generated in the reaction.

Benzene has been identified as a carcinogen; OSHA has issued emergency standards for its use. All procedures involving benzene should be carried out in a well-ventilated hood, and glove protection is required.

A. *(S)-$(-)$-Menthyl p-toluenesulfinate.* In a dry, 250-mL, three-necked, round-bottomed flask equipped with a nitrogen inlet are placed a magnetic stirring bar and 65 g (40 mL, 0.55 mol) of thionyl chloride (Note 1). The liquid is stirred under a nitrogen atmosphere as 35.6 g (0.200 mol) of anhydrous sodium *p*-toluenesulfinate (Note 2) is added in portions over about 1 hr (Note 3). The solution immediately develops a yellow-green tinge as sulfur dioxide is liberated. After about three-fourths of the sulfinate has been added, 30 mL of benzene is added to facilitate stirring. The greenish slurry is stirred for another 1.5 hr, after which time 75 mL of benzene is added. The mixture is transferred to a 500-mL, round-bottomed flask, along with 75 mL of benzene used to rinse the flask. Excess thionyl chloride and benzene are removed by rotary evaporation and gentle heating. Four 150-mL portions of benzene are added to the residue, and each portion is evaporated to complete the removal of the thionyl chloride. The flask is equipped with a magnetic stirring bar and a 125-mL, pressure-equalizing dropping funnel. The crude *p*-toluenesulfinyl chloride, sodium chloride, and residual benzene are dissolved in 150 mL of anhydrous diethyl ether. The resulting ethereal suspension is stirred and cooled in an ice bath as 31.3 g (0.200 mol) of (−)-menthol (Note 1) in 25 mL of pyridine is added over ca. 2 min. The mixture is allowed to stir overnight, after which 70 g of ice is added. The layers are separated and the aqueous layer is extracted with one 100-mL portion of ether. The ethereal solutions are combined, washed three times with 50-mL portions of 20% aqueous hydrochloric acid, and dried with a mixture of anhydrous sodium sulfate and potassium carbonate. Filtration to separate the drying agents and rotary evaporation until a pressure of 3 mm is sustained leaves 57.5 g of crude methyl *p*-toluenesulfinate as a clear liquid admixed with white crystals. The less soluble (*S*)-(−) diastereomer (**1**) is isolated in several crops by crystallization from 1.2 volumes of reagent-grade acetone at −20°C. After the first crop has been collected, 3 drops of concd hydrochloric acid is added to the acetone mother liquor to effect equilibration of the sulfinate diastereomers. A total of 40.9–42.2 g of crystalline sulfinate is obtained in six crops. Recrystallization from acetone affords two crops of (*S*)-(−)-methyl *p*-toluenesulfinate, mp 105–106°C, $[\alpha]_D^{25}$ −199.4 (acetone, *c* 1.5), weighing 36.9–38.2 g (63–65%) (Note 4).

B. *(S)-(+)-2-(p-Toluenesulfinyl)-2-cyclopentenone ethylene ketal.* A 250-mL, three-necked, round-bottomed flask equipped with two rubber septa, a nitrogen inlet, 125-mL pressure-equalizing dropping funnel, and a magnetic stirring bar is flame-dried under nitrogen. After the apparatus cools to room temperature, the flask is charged with 70 mL of anhydrous tetrahydrofuran (Note 5) and cooled in an isopropyl alcohol–dry ice bath. Stirring is begun as 42 mL (60.8 mmol) of 1.45 *M* butyllithium in hexane (Note 6) is added slowly through the dropping funnel over 10–30 min. After another 10 min. a solution of 11.3 g (55.1 mmol) of 2-bromo-2-cyclopentenone ethylene ketal (Note 7) is added from the dropping funnel over 30 min. The colorless or pale-yellow solution is stirred and cooled at −78°C for 1.5 hr. A 1-L, three-necked, round-bottomed flask equipped with a magnetic stirring bar, two rubber septa, and a stopcock connected to a bubbler gas exit is flushed with nitrogen and charged with 24.4 g (82.9 mmol) of (*S*)-(−)-menthyl *p*-toluenesulfinate and 460 mL of anhydrous tetrahydrofuran. The sulfinate suspension is stirred vigorously (Note 8) and cooled at −78°C as the vinyllithium reagent (**2**) in the first flask is then transferred into the second flask through a cooled cannula by means of nitrogen pressure (Note 9). As the 50-min transfer proceeds, the sulfinate suspension becomes yellow. The mixture is stirred for another 15 min at −78°C, the cooling bath is removed, and 125 mL of saturated aqueous sodium dihydrogen phosphate is added. When the contents have warmed to room temperature, the tetrahydrofuran is removed by rotary evaporation. The residue is partitioned between 300 mL of water and 200 mL of chloroform. The aqueous layer extracted with three 100-mL portions of chloroform. The chloroform extracts are combined and dried over anhydrous potassium carbonate. Filtration of the drying agent and evaporation of the chloroform gives 40–55 g of a viscous brown oil consisting of the sulfinyl ketal, menthol, menthyl sulfinate, minor by-products, and residual chloroform. The sulfinyl ketal is isolated by modified flash chromatography on 500 g of Woelm silica gel (32–64 μm) packed in dry diethyl ether in a 6.5-cm × 45-cm column (Note 10). The crude product is applied to the column in 25 mL of chloroform and the column is eluted with ether under sufficient compressed air pressure to achieve a flow of 60 mL per min. After thirty 60-mL fractions are collected, the solvent is changed to ethyl acetate, and another forty 60-mL fractions are collected and analyzed by thin-layer chromatography (Note 11). Combination and evaporation of fractions 40–60 provides 9.05–9.75 g (62–67%) of crude (S)-$(+)$-2-$(p$-toluenesulfinyl)-2-cyclopentenone ethylene ketal as a pale-yellow oil, $[\alpha]_D^{25}$ +78° (CHCl$_3$, c 0.25) (Note 12).

C. *(S)-($+$)-2-(p-Toluenesulfinyl)-2-cyclopentenone.* A magnetic stirring bar, 100 g of anhydrous copper(II) sulfate, and a solution of 9.05–9.75 g of the sulfinyl ketal in 300 mL of acetone are placed in a 500-mL Erlenmeyer flask. The flask is flushed with nitrogen and stoppered. The suspension is stirred vigorously overnight, the copper sulfate is separated by filtration, and the filtercake is washed thoroughly with 500–700 mL of acetone. Concentration of the combined filtrates by rotary evaporation gives 7.36–7.58 g of tan crystals. Recrystallization is carried out by dissolving the product in a minimum volume of ethyl acetate (ca. 80 mL) at room temperature, treating with Norite, diluting with an equal volume of diethyl ether, and cooling to −20°C. After the resulting crystals are collected, the mother liquor is evaporated under reduced pressure at room temperature, and the procedure is repeated twice. The mother liquor is again evaporated and the residue (1.4–1.8 g) is purified by flash chromatography on 110 g of Woelm silica gel using ethyl acetate as eluant (Note 13). Combination of appropriate fractions, evaporation, and recrystallization affords two additional crops of crystalline product (0.4–0.7 g). The yield of (S)-$(+)$-2-(p-toluenesulfinyl)-2-cyclopentenone, mp 125–126°C, $[\alpha]_D^{25}$ +148° (CHCl$_3$, c 0.11), is 6.02–6.60 g (50–54% based on bromo ketal) (Notes 14 and 15).

2. Notes

1. This reagent was purchased from Aldrich Chemical Company, Inc.

2. Sodium p-toluenesulfinate hydrate, purchased from Aldrich Chemical Company, Inc., was dried overnight in a vacuum oven at 140°C to remove the water of hydration. The weight loss amounts to 19–21%.

3. The checkers added the sodium sulfinate from a 100-mL, three-necked flask via a bent sidearm fitted to the reaction vessel. A stream of nitrogen flowing through the 100-mL flask prevented backflow of fumes from the reaction and caking of the sodium sulfinate powder.

4. The spectral properties of the (S)-$(-)$ sulfinate are as follows: IR (CCl$_4$) cm^{-1}: 2958 (s), 2924 (s), 2870 (s), 1455 (m), 1135 (s), 961 (s), 919 (s), 853 (s); ^{1}H NMR (90 MHz, CDCl$_3$) δ: 0.72 (d, 3 H, J = 6, CHCH$_3$), 0.94 and 0.86 [2 d, 6 H, J = 7, CH(CH$_3$)$_2$], 2.37 (s, 3 H, ArCH$_3$), 4.08 (t of d, 1 H, J = 5, 10, CHOSO$_2$), 7.26 and 7.56 (2 d, 4 H, J = 8, ArH).

5. Tetrahydrofuran was dried by distillation from sodium–benzophenone ketyl before use.

6. Butyllithium in hexane is available from Aldrich Chemical Company, Inc. and Alfa Products, Morton Thiokol, Inc. The reagent was titrated with anhydrous diphenylacetic acid as described in the literature.[2]

7. 2-Bromo-2-cyclopentenone ethylene ketal was prepared according to a published procedure.[3] The compound is quite unstable and should be purified by distillation before use to remove impurities. The submitters stored the bromo ketal at $-20°C$ over Linde 3A molecular sieves and redistilled a portion in a Kugelrohr apparatus with an oven temperature of 38°C (0.1 mm) immediately before use. The checkers found it necessary to distill the bromo ketal a second time to increase its purity. The compound was stored at $-20°C$ and used in Step B the next day.

8. The submitters caution that rapid stirring is essential to avoid local heating from the exothermic reaction and, as a consequence, diminished yields.

9. The checkers used a 61-cm, 16-gauge cannula with a single loop ca. 6 cm in diameter immersed in an isopropyl alcohol–dry ice bath. The submitters report that lower yields were obtained when the vinyllithium reagent was allowed to warm above $-78°C$ briefly during the transfer.

10. The submitters purified the product by medium-pressure liquid chromatography on a 60-cm × 5-cm column packed with 230–400-mesh silica gel 60 purchased from E. Merck. Ethyl acetate was used as eluant at a flow rate of 4.0 mL per min. Fractions (20 mL) were collected and analyzed by thin-layer chromatography.

11. Thin-layer chromatograms were obtained with silica gel as absorbent and ethyl acetate as developing solvent. The order of elution and R_f values of the major components are as follows: menthyl sulfinate (0.65). menthol (0.59), sulfinyl ketal (0.30).

12. The ^{1}H NMR spectral characteristics of the ketal are as follows (CDCl$_3$) δ: 2.0–2.2 (m, 2 H, CH$_2$), 2.3–2.6 (m, 2 H, C=CCH$_2$), 2.37 (s, 3 H, CH$_3$), 3.7–3.9 (m, 4 H, OCH$_2$CH$_2$O), 6.67 (t, 1 H, $J = 2$, C=CH), 7.24 (2 d, 4 H, $J = 8$, aryl H).

13. Flash chromatography was carried out according to a procedure in the literature.[4]

14. The spectral properties of the sulfinyl enone are as follows: IR (CCl$_4$) cm^{-1}: 2924 (m), 1715 (s), 1287 (m), 1152 (s), 1083 (s), 1054 (s), 728 (m), ^{1}H NMR (CDCl$_3$) δ: 2.2–2.5 (m, 2 H, CH$_2$), 2.30 (s, 3 H, CH$_3$), 2.6–2.8 (m, 2 H, C=CCH$_2$), 7.19 and 7.58 (2 d, 4 H, $J = 8$, aryl H), 8.03 (t, 1 H, $J = 2$, C=CH); mass spectrum (70 eV), m/z (relative intensity): 220 (M$^+$, 30), 172 (100), 139 (48), 129 (72). The product was analyzed by the submitters: Anal. calcd. for C$_{12}$H$_{12}$SO$_2$: C, 65.43; H, 5.49; S, 14.56. Found: C, 65.53; H, 5.51; S, 14.72.

15. The submitters report that the sulfinyl ketone may be stored in vials in a desiccator at 0°C for more than 1 year without evidence of decomposition. Although storage under an inert atmosphere is not necessary, the checkers found that product exposed to the atmosphere at room temperature became discolored after several weeks.

3. Discussion

Enantiomerically pure β-substituted carbonyl compounds serve as useful intermediates in the synthesis of many chiral organic compounds. The enantioselective synthesis

TABLE I
ENANTIOMERICALLY PURE α-SULFINYL-α,β-ENONES PREPARED FROM ETHYLENE KETALS OF α-BROMO-α,β-ENONES

Sulfinyl Enone	R	Yield (%)	mp (°C)	$[\alpha]_D^{25}$
	p-MeC$_6$H$_4$	50–54	125–126	+ 142°
	1-Naphthyl	65	96.5–97.0	+ 292°
	p-MeOC$_6$H$_4$	76	120.5–121.5	+ 141°
	p-MeC$_6$H$_4$	66	101–102	+ 210°
	Me	38	90.5–91.0	+ 21.0°
	p-MeC$_6$H$_4$	35	132–133	− 322°

of acyclic β-substituted carboxylic acids has been reported by Meyers,[5] Mukaiyama,[6] and Koga.[7] However, no effective, general method for the enantio-controlled preparation of β-substituted cycloalkanones was available prior to the investigations by the submitters.[8] For example, poor enantioselectivity was observed in conjugate additions of organometallic reagents to cyclic α,β-enones in the presence of optically active solvents[9] or chiral ligands.[10] In contrast, the submitters have found that conjugate addition to chiral cyclic α-sulfinyl α,β-enones occurs with high enantioselectivity.[11] Thus, the title compound is a useful intermediate for the synthesis of a variety of β-substituted cyclopentanones.

The preparation of (S)-(−)menthyl p-toluenesulfinate described in Step A is based upon the procedure reported by Solladié.[12] 2-Bromo-2-cyclopentenone ethylene ketal is available from 2-cyclopentenone by the procedure of Smith and co-workers.[3] The present procedure has been used by the submitters to prepared analogous chiral α-sulfinyl α,β-enones (Table I).[11] The utility of these chiral synthons is enhanced by their stability, the facility of their conjugate addition reactions, and the capability of producing either enantiomeric β-substituted adduct by varying the reaction conditions.[13] Similar methodology has allowed conversion of some enantiomerically pure butenolide sulfoxides into the corresponding β-substituted butyrolactones.[14]

Both (S)-(−)- and (R)-(+)-menthyl 4-toluenesulfinates are now available from the Aldrich Chemical Company, Inc.

1. Department of Chemistry, The Johns Hopkins University, Baltimore, MD 21218. Financial support from the National Science Foundation (CHE 7915161) is gratefully acknowledged.
2. Kofron, W. G.; Baclawski, L. M. *J. Org. Chem.* **1976**, *41*, 1879–1880.
3. Smith, A. B., III; Branca, S. J.; Guaciaro, M. A.; Wovkulich, P. M.; Korn, A. *Org. Synth.*, *Coll. Vol. VII* **1990**, 271.

4. Still, W. C.; Kahn, M.; Mitra, A. *J. Org. Chem.* **1978,** *43*, 2923–2925.

5. (a) Meyers, A. I. *Acc. Chem. Res.* **1978,** *11*, 375–381; (b) Meyers, A. I.; Smith, R. K.; Whitten, C. E. *J. Org. Chem.* **1979,** *44*, 2250–2256.

6. Mukaiyama, T.; Takeda, T.; Osaki, M. *Chem. Lett.* **1977,** 1165–1168.

7. (a) Hashimoto, S.-i.; Komeshima, N.; Yamada, S.-i.; Koga, K. *Chem. Pharm. Bull.* **1979,** *27*, 2437–2441; (b) Hashimoto, S.-i.; Yamada, S.-i.; Koga, K. *J. Am. Chem. Soc.* **1976,** *98*, 7450–7452.

8. For recently reported syntheses of optically active β-alkylcycloalkanones, see Taber, D. F.; Saleh, S. A.; Korsmeyer, R. W. *J. Org. Chem.* **1980,** *45*, 4699–4702; Taber, D. F.; Raman, K. *J. Am. Chem. Soc.* **1983,** *105*, 5935–5937.

9. Langer, W.; Seebach, D. *Helv. Chim. Acta* **1979,** *62*, 1710–1722.

10. Colonna, S.; Re, A.; Wynberg, H. *J. Chem. Soc., Perkin Trans. 1* **1981,** 547–552 and references cited therein.

11. (a) Posner, G. H.; Frye, L. L.; Hulce, M. *Tetrahedron* **1984,** *40*, 1401–1405; (b) Posner, G. H.; Frye, L. L.; Hulce, M. *Israel J. Chem.* **1984,** *24*, 88–92; (c) Posner, G. H.; Kogan, T. P.; Hulce, M. *Tetrahedron Lett.* **1984,** *25*, 383–386; (d) Posner, G. H. In "Asymmetric Synthesis," Morrison, J., Ed.; Academic Press: New York, 1983; Vol. 2, p. 225–241; (e) Posner, G. H. *Acc. Chem. Res.* **1987,** *20*, 72; (f) Posner, G. H. In "The Chemistry of Sulfones and Sulfoxides," Patai, S.; Rappaport, Z.; Stirling, C. M. J., Eds.; Wiley: New York, 1987, p. 157.

12. Solladié, G. *Synthesis* **1981,** 185–196.

13. Posner, G. H.; Hulce M. *Tetrahedron Lett.* **1984,** *25*, 379–382.

14. Posner, G. H.; Kogan, T. P.; Haines, S. R.; Frye, L. L. *Tetrahedron Lett.* **1984,** *25*, 2627–2630.

cis-*N*-TOSYL-3-METHYL-2-AZABICYCLO[3.3.0]OCT-3-ENE

(Cyclopenta[*b*]pyrrole, 1,3a,4,5,6,6a-hexahydro-2-methyl-1-[4-methylphenyl)sulfonyl]-, *cis*-)

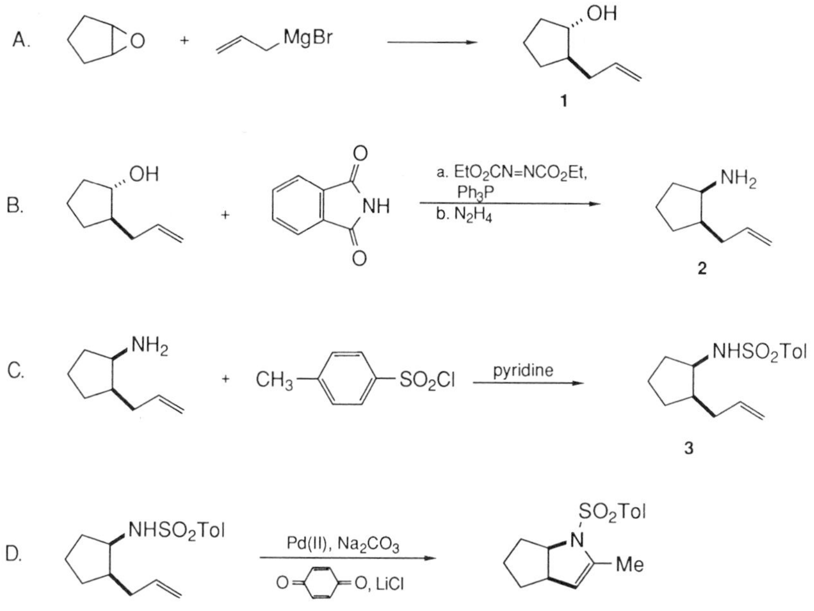

A.

B.

C.

D.

Submitted by Louis S. Hegedus, Michael S. Holden, and James M. McKearin[1]
Checked by Christoph Nubling and Ian Fleming

1. Procedure

A. *trans*-2-*(2-Propenyl)cyclopentanol*. A 500-mL, three-necked, round-bottomed flask equipped with a magnetic stirring bar, reflux condenser with a stopcock, and a 250-mL addition funnel is charged with 18.3 g (750 mmol) of magnesium turnings (Note 1). The system is evacuated and placed under argon, then 100 mL of ethyl ether (Note 2) is added to the system via cannula. The system is placed in an ice–water bath, and 2 mL of allyl bromide (Note 3) is added via syringe to the magnesium suspension to initiate Grignard formation. The addition funnel is charged with 45.5 g (375 mmol) of allyl bromide and 30 mL of ethyl ether. Another 100 mL of ethyl ether is added to the reaction flask. Stirring is begun, and the allyl bromide–ethyl ether mixture is added dropwise to the cooled reaction flask over a period of about 2 hr. After the addition is complete, the dark-gray solution is stirred for several hours at ambient temperature (Note 4). Meanwhile, a 500-mL, three-necked, round-bottomed flask equipped with a magnetic stirring bar, a reflux condenser with a stopcock, and a 60-mL addition funnel is evacuated and placed under argon. The Grignard solution is transferred, via a cannula, into the flask, and 16.8 g (200 mmol) of cyclopentene oxide (Note 5) is placed in the addition funnel. While the solution is stirred, the epoxide is added dropwise to the Grignard reagent at a rate sufficient to maintain a mild reflux. After the solution is stirred for several hours or overnight, the flask containing the dark gray reaction mixture is placed in an ice–water bath, and excess Grignard reagent is hydrolyzed with 40 mL of a saturated aqueous ammonium chloride solution. The fine white precipitate is allowed to settle (Note 6), and the liquid is decanted into a 500-mL separatory funnel. The precipitate is washed with ethyl ether (4 × 50 mL) (Note 7), and all the ethyl ether solutions are combined, washed with saturated aqueous sodium bicarbonate solution (3 × 20 mL), then with saturated aqueous sodium chloride (2 × 20 mL). The aqueous layers are combined and washed with ethyl ether (2 × 20 mL). The ether layers are combined and dried over anhydrous potassium carbonate. The desiccant is removed by gravity filtration, and the solvent removed under reduced pressure to give 24.1–26.4 g (96– 105%) of a yellow oil. Distillation (43°C, 0.250 mm) yields **1** (19.8–23.0 g, 78–91%) as a clear, colorless oil (Note 8).

B. *cis-2-(2-Propenyl)cyclopentylamine.* A 1000-mL, three-necked, round-bottomed flask equipped with a magnetic stirring bar, two addition funnels, and a stopcock is charged with 41.5 g (158 mmol) of triphenylphosphine (Note 9) and 23.3 g (158 mmol) of phthalimide (Note 10). The system is evacuated and placed under argon. To one addition funnel, 20.0 g (158 mmol) of *trans*-2-(2-propenyl)cyclopentanol is added; 27.5 g (158 mmol) of diethyl azodicarboxylate (Note 11) is added to the other. Tetrahydrofuran, 500 mL (Note 12), is added to the flask via cannula, and stirring is begun. The substrate and diethyl azodicarboxylate are simultaneously added dropwise, slowly over about 30 min (Note 13), with stirring; the solution turns clear and yellow (Note 14). The reaction is permitted to proceed for 2 days at room temperature; the solution is then transferred to a 1000-mL, one-neck, round-bottomed flask, and the solvent is removed under reduced pressure, to leave a yellow–white semisolid. A magnetic stirring bar is added to the flask and the semisolid is taken up in 250 mL of reagent-grade methyl alcohol. To this, 10.1 g (316 mmol) of hydrazine (Note 15) is added. A reflux condenser is attached to the flask, stirring is begun, and the system is brought to reflux (Note 16). A large amount of clumpy white solid forms in a yellow-to-orange solution. After 4 hr at reflux, the solution is allowed to cool to room temperature; a mixture of 20 mL of hydrochloric acid (Note 17) and 65 mL of methyl alcohol is added, and the system is refluxed overnight. The resulting reaction mixture is filtered to remove the precipitate, and the solvent is removed under reduced pressure to yield a white-to-pink solid, which is taken up in 800 mL of water and 28 mL of hydrochloric acid. The solution is filtered, and the solid washed with water (2 × 200 mL) and hydrochloric acid (20 mL). The liquids are combined, placed in a 2000-mL separatory funnel, and washed with chloroform (3 × 250 mL), and ethyl ether (1 × 250 mL). The aqueous layer is transferred to a 2000-mL Erlenmeyer flask and cooled in an ice–water bath. A saturated aqueous sodium hydroxide solution is used to make the solution basic, to approximately pH 14, whereupon the solution turns dark olive green. The basic solution is extracted with ethyl ether (10 × 250 mL or by continuous extraction overnight) and the combined organic layers are dried over a mixture of anhydrous sodium sulfate and anhydrous potassium carbonate. Filtration and solvent removal at atmospheric pressure yields a green–yellow oil. Distillation (52–58°C, 8–11 mm) gives **2** (11.8–12.5 g, 60–63%) as a clear, colorless oil (Note 18).

C. *cis-1-N-Tosyl-2-(2-propenyl)cyclopentylamine.* A 100-mL, one-necked, round-bottomed flask equipped with a sidearm, a magnetic stirring bat, stopcock, and a serum cap on the sidearm, is charged with 8.00 g (64 mmol) of *cis*-2-(2-propenyl)cyclopentylamine. The system is evacuated and placed under argon. Via cannula, 50 mL of pyridine (Note 19) is added. The flask is cooled in an ice–water bath, the stopcock removed, 12.58 g (66 mmol) of *p*-toluenesulfonyl chloride (Note 20) is added to the reaction mixture, and the stopcock replaced. The reaction mixture immediately turns orange; it is allowed to stir at 0°C overnight, during which time the reaction mixture turns deep purple. The reaction mixture is then poured into a separatory funnel, 60 mL of distilled technical grade ethyl acetate is added, and the solution is washed with 100-mL portions of 1 : 1 2 N HCl: saturated aqueous sodium chloride until the washings are acidic. The organic layer is washed with saturated aqueous sodium chloride (2 × 60 mL), and dried over anhydrous magnesium sulfate. Gravity filtration and solvent removal under reduced pressure yield a dark red–brown solid. This is purified by recrystallization from 250 mL of ethyl alcohol : water (4 : 1); crystallization is completed in the refrigerator to give **3** (12.7–13.9 g, 71–78%) as off-white plates, mp 109–110°C (Note 21).

D. *cis-N-Tosyl-3-methyl-2-azabicyclo[3.3.0]oct-3-ene.* A 500-mL, one-necked, round-bottomed flask equipped with a magnetic stirring bar and reflux condenser is charged with 5.159 g (18.49 mmol) of *cis*-1-*N*-tosyl-2-(2-propenyl)cyclopentylamine, 1.998 g (18.49 mmol) of *p*-benzoquinone (Note 22), 0.096 g (0.370 mmol, 2 mol%) of PdCl$_2$(CH$_3$CN)$_2$ (Note 23), 3.920 g (92.46 mmol, 500 mol%) of lithium chloride (Note 24), and 1.960 g (18.49 mmol) of sodium carbonate (Note 25). Tetrahydrofuran (100 mL) (Note 12) is added and stirring is begun. The yellow–orange solution is heated at reflux until thin-layer chromatography (3 : 1 hexane : ethyl acetate, SiO$_2$) shows that no starting material remains (about 3–4 hr); it is then poured into a 500-mL separatory funnel and 100 mL of ethyl acetate is added. This is washed with 100-mL portions of 1 : 1 saturated aqueous sodium chloride : sodium hydroxide (1%) until the aqueous layer is clear; then the yellow–green organic layer is washed with saturated aqueous sodium chloride (2 × 50 mL). The organic layer is dried over anhydrous magnesium sulfate, filtered by gravity, and passed through a short column (approximately 5 cm) of neutral alumina, and the column is washed with 100 mL of ethyl acetate. The combined solvents are removed under reduced pressure to give 4.9–5.1 g (94–99%) of a tan solid. The product is recrystallized from 100 mL of methyl alcohol : water (4 : 1) to yield **4** (3.9–4.45 g, 76–87%) as white needles, mp 91–92°C (Notes 26 and 27).

2. Notes

1. Magnesium turnings, purified for Grignard reactions, are purchased from J. T. Baker Chemical Company and used without further purification.

2. Ethyl ether is freshly distilled from sodium/benzphenone ketyl at atmospheric pressure under nitrogen.

3. Allyl bromide, purchased from Aldrich Chemical Company, Inc., is distilled and stored in a brown bottle away from light.

4. Successful reactions have been run with this induction period lasting from 1 hr to overnight.

5. Cyclopentene oxide is purchased from Arapahoe Chemicals, Boulder, CO, and used without purification. The checkers bought cyclopentene oxide from Lancaster Synthesis.

6. The fine precipitate may take several hours to settle. Filtration is often ineffective, but settling can be accelerated by centrifuging.

7. Since the efficiency of this washing is dependent on the degree of settling, the checkers recommend that washing with 50-mL batches of ether be continued until the smell of the alcohol is no longer detectable on a sample of the dry salts.

8. The spectral properties are as follows: ^{1}H NMR (CDCl$_3$) δ: 1.0–2.4 (m, 9 H); 3.0–3.3 (br s, 1 H, O—H); 3.7–4.1 (m, 1 H, CH—O); 4.8–5.3 (m, 2 H, =CH$_2$); 5.5–6.2 (m, 1 H, —CH=).

9. Anhydrous triphenylphosphine is purchased from Sigma Chemical Company and is used without further purification

10. Phthalimide, 98%, is purchased from Aldrich Chemical Company, Inc. and is used without further purification.

11. Diethyl azodicarboxylate is purchased from Aldrich Chemical Company, Inc. and is used without further purification.

12. Tetrahydrofuran is freshly distilled from sodium/benzophenone ketyl at atmospheric pressure under nitrogen.

13. Too rapid a rate of addition may cause the solution to boil.

14. The solution does not become homogeneous until it is warmed by the heat of the reaction.

15. Anhydrous hydrazine, ≥97%, is purchased from Matheson, Coleman and Bell, Norwood, OH 45212 and is used without further purification.

16. *Caution! Because of the dangerous nature of hydrazine, a safety shield should always be in place during this reaction.*

17. ACS reagent hydrochloric acid is purchased from Fisher Scientific Company and used without further purification.

18. The spectral properties are as follows: ^{1}H NMR (CDCl$_3$) δ: 0.8 (s, 2 H, NH$_2$); 1.3–2.4 (m, 9 H, CH$_2$, CH); 3.1–3.4 (m, 1 H, HC—N); 4.8–5.2 (m, 2 H, =CH$_2$); 5.4–6.1 (m, 1 H, HC=).

19. Pyridine is distilled from CaH$_2$ and stored over CaH$_2$ under argon.

20. *p*-Toluenesulfonyl chloride is purchased from J. T. Baker Chemical Company and purified by dissolving 20 g in 50 mL of chloroform, adding 250 mL of hexane, filtering, and removing solvent under reduced pressure.[2]

21. The spectral properties are as follows: ^{1}H NMR (CDCl$_3$) δ: 1.0–2.3 (m, 9 H, CH$_2$, CH); 2.41 (s, 3 H, CH$_3$); 3.4–3.8 (m, 1 H, CHN); 4.7–5.1 (m, 3 H, =CH$_2$, NH); 5.2–6.1 (m, 1 H, =CH); 7.25 (d, 2 H, J = 8, ArH); 7.8 (d, 2 H, J = 8 ArH).

22. *p*-Benzoquinone, ≥98%, is purchased from the Aldrich Chemical Company, Inc., sublimed at 60°C/15 mm, and stored under argon. The checkers used it as supplied.

23. Palladium(II) chloride–acetonitrile complex is formed by placing 8.00 g of $PdCl_2$ in 200 mL of acetonitrile and stirring for 2 days or refluxing for 3 hr. The complex (11.43 g, 97.8%) is collected by filtration, washed, and dried.

24. Lithium chloride is purchased from Fisher Scientific Company and used without further purification.

25. Sodium carbonate is purchased from Aldrich Chemical Company, Inc. and used without further purification.

26. The spectral properties are as follows: ^{1}H NMR ($CDCl_3$) δ: 1.40–2.00 (m, 6 H, CH_2); 2.10 (m, 3 H, $CH_3C=$); 2.40 (s, 3 H, $ArCH_3$); 2.80–3.20 (m, 1 H) 4.20–4.50 (m, 1 H, CHN); 4.70 (m, 1 H, CH=); 7.30 (d, 2 H, $J = 8$, ArH); 7.70 (d, 2 H, $J = 8$, ArH).

27. The checkers also carried out the entire sequence on three times the scale with slightly better yields.

3. Discussion

Synthesis of the title compound is representative of a number of syntheses of non-aromatic nitrogen heterocycles via Pd(II)-catalyzed amination of olefins.[3] These tosylated enamines are not readily available by standard synthetic methods and show potential for further functionalization of the heterocycle.[4] The saturated amine can be synthesized from the title compound by hydrogenation of the double bond followed by photolytic deprotection.[3]

In terms of cost, the effectiveness of the catalytic cycle in the ring closure makes this process economical in palladium. The first three steps in the reaction sequence—ring opening of an epoxide by a Grignard reagent,[5] conversion of an alcohol to an amine with inversion,[6] and sulfonamide formation from the amine[7]—are all standard synthetic processes.

1. Department of Chemistry, Colorado State University, Fort Collins, CO 80523.
2. Fieser, L. F.; Fieser, M. In "Reagents for Organic Synthesis"; Wiley: New York, 1967; Vol. 1, p. 1180.
3. Hegedus, L. S.; McKearin, J. M. *J. Am. Chem. Soc.* **1982**, *104*, 2444.
4. Unpublished observations, these laboratories.
5. Speziale, V.; Amat, M. M.; Lattes, A. *J. Heterocycl. Chem.* **1976**, *13*, 349.
6. Mitsunobu, O.; Wada, M.; Sano, T. *J. Am. Chem. Soc.* **1972**, *94*, 679.
7. Gold, E. H.; Babad, E. *J. Org. Chem.* **1972**, *37*, 2208.

TRIFLOROACETYL TRIFLATE

(Acetic acid, trifluoro-, anhydride with trifluoromethanesulfonic acid)

$$CF_3COOH \quad + \quad CF_3SO_3H \quad \xrightarrow{P_2O_5} \quad CF_3COOSO_2CF_3$$

Submitted by Stephen L. Taylor, T. R. Forbus, Jr., and J. C. Martin[1]
Checked by Thomas W. Panunto and Edwin Vedejs

1. Procedure

Caution! The volatile product reacts rapidly with water to give corrosive strong acids. It also reacts rapidly with other nucleophiles. Care should therefore be exercised to avoid inhalation of its vapors. It should be handled in a well-vented fume hood.

To a 1-L flask containing 160 g (1.13 mol) of powdered phosphorus oxide (P_2O_5), thoroughly mixed with an equal volume of dried fine sand (Note 1), is added a mixture of 85.5 g (0.75 mol) of trifluoroacetic acid (TFA) (Note 2) and 56.5 g (0.750 mol) of triflic acid (TfOH) at $-20°C$ (Note 3). The stoppered flask (Note 4) is vigorously shaken for 5 min and then fitted for simple distillation, with the receiving flask cooled to $-78°C$, and allowed to stand at room temperature under a dry nitrogen atmosphere for 2.5 hr. The liquid is removed from the solid mixture by simple distillation at a bath temperature of 240°C (Note 5) for 3.5 hr (Note 6). The distillate is then carefully fractionally distilled (Note 7) from 5 g of powdered P_2O_5 (Note 8) with the receiving flasks cooled at $-78°C$. The colorless liquid collected at 62.5–63°C (760 mm) (Note 9), 69 g (75%) of trifluoroacetyl triflate (TFAT), is of 99% purity (Note 10), as determined by fluorine magnetic resonance (Note 11).

2. Notes

1. In recent synthetic applications of this method the powdered phosphorus oxide was mixed with *twice* as much volume of dried fine sand. The increased volume of sand makes it easier to remove the TFAT by distillation and reduces the probability that the flask will be broken after the distillation.

2. The 99% TFA obtained from Aldrich Chemical Company, Inc. was used without further purification.

3. Triflic acid (TfOH), obtained from Minnesota Mining & Manufacturing Company, (3M), in kilogram quantities was used without further purification.

4. Ground-glass joints were connected using Teflon sleeves or a chlorofluorocarbon stopcock grease.

5. High temperatures are needed to distill the products from P_2O_5. The use of temperatures higher than 250°C, however, causes the round-bottomed flask to break when the temperature is lowered to near room temperature. On completion of the reaction, the P_2O_5 sand mixture can be removed from the flask by careful, slow addition of water. The checkers used an equilibrated bath of sand in a large heating mantle; the flask always broke after distillation (see Note 1).

6. The nitrogen outlet from the distillation apparatus should be well vented.

7. An 8-mm × 1-m jacketed column packed with a coiled tantalum wire was used by the submitters. The checkers used a Vigreux column of similar size.

8. Since the distillate contains 1–3% of the starting acids, P_2O_5 is added to prevent the reaction of TFA and TFAT, which gives trifluoroacetic anhydride (TFAA) and TfOH.

9. The first fraction is TFAA, bp 38.5–41°C (760 mm).

10. The impurity is TFAA.

11. The reactants and products show only singlets in their fluorine magnetic resonance spectra with the following chemical shifts (downfield from fluorotrichloromethane internal standard) δ: TFA, −76.3; TfOH, −77.3; TFAT, −73.3 and −74.8; TFAA, −75.9; triflic anhydride, −72.6 ppm.

3. Discussion

Trifluoroacetyl triflate is probably the most powerful trifluoroacetylating agent known, as evidenced by its reactivity toward several types of nucleophiles under mild conditions. A sterically hindered base, 2,6-di-*tert*-butyl-4-methylpyridine,[2] may be used to scavenge the triflic acid produced in the reactions, since it does not react with TFAT under these conditions.

Trifluoroacetylation occurs at carbon in activated arenes such as anthracene[3] under milder conditions using TFAT than when using TFAA. Trifluoroacetate esters are formed from alcohols and phenols,[4] while ketones are acylated at oxygen to yield enol trifluoroacetates.[3] Amines[4] give the corresponding amides on reaction with 1 equiv of TFAT or imides on reaction with 2 equiv. Some covalent halides (fluorides[5] and chlorides[3]) are acylated at halogen by TFAT to yield the very volatile trifluoroacetyl halides and ionic triflates. It was recently reported that TFAT reacts with a thioketone to give a stable cation.[6] Reaction of TFAT with the methyl ester of glutaconic acid gives 2,6-dimethoxypyrylium triflate, the first member of a new class of pyrylium salts[4] with alkoxy groups at positions-2 and -6.

The high reactivity of TFAT limits the number of solvents that can be used for its reactions. We have found that TFAT is unreactive towards saturated hydocarbons, benzene, and common halogenated solvents. It reacts only very slowly with nitromethane, but reacts relatively rapidly with ether, tetrahydrofuran, ethyl acetate, and acetonitrile.

1. Roger Adams Laboratory, University of Illinois, Urbana, IL 61801.
2. Anderson, A. G.; Stang, P. J. *Org. Synth., Coll. Vol. VII* **1990**, 144.
3. Forbus, T. R., Jr.; Martin, J. C. *J. Org. Chem.* **1979**, *44*, 313.
4. Taylor, S. L; Forbus, T. R., Jr.; Martin, J. C., *J. Org. Chem.* **1987**, *52*, 4156.
5. Michalak, R. S.; Martin, J. C. *J. Am. Chem. Soc.* **1980**, *102*, 5921.
6. Mass, G.; Stang, P. J. *J. Org. Chem.* **1981**, *46*, 1606.

m-TRIFLUOROMETHYLBENZENESULFONYL CHLORIDE

(Benzenesulfonyl chloride, *m*-(trifluoromethyl)-)

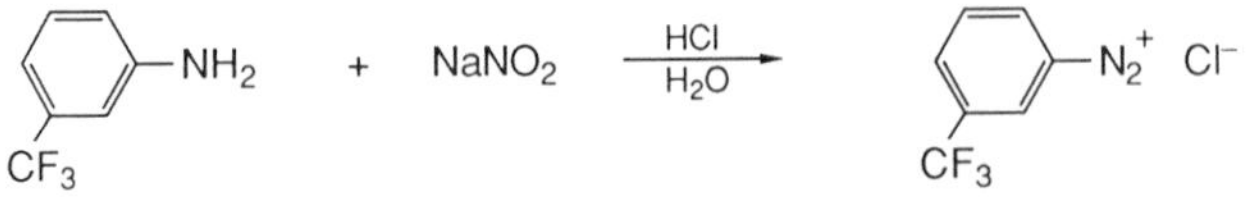

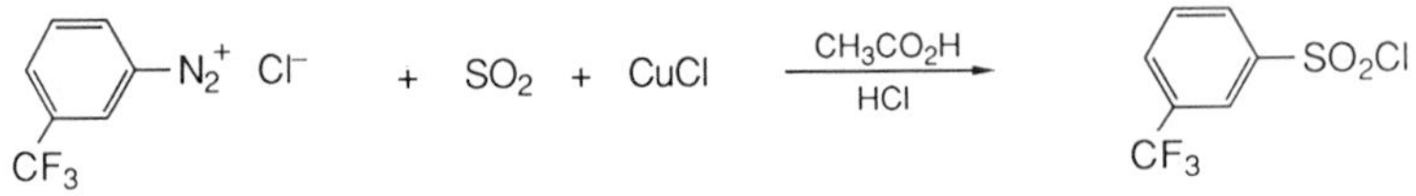

Submitted by R. V. Hoffman[1]
Checked by G. Saucy, G. P. Roth, and J. W. Scott

1. Procedure

Caution! All operations should be carried out in a hood! m-*Trifluoromethylben-zenesulfonyl chloride is a lachrymator. Spills should be treated with saturated sodium carbonate.*

α,α,α-Trifluoro-*m*-toluidine (*m*-aminobenzotrifluoride) (96.7 g, 0.6 mol) (Note 1) is added in one portion to a mixture of concentrated hydrochloric acid (200 mL) and glacial acetic acid (60 mL) in a 1000-mL beaker arranged for efficient mechanical stirring (Note 2). The white hydrochloride salt precipitates (Note 3). The beaker is placed in a dry ice-ethanol bath and, when the temperature of the stirred mixture has reached $-10°C$, a solution of sodium nitrite (44.8 g, 0.65 mol) in water (65 mL) is added dropwise at such a rate that the temperature does not exceed $-5°C$ (Note 4). After all the sodium nitrite solution has been added, the mixture is stirred for 45 min while the temperature is maintained between $-10°C$ and $-5°C$ (Note 5).

While the diazotization is being completed, glacial acetic acid (600 mL) is placed in a 4000-mL beaker and stirred magnetically. Sulfur dioxide is introduced by a bubbler tube with a fritted end immersed below the surface of the acetic acid until saturation is evident (Note 6). Cuprous chloride (15 g) (Note 7) is added to the solution. The introduction of sulfur dioxide is continued until the yellow–green suspension becomes blue-green. Most of the solids dissolve during this time (20–30 min). The mixture is then placed in an ice bath and cooled with stirring. When the temperature approaches $10°C$, the diazotization reaction mixture (Note 8) is added in portions over a 30-min period to the sulfur dioxide solution. Considerable foaming occurs after each addition, and this can be disrupted with a few drops of ether. The temperature rises during the addition, but it should not exceed $30°C$. After all the diazonium salt mixture has been added, the mixture is poured into ice water (1 : 1, 2000 mL), stirred magnetically until the ice has melted, and added to a 4000-mL separatory funnel. The product separates as a yellow oil that is drawn off. The reaction mixture is extracted with 200-mL portions of ether until the ether washings are colorless (Note 9), and these washings are added to the initial product. The combined organic fraction is washed with saturated aqueous sodium bicarbonate until neutral (Note 10), then with water, and is then dried with magnesium sulfate. The solvent is removed with a rotary evaporator, and the residue is distilled (bp 54–55°C, 0.1 mm) through a 10-cm vacuum-jacketed Vigreux column to give *m*-trifluoromethylbenzenesulfonyl chloride (100–115 g, 68–79%) as a colorless or slightly yellow, clear liquid (Note 11, 12).

2. Notes

1. α,α,α-Trifluoro-*m*-toluidine was obtained from Aldrich Chemical Company, Inc. The checkers distilled this material prior to use (bp 187–189°C).

2. A chain beaker clamp is very satisfactory for supporting the beaker, as it can later be used as a handle to pour the diazonium solution. For efficient stirring the blade of the stirrer was made by trimming the ends of a large Teflon stirring paddle to the diameter of the beaker. The paddle was inverted (straight edge on bottom) and should rotate 1–1.5 cm from the bottom of the beaker.

3. If solid amines are used, they should be thoroughly crushed in a mortar and pestle before adding to the acid mixture.

4. Temperature control during the sodium nitrite addition is essential to the success of the preparation. The temperature can go as low as −15°C but must not exceed −5°C. The addition takes ca. 1 hr. At temperatures greater than −5°C, dark-red by-products form which lower the yield.

5. Temperature control is conveniently accomplished by raising and lowering the dry-ice bath. It does not seem to matter if longer reaction times are employed, but the temperature should be lowered to −10°C or below after 45 min.

6. Saturation, which requires 15–30 min, is conveniently noted by observing that most sulfur dioxide bubbles reach the surface of the acetic acid.

7. The original literature[2] suggests that copper(II) chloride dihydrate can be used as a catalyst, since it is reduced by the sulfur dioxide to copper(I). It has been noted on several occasions that catalytically inactive mixtures result. If copper(II) chloride dihydrate is used, it is expedient to add copper(I) chloride (1 g) to ensure efficient catalysis in the early stages of reaction.

8. This mixture should be a pale tan suspension, and it should be cooled between additions.

9. The first portion of ether may be larger (400 mL) since much dissolves in the aqueous mixture. A total of 1000 mL of ether is usually sufficient.

10. A considerable amount of acid is present in the ether extracts, so vigorous gas evolution occurs during the sodium bicarbonate extraction. Caution must be exercised at this point.

11. The product is sufficiently pure for most purposes. A second distillation affords a colorless product (lit.[3] bp 88–90°C, 6 mm).

12. With many anilines used as precursors in this reaction, the sulfonyl chloride product is a solid and an alternate workup procedure is used. After the reaction is quenched with ice water, the solid product is filtered with suction and washed copiously with cold water. The crude product tends to occlude water and copper

salts, which may be detrimental in later reactions. A good washing protocol involves rinsing the solid on the filter with water (200 mL), then suspending the solid in cold water (1000 mL), stirring briskly, and filtering with suction. The latter process should be repeated three times. The final water wash should be only very slightly yellow. After air drying the product can be recrystallized from an appropriate solvent.

3. Discussion

m-Trifluoromethylbenzenesulfonyl chloride has been prepared by treatment of *m*-trifluoromethylbenzenediazonium chloride with sulfur dioxide and hydrochloric acid[4] and by conversion of benzotrifluoride to *m*-trifluoromethylbenzenesulfonic acid with oleum, followed by chlorination with phosphorus pentachloride.[3] Derivatives of this compound, such as esters and amides, are quite useful in that they display reactivities similar to *p*- and *m*-nitrobenzenesulfonyl compounds but have greatly improved solubilities.

The described procedure essentially follows that described by Meerwein et al.[2] as modified slightly by Yale and Sowinski.[4] This same method can be used for a great variety of substituted anilines with good results. As evident in Table I, good yields are obtained in most cases, and the reaction works better for anilines with electron-withdrawing substituents. The identical procedure has been used to prepare many other examples, such as *m*-F, *o*-F, 3,5-di-CF$_3$.[5] This method readily provides many unavailable arylsulfonyl chlorides; it is experimentally straightforward, and the products are isolated without complications.

There are two general routes to arylsulfonyl chlorides. The first involves the conversion of an already sulfur-substituted aromatic compound to the sulfonyl chloride. Thus arylsulfonic acids or their alkali metal salts yield sulfonyl chlorides by treatment with a variety of chlorinating agents such as phosphorus pentachloride, thionyl chloride, phosgene, and chlorosulfonic acid. Alternatively, substituted thiophenols or aryl disulfides can be oxidized by chlorine-water to the sulfonyl chloride.[6]

TABLE I

CONVERSION OF ARYLAMINES TO ARYLSULFONYL
CHLORIDES

Amine, XC$_6$H$_4$NH$_2$, X =	Yield (%) of XC$_6$H$_4$SO$_2$Cl
m-CF$_3$	72
p-NO$_2$	68
m-NO$_2$	86
p-Cl	90
p-CO$_2$CH$_3$	90
3,5-di-NO$_2$	81
m-CH$_3$	71
H	53
p-OCH$_3$	27

The second route utilizes the introduction of the chlorosulfonyl substituent directly onto the aromatic nucleus. The reaction of substituted benzenes with chlorosulfonic acid gives good yields of arylsulfonyl chlorides; however, the aryl substituent dictates the position of attachment of the chlorosulfonyl function in this electrophilic aromatic substitution.[7] The method described herein allows replacement of a diazotized amine function by the chlorosulfonyl group. The ready availability of substituted anilines makes this the method of choice for the preparation of arylsulfonyl chlorides.

Arylsulfonyl chlorides are pivotal precursors for the preparation of many diverse functional types including sulfonate esters,[8] amides,[4] sulfones,[9] sulfinic acids,[10] and others.[11] Furthermore, sulfonyl fluorides are best prepared from sulfonyl chlorides.[12] The sulfonyl fluorides have many uses, among which is their utilization as active site probes of chymotrypsin and other esterases.[13] The trifluoromethyl group also plays valuable roles in medicinal chemistry.[14]

1. Department of Chemistry, Box 3C, New Mexico State University, Las Cruces, NM 88003.
2. Meerwein, H.; Dittmar, G.; Gollner, R.; Hafner, K.; Mensch, F.; Steinfort, O. *Chem. Ber.* **1957**, *90*, 841–852.
3. Yagupolskii, L. M.; Troitskaya, V. I. *Zh. Obsch. Khim.* **1959**, *29*, 552–556; *Chem. Abstr.* **1960**, *54*, 356f.
4. Yale, H. L.; Sowinski, F. *J. Org. Chem.* **1960**, *25*, 1824–1826.
5. Gerig. J. T.; Roe, D. C. *J. Am. Chem. Soc.* **1974**, *96*, 233–238.
6. Muth. F., In Houben-Weyl, "Methoden der Organischen Chemie," Müller, E., Ed.; Georg Thieme Verlag: Stuttgart, Germany, 1955; Vol. 9 pp. 563–585.
7. Gilbert, E. E. "Sulfonation and Related Reactions"; Interscience: New York, 1965; pp. 84–87.
8. See: Coates, R. M.; Chen, J. P. *Tetrahedron Lett.* **1969**, 2705–2708 and references cited therein.
9. Truce, W. E. In "Organic Chemistry of Sulfur," Oae, S., Ed.; Plenum Press: New York, 1977; pp. 532–536.
10. Truce, W. E.; Murphy, A. M. *Chem. Rev.* **1951**, *48*, 68–124.
11. Trost, B. M. *Acc. Chem. Res.* **1978**, *11*, 453–461.
12. Davis W.; Dick, J. H. *J. Chem. Soc.* **1932**, 483–484; DeCat, A., Van Poucke, R.; Verbrugghe, M. *J. Org. Chem.* **1965**, *30*, 1498–1502.
13. See Ref. 5 for a good introduction.
14. Yale, H. L. *J. Med. Pharm. Chem.* **1959**, *1*, 121.

SILYLATION OF KETONES WITH ETHYL TRIMETHYLSILYLACETATE: (Z)-3-TRIMETHYLSILOXY-2-PENTENE

(Silane, [(1-ethyl-1-propenyl)oxy]trimethyl-, (Z)-)

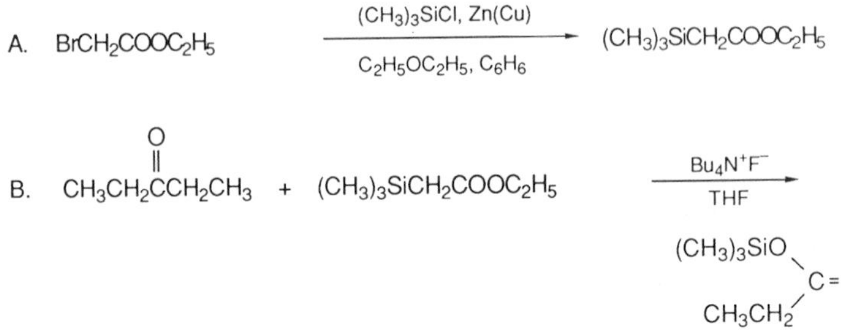

A. $BrCH_2COOC_2H_5$ $\xrightarrow[\text{C}_2\text{H}_5\text{OC}_2\text{H}_5, \text{ C}_6\text{H}_6}{\text{(CH}_3\text{)}_3\text{SiCl, Zn(Cu)}}$ $(CH_3)_3SiCH_2COOC_2H_5$

B. $CH_3CH_2\overset{\overset{\displaystyle O}{\|}}{C}CH_2CH_3$ + $(CH_3)_3SiCH_2COOC_2H_5$ $\xrightarrow[\text{THF}]{\text{Bu}_4\text{N}^+\text{F}^-}$

$$\underset{CH_3CH_2}{\overset{(CH_3)_3SiO}{}}C=C\underset{H}{\overset{CH_3}{}}$$

Submitted by Isao Kuwajima, Eiichi Nakamura, and Koichi Hashimoto[1]
Checked by Peter J. Card and Richard E. Benson

1. Procedure

Caution! Ethyl bromoacetate is intensely irritating to eyes and skin. The preparation of this ester should be carried out in an efficient hood.

Benzene has been identified as a carcinogen; OSHA has issued emergency standards on its use. All procedures involving benzene should be carried out in a well-ventilated hood, and glove protection is required.

A. *Ethyl trimethylsilylacetate (Note 1).* In a 3-L, three-necked flask fitted with a 1-L, pressure-equalizing dropping funnel, mechanical stirrer, and efficient condenser that is connected to a nitrogen source are placed 97.5 g (1.5 mol) of zinc powder (Note 2) and 14.9 g (0.15 mol) of cuprous chloride (Note 3). After the reaction vessel is flushed with nitrogen, a static nitrogen atmosphere is maintained for the remainder of the reaction. A mixture of 150 mL of benzene (Note 5) is added to the flask, and the resulting mixture is refluxed with stirring for 30 min with the aid of an electric heating mantle. Heating is discontinued and a solution of 109 g (128 mL, 1.0 mol) of chloro-trimethylsilane (Note 6) and 184 g (123 mL, 1.1 mol) of ethyl bromoacetate (Note 7) in a mixture of 90 mL of ether and 350 mL of benzene is promptly added through the dropping funnel at such a rate as to maintain the reaction at gentle reflux. The addition takes about 1 hr. After the addition is complete, the mixture is heated at reflux for 1 hr and then cooled in an ice bath. While the mixture is stirred, 300 mL of aqueous 5% hydrochloric acid is added through the dropping funnel over a 10-min period. The liquid layer is decanted into a 3-L separatory funnel and the flask is washed with two 100-mL portions of ether. The ether solutions are added to the separatory funnel, the organic layer is separated, and the aqueous layer is extracted with two 200-mL portions of ether. The organic phases are combined and washed twice with 200-mL portions of saturated aqueous sodium chloride, twice with 200-mL portions of saturated aqueous sodium bicarbonate, and finally with 200 mL of saturated aqueous sodium chloride. The organic layer is dried over anhydrous magnesium sulfate, the mixture is filtered, and the filtrate is concentrated on a rotary evaporator to a volume of about 400 mL. The residual yellow liquid is distilled in a 30-cm vacuum-jacketed Vigreux column at atmospheric pressure until the boiling point is 90°C. The remaining liquid is distilled at reduced pressure to give, after a small forerun, 101–108 g (63–74%, Note 8) of ethyl trimethylsilylacetate, bp 93–94°C (104 mm), n_D^{20} 1.4152–1.4154 (Note 9).

B. *(Z)-3-Trimethylsiloxy-2-pentene.* In a dry, 200-mL flask (Note 10) equipped with a Teflon-coated magnetic stirring bar and a three-way stopcock, one exit of which is capped with a small rubber septum, is quickly placed 1.5 g (ca. 6 mmol) of dried tetrabutylammonium fluoride hydrate (Note 11). With the aid of a hypodermic syringe, 50 mL of dry tetrahydrofuran (THF, Note 12) is added through the septum, and the clear solution is stirred. After 5 min, the reaction vessel is immersed in a hexane/dry ice bath, and 38.4 g (0.240 mol) of ethyl trimethylsilylacetate is added during 10 min through a syringe that is rinsed with 15 mL of dry THF. After 10 min a solution of 17.2 g (0.200 mol) of 3-pentanone (Note 13) in 15 mL of dry THF is introduced during 10 min to the stirred solution with the aid of a syringe, which is then rinsed with 5 mL of dry THF. The clear solution is stirred for 3 hr, then warmed gradually to 0°C over about 1 hr and finally the temperature is held at 0°C for 2–4 hr (Note 14). Meanwhile, 400 mL of pentane (Note 15) in a dry, nitrogen-filled, 1-L flask equipped with a drying tube and a magnetic stirring bar is cooled with stirring in a hexane/dry ice bath, and the dark-orange reaction mixture is poured into it. The reaction vessel is rinsed with three 50-mL portions of pentane. The pentane rinses are added to the reaction solution and the resulting mixture is filtered through a pad of Hyflo Super Cell on a sintered-glass filter, and the filtrate is washed with 100 mL of saturated aqueous sodium bicarbonate and 100 mL of saturated aqueous sodium chloride. The organic layer is dried over magnesium sulfate, the drying agent is remove by filtration, and the resulting solution is concentrated on a rotary evaporator at room temperature to a volume of 150 mL. The remaining liquid is distilled through a 10-cm Vigreux column. After a very small amount of forerun (< 1 g), 21.9–24.1 g (69–76%) of 3-trimethylsiloxy-2-pentene is obtained, bp 139–142°C; n_D^{20} 1.4133–1.4135 (Note 16).

2. Notes

1. This procedure is based on a report by Fessenden and Fessenden.[2a] Cuprous chloride[3] is a more efficient initiator than iodine as specified in the original procedure.

2. The submitters used zinc powder purchased from Koso Chemical (Japan) without any purification. The checkers used product available from Fisher Scientific Company. It is essential to use excess zinc to ensure complete consumption of ethyl bromoacetate, which interrupts the catalytic cycle in Step B of the present silylation reaction.

3. The submitters used cuprous chloride purchased from Koso Chemical Co. Ltd. without purification. The checkers used cuprous chloride available from Fisher Scientific Company.

4. The submitters used diethyl ether, obtained from Showa Ether, after distillation from sodium wire. The checkers distilled the product obtained from Fisher Scientific Company from lithium aluminum hydride.

5. Benzene was distilled over sodium wire before use

6. The submitters used chlorotrimethylsilane obtained from Nakarai Chemical. The material was distilled from calcium hydride or sodium wire before use. The checkers used product available from Aldrich Chemical Company, Inc.

7. The submitters used ethyl bromoacetate (GR grade) obtained from Tokyo Kasei and distilled it before use in an efficient hood. The checkers used product available from Aldrich Chemical Company, Inc.

8. The submitters state that the yield ranged from 68 to 70% for runs made on a 1.5-mol scale.

9. Ethyl trimethylsilylacetate is stable to the usual manipulations, and can be stored in glass containers for years without change of physical and spectral properties. IR (liquid film) cm^{-1}: 1720, characteristic of α-silyl esters. The reported physical constants are bp 76–77°C (40 mm), n_D^{25} 1.4136,[2a] n_D^{20} 1.4149.[2b] ^{1}H NMR (CCl$_4$) δ: 0.17 (s, 9 H, CH_3Si), 1.31 (t, 3 H, $J = 7$, CH_3CH$_2$), 1.88 (s, 2 H, SiCH_2), and 4.14 (q, 2 H, $J = 7$, CH_2O).

10. Tetrabutylammonium fluoride is very hygroscopic. A drybox may be used to avoid rapid manipulation of the fluoride in the atmosphere and exposure of the reagent in the storage vessel to moisture. Alternatively, hydrated tetrabutylammonium fluoride (Note 11) can be dried in the reaction vessel and used directly.

11. Tetrabutylammonium fluoride trihydrate obtained from Fluka AG was dried over phosphorus pentoxide for 48 hr at a pressure of ~0.1 mm. The hygroscopic fluoride was pulverized with the aid of a spatula in a dry atmosphere. The checkers prepared the dry salt by this method using material obtained from Tridom Chemical, Inc.

Alternatively, the fluoride can be prepared as follows: A 10–40% aqueous or alcoholic solution of tetrabutylammonium hydroxide available from several sources is placed in a glass flask fitted with a Teflon-coated magnetic stirring bar and stirred gently. The pH of the solution is adjusted to about 8 by rapid addition of an almost theoretical amount of 48% aqueous hydrofluoric acid with the aid of a plastic pipette. *Caution: Hydrofluoric acid in contact with the skin produces extremely painful burns. Long, acid-resistant gloves should be worn.* Final adjustment of the pH to 7–8, measured with a pH meter, is achieved by addition of 5% aqueous acid. The bulk of the solvent is removed by distillation on a rotary evaporator at ~30°C (1 mm). The resulting white paste is further dried as described above to give the salt as a white mass.

The submitters state that in some cases, probably depending on the source of the hydroxide, the dried salt did not solidify. On such an occasion, the aqueous solution was diluted with deionized water to obtain a ~0.5 M aqueous solution. The resulting solution was cooled to 5–10°C and allowed to stand to give a white clathrate. The supernatant liquid was removed by a pipette and the clathrate was washed once with cold water. When the clathrate was dried as described above the fluoride was obtained as a solid.[4]

12. Tetrahydrofuran was distilled successively from cuprous chloride and sodium wire,[5] and further purified by distillation from sodium benzophenone ketyl in a recycling still. The checkers used product obtained from Fisher Chemical Company that was distilled from lithium aluminum hydride prior to use.

13. 3-Pentanone obtained form Tokyo Kasei (GR grade) was distilled before use. The checkers used product available from Aldrich Chemical Company, Inc.

14. The reaction is normally complete at $-78°C$, affording a product of 99.5% isomeric purity. It is advisable, however, to raise the reaction temperature finally to $0°C$, since some unknown factors occasionally retard this catalyzed reaction. Development of an orange-to-red color of the mixture usually indicates the progress of the reaction.

15. Pentane was stored over sodium wire. The checkers used product available from Eastman Organic Chemicals.

16. The spectral properties of 3-trimethylsiloxy-2-pentene are as follows: ^{1}H NMR (CCl_4) δ: 0.18 (s, 9 H, $SiCH_3$), 1.03 (t, 3 H, $J = 7$, CH_3CH_2), 1.48 (d of t, 3 H, $CH_3C=CH$, $J = 1$ and 6.5), 2.02 (unresolved quartet, 2 H, CH_2CH_3, $J = 7$), 4.47 (q, 1 H, $J = 7$, $CH_3CH=C$). IR spectrum (liquid film) cm^{-1}: 1678, 1250, and 835. The isomeric purity was 96–99.5% of Z isomer as determined by the submitters by GLC comparison with an authentic E-[6,7] or Z-enriched[7] mixture. The GLC analysis was carried out using the following column and conditions: 3-mm × 6-m stainless steel column, 5% XE-60 on 60–80-mesh Chromosorb P(AW), $80°C$, 45 mL of nitrogen per min. The retention times for the E-isomer, the Z-isomer, 3-pentanone, and ethyl trimethylsilylacetate are 5.2, 5.6, 6.2, and 15.9 min, respectively.

3. Discussion

Enol trimethylsilyl ethers belong to a most important class of enol derivatives[8] and serve as good precursors of isomerically pure enolate anions.[7,9] The double bond also resembles that of electron-rich olefins in reactions with electrophiles and sometimes is reactive in electrocyclic reactions.

Among the methods for their preparations, two reactions described by House have been employed widely:[7] a thermodynamically controlled silylation with chlorotrimethylsilane/triethylamine in hot dimethylformamide or a kinetically controlled reaction that involves lithiation with a lithium dialkylamide followed by quenching with the chlorosilane. Each method has its own merits and drawbacks with respect to three important factors: regio-, stereo-, and chemoselectivities.

The present silylation reaction[10] represents a new procedure based on metathetical generation of reactive enolate species,[11] and some characteristic features described below make this reaction complementary to the previous methods.

The excellent stereoselectivity as described in the present example is one of the advantages that merits attention.[10] The reaction affords only Z-enol silyl ethers when applied to acyclic ketones. For instance, silylation of 5-nonanone and 2-octanone gave (Z)-5-trimethylsiloxy-4-nonene and (Z)-2-trimethylsiloxy-2-octene (together with 14% of its regio isomer), both in 91% yield.

Chemoselectivity of the reaction constitutes another point of interest. Ketones can be silylated in the presence of functional groups that include oxiranes, esters, nitriles,[10] and even ketones. Thus silylation of one ketone can be performed in the presence of another. The equation shown below illustrates this selectivity.[12]

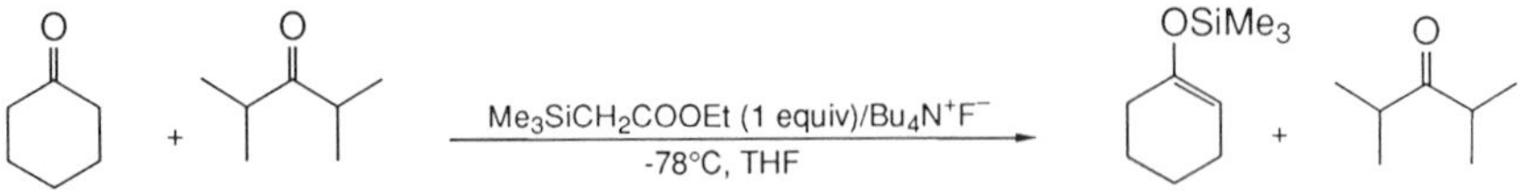

Alkyl halides[11] and aldehydes[13] are not compatible with the present silylation reaction.

Kinetic selectivity of the silylation reaction is high with methyl isopropyl ketone (99.5% of the less highly substituted isomer),[12] and methyl isobutyl ketone (~ 90%), and fair with 2-methylcyclohexanone (~ 80%).[10] The nature of the regioselectivity of this reaction appears different from that with lithium dialkylamide for which steric factors may influence the regioselectivity. In fact, silylation of 3-phenylthio-2-butanone with ethyl trimethylsilylacetate at 0°C produced 2-phenylthio-3-trimethylsiloxy-2-butene, whereas treatment with lithium diisopropylamide followed by quenching with chlorortrimethylsilane gave mainly the less highly substituted regioisomer.[12]

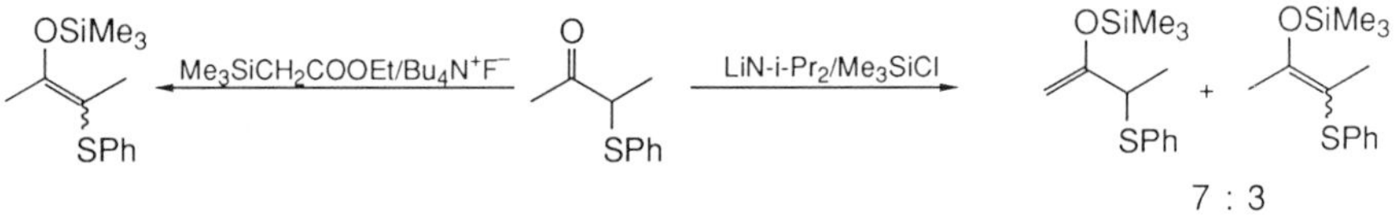

Since the only by-product of the reaction is ethyl acetate, the silylated product can be employed for further reactions without purification. Examples include the fluoride-catalyzed aldol reaction[14] and bromination with N-bromosuccinimide.[10]

The present reaction can be applied to a variety of ketones including four- to eight-membered and twelve-membered cycloalkanones and acyclic and α,β-unsaturated ketones.[10] It has also been used for primary, secondary, and tertiary alcohols,[15] alkanethiols,[15] phenols,[15] and arylacetylenes.[10]

Ethyl trimethylsilylacetate has also been used for the synthesis of α,β-unsaturated esters.[16] The chemistry of tetrabutylammonium fluoride as a base with mild reactivity has been reviewed.[17]

1. Department of Chemistry, Tokyo Institute of Technology, Tokyo 152, Japan.
2. (a) Fessenden, R. J.; Fessenden, J. S. *J. Org. Chem.* **1967**, *32*, 3535; (b) Gold, J. R.; Sommer, L. H.; Whitmore, F. C. *J. Am. Chem. Soc.* **1948**, *70*, 2874.
3. Rawson, R. J.; Harrison, I. T. *J. Org. Chem.* **1970**, *35*, 2057.
4. McMullan, R.; Jeffrey, G. A. *J. Chem. Phys.* **1959**, *31*, 1231.
5. See "WARNING," *Org. Synth., Coll. Vol. V* **1973**, 976.
6. Ireland, R. E.; Mueller, R. H.; Willard, A. K. *J. Am. Chem. Soc.* **1976**, *98*, 2868.
7. House, H. O.; Czuba, L. J.; Gall, M.; Olmstead, H. D. *J. Org. Chem.* **1969**, *34*, 2324.
8. Review: Rasmussen, J. K. *Synthesis* **1977**, 91.
9. Stork, G.; Hudrlik, P. F. *J. Am. Chem. Soc.* **1968**, *90*, 4462, 4464.
10. Nakamura, E.; Murofushi, T.; Shimizu, M.; Kuwajima, I. *J. Am. Chem. Soc.* **1976**, *98*, 2346; Nakamura, E.; Hashimoto, K.; Kuwajima, I. *Tetrahedron Lett.* **1978**, 2079.
11. Kuwajima, I.; Nakamura, E. *J. Am. Chem. Soc.* **1975**, *97*, 3257.

12. Kuwajima, I.; Nakamura, E. *Acc. Chem. Res.* **1985,** *18,* 181.
13. Nakamura, E.; Shimizu, M.; Kuwajima, I. *Tetrahedron Lett.* **1976,** 1699.
14. Noyori, R.; Yokoyama, K.; Sakata, J.; Kuwajima, I.; Nakamura, E.; Shimizu, M. *J. Am. Chem. Soc.* **1977,** *99,* 1265.
15. Nakamura, E.; Hashimoto, K.; Kuwajima, I. *Bull. Chem. Soc. Jpn.* **1980,** *54,* 804.
16. Taguchi, H.; Shimoji, K.; Yamamoto, H.; Nozaki, H. *Bull. Chem. Soc. Jpn.* **1974,** *47,* 2529.
17. Review: Kuwajima, I. *J. Synth. Org. Chem. Jpn.* **1976,** *34,* 964; *Chem. Abstr.* **1977,** *86,* 106694v.

TRIMETHYLSILYL CYANIDE: CYANOSILATION OF *p*-BENZOQUINONE

(Silanecarbonitrile, trimethyl)

A. $(CH_3)_2C\overset{OH}{\underset{CN}{\big<}}$ + LiH $\xrightarrow[20\text{-}30°C]{THF}$ LiCN + $(CH_3)_2CO$ + H_2

$(CH_3)_3SiCl$ + LiCN $\xrightarrow[25°C]{CH_3O(CH_2CH_2O)_4CH_3}$ $(CH_3)_3SiCN$ + LiCl

B.

Submitted by TOM LIVINGHOUSE[1]
Checked by TOD HOLLER, KEVIN J. CARLIN, and G. BÜCHI

1. Procedure

Caution! Trimethylsilyl cyanide is very toxic. All reactions in this sequence should be carried out in a hood.

A. *Trimethylsilyl cyanide.* A 1-L, round-bottomed flask equipped with a magnetic stirrer, nitrogen inlet, and a 60-mL addition funnel is charged with 5.0 g (0.624 mol) of lithium hydride (Note 1) and 500 mL of anhydrous tetrahydrofuran (Note 2). The stirred suspension is cooled in an ice bath and 42.6 g of acetone cyanohydrin (45.7 mL, 0.501 mol) (Note 3) is added dropwise over 15 min. After the addition is complete, the ice bath is removed and the mixture stirred for 2 hr at room temperature (Note 4). The magnetic stirring bar is removed and the solvent evaporated as completely as possible on a rotary evaporator. The white lithium cyanide is then dried in vacuo for 3 hr (Notes 5 and 6). The lithium cyanide is freed from the sides of the flask and broken up with a spatula (Note 7). A 250-mL round-bottomed flask equipped with an ice bath, magnetic stirrer, thermometer, and nitrogen inlet is charged with 54.32 g (63.46 ml, 0.500 mol) of trimethylchlorosilane (Note 8) and 100 mL of bis[2-(2-methoxyethoxy)ethyl] ether (Note 9). The lithium cyanide is added to this stirred solution over 15 min through Gooch tubing (Note 10). After the addition is complete, the ice bath is removed and the milky suspension stirred overnight at room temperature. The Gooch tubing and the thermometer are then removed from the reaction flask and a stillhead equipped for downward vacuum distillation is attached. A 100-mL, round-bottomed flask immersed halfway in an acetone-dry ice slush bath (Note 11) is employed as the receiver. The volatile compounds are distilled under a pressure of 50 mm (bp 25–55°C) by heating the contents of the pot using an oil bath (Note 12). The distillate is carefully redistilled through a well-insulated 15-cm column packed with glass helices under an inert atmosphere. A 25–40 mL forerun (bp 66–113°C), consisting primarily of tetrahydrofuran and hexamethyldisiloxane, is first collected. The second fraction, containing 29–41 g (59–82%) of trimethylsilyl cyanide, bp 114–117°C, n_D^{25} 1.3902 (Note 13), then distills. A purity of ca. 97% was established by GC analysis (Notes 14 and 15); the product is suitable for synthetic use without further purification.

B. *Cyanosilylation of* p-*benzoquinone.* A 100-mL, round-bottomed flask equipped with a magnetic stirrer, West condenser, and a nitrogen inlet is charged with 6.30 g (58.2 mmol) of p-benzoquinone (Note 16), 10 mL of dry carbon tetrachloride, and 8 mL (63.03 mmol) of trimethylsilyl cyanide. The stirred suspension is heated to a gentle reflux by means of a heat gun to dissolve all the p-benzoquinone. It is then allowed to cool slowly until the crystallization of the p-benzoquinone starts (Note 17), at which time 5 mg of the 1 : 1 complex between potassium cyanide and 18-*crown*-6 (Note 18) is added through the top of the condenser. An immediate vigorous reflux sets in and continues for 1–2 min (Note 19). The stirred reaction mixture is permitted to cool slowly to room temperature, whereupon the condenser is removed and 3 g of Florisil (Note 20) is added. After stirring for an additional 15 min, 10 mL of dry carbon tetrachloride is added. The suspension is then filtered and the filtercake leached with three 5-mL portions of carbon tetrachloride. The solvent is evaporated from the filtrate as completely as possible on a rotary evaporator, at which point crystallization of the residue usually begins (Note 21). The last traces of solvent and trimethylsilyl cyanide are then removed in vacuo over 20 hr at 50 μ to afford 12.0–12.2 g of crude product. The trimethylsilyl cyanohydrin is recrystallized by dissolving the crude material in 25 mL of hot hexane and allowing the resulting solution to cool slowly to room temperature (Note 22). After collection by filtration the product is rinsed with two 5-mL portions of hexane and air-dried to yield 7.54–9.77 g (63%–81%) of white to buff-colored needles, mp 65–67°C (Notes 23 and 24).

2. Notes

1. Commercial lithium hydride (Alfa Products, Morton Thiokol, Inc.) was used.
2. Commercial tetrahydrofuran was distilled from sodium benzophenone ketyl immediately before use.
3. Commercial acetone cyanohydrin (Aldrich Chemical Company, Inc.) was used without further purification.

4. A vigorous evolution of hydrogen gas occurs during the addition of the acetone cyanohydrin. Hydrogen evolution virtually ceases after stirring at room temperature for 2 hr.

5. It is essential to exclude atmospheric moisture as much as possible during this operation.

6. A small quantity of tetrahydrofuran remains complexed in the solid lithium cyanide and is separated later in the preparation.

7. This operation must be performed rapidly to avoid water absorption by the hygroscopic lithium cyanide.

8. Commercial trimethylchlorosilane (Silar Laboratories, Inc.) was distilled from calcium hydride immediately before use.

9. Commercial bis[2-(2-methoxyethoxy)ethyl] ether, "tetraglyme" (Eastman Organic Chemicals), was dried over Linde 4A molecular sieves for 24 hr before use.

10. The internal temperature is maintained at or below 35°C during this operation by periodic cooling with an ice bath.

11. Trimethylsilyl cyanide solidifies in the receiver during the course of the distillation. *It is absolutely necessary that the receiver be immersed no more than halfway in the slush bath.* Further immersion may cause the product to solidify in the end of the condenser. This necessitates cessation of the distillation to unclog the apparatus.

12. The temperature of the oil bath is raised from 25 to 110°C over 45 min and then maintained at the upper temperature until no more product distills.

13. The product exhibits the following properties: ^{1}H NMR (CCl$_4$ with CHCl$_3$ internal standard) δ: 0.4 [s, Si(CH$_3$)$_3$]; IR (neat) cm^{-1}: 2200 ($-$CN).

14. The GC analysis was performed on an 8-ft column packed with 5% OV-17 on Anachrome ABS.

15. Trimethylsilyl cyanide hydrolyzes rapidly in moist air and is best stored under an inert atmosphere.

16. Commercial *p*-benzoquinone (Matheson, Coleman, and Bell, Inc.) was recrystallized from 95% ethanol before use.

17. The initiation of crystallization indicates the optimum reaction temperature for the catalyzed cyanosilylation of *p*-benzoquinone. The use of higher temperatures results in excessive darkening of the product and a decrease in yield.

18. The 1 : 1 complex is conveniently prepared by dissolving 0.652 g (10 mmol) of pulverized potassium cyanide and 2.640 g (10 mmol) of commercial 18-*crown*-6 (Aldrich Chemical Company, Inc.) in 45 mL of anhydrous methanol by swirling and warming. The methanol is then evaporated at a rotary evaporator and the white complex dried in vacuo over night.

19. *Caution! Extreme care must be taken during the addition of the catalyst. The addition of too much catalyst or the use of higher reaction temperatures may result in the reaction mixture boiling over.*

20. Florisil obtained from Matheson, Coleman, and Bell, Inc. was used.

TABLE I
CYANOSILYLATION OF KETONES AND ALDEHYDES

Substrate	Silylcyanohydrin	Yield (%)	Ref.
Benzophenone	(78)[a]	98	2
Crotonaldehyde	(88)[a]	98	2
Furfural		99	2
Cyclooctanone		94	7
Cyclododecanone		94	7
Camphor		>95[b]	7
α-Tetralone		>95[b]	7
3-Methyl-3-penten-2-one		91	7

[a] No catalyst employed; zinc iodide catalyst used in all other cases.
[b] Yield determined by GLC analysis.

21. Crystallization of the residue may also be induced by the addition of a seed crystal or scratching with glass rod.

22. In some instances addition of a seed crystal during cooling is necessary.

23. p-Benzoquinone monotrimethylsilyl cyanohydrin darkens on prolonged exposure to light and air. It is best stored under nitrogen in the dark.

24. An analytically pure sample, mp 67–67.5°C, may be obtained by a second recrystallization from cyclohexane: ^{1}H NMR (CCl$_4$) δ: 0.30 (s, 9, CH$_3$), 6.30 (d, 1, J = 10, C=CH), 6.83 (d, 1, J = 10, C=CH); IR (CCl$_4$) cm^{-1}: 1678 (C=O), 1252, 845 (Si–CH$_3$).

3. Discussion

Trimethylsilyl cyanide is useful reagent for the preparation of β-amino alcohols,[2] α-amino nitriles,[3] and α-trimethylsiloxyacrylonitriles[4] from the corresponding ketones, imines, and ketenes. The reagent adds rapidly to the carbonyl of aldehydes at 25°C,[2] and the resulting adducts have proved useful precursors for the preparation of carbonyl anion synthons.[5] Enones give exclusively the products derived from 1,2-addition.[2]

Trimethylsilyl cyanide has been prepared in modest yield by the action of hexamethyldisilazane on hydrogen cyanide[6] and the reaction of silver cyanide with trimethylchlorosilane.[6,7] It has been prepared in good yield by the treatment of preformed lithium cyanide (from LiH and HCN) with trimethylchlorosilane in ether.[7] The procedure described here not only affords trimethylsilyl cyanide in good yield, but also avoids the use of hydrogen cyanide and the need for Schlenk ware.

Table I illustrates the cyanosilylation of several representative ketones and aldehydes.

1. Department of Chemistry, University of California, Los Angeles, 405 Hilgard Ave., Los Angeles, CA 90024. Present address: Department of Chemistry, Montana State University, Bozeman, MT 59717.

2. Evans, D. A.; Truesdale, L. K.; Carroll, G. L. J. Chem Soc., Chem. Commun. 1973, 55–56.

3. Ojima, I.; Inabe, S.; Nakatsugawa, K.; Nagai, Y. Chem. Lett. 1975, 331–334.

4. Hertenstein, U.; Hünig, S. *Angew. Chem. Int. Ed. Engl.* **1975,** *14,* 179–180.
5. Hünig, S.; Wehner, G. *Synthesis* **1975,** 180–182.
6. Bither, T. A.; Knoth, W. H.; Lindsey, R. V.; Sharkey, W. H. *J. Am. Chem. Soc.* **1958,** *80,* 4151–4153.
7. Evans, D. A.; Carroll, G. L.; Truesdale, L. K. *J. Org. Chem.* **1974,** *39,* 914–917.

in situ CYANOSILYLATION OF CARBONYL COMPOUNDS: *O*-TRIMETHYLSILYL-4-METHOXYMANDELONITRILE

(Benzeneacetonitrile, 4-methoxy-α-[(trimethylsilyl)oxy]-)

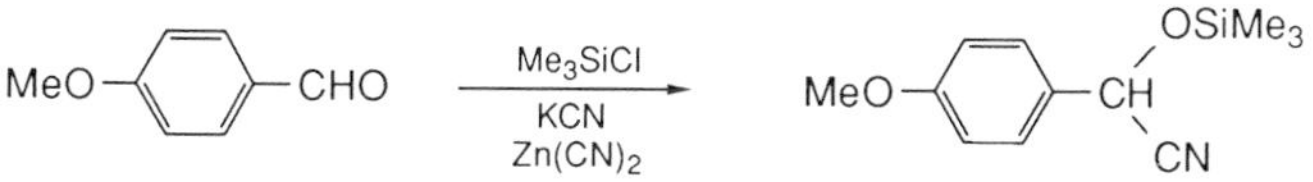

Submitted by J. K. RASMUSSEN and S. M. HEILMANN[1]
Checked M. F. SEMMELHACK and RAJ N. MISRA

1. Procedure

Caution! Potassium cyanide is highly toxic. Care should be taken to avoid direct contact of the chemical or its solutions with the skin, and impervious gloves should be worn to handle the reagent.

In a 1-L, three-necked, round-bottomed flask equipped with a mechanical stirrer, a reflux condenser fitted with a nitrogen-inlet tube, and a rubber septum (Note 1) are placed 97.5 g (1.5 mol) of finely ground potassium cyanide (Note 2), 81.4 g (0.75 mol, 95.2 mL) of chlorotrimethylsilane (Note 3), 68 g (0.5 mol) of *p*-anisaldehyde (Note 4), 100 mL of dry acetonitrile (Note 5), and 0.5 g (4.25 mmol) of zinc cyanide (Note 6). The reaction mixture is blanketed with dry nitrogen (Note 7), stirring is begun, and the temperature is raised (heating mantle) to maintain gentle reflux. Heating is continued under these conditions for 30 hr (Note 8), with the occasional removal of small samples by syringe for monitoring by GLC (Note 9). On completion of the reaction, the mixture is cooled to ambient temperature and filtered. The filtercake is washed twice with 50 mL of dry acetonitrile and the combined filtrates are concentrated on a rotary evaporator. The residue is distilled at reduced pressure (Note 10). The yield of the colorless liquid (Note 10), which boils at 93–98°C (0.15 mm), amounts to 105–115 g (90–98% based on *p*-anisaldehyde).

2. Notes

1. All glassware was oven-dried overnight at 130°C, assembled hot, and allowed to cool under a flow of dry nitrogen.
2. Reagent-grade potassium cyanide was purchased from Matheson, Coleman and Bell, and dried at 115°C (0.5 mm) for 24 hr. The checkers found it necessary to use newly purchased potassium cyanide. The use of potassium cyanide which

was several years old gave incomplete reaction even at extended reaction times. The large excess of potassium cyanide is used simply to obtain convenient reaction times. For comparison, use of 1.5 equiv of KCN gave 38% conversion under conditions where 3 equiv produced 100% conversion.

3. Chlorotrimethylsilane was supplied by Petrarch Systems, Inc. and used without further purification.

4. p-Anisaldehyde (4-methoxybenzaldehyde), 95%, was used as supplied by Aldrich Chemical Co.

5. Acetonitrile, 99%, supplied by Aldrich Chemical Co., was dried over Linde 4A molecular sieves for 12 hr and decanted.

6. Technical-grade zinc cyanide was used as supplied by MCB, Inc. Other Lewis acids, notably aluminum chloride, zinc bromide, and zinc iodide, may be used as catalysts for the reaction.

7. To "blanket with nitrogen," the checkers simply prepared the reaction mixture with the flask open, introduced a flow of nitrogen over the surface for a few minutes, and then closed the system with an exit through a mercury bubbler to maintain a positive pressure.

8. The reaction time required depends on the catalyst. Zinc iodide, zinc cyanide, and zinc bromide produce essentially complete conversion under these conditions in approximately 16.5, 28, and 30 hr, respectively, probably reflecting solubility differences. When zinc iodide is used, the distilled product is often colored because of the formation of small amounts of iodine.

9. This may be done using a simple boiling-point column. We have employed either 10% UCW-98 on Chromasorb W or SP-2100 on 80/100 Supelcoport G2642. The checkers did not monitor the reaction except to extract a small sample after 30 hr in order to verify the absence of starting aldehyde by ^{1}H NMR spectroscopy.

10. Distillation should be below 100°C. In some instances, at distillation temperatures in excess of 100°C, reversion to the starting aldehyde and trimethylsilyl cyanide has been observed. The pure compound shows the following spectral data: ^{1}H NMR (CCl$_4$): δ 0.28 (s, 9 H), 3.86 (s, 3 H), 5.35 (s, 1 H), 6.83 (d, $J = 9$, 2 H), 7.35 (d, $J = 9$, 2 H); IR (film) cm^{-1}: 2965, 1614, 1512, 1258, 1180, 1089, 878, 850. The purity of the crude product is generally such that a distillation forecut need not be taken.

3. Discussion

Cyanosilylations have generally been accomplished by addition of a trialkylsilyl cyanide to the corresponding aldehyde or ketone.[2-5] Although this method is straightforward and proceeds in good to excellent yield, use of preformed trialkylsilyl cyanides has a number of disadvantages, particularly when one considers larger-scale preparations. Trialkylsilyl cyanides can be prepared[6] by treatment of the corresponding silyl chlorides with either silver cyanide or lithium cyanide generated in situ by reaction of lithium hydride with hydrogen cyanide. The former procedure involves the use of stoichiometric quantities of a rather expensive reagent, while the latter involves handling

TABLE I
IN SITU CYANOSILYLATION OF CARBONYL COMPOUNDS

$\begin{array}{c} OSi(CH_3)_3 \\ | \\ R^1{-}C{-}R^2 \\ | \\ CN \end{array}$	Distilled Yield (%)	bp (°C) (pressure, mm)
$R^1 = C_6H_5$, $R^2 = H$	95–98	93–95 (1.75)
$R^1 = 4\text{-}CH_3C_6H_4$, $R^2 = H$	91	87 (0.45)
$R^1 = 2\text{-}ClC_6H_4$, $R^2 = H$	99	92–93 (0.45)
$R^1 = 4\text{-}ClC_6H_4$, $R^2 = H$	93	100 (0.45)
$R^1 = C_6H_5$, $R^2 = CH_3$	93	73–75 (0.9)
$R^1, R^2 = ({-}CH_2{-})_5$	89	96 (15)
$R^1 = c\text{-}C_6H_{11}$, $R^2 = H$	87	106–108 (6.5)

fairly large quantities of hydrogen cyanide gas. In addition, both procedures require relatively long reaction times and distillation of the silyl cyanide, and produce only moderate to good yields. More recently, improved syntheses of trimethylsilyl cyanide have appeared.[7,8] Commercially available tremethylsilyl cyanide is also rather expensive.

Silylated cyanohydrins have also been prepared via silylation of cyanohydrins themselves[9] and by the addition of hydrogen cyanide to silyl enol ethers.[10] Silylated cyanohydrins have proved to be quite useful in a variety of synthetic transformations, including the regiospecific protection of p-quinones,[11] as intermediates in an efficient synthesis of α-aminomethyl alcohols,[6] and for the preparation of ketone cyanohydrins themselves.[12] The silylated cyanohydrins of heteroaromatic aldehydes have found extensive use as acyl anion equivalents, providing general syntheses of ketones[13] and acyloins.[14] Acyloins are also readily prepared via addition of Grignard reagents to silylated cyanohydrins followed by hydrolysis of the magnesium imine intermediate.[15] Alternatively, reduction of this same intermediate with borohydride provides a general synthesis of aminoalcohols.[16] Tetronic acids may be produced on reaction of silylated cyanohydrins with Reformatsky reagents.[17]

The in situ cyanosilylation of p-anisaldehyde is only one example of the reaction that can be applied to aldehydes and ketones in general.[18] The simplicity of this one-pot procedure, coupled with the use of inexpensive reagents, are important advantages over previous methods. The silylated cyanohydrins shown in Table I were prepared under conditions similar to those described here. Enolizable ketones and aldehydes have a tendency to produce silyl enol ethers as by-products in addition to the desired cyanohydrins. The problem can be overcome by using a modified procedure in which dimethylformamide is employed as solvent.[18]

1. Central Research Laboratories, 3M Company, 3M Center, St. Paul, MN 55144.
2. Evans, D. A.; Truesdale, L. K.; Carroll, G. L. *J. Chem. Soc., Chem. Commun.* **1973**, 55–56.

3. Evans, D. A.; Truesdale, L. K. *Tetrahedron Lett.* **1973**, 4929–4932.
4. Lidy, W.; Sundermeyer, W. *Chem. Ber.* **1973**, *106*, 587–593.
5. Neef, H.; Muller, R. *J. Prakt. Chem.* **1973**, *315*, 367–374.
6. Evans, D. A.; Carroll, G. L.; Truesdale, L. K. *J. Org. Chem.* **1974**, *39*, 914–917.
7. Rasmussen, J. K.; Heilmann, S. M. *Synthesis* **1979**, 523–524.
8. Hünig, S.; Wehner, G. *Synthesis* **1979**, 522–523.
9. Frisch, K. C.; Wolf, M. *J. Org. Chem.* **1953**, *18*, 657–660.
10. Parham, W. E.; Roosevelt, C. S. *Tetrahedron Lett.* **1971**, 923–926.
11. Evans, D. A.; Hoffman, J. M.; Truesdale, L. K. *J. Am. Chem. Soc.* **1973**, *95*, 5822–5823.
12. Gassman, P. G.; Talley, J. J. *Tetrahedron Lett.* **1978**, 3773–3776.
13. Deuchert, K.; Hertenstein, U.; Hünig, S.; Wehner, G. *Chem. Ber.* **1979**, *112*, 2045–2061.
14. Hünig, S.; Wehner, G. *Chem. Ber.* **1979**, *112*, 2062–2067.
15. Krepski, L. R.; Heilmann, S. M.; Rasmussen, J. K. *Tetrahedron Lett.* **1983**, *24*, 4075–4078.
16. Krepski, L. R.; Jensen, K. M.; Heilmann, S. M.; Rasmussen, J. K. *Synthesis* **1986**, 301–303.
17. Krepski, L. R.; Lynch, L. E.; Heilmann, S. M.; Rasmussen, J. K.; *Tetrahedron Lett.* **1985**, *26*, 981–984.
18. Rasmussen, J. K.; Heilmann, S. M. *Synthesis* **1978**, 219–221.

STEREOSPECIFIC REDUCTION OF PROPARGYL ALCOHOLS: (*E*)-3-TRIMETHYLSILYL-2-PROPEN-1-OL

[2-Propen-1-ol, 3-(trimethylsilyl)-, (*E*)-]

A. $HC \equiv C - CH_2OH \xrightarrow[\text{THF, 10°C}]{\text{EtMgBr}} \xrightarrow[5° \to 70°C]{\text{Me}_3\text{SiCl}} \xrightarrow[45°C]{1.4 \text{ M } H_2SO_4} Me_3Si - C \equiv C - CH_2OH$

B. $Me_3Si - C \equiv C - CH_2OH \xrightarrow[\text{ether/toluene, 20°C}]{\text{NaAlH}_2(\text{OCH}_2\text{CH}_2\text{OMe})_2}$

Structure showing (E)-alkene with Me₃Si and H on one carbon, H and CH₂OH on the other.

Submitted by TODD K. JONES and SCOTT E. DENMARK[1]
Checked by STEVEN M. VITI and K. BARRY SHARPLESS

1. Procedure

A. *3-Trimethylsilyl-2-propyn-1-ol.* A 3-L, three-necked, round-bottomed flask (equipped with a mechanical stirrer and a thermometer) is fitted with a Claisen adapter on which is mounted a 250-mL pressure-equalizing addition funnel and a reflux condenser (Note 1). The apparatus is flushed with nitrogen and then charged with 48.7 g (2.0 mol) of magnesium turnings and 1 L of dry tetrahydrofuran (Note 2). To the stirred suspension is added dropwise 149.5 mL (218.3 g, 2.0 mol) of bromoethane over 3 hr while maintaining the temperature at 50°C or less. After complete addition, the gray–green solution is heated at 50°C for 1 hr and then cooled to 5°C on ice. A solution of 41.6 mL (40.5 g, 0.72 mol) of propargyl alcohol (Note 3) in 42 mL of tetrahydrofuran is cautiously added dropwise to the gray suspension over 2.25 hr while maintaining the temperature at 10°C or less (Note 4). The addition funnel is rinsed with 25 mL of tetrahydrofuran and the gray–green suspension is stirred overnight. The resulting solution is cooled to 5°C on ice and the addition funnel is charged with 254 mL (217 g, 2.0 mol) of chlorotrimethylsilane (Note 5). This is added dropwise to the stirred solution over 1 hr while maintaining the temperature at 25°C or less by external cooling with ice. After complete addition, the mixture is heated to reflux for 2 hr with a heating mantle (Note 6). The suspension is cooled to 20°C on ice and then 800 mL of 1.4 M aqueous sulfuric acid is cautiously added over 0.75 hr so that the temperature remains below 45°C. The resulting solution is stirred for 5 min and then 600 mL of ether is added. Both phases are transferred to a 4-L separatory funnel and the layers are separated. The aqueous phase is extracted twice with 400-mL portions of ether and all ether layers are individually washed in series with two 1-L portions of water and once with 800 mL of saturated sodium chloride solution. The combined organic extracts are dried over magnesium sulfate and concentrated by rotary evaporation. The yellow–brown residue is purified by short path distillation to afford 82–86 g (91–94% yield) of 3-trimethylsilyl-2-propyn-1-ol as a clear, colorless liquid (Note 7), bp 76°C (20 mm) (Note 8).

B. *(E)-3-Trimethylsilyl-2-propen-1-ol.* A three-necked, 2-L, round-bottomed flask fitted with a thermometer, nitrogen inlet, 250-mL pressure-equalizing addition funnel, and magnetic stirring bar is charged with 147 mL of a 3.4 M solution of sodium bis(2-methoxyethoxy)aluminum hydride (SMEAH, Note 9) and 200 mL of anhydrous ether (Note 10). The SMEAH solution is cooled to 3°C on ice and then treated dropwise from the addition funnel with a solution of 40 g (0.31 mol) of 3-trimethylsilyl-2-propyn-1-ol in 180 mL of ether over 1.25 hr, while maintaining the temperature at 5°C or less. Then 10 min after complete addition, the ice bath is removed and the reaction is complete within 1 hr (Note 11). The mixture is cooled to 0°C and then quenched by the addition of 1 L of 3.6 M aqueous sulfuric acid (Note 12). The layers are separated in a separatory funnel and the aqueous phase is extracted twice with 200-mL portions of ether. All ether layers are individually washed in series with two 200-mL portions of water and once with 200 mL of saturated sodium chloride. The combined organic extracts are dried over magnesium sulfate and concentrated by rotary evaporation. Distillation of the yellow residue with a capillary bleed affords 27.7–29.0 g (68–71%) of (*E*)-3-trimethylsilyl-2-propen-1-ol (Note 13) as a clear, colorless liquid, bp 73–75°C (20 mm) (Note 14).

2. Notes

1. It is not necessary to flame or oven-dry this apparatus, but a nitrogen inlet on the reflux condenser is desirable. The size of the stirring paddle is critical because of the viscous nature of the solution during this protection step. A paddle at least 11 cm in length is recommended to ensure complete mixing.

2. Magnesium turnings and bromoethane are Mallinckrodt AR grade and are used without purification. Tetrahydrofuran is Aldrich Gold Label and is distilled from sodium benzophenone ketyl prior to use.

3. Propargyl alcohol is obtained from Aldrich Chemical Company, Inc. and is distilled from potassium carbonate.

4. Evolution of ethane can conveniently be monitored with a Nujol bubbler in the nitrogen line by turning off the nitrogen flow.

5. Chlorotrimethylsilane is purchased from Silar and used as received.

6. The progress of the reaction can be monitored by gas chromatography. Column: 5% Carbowax 12 M on acid-washed Chromosorb W, 6 ft × one-eighth in; temperature program: 70°C (2 min), 20°C/min, 200°C (5 min). Retention times: propargyl alcohol, 2.4 min; 3-trimethyl-2-propyn-1-ol, 4.8 min.

7. *Caution: The distillation pot may ignite if it is exposed to air before it is allowed to cool.* The product thus obtained is 94–98% pure by GC analysis and is of suitable purity for reduction. Further purification can be effected by distillation through a 6-in. Vigreux column.

8. The product has the following spectral characteristics: ^{1}H NMR (90 MHz, CDCl$_3$) δ: 0.27 [s, 9 H, (CH$_3$)$_3$Si], 1.65 (s, 1 H, OH), 4.28 [s, 2 H, 2 H—C(1)].

9. Sodium bis(2-methoxyethoxy)aluminum hydride is obtained as a 70% solution in toluene from Aldrich Chemical Company, Inc. (Red-Al). Iodometric titration gives a 3.6 M concentration.

10. Anhydrous ether is obtained from Mallinckrodt, Inc. (AR grade) and used without purification.

11. The reaction can be monitored by gas chromatography (Note 6), temperature program: 70°C (2 min), 20°C/min, 150°C (2 min). Retention times: (E)-3-trimethylsilyl-2-propen-1-ol, 4.2 min; 3-trimethylsilyl-2-propyn-1-ol, 6.1 min.

12. A vigorous evolution of hydrogen accompanies the addition of the first milliliters of sulfuric acid. The reaction mixture becomes gelatinous and unstirable but clarifies on further addition of acid.

13. The product is 100% E geometry by GC analysis.

14. The product has the following spectral characteristics: ^{1}H NMR (90 MHz, CDCl$_3$) δ: 0.23 [s, 9 H, (CH$_3$)$_3$Si], 1.5 (t, 1 H, J = 6, OH), 4.22 [d of d, 2 H, J = 6 and 4, 2 H—C(1)], 5.93 [d, 1 H, J = 18, H—C(3)], 6.23 [d of t, 1 H, J = 18 and 4, H—C(2)].

3. Discussion

The silylation of propargyl alcohol dianion[2a] described here is a further modification of the procedure recently reported.[2b] By replacing ether with tetrahydrofuran the reaction mixture is more manageable and the silyl ether can be hydrolyzed in situ obviating an unnecessary workup and distillation. The yield correspondingly improves by up to 91–94%. Silylation of the dilithium salt in ether is reported[3] to proceed in 86% yield.

Reduction of 3-trimethylsilyl-2-propyn-1-ol exemplifies the problem of stereoselectivity in hydride reduction of acetylenic alcohols to E-allyl alcohols.[4] Early reports[5] that lithium aluminum hydride stereoselectively reduced acetylenic alcohols gave way to closer scrutiny, which revealed a striking solvent dependence of the stereochemistry.[6] Specifically, the percentage of trans reduction is seen to increase with increasing Lewis basicity of solvent. Similarly, the addition of less Lewis acidic cations to the reducing mixture leads to improved trans : cis ratios.[7] Sodium bis(2-methoxyethoxy)aluminum hydride (SMEAH)[8] makes use of these phenomena simultaneously (even in ether–toluene mixtures) and leads to completely stereospecific trans reduction where lithium aluminum hydride in various solvents or with sodium methoxide is less selective.[2b,9,10a,b] The use of SMEAH to reduce stereospecifically other acetylenic alcohols has been reported.[11]

(E)-3-Trimethylsilyl-2-propen-1-ol is a versatile intermediate used to introduce organosilicon functional groups into organic molecules.[9,12] The corresponding aldehyde has found use in the preparation of β-silyl divinyl ketones[13] and as a precursor for 1-trimethylsilyl-substituted dienes.[10]

1. Department of Chemistry, School of Chemical Sciences, University of Illinois, Urbana, IL 61801.

2. (a) Mironov, V. F.; Maksimova, N. G. *Bull. Acad. Sci. USSR, Div. Chem. Sci. (Engl. transl.)* **1960**, 1911; (b) Denmark, S. E.; Jones, T. K. *J. Org. Chem.* **1982**, *47*, 4595.

3. Brandsma, L.; Verkruijsse, H. D. "Synthesis of Acetylenes, Allenes and Cumulenes: A Laboratory Manual"; Elsevier: Amsterdam, 1981; p. 58.

4. (a) House, H. O. "Modern Synthetic Reactions," 2 ed.; W. A. Benjamin: Menlo Park, CA, 1972; p. 91; (b) Hajos, A. "Complex Hydrides and Related Reducing Agents in Organic Synthesis"; Elsevier: New York, 1979.

5. (a) Bates, E. B.; Jones, E. R. H.; Whiting, M. C. *J. Chem. Soc.* **1954**, 1854; (b) Snyder, E. I. *J. Am. Chem. Soc.* **1969**, *91*, 2579.

6. Grant, B.; Djerassi, C. *J. Org. Chem.* **1974**, *39*, 968.

7. (a) Molloy, B. B.; Hauser, K. L. *J. Chem. Soc., Chem. Commun.* **1968**, 1017; (b) Corey, E. J.; Katzenellenbogen, J. A.; Posner, G. H. *J. Am. Chem. Soc.* **1967**, *89*, 4245.

8. The uses of SMEAH have been reviewed: (a) Vit, J. *Org. Chem. Bull.* **1970**, *42(3)*, 1–9; *Chem. Abstr.* **1971**, *74*, 99073p; (b) Vit, J.; Papaionnou, C.; Cohen, H.; Batesky, D. *Eastman Org. Chem. Bull.* **1974**, *46(1)*, 1–6; *Chem. Abstr.* **1974**, *80*, 120098m; (c) Hajós, A. "Complex Hydrides and Related Reducing Agents in Organic Synthesis"; Elsevier: New York, 1979; pp. 159–167.

9. Stork, G.; Jung, M. E.; Colvin, E.; Noel, Y. *J. Am. Chem. Soc.* **1974**, *96*, 3684.

10. (a) Jung, M. E.; Gaede, B. *Tetrahedron* **1979**, *35*, 621; (b) Carter, M. J.; Fleming, I.; Percival, A. *J. Chem. Soc., Perkin Trans. 1* **1981**, 2415; (c) Petrzilka, M.; Grayson, J. I. *Synthesis* **1981**, 753.

11. (a) Chan, K.-K.; Specian, A. C., Jr.; Saucy, G. *J. Org. Chem.* **1978**, *43*, 3435; (b) Chan, K.-K.; Cohen, N.; De Noble, J. P.; Specian, A. C., Jr.; Saucy, G. *J. Org. Chem.* **1976**, *41*, 3497.

12. Kuwajima, I.; Tanaka, T.; Atsumi, K. *Chem. Lett.* **1979**, 779.

13. (a) Denmark, S. E.; Jones, T. K. *J. Am. Chem. Soc.* **1982**, *104*, 2642; (b) Jones, T. K.; Denmark, S. E. *Helv. Chim. Acta* **1983**, *66*, 2377, 2397; (c) Denmark, S. E.; Habermos, K. L.; Hite, G. A.; Jones, T. K. *Tetrahedron* **1986**, *42*, 2821; (d) Denmark, S. E.; Habermos, K. L.; Hite, G. A. *Helv. Chim. Acta* **1988**, *71*, 161.

TRIS(DIMETHYLAMINO)SULFONIUM DIFLUOROTRIMETHYLSILICATE

[Sulfur(1 +), tris(N-methylmethanaminato)-, difluorotrimethylsilicate(1-)]

$$3 \text{ Me}_3\text{SiNMe}_2 \quad + \quad \text{SF}_4 \quad \xrightarrow[25°C]{ether} \quad (\text{Me}_2\text{N})_3\text{S}^+ \text{ F}_2\text{SiMe}_3^- \quad + \quad 2 \text{ FSiMe}_3$$

Submitted by WILLIAM J. MIDDLETON[1]
Checked by FRED G. WEST and EDWIN VEDEJS

1. Procedure

Caution! This procedure should be conducted in an efficient hood to avoid exposure to the toxic gas sulfur tetrafluoride.

A dry, 500-mL, four-necked flask equipped with a magnetic stirrer, dry ice condenser, thermometer (−100°– 50°C) and a gas inlet tube is assembled as shown in Figure 1 (connections were all-glass or polyethylene tubing). The system is flushed with nitrogen through three-way stopcocks A and B, the four-necked flask is charged with 150 mL of dry ether (Note 1), and the dropping funnel is charged with 46.9 g (0.40 mol) of N,N-dimethylaminotrimethylsilane (Note 2). The reaction vessel is maintained under a positive nitrogen pressure using a bypass nitrogen stream and bubbler. Stopcock A is connected to the sulfur tetrafluoride (SF₄) tank (Note 3) and stopcock B is turned to vent directly into a nitrogen bypass line and bubbler. While the graduated cylinder C is cooled in acetone-dry ice, SF₄ is slowly passed into the cylinder until 7 mL (13 g at −70°C, 0.12 mol) of liquid SF₄ have condensed. Stopcock A is closed and B is vented directly into the three-necked flask. Removal of the cooling bath from graduated cylinder C allows distillation of SF₄ into the cooled reaction vessel.

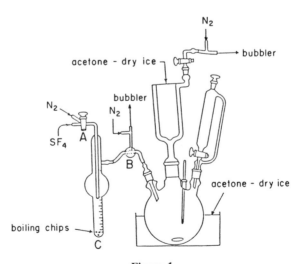

Figure 1

A slow stream of nitrogen is passed into the reaction vessel through stopcock B and the N,N-dimethylaminotrimethylsilane is added to the stirred SF_4 solution at a rate sufficiently slow to keep the temperature below $-60°C$ (about 30 min). The cooling bath is removed, the mixture is allowed to warm to room temperature, and the entire system is placed inside a nitrogen-flushed glove bag. The dropping funnel and condenser are replaced by stoppers, stopcock B is closed, and the closed system is stirred for 3 days with constant nitrogen flow through the glove bag (Note 4). During this time, the product separates as fine crystals. The crystals are collected in a nitrogen pressure filter, washed with 50–100 mL of dry ether, and dried by passing a stream of dry nitrogen through them to give 23–26 g (71–78% yield) of tris(dimethylamino)sulfonium difluorotrimethylsilicate as hygroscopic (Note 5), colorless needles, mp 98–101°C (Note 6).

2. Notes

1. It is important that the ether be very dry (distilled from Na/benzophenone). Otherwise, the quality of the product and the yield will be substantially lower.

2. N,N-Dimethylaminotrimethylsilane is available from Petrarch Systems, Inc. Care should be taken to assure that there is no free dimethylamine present. Commercial samples can be purified by distillation through a 6-in. Vigreux column, bp 86–87°C. The submitters used a spinning band column for removal of hexamethyldisiloxane, bp 99–100°C, which is present as a contaminant.

3. Sulfur tetrafluoride is available from Air Products and Chemicals, Inc. or Matheson Gas Products. Commercial SF_4 was used without purification. In a more convenient modification of this procedure, dimethylaminosulfur trifluoride (methyl DAST), available from Carbolabs, Inc., can be substituted for SF_4. Dimethylaminosulfur trifluoride (1 mol) is mixed with dry ether and dimethylaminotrimethylsilane (2.1 mol) is added over a one-half–2 hr period with the temperature held below 20°C. After this addition, all other aspects of the run are carried out in the same manner as those of the procedure using SF_4.

4. The submitters obtained good yields without a glove bag, but the checkers encountered 30–40% yield reduction without this precaution. A drybox is also suitable. More than 3 days may be required for the reaction to go to completion if the laboratory temperature drops below 20°C.

5. Because tris(dimethylamino)sulfonium difluorotrimethylsilicate is very hygroscopic, it is best transferred in a dry atmosphere of nitrogen or argon (dry box or glove bag).

6. Even a brief exposure to moist air will cause the product to react with the available water vapor to give $(Me_2N)_3S^+ HF_2^-$ and $(Me_3Si)_2O$, and the presence of these products will appreciably lower the melting point. A melting point as low as 58–62°C can be obtained after a brief exposure.

3. Discussion

Tris(dimethylamino)sulfonium difluorotrimethylsilicate is a source of soluble organic fluoride ion of high anionic reactivity. Fluoride ion from this salt and other tris(dialkylamino)sulfonium difluorotrimethylsilicates has been used to displace halogen from carbon[2] and to cleave $Si-O^{3-7}$ and $Si-C=O^{7,8}$ bonds. Since these salts can be prepared in a rigorously anhydrous state, they have an advantage over quaternary ammonium fluorides, which usually contain some water. Tris(dialkylamino)sulfonium difluorotrimethylsilicates have also been used to prepare other sulfonium salts with high nucleophilic reactivity, including $(R_2N)_3S^+$ enolates,[6] phenoxide,[5] cyanide, azides, and cyanates.[2]

This method has been used to prepare several different tris(dialkylamino)sulfonium difluorotrimethylsilicates, including salts with greater organic solubility such as the tris(diethylamino)sulfonium[2,3] and tris(pyrrolidino)sulfonium[2] difluorotrimethylsilicates. The tris(dimethylamino)sulfonium salt, however, is highly crystalline and thus has an advantage in ease of preparation and purification over these other salts.

1. E. I. du Pont de Nemours and Company, Central Research and Development Department, Experimental Station, Wilmington, DE 19898. Present address: Department of Chemistry, Ursinus College, Collegeville, PA 19426.
2. Middleton, W. J. U.S. Patent 3940402, 1976; *Chem. Abstr.* **1976,** *85,* P6388j.
3. Noyori, R.; Nishida, I.; Sakata, J. *J. Am. Chem. Soc.* **1981,** *103,* 2106.
4. Noyori, R.; Nishida, I.; Sakata, J. *Tetrahedron Lett.* **1980,** *21,* 2085.
5. Noyori, R.; Nishida, I.; Sakata, J. *Tetrahedron Lett.* **1981,** *22,* 3993.
6. Noyori, R.; Nishida, I.; Sakata, J.; Nishizawa, M. *J. Am. Chem. Soc.* **1980,** *102,* 1223.
7. Brinkman, K. C.; Gladysz, J. A. *J. Chem. Soc., Chem. Commun.* **1980,** 1260.
8. Blakeney, A. J.; Johnson, D. J.; Donovan, P. W.; Gladysz, J. A. *Inorg. Chem.* **1981,** *20,* 4415.
9. Farnham, W. B.; Harlow, R. L. *J. Am. Chem. Soc.* **1981,** *103,* 4608.

REDUCTION OF α-AMINO ACIDS: L-VALINOL

(1-Butanol, 2-amino-3-methyl-, (S)-)

Submitted by D. A. DICKMAN,[1a] A. I. MEYERS,[1a] G. A. SMITH[1b] and R. E. GAWLEY[1b]
Checked by (A) KARL M. SMITH and CLAYTON H. HEATHCOCK; (B) ALAN T. JOHNSON, KRAIG M. YAGER, and JAMES D. WHITE

1. Procedure

A. *Caution! Because of the hydrogen gas evolved during this reaction, this procedure should be carried out in an efficient fume hood.*

An oven-dried, 3-L, three-necked flask equipped with a mechanical stirrer, a Friedrich condenser, and a nitrogen-inlet tube is flushed with nitrogen, and then charged with a suspension of lithium aluminum hydride (47.9 g, 1.26 mol) in 1200 mL of tetrahydrofuran (THF) (Note 1). The mixture is cooled (10°C, ice bath) and L-valine (100 g, 0.85 mol) is added in portions over a 30-min period from a 200-mL round-bottomed flask connected to the reaction flask via a flexible plastic sleeve so as not to produce too vigorous an evolution of hydrogen (Note 2). After the addition is complete, the plastic sleeve is replaced by a stopper, the ice bath is removed, and the reaction mixture is warmed to room temperature and then refluxed for 16 hr. The reaction mixture is then cooled again (10°C, ice bath) and diluted with ethyl ether (1000 mL) (Note 3). The reaction is quenched over a 30-min period with water (47 mL) (*Caution! See Note 4*), aqueous 15% sodium hydroxide (47 mL, over 20 min), and water (141 mL, over 30 min). The solution is stirred for 30 min and the white precipitate is filtered. The filter cake is washed with ethyl ether (3 × 150 mL) and the organic filtrates are combined, dried with anhydrous sodium sulfate, and concentrated under reduced pressure. Distillation of the residue under vacuum affords L-valinol (63.9–65.7 g, 73–75%) (Note 5) as a clear liquid: bp 63–65°C (0.9 mm) (Note 6): $[\alpha]_D^{20}$ +14.6° (neat); n_D^{20} 1.455; IR (neat) cm^1: 3300, 1590; ^{1}H NMR (CDCl$_3$) δ: 0.92 (d, 6 H), 2.38–2.74 (m, 4 H), 3.13–3.78 (m, 2 H).

B. *Caution! Because of the foul odor of the methyl sulfide given off, this procedure, up to the methanol quench, should be carried out in a hood.*

A 2-L, three-necked, round-bottomed flask is equipped with a mechanical stirrer, heating mantel, 250-mL graduated addition funnel, and an 8-in., air-cooled reflux condenser (West type) topped with a water-cooled distillation head and a 1-L receiving flask. It is connected to a nitrogen line through the still head. The glassware is either oven-dried and cooled in a desiccator or flame-dried and assembled while still hot. The assembly is flushed with nitrogen and charged with 200 g of L-valine (1.7 mol), 400 mL of tetrahydrofuran (THF) (Note 1), and 210 mL of freshly distilled boron trifluoride etherate (242 g, 1.7 mol). The mixture is heated at a rate sufficient to cause the THF to reflux gently (Note 7) and 188 mL (1.88 mol) of borane–methyl sulfide complex (BMS) (Note 8) is added dropwise over the course of 2 hr (Note 9). The solution is then refluxed for 18 hr. The methyl sulfide that has collected at the stillhead is discarded (Note 10), and the reaction mixture is cooled to 0°C and quenched by the slow addition of 200 mL of methanol. The addition funnel is replaced by a glass stopper, and the air-cooled condenser is removed, leaving the flask equipped for distillation of solvent through the distillation head. The reaction mixture is concentrated under reduced pressure with heating and stirring. The distillation head is replaced by a water-cooled reflux condenser, and the residue is dissolved in 1 L of 6 M sodium hydroxide and refluxed for 4 hr. The mixture is saturated with potassium carbonate (ca. 400 g); cooled; filtered through a Celite pad on a coarse, fritted funnel; and extracted with three 1-L portions of chloroform. The combined extracts are washed with three portions of saturated sodium chloride (500 mL each), stirred over anhydrous potassium carbonate for 24 hr, and concentrated under reduced pressure to give a yellow oil. The crude material is vacuum distilled to give 77.5 g (44%) of purified L-valinol, bp 62–67°C/2.5 mm (Note 6): $[\alpha]_D^{20}$ +14.6° (neat), n_D^{20} 1.455; IR (neat film) cm^{-1}: 3300 (OH), and 1590 (NH$_2$); NMR δ: 0.92 (d, 6 H), 1.54 (m, 1 H), 2.38–2.74 (m, 4 H), 3.13–3.78 (m, 2 H).

2. Notes

1. Tetrahydrofuran is dried by distillation from sodium/benzophenone ketyl.
2. The hydrogen gas released during the addition of the amino acid should be vented through the nitrogen inlet to a bubbler at the back of the fume hood, well away from the mechanical stirrer motor and any other source of electrical spark.
3. Dilution is necessary to keep the reaction mixture from becoming too thick during the quench. Dilution with THF results in significantly lower yields.
4. An addition funnel was used for the dropwise addition of water. Care *must* be taken during this quench to ensure that all the escaping hydrogen gas is vented to the back of the fume hood through the nitrogen bubbler.
5. The submitters reported a yield of 73.7 g (84%).
6. The checkers found that L-valinol (mp 29–31°C) solidifies on distillation and will clog a water-cooled condenser. The use of a heat gun is recommended to avoid obstruction of the distillation pathway.
7. The temperature is maintained sufficiently high so that THF refluxes in the air-cooled condenser while ether and methyl sulfide distill through the short-path distillation head.
8. The borane–methyl sulfide complex is available from Aldrich Chemical Company, Inc.
9. It is important that gentle reflux be maintained throughout the addition. If the solution is not heated during this period, an exothermic reaction occurs when the solution is refluxed.
10. Methyl sulfide should be destroyed by slowly pouring the volatile distillate into 1 gallon of household bleach (5% sodium hypochlorite). After 30 min, the bleach solution may be discarded in the drain.

3. Discussion

The reduction of amino acids to the corresponding amino alcohols via their ethyl ester hydrochlorides has been reported using lithium aluminum hydride[2] and sodium borohydride.[3] The reduction of several amino acids with borane–methyl sulfide (BMS) has also been reported.[4] The reduction of proline to prolinol with lithium aluminum hydride in THF was reported by Enders[5] and of valine to valinol by Meyers.[6] Procedure A (see Section A) is adapted from the latter work. The submitters have used the same procedure to reduce alanine to alaninol (70% yield), phenylglycine to phenylglycinol (76% yield), phenylalanine to phenylalaninol (87% yield), and N-benzoylvaline to N-benzylvalinol (76% yield).

Procedure A using LAH is faster and more convenient than the BMS procedure (B). Both procedures are general for the reduction of amino acids to amino alcohols. If functional group incompatibility is precluded, lithium aluminum hydride reduction is preferable.

The borane procedure (B) is a hybrid of two methods, Lane's procedure for BMS/trimethyl borate reduction of anthranilic acid,[7] and Brown's procedure for enhanced-rate reductions of several functional groups with BMS by distilling off the methyl sulfide during the course of the reaction.[8] The submitters have obtained a 97% crude yield (44–51% yield after distillation) of prolinol using this procedure. Lane reports that the following additional amino acids can be reduced using BMS/BF₃ etherate: leucine, phenylalanine, and 6-aminocaproic acid.[4] Meyers has added phenylglycine to the list, and has confirmed the optical purity of the amino alcohols obtained by preparation of the Mosher amides.[9]

1. (a) Procedure A. Department of Chemistry, Colorado State University, Ft. Collins, CO 80523; (b) Procedure B, Department of Chemistry, University of Miami, Coral Gables, FL 33124.
2. Karrer, P.; Portmann, P.; Suter, M. *Helv. Chim. Acta* **1949**, *32*, 1156; Karrer, P.; Naik, A. R. *Helv. Chim. Acta* **1948**, *31*, 1617.
3. Seki, H.; Koga, K.; Matsuo, H.; Ohki, S.; Matsuo, I.; Yamada, S. *Chem. Pharm. Bull.* **1965**, *13*, 995.
4. Lane, C. F., U.S. Patent 3935280, 1976; *Chem. Abstr.* **1976**, *84*, 135101p.
5. Enders, D.; Eichenauer, H. *Chem. Ber.* **1979**, *112*, 2933.
6. Meyers, A. I.; Dickman, D. A.; Bailey, T. R. *J. Am. Chem. Soc.* **1985**, *107*, 7974.
7. Lane, C. F.; Myatt, H. L.; Daniels, J.; Hopps, H. B. *J. Org. Chem.* **1974**, *39*, 3052.
8. Brown, H. C.; Choi, Y. M.; Narasimhan, S. *J. Org. Chem.* **1982**, *47*, 3153.
9. Poindexter, G. S.; Meyers, A. I. *Tetrahedron Lett.* **1977**, 3527.

TETRAMETHYLBIPHOSPHINE DISULFIDE

Warning: It has been reported[1,2] *that serious explosions have occurred during the preparation of tetramethylbiphosphine disulfide by the method described in* Inorganic Syntheses.[3] No such incidents have been reported in the synthesis of the compound published in this series,[4] but the two procedures are sufficiently similar that caution is indicated. The following precautions are strongly urged:

1. The phosphorus trichloride sulfide (PSCl₃) should be distilled before use.

2. The reaction vessel should be cooled with an ice–salt bath (rather than an acetone–dry ice bath as specified in the published procedure) during the addition of the PSCl₃ solution to the Grignard reagent. The reaction temperature should be monitored carefully. If it falls below $-5°$, the addition should be stopped and the reaction mixture cautiously rewarmed to $0-5°C$ before addition is resumed.

3. The reaction apparatus should be shielded throughout the addition of the PSCl₃ solution and the subsequent warming of the reaction mixture.

1. Bercaw, J. E. *Chem. Eng. News* **1984**, *62*(18), 4, April 30.
2. Davies, S. G. *Chem. Britain* **1984**, *20*(5), 403, May.
3. Butter, S. A.; Chatt, J. *Inorg. Synth.* **1974**, *15*, 186.
4. Parshall, G. W. *Org. Synth.*, **1965**, *45*, 102; *Org. Synth., Coll. Vol. V* **1973**, 1016.

TYPE OF REACTION INDEX

This index lists the preparations contained in this volume in accordance with general types of reactions. Only those preparations that can be classified under the selected heading with some definiteness are included. The arrangement of types and of preparations is alphabetical. The reference page is to the title page of the procedure wherever the preparations may be found.

TYPE OF COMPOUND INDEX

This index lists the preparations contained in this volume by functional groups or by ring systems. Compounds are listed as unsubstituted if they contain phenyl, alkenyl, allenyl or alkynyl groups. Salts are included with the corresponding acids and bases.

FORMULA INDEX

All preparations listed in the Contents are recorded in this index. The system of indexing is that used by *Chemical Abstracts*. The essential principles involved are as follows: (1) The arrangement of symbols in formulas is alphabetical except that in carbon compounds C always comes first, followed immediately by H if hydrogen is also present. (2) The arrangement of formulas is also alphabetical except that the number of atoms of any specific kind influences the order of compounds: e.g., all formulas with one carbon atom precede those with two carbon atoms, thus: CH_2I_2, CH_3NO_2, CH_5N, C_2H_2O. (3) The arrangement of entries under any heading is strictly alphabetical according to the names of the isomers. (4) Inorganic salts of organic acids and inorganic addition compounds of organic compounds are listed under the formulas of the compounds from which they are derived.

555

AUTHOR INDEX

Aebi, J., 153
Albini, A., 23
Alderdice, M., 351
Alexakis, A., 290
Anderson, A. G., 144
Andrews, S. A., 197
Anello, L. B., 251
Arduengo, A. J., III, 195
Arora, S. K., 263
Arvanaghi, M., 451

Bach, R. D., 126
Bal, B., 185
Balderman, D., 433
Ballesteros, P., 142
Banner, B. L., 297
Bartlett, P. A., 164
Barnier, J. P., 129
Bastiaansen, L. A. M., 287
Batcho, A. D., 34
Bauer, M., 435
Beadle, J. R., 108
Benac, B. L., 195
Berlin, K. D., 210
Bertz, S. H., 50
Bettinetti, G. F., 23
Blacklock, T. J., 203
Böhm, I., 485
Bond, F. T., 77
Branca, S. J., 271
Brown, H. C., 427
Buchschacher, P., 368
Buntin, S. A., 361
Burgess, E. M., 195
Buse, C. T., 185, 381

Cahiez, G., 290
Caldwell, W. E., 315
Cargill, R. L., 315
Carozza, L., 297
Carr, R. V. C., 453
Cassady, J. M., 319
Chadha, N. K., 339

Chamberlin, A. R., 77
Champion, J., 112, 129
Chan, D. M. T., 266
Chan, W. K., 81, 87
Chang, V. S., 397
Charleson, D. A., 282
Chatziiosifidis, I., 424
Chiu, I-C., 249
Citterio, A., 105
Claus, R. E., 168
Clausen, K., 372
Cohen, N., 297
Conia, J. M., 112, 129
Cook, J. M., 50
Crabbé, P., 276
Crass, G., 41
Creary, X., 438
Crossland, I., 12

Dalton, J. R., 315
Danishefsky, S., 312, 411
Denis, J. M., 112
Denmark, S. E., 524
Dessau, R. M., 400
DiBiase, S. A., 108
Dickman, D. A., 530
Dittmer, D. C., 491

Earl, R. A., 334
Ellis, M. K., 356
Exon, C. M., 461

Fadel, A., 117
Fahrni, H. P., 30
Farrand, R., 302
Fishel, D. I., 420
Fleming, M. P., 1
Fludzinski, P., 208, 221, 241
Forbus, T. R., Jr., 506
Fry, J. L., 393
Frye, L. L., 495
Fung, A. P., 254
Fürst, A., 368

GENERAL INDEX

This index includes references to and registry numbers of the reagents, chemicals, and catalysts used in the various preparations as well as to the intermediates and final products obtained from the preparations. Special apparatus is also included. It also includes selected references to reaction types and compound types involved in or referred to in the preparations, but the Reaction Index or Compound Index should also be referred to for possible additional entries in these categories. Most entries in this index refer to the preparative parts of the procedures or to extensions thereof; thus the discussion sections of the procedures have not been extensively indexed.

The name of a compound in capital letters together with a number in **boldface** type indicates complete preparative directions for the substance named. A name in regular type together with a number in regular type indicates a compound or an item, usually starting materials, mentioned in connection with a preparation.

In almost all cases, the page reference is to the title page of the procedure, but references to tables will be to the specific page.

563

HAZARD INDEX

CONCORDANCE INDEX

The annual volume (in bold-faced type) and page number where each procedure first appeared is given on the left; the page number for the revised procedure is given on the right.

Annual Volume		This Volume	Annual Volume		This Volume
Number	**Page**	**Page**	**Number**	**Page**	**Page**
60	1	181		35	435
	6	12		39	200
	11	397		42	447
	14	20		48	87
	18	112		56	210
	20	114		59	249
	25	129		62	256
	29	121		65	271
	34	144		71	302
	41	485		74	124
	49	162		77	319
	53	203		82	361
	58	479		85	375
	63	126		93	223
	66	411		98	23
	72	287		103	427
	78	482		112	304
	81	334		116	66
	88	332		122	512
	92	386		129	473
	101	396		134	81
	104	433		141	77
	108	393		147	312
	113	1			
	117	414	**62**	1	290
	121	508		9	137
	126	517		14	351
				24	229
61	1	467		31	245
	5	213		39	172
	8	420		48	501
	14	27		58	266
	17	56		67	105
	22	400		74	326
	24	41		86	443